THE CLUSTER AND PHOENIX MISSIONS

THE CLUSTER AND PHOENIX MISSIONS

edited by

C.P. ESCOUBET

Solar System Division, Space Science Department, ESA/ESTEC,
Noordwijk, the Netherlands

C.T. RUSSELL

Institute of Geophysics and Planetary Physics,
University of Los Angeles, USA

and

R. SCHMIDT

Solar System Division, Space Science Department, ESA/ESTEC,
Noordwijk, the Netherlands

Reprinted from *Space Science Reviews*, Vol. 79, Nos. 1–2, 1997.

KLUWER ACADEMIC PUBLISHERS
DORDRECHT / BOSTON / LONDON

A C.I.P. Catalogue for this book is Available from the Library of Congress

ISBN 0-7923-4411-1

Published by Kluwer Academic Publishers,
P.O. Box 17, 3300 AA Dordrecht, The Netherlands

Kluwer Academic Publishers incorporates
the publishing programmes of
D. Reidel, Martinus Nijhoff, Dr W. Junk and MTP Press.

Sold and distributed in the U.S.A. and Canada
by Kluwer Academic Publishers,
101 Philip Drive, Norwell, MA 02061, U.S.A.

In all other countries, sold and distributed
by Kluwer Academic Publishers Group,
P.O. Box 322, 3300 AH Dordrecht, The Netherlands

Printed on acid-free paper

Printed in Belgium

Table of Contents

Space Science Reviews **79:** v–vii, 1997.

PREFACE

On June 4 1996, the catastrophic failure of the first flight of Ariane-5 with the four Cluster spacecraft onboard devastated the whole Cluster community. Ten years of work was annihilated in less than 1 minute.

Figure 1. Explosion of the first Ariane-5 rocket with the four Cluster spacecraft onboard.

Space Science Reviews **79:** 1–5, 1997.

Figure 2. Parts of the Cluster spacecraft in the swampy area around the launch pad.

After a short break in the countdown due to bad weather on the launch pad, Ariane-5 rose flawlessly, producing a tremendous cloud of smoke and vapour. The flight trajectory was nominal up to an altitude of 3.5 km, when the sudden swivelling of both solid booster nozzles forced the launcher to tilt sharply. The intense aerodynamic loads on the launcher structure resulted in the breakup of the launcher which caused the self-destruction of all launcher elements (Figure 1). Parts of the Cluster spacecraft were dispersed throughout the swampy area around the launch pad (Figure 2). An Enquiry Board has now determined the accident to be attributable to the flight software and has made recommendations to correct the anomaly.

Shortly after the failure, the Cluster project proposed to the Principal Investigators (PI) to refurbish quickly the Cluster structural model using the spare experiments and make it ready for flight. This spacecraft, considered to be the first of a new fleet, was to be called Phoenix. The PI's, however, stressed that an early launch of Phoenix could not achieve the scientific objectives of the four Cluster spacecraft (Figure 3).

On 17 and 18 June a Science Working Team (SWT) meeting was convened which re-confirmed the Cluster scientific objectives. After extensive discussion on the recovery scenario, a resolution was prepared listing two possible options which would fulfill the Cluster scientific objectives:

(1) Refly the Cluster mission: this option would use four re-built Cluster spacecraft and would totally recover the scientific objectives of Cluster and a

Figure 3. The four Cluster spacecraft before launch

substantial portion of the ESA Solar Terrestrial Science Programme (STSP). This is the highest priority option and is the one which best exploits previous investments.

(2) Re-build one Cluster spacecraft, Phoenix, and launch it together with three potentially smaller spacecraft provided in a special programme with national agencies.

Then, on 2 and 3 July, the ESA Science Programme Committee (SPC) approved the preparation for flight of Phoenix and requested a study on how to launch Phoenix and of the various options on how to obtain the three other spacecraft. The result of the study will be presented at the next SPC meeting in November 1996.

The case for the launch of a single Phoenix spacecraft into a 'Cluster-type' orbit in the immediate future rests upon two central arguments: (a) primary investigation using modern high-resolution plasma and field instrumentation of key magnetospheric regions, principally the high-altitude dayside cusp, and (b) coordinated observations with a number of key spacecraft which are currently operational or which are due to be launched in the immediate future, and with the new ground-based infrastructure which was timed for Cluster.

The primary driver in the choice of orbit intended for the original Cluster mission was that it should pass through the high-latitude boundary of the Earth's magnetosphere in the vicinity of the dayside cusp. Most of the information on this region which is presently available was derived from measurements made by the ESRO HEOS- II spacecraft during 1972–1974. These data were sufficient to provide the first evidence of time-dependent solar wind-magnetosphere coupling processes at the magnetopause boundary (then termed 'flux-erosion events'), and also indicated, via the discovery of the 'entry layer', that this region was probably a primary site of solar wind plasma entry into the Earth's plasma environment. Nevertheless, the instrumentation available $\sim$ 25 years ago was relatively crude, and sufficient only to delineate the basic plasma structures and the basic nature of the physical processes involved. For example, the plasma instrumentation, though advanced for its day, was capable only of observations with $\sim$ 4-min resolution. By comparison the plasma structures involved at the boundary are expected to have scales of 100–1000 km (current sheets and flux ropes due to boundary interaction processes), and with boundary motions of typically several tens of km s^{-1}, require instrumentation capable of resolution down to a few seconds to investigate the physics involved. There is therefore no doubt that a single spacecraft with Cluster-class instrumentation will provide major new insights and information on this key region, though not, of course, of the same order as would have been provided by the four-point information from Cluster itself.

In addition to the above 'high-altitude' investigations, there will also be considerable interest in the Phoenix perigee passes, when the spacecraft will cut rapidly across the dayside cusp, boundary layer, and plasma sheet field lines in the middle-altitude ($\sim$ 4 Earth radii) distance regime. Such data will also be capable of providing new insights into dayside boundary and tail processes, when coordinated with data from ground-based facilities.

A second major area of activity which will uniquely become possible with the Phoenix spacecraft launched in the immediate future will be coordinated observations with the large number of magnetospheric-related spacecraft which are either currently operational or which are planned to become so over the next 1–2 years. Phoenix would fill a gap in the global coverage of the magnetosphere created by the loss of the Cluster mission. Four major agencies (ESA, NASA, the Russian Institute of Space Research, IKI, and the Japanese Institute of Space and Astronautical Science, ISAS) have coordinated their missions in the field of solar terrestrial science in order to study large-scale magnetospheric processes. This Programme has been augmented by an unprecedented network of ground based observatories. The four-point Cluster measurements were an important element in this global programme. A single spacecraft would be able to recover some, but not all, of these objectives.

Since Phoenix, as fifth Cluster spacecraft, will be equipped with the spare Cluster experiments, the instrumentation articles are still quite appropriate to the new mission. Furthermore, the objectives of the recovery mission, the ground systems, the ground observation program and the theory and modelling efforts all remain unchanged. Thus this series of articles will continue to be essential to the Cluster team as the recovery mission is implemented.

C. P. ESCOUBET R. SCHMIDT J. CREDLAND

FOREWORD

Since the earliest reports around 2500 B.C. by the Chinese, the auroras borealis have always been of great interest to mankind. Their usual manifestation is close to the geographic pole. Very active auroras, however, are sometimes visible even close to the equator. This unpredictable behaviour has attracted the interest of some of the greatest scientists in the world. Early theories explained the auroras as an evaporation of air which then became incandescent at high altitude or, alternatively, as air clouds illuminated by sunlight.

It was not until the eighteenth century that magnetic disturbances were found to be associated with auroras. This discovery marked an important step in the physics of auroras. Magnetic disturbances remain one of the main tools to characterize auroral activity today. The second major step forward in understanding auroras concerned the effect of solar activity. In the nineteenth century it was realized that there was a clear correlation between the number of sunspots and the intensity and number of geomagnetic disturbances.

Then in the twentieth century with the International Geophysical Year, the first spacecraft, carrying scientific experiments to measure electric and magnetic fields, were launched into space. These *in-situ* measurements gave an unprecedentedly detailed view of the plasma surrounding the Earth, namely the magnetosphere. All the various plasma regions, from the solar wind (the extension of the solar corona), down to the plasmasphere much closer to the Earth, were identified. The direct influence of the solar wind was also observed, with multi-point measurements made simultaneously in the solar wind and in the Earth's ionosphere. Although the global view of the magnetosphere is now quite well known, many questions remain concerning the microphysics that governs the transfer of energy from the solar wind to the Earth's magnetosphere and from the magnetosphere to the ionosphere. One major unknown in this Sun-Earth system concerns the size and shape of the small-scale plasma structures and their role in the transfer of energy. ESA's Cluster, with four spacecraft travelling together through the magnetosphere, will be the first mission to study these plasma structures in three dimensions.

Cluster is one of the two missions – the other being the Solar and Heliospheric Observatory (SOHO) – constituting the Solar Terrestrial Science Programme (STSP), the first 'Cornerstone' of ESA's Horizon 2000 Programme. The Cluster mission was first proposed in November 1982 in response to an ESA Call for Proposals for the new scientific mission. The original idea of four spacecraft in a tetrahedral configuration came from a French study (ESSAIM), the prime objective of which was to visit the magnetotail in the equatorial region. The tetrahedron idea was subsequently adopted in the proposal and a polar orbital plane was chosen in order to be able to visit both the polar cusp and the magnetotail region. Following

Space Science Reviews **79:** 7–9, 1997.

an Assessment and a Phase-A Study, the Cluster mission was presented to the scientific community at the end of 1985. In February 1986 the STSP programme, combining both Cluster and Soho, was selected by the ESA Science Programme Committee. Following a joint ESA/NASA Announcement of Opportunity issued in March 1987, the eleven instruments making up the scientific payload were selected in March 1988. The launch of the four Cluster spacecraft is currently scheduled for May 1996 on the first flight of Ariane-5.

This book describes the Cluster mission and its scientific objectives, the instrumentation, which is identical on the four spacecraft, the operations carried out by the two operations centres and the data dissemination to both the Cluster and the wider scientific community. The volume begins with an overview of the Cluster mission and its scientific objectives, followed by a description of the spacecraft design and the engineering challenges of building four identical spacecraft. Each of the eleven instruments is then described in detail by their respective Principal Investigators. Seven of the instruments will measure electric and magnetic fields from static fields to high frequency fields. A further three instruments will measure electron and ion distribution functions with high time resolution. Finally, one instrument will control the potential of the spacecraft by emitting a beam of indium ions. These instrument papers are followed by a description of the mission operations undertaken by the European Space Operations Centre and the scientific operations undertaken by the Joint Science Operations Centre (JSOC). One further contribution describes the raw data dissemination, first to the principal investigators via an on-line service and later to the Cluster community on CDroms. The Cluster community, which includes the Principal Investigators and Co-Investigators, and also the wider scientific community, will have access to processed physical parameters from all instruments. The processing of the parameters and their distribution is the task of the Cluster Science Data System (CSDS). Finally, a directory of the Cluster community members is included as an Annex. A description of the Cluster mission and the latest news can be found in a World Wide Web page at http:://www.estec.esa.nl/spdwww/cluster/html/.

All contributions in this volume have been refereed by two experts, one from within and the other from outside the Cluster community. We would like to thank all the referees for their fruitful comments which will make these articles easy to read and informative for both the Cluster community and the general scientific community. We gratefully acknowledge the help of Erica Rolfe in editing the manuscripts.

Many individuals have made the challenging Cluster programme a reality. They all should be specially acknowledged for their key role. First, R. M. Bonnet, ESA director of scientific programmes, who directed the Cluster programme through all phases of the development. J. Credland, ESA Cluster project manager, and his team conducted the production of the 4 Cluster spacecraft with unrelenting efforts. M. Warhaut, ESA Cluster ground segment manager, P. Ferri, ESA Cluster spacecraft operations manager, and their team made outstanding efforts to achieve the preparation of the ground segment. T. Dimbylow, JSOC project manager, and M. Hapgood, JSOC project scientist, together with their team worked remarkably to prepare the coordination of the science operations. The eight CSDS data centre managers and the members of the CSDS steering committee and implementation working group achieved successfully the readiness of the Cluster data system. Finally, G. Paschmann, chairman of the science operation working group, together with N. Sckopke conducted excellently the preparation of the science operations.

During the process of editing this volume, we were deeply grieved to hear of the tragic death of Les Woolliscroft, the Principal Investigator of the Digital Wave Processing instrument. He was an enthusiastic scientist and a key figure in the preparation of the Cluster mission. He was actively participating in all Cluster working groups from the Cluster Science Data System Steering Committee to the Wave Experiment Consortium Operation Working Group, of which he was Chairman. We will always remember his fundamental role in the definition of the Cluster Science Data System and his active support of the Joint Science Operation Centre. Les Woolliscroft was Reader in Physics at the University of Sheffield. He was an excellent scientist and a good friend; he will be greatly missed.

C. P. ESCOUBET C. T. RUSSELL R. SCHMIDT

CLUSTER – SCIENCE AND MISSION OVERVIEW

C.P. ESCOUBET and R. SCHMIDT
Space Science Department of ESA/ESTEC, P.O. Box 299, 2200 AG Noordwijk, The Netherlands
M.L. GOLDSTEIN
Laboratory for Extraterrestrial Physics, NASA/GSFC Code 692, Greenbelt MD 20771, U.S.A.

Abstract. The European Space Agency's Cluster programme is designed to study the small-scale spatial and temporal characteristics of the magnetospheric and near-Earth solar wind plasma. The programme is composed of four identical spacecraft which will be able to make physical measurements in three dimensions. The relative distance between the four spacecraft will be varied between 200 and 18 000 km during the course of the mission. This paper provides a general overview of the scientific objectives, the configuration and the orbit of the four spacecraft and the relation of Cluster to other missions.

1. Introduction

Cluster is one of the two missions – the other being the Solar and Heliospheric Observatory (SOHO) – constituting the Solar Terrestrial Science Programme (STSP), the first 'Cornerstone' of ESA's Horizon 2000 Programme. The Cluster mission was first proposed in November 1982 in response to an ESA Call for proposals for the next series of scientific missions. The assessment study was conducted in 1983 to prove the feasibility of the mission concept. Subsequently, the Phase-A study was conducted jointly with NASA during 1984-1985. At the end of 1985 the Cluster mission was presented to the scientific community and in February 1986, the STSP programme, combining both Cluster and Soho missions, was selected by ESA Science Programme Committee. After a joint ESA/NASA Announcement of Opportunity issued in March 1987, the 11 instruments making up the scientific payload were selected in March 1988. The launch of the four Cluster spacecraft is currently scheduled for May 1996 on the first Ariane-5 flight.

The four Cluster spacecraft will be launched together as two pairs in the two launcher payload compartments of an Ariane-5 vehicle. As this is the first qualification flight of Ariane-5, the spacecraft will be placed into Geostationary Transfer Orbit (GTO). Then a series of complex manoeuvres will be conducted for each spacecraft, over about three weeks until the spacecraft reach their final polar orbit with a perigee of 4 R_E and an apogee of 19.6 R_E. After the final orbit is reached, the commissioning phase, consisting of boom deployments and checking of the instruments, will be conducted over approximately 2.5 months. After this extensive checkout period, the mission phase itself will begin, lasting a nominal period of two years.

Space Science Reviews **79:** 11–32, 1997.

Each Cluster spacecraft is cylindrical with a diameter of 2.9 m and a height of 1.3 m and will be spin-stabilised at 15 rpm (see Credland *et al.*, this issue for details on the spacecraft). The spin axis will point toward the North ecliptic pole. The total mass of each spacecraft is 1200 kg, 72 kg for the 11 experiments, 478 kg for the spacecraft dry mass and 650 kg for the propellant. The later is necessary for the extensive series of manoeuvres to be performed to reach the operational orbit and, during the course of the mission, to change the configuration of the spacecraft. The power allocated to the instruments is 47 W out of a total of 224 W available from the solar arrays. The in-orbit spacecraft configuration is characterised by four 50 m experiment wire booms, two 5 m experimental radial booms and two axial telecommunication antenna booms.

The orbital parameters of the four spacecraft will be slightly different to obtain a tetrahedral configuration in the regions of scientific interest. The size of this tetrahedron will be varied from 200 km up 18 000 km during the course of the two-year mission. As the Cluster orbit is fixed in the inertial system, the rotation of the Earth around the Sun will cause the spacecraft to cross the various near-Earth plasma regions, for example the Earth magnetotail after launch and the polar cusp and solar wind six months later. All regions of interest should be crossed again during the second year of the mission which will allow the possibility to revise the configuration strategy in view of the experience gained during the first year.

The interaction between the solar wind, the extension of the solar atmosphere, and the magnetosphere, the cavity which contains the Earth magnetic field, is a key element in the Solar Terrestrial Science Programme. One example of this interaction is the direct entry of solar wind particles into the magnetosphere through the polar cusps. The polar cusps are two magnetic funnels, one in each hemisphere, which focus the solar wind particles, rather like an astronomical telescope focuses photons. The solar wind particles enter the exterior cusp, which has a diameter of approximately 50 000 km, and then follow the converging magnetic field down to the ionosphere where the cusp size is around 500 km. This converging magnetic field allows the study of a very large area of the magnetopause through a limited region of space inside the cusps.

Another example of the interaction of the solar wind with the magnetosphere is the acceleration of plasma in the magnetotail during substorms. The magnetotail is a large reservoir of both solar wind and ionospheric particles which under some circumstances in the solar wind, for instance the interplanetary magnetic field reversal from north to south, releases a large amount of particles toward the Earth. Both mechanisms, particles entering in the polar cusps and the substorms, produce aurorae when the precipitating particles, electrons and ions, hit the neutral gas of the atmosphere. Sometimes these particles are very energetic and can have dramatic effects on human activities, for example disruption of electrical power and telecommunications and also serious anomalies in the operation of satellites, particularly at geostationary orbit.

Cluster will determine the physical processes involved in the interaction between the solar wind and the magnetosphere by visiting the key regions like the polar cusps and the magnetotail. The four Cluster spacecraft will map in three dimensions the plasma structures contained in these regions. In addition, the simultaneous four-point measurements will permit to derive for the first time differential plasma quantities. For example, the current density near the four spacecraft can, under certain conditions, be derived from the magnetic field measurements using Ampère's law. The inter-spacecraft comparison requires identical experiments. This was controlled during extensive calibrations performed on ground and will finally be verified in orbit in selected plasma regions.

2. Scientific Objectives

The main goal of the Cluster mission is to study the small-scale plasma structures in space and time in the key plasma regions (Figure 1):

- solar wind and bow shock,
- magnetopause,
- polar cusp,
- magnetotail,
- auroral zone.

In this paper we will briefly list some of the scientific questions that the four Cluster spacecraft will answer. More information on the subject can be found in the Proceedings of the Cluster workshop on Physical Measurements and Mission Oriented Theory (Mattok, 1995) and in the supplement of Reviews of Geophysics (Shea, 1991).

The bow shock is the region where the solar wind becomes decelerated from a supersonic to a subsonic speed before being deflected around the Earth (Figure 1). The length scales of various processes occurring at the bow shock, like the reflection and thermalisation of the ions, are not yet known (Sckopke *et al.*, 1990) and a better knowledge requires multi-point measurements on various scales. Further upstream of the bow shock, the solar wind is the place where electromagnetic waves play a major role and a further understanding of these waves is needed, in particular their transmission through the bow shock and the magnetosheath. Cluster, which carries five instruments to measure the characteristics of electromagnetic waves in various frequency domains (see papers by Cornilleau-Wehrlin *et al.*; Gustafsson *et al.*; Décréau *et al.*; Gurnett *et al.*; and Woolliscroft *et al.*, this issue), would be ideally located for this study with, for instance, two spacecraft downstream and two spacecraft upstream of the bow shock. In addition, hot flow anomalies (Schwartz *et al.*, 1985; Thomsen *et al.*, 1986; Paschmann *et al.*, 1988; Onsager *et al.*, 1990), which are extreme disturbances observed near the bow shock, are still an enigma. Multi-point measurements would help to define their shape and structure and understand their origin.

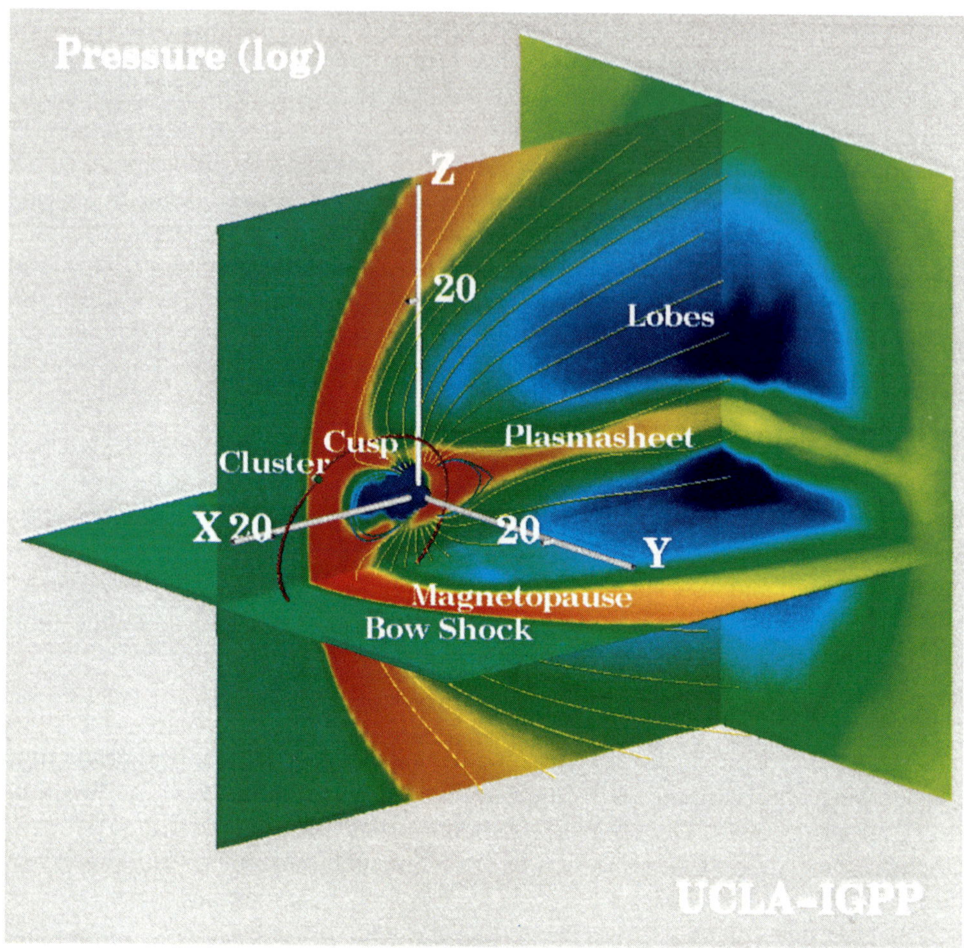

Figure 1. Results of three-dimensional global magnetohydrodynamic simulation for westward interplanetary magnetic field (Courtesy of Jean Berchem, UCLA). Only a small part ($75 \times 60 \times 60\ R_E$) of the simulation system is represented by a 3D perspective view of 2D colour-coded contour of the logarithm of plasma pressure. Three cuts are obtained by using the standard Advanced Visual Systems orthogonal slicers. One cut is taken along the meridian geocentric solar ecliptic (GSE) $X - Z$ plane, the second plane is around the equatorial $X - Y$ plane and the last one is taken in the $Y - Z$ plane around 30 R_E in the tail. The magnetospheric field lines are also shown in the meridional plane (closed field lines in blue and open field lines in yellow). The Cluster orbit is shown in red, crossing the various magnetospheric regions.

The magnetopause is the thin plasma layer that separates the solar wind magnetic field from the Earth magnetic field. It is the locus where the plasma pressure of the solar wind is in equilibrium with the magnetic pressure inside the magnetosphere. Due to a continuous variation of the solar wind pressure, this boundary moves continuously. Cluster, with its four-point measurements, will help to characterise this motion. Although to first order, the magnetopause is considered as an impenetrable boundary, some plasma from the solar wind can enter the magnetosphere.

Various processes have been proposed to account for this penetration of plasma: (1) pressure pulses, when the solar wind dynamic pressure suddenly increases leading to an indentation of the magnetopause, (2) reconnection between the interplanetary magnetic field and the Earth magnetic field, (3) viscous interaction and (4) impulsive penetration where plasma filaments which have a higher momentum than the surrounding solar wind plasma hit and possibly penetrate through the magnetosphere. All these mechanisms have been proposed to explain the penetration of solar wind plasma into the magnetosphere, but there is still no consensus on which plays the major role. The Cluster spacecraft will measure the three-dimensional size and motion of the structures observed at the magnetopause and will also be able to determine the local geometry of the magnetopause (Mottez and Chanteur, 1994) which will make it possible to distinguish between these mechanisms. In addition, the Electron Drift Instrument (EDI) (Paschmann *et al.*, this issue) and the Electric Field and Wave experiment (EFW) (Gustafsson *et al.*, this issue) will measure the electric field at the magnetopause which is a key element in the reconnection process.

Almost anywhere near the surface of the magnetopause, the Earth's magnetic field is tangential to the magnetopause, acting like a natural barrier to the solar wind particles. There are only two regions, one in each hemisphere, where the magnetic field is approximately perpendicular to the magnetopause, namely the polar cusps, allowing a more direct entry of solar wind particles into the magnetosphere. Reconnection of the interplanetary magnetic field with the geomagnetic field appears to occur around these regions, southward of the cusp in the entry layer (Haerendel *et al.*, 1978) and northward of it on the magnetotail lobes (Berchem *et al.*, 1995). However only a few examples have so far been found and only with limited resolution measurements. Cluster will make measurements with much higher resolution, especially in burst mode where the spacecraft operates at 107 kbit s^{-1}, allowing full three-dimensional ion and electron distribution functions to be measured every spin (4 s) (see Rème *et al.*; Johnstone *et al.*, this issue). The exterior cusp is also seen as a very turbulent region with vortices in the plasma flow. Cluster will measure the flow with high resolution in time and at four locations in space to determine the characteristics of these vortices. An optimisation of the orbital parameters was also conducted to cross the cusps as many times as possible with the required spacecraft configuration (Rodriguez-Canabal *et al.*, 1993).

The magnetotail is characterised by magnetic field lines stretched by the solar wind flow in the anti-sunward direction. The outer region consists of two magnetotail lobes and the inner region of the plasma sheet boundary layer and the central plasma sheet (Figure 1). The lobes are large reservoirs of magnetic energy which contain plasma of densities much less than one particle per cubic centimeter. A spacecraft in this very tenuous plasma will charge to a high positive potentials (+30 to +50 V). The effect of this potential on low-energy-particle measurements is dramatic since the electron measurements are saturated by photo-electrons coming back to the spacecraft and ions are repelled. The technique of potential control by ion emitters has already been used on Geotail and has proven to be very valuable (Schmidt *et al.*, 1995), particularly in the far tail. The Cluster potential will be controlled by the Active Spacecraft Potential Control instrument (ASPOC) (see Riedler *et al.*, this issue) and hence will allow measurement of the full distribution functions of the ions and electrons down to around 2 eV in the lobes and the polar cap. These measurements will determine the role of the lobes as a transit region for the particles originating from the ionosphere and the solar wind.

The plasma sheet boundary layer is often the most active plasma region of the magnetotail. Ion beams are observed coming from and toward the Earth. Multi-point measurements are essential to derive the exact origin of these beams and to learn more about the generating mechanism. Magnetic field-aligned currents flowing toward and away from the Earth are also often observed in this region. The Cluster Fluxgate Magnetometer (FGM) (see Balogh *et al.*, this issue) will be able, for the first time, to calculate the current flowing in the vicinity of the spacecraft without having to make any assumption about the shape and orientation of the current sheet. In the centre of the plasma sheet lies the neutral sheet where the magnetic field is weak. The neutral sheet is expected to be the ideal site for magnetic reconnection which accelerates ions toward the Earth and downstream. The Cluster Research with Adaptative Particle Imaging Detectors (RAPID) (Wilken *et al.*, this issue) will make multi-point measurements of these accelerated particles and allow to derive the characteristics of the acceleration process. The central plasma sheet is the location of the cross-tail current flowing from dawn to dusk. Sometimes a disruption of this current occurs and part of it is directed toward the ionosphere. This disruption initiates in the near-Earth tail and then propagates further downstream at high speed (Jacquey *et al.*, 1993). Furthermore these disruptions seem to be associated with substorms, one of the most intriguing phenomena in the magnetosphere. Cluster will be able to sense remotely the beginning of the disruption when the spacecraft are in the lobes and then observe the consequences of the disruption further downstream in the plasma sheet.

The auroral zone is a ring of light emission, around the magnetic pole, created by the precipitation of particles in the atmosphere. The cusp and boundary layers on the dayside and the plasma sheet and plasma sheet boundary layer on the nightside are the sources of these precipitations. Transient plasma flows and particle precipitation into the low-altitude cusp are of great interest, as they appear to be produced by phenomena occurring on the magnetopause, like Flux Transfer Events (FTEs) (Lockwood and Smith, 1989) or solar wind dynamic pressure variations (Sibeck and Croley, 1991). The spatial scales and time variations of these transients are very important to distinguish between these two mechanisms. The four Cluster spacecraft will cross the cusp and the auroral field lines between 4 and 6 R_E like a 'string of pearls' with a time delay between the spacecraft ranging from a few minutes to up to 40 min, thus allowing the time variations of the transient events to be studied. The ionosphere is now believed to be an important source of plasma for the magnetosphere and in particular for the magnetotail (Moore and

Table I
Instruments on cluster

Acronym	Instrument	Principal investigator
FGM	Fluxgate magnetometer	A. Balogh (IC, U.K.)
STAFF[a]	Spatio-temporal analysis of Field fluctuation experiment	N. Cornilleau-Wehrlin (CETP, France)
EFW[a]	Electric field and wave experiment	G. Gustafsson (IRFU, Sweden)
WHISPER[a]	Waves of high frequency and sounder for probing of electron density by relaxation	P. M. E. Décréau (LPCE, France)
WBD[a]	Wide band data	D. A. Gurnett (Iowa U., U.S.A.)
DWP[a]	Digital wave processing experiment	L. J. C. Woolliscrooft[b], H. Alleyne (Sheffield U., U.K.)
EDI	Electron drift instrument	G. Paschmann (MPE, Germany)
ASPOC	Active spacecraft potential Control	W. Riedler (IWF, Austria)
CIS	Cluster Ion spectrometry	H. Rème (CESR, France)
PEACE	Plasma electron and current Experiment	A. D. Johnstone (MSSL, U.K.)
RAPID	Research with adaptive particle Imaging detectors	B. Wilken (MPA, Germany)

[a] Members of the wave experiment consortium (WEC).
[b] We regret to announce the death of Les Woolliscroft, the DWP Principal Investigator. His successor is Hugo Alleyne.

Delcourt, 1995). The polar wind, the cleft ion fountain, and the nightside auroral zone contribute to the filling of the magnetosphere with plasma. However, their exact contribution may have been underestimated due to the limited measurements at low energy (<10 eV) (Chappell *et al.*, 1987). Cluster should help to refine these results with ion measurements (with distinction of mass) down to around 2 eV.

3. Cluster Instrumentation

The Cluster instrumentation has state-of-the-art capability to measure the electric and magnetic fields and electron and ion distribution functions. The four Cluster spacecraft are identical and each contain 11 instruments, giving a total of 44 instruments built by the Principal Investigators (Table I). The location of the 11 instruments on the spacecraft is show in Figure 2.

The fluxgate magnetometer FGM (Figure 2(b) number 1) is designed to provide inter-calibrated measurements of the magnetic-field vector at the location of the four Cluster spacecraft. The instrument consists of two tri-axial fluxgate sensors and a failure-tolerant data-processing unit. The combined analysis of the data from the four spacecraft will yield parameters such as the current density vector, wave vectors and the geometry and structure of discontinuities. FGM provides to all other instruments, except ASPOC, the magnetic-field vector and the event memory trigger which is the time when a given scientific event (e.g., crossing of the magnetopause) is occurring.

A suite of five instruments, EFW, STAFF, WHISPER, WBD, and DWP forms the Wave Experiment Consortium (WEC) (Pedersen *et al.*, 1995).

The Electric Field and Wave (EFW) experiment (Figure 2(a) number 2) has been specifically designed to study the fast time- and space-varying vectorial electric fields including the DC to low-frequency range. In addition, the instrument will measure the spacecraft potential and the electron density and temperature when operated in Langmuir mode. EFW consists of four spherical sensors orthogonally deployed with 50 m cables in the spin plane of the spacecraft. The dynamic range of the instrument spans over more than four decades which allows the measurement of electrostatic shocks and double layers up to 700 mV m^{-1} as well as large-scale convection electric fields down to 0.1 mV m^{-1}.

The Spatio-Temporal Analysis of Field Fluctuation (STAFF) experiment (Figure 2(a) number 1) has two main parts: the search coil magnetometer which measures the magnetic component of the electromagnetic fluctuations and the spectrum analyser which performs auto- and cross-correlation between electric and magnetic components. Such measurements will determine the shape, current density and motion of small-scale current structures and identify the source of plasma waves and turbulence. To this end, STAFF covers the frequency range up to 4 kHz.

The Waves of High frequency and Sounder for Probing of Electron density by Relaxation (WHISPER) (Figure 2(a) number 4) is an intermittent transmitter/receiver instrument that can also be operated in a passive (receive-only) mode. The transmitter emits a short pulse to stimulate plasma resonances. After each emission, the receiver part is activated to detect plasma echoes in the frequency range 4–80 kHz. Echoes at characteristic frequencies let the investigators determine the plasma density in the range 0.2–80 cm^{-3}. It should be pointed out that the measurement of the plasma density with this technique is not influenced by the spacecraft potential fluctuations.

The objective of the Wide Band Data (WBD) receiver (Figure 2(a) number 5) system is to provide high resolution electric field waveforms and frequency-time spectrograms of terrestrial plasma waves and radio emissions. WBD will also perform unique measurements which are the Very-Long-Baseline-Interferometer (VLBI) measurements with up to four spacecraft. VLBI will give new information on the angular size and motion of the sources of terrestrial radio emissions such as auroral kilometric radiation. The electric field measured by WBD is in the range 25 Hz–577 kHz and the magnetic field in the range 25 Hz–4 kHz.

The Digital Wave Processing (DWP) instrument (Figure 2(a) number 3) coordinates WEC measurements and performs particle correlations. These correlations consist of auto-correlation functions of the time series of particle detector counts

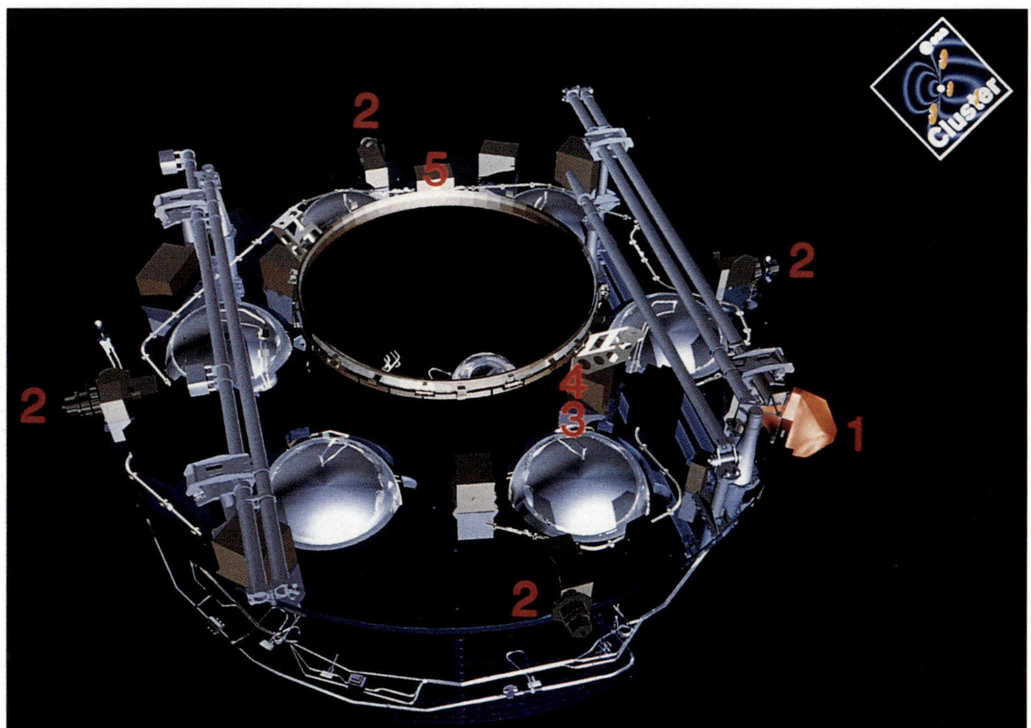

Figure 2a. Cluster instruments accommodation on the spacecraft. 1 STAFF, 2 EFW, 3 DWP, 4 WHISPER, 5 WBD.

as a function of energy and pitch angle. A hardwired link with PEACE provides the electron count measurements. These correlation functions will allow the investigation of nonlinear wave-particle interactions which are believed to be the source process of many plasma transport mechanisms.

The five experiment that together form the WEC consortium provide another dimension to the scientific return. DWP is the 'heart' of the WEC's on-board data-processing. Its tasks are to perform event selection, control the modes, compact/compress data for WHISPER and STAFF. Furthermore, the interconnection between EFW and STAFF permits time- and phase-coordinated measurements that will help to both localise and characterise electrostatic (e.g., double layers) and electromagnetic structures (e.g., field-aligned current).

Cluster carries another instrument to measure electric fields. The Electron Drift Instrument (EDI) (Figure 2(b) number 2) is based on the emission and subsequent detection of tracer electrons to derive an ambient electric field. The instrument consists of two sets of electron guns/detectors at 180° to each other. EDI employs two different methods to measure the electric field. In the case of strong ambient magnetic fields, the displacement of the electron over one gyration can be measured by a triangulation method using the directions of emission of the two electron

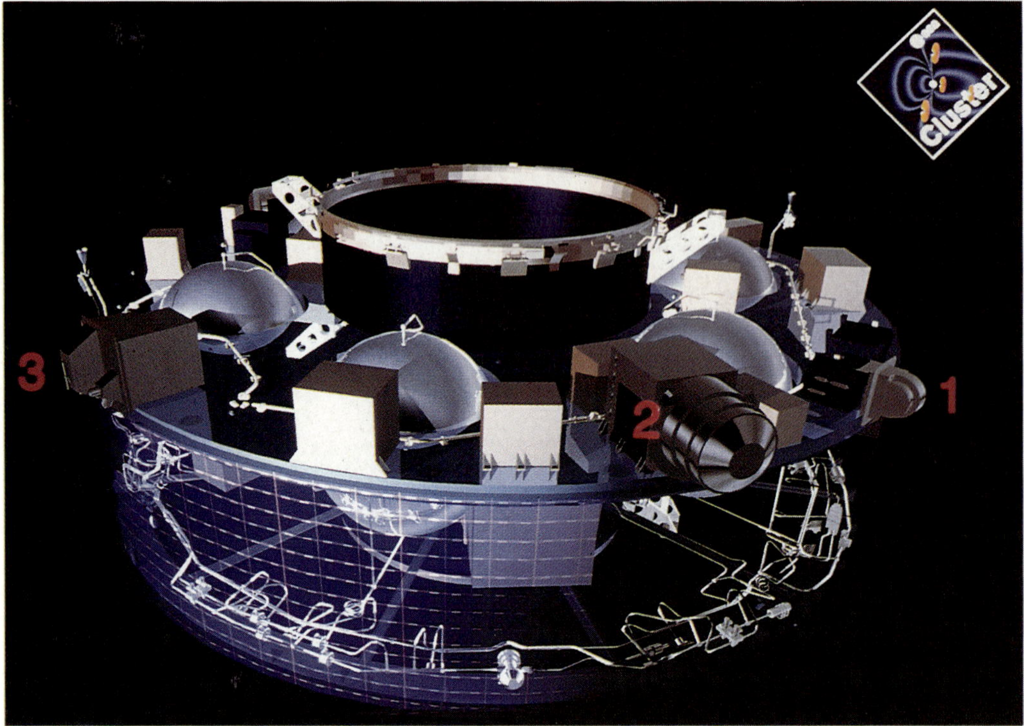

Figure 2b. Cluster instruments accommodation on the spacecraft. 1 FGM, 2 EDI, 3 ASPOC.

beams. When the ambient magnetic field is small, the instrument works in a mode whereby two beams are emitted in opposite directions and the time-of-flight is then measured. In addition, by varying the electron energy, the instrument can determine gradients in the local magnetic field.

The accurate measurement of the cold plasma population demands that the electrostatic potential of the spacecraft with respect to the ambient plasma be maintained at a very low level. Cluster will be equipped with an ion emitter, the Active Spacecraft Potential Control (ASPOC) experiment (Figure 2(b) number 3) to routinely control this potential. ASPOC is specifically designed to stabilise the fluctuating potential by emission of indium ions with an energy between 5 and 8 keV and a total current up to 50 μA. A hardwire link with PEACE and EFW provides the measured surface potential against which the emission is controlled. ASPOC and WEC will also undertake investigations of the interaction between a weak ion beam and the ambient plasma.

The Cluster Ion Spectrometry (CIS) experiment (Figure 2(c) number 2) employs two sensors to obtain the full three-dimensional ion distribution of the major species with high time resolution and mass per charge plasma composition. One sensor, the time-of-flight ion Composition and Distribution Function analyser (CODIF) will measure the distribution of the major ion species from $\sim$ 0 to 40 keV q^{-1} with

Figure 2c. Cluster instruments accommodation on the spacecraft. 1 PEACE, 2 CIS, 3 RAPID.

an angular resolution of 22.5° × 10.25° and two different sensitivities. CODIF also uses a retarding potential analyser to make more accurate measurements below 15 eV q^{-1}. The other sensor, the Hot Ion Analyser (HIA), will measure the distribution of the ions without distinction of mass from ~ 5 eV q^{-1} to 32 keV q^{-1} with an angular resolution of 5.6° × 5.6° and two different sensitivities. HIA is specifically designed for the highly directional ion beams observed in the solar wind. A specific feature of CIS is the double sensitivity of the sensors which will allow the precise measurement of the large flux of ion beams in the solar wind as well as the low flux of ions in the lobes of the magnetosphere.

The plasma electron measurements will be performed by the Plasma Electron and Current Experiment (PEACE) (Figure 2(c) number 1). The instrument consists of two sensors: the Low Energy Electron Analyser (LEEA) and the High Energy Electron Analyser (HEEA). The detection of cold electrons requires a very careful design to eliminate spurious effects introduced by the photo-electrons which are known to be abundant near the spacecraft skin. LEEA is designed to measure the low-energy electrons from 0.7 to 10 eV but is also capable of covering the full energy range up to 30 keV. HEEA, with a geometric factor five times higher than LEEA, covers the full energy range from 0.7 eV to 30 keV. The two sensors are mounted opposite to each other on the spacecraft which allows the three-dimensional distribution function to be measured every half of spacecraft spin in the energy part common to both sensors.

Energetic particles are sensitive probes for remote-sensing techniques in nearly all plasma regions of geospace. These particles help to identify distant acceleration regions and they can be used to trace plasma flows and magnetic field line topologies. The Research with Adaptative Particle Imaging Detectors (RAPID) (Figure 2(c) number 3) consists of two spectrometers, each containing position-sensitive solid-state detectors: the Imaging Ion Mass Spectrometer (IIMS) which measures the ion distribution function from 30 to 1500 keV q^{-1} with distinction of mass and the Imaging Electron Spectrometer (IES) which measures the electron from 20 to 450 keV. A special mode of RAPID allows also the measurement of the energetic neutral atoms.

The payload can be commanded to generate alternately two bit streams: 16.8 (normal mode) and 107 kbit s^{-1} (burst mode). If the spacecraft is in real-time contact with the ground upon entering one of the regions of scientific interest, then the science data are directly downlinked to the station; otherwise all data are stored, on all four spacecraft, in one of the two on-board solid-state recorders of the capacity of 2.25 Gbit each. This large capacity and the ground segment system consisting of two ground stations allows a coverage of the full orbit under normal mode or around 50% of the orbit if the burst mode is used for 1.5 hour together with 27 hours of normal mode. As burst mode is necessary for some experiments, for example the particle experiments, to get the full three-dimensional distribution function every spin, a well-balanced distribution between normal and burst mode has been established. A Master Science Plan, which has been produced by the Science Operation Working Group, defines the data-taking and the instrument modes for all orbits ($\sim$ 144) during the first year of the mission. All regions of scientific interest are covered many times in both normal and burst modes and the average data coverage for the first year is around 47% per orbit. The second year of operations will be defined in the light of experience gained during the first year.

The spacecraft commanding involves the Principal Investigators which select the modes of operation of their respective instruments, the Joint Science Operations Centre (JSOC) (see Hapgood *et al.*, this issue) which coordinates the command schedule and the mission planning system at the European Space Operations Centre (ESOC) (see Ferri and Warhaut, this issue) which verify and upload the commands to the spacecraft. The telemetry is acquired and processed by ESOC and the science data are distributed to the PIs via both an on-line data disposition system for quick access and limited volume and later on CDroms containing the full data base (Soerensen *et al.*, this issue). Selected physical parameters are then processed and made available to the Cluster scientific community and to the scientific community in general by the Cluster Science Data System (CSDS) (Schmidt *et al.*, this issue).

The desire for highly accurate measurements imposes requirements on the spacecraft design concerning attitude reconstitution, electromagnetic cleanliness and timing accuracy ($\pm$2 ms) of the data acquisition. In addition, to derive three-dimensional quantities like the current density, accurately, the separation distances between the satellites should be measured with high precision: the limit in the accuracy is 10 km for separations of less than 1000 km and 1% when separations are greater than 1000 km.

4. Orbit and Separation Strategy

In order to meet the scientific objectives of the mission, the orbit was chosen with a perigee at 4 R_E, an apogee at 19.6 R_E, an inclination of 90° and a line of apsides around the ecliptic plane. The orbital period is 57 hours. After launch the spacecraft will take around three weeks to reach their final orbit after many manoeuvres (see Credland *et al.*, this issue) and then the commissioning and verification phase will take about 2.5 months. Therefore the mission operation phase will start around 3 months after launch. At that time, the orbit will initially be in the dawn-dusk meridian with the apogee at dusk (Figure 3). Then, as the Earth is rotating around the Sun, the apogee will be in the solar wind around local noon after 3 months, around the flank of the magnetosphere at dawn after 6 months, in the plasma sheet around midnight local time after 9 months and then again on the flank of the magnetosphere at dusk one year later. The same scenario will be repeated for the second year of Cluster operations.

The Cluster orbit will cross all regions of scientific interest during the course of the mission. Figures 4(a) and 4(b) show the orbit when the apogee is around local noon and local midnight, respectively. When the apogee is around local noon, the regions successively crossed are the nightside auroral zone, the northern cusp, the magnetopause, the bow-shock, the solar wind and then again the same regions in the southern hemisphere. Note that the spacecraft configuration has been enlarged by a factor of 30 in order to be visible on the figure. Special emphasis in terms of separation has been put on the northern cusp and the southern magnetopause/bow shock. In these two areas a perfect tetrahedron will be constituted. An advantage of this scenario is that the configuration stays quite close to a tetrahedron throughout the magnetosheath and solar wind. Near perigee, the configuration becomes elongated and the spacecraft will cross the auroral zone as a ‘string-of-pearls’. The inter-spacecraft distance has been fixed at around 600 km for the cusp crossing during the first year (Table II). For the second year, the separation distance is planned to be between 200 and 2000 km, however the exact distance will be fixed around six months before the event. When apogee is in the tail (Figure 4(b)), the regions crossed by Cluster will be the mid-altitude cusp (between 4 and 6 R_E), the polar cap and tail lobes and the plasma sheet. The spacecraft will form a tetrahedron in the plasma sheet. Note that the spacecraft configuration has been enlarged by a factor of 5 in Figure 4(b). The separation distance in the plasma sheet is planned to be between 2000 and 5000 km (exact distance to be fixed 6 months before). For the second year of operations the separation distance will be modified with a

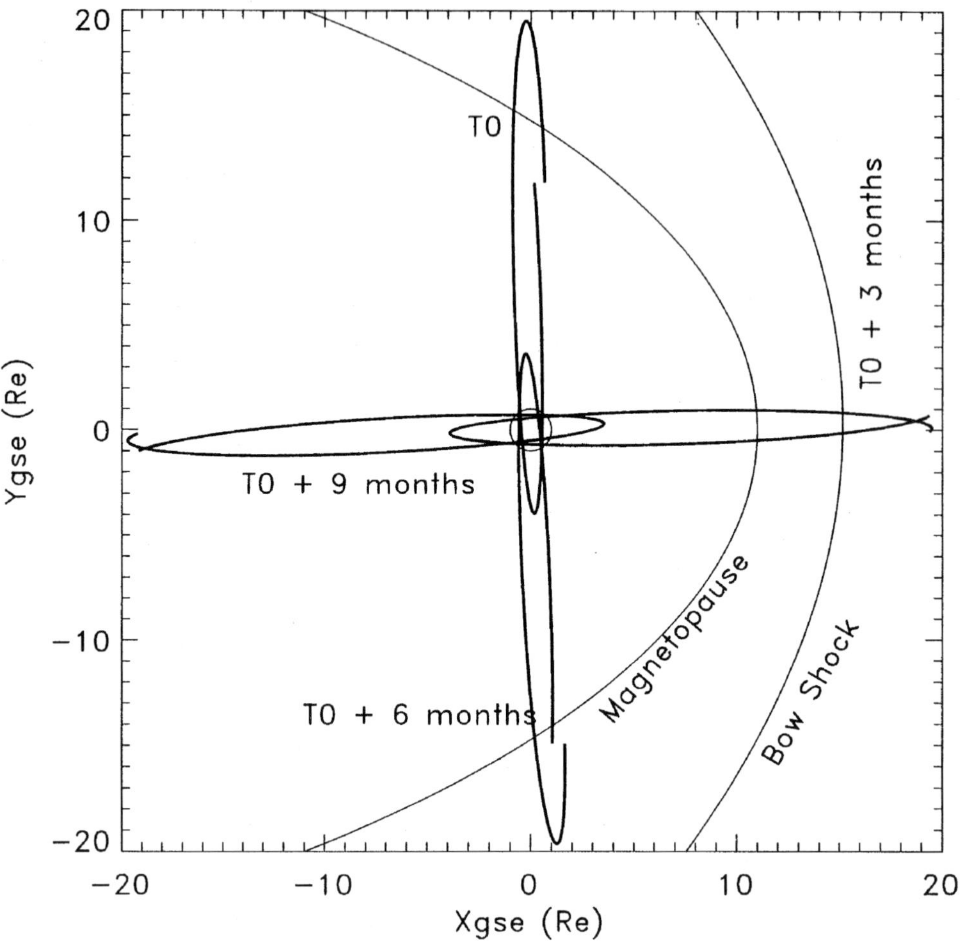

Figure 3. Orbits of Cluster spacecraft at three month intervals in the GSE equatorial plane. The operational phase starts when the apogee is at dusk (T0) and then the orbital plane is rotating with respect to the GSE Z axis. The complete rotation takes one year.

possible separation distance in the tail of up to 3 R_E. However this distance will be fixed once experience has been gained during the first year and also once the fuel reserves are known.

The four Cluster will cross the mid-altitude cusp successively like a 'string-of-pearls'. The footprint of the field lines where the satellites will be located is shown in Figure 4(c). The spacecraft move toward the pole in the same meridional plane. They will cross successively the cusp around 75° of latitude with a delay varying from a few min up to 40 min. Note that 6 months later, the spacecraft will cross the nightside auroral zone with a time delay from 30 s up to 5 min. These delays are shorter than the delay for the mid-altitude cusp crossings since the separation distance at apogee is also smaller (600 km against 1 R_E).

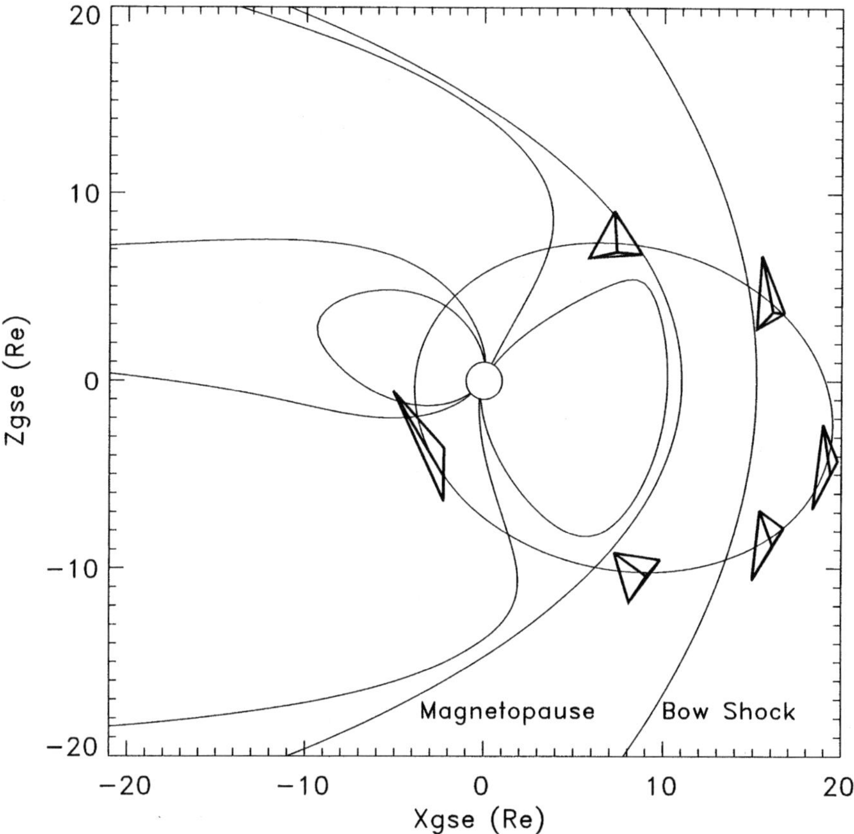

Figure 4a-c. Cluster orbit and tetrahedron configuration in the GSE $X - Z$ plane when the apogee is in the solar wind (a) and in the magnetotail (b). The inter-spacecraft distances have been enlarged by a factor of 30 in (a) and by a factor of 5 in (b). The magnetic field model is from Tsyganenko (1987) with $Kp = 2$, the magnetopause from Sibeck *et al.* (1991) with solar wind pressure equal to 2 nPa and the bow shock from Peredo *et al.* (1995) with the Alfvén mach number in the range (2–20). (c) shows the footprint of the field lines crossed by the four Cluster in the mid-altitude cusp region (near perigee on (b)) in a diagram corrected geomagnetic latitude (CMLAT) – magnetic local time (MLT).

Table II
Separation distances for primary target

Phase	Primary target	Separation	Comment
1	cusp	$600 \pm 20\%$ km	
2	tail	2000–5000 km	target value to be specified later
3	cusp	200–2000 km	target value to be specified after 1st cusp experience
4	tail	1–3	target value to be specified once fuel reserves known

Zgse (Re)
Xgse (Re)
Magnetopause
Bow Shock
20
10
0
−10
−20
−20
−10
0
10
20

12H CMLT
C4
C1
10 min
C2
C3
18H
06H
60°
70°
80°
90° CMLAT

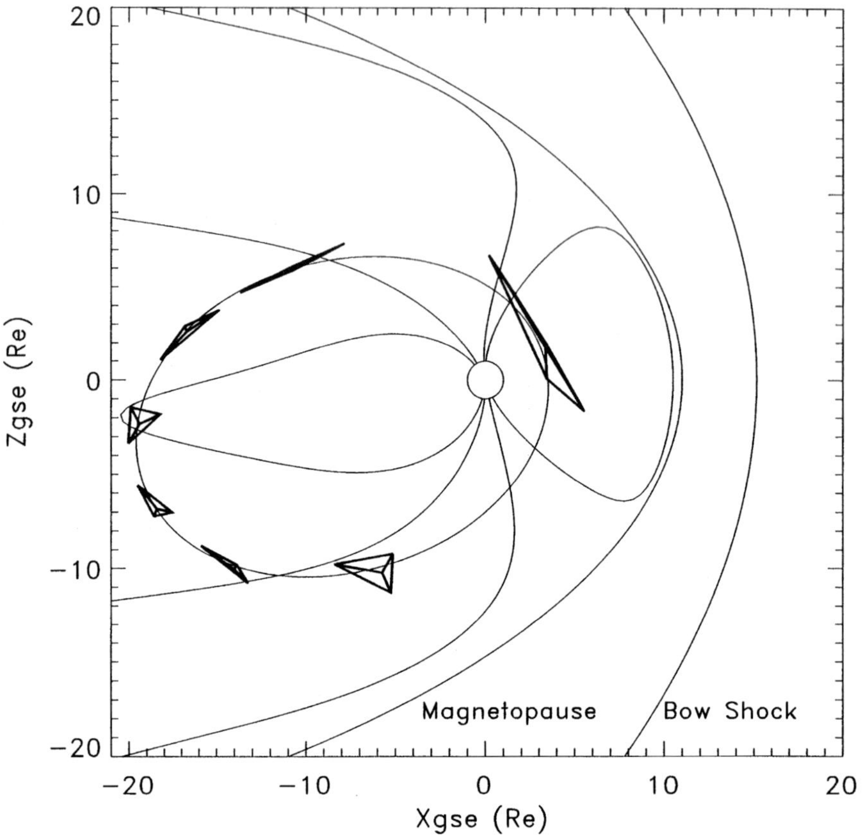

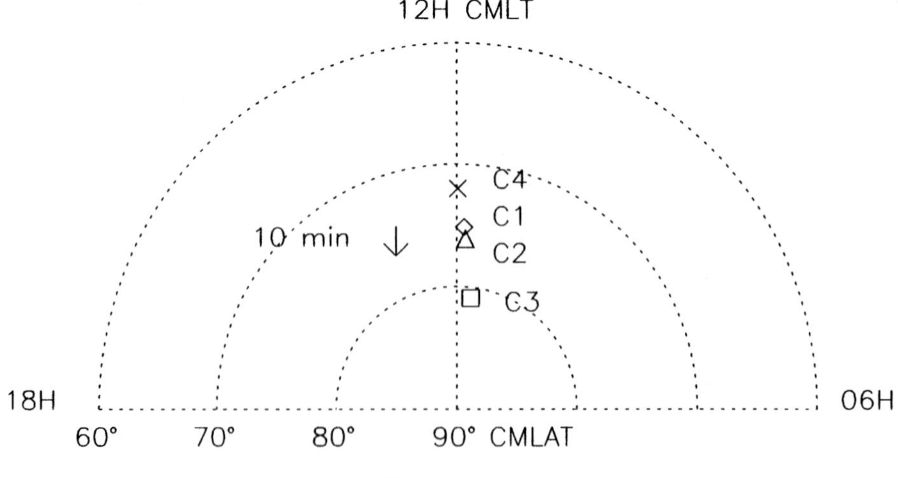

Figure 4b-c.

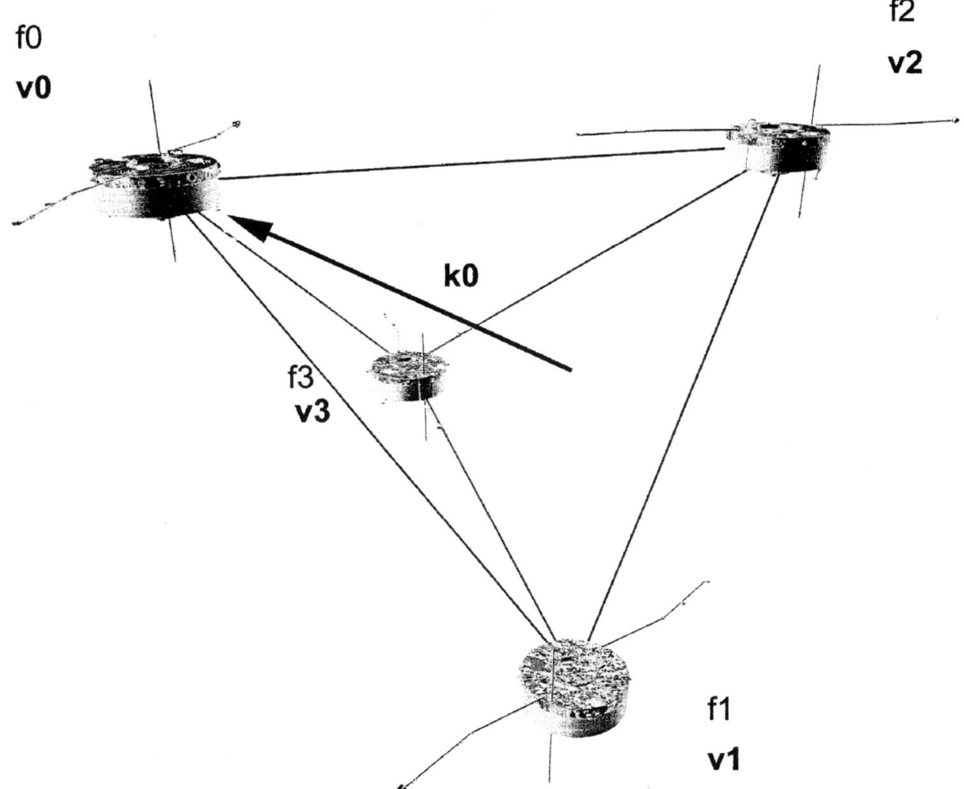

Figure 5. The four Cluster spacecraft in a tetrahedron configuration. The vector **v** and the scalar f is measured simultaneously at each spacecraft. The vector **k** is the reciprocal vector, joining one spacecraft to the plane defined by the three other (**k** being perpendicular to that plane).

5. Specific Capabilities of Cluster

Cluster, with four spacecraft in a tetrahedron configuration, is the first mission to study the small-scale space plasma structures in three dimensions (Figure 5). The main studies which will be performed using four spacecraft and which have already been tested using simulations or empirical models are the curlometer technique, the discontinuity analysis and the k-filtering technique.

A simple method to compute the gradient of any physical quantity which can be a scalar or a vector has been described in Chanteur and Mottez (1993). It should be noted that this method assumes that the variation of the measured quantity between two measurement points (two spacecraft) is linear. To summarise the method: if we have a measurement of the vector $\mathbf{v}_\alpha$ or of the scalar f_α ($\alpha = 0, 1, 2, 3$) performed at each vertices α of the tetrahedron (Figure 5), the differential operators of fields $\mathbf{V}$ and F can be calculated with the following equations:

$$\mathbf{grad}\ F = \Sigma\ \mathbf{k}_\alpha\ f_\alpha \qquad (1)$$

$$\text{div}\,\mathbf{V} = \Sigma\,\mathbf{k}_\alpha \cdot \mathbf{v}_\alpha \tag{2}$$

$$\mathbf{curl}\,\mathbf{V} = \Sigma\,\mathbf{k}_\alpha \times \mathbf{v}_\alpha \tag{3}$$

where $\mathbf{k}_\alpha$ are the reciprocal vectors of the tetrahedron given by

$$\mathbf{k}_0 = \frac{\mathbf{r}_{12} \times \mathbf{r}_{13}}{\mathbf{r}_{10}(\mathbf{r}_{12} \times \mathbf{r}_{13})} \quad \mathbf{k}_1 = \frac{\mathbf{r}_{20} \times \mathbf{r}_{23}}{\mathbf{r}_{21}(\mathbf{r}_{20} \times \mathbf{r}_{23})},$$

$$\mathbf{k}_2 = \frac{\mathbf{r}_{13} \times \mathbf{r}_{10}}{\mathbf{r}_{12}(\mathbf{r}_{13} \times \mathbf{r}_{10})} \quad \mathbf{k}_3 = \frac{\mathbf{r}_{21} \times \mathbf{r}_{20}}{\mathbf{r}_{23}(\mathbf{r}_{21} \times \mathbf{r}_{20})},$$

where $\mathbf{r}_{12} = \mathbf{r}_2 - \mathbf{r}_1$ is the vector joining spacecraft 1 to spacecraft 2. Note that these equations are also valid for irregular tetrahedra.

This formalism can be applied to calculate the current density **J** flowing through the Cluster area and given by the Ampère's law $\mathbf{curl}\,\mathbf{B} = \mu_0\,\mathbf{J}$. In practice, using the vector of the magnetic field **B** measured at the four Cluster spacecraft and Equation (3), the current density **J** can be calculated. In addition Equation (2) can be used to check that the Maxwell's equation $\text{div}\,\mathbf{B} = 0$ is verified. Coeur-Joly *et al.* (1995) calculate curl **B** and div **B** along the orbit of Cluster spacecraft using the Tsyganenko magnetic field model. Their results showed that the current density was very accurately estimated all along the orbit except at perigee where the deformation of the tetrahedron is too large (the spacecraft are almost on the same line, see Figure 4(b)). It should be noted that the curlometer technique can only be applied when the spacecraft separation distances are smaller than the current sheet size (Balogh *et al.*, this issue). When the separation distances are larger than the current sheet size, e.g., the magnetopause, then the discontinuity analysis can be applied.

Other quantities can be derived from this formalism, for instance the vorticity of the flow $\boldsymbol{\Omega} = \text{curl}\,\mathbf{v}$ where $\mathbf{v}$ is the plasma velocity measured at the four spacecraft. It is important to note that the accuracy of the measured plasma velocity depends strongly on the density and temperature of the plasma (Martz *et al.*, 1993), therefore before calculating $\mathbf{curl}\,\mathbf{v}$ the accuracy of the measurement of $\mathbf{v}$ should be verified.

The discontinuity analysis investigates the local geometry of a surface crossed by Cluster and determines its velocity in the frame of reference of Cluster (Mottez and Chanteur, 1994). This method uses the magnetic field measurements and the minimum variance analysis to calculate the vector normal to the discontinuity at the four Cluster crossings. Then, the difference between the normal vectors at the four positions of the surface is used to derive the curvature of the surface. This method also provides the velocity of the surface with respect to Cluster. Note the assumption that the surface does not deform or rotate during the time interval needed for the four Cluster to cross the surface. This method should help to distinguish between the back-and-forth motion of the magnetopause and a surface wave propagation on its surface. In addition, the four Cluster spacecraft will also permit the investigation of non-planar structures like flux transfer events or field line resonance (Dunlop *et al.*, 1995).

Finally, the k-filtering technique uses magnetic field vector **B** and the electric field vector **E** to identify the various three-dimensional electromagnetic structures in the wave field (Pinçon, 1995). This technique is valid if two assumptions are met: firstly, the electromagnetic fields must be homogeneous in space and stationary in time and secondly, the characteristic length of the field energy should be larger than the mean inter-spacecraft distance. The k-filtering technique has been applied to real and simulated data and the results show that all initial wave modes can very often be retrieved.

6. Cluster and the Other Missions

Cluster, together with SOHO, forms the first 'Cornerstone' of ESA's Horizon 2000 programme. These two missions were selected at the same time to investigate the relation between the Sun and the Earth's environment. SOHO, located at the Lagrangian point at around 240 R_E from the Earth, successfully launched in December 1995, will be recording continuously the activity of the Sun and the flux of high-energy particles in the solar wind (Fleck *et al.*, 1995). At the same time Cluster, much closer to the Earth, between 4 and 19.6 R_E, will record the effect of the Sun and the solar wind on the Earth's magnetosphere.

Cluster is also part of an international activity organised by the Inter-Agency Consultative Group (IACG) consisting of ESA, ISAS (Japan), NASA and the Russian Space Agency. During an IACG workshop held in Graz (Austria) in April 1993, various campaigns were defined to investigate the role of current sheets in space plasmas. Cluster and SOHO, together with the NASA Global Geospace Science (GGS) spacecraft Wind and Polar, the NASA/ISAS spacecraft Geotail, the ISAS Yohkoh spacecraft, the two Russian Interball spacecraft and the German spacecraft Equator-S, form the core mission of the campaigns. As soon as Cluster is launched, its orbit will be analysed in conjunction with the other spacecraft and the campaigns will be planned in terms of time of data acquisition and eventually in terms of modes of operation.

7. Conclusions

The ESA Cluster mission consists of four identical spacecraft which will study in three dimensions the small-scale plasma structures in the near-Earth environment. During the two-year mission, the spacecraft will observe the key regions of scientific interest, namely the solar wind, bow shock, magnetopause, cusps, plasma sheet and auroral acceleration region. The Cluster instrumentation is composed of 11 experiments, on each of the four spacecraft, to measure with very high accuracy the electric and magnetic field, the ion and electron distribution function and the electromagnetic waves. The inter-spacecraft separation will be varied from 200 km up to 18 000 km with respect to the plasma phenomena being studied. Furthermore, the four-point measurements in space will allow for the first time to derive differential plasma quantities like the current density flowing around the spacecraft.

This volume is composed of a description of the 11 experiments by the investigators. This is followed by an overview of the spacecraft by Credland *et al.* Then the mission operations and ground segment system are described by Ferri and Warhaut and the data distribution by Sorensen *et al.*. Furthermore the volume has a review of the Cluster Science Data System (CSDS) by Schmidt *et al.* and the Joint Science Operation Centre (JSOC) by Hapgood *et al.* The theoretical studies in supprt of the Cluster mission are described by Chanteur and Roux and finally, the Ground-Based investigations are reviewed by Opgenoorth and Lockwood. Finally, a directory of the Cluster community members is included as an Annex.

Acknowledgements

We gratefully acknowledge the Cluster Project Manager J. Credland and his team and the Cluster Ground Segment Manager M. Warhaut and his team for their unfailling support. We thank G. Paschmann as chairman of the Science Operation Working Group and M. A. Hapgood, P. Ferri, J. Rodriguez-Canabal and J. Schoenmaekers for their support. We would like to acknowledge M. Ashour-Abdalla, J. Berchem and J. Raeder for providing the MHD model of the magnetosphere.

References

Balogh, A. *et al.*: 1995, 'The Cluster Magnetic Field Investigation', *Space Sci. Rev.*, this issue.

Berchem, J., Raeder, J., and Ashour-Abdalla, M.: 1995, 'Magnetic Flux Ropes at the High-Latitude Magnetopause', *Geophys. Res. Letters* **22**, 1189.

Chanteur, G. and Mottez, F.: 1993, 'Geometrical Tools for Cluster Data Analysis', *Proceedings of the International Conference on Spatio-Temporal Analysis for Resolving Plasma Turbulence (START)*, ESA WPP-07, p. 341.

Chanteur, G. and Roux, A.: 1995, 'European Network for the Numerical Simulation of Space Plasmas', *Space Sci. Rev.*, this issue.

Chappell, C. R., Moore, T. E., and Waite, J. H.: 1987, 'The Ionosphere as a Fully Adequate Source of Plasma for the Earth's Magnetosphere', *J. Geophys. Res.* **92**, 5896.

Coeur-Joly, O., Robert, P., Chanteur, G., and Roux, A.: 1995, 'Simulated Daily Summaries of Cluster Four-Point Magnetic Field Measurements', *Proceedings of the Cluster Workshop on Physical Measurements and Mission-Oriented Theory*, ESA SP-371, p. 223.

Cornilleau-Wehrlin, N. *et al.*: 1995, 'The Cluster Spatio-Temporal Analysis of Field Fluctuations (STAFF) Experiment', *Space Sci. Rev.*, this issue.

Credland, J. et al.: 1995, 'The Cluster Mission: ESA's Spacefleet to the Magnetosphere', *Space Sci. Rev.*, this issue.

Décréau, P. M. E. *et al.*: 1995, 'WHISPER, A Resonance Sounder and Wave Analyser: Performances and Perspectives for the Cluster Mission', *Space Sci. Rev.*, this issue.

Dunlop, M. W., Woodward, T. I., Motschmann, U., Glassmeier, K.-H., Southwood, D. J., and Balogh, A.: 1995, 'Analysis of Non-Planar Structures with Multipoint Techniques: A Roadmap for Cluster', *Proceedings of the Cluster Workshop on Physical Measurements and Mission Oriented Theory*, ESA SP-371, p. 267.

Ferri, P., and Warhaut, M.: 1995, 'Cluster Mission Operations', *Space Sci. Rev.*, this issue.

Fleck, B., Domingo, V., and Polland A. I. (eds.): 1995, *Solar Phys.* **162**.

Gurnett, D. A. *et al.*: 1995, 'The Wide-Band Plasma Wave Investigation', *Space Sci. Rev.*, this issue.

Gustafsson, G. *et al.*: 1995, 'The Electric Field and Wave Experiment for the Cluster Mission', *Space Sci. Rev.*, this issue.

Haerendel, G., Paschmann, G., Scokpke, N., Rosenbauer, H., and Hedgecock, P. C.: 1978, 'The Frontside Boundary Layer of the Magnetosphere and Problem of Reconnection', *J. Geophys. Res.* **83**, 3195.

Hapgood, M. A. *et al.*: 1995, 'The Joint Science Operations Centre', *Space Sci. Rev.*, this issue.

Jacquey, C., Sauvaud J. A., and Dandouras, J.: 1993, 'Tailward Propagating Tail Current Disruption and Dynamics of the Near-Earth Tail: a Multi-Point Measurement Analysis', *Geophys. Res. Letters* **20**, 983.

Johnstone, A. D. *et al.*: 1995, 'PEACE: A Plasma Electron and Current Experiment', *Space Sci. Rev.*, this issue.

Lockwood, M. and Smith, M. F.: 1989, 'Low-Altitude Signature of the Cusp and Flux Transfer Events', *Geophys. Res. Letters* **16**, 879.

Martz, C., Sauvaud, J.-A., and Rème H.: 1993, 'Accuracy on Ion Distribution Measurements and Related Parameters Using the Cluster CIS Experiment', *Proceedings of the International Conference on Spatio-Temporal Analysis for Resolving Plasma Turbulence (START)*, ESA WPP-07, p. 229.

Mattok, C. (ed.): 1995, *Proceedings of the Cluster Workshop on Physical Measurements and Mission Oriented Theory*, ESA SP-371.

Moore, T. E. and Delcourt, D. C.: 1995, 'The Geopause', *Rev. Geophys.* **33**, 175.

Mottez, F. and Chanteur, G.: 1994, 'Surface Crossing by a Group of Satellites: A Theoretical Study', *J. Geophys. Res.* **99**, 13 499.

Onsager, T. G., Thomsen, M. F., Gosling, J. T., and Bame, S. J.: 1990, 'Observational Test of a Hot-Flow Anomaly Formation Mechanism', *J. Geophys. Res.* **95**, 11 967.

Opgenoorth, H. J. and Lockwood, M.: 1995, 'Opportunities for Magnetospheric Research with Coordinated Cluster and Ground-Based Observations', *Space Sci. Rev.*, this issue.

Paschmann, G., Haerendel, G., Scokpke, N., Möbius, E., Lühr, H., and Carlson, C. W.: 1988, 'Three-Dimensional Plasma Structures with Anomalous Flow Directions Near the Earth's Bow Shock', *J. Geophys. Res.* **93**, 11 279.

Paschmann, G. *et al.*: 1995, 'The Electron Drift Instrument for Cluster', *Space Sci. Rev.*, this issue.

Pedersen, A. *et al.*: 1995, 'The Wave Experiment Consortium (WEC)', *Space Sci. Rev.*, this issue.

Peredo, M., Slavin, J. A., Mazur, E., and Curis, S. A.: 1995, 'Three-Dimensional Position and Shape of the Bow Shock and Their Variation with Alfvénic, Sonic and Magnetosonic Mach Numbers and Interplanetary Magnetic Field Orientation', *J. Geophys. Res.* **100**, 7907.

Pinçon, J. L.: 1995, 'Cluster and the k-Filtering', *Proceedings of the Cluster Workshop on Physical Measurements and Mission Oriented Theory*, ESA SP-371, p. 87.

Rème, H. *et al.*: 1995, 'The Cluster Ion Spectrometry (CIS) Experiment', *Space Sci. Rev.*, this issue.

Riedler, W. *et al.*: 1995, 'Active Spacecraft Potential Control', *Space Sci. Rev.*, this issue.

Rodriguez-Canabal, J., Warhaut, M., Schmidt, R., and Bello-Mora, M.: 1993, 'The Cluster Orbit and Mission Scenario', *Cluster: Mission, Payload and Supporting Activities*, ESA SP-371, p. 259.

Schmidt, R., Arends, H., Pedersen, A., Rudenauer, F., Fehringer, M., Narheim, B. T., Svenes, R., Kvernsveen, K., Tsuruda, K., Mukai, T., Hayakawa, H., and Nakamura, M.: 1995, 'Results from Active Spacecraft Potential Control on the Geotail Spacecraft', *J. Geophys. Res.* **100**, 17 253.

Schmidt, R. *et al.*: 1995, 'The Cluster Science Data System (CSDS) – a New Approach to the Distribution of Scientific Data', *Space Sci. Rev.*, this issue.

Schwartz, S. J., Chaloner, C. P., Hall, D. S., Johnstone, A. D., Rijnbeek, R. P., Southwood, D. J., and Woolliscroft, L. J. C.: 1985, 'An Active Current Sheet in the Solar Wind', *Nature* **318**, 269.

Sckopke, N., Paschmann, G., Brinca, A. L., Carlson, C. W., and Lühr, H.: 1990, 'Ion Thermalization in Quasi-Perpendicular Shocks Involving Reflected Ions', *J. Geophys. Res.* **95**, 6337.

Shea, M. A. (ed.): 1991, 'U.S. National Report to International Union of Geodesy and Geophysics 1987–1990', *Rev. Geophys. Suppl.*

Sibeck, D. G. and Croley, D. J.: 1991, 'Solar Wind Dynamic Pressure Variations and Possible Ground Signatures of Flux Transfer Events', *J. Geophys. Res.* **96**, 5489.

Sibeck, D. G., Lopez, R. E., and Roelof, E. C.: 1991, 'Solar Wind Control of the Magnetopause Shape, Location and Motion', *J. Geophys. Res.* **96**, 1669.

Soerensen, E. M. *et al.*: 1995, 'The Cluster Data Processing System: A Distributed System in Support of a Challenging Scientific Mission', *Space Sci. Rev.*, this issue.

Thomsen, M. F., Gosling, J. T., Fuselier, S. A., Bame, S. J., and Russel, C. T.: 1986, 'Hot Diamagnetic Cavities Upstream from the Earth's Bow Shock', *J. Geophys. Res.* **91**, 2961.

Tsyganenko, N. A.: 1987, 'Global Quantitative Models of the Geomagnetic Field in the Cislunar Magnetosphere for Different Disturbance Levels', *Planetary Space Sci.* **35**, 1347.

Wilken, B. *et al.*: 1995, 'RAPID, The Imaging Energetic Particle Spectrometer on Cluster', *Space Sci. Rev.*, this issue.

Woolliscroft, L. J. C. *et al.*: 1995, 'The Digital Wave-Processing Experiment on Cluster', *Space Sci. Rev.*, this issue.

THE CLUSTER MISSION: ESA'S SPACEFLEET TO THE MAGNETOSPHERE

J. CREDLAND, G. MECKE and J. ELLWOOD
Cluster Project Division, ESA Directorate of the Scientific Programme, ESTEC, Noordwijk, The Netherlands

Abstract. During the first half of 1996, the European Space Agency (ESA) will launch a unique flotilla of spacecraft to study the interaction between the solar wind and the Earth's magnetosphere in unprecedented detail. The Cluster mission was first proposed to the Agency in late 1982 and was selected, together with SOHO, as the Solar Terrestrial Science Programme (STSP), the first cornerstone of ESA's Horizon 2000 Programme. It is a complex four-spacecraft mission designed to carry out three-dimensional measurements of the magnetosphere, covering both large- and small-scale phenomena in the sunward and tail regions. The mission is a 'first' for ESA in a number of ways:

– the first time that four identical spacecraft have been launched on a single launch vehicle,
– the first time that ESA has built spacecraft in true series production and operated them as a single group,
– the first time that European scientific institutes have produced a series of up to five instruments with full intercalibration, and
– the first launch of the Agency's new heavy launch vehicle Ariane-5.

The article gives an overview of this unique mission and the requirements that governed the spacecraft design. It then describes in detail the resulting design and how the particular engineering challenges posed by the series production of four identical spacecraft and sets of scientific instruments were met by the combined efforts of the ESA Project Team, Industry and the experiment teams.

1. Mission and System Design

1.1. Mission overview

The Cluster mission is an *in-situ* investigation of plasma processes in the Earth's magnetosphere using four identical spacecraft simultaneously (Figure 1). It will permit the accurate determination of three-dimensional and time-varying phenomena and will make it possible to distinguish between spatial and temporal variations.

The four Cluster spacecraft will be placed in nearly identical, highly eccentric polar orbits, with a nominal apogee of 19.6 R_E and a perigee of 4 R_E (Figure 2). This nominal orbit is essentially inertially fixed, so that in the course of the two-year mission a detailed examination may be made of all significant regions of the magnetosphere. The plane of this orbit bisects the geomagnetic tail at apogee during the northern-hemisphere summer, and passes through the northern cusp region of the magnetosphere six months later. The operational orbit finally selected for each spacecraft will depend upon the actual launch date and orbit-transfer scenario, but will ensure that each is located at a vertex of a pre-determined tetrahedron (Figure 3) when crossing the regions of interest within the magnetosphere. The relative

Space Science Reviews **79:** 33–64, 1997.

Figure 1. The four Cluster flight-model spacecraft together for the first time in the cleanroom at IABG, Ottobrunn (D). Two spacecraft are shown complete with thermal hardware and stacked in launch configuration, whilst the other two are 'open' with the payload platform visible.

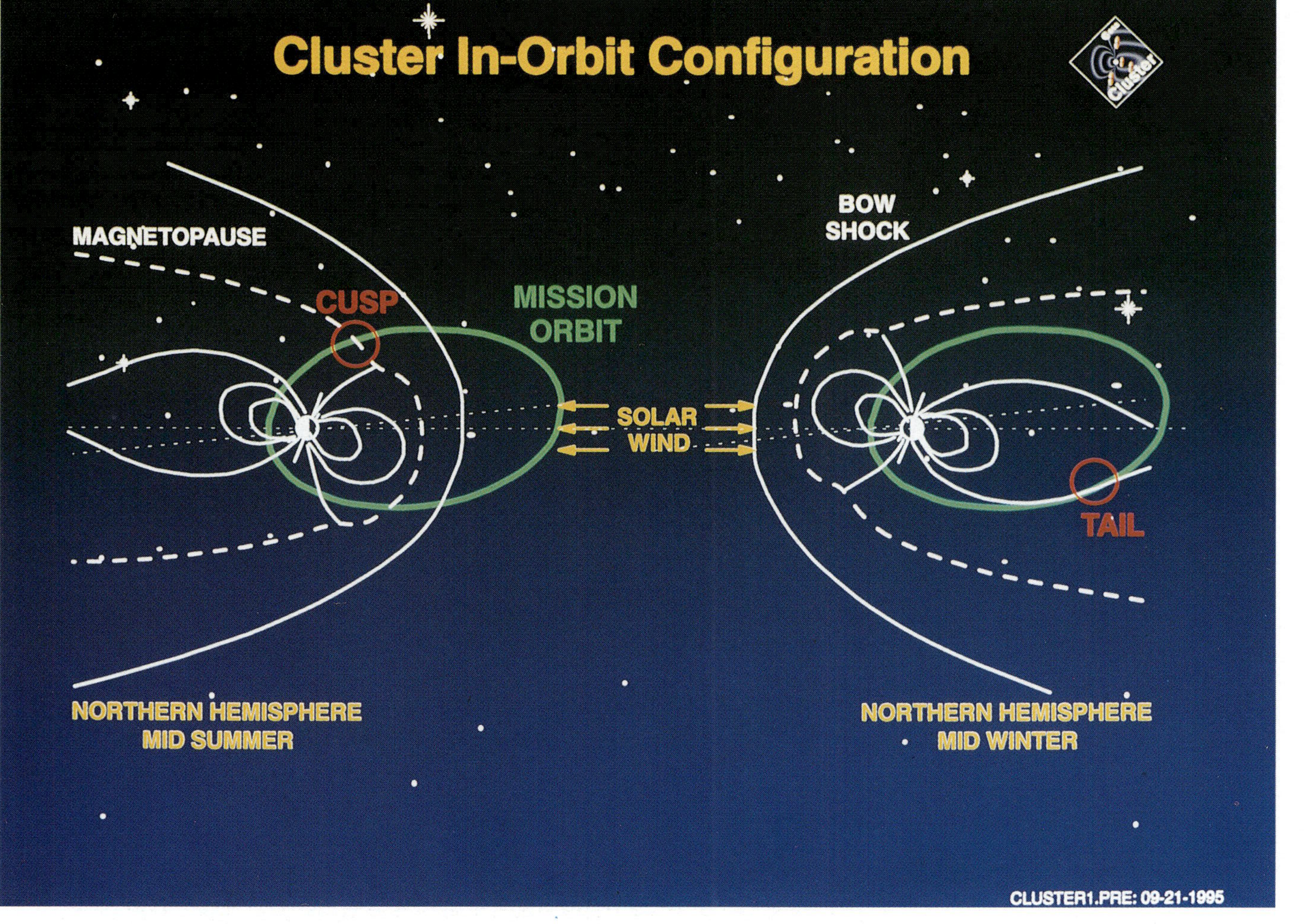

Figure 2. The Cluster operational orbit. This highly elliptical polar orbit of $4 \times 19.6\ R_E$ is essentially inertially fixed, so that in the course of a year the apogee moves from the tail region into the solar-wind region, thereby allowing all areas of the magnetosphere to be addressed. (Note: this figure reflects the original November 1995 launch date, hence the orbit bisects the cusp in northern-hemisphere winter.)

separations will be adjusted during the mission to correspond with the spatial scales of the structures to be studied, and will vary from a few hundred kilometres to a few Earth radii. The separation manoeuvres will be performed at intervals of approximately six months, synchronised with normal orbit-maintenance manoeuvres.

In the programme baseline, all four spacecraft will be injected into Geostationary Transfer Orbit (GTO) on a single Ariane-5 launch vehicle. The spacecraft will then be transferred in pairs to their mission orbits through a series of spacecraft propulsive manoeuvres.

In orbit, the spacecraft will be spin-stabilised at all times. Their attitudes will be selected to ensure a solar-aspect angle of approximately 90°, which optimises the performance of the spacecraft's solar-power generator and thermal-control subsystem throughout the mission. This attitude will be maintained during manoeuvres in the operational phase of the mission, but it will be necessary to reorient the spacecraft temporarily during the transfer-orbit phase in order to permit specific manoeuvres to be carried out, as discussed in Section 1.4.

1.2. The Scientific Payload

The Cluster payload experiments gave rise to some fundamental spacecraft design requirements, described in more detail in Section 3. In particular, the mission demands that the spacecraft have a very high degree of electromagnetic and electrostatic cleanliness to avoid disturbing the local plasma field or otherwise interfering with the sensor measurements. The particle experiments all require unobstructed fields of view and a high level of chemical cleanliness to preserve the sensitivity of their detectors.

In order to measure the undisturbed local magnetic field to the required sensitivity, the fluxgate magnetometer sensors are mounted on booms as far as possible from the main body of the spacecraft.

Each Cluster scientific payload produces data at different rates, depending on its operating mode. In nominal mode, it will deliver data at a rate of 17 kbit/s, but at certain locations within the orbit it will be operated in a 'burst mode' of 105 kbit s^{-1}. In addition, the Wide Band Data (WBD) experiment produces data at the very high rate of some 220 kbit s^{-1}, but this experiment will only be operational for approximately 30 min of each orbital revolution.

1.3. The Ground Segment

The European Space Operations Centre (ESOC) in Darmstadt, Germany, is responsible for the Cluster Operations Control Centre (OCC). It will control the four spacecraft in their mission orbits via ESA's Odenwald and Redu ground stations (Ferri and Warhaut, this issue). Additional ESA ground stations will be used during the mission's Launch and Early Orbit Phase (LEOP).

A ground station belonging to NASA's Deep-Space Network (DSN) will support the mission during certain specific scientific operations, when data from the WBD

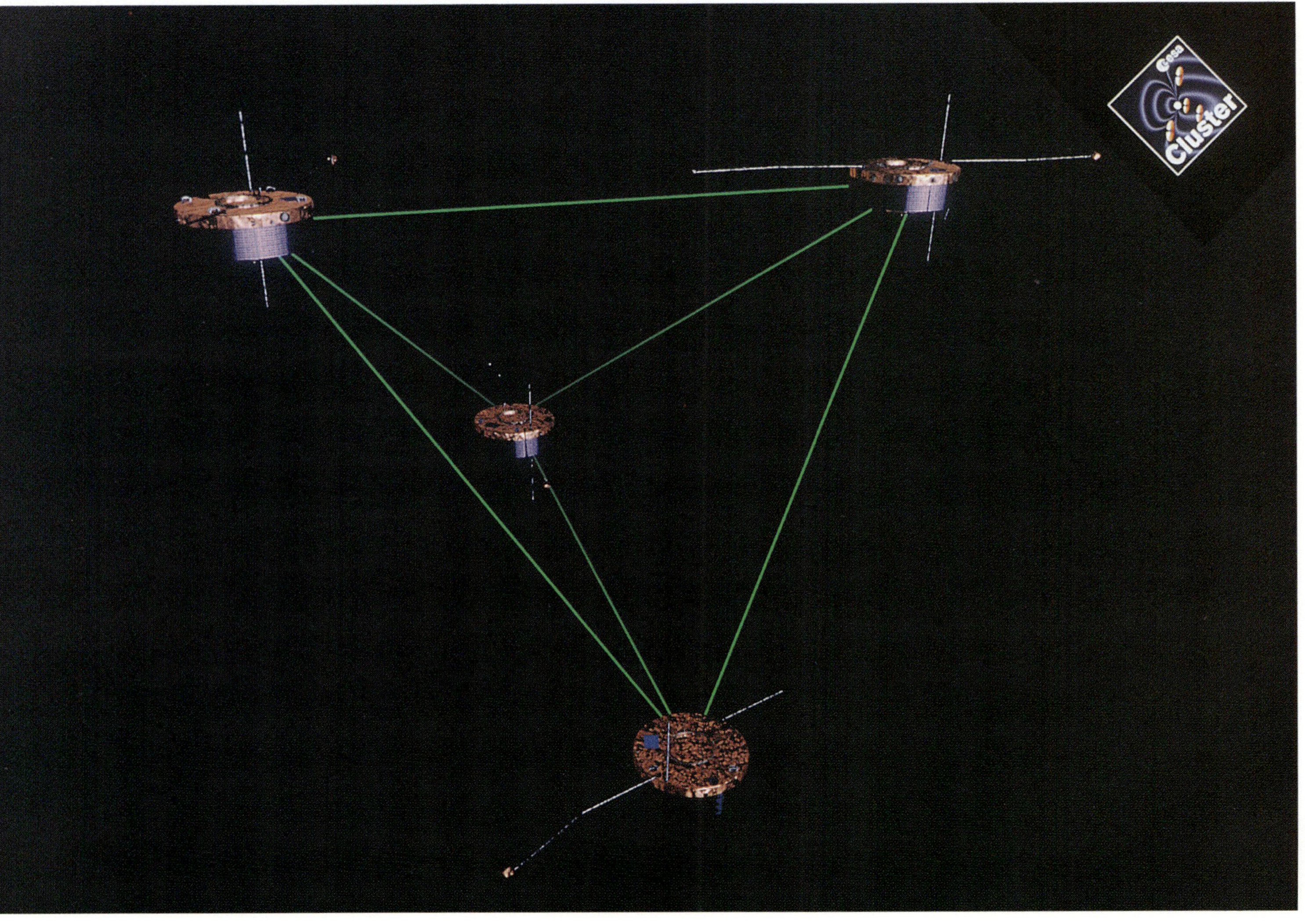

Figure 3. The tetrahedral formation in orbit.

experiment is being transmitted at high bit rate. Further NASA/DSN support will used to back up the ESA network, including ranging, telemetry acquisition and telecommand back-up functions during the critical transfer-orbit phase.

Orbit determination for all mission phases will also be performed by ESOC. The Cluster mission requires that the relative separations of the four satellites be determined to within 1% or 10 km, whichever is the smaller. This will be achieved by determining the orbit of each spacecraft individually from the ground.

Simultaneous acquisition of two or more spacecraft from one ground station is not planned. The OCC, however, is able to monitor and control two spacecraft simultaneously by using two ground stations. All communications with the spacecraft will be in S-band, and the up- and down-links of each individual satellite are assigned different frequencies.

1.4. Mission operations

The mission is divided into several distinct phases, each with specific operational objectives. One special feature of this mission is that the four spacecraft will be controlled in a time-sharing mode, presenting the OCC with an unusually complex task. This will be particularly true during the early mission phases, when many critical operations will have to be performed under strict time constraints. The spacecraft design therefore includes a degree of autonomy and a flexible on-board software concept to make it easier to meet these operational requirements.

In the Launch and Early Orbit Phase (LEOP), following injection into GTO by Ariane, ground contact will be established with all four spacecraft at the earliest possible opportunity. This will permit a quick-look status verification of essential parameters, which will be followed by less time-critical checkout and orbit-determination activities.

The Transfer-Orbit Phase (TOP) is characterised by a number of large orbit-profile and inclination-change manoeuvres, to target the spacecraft into the desired mission orbits. The spacecraft attitude will be temporarily adjusted for each of these manoeuvres to align the single main engine with the desired thrust direction. Throughout the TOP, the four spacecraft will be treated as two pairs, and injected with an interval of approximately 40 hr between spacecraft pairs.

Once in mission orbit, a Commissioning and Verification Phase (CVP) will commence, which will be devoted mainly to payload operations. These include the deployment of the two rigid radial booms, followed by the deployment of the flexible wire booms, the experiment check-out, and the calibration activities.

During the main Mission Operations Phase (MOP), the primary objective is to maximise the scientific data return from the payload. The areas of scientific interest within the orbit will vary seasonally, and do not generally coincide with real-time ground contacts. To accommodate these, in the nominal operating mode the scientific data will be stored on on-board solid- state memory units, which will then dump the data, together with the real-time telemetry, during subsequent ground-contact periods (Sørensen *et al.* this issue) .

1.5. Launch Campaign

The Cluster launch campaign started in late August 1995 to meet a by then declared, slightly delayed launch date of 17 January, 1996 for Ariane-5's first flight (V501). A natural break point occurred in the programme in November 1995 with the completion of all electrical checks and prior to spacecraft fuelling. The launch campaign was suspended at this point and the four spacecraft were sealed in their containers for temporary storage. Following declaration of 15 May, 1996 as the earliest expected launch date, the campaign recommenced in early February. The transfer of 650 kg of fuel to each spacecraft was completed at the end of March. The launch date has now been confirmed for late May and the batteries have been integrated in preparation for the mating of the four spacecraft with the launch vehicle.

1.6. The Industrial Team

The Invitation to Tender for the Cluster mission was issued by the Agency in mid-1988. Following a phase of industrial competition, the Prime Contractorship was awarded to Dornier in Summer 1989.

Phase-B, the design phase, commenced in October 1989 with a core team of contractors already selected with the Dornier proposal. The full industrial team (Figure 4) was built up during this phase, with lower-tier subcontractors being selected on a technical-competence and price basis to achieve the requisite geographical distribution targets set for the project.

The achievement of those geographical-return targets was made easier by an agreement with the ESA Industrial Policy Committee (IPC) that the targets could be met for the STSP Cornerstone as a whole, rather than at individual project (SOHO and Cluster) level. Several pieces of equipment (power-distribution units and transponders) were in fact procured as common items between the two projects as a pragmatic way of achieving the required targets.

The main Cluster development programme (Phase-C/D) commenced in April 1991 with delivery of the four spacecraft to the Agency in April 1995 as a target. In reality, despite a number of critical setbacks during the programme, delivery by Dornier was completed in July 1995, still on schedule for the originally foreseen 1 December, 1995 launch date.

The flight-model system-level environmental test programme was conducted entirely at IABG in Ottobrunn (D) and lasted for two years. During this period, all four Cluster spacecraft successfully underwent sine-vibration, acoustic noise, thermal-balance/vacuum and DC- magnetic testing. The flight model programme, together with the structural-model testing also conducted at IABG in 1992, con-

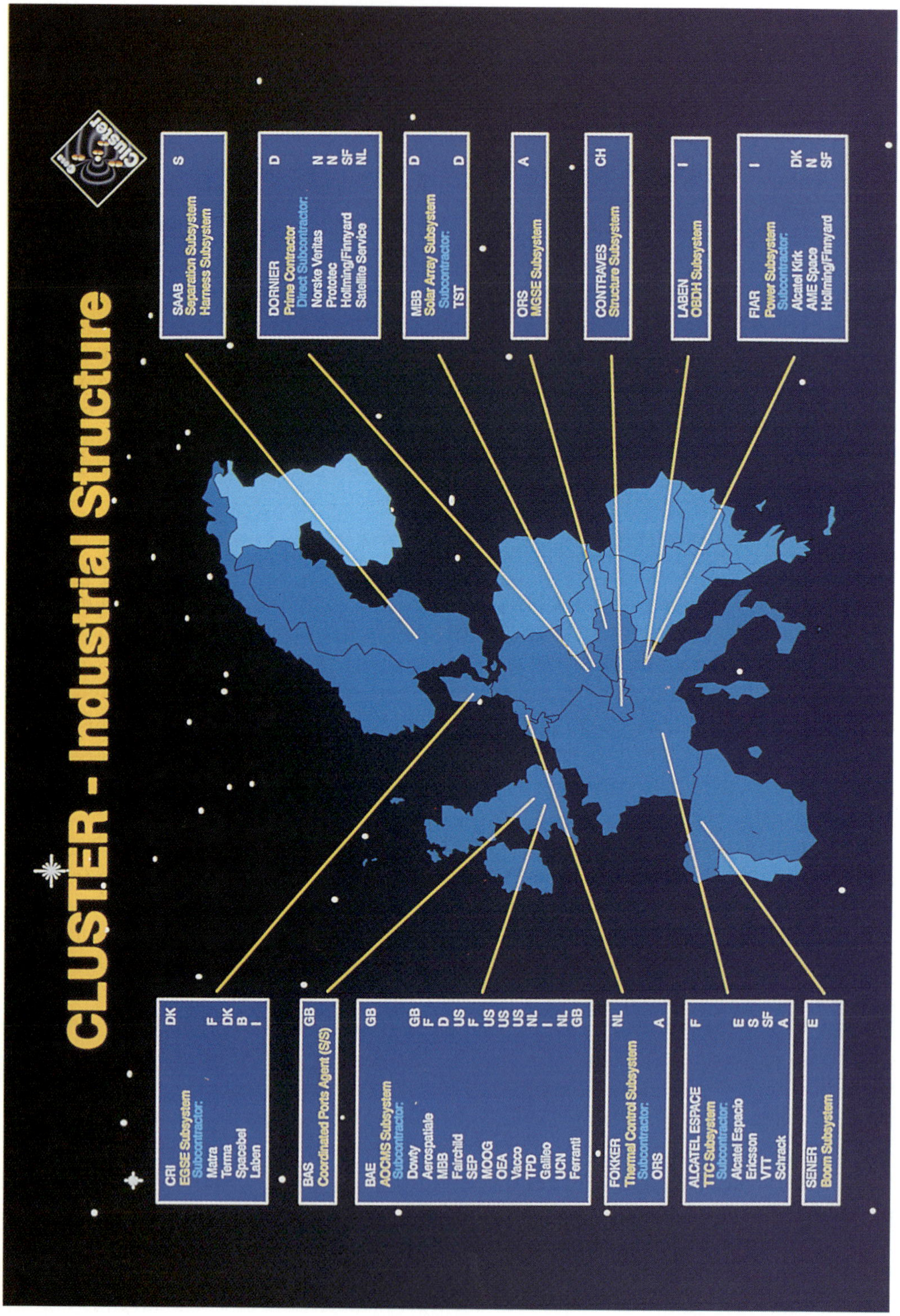

Figure 4. The Cluster Industrial Team.

stituted the most extensive test programme of this nature ever undertaken for any ESA project.

2. Spacecraft Design

2.1. Mission Requirements

Long before the start of the Cluster design phase (Phase-B) activities in October 1989, a launch date towards the end of 1995 had been fixed. This limited the time available for development, manufacture and testing of the four spacecraft to that which other satellite projects would normally need to deliver only one flight model. Consequently, the logistics for Cluster hardware and software production and verification were crucial, leading to such requirements as modular design at system level, allowing parallel pre-integration and late exchange of the Programmable Read-Only Memory (PROMs) in the Central Data Management Units.

Since the Cluster mission entails scientific measurements on four spacecraft simultaneously, maximum similarity of the various spacecraft functions on the four flight models is clearly necessary. Reproducibility of the Cluster hardware at all levels has therefore been of major importance.

The mission orbits selected impose eclipse durations of up to 4 hr and this has been a major design driver for the whole spacecraft system, and for the thermal-control, power and structural subsystems in particular.

Because the development of Cluster and of Ariane-5 have been proceeding in parallel, specific and sometimes severe requirements were imposed on Cluster to cover uncertainties in the launcher development programme. For example, conservative loads were specified for sine, random and acoustic vibration testing. Nevertheless, very late in the Cluster development programme, in fact after acceptance of the Cluster hardware, unusually high predicted shock loads originating from launcher separations were identified by Ariane, necessitating additional, unforeseen verification activities at unit and system level for Cluster.

A further constraint is that, as this will be the first flight of Ariane-5, launch must occur within a daily 2 hr launch window during Kourou daylight time.

The four spacecraft will be accommodated in two pairs within the two launcher payload compartments (Figure 5). This requires that each of the lower spacecraft must have a dedicated separation system on top to carry the upper spacecraft of the pair. Since they will be injected into GTO rather than their mission operational orbits, each spacecraft will have to perform a complex series of orbit transfer manoeuvres, requiring large quantities of propellant. The strategy of handling the spacecraft as two pairs during the early mission phases, leaving one pair in GTO twice as long as the other, was adopted to make this transfer scenario manageable from the operational point of view. The design of on-board equipment, including the solar arrays, has had to take the associated increased radiation environment

Figure 5. The Cluster launch configuration aboard Ariane-5.

Table Ia

Nominal dry mass breakdown (kg) without balance mass

Payload	69.3
Solid state recorders	18.6
High power amplifiers	4.0
Structure	139.6
Thermal control	19.0
Booms	27.7
Attitude and orbit control and meas. system (AOCMS)	89.4
Data handling	27.0
Telecommunications	9.8
Power supply	67.6
Pyro unit	3.1
Solar array	10.8
Harness	33.4
Separation system	11.1
Total spacecraft subsystems	438.5

into account. Operability requirements in general have played a major role during Cluster's development, to facilitate safe parallel operation of the four spacecraft by ESOC.

2.2. The Resulting Spacecraft Design

The final design resulting from the synthesis of the above constraints is a spin-stablised spacecraft rotating at 15 rpm in nominal configuration with four 50 m experiment wire booms, two 5 m experiment radial rigid booms and two axial telecommunication antenna booms (Figure 6).

The nominal mass breakdown and power allocation for each spacecraft is shown in Table I.

2.2.1. *Structural Design*

The spacecraft's cylindrical design is driven by the body-mounted solar array and also optimises the fields of view available to the experiments, which are accommodated around the rim of the main equipment platform on the upper side of the spacecraft. The height of the spacecraft body has been minimised to make optimal use of the fairing volume offered by the launch vehicle. Each spacecraft is 1.3 m high and 2.9 m in diameter.

This compact spacecraft primary structure provides mass-efficient load paths to its mechanical interfaces. It consists of the central cylinder, the Main Equipment Platform (MEP), a tank support structure, a platform internal to the central cylinder and a Reaction Control System (RCS) support ring. The central cylinder is fabricated as a carbon-fibre-reinforced plastic (CFRP)-skinned aluminium honeycomb

Figure 6. The Cluster flight configuration.

Table Ib
Nominal power allocation in mission orbit (W)

Payload	39.3
Spacecraft bus	
Thermal control excl. IPD/EPD	1.0
Internal Power Dumpers (IPD) total	22.0
External Power Dumpers (EPD) total	30.0
AOCMS	3.8
Telecommunication	60.1
On-board data handling (OBDH)	51.5
Power	27.3
Harness	1.7
Subtotal	*236.7*
Total with 5% margin for lifetime variation	245.9
Battery	0.0
Solar array (end-of-life 247 W beginning of life 315 W)	245.9
Thermal conditions	
Spacecraft internal power dissipation[a] (W)	196.2
Resulting average MEP temp. (°C)	21.2

[a] Subtotal with margin.

sandwich, and the MEP as an aluminium-skinned honeycomb panel reinforced by an outer aluminium ring. The MEP is supported by symmetrically arranged CFRP struts connected to the central cylinder.

The overall design allows for parallel integration of all equipment with the MEP on one side, and the central cylinder with the RCS components on the other. This feature proved essential in maintaining the imposed schedule. The structural model spacecraft is shown in Figure 7.

Six cylindrical titanium propellant tanks with hemispherical ends are each mounted to the central cylinder via four CFRP struts and a boss. The propellant carried in these tanks constitutes more than half of each spacecraft's launch mass of around 1180 kg.

Six curved solar-array panels together form the outer cylindrical shape of the spacecraft body and are attached to the MEP. This platform provides the mounting area for most of the spacecraft units, the payload units being accommodated on the upper surface and the subsystem units, in general, on the lower surface. The five batteries and their associated regulator units needed to provide energy to the spacecraft during the 4 hr eclipses in mission orbit are mounted directly on the central cylinder.

Figure 7. The Cluster Structural Model during integration. The six propellant tanks, booms, Main Equipment Platform and central cylinder are visible, together with the harness and mass dummies of the various units.

At their lower ends, the solar-array panels support a ring accommodating many of the RCS components, including four radial 10 Newton (N) thrusters. Four axial 10 N thrusters are mounted on studs on the upper and lower faces of the spacecraft. All thruster positions were carefully chosen to minimise the chances of contamination reaching the sensitive experiments.

Because the solar-array panels will experience extremely low temperatures during the eclipses, special care has had to be taken in designing their attachments to the structure and the thermal insulation of their inner faces, in order to minimise the on-board heating requirements in eclipse.

The inner equipment panel inside the central cylinder supports the single main engine, two high-pressure tanks and associated propellant-management hardware.

The central cylinder carries aluminium interface rings at both its upper and lower ends. The lower ring is compatible with the Ariane 1194 mm diameter adaptor and separation mechanism. The upper ring simulates the interface offered by the adaptor and is equipped with a separation mechanism. This allows two spacecraft to be stacked on top of one another, whilst themselves still remaining identical to the maximum extent possible.

Two rigid, double-hinged radial booms on the upper surface of the MEP carry payload sensors. These booms are stowed for launch, as are the four payload wire booms and the two rigid, single-hinged antenna booms carrying the S-band antennas. The rigid booms consist of CFRP tubes with titanium-alloy end fittings and deployment mechanisms. The radial booms will be deployed mainly by the centrifugal force developed by the spinning spacecraft, while the antenna booms are driven by redundant springs.

2.2.2. *Propulsion Design*

The Reaction Control Subsystem (RCS) is configured as a conventional bi-propellant system, based on a single 400 N main engine and eight 10 N thrusters (Figure 8). It is arranged in two redundant branches (with the exception of the main engine), each of which is capable of performing a complex mission profile. Electrical cross-coupling permits the operation of either of the two branches from either redundant half of the Attitude Determination and Control Electronics (ADCE). The propellant is stored in the six tanks pressurised by helium stored in two smaller spherical tanks. Pressure-regulation and propellant-delivery systems manage the pressurant and propellant conditioning and distribution functions. During launch, the pressurant, fuel, oxidiser and the manifold will be isolated from each other by pyrotechnically operated valves, to comply with launch-vehicle safety requirements. After each manoeuvre, the main engine and thrusters will also be isolated by additional latching valves, thereby increasing reliability by eliminating potential leak paths.

2.2.3. *Thermal Design*

The passive thermal control of the Cluster spacecraft is based on a low-emissivity concept, insulating the spacecraft from the exterior environment enough to survive the eclipses, whilst still allowing the internally generated heat to be rejected (Figure 9). The thermal closeout is provided by three types of hardware: low-emissivity double foil shields on the upper and lower surfaces of the spacecraft; multi-layer insulation (MLI) on the top and bottom of the central cylinder, below the RCS ring and around the upper part of the satellite, enclosing the experiments; and thermal insulation of the inner sides of the solar-array panels and of the main engine. An Optical Surface Reflector (OSR) radiator is integrated into the top surface to allow for the high dissipation of the RF power amplifiers. An External Power Dumper (EPD) radiator located in the upper thermal shield within the central cylinder dissipates excess power generated by the solar arrays. Heaters are used to keep equipment within specified temperature ranges throughout all mission

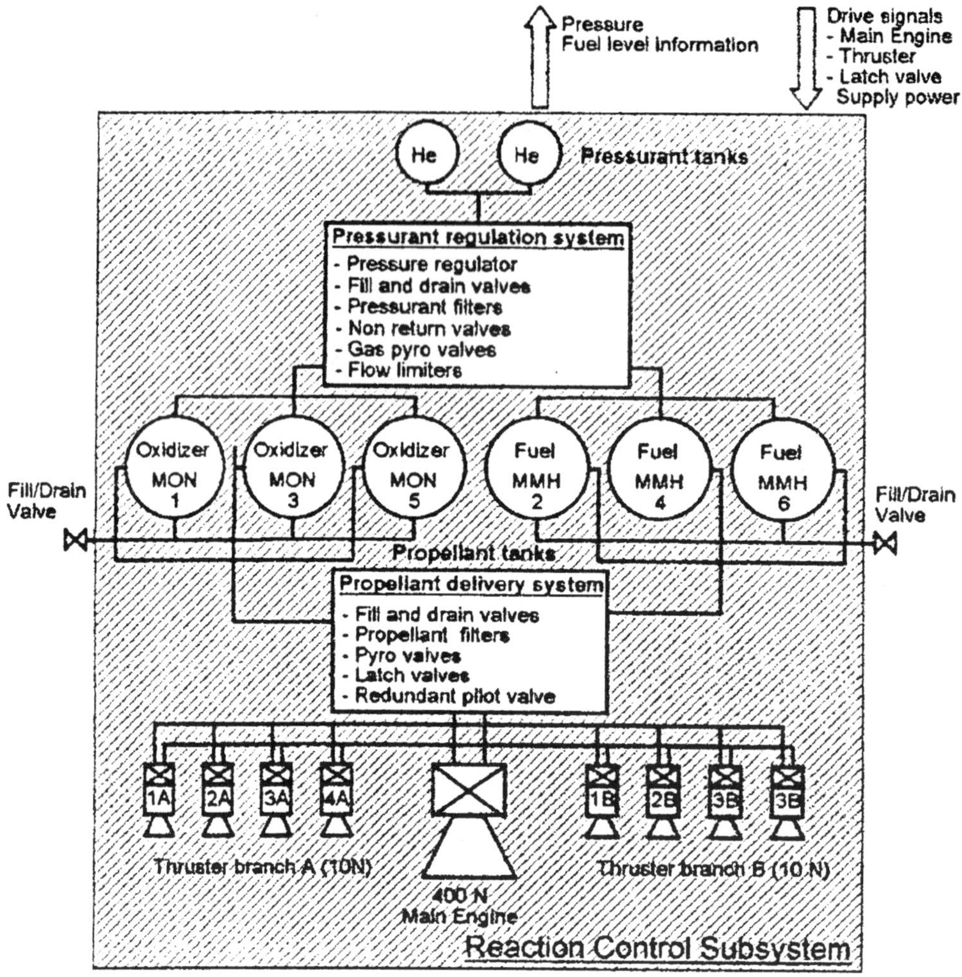

Figure 8. The Reaction Control Subsystem (RCS).

phases, including eclipses. Temperature control is achieved by a combination of thermostats with thermistor surveillance and of thermistor-guided software control.

The thermal design has been optimised for the almost constant solar-aspect-angle (SAA) range ($90° < \mathrm{SAA} < 96°$) that will apply throughout the nominal mission phase. During the orbit transfer manoeuvres, however, the spacecraft may experience a much wider SAA range ($65° < \mathrm{SAA} < 115°$). The heat-rejection concept selected therefore permits the satellite to dissipate heat through either the upper or the lower thermal shield. With these precautions, Cluster can safely withstand the complete range of solar aspect angles that will be encountered.

A heated-environment concept has been chosen for the lower spacecraft compartment, comprising RCS equipment, batteries and battery regulators. The com-

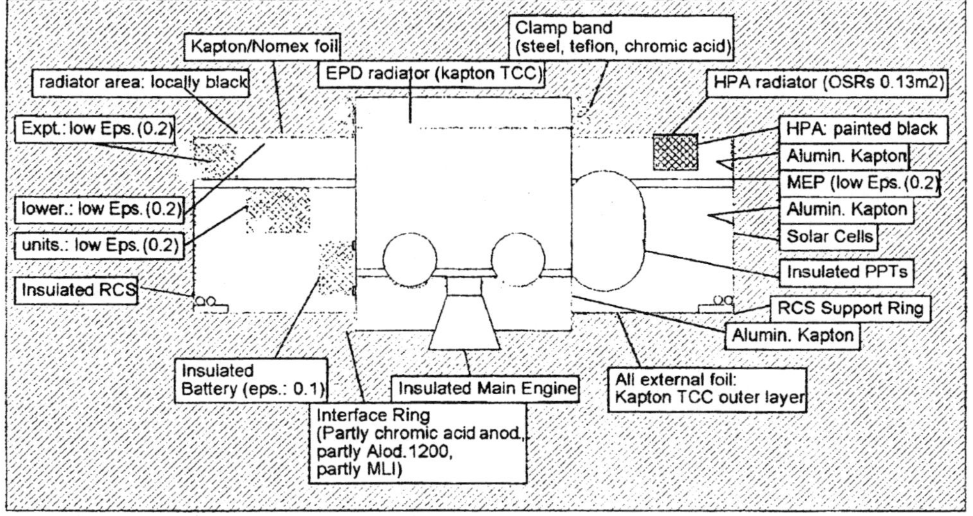

Figure 9. The thermal-control concept.

plexity and duration of the assembly and integration activities were greatly reduced by this approach compared to a solution with insulated components, but it does require somewhat more heater power during eclipses.

All external surfaces, including the solar cells, blankets, double foils and radiator have been finished with an electrically conductive Indium Tin Oxide (ITO) coating to comply with the electrostatic requirements imposed by the experiments.

2.2.4. *Electrical Design*

Electrically, the spacecraft is configured in four major functional areas (Figure 10):

- a power-supply subsystem including a pyrotechnics unit,
- an on-board data-handling subsystem,
- an attitude and orbit control and measurement subsystem,
- a telecommunications subsystem.

Dedicated, physically separated and carefully shielded harnesses interconnect the various subsystem and payload units. The payload includes a separate experiment interconnection harness.

– *Power*

Spacecraft power demands will be met by the body-mounted solar array and five silver/cadmium batteries, chosen for their non-magnetic characteristics. The batteries will power the spacecraft during eclipses and supplement the solar-array output during periods of peak power demand. Full payload operation will be supported throughout the entire mission orbit phase, except during eclipses. A block diagram of the Power Subsystem is shown in Figure 11.

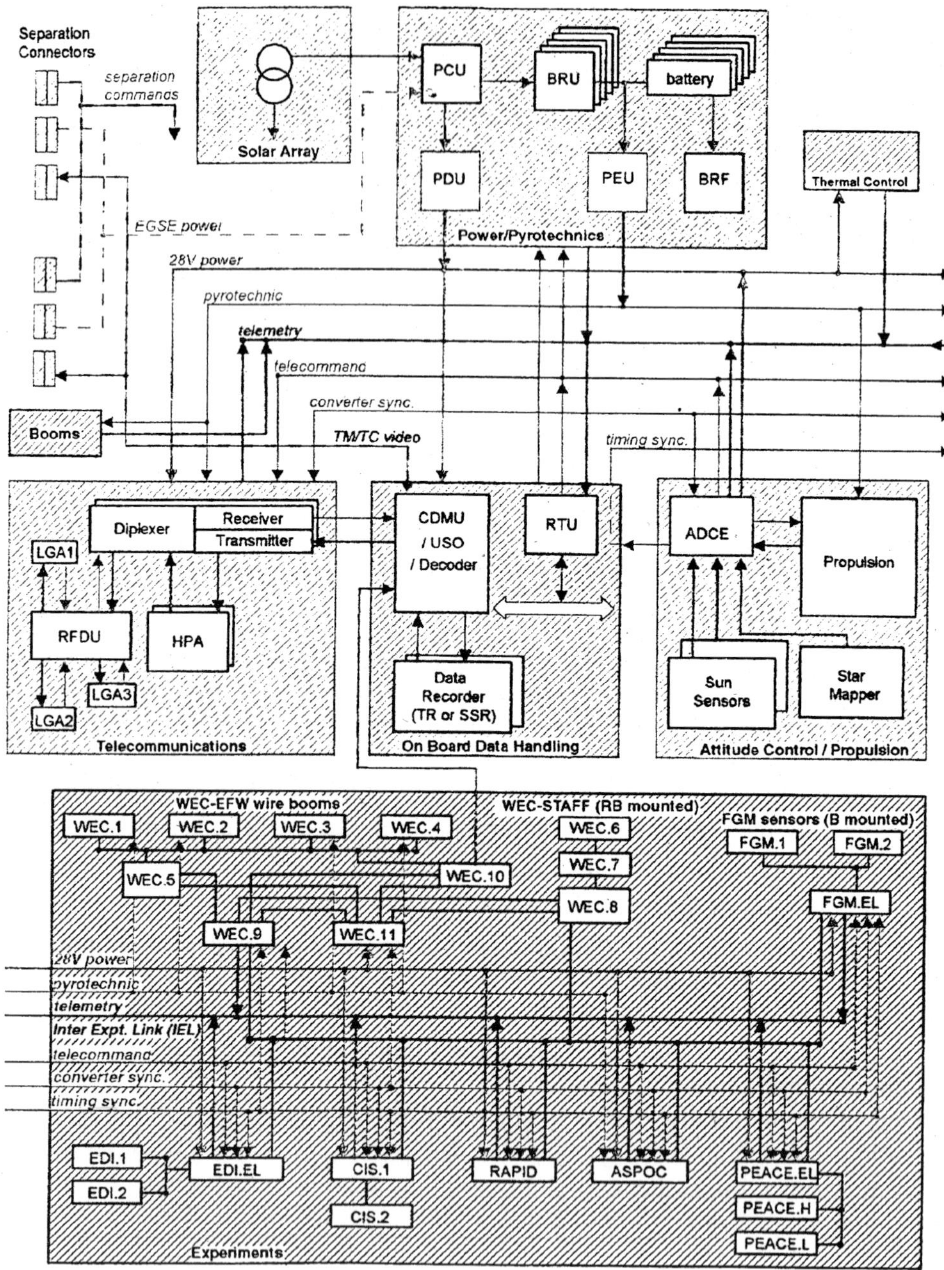

Figure 10. The electrical system.

The solar arrays consist of Back-Surface-Reflection (BSR) cells, arranged in self-compensating formations to minimise the generation of DC magnetic fields.

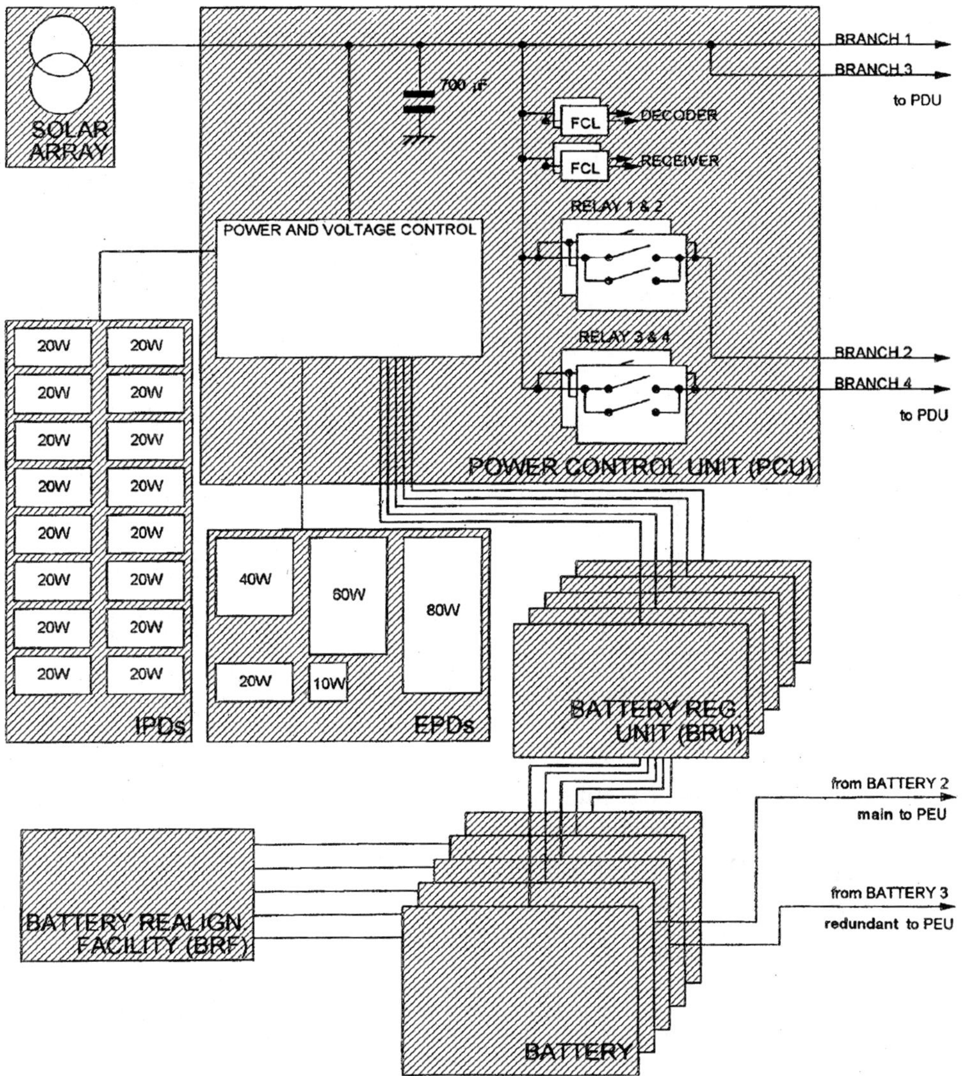

Figure 11. The Cluster power subsystem.

The conductive coating on the cell cover glass minimises the build-up of differential charge potentials.

Spacecraft power is conditioned and distributed via a voltage-regulated bus and redundant current-limiting switches. Where required, permanent 'keep alive' lines are provided to payload and subsystem units. Full protection against short-circuit or overload is provided by limiting the maximum current in any supply line. Excess solar-array output power is automatically routed to Internal Power Dumpers or shunted by commandable switching to an External Power Dumper. The power bus

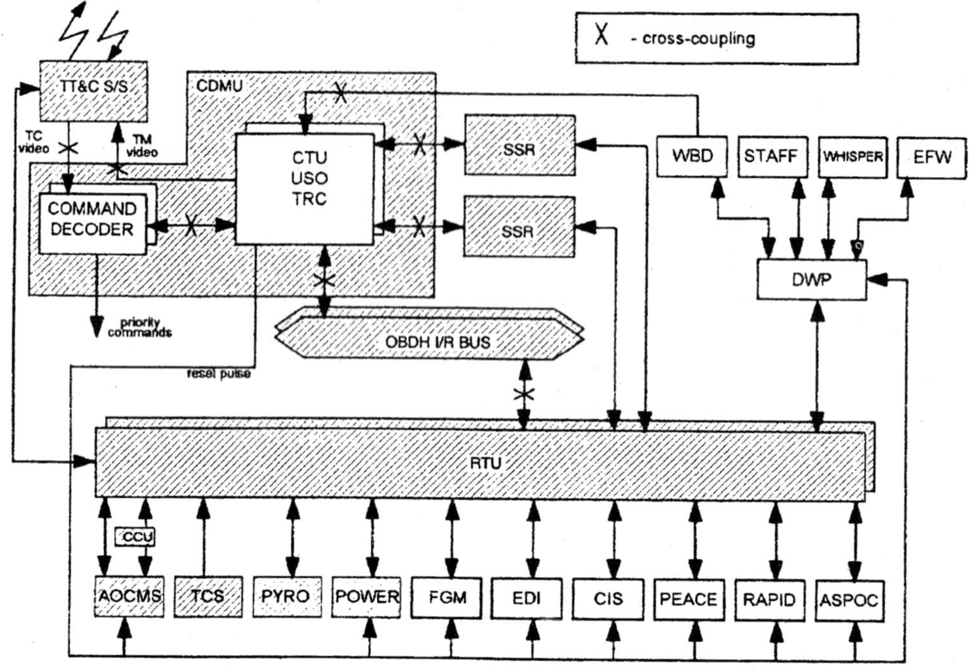

Figure 12. The Cluster On-Board Data-Handling (OBDH) subsystem.

operates on a linear shunt regulation approach, rendering the main bus voltage extremely 'clean' during payload measurement periods. This significantly reduces any potential electromagnetic disturbances due to the power subsystem.

– *On-Board Data Handling*

The On-Board Data Handling (OBDH) subsystem, which will perform the primary spacecraft control functions, is based on an ESA standard approach (Figure 12). It consists of a Central Data Management Unit (CDMU), a Remote Terminal Unit (RTU), and two Solid-State Recorders (SSRs). The CDMU and RTU are internally redundant; each SSR provides storage for about 2.2 Gbit of data at beginning of life.

The OBDH decodes and distributes commands, received by the telecommunication subsystem at a command bit rate of 2 kbit s^{-1}, and acquires and encodes telemetry from payload and subsystem units. This telemetry is delivered either to the telecommunications subsystem for real-time transmission to ground, and/or to the SSRs for later transmission. Dedicated high-data-rate interfaces are provided to the Wide-Band Data (WBD) experiment and the SSRs. Stored data will be played back at a much higher rate than real-time data, in order to reduce the duration of the downlink during the limited ground-station visibility periods. WBD telemetry will only be transmitted in real-time.

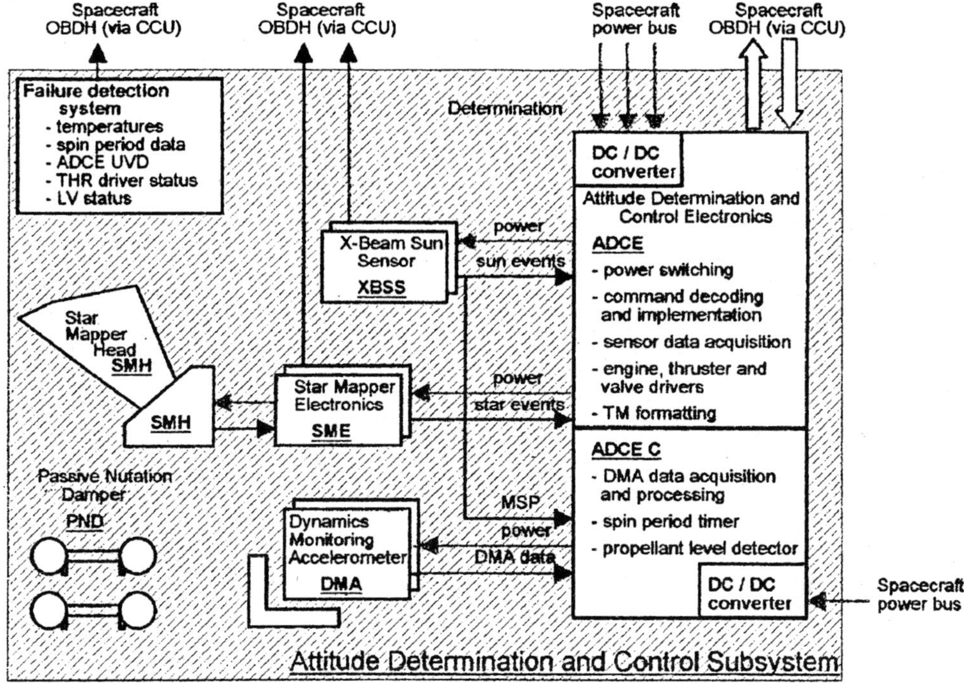

Figure 13. The Attitude-and-Orbit Control and Measurement Subsystem (AOCMS).

Telemetry-stream bit rates are fixed at about 2 kbit s^{-1} for housekeeping telemetry, 22 kbit s^{-1} for nominal science telemetry, 131 kbit s^{-1} for burst science-data/recorder playback, and 262 kbit s^{-1} for WBD transmission/recorder playback.

The OBDH also provides timing and synchronisation signals to payload and subsystem units, as well as AOCMS data to the payload. It will perform a surveillance function using on-board software to provide the autonomy required because of the extended non-visibility periods.

– *Attitude and Orbit Control and Measurement*

Maintenance of the orbit and attitude of the spacecraft is performed by the Attitude and Orbit Control and Measurement Subsystem (AOCMS), a block diagram of which is shown in Figure 13. Spacecraft attitude and spin data are provided by an internally redundant star mapper and an internally redundant X-beam Sun sensor. The reconstitution of attitude data, such as inertial attitude, spin rate and spin phase, will be performed on the ground. This information is essential for the interpretation of the payload science data. The necessary accuracies for these attitude data are comfortably met by the subsystem.

Orbit and attitude maintenance will be performed by using control thrusters, both semi-radial and axial, together with the main engine, which will be used to

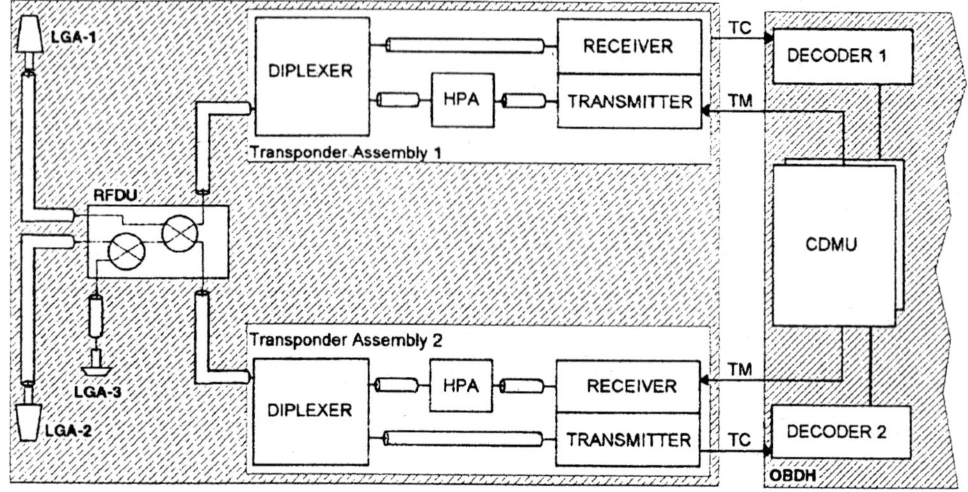

Figure 14. The telecommunication subsystem.

perform the large orbital change manoeuvres required to reach the polar mission orbit from GTO.

Telecommunications

Communications with the spacecraft will be established through the Telecommunications Subsystem (Figure 14), which includes uplink and downlink capabilities to support the telecommand, telemetry and tracking functions. It will interface with the ESA Ground Segment and the NASA Deep-Space Network at S-band frequencies (uplink 2025–2110 MHz; downlink 2200–2290 Mhz).

The subsystem includes three low-gain antennas, a redundant set of transponders (including a NASA-supplied 10 W RF amplifier), an RF distribution unit, and associated RF harnesses. Two low-gain antennas are mounted on deployable booms attached to the upper and lower faces of the spacecraft. They will ensure full spherical coverage for uplinking and hemispherical coverage for downlinking. A third antenna mounted on the lower side of the spacecraft will be used until the in-orbit deployment of the lower antenna boom.

2.3. THE CLUSTER PRODUCTION LINE

The four Cluster spacecraft have constituted a low-volume, process-oriented production run of a custom design. Rather like custom-built homes designed by architects and built by contractors to customer's specifications, they are characterised by their uniqueness and the need for both high quality and on-time delivery. This combination of characteristics demanded considerable flexibility in the production process in order to be able to cope with unforeseen difficulties and yet still achieve the original design goals.

Whilst such characteristics also apply to other spacecraft programmes, the situation with Cluster differed considerably in that the four flight models had to be produced within the time normally allowed for one. Although Cluster was of course not a series production in the full commercial sense, at system level it stands somewhere between the 'one-off' product of other scientific projects and the 'mass production' in the telecommunications satellite sector, for example of satellites for mobile services.

The picture changes completely at the unit and component levels: the hardware of the four flight spacecraft alone is comprised of a total of about 360 units, 16 rigid booms (without the 16 wire booms), 36 propellant tanks, 8 pressure tanks, 32 thrusters, 320 m of RCS pipework, almost 17 km of harness, 1440 connectors and more than 57 000 electrical contacts. Here, Cluster is indeed much closer to a series production product.

A mixture of strategies originating from the two extremes of a 'one-off' and a series production was therefore employed in the Cluster programme. 'Standardised' or 'off-the-shelf' items are typically used in mass production because they represent readily available identical units at reasonable cost. Such items have therefore been used in many areas on Cluster; e.g. RCS equipment, silver-cadmium batteries, battery regulator units, sensors, booms and pressure tanks. Because such items are considered 'flight-proven' and usually have a long history of successful application in space, both development time and risk can be reduced by using them.

The designs of some 'existing items' like booms and battery regulator units had to be slightly changed to cope with Cluster's specific requirements, which required the initiation of a full new space-qualification process. In other cases such as the pressure tanks, waivers against the Cluster requirements were granted after careful evaluation of the item's acceptability.

Parallel work flows are a typical mass-production technique for reducing overall production time. For Cluster, the modular design at system level allowed the parallel integration of the RCS system with the central cylinder, and the various equipment items with the MEP. Mating of these two modules was performed after their pre-integration. Without this approach, the overall schedule could not have been met.

Parallel testing at unit and system level also became necessary to meet the overall schedule. This situation was not ideal in that the first system-level tests had to be performed in parallel with the production of the next spacecraft model's units. Compromises between requirements and actual performances and between unit/subsystem and system verification became necessary, tending sometimes to increase the risk. Further on, this situation resulted in extremely high workloads for longer periods than usual for those involved simultaneously in the unit, subsystem and system activities.

The system-level testing periods were extremely long, embracing static load testing of the Structural Model/Spacecraft Mass Dummy (SM/SMD) stack, electrical testing on the Engineering Model (EM), sine/random/acoustic vibration campaigns on the SM/SMD stack, two sine/acoustic vibration campaigns for Flight Model stacks, and thermal-balance/thermal-vacuum tests on each individual Flight Model. Most of these tests were performed at IABG in Munich (D) between March 1992 and March 1995, except for the period from mid- to end-1993 during which the EM tests took place at the Prime Contractor's site. The electrical system-level tests ran almost continuously on the various flight models.

While this situation presented logistic difficulties, the positive effect was that the various specialists worked with high efficiency because they could move almost immediately from one spacecraft to the next. A pronounced learning-curve effect due to repeated integration and testing activities also became visible at all levels, helping to achieve the prescribed overall schedule.

The flexibility during the production process that resulted from the production-line approach because more than one model per unit was available for much of the time, provided more possibilities for work-arounds in the event of a failed unit.

The high motivation of all personnel involved in the activities was another prerequisite for success, allowing the timely solution of many unforeseen difficulties and problems. This included, for example, three-shift working at the Prime Contractor for an extended period during the early electrical-testing campaigns, to compensate for unit delivery delays of up to six months.

The timely availability of hardware and software was of the utmost importance to achieve the prescribed delivery date for the Cluster flight models.

Problems with long-lead-time items and some specific items like CMOS devices, hybrids, specific heaters, double foils for the thermal top and bottom shields, etc. had to be resolved. Sometimes only limited quantities of particular parts or materials were available because of production stops in industry.

Traceability of the hardware and software has been another important element in Cluster's four-model programme. Each unit model has its own performance characteristics and its own calibration curves for the conversion of telemetry data into meaningful physical measurements, with slight nuances occurring from one model to another. It was therefore important to keep track of which model of a given unit was finally integrated to which spacecraft and in which position, not only for the ground testing, but even more so for later in-orbit operations. Databases containing this information are therefore being continuously updated until launch.

Identical functioning of the four Cluster spacecraft is extremely important, as noted earlier. This goal was achieved at unit level by applying common specification limits for all models of any given unit. Proof of specification compliance, and hence of 'identicality' between models, was provided by the individual unit acceptance tests.

Identical functionality at system level could be checked via system-parameter measurements on each of the Cluster flight models. Only a very small scatter was found between the four spacecraft models, as evident from the example of launch masses given in Table II.

Table II
Launch masses of the four Cluster spacecraft

	FM1	FM2	FM3	FM4	ALL 4	
Total spacecraft dry mass	531.2	519.9	519.8	532.5	2104.4	
Maximum propellant	651.4	651.4	651.4	651.4		
Total spacecraft launch mass[a]	1182.6	1169.3	1171.2	1183.9	4707.0	
				Ariane 5 limit	4800.0	
				Margin	93.0	1.9%

[a] Masses of FM1 and FM4 include separation systems of 12 kg.

3. The Cluster Payload: a Unique Engineering Challenge

Cluster's payload is an advanced set of experiments to measure electric and magnetic fields, plasmas and energetic particles. In addition to the sophisticated measurement techniques that have been incorporated, the engineering challenges of manufacturing the state-of-the-art instruments and accommodating them successfully on board the spacecraft have been immense. The unique opportunity to build, integrate and test four identical sets of eleven instruments has also created problems not only of a technical, but also of a managerial nature, with a complexity of interfaces and logistics never previously experienced in an ESA scientific programme.

3.1. The Overall Payload Complement

Each of the four Cluster spacecraft is equipped with the same state-of-the-art electrical and magnetic instruments and particle detectors. Table III lists all of the experiments (with their acronyms), the basic technical parameters of the instruments, and the names of their respective Principal Investigators (see also Escoubet and Schmidt, this issue).

3.2. The Payload's Demands on the Spacecraft

Installation of the payload has involved accommodating a total of 24 units on each of the four spacecraft. More than half of these units require unobstructed fields of view in space, something that has been difficult to ensure given the presence of the many other experiments and the deployable rigid and the ultra-long wire booms (Table IV).

In addition, the detectors are highly sensitive to contamination, which could have occurred on the ground during the very long integration and test programme or could come from the thrusters and main engine of the spacecraft itself after launch. The first problem was solved by imposing a rigorous cleanliness programme during the integration and test phases, and only integrating the flight sensors at the last possible moment. In designing the spacecraft, the thrusters and main engine have

Table III
Investigations to be performed on Cluster

Instrument	Principal Investigator	Measurement	Technique
Fluxgate Magnetometer (FGM)	A. Balogh, Imperial College, London, UK	**B**, wave form DC to ~10 Hz; resolution ≥6 pT	Two three-axis fluxgate sensors on 5 m boom
Spatio-Temporal Analysis of Field Fluctuations (STAFF)*	N. Cornilleau-Wehrlin, Centre de Recherche en Physique de l'Environnement Terrestre et Planetaire, Paris, F	**B**., wave form up to 10 Hz, compressed data up to 4 kHz. Cross-correlator for <**E**, **B**>	Three-axis search-coil sensor on 5 m boom
Electric Fields and Waves (EFW)*	G. Gustafsson, Swedish Institute of Space Physics, Uppsala, S	**E**, wave form up to 10 Hz, compressed data up to 100 kHz, sensitivity <50 nV/m $(Hz)^{1/2}$	Double probes, two pairs wire booms, each 100 m tip-to-tip
Waves of High Frequency and Sounder for Probing of Density by Relaxation (WHISPER) *	P.M.E. Décréau, Laboratoire de Physique et Chimie de l'Environnement, Orléans, F	Active: Total electron density Passive: Natural plasma waves up to 400 kHz	Sounding, using parts of EFW wire booms Filter banks
Wide Band Data (WBD)*	D.A. Gurnett, Univ. of Iowa, USA	Transmissions of E-field wave form up to ~100 kHz, variable centre frequency	Using sensors of EFW
Digital Wave Processor (DWP)*	L.J.C. Woolliscroft, Univ. of Sheffield, UK	Data compaction & compression, event selection, particle/wave correlation, control of WHISPER.	CMOS multiprocessor unit
Electron Drift Instrument (EDI)	G. Paschmann, MPI für Extraterrestrische Physik, Garching, FRG	**E**, (0.1 – 10 mV/m, <100 Hz), ∇**B**, |**B**| (5 – 1000 nT), emission and tracking of two electron beams	Two emitter/detector assemblies, each with 2π field of view (FOV)
Cluster Ion Spectrometry (CIS)	H. Rème, Centre d'Etude Spatial des Rayonnement, Toulouse, F	CODIF: Composition and Distribution Functions Analyser, 0 – 40 keV/q	Symmetric hemispherical analyser with RPA and TOF, $2\pi \times 8°$ FOV, split geometric factor
		HIA: Hot Ion Analyser for high time resolution (e.g. solar wind), 3 eV/q – 40 keV/q	Symmetric quadrispherical analyser, $2\pi \times 8°$ FOV with high resolution (≥2.8°).
Plasma Electron and Current Analyser (PEACE)	A.D. Johnstone, Mullard Space Science Laboratory, Holmbury St. Mary, UK	LEEA: Low-Energy Electron Analyser, 0 – 100 eV	Spherical electrostatic analyser, $\pi \times 3.8°$ radial FOV.
		HEEA: High-Energy Electron Analyser, 0.1 – 30 keV.	Toroidal electrostatic analyser, $2\pi \times 4.6°$ FOV
Research with Adaptive Particle Imaging Detectors (RAPID)	B. Wilken, MPI für Aeronomie, Lindau/Harz, FRG	IIMS: Imaging Ion Mass Spectrometer, ion distribution and species, energy 2 – 1500 keV/nuc.	Position-sensitive solid-state detectors with TOF section.
		IES: Imaging Electron Spectrometer, distribution of energetic electrons, energy 20 – 400 keV	Position-sensitive solid-state detector
Active Spacecraft Potential Control (ASPOC)	W. Riedler, Institut für Weltraumforschung, Graz, A	Spacecraft potential control, emission current of oder 20 μA, indium ions	Field-ionisation, liquid-metal ion emitter

* Members of the Wave Consortium
\+ We regret to announce the death of Les Woolliscroft. His role as PI has been taken over by H. Alleyne

been kept as far away as possible from the sensitive instruments, which are mounted around Cluster' s upper periphery. The thrusters and main engine are thus on the opposite end of the spacecraft.

Table IV
Major payload requirements on the Cluster spacecraft

Number of units per spacecraft	$\sim$24
Total mass of payload	$\leq$72 kg
Total power available for payload (end-of-life)	$\leq$47 W
Residual magnetism of spacecraft at magnetic sensors	$\leq$0.25 nT
Stability of residual magnetic field	$\leq$0.1 nT
Spacecraft skin potential point-to-point	$\leq$1 V
Accomodation of 100 m tip-to-tip wire booms	
High EMV cleanliness	
High cleanliness for AIV programme and thruster positions	

Trying to keep within the mass and power allocations (Table I) in the system budgets has been extremely demanding. Cluster' s various operating modes and the areas of operation within the elliptic orbit mean that the payload has a requirement for data rates varying between 17 and 220 kbit s^{-1}, whilst the overall volume of data to be handled per orbit can be as high as 2 Gbytes.

One of the major technical requirements on both the payload and the spacecraft has been to maintain electromagnetic cleanliness (EMC), as electric-wave and magnetic-field measurements are both adversely affected by spacecraft residual values. This resulted in a unique EMC programme for Cluster.

The planned detection of cold electrons by PEACE (Johnstone *et al.*, this issue) called for very special design efforts to eliminate the possibility of spurious effects being introduced by photoelectrons, which it is known will be abundant near the spacecraft' s skin. The elimination of voltage potential variations over the surface has meant that the outer skin of the spacecraft must be as conductive as possible.

3.3. ACHIEVING UNIFORMITY

The unique requirement for four identical sets of eleven instruments on four identical spacecraft operating in closely coordinated orbits imposes more stringent constraints on the accuracy of the individual instrument measurements than in a normal single-spacecraft mission.

The main error sources had to be quantified for each instrument and major efforts were required to reduce such errors by careful design, stringent control of materials and manufacturing processes, and extensive pre-flight (and later in-flight) calibration of the individual instruments, as well as careful intercomparison of different measurements from the different instruments.

For the particle detectors, CIS, PEACE, and RAPID, proper sensor calibration is especially important. It is also known that some of the long-term drifts in such instruments, mainly involving the micro-channel plates, are difficult to correct entirely.

PEACE is a good example of how differences between the instruments on the four spacecraft were assessed during the development programme. It is necessary to maintain comparability between the analysers on the four flight spacecraft to within 1%, including the in-flight intercalibration.

The PEACE analyser consists of two concentric hemispherical shells between which the electrons are deflected. In order to maintain the 1% goal, the concentricity of the hemispheres had to be maintained to better than 40 microns. When tested, the initial design was found to have a value of 150 microns. Even by improving the manufacturing tolerances to state-of-the-art values using specialist facilities, it could only be reduced to 80 microns. A major redesign was therefore undertaken both to increase the rigidity of the hemispheres and to connect them by a shorter load path using materials with similar coefficients of expansion (thereby reducing the effects of minor temperature differences between detectors). By these means, and by using a Kapton instead of a ceramic anode and changing the mechanical design, the tolerances could be reduced to 37 microns even using the normal flight-standard manufacturing tolerances.

For the wave experiments, the question of similarity on the four spacecraft was not a central issue. The stability of the oscillators is far better than the required frequency resolution, which is mainly dictated by telemetry constraints. In order to ensure correlation between measurements on the four spacecraft, the relative timing between the measurements on each spacecraft is very important, which has meant that operational constraints have had to be rigorously defined.

The Fluxgate Magnetometer (FGM) requires an overall single instrument accuracy of 0.1%. This has been demonstrated to be possible, and can also be checked in orbit using a four-point intercalibration technique (Div **B** must always be zero).

3.4. Optimisation of Resources

Payload resources were tight from the outset, with the baseline 72 kg/47 W/17 kbit s^{-1} already being oversubscribed at the proposal stage. A de-scoping and rationalisation exercise had to take place immediately, which resulted in the formation of the Wave Experiment Consortium (WEC) and the provision of the power-supply and data-handling functions via boxes common to all five of those experiments (Pedersen *et al.*, this issue).

During the development phase, the payload mass again began to grow and every additional gramme had to be accounted for. Very thin walled structures were developed for some electronic boxes (e.g., FGM) and some instruments switched to using magnesium as their primary structural material (e.g., WBD).

Power for driving the sophisticated instruments and their computers was also critical from an early stage. Various operating modes were therefore developed for each instrument to conserve power by, for example, running the processors at different speeds during the two-year mission. The data rates have also been optimised by defining different operational modes and time-sharing between the individual instruments.

The payloads to be launched on the flight spacecraft are still within their allocated resources, thanks to the continual attention that has been paid to the problem by both the Experimenters and the ESA Project Team.

3.5. SURVIVING THE LAUNCH AND IN-ORBIT ENVIRONMENT

Before making any scientific measurements, Cluster's instruments will have to survive the launch-induced vibrations. Once in orbit, they must work in an environment varying between cold deep space and hot sunlight, and also endure a constant shower of strong radiation.

The Cluster launch vibration environment has been assumed to be slightly harsher than usual, to take into account the inherent uncertainties due to it being the first Ariane-5 launch. This has had design repercussions for some instrument units. The CIS experiment, for example, contains very thin carbon foils (density of order 3 μg cm^{-2}). It was very soon realised that these would have problems surviving the vibration environment and so the experiment was mounted on rubber anti-vibration mounts. During testing, however, these foils still broke due to the high induced displacements causing the foil to hit the surrounding structure. A design modification and further testing has ensured that CIS will survive the launch.

The in-orbit thermal environment, varying from hot sunlight to long cold eclipses with a minimum of heater power provided by the spacecraft, has presented a severe challenge for all protruding sensors. These need an environment close to room temperature for their sensitive components, like the micro-channel plates, when operating, and no colder than a freezer when non-operational.

A few sensors required further design optimisation after the spacecraft thermal-balance test. In particular, their coatings and thermal blanket interfaces had to be changed. The thermal blankets are now 'hand-tailored' around the protruding instruments.

The predicted in-orbit radiation environment of 20 krad total dose that Cluster must endure has required the use of radiation-hardened or shielded components. Special circuitry to survive the latch-up that could be caused by particle intrusion has also been necessary, especially for the CMOS technologies.

3.6. THE ELECTROMAGNETIC-CLEANLINESS PROBLEM

The instruments aboard each new generation of scientific satellite endeavour to exploit the latest sensor and electronic technology to achieve the highest possible sensitivity and resolution. For a plasma mission like Cluster, this ideally means that the spacecraft itself should be 'invisible' , with no apparent interaction with the space environment.

Reality is rather different; the spacecraft does become charged under the cyclic influence of sunlight and shadow. In addition, certain of the spacecraft' s subsystems contain relays and latch valves, which generate DC or slowly-varying magnetic fields. It has an electrical support system for power and data-handling, which also constitutes a source of electromagnetic interference. Last but not least, the scientific payload itself may generate signals that could influence the performance of other instruments on board. Most of these phenomena are either impossible to simulate, or the levels involved are orders of magnitude too small to be measured on the ground.

How then can a suitable spacecraft and its payload be built and tested to the satisfaction of the scientific community? The process begins with a careful selection of design concepts and materials, which are then translated into an electromagnetic design and test specification. The necessary electrostatic cleanliness is achieved by using a conductive coating on the spacecraft's external surfaces, including the solar arrays. Thermal insulation blankets and foils are coated with indium tin oxide and, along with all other external parts, locally grounded to the spacecraft structure to avoid the build-up of electrostatic potential. DC magnetic cleanliness imposes the selection of non-magnetic materials wherever possible. In addition, the magnetic sensors are mounted on deployable booms, as far as possible from the spacecraft body.

Electromagnetic interference is usually controlled by the synchronisation of clock signals on-board the spacecraft. This, together with an optimised electrical harness configuration and grounding scheme, ensures minimum disturbance of the frequency bands being observed by the scientific instruments.

However, when these design optimisations came to be translated into hardware for Cluster, some of the above concepts could not be realised: either the ideal technology could not support the original purpose of a certain element or instrument, or it was not available within the given constraints of mass, power or schedule. An EMC Review Board was therefore established, made up of scientists from each particular area of Cluster plasma science, from Industry and from ESA, to analyse the problems and to agree on acceptable solutions.

The verification phase involved various EMC analyses and tests, not only state-of-the-art conductivity and susceptibility tests, but also dedicated experiment tests for the particularly sensitive WEC and PEACE instruments. After several iterations and optimisation of the grounding schemes of these experiments and the spacecraft interface, these two experiment teams were able to confirm a satifactorily low noise environment on the spacecraft. Radiated emission and susceptibility testing at IABG in Munich, with the EFW wire booms partially deployed, finally confirmed that the Cluster spacecraft generates levels very close to the instruments' own background noise.

In summary, the way to EMC cleanliness was not straightforward, but the close cooperation of all parties involved in addressing the problem has resulted in Cluster being one of the most electromagnetically clean spacecraft that ESA has ever launched.

3.7. Seven PCs on Each Spacecraft

A feature of modern-day instrumentation and the result of the great advances in microtechnology is the fact that each experiment has almost as much on-board computational capability as the spacecraft' s own on-board computer. In fact, the Cluster payload has the processing capability of almost seven personal computers (PCs), the Wave Consortium being served by one processing unit (DWP).

This on-board computational power not only provides the interfaces with the spacecraft bus in terms of telemetry and telecommand handling, but also the interface with the various experiment units. It also performs a certain amount of on-board scientific processing, which reduces the data rates needed to the ground.

The Digital Wave Processor (DWP), perhaps the most performant computer, consists of three T222 transputers, each with 32 kbytes of external RAM (internal memory disabled to provide increased radiation tolerance) and 32 kbytes of PROM. DWP's design permits the transputers to be operated at input clock frequencies of 2.5 or 5 MHz, the slower rate requiring less power.

DWP (Woolliscroft *et al.*, this issue) performs two major software-driven tasks. The first is particle correlation, with a novel diagnostic technique based on forming auto-correlation functions of the time series of particle-detector counts as a function of energy and pitch angle. Secondly, it performs data compression in order to cope optimally with the restrictions on the available telemetry bandwidth. Various data-compression methods are implemented within the DWP to remove any redundant information from the data stream.

3.8. Experiment Management

Management of the Cluster experiments has provided some unique challenges for the scientific community. To begin with, getting together the best payload suite in Europe to satisfy the mission requirements meant the involvement of 11 Principal Investigators and interfacing with the Co-Investigators who will study the Cluster data. This is not a unique scenario in terms of ESA missions, but it did mean that rationalisation of the instrument groups was required early in the programme. One result of this was the setting up of the WEC, mentioned earlier.

The next major problem was that, because the integration, testing and preparation for launch of the four flight spacecraft would take place over two years, engineering breadboards were required very early in the programme compared to the launch date. The experimenter groups performed this difficult task in a very satisfactory way for the hardware, but the software for ground testing and on-board processing always lagged behind.

For a nominal one-off spacecraft, once the flight hardware has been built and tested at the home institute, the payload team is available to assist the ESA project team with the system-level Assembly, Integration and Verification (AIV) programme. In Cluster' s case, once the first flight instruments had been delivered, another four (including flight spare) were needed. In addition, system-level testing had to be performed simultaneously on up to three flight spacecraft (e.g. functional testing, EMC, and thermal vacuum), all of which required payload support. The short-term schedule for such system testing was also varying on a day-to-day basis. The logistical problems for the experimenter teams have therefore been immense and a successful outcome hinged on the trust built up between the ESA project team and the payload groups in the early days of the programme, and on the flexibility shown by both sides. This has been supplemented by technological advances such as the remote links established to both the integration and test sites (Dornier and IABG), which gave the payload teams remote access to their test data from their home institutes.

4. Conclusion

The engineering challenges faced with the Cluster mission have been some of the most demanding of any spacecraft ever launched. They have been faced together by a joint team of the ESA project, European Industry and the scientific community, resulting in the production of four near-identical spacecraft carrying an advanced plasma payload. Once launched, the intercalibration of the scientific payload will achieve the objective of having four near-identical instruments sets in orbit to perform the first truly three-dimensional measurements of the Earth's magnetospheric environment. The efforts of the combined team over the past eight years will ensure that Cluster provides the worldwide scientific community with a wealth of data on the magnetosphere which, when combined with data from other missions such as SOHO, Wind, Polar and Geotail, will form a unique data set on the Earth/Sun interaction.

References

Escoubet, C. P. and Schmidt, R.: 1996, 'Cluster-Science and Mission Overview', *Space Sci. Rev.*, this issue.

Ferri, P. and Warhaut, M.: 1996, 'Cluster Mission Operations', *Space Sci. Rev.*, this issue.

Johnstone, A. *et al.*: 1996, 'PEACE: a Plasma Electron and Current Experiment', *Space Sci. Rev.*, this issue.

Pedersen, A. *et al.*: 1996, 'The Wave Experiment Consortium (WEC)', *Space Sci. Rev.*, this issue.

Sørensen, E. M., Merri, M., and Di Girolamo, G.: 1996, 'The Cluster Data Processing System: a Distributed System in Support of a Challenging Mission', *Space Sci. Rev.*, this issue.

Woolliscroft, L. J. C. *et al.*: 1996: 'The Digital Wave-Processing Experiment on Cluster', *Space Sci. Rev.*, this issue.

THE CLUSTER MAGNETIC FIELD INVESTIGATION

A. BALOGH, M. W. DUNLOP, S. W. H. COWLEY, D. J. SOUTHWOOD and J. G. THOMLINSON
The Blackett Laboratory, Imperial College, London, U.K.

K. H. GLASSMEIER, G. MUSMANN, H. LÜHR and S. BUCHERT
Institut für Geophysik und Meteorologie, Technische Universität Braunschweig, Germany

M. H. ACUÑA, D. H. FAIRFIELD and J. A. SLAVIN
Goddard Space Flight Center, Greenbelt, MD., U.S.A.

W. RIEDLER and K. SCHWINGENSCHUH
Institut für Weltraumforschung, Graz, Austria

M. G. KIVELSON
Institute for Geophysics and Planetary Physics, University of California, Los Angeles, CA., U.S.A.

THE CLUSTER MAGNETOMETER TEAM

Abstract. The Cluster mission provides a new opportunity to study plasma processes and structures in the near-Earth plasma environment. Four-point measurements of the magnetic field will enable the analysis of the three dimensional structure and dynamics of a range of phenomena which shape the macroscopic properties of the magnetosphere. Difference measurements of the magnetic field data will be combined to derive a range of parameters, such as the current density vector, wave vectors, and discontinuity normals and curvatures, using classical time series analysis techniques iteratively with physical models and simulation of the phenomena encountered along the Cluster orbit. The control and understanding of error sources which affect the four-point measurements are integral parts of the analysis techniques to be used. The flight instrumentation consists of two, tri-axial fluxgate magnetometers and an on-board data-processing unit on each spacecraft, built using a highly fault-tolerant architecture. High vector sample rates (up to 67 vectors s^{-1}) at high resolution (up to 8 pT) are combined with on-board event detection software and a burst memory to capture the signature of a range of dynamic phenomena. Data-processing plans are designed to ensure rapid dissemination of magnetic-field data to underpin the collaborative analysis of magnetospheric phenomena encountered by Cluster.

1. Introduction: Overview of Objectives

The four-spacecraft Cluster mission (Escoubet *et al.*; Credland *et al.*, this issue) will provide the first opportunity to determine the three-dimensional, time-dependent characteristics of small-scale processes and structures in the near-Earth space plasma, both in the magnetosphere and in the nearby interplanetary medium. Small-scale phenomena, such as localised, transient magnetic reconnection (flux transfer events) or turbulent diffusion, which operate at the boundaries between plasmas of different origin, are largely responsible for determining the nature and geometry of the interactions. The objectives of the Cluster mission and of the magnetic-field

Space Science Reviews **79:** 65–91, 1997.

investigation are the study of phenomena on spatial scales corresponding to the proposed separation distances of the four spacecraft and their interpretation in terms of the role they play on a more global scale to shape the properties of different regions of the magnetosphere and the upstream solar wind.

Cluster will sample most of the key regions of the magnetosphere: the dayside magnetospheric boundary, both at mid-latitudes and in the cusp, where processes associated with magnetic reconnection and turbulence are believed to occur; the near-Earth magnetospheric tail on the nightside which undergoes frequent large-scale magnetic reconfigurations during substorms; and the upstream solar wind, bow shock and magnetosheath. The anisotropy which is inherent in all magnetised plasma processes, introduced by the magnetic field, makes the accurate determination of the magnetic field at high time resolution a critical contribution to the probing of small-scale structures and dynamics encountered in these regions.

The Cluster mission was conceived to overcome some of the inherent limitations and difficulties in interpreting previous magnetospheric observations, which were made overwhelmingly by a single spacecraft. The range of phenomena in the magnetosphere and its boundaries that can be uniquely addressed by multi-point magnetic field measurements is considerable. The following examples are used to illustrate, rather than exhaustively to describe, the potential advantages of the four-spacecraft observations.

The current understanding of the structure and dynamics of the main boundaries on the sunward side of the magnetosphere, the bow shock and the magnetopause, is based to a large extent on observations of single crossings and numerical simulations. However, as shown by dual spacecraft observations, these boundaries have complex, evolving structures and are non-stationary on several time- and length scales (see, for example, the reviews of upstream waves, bow shock and magnetopause observations by Russell (1989), Thomsen (1989), and Elphic (1989), respectively). Four-point magnetic-field observations are expected to provide information on the immediate neighbourhood of the boundaries which will clarify the phenomenology of boundary-associated phenomena simultaneously at the four locations. A particularly interesting objective of Cluster on this topic is the exploration of processes occuring at quasi-parallel bow shock geometries, where the generally accepted shock re-formation occurs (Burgess, 1989; Scholer and Burgess, 1992), as has been shown by increasingly sophisticated simulations (e.g., Scholer *et al.*, 1993; Giacalone *et al.*, 1994). Boundary crossing phenomena include complex wave fields that are an integral part of the MHD processes that form the wider environment of the boundaries; inherent non-stationarity prevents the resolution of these wave fields by single- or even two-point measurements. During a two-year mission, Cluster will cross the dayside bow shock and magnetopause several hundred times. This will allow the observation of both boundaries under a range of conditions which will provide the basis for a separation of different phenomena and causal processes. The magnetometers on Cluster, with their high time resolution and their event-driven Microstructure Analyser (see Section 3), will generate the necessary data sets to determine the context and processes at these boundaries.

Other specific phenomena for four-point analysis to which the observations of the magnetic field on Cluster will contribute include dayside magnetic reconnection and the nature of flux transfer events in particular; the role of the Kelvin–Helmholtz instability in magnetopause processes; impulsive events, possibly driven by discontinuities in the solar wind; and, as a prime objective of Cluster, the nature, structure and dynamics of the outer cusp (as summarised for example by Paschmann, 1995; Scholer, 1995).

In the magnetospheric tail, Cluster observations will target both phenomena at small scales, such as the structure and time evolution of current sheets associated with the growth phase of substorms and, on larger scales, to establish the magnetic topology of the different regions in the tail and their dynamics. The role, in particular, of a range of current structures and their magnetic signatures, all on scales which require high time resolution magnetic measurements, simultaneously at several locations, will contribute to resolving the time evolution of the magnetospheric tail prior to and during substorms (as summarised for example by Cowley, 1995; Roux and Sauvaud, 1995).

A significant subset of the objectives of the magnetic field investigation has close connections with the wider field of solar-terrestrial relations. In this context, collaboration with simultaneous ground-based observations will play a key role in relating magnetospheric responses with signatures observed by remote-sensing, ground-based instruments and facilities, such as magnetometer chains and ionospheric radars (Opgenoorth *et al.*, this issue).

The analysis of four-point magnetic field measurements presents formidable conceptual and technical challenges. In Section 2, we outline the main strands of scientific data analysis techniques which have been developed or considered for the Cluster magnetic-field investigation and comment on their expected applicability. Some of these techniques, together with details of their applications and the complexities of the expected results are summarised in this paper. The evolution of the Cluster orbit, the configuration of the tetrahedron and the separation distances are discussed briefly, in the light of the importance of the orientation and shape of the tetrahedron along the orbit with respect to the magnetic field in the different regions of the magnetosphere.

While magnetometers, in particular of the fluxgate type, have been the mainstay of magnetospheric missions, the design of the Cluster magnetometers presents specificities which are described in Section 3. More details of the instrument functions can be found in Balogh *et al.* (1993). The intercalibration of the magnetometers on the four spacecraft is a major requirement, as the combined analysis of the data is largely based on the measurement of differences between spacecraft. Error analysis, ground calibration and plans for in-flight calibration procedures have therefore played a greater than usual role in the pre-launch development of the instrument; these are summarised in Section 4.

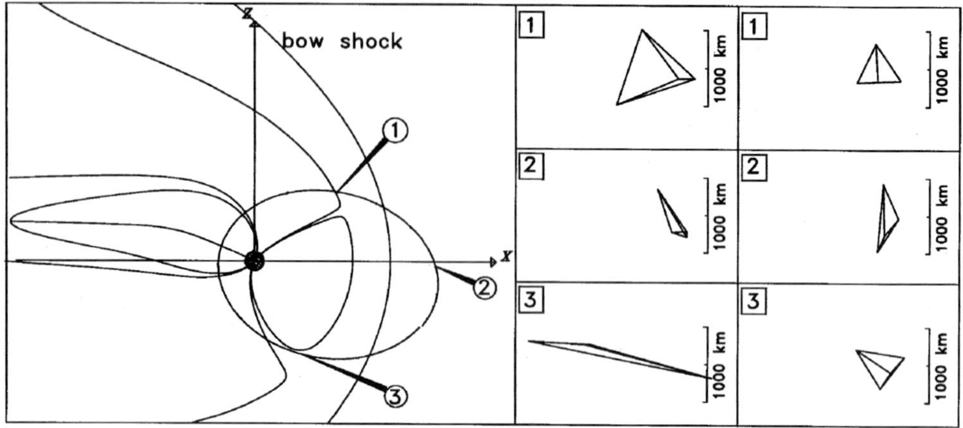

Figure 1. Model orbit of Cluster in the noon-midnight meridian, with representative shapes of the Cluster tetrahedron corresponding to different separation strategies.

The orbit selected for Cluster is polar, with a perigee of 4 R_E and an apogee of 19.6 R_E. The shape of the four-spacecraft Cluster configuration is basically a tetrahedron which, once set up at a particular point, evolves around the orbit in a deterministic way (for example, Dunlop, 1990). The orbital evolution of the tetrahedral shape, its orientation and its degree of deformation with respect to both the main local anisotropies and the reference regular tetrahedron in the different regions of the magnetosphere, as well as the separation distances, all play a key role in constraining the applicable analysis techniques. While some of the parameters of the orbit and of the configuration are relatively free and will be tuned during the mission, others are more constrained by both operational and dynamic considerations.

For the purpose of illustrating the context of the measurements, Figure 1 represents one of the model orbits of Cluster, with the shape of the tetrahedron magnified at three locations. The orbit represents the epoch when the apogee of the spacecraft is on the sunward side of the magnetosphere. The two sets of tetrahedra correspond to two possible constellation strategies. The first is the baseline mission (although the exact details are dependent on launch date), with a regular tetrahedron set up at the location of the northern cusp. The second constellation strategy was developed by recognising inherent flexibilities in orienting the tetrahedron which allow, as a secondary constraint, the setting up of a regular tetrahedron at a second location (in this case, at the southern cusp) along the orbit. The deformation of the constellation at the location of the southern cusp in the first case illustrates well the evolution of the tetrahedron; the fact that both strategies are possible (and a significant range in between) illustrates the complexity involved in the planning of the different orbit strategies. Other significant and unforeseeable complications arise from the dynamic reactions of the magnetosphere to changing solar wind and

Table I
Cluster magnetic field investigation (FGM) investigator team

A. Balogh (PI)	
	The Blackett Laboratory, Imperial College, London, U.K. D.J. Southwood
S.W.H. Cowley	
	Dept. of Physics and Astronomy, University of Leicester,
Leicester, U.K.	
K.-H. Glassmeier	Institut für Geophysik und Meteorologie,
H. Lühr	Technische Universität Braunschweig, Germany
G. Musmann	
M. H. Acuña	NASA/GSFC, Greenbelt, MD., U.S.A.
D. H. Fairfield	
J. A. Slavin	
W. Riedler	Institut für Weltraumforschung, Graz, Austria
K. Schwingenshuh	
F. M. Neubauer	Institut für Geophysik und Meteorologie, Universität zu Köln, Germany
M. G. Kivelson	Institute for Geophysics and Planetary Physics, UCLA, Los Angeles, CA., U.S.A.
M. Tatrallyay	RMKI/KFKI, Budapest, Hungary
R. C. Elphic	Los Alamos National Laboratory, NM, U.S.A.
F. Primdahl	Dansk Rumforkningsinstitut, Lyngby, Denmark
A. Roux	CETP/USQV, Velizy, France
B. T. Tsurutani	Jet Propulsion Laboratory, Pasadena, CA, U.S.A.

interplanetary magnetic-field conditions; the magnetosphere illustrated is based on the Tsyganenko model, assuming moderately disturbed conditions. An early, but thorough discussion of the effects of the potential range of orbit and constellation parameters on magnetic field measurements and analysis can be found in Dunlop (1990).

The Cluster magnetic-field investigation is directly supported by a large scientific team, representing eleven institutions in seven countries. The investigator team is listed in Table I. Extensive plans for post-launch operations, data processing and exploitation have been drawn up; these are summarised in Section 5. The coordination of the data dissemination, through the Cluster Science Data System in particular will enable the establishment of early initiation of scientific cooperation between the magnetic field investigators, the Cluster community, and the wider interested science community.

2. Four-spacecraft Data Analysis Techniques

The most important attribute of the mission and of the magnetic-field investigation is the possibility that is provided to compare observations that are relatively closely spaced compared to the principal dimensions of the magnetosphere. It is anticipated that many of the processes on the scales to be sampled by Cluster are in fact those which shape the larger-scale structure and properties of the system. While much is known in terms of such larger-scale processes and structures, the small scale is often assumed rather than understood; in this sense, Cluster is an exploratory mission. The full phenomenology of small-scale processes is likely to be discovered during the mission. This necessitates the preparation of the data analysis of the forthcoming magnetic-field observations in terms of tools that recognise basic physical processes and structures. However, the likely complexity of phenomena on these scales, both temporally and spatially, implies the need for a flexible methodology to discern the underlying component processes. Therefore a dual approach, based on both the underlying physics and judicious modelling of more complex superpositions of temporal and spatial dependences, is required for data analysis.

In general terms, the analysis and interpretation of the Cluster magnetic-field data has to proceed by successively iterating assumed models, based on expectations derived from knowledge or assumptions about magnetospheric structures and processes, and their goodness of fit to the data. This general analysis procedure is illustrated in Figure 2. Despite the apparently self-evident generality of this procedure, a detailed analysis of the methodology in these terms is essential to identify the specific problems and the applicability of any data-analysis technique to the case of the Cluster magnetic-field data. Several studies based on this general methodology, applied to the expected Cluster magnetic-field data, have explored specific aspects of the problem of interpreting observed signatures in terms of a known input model.

Given that the Cluster magnetic-field data set consists of the vector measurements of the magnetic field at the location of the four Cluster spacecraft, combined with the location and time of the measurements, three different data analysis techniques have been considered. These are the curlometer, the wave telescope and the discontinuity analyser, as summarised in Table II. As shown in the Table, the main distinguishing feature of the three techniques is their applicability relative to the scale sizes of the basic phenomena to be analysed. This interaction between scale size and spacecraft separation is one of the main characteristics of the expected magnetic-field observations.

The next level of complexity in the magnetometer data analysis arises as a result of fundamental anisotropies in the magnetosphere. While it is assumed that the best basic configuration of the Cluster tetrahedron is a regular one, the spacecraft constellation naturally evolves through the orbit. This leads to the need to consider the dominant orientation (e.g., the elongation) of the four-spacecraft configuration with respect to the natural anisotropies in the different magnetospheric regions

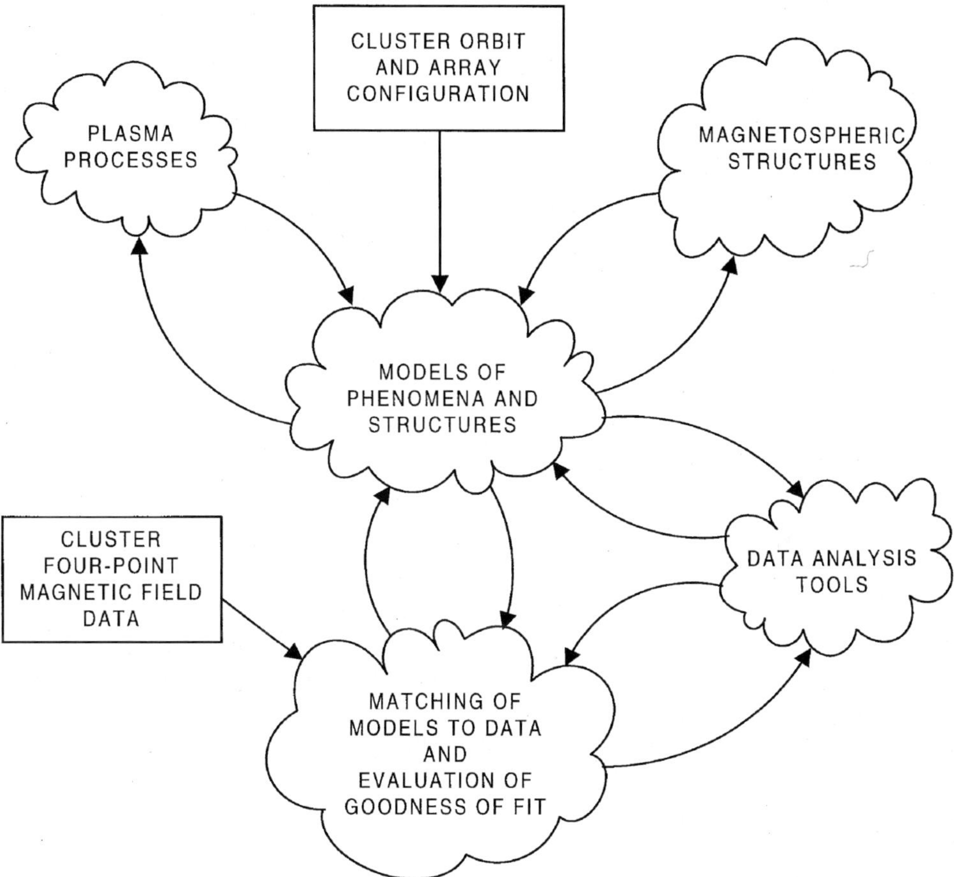

Figure 2. Conceptual methodology for the analysis of four-point magnetic field data from Cluster, representing the need for multiple iteration between data analysis techniques and physical modelling of magnetospheric phenomena.

Table II

Relative scales	Analysis technique	Representative expression
$L > \mathbf{R}_{ij}$	Curlometer	$\mu_0\mathbf{J}\cdot(\Delta\mathbf{R}_i \times \Delta\mathbf{R}_j) = \Delta\mathbf{B}_i \cdot \Delta\mathbf{R}_j - \Delta\mathbf{B}_j \cdot \Delta\mathbf{R}_i$ $\langle \mathrm{div}\,\mathbf{B}\rangle_{\mathrm{average}} = 0$
$L \approx \mathbf{R}_{ij}$	Wave telescope	$\phi_i = \mathbf{k}\cdot\Delta\mathbf{R}_i - \omega\Delta t_{ij}$ $\mathbf{k} = \mathbf{k}_0 + \mathbf{G}$
$L < \mathbf{R}_{ij}$	Discontinuity analyser	$\langle n\rangle, n_j, v_{\parallel}, a_{\parallel}$

encountered by Cluster. The orientation and scale length of the tetrahedron with respect to the background magnetic field, the background flow and the normals to discontinuity surfaces (bow shock, magnetopause) strongly condition the expected differential magnetic signatures, and therefore the most appropriate analysis tools to be used for the identification of specific small-scale properties of the plasma phenomena. An overview of the relationship between phenomena expected in and near the magnetosheath, their anisotropic properties and the applicability of data-analysis tools has been given in Dunlop *et al.* (1993). In the magnetosheath, the interplay between the orientation of the Cluster tetrahedron with respect to the dominant flow and field directions on the one hand, and the applicability of analysis tools on the other, results in the sampling of extended structures along the background field, short scales transverse to the field and intermediate scales roughly along the direction of the flow component transverse to the magnetic field. Other regions of the magnetosphere will present other combinations between the measurement anisotropies (due to the shape of the tetrahedron and the use of different data analysis methods) and the preferred orientation and scale sizes of static and dynamic magnetospheric phenomena.

The combination of vector differences between simultaneous measurements at the four locations, using Maxwell's equations, can be interpreted in terms of the average current density of the plasma. This technique (the curlometer) yields, assuming stationarity, a measure of the components of the current density vector. However, simple modelling of curlometer measurements in a model magnetosphere have shown (Dunlop *et al.*, 1990) that field gradients on the scale of, or shorter than the Cluster separation can bias the estimate of the current density vector, indicated by the implied non-zero divergence of the vector differences. It is important to recognise that natural gradients in a static model magnetosphere already introduce such distortions in the estimates based on four-spacecraft data; in other words, the technique is sensitive to the global current systems which shape the magnetosphere. Several model studies (Dunlop *et al.*, 1990; Robert and Roux, 1993; Dunlop and Balogh, 1993; Robert *et al.*, 1995) along the Cluster orbit have explored the sensitivity of the current density estimates based on difference measurements to the shape of the Cluster tetrahedron.

Dynamic effects, on both large- and small-scales, are additional sources of current which affect the curlometer output. Nevertheless, a routine measure of the current density, implied by the difference measurements of the magnetic field, can detect the presence and direction of local current systems.

The variability of the magnetic field on scales comparable to the spacecraft separation vectors can be examined in terms of 'waves'. The 'wave telescope' method of data analysis has already been extensively investigated in connection with Cluster (for example Neubauer and Glassmeier, 1990, Neubauer et al., 1990, Motschmann *et al.*, 1995a, b). The complexity of the possible wave fields and and the task to describe them in terms of self-consistent sets of quantitative parameters require the use, not just of advanced signal analysis techniques, but also a range of physical assumptions (Pinçon and Lefeuvre, 1991, Motschmann and Glassmeier, 1995, Motschmann *et al.*, 1995b, Pinçon, 1995). Temporal and spatial stationarity can be assumed only as a first approximation; the small-scale dynamics of magnetospheric processes will invariably lead to the need to consider, almost on a case-by-case basis, an iterative approach of data analysis as outlined above.

The third basic tool to be used for the analysis of the magnetic-field data from the four Cluster spacecraft is the discontinuity analyser. Recognising that boundaries between plasma regimes are almost always in motion, a simple fitting of the times of boundary crossings detected at the four locations is unlikely to be sufficient. The minimum-variance technique (Sonnerup and Cahill, 1967), originally developed for the study of boundary crossings by single spacecraft, has been extensively developed and applied to the determination of boundary normals in space plasmas. Application of minimum-variance analysis to the Cluster magnetic field data has required a re-evaluation of this technique (Dunlop *et al.*, 1995a) to explore its limitations and to delineate the conditions for its use. Other techniques also need to be investigated (Chapman and Dunlop, 1993; Watkins *et al.*, 1995). The starting point may be the application of the technique to single-spacecraft data, followed by a comparison of the properties (normals and velocities) derived at the four points. An ideal set would provide at most small differences between the normal directions, thus satisfying the condition for (near) planarity of the boundary. It is expected that an important class of boundary crossings will satisfy the near-planarity condition, as well that of near-stationarity (i.e., when the boundary crossing is fast compared to the evolution of the Cluster tetrahedron).

However, if the normals determined at the four locations are significantly different, more complex assumptions need to be built into the data analysis (Dunlop *et al.*, 1995b). Some important non-planar structures expected are flux transfer events and field-line resonance structures. A hierarchical data analysis approach, iteratively feeding back data on the event parameters to improve a modelled boundary structure, is needed to resolve the overdetermined, but ambiguous four-spacecraft data. Complex boundaries where time stationarity breaks down significantly, such as quasi-parallel bow shock crossings, represent an important target for Cluster where extremes of temporal and spatial variability are expected (Giacalone *et al.*, 1994; Burgess, 1995). Even one-dimensional simulations of the shock reformation process which characterises the dynamics of the high Mach number, quasi-parallel bow shock illustrate well the challenge posed to Cluster to detect the spatial and temporal evolution of such shock waves.

An illustration of Cluster magnetic field measurements through a simple non-planar two dimensional spatial structure is given in Figure 3. The magnetic field is defined by $b_x = B_0 e^{-ikx} e^{-k|y|}$, $b_y = ib_x$ for $y < 0$ and $b_y = -ib_x$ for $y > 0$, with b_z an arbitrary constant. Ignoring this last component, the structure is that of a stationary sinusoidal wave along the x direction, with wave normal k, but with an exponentially decaying magnitude on either side of the $y = 0$ axis. Two fly-throughs were considered (Dunlop et al., 1995b), one parallel to

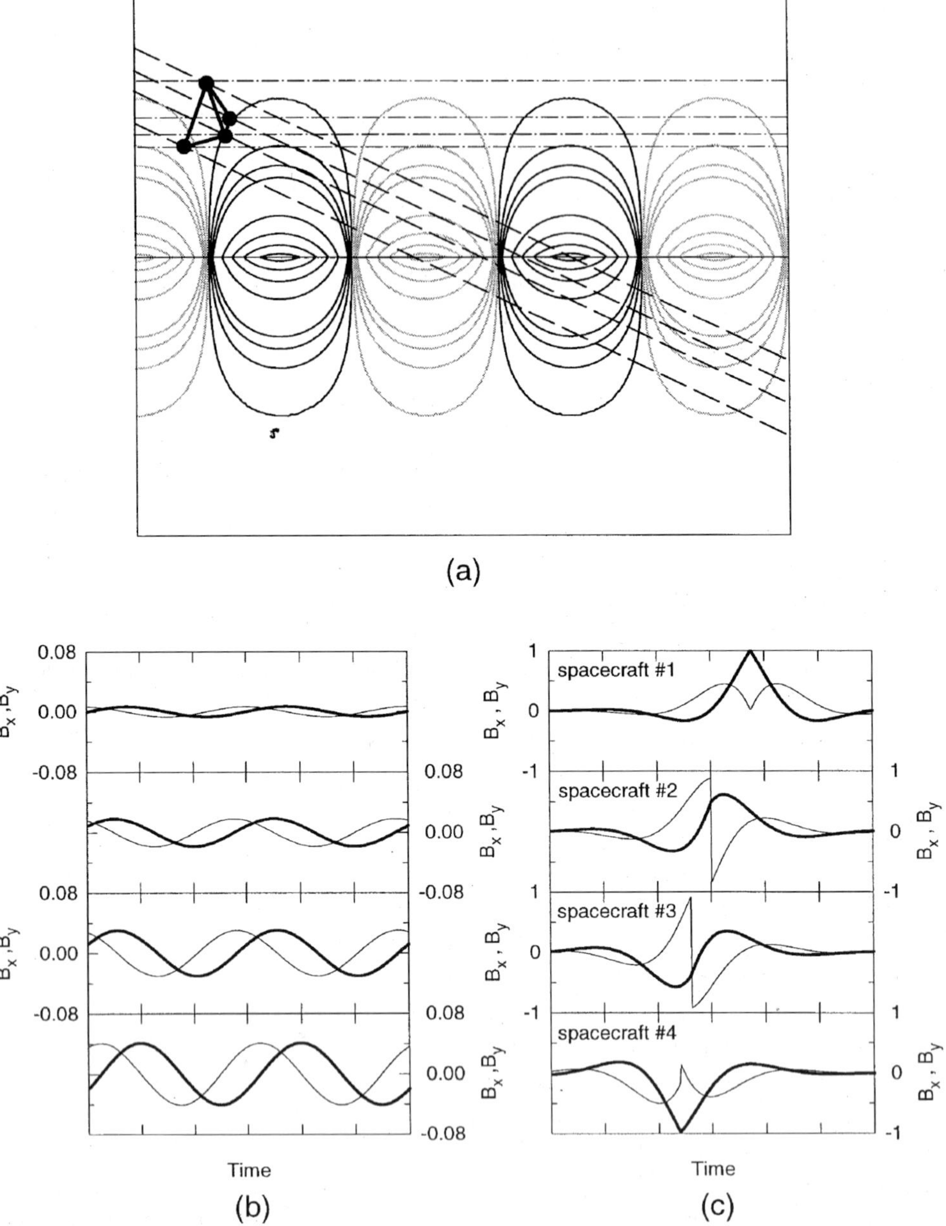

Figure 3. (a) Model of surface wave of the magnetic field and two possible Cluster trajectories through it. (The contour plot of b_x is shown.) (b) and (c) Signatures of the surface wave in (a) for the two sample trajectories of Cluster in the magnetic-field data. (Thick and thin lines show the b_x and b_y components, respectively.)

the x axis, and one obliquely across the structure, as illustrated in Figure 3(a). The 'measured' b_x and b_y components for the two trajectories are illustrated in Figures 3(b) and 3(c), respectively. For the trajectory along the x axis, the two components are in quadrature, with different amplitudes measured at the location of the four spacecraft. A variance analysis yields a minimum direction along the z axis, clearly in contradiction with the geometry of the model. However, the non-planar nature of the structure can be deduced from the two components. The oblique trajectory yields, in the b_x component, the apparent signature of a wave packet, with a propagation vector approximately along the fly-through trajectory. The behaviour of the b_y component, with abrupt phase changes at the crossing of the $y = 0$ surface, indicates, however, the true nature of the field configuration. The conclusions to be drawn from this simple example is that the signature of the magnetic field configuration in the Cluster data set will depend heavily on the respective orientations and scale sizes of the Cluster tetrahedron and the actual field geometry.

3. Instrumentation

Each Cluster spacecraft carries an identical instrument to measure the magnetic field. Each instrument, in turn, consists of two triaxial fluxgate magnetometers and an on-board Data-Processing Unit (DPU). The block diagram of the instrument is shown in Figure 4. The mass of each of the triaxial fluxgate sensors is 290 g, with an additional 48 g for the thermal cover. The mass of the electronics box is 2060 g. The instrument power consumption in normal operations is 2460 mW.

The fluxgate magnetometers are similar to many previous instruments flown in Earth-orbit and on other, planetary and interplanetary missions. In order to minimise the magnetic background of the spacecraft, one of the magnetometer sensors (the outboard, or OB sensor) is located at the end of one of the two 5.2 m radial booms of the spacecraft, the other (the inboard, or IB sensor) at 1.5 m inboard from the end of the boom. In flight, either sensor can be designated as the Primary Sensor, for acquiring the main data stream of magnetic-field vectors. Selection of the sensors as Primary or Secondary is made by ground command; in the default configuration, the OB sensor is used as the Primary Sensor. Data are also acquired simultaneously from the other, Secondary Sensor, albeit at a lower rate.

The magnetometers have eight possible operating ranges; of these, five are used on the Cluster magnetometers, shown in Table III. These ranges were selected to provide good resolution in the solar wind (with expected field magnitudes between 3 and 30 nT), and up to the highest field values expected in the magnetosphere along the Cluster orbit (up to about 1000 nT). The highest range (65 500 nT) is used only to facilitate ground testing. Range selection can be automatic (controlled by the instrument DPU) or commanded from the ground. When in the automatic mode, a range selection algorithm running in the DPU continuously monitors each

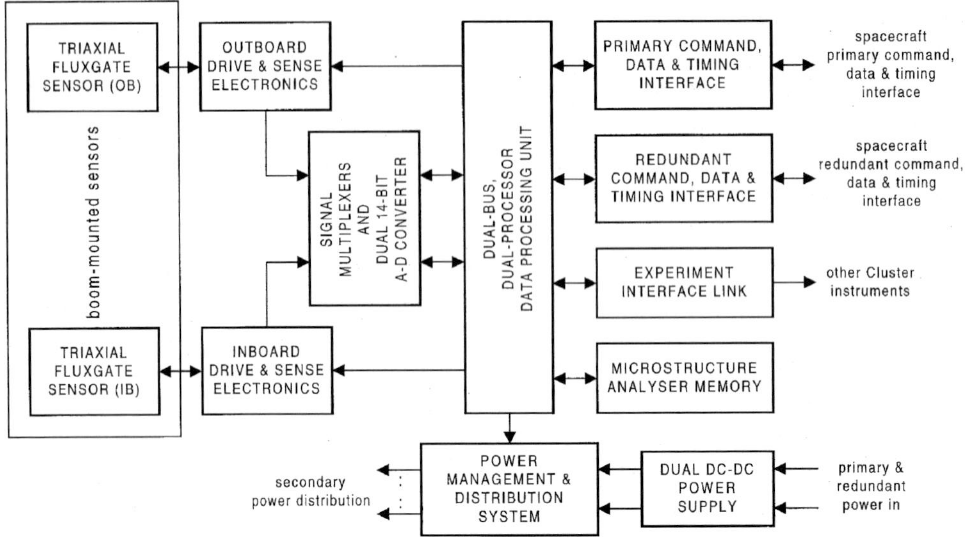

Figure 4. Block diagram of the Cluster magnetometer (FGM). The instrument has a high level of redundancy and fault tolerance.

Table III

Range (nT)	Digital resolution (nT)
−64 to +63.992	7.813×10^{-3}
−265 to +255.97	3.125×10^{-2}
−1024 to +1023.9	0.125
−4096 to +4095.5	0.5
−65536 to +65528	8

component of the measured field vector. If any component exceeds a fraction (set at 90%) of the range, an up-range command is generated and transmitted to the magnetometer. If all three components are smaller than 10% of the range for more than a complete spin period (in fact for more than the telemetry frame period of 5.15222 s), an automatic down-range command is generated to the magnetometer.

The instrument uses two 16-bit Analogue-to-Digital Converters (ADC), normally in cold redundancy. The most significant 14 bits of the converted field-component values are used for generating the output of the instrument. The digital resolution, corresponding to the five ranges of the magnetometers, is shown in Table III.

The DPU of the instrument contains two, redundant, Central Processing Units (CPU), based on the Low-Power Processor System developed by British Aerospace, using the MAS281 microprocessor, implementing the MIL-STD-1750A instruction set. The two CPUs are linked to each other and to other functions within the DPU by a dual, redundant, data and address bus. Two, redundant spacecraft interface units are used to receive telecommands and to transmit data to the telemetry, via the two (primary and redundant) spacecraft command and telemetry interfaces. An internal power management system, controlled by ground commands, allows the selective switching of functional units within the DPU. The power switches can also automatically switch off functions in the DPU when an excessive current drain is detected. The instrument is powered by two, redundant, power converter units, which receive power independently from the spacecraft. The overall design of the instrument has a highly failure-tolerant architecture.

The data-acquisition process is controlled by the DPU. The three components of the magnetic- field vector measured by the Primary Sensor are sampled at a constant rate of 201.75 Hz, independently of the finally transmitted vector rate. Data acquired at this rate are digitally filtered to match the transmitted rate, according to the operating modes of the instrument and the telemetry data rate allocated to it.

The spacecraft telemetry operates in different modes, corresponding to different data acquisition and transmission rates for the different instruments. The magnetometer has been allocated four telemetry data rates, corresponding to the Nominal Mode (at 1211.13 bits s^{-1}), Burst Mode 1 (3465.69 bits s^{-1}), Burst Mode 2 (1347.77 bits s^{-1}) and Burst Mode 3 (5583.61 bits s^{-1}). The spacecraft telemetry is based on a packetised protocol, with a constant frame length of 5.15222 s, independent of telemetry rate. The vector data rates from the magnetometer associated with the different telemetry modes of the spacecraft are shown in Table IV. In the Nominal and Burst Mode 2 telemetry modes, the instrument uses three, commandable vector transmission rates (FGM Telemetry Options A, B, and C), distinguished by different rates for the Primary and Secondary Sensors, and the data rate read out from the Microstructure Analyser memory. In Burst Mode 1, no data are read out from the Microstructure Analyser memory. In Burst Mode 3, scheduled collectively for all the instruments with burst memory capabilities, data are dumped fast from the Microstructure Analyser memory. The magnetometer DPU uses Gaussian-windowed digital filtering to reduce the stream of primary vector samples to match the bandwidth of the telemetry. Figure 5 shows (in the upper panel) the (unnormalised) weights associated with the transmitted vector data rates and (in the lower panel) the weighting of successive transmitted vector samples at the 15.5 vectors s^{-1} rate.

The instrument incorporates a Microstructure Analyser (MSA) which consists of a burst memory capable of storing about 32 000 magnetic field vectors, and the associated event recognition software in the DPU. The objective of the MSA is to record events of short duration, at acquisition rates not available at normal or burst telemetry rates, when the magnetic signature of the event satisfies programmable preset criteria. Data recorded in the MSA are read out in the telemetry either in a background mode, in parallel with the normal data acquisition, or in a short burst, in Burst Mode 3 of the spacecraft telemetry. Considerable flexibility has been built

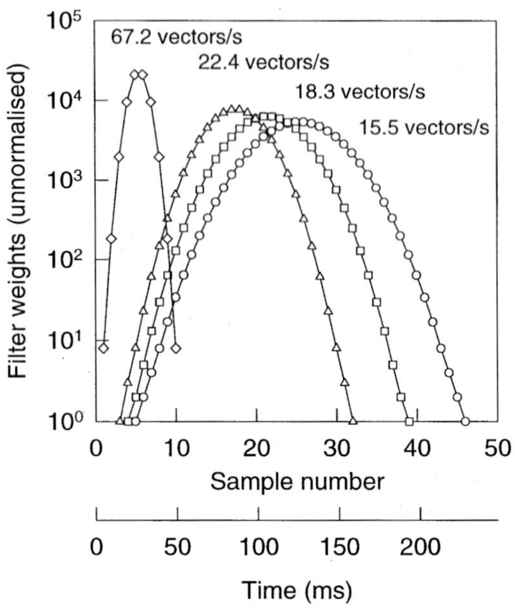

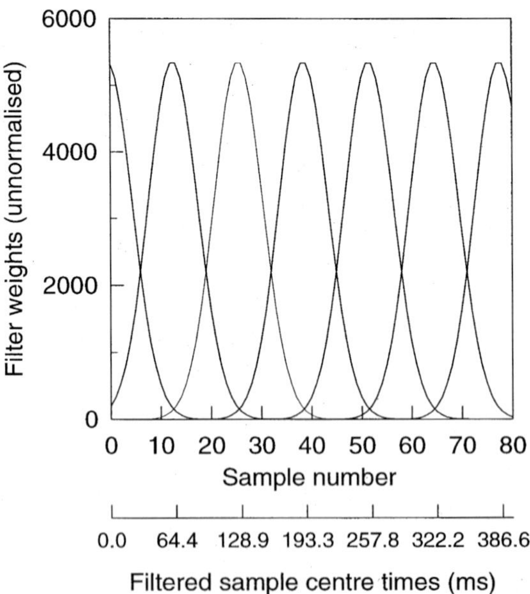

Figure 5. (a) Gaussian filter weights used in the FGM Data Processing Unit to match the internally sampled vector rate to the transmitted telemetry rate. (b) Overlap between the digitally filtered individual telemetry samples of the magnetic-field vector at the 15.5 vector s^{-1} rate.

Table IV

Spacecraft telemetry modes	FGM telemetry option	Primary sensor vector rate (vector s^{-1})	Secondary sensor vector rate (vector s^{-1})
Nominal modes 1, 2, 3	A	15.519	1.091
Burst mode 2	B	18.341	6.957
	C	22.416	3.011
Burst mode 1	D	67.249	7.759

into the operation of the MSA, to allow in-flight optimisation of its operation and to program it to recognise different classes of short-duration events.

The DPU continuously calculates the long- and short-term averages of five parameters which are used in the event detection algorithm. These are as follows:

– The 'pseudo-variance', $\sigma_x = \langle|B_x - b_x|\rangle$, representing a measure of the average scatter of the magnetic field component along the spin axis sampled at the full rate (b_x) around the filtered and transmitted value (B_x) of this component. The calculated pseudo-variance is transmitted regularly in the telemetry and is also used as an input to the event detection algorithm.

– The spin axis component (B_x) of the magnetic field, corrected for offset.

– The squared magnitude of the magnetic field, $B^2 = B_x^2 + B_y^2 + B_z^2$, with all components corrected for offsets.

– The squared magnitude of magnetic field in the spin plane, $B^2 = B_y^2 + B_z^2$, corrected for offsets.

– A measure of the angular change, $B_L^2 = (\delta B_y)^2 + (\delta B_z)^2$, in the spin plane component of the magnetic field.

Running long- and short-term averages of the above quantities are calculated over programmable intervals. For each parameter, the difference between the long term average (L) and the short term average (S) is continuously compared to a programmable threshold factor (T) multiplied by the long-term average, and the whole offset by a programmable parameter (O). If the condition $|L-S| > |TL|+O$ is met, a flag is set for that parameter. The instrument Event Flag is set if a programmable combination of these individual flags is detected.

Sampled vectors, at rates which can be programmed up to the full sampling rate of the instrument, are continuously written into the MSA memory. If the MSA event detection mode is enabled, the memory is frozen after a preset interval following the detection of the Event Flag in such a way that the occurrence of the event is approximately half-way through the contents of the memory. In this way, data are recorded both before and after the occurrence of the event which triggered the MSA memory. Both the triggering and readout of the MSA memory are programmable in such a manner that its operation is optimised for the acquisition of physically significant events around the Cluster orbit.

As the data coverage around the Cluster orbit is not complete, but subject to careful planning to cover regions of greatest interest for the four-spacecraft capabilities of the mission, the overall context of the observations will not always be apparent. This can be alleviated to some extent by the magnetometer using a unique flight operations mode developed late in the pre-launch programme. In this mode, the magnetometer records spin-averaged magnetic field vectors in the MSA memory during periods without telemetry coverage and then dumps the data thus acquired when telemetry coverage is resumed. It is possible to record data in this way for up to 27 hours, thus providing the context for the intervals when all instruments acquire data. The implementation of this mode is dependent on details of the operations plans, but provides a potentially important tool for enhancing the return for the mission as a whole. The operation of the mode does not affect the use of the MSA for its original purpose of acquiring snapshots of very high resolution magnetic field data during periods of normal telemetry coverage.

The magnetometer also provides, on board, a stream of sampled vectors to other instruments through the Inter-Experiment Link (IEL), to coordinate measurement sequences and to assist on-board data reduction in real time, as a function of the direction and magnitude of the magnetic field. Vectors sampled at intervals of 64.35 ms are transmitted to the plasma ion and electron detectors (CIS and PEACE), the energetic particle detector (RAPID), to the instruments in the Wave Consortium Experiment and, at a higher clock speed, to the electron gun experiment (EDI). Additionally, the trigger signal derived in the magnetometer Microstructure Analyser, signalling the occurence of an 'event' (as defined above), is also transmitted on the Inter-Experiment Link.

The magnetometer flight instrumentation was provided by the Goddard Space Flight Center (sensors and analogue electronics), the Institute für Weltraumforschung, Graz (Analogue-to-Digital Converters), Imperial College, London (Data Processing Unit, Spacecraft Interfaces and Power Converter), and the Technische Universität Braunschweig (Microstructure Analyser). Instrument management and integration, the spacecraft-level test programme and the Ground Support Equipment were the responsibility of Imperial College. Instrument calibration and a significant contribution to the overall magnetic cleanliness programme were the responsibility of the Technische Universität Braunschweig.

4. Measurement Error Sources and Instrument Calibration

The objectives of the magnetic field investigation on Cluster rely on the accurate measurement of vector differences between spacecraft, together with accurate knowledge of the vector differences between spacecraft locations. The correlation of the measurements also requires accurate timing. The quantitative evaluation of the effects of errors on the four-point parameters was first given by Dunlop *et al.* (1990). This is an essential task, intended to ensure that the effect of measurement errors and uncertainties is minimised in the intrinsically complex data analysis procedures outlined in Section 2. To illustrate the necessary error analysis, we present the considerations which apply to the determination of the current density vector, using the difference approximation to curl **B**.

The estimate of the current density vector **J** depends on the vector differences between spacecraft locations, defined as $\mathsf{R}_{ij} = \mathsf{R}_i - \mathsf{R}_j$, where R_i and R_j are the locations of spacecraft i and j, respectively; and on the vector differences $\mathsf{B}_{ij} = \mathsf{B}_i - \mathsf{B}_j$ between the magnetic field vectors B_i and B_j measured by the magnetometers on those spacecraft. The upper limit of the error in the differences in spacecraft locations can be taken as $\mathsf{R}_{ij} = \delta\mathsf{R}$, specified for Cluster as 10 km for each component of the difference. The total error in the differences in the magnetic field between spacecraft is $\delta\mathsf{B}_{ij} = (\delta\mathsf{B}_i + \delta\mathsf{B}_j + \mathsf{B}^t_{ij}$, where $\delta\mathsf{B}_i = \Sigma_k\, \delta\mathsf{B}^k_i$ at each spacecraft, due to frame transformation (including orthogonality) errors and offsets, and $\delta\mathsf{B}^t_{ij} = (\delta|\mathsf{B}|/\delta t)_{ij}\delta t_{ij}$ is the error component due to inter-spacecraft timing errors.

The errors in B_i and B_j arise in the overall error budget through the transformations which are needed to go from the measurement frame to a geophysical frame:

$$\begin{pmatrix} \mathsf{B}_x \\ \mathsf{B}_y \\ \mathsf{B}_z \end{pmatrix} = \mathsf{G}(\rho, \psi)\mathsf{R}(\varphi)\mathsf{T}(\alpha\beta) \begin{pmatrix} \mathsf{S}_1/\lambda_1 - \mathsf{b}_1 \\ \mathsf{S}_2/\lambda_2 - \mathsf{b}_2 \\ \mathsf{S}_3/\lambda_3 - \mathsf{b}_3 \end{pmatrix},$$

where S_l is the sensor output along its l axis (l = 1, 2, or 3); b_l are the offsets, λ_l are the sensitivities; T is the tranformation matrix from instrument to (spinning) spacecraft coordinates, R is the despin matrix and G is the transformation from the spacecraft frame into a geophysical coordinate system.

Under some assumptions the effects of the errors on the estimate of the current density can be written as

$$\frac{\delta\mathsf{J}}{|\mathsf{J}|} = \mathsf{S}_R\frac{\delta\mathsf{R}}{\mathsf{R}_{ij}} + \mathsf{S}_B\frac{\delta\mathsf{B}}{\mathsf{B}_{ij}},$$

where S_R and B_B are coefficients which represent the effect of the shape of the Cluster tetrahedron on the estimate. In the best case (for a regular tetrahedron) S_R and S_B are of order 1, but for deformed configurations, there is a magnifying effect, with the two coefficients of order 5 to 10. Essentially similar results were found by Robert and Roux (1993), and Robert *et al.* (1995), using a range of criteria for characterising the 'quality' of the tetrahedron, as it evolves around each orbit from a regular to an elongated, finally planar and even, occasionally, nearly aligned configuration (before repeating the cycle on the next orbit). In terms of errors in the determination of the current density, typical fractional errors of order 10% are expected for spacecraft separation uncertainties of 5 to 10% for the case of the regular tetrahedron, but 50% or more when the tetrahedron is significantly distorted.

Inter-spacecraft timing errors are limited to 4 ms by specification of the Cluster mission. This introduces an uncertainty in the estimated current density which, in general, is less than that due to the inter-spacecraft distance errors.

Other four-point parameters which can be derived by the analysis tools discussed in Section 2 are similarly affected by the uncertainties in the magnetic field measurements, the spacecraft separation vectors and inter-spacecraft timing. However, the error sources enter into the analysis in a more complex manner and their effects need to be evaluated on a case-by-case basis. For instance, uncertainties associated with the determination of normal directions of discontinuity surfaces using minimum variance analysis (even in the case of single spacecraft data) are more often due to the complexity associated with phenomena near the discontinuity and the selection of the intervals used than with inherent errors in the data set.

Given that uncertainties arise not only from intrinsic instrument performance (including cross-calibration of the instruments on the four spacecraft), but from mission-associated performance and constraints (spacecraft attitude, inter-spacecraft distances, timing), error control by the magnetometer is limited to comprehensive instrument calibration, both pre-launch and in flight. Extensive instrument-level calibrations were carried out in the Magnetsrode facility of the Technische Universität, Braunschweig, with additional calibration activities at the large magnetic facility (MFSA) of the Industrieanlagen Betriebsgesellschaft (IABG) where spacecraft-level magnetic tests were also carried out.

Instrument calibration has covered the determination of the magnetic alignment of the sensors with respect to their reference (geometric) axes; offsets and sensitivity factors for all ranges; temperature dependences of these parameters; frequency and transient responses. The frequency response of the sensors is adapted to the basic sampling rate of the instrument; the results of calibration of one of the tri-axial sensors is shown in Figure 6. These pre-launch calibration parameters will serve as a first input to a continuous monitoring programme in flight. For measuring accurately the sensitivity of the magnetometers, a calibration field was switched on and off periodically, with a period of 2 s. The response of one of the sensors to a calibration field of 38 nT is shown in Figure 7(a). Given that the complete calibration sequence consists of 64 such on-off cycles of the calibation field, the response can be analysed in the frequency domain. The result of such an anlysis for the frequency range 0 to 1 Hz is shown in Figure 7(b). The signal peak of 38.071 nT at 0.5 Hz shows the magnitude of the sensor response to the applied calibration field. The signal-to-noise ratio of the calibration step is of order 60 db. This result was obtained during tests at IABG, using calibrated magnetic field steps; in space a calibration voltage is applied in the magnetometer feedback loop to provide, using the same analysis techniques, an accurate check on the sensitivity of the sensors.

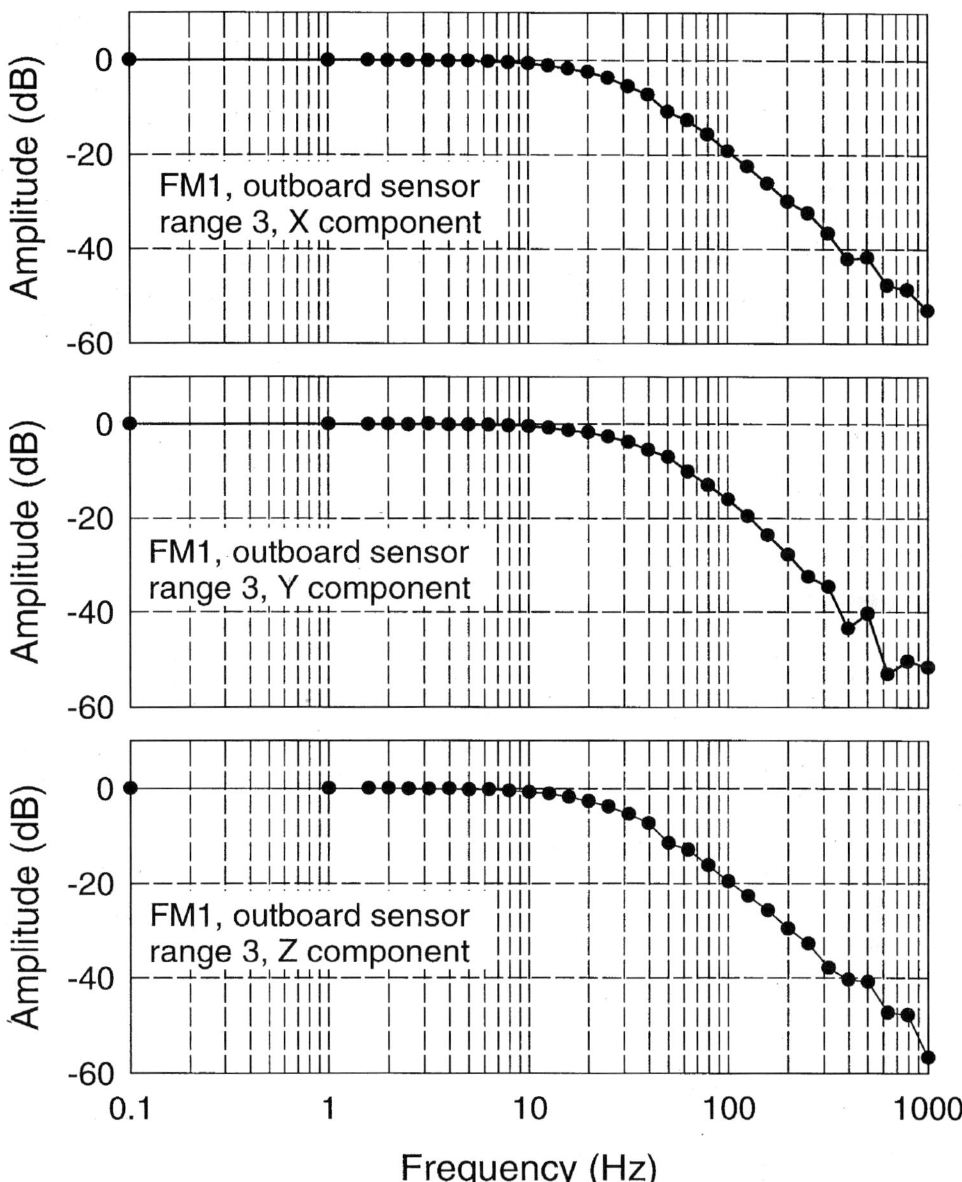

Figure 6. Analogue response of the three sensors of the one of the vector fluxgate magnetometers.

Many calibration parameters, such as alignment with respect to spacecraft axes, are susceptible to be refined in space; for some, such as offsets, continuous monitoring is essential. In-flight calibration techniques are routinely applied to magnetic-field data from single spacecraft. However, on Cluster, there is the additional opportunity to compare closely related data from four spacecraft. While such com-

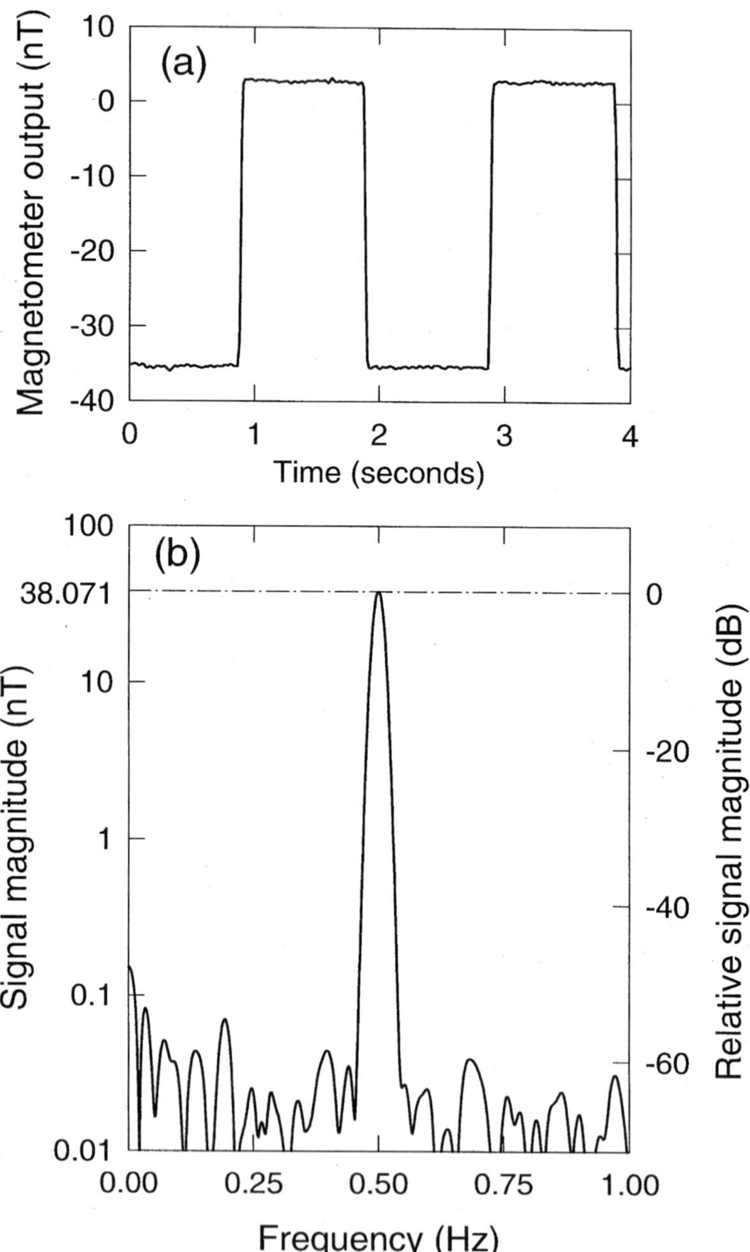

Figure 7. (a) Time domain response of one of the magnetometer sensors to the repeated application of a −38 nT magnetic field. (b) Frequency domain analysis of a magnetometer sensor to a 38 nT magnitude, 2 s period, 64-cycle calibration signal.

parisons are made, in the first instance, to derive differential quantities required by the basic objectives of Cluster, a systematic study of the residuals provides an opportunity directly to inter-calibrate the instruments in space.

A major potential source of uncertainties for magnetic-field measurements in space is background fields due to the spacecraft at the location of the magnetometer sensors. For the Cluster mission, as for previous ESA spacecraft with similar requirements, a comprehensive magnetic cleanliness programme was implemented, to ensure that any disturbance caused by the spacecraft is minimised. The sensors are mounted on a 5 m boom, which locates them at about 4.7 m and 6.2 m from the centre-line of the spacecraft, or at 3.2 m and 4.7 m from the surface of the spacecraft, respectively. The maximum allowable DC magnetic field at the location of the outboard magnetometer sensor is 0.25 nT; the field should not vary by more than 0.1 nT in 100 s. Strict magnetic cleanliness requirements were enforced, with all elements of the spacecraft allocated a maximum magnetic budget. All units were individually mapped and, if necessary, compensated. The magnetic signatures of all units were incorporated into a synthetic magnetic model for each spacecraft, maintained by the Technische Universität, Braunschweig. The overall magnetic signatures of all four spacecraft were measured in the MFSA-IABG facility. Three of the spacecraft were found to be outside the 0.25 nT specification, but were compensated. Results of the measured and modelled fall-off of the F3 spacecraft's magnetic field is shown in Figure 8 (taken from the report on these tests prepared by K. Mehlem); as a result of compensation, the background field at both inboard and outboard sensors is 0.12 nT, a very satisfactory value for the magnetic field investigation. The other three spacecraft have, similarly, background fields significantly better than the specification.

5. Operations and Data Processing Plans

The in-flight operation of the magnetometer is, in principle, simple. It basically measures a single quantity, the three components of the magnetic field vector, and transmits a time series of vectors at rates which are closely related to the operating mode of the spacecraft telemetry. The time resolution of the magnetometer in fact always exceeds by up to two orders of magnitude the time resolution of the plasma instruments, which have considerably more complex measurement tasks. As a result, the scientific operation of the mission, which includes the selection of parts of the orbit for higher time resolution, burst mode, data taking, is largely dependent on the requirements of instruments other than the magnetometer.

Science operations, in particular the planned data coverage, can be looked at from two related, but somewhat conflicting points of view. The complex interplay between the data coverage along the orbit and ground station visibility, combined with the significant, but finite on-board data storage leads to trade-offs between burst mode (high data rate) and nominal mode (standard data rate) operations.

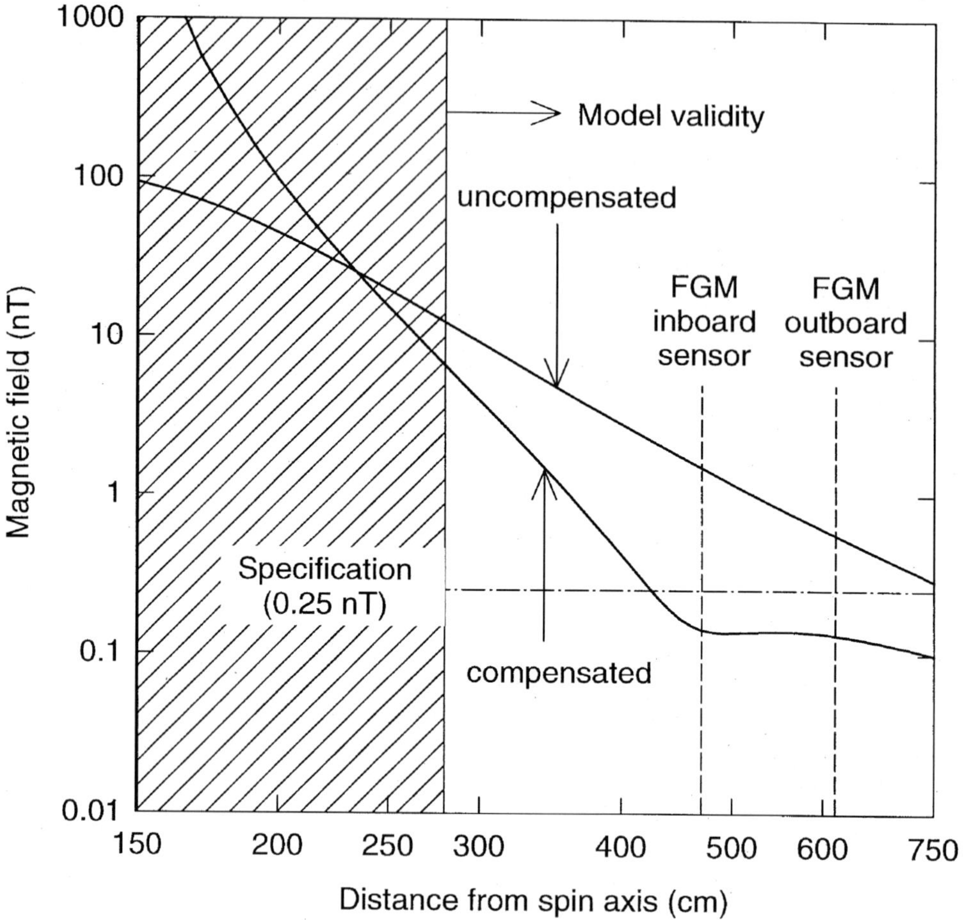

Figure 8. Measured fall-off of the spacecraft background magnetic field (compensated and uncompensated) along the axis of the magnetometer boom, showing the final low value of the magnetic disturbance caused by the spacecraft at the location of the magnetometer sensors. (The results of the modelling data are shown courtesy of K. Mehlem, ESTEC.)

While 'magnetospheric' objectives would need maximal, preferably complete coverage along the orbit, objectives associated with 'small-scale plasma processes' require operations in the high-data-rate mode, even at the expense of complete coverage. These objectives can be regarded as complementary; however, for the magnetometer (because of its inherently high sample rate) maximising the coverage along the orbit would normally be preferable.

The operation of the magnetometer will be fully coordinated through the Cluster Joint Science Operations Centre (JSOC) (Hapgood *et al.*, this issue). Inputs to science operations planning will be done through the Cluster Science Working Team and the Science Operations Working Group. Routine command generation and monitoring of the status, health and safe operation of the instrument through access to quick-look data will be carried out from Imperial College.

Data processing development for the magnetic field investigation has been led by the Technische Universtät Braunschweig, with support by Imperial College. In common with other Cluster investigations, full support has been given to the Cluster Science Data System (Schmidt and Escoubet, this issue), in particular to the UK Cluster Data Handling Facility at the Rutherford Appleton Laboratory, to generate the magnetic field Prime and Summary Parameter Data Bases, intended for use by the wider investigator community as the basis for collaborative studies. Continuous effort will be made to ensure the quality of the magnetic field Prime Parameter Data Base (PPDB) to make it suitable to serve as the basis of multi-instrument investigations. Furthermore, the need for the magnetic field data by other investigations to generate their own Prime Parameter Data will be supported by the distribution and maintenance of software (by the Hungarian Data Centre) at the national CSDS centres which will generate the magnetic-field Processing Support Data Set.

Internal to the magnetic field investigator team, software to generate full time resolution data will be used at most sites. A 1 s averaged Reference Magnetic Field data set will be generated and maintained by the team. In-flight calibration of the instrument and the generation of calibration files will be coordinated by Imperial College. An important architectural aspect of the data processing is the handling of the calibration files. Several versions of these will be generated, with a strict account maintained of their use. The routine processing of the Prime and Summary Parameter Data requires an early estimate of the calibration parameters. These needs will be covered by calibration files generated for operational use, based on an early evaluation of the calibration parameters using the results of ground calibration, passes through the quick-look data and a first pass through the complete daily data set. These operational files will be recorded but will not be re-used. A 'current best' calibration file will be generated and maintained at Imperial College which will cover the best estimates of calibration parameters covering the complete period from launch. This file will be updated as refinements are made to the calibration parameters, both at single spacecraft and four-spacecraft level.

It is expected that data gathered by the magnetometers on Cluster will provide a basic input to magnetospheric research for a duration well beyond the lifetime of the mission. Archiving plans, although at a preliminary stage, will be implemented to ensure the continued availability of the data. These plans concern specifically the Prime Parameter Data sets generated at the Cluster Data Centres, the 1 second Reference Data Set generated by the Cluster FGM team and access to high resolution data from the Cluster Raw Data Medium. For the processed data sets, archiving will consist in maintaining the data in their final calibrated state. For access to higher resolution data, archiving will consist not only in the maintenance of the raw data, but also in the maintenance of related orbit and calibration files, the processing software, and coordinate transformation routines needed to generate calibrated data sets in geophysical coordinate systems.

6. Summary

– The magnetic field investigation (FGM) on Cluster will provide accurate, high time-resolution, intercalibrated four-point measurements of the magnetic-field vector in the magnetosphere and in the upstream solar wind along the Cluster orbit.
– These measurements, in conjunction with other plasma parameters measured on Cluster will enable the analysis of complex, three-dimensional processes, boundaries and their dynamics. The applicability of these techniques depends strongly on the relative size and orientation of the Cluster tetrahedron as it evolves around the orbit with respect to anisotropies and scale sizes of plasma phenomena and structures in the different regions of the magnetosphere.
– The four-point magnetic field measurements require a range of new analysis techniques which integrate classical time series analysis with modelling and simulation of the observed phenomena. The different analysis techniques rely on the detailed analysis of the error sources which arise not only from uncertainties in magnetic-field measurements but also from uncertainties in the knowledge of the Cluster geometry. Pre-flight and in-flight calibration of the magnetometers, including four-spacecraft intercalibration techniques in flight, will minimise the errors in the magnetic field measurements. The intensive magnetic-cleanliness programme has ensured that the spacecraft background will not disturb the measurements.
– The coordination of in-flight operations and the extensive data-processing programme will ensure a timely dissemination of the magnetic field data to underpin the expected cooperative data analysis programmes necessary to exploit the opportunities offered by the Cluster mission.

Acknowledgements

We acknowledge our indebtedness to many people at the different institutions who contributed to the successful implementation of the Cluster magnetic field investigation. These have included T. Beek, C. Carr, H. Larbie, E. Serpell, B. Wingfield, T. Woodward, and S. Balogh at Imperial College; R. Kempen, H. Kügler, M. Rahm, I. Richter, J. Warnecke at the Technische Universität Braunschweig; J. Scheifele at the Goddard Space Flight Center; O. Aydogar at the Institut für Weltraumforschung, Graz; K. Khurana at UCLA. We are grateful for the support of the Cluster Project Scientist, Dr R. Schmidt and his Deputy, Dr P. Escoubet; the support and unfailing cooperation of the ESA Cluster Project Team: J. Credland, J. Ellwood, B. Gramkow, H. Bachmann, and their colleagues; the support provided for mission analysis and operations by the ESOC Team: M. Warhaut, J. Rodriguez-Canabal, P. Ferri and their colleagues; and the professionalism of the Dornier Cluster Project Team.

Financial support for the Cluster magnetic field investigation has been provided by the Particle Physics and Astronomy Research Council in the United Kingdom, DARA in Germany, the Osterreichisches Akademie der Wissenschaften in Austria and NASA in the United States.

References

Balogh, A., Cowley, S. W. H., Dunlop, M. W., Southwood, D. J., Thomlinson, J. G., Glassmeier, K.-H., Musmann, G., Lühr, H., Acuña, M. H., Fairfield, D. H., Slavin, J. A., Riedler, W., Schwingenschuh, K., Neubauer, F. M., Kivelson, M. G., Elphic, R. C., Primdahl, F., Roux, A., and Tsurutani, B. T.: 1993, 'The Cluster Magnetic Field Investigation: Scientific Objectives and Instrumentation', *Cluster: Mission, Payload and Supporting Activities*, ESA SP-1159, 95.

Burgess, D.: 1989, 'Cyclic Behaviour at Quasi-Parallel Collisionless Shocks', *Geophys. Res. Letters* **16**, 345.

Burgess, D.: 1995, 'Virtual Instruments for Space Plasmas', *Proc. Cluster Workshop on Physical Measurements and Mission Oriented Theory*, ESA SP-371, 179.

Chapman, S. C. and Dunlop, M. W.: 1993, 'Some Consequences of the Shift Theorem for Multispacecraft Measurements', *Geophys. Res. Letters* **20**, 1033.

Coeur-Joly, O., Robert, P., Chanteur, G., and Roux, A.: 1995, 'Simulated Daily Summaries of Cluster Four-Point Magnetic Field Measurements', *Proc. Cluster Workshop on Physical Measurements and Mission Oriented Theory*, ESA SP-371, 223.

Cowley, S. W. H.: 1995, 'Prospects for Progress in Geotail and Substorm Studies with Cluster', *Proc. Cluster Workshop on Physical Measurements and Mission Oriented Theory*, ESA SP-371, 253.

Credland, J. *et al.*: 1996, *Space Sci. Rev.*, this issue.

Dunlop, M. W.: 1990, 'Review of the Cluster Orbit and Separation Strategy: Consequences for Measurements', *Proc. Int. Workshop on Space Plasma Physics Investigations by Cluster and Regatta*, ESA SP-306, 17.

Dunlop, M. W. and Balogh, A.: 1993, 'On the Analysis and Interpretation of Four-Spacecraft Magnetic Fields Measurements in Terms of Small Scale Plasma Processes', *Spatio-Temporal Analysis for Resolving Plasma Turbelence*, ESA WPP-047, 223.

Dunlop, M. W., Southwood, D. J., Glassmeier, K.-H., and Neubauer, F. M.: 1988, 'Analysis of Multi-Point Magnetometer Data', *Adv. Space Res.* **8** (9), 273.

Dunlop, M. W., Balogh, A., Southwood, D. J., Elphic, R. C., Glassmeier, K.-H., and Neubauer, F. M.: 1990, 'Configurational Sensitivity of Multipoint Magnetic Field Measurements', *Proc. Int. Workshop on Space Plasma Physics Investigations by Cluster and Regatta*, ESA SP-306, 23.

Dunlop, M. W., Southwood, D. J., and Balogh, A.: 1993, 'The Cluster Configuration and the Directional Dependence of Coherence Lengths in the Magnetosheath', *Spatio-Temporal Analysis for Resolving Plasma Turbulence*, ESA WPP-047, 295.

Dunlop, M. W., Woodward, T. I., Motschmann, U., Southwood, D. J., and Balogh, A.: 1996, 'Analysis of Non-Planar Structures with Multipoint Measurements', *Adv. Space Res.* **18** (8), 309.

Dunlop, M. W., Woodward, T. I., and Farrugia, C. J.: 1995a, Minimum Variance Analysis: Cluster Themes', *Proc. Cluster Workshop on Data Analysis Tools*, ESA SP-371, 33.

Dunlop, M. W., Woodward, T. I., Motschmann, U., Glassmeier, K.-H., Southwood, D. J., and Balogh, a.: 1995b, 'Analysis of Non-Planar Structures with Multi-Point Techniques: a Roadmap for Cluster?', *Proc. Cluster Workshop on Physical Measurements and Mission Oriented Theory*, ESA SP-371, 267.

Elphic, R. C.: 1988, 'Multipoint Observations of the Magnetopause: Results from ISEE and AMPTE', *Adv. Space Res.* **8** (9), 223.

Escoubet, P. and Schmidt, R.: 1996, 'Cluster – Science and Mission Overview', *Space Sci. Rev.*, this issue.

Ferri, P. and Warhaut, M.: 1996, 'Cluster Mission Operations', *Space Sci. Rev.*, this issue.

Giacalone, J., Schwartz, S. J., and Burgess, D.: 1994, 'Artificial Spacecraft in Hybrid Simulations of the Quasi-Parallel Earth's Bow Shock: Analysis of Time Series Versus Spatial Profiles and a Separation Strategy for Cluster', *Ann. Geophys.* **12**, 591.

Glassmeier, K.-H., Motschmann, U., and Stein, R. von: 1995, 'Mode Recognition of MHD Wave Fields at Incomplete Dispersion Measurements', *Ann. Geophys.* **13**, 76.

Hapgood, M. A., Trimbylow, T. G., Sutcliffe, D. C., Chaizy, P. A., Ferran, P. S., Hill, P. M., and Tiratay, X. T.: 1996, 'The Joint Operations Centre', *Space Sci. Rev.*, this issue.

Motschmann, U. and Glassmeier, K.-H.: 1995, 'Mode Recognition of MHD Wave Fields', *Proc. Cluster Workshop on Data Analysis Tools*, ESA SP-371, 103.

Motschmann, U., Woodward, T. O., Glassmeier, K.-H., and Dunlop, M. W.: 1995a, 'Array Signal Processing Techniques', *Proc. Cluster Workshop on Data Analysis Tools*, ESA SP-371, 79.

Motschmann, U., Woodward, T. I., Glassmeier, K.-H., and Southwood, D. J.: 1995b, 'Wavelength and Direction Filtering by Magnetic Measurements of Satellite Arrays: Generalised Minimum Variance Analysis', *J. Geophys. Res.*, submitted.

Neubauer, F. M. and Glassmeier, K.-H.: 1990, 'Use of an Array of Satellites as Wave Telescope', *J. Geophys. Res.* **95**, 19 115.

Neubauer, F. M., Glassmeier, K.-H., Walter, R., and Dunlop, M. W.: 1990, 'Cluster as a Wave Telescope', *Proc. Int. Workshop on Space Plasma Physics Investigations by Cluster and Regatta*, ESA SP-306, 51.

Opgenoorth, H. J. *et al.*: 1996, 'Opportunities for Magnetospheric Research with Coordinated Cluster and Ground-Based Measurements', *Space Sci. Rev.*, this issue.

Paschmann, G.: 1995, 'Magnetopause, Cusp and Dayside Boundary Layer: Experimental Point of View', *Proc. Cluster Workshop on Physical Measurements and Mission Oriented Theory*, ESA SP-371, 149.

Pinçon, J. L.: 1995, 'Cluster and the k-filtering', *Proc. Cluster Workshop on Physical Measurements and Mission Oriented Theory*, ESA SP-371, 87.

Pinçon, J. L. and Lefeuvre, F.: 1991, 'Local Characterization of Homogeneous Turbulence in a Space Plasma from Simultaneous Measurement of Field Components at Several Points in Space', *J. Geopys. Res.* **96**, 1789.

Robert, P. and Roux, A.: 1990, 'Accuracy of the Estimate of J via Multipoint Measurements', *Proc. Int. Workshop on Space Plasma Physics Investigations by Cluster and Regatta*, ESA SP-306, 23.

Robert, P. and Roux, A.: 1993, 'Dependence of the Shape of the Tetrahedron on the Accuracy of the Estimate of the Current Density', *Spatio-Temporal Analysis for Resolving Plasma Turbulence*, ESA WPP-047, 289.

Robert, P., Roux, A., and Coeur-Joly, O.: 1995, 'Validity of the Estimate of the Current Density Along the Cluster Orbit with Simulated Magnetic Field Data', *Proc. Cluster Workshop on Physical Measurements and Mission Oriented Theory*, ESA SP-371, 229.

Roux, A. and Sauvaud, J. A.: 1995, 'Splinter Session on Magnetotail and Substorms', *Proc. Cluster Workshop on Physical Measurements and Mission Oriented Theory*, ESA SP-371, 205.

Russell, C. T.: 1988, Multipoint measurements of upstream waves, *Adv. Space Res.* **8** (9), 147.

Schmidt, R. and Escoubet, C. P.: 1996, 'The Cluster Science Data System (CSDS) – a New Approach to the Distribution of Scientific Data', *Space Sci. Rev.*, this issue.

Scholer, M.: 1995, 'Magnetopause, Cusp and Dayside Boundary Layer: Theoretical Point of View', *Proc. Cluster Workshop on Physical Measurements and Mission Oriented Theory*, ESA SP-371, 153.

Scholer, M. and Burgess, D.: 1992, 'The Role of Upstream Waves in Supercritical Quasi-Parallel Shock Re-Formation', *J. Geophys. Res.* **97**, 8319.

Scholer, M., Fujimoto, M., and Kucharek, H.: 1993, 'Two-Dimensional Simulations of Supercritical, Quasi-Parallel Shocks: Upstream Waves, Downstream Waves and Shock Re-Formation', *J. Geophys. Res.* **98**, 18 971.

Sonnerup, B. U. O. and Cahill, L. J.: 1967, 'Magnetopause Structure and Attitude from Explorer 12 Observations', *J. Geophys. Res.* **72**, 171.

Southwood, D. J.: 1990, 'Multispacecraft Measurements at Varying Scalelengths – the Cluster/Regatta Opportunity', *Proc. Int. Workshop on Space Plasma Physics Investigations by Cluster and Regatta*, ESA SP-306, 1.

Thomsen, M. F.: 1988, 'Multi-Spacecraft Observations of Collisionless Shocks', *Adv. Space Res.* **8** (9), 157.

Tsyganenko, N. A.: 1987, 'Global Quantitative Models of the Geomagnetic Field in the Cislunar Magnetosphere for Different Disturbance Levels', *Planetary Space Sci.* **35**, 1347.

von Stein, R., Glassmeier, K.-H., and Motschmann, U.: 1993, 'Cluster as a Wave Telescope and Mode Filter', *Spatio-Temporal Analysis for Resolving Plasma Turbulence*, ESA WPP-047, 211.

Watkins, N. W., Chapman, S. C., and Dunlop, M. W.: 1995, 'Fourier Techniques in Space Plasma Measurements', *Proc. Cluster Workshop on Physical Measurements and Mission Oriented Theory*, ESA SP-371, 235.

THE WAVE EXPERIMENT CONSORTIUM (WEC)

A. PEDERSEN[1]
Space Science Department, ESTEC, Keplerlaan 1, Noordwijk, The Netherlands
N. CORNILLEAU-WEHRLIN[3], B. DE LA PORTE[4], A. ROUX[2] and A. BOUABDELLAH[5]
CETP/UVSQ, 10/12 Ave. De l'Europe, Velizy, France
P. M. E. DÉCRÉAU[3], F. LEFEUVRE[2] and F. X. SÈNE[5]
LPCE/CNRS, 3A Ave. de Recherche Scientifique Orléans, France
D. GURNETT[3] and R. HUFF[5]
University Iowa, Department Physics and Astronomy, Iowa City, U.S.A.
G. GUSTAFSSON[3] and G. HOLMGREN[6]
Swedish Institute of Space Physics, Uppsala, Sweden
L. WOOLLISCROFT[3+7], H. ST. C. ALLEYNE[3] and J. A. THOMPSON[5]
University Sheffield, Department Automatic Control and Systems Engineering, Sheffield, England
P. H. N. DAVIES[5]
Sussex Space Centre, ENGG2, Falmer, Brighton, U.K.

Abstract. In order to get the maximum scientific return from available resources, the wave experimenters on Cluster established the Wave Experiment Consortium (WEC). The WEC's scientific objectives are described, together with its capability to achieve them in the course of the mission. The five experiments and the interfaces between them are shown in a general block diagram (Figure 1). WEC has organised technical coordination for experiment pre-delivery tests and spacecraft integration, and has also established associated working groups for data analysis and operations in orbit. All science operations aspects of WEC have been worked out in meetings with wide participation of investigators from the five WEC teams.

1. Introduction

The five teams proposing field and wave experiments on the Cluster mission have decided to coordinate their experiment proposals in order to be able to implement modern technology and to optimise the use of limited spacecraft resources. This has led to the formation of the Wave Experiment Consortium (WEC). The five WEC experiments are designed to measure quasi-static electric fields, electric and magnetic fluctuations, and small-scale plasma density structures in the Earth's magnetosphere and the solar wind. These experiments are:

[1] Present WEC Chair.
[2] Previous WEC Chair.
[3] Principal Investigator.
[4] Technical Coordinator.
[5] Technical Coordination Office.
[6] Data Working Group Chair.
[7] Operations Working Group Chair.

Space Science Reviews **79:** 93–105, 1997.

(A) A spherical double-probe electric-field experiment, known as the Electric Field and Wave (EFW) experiment, which will also be used to measure density fluctuations. PI: G. Gustafsson (Swedish Institute of Space Physics, Sweden).

(B) A triaxial search-coil magnetometer and a five-component (2 electric, 3 magnetic) spectrum analyser to perform spatio-temporal analysis of electric and magnetic field fluctuations (STAFF). PI: N. Cornilleau-Wehrlin (CETP/UVSQ, France).

(C) A relaxation sounder for probing eletron density and also used as a natural wave receiver to measure high frequency waves (WHISPER). PI: P. Décréau (LPCE/CNRS, France).

(D) A wide-band receiver system providing wideband waveform data (WBD). PI: D. A. Gurnett (University Iowa, USA)

(E) A digital wave-processing experiment (DWP) charged with on-board control of the WEC instruments and including a particle and wave/particle correlator. PI: H. St. C. Alleyne (University Sheffield, U.K.).

More detailed descriptions of each experiment, in terms of performance and scientific output, are given in companion papers (Gustafsson *et al;* Cornilleau-Wehrlin *et al.*; Décréau *et al*; Gurnett *et al.*; Woolliscroft *et al.*; this issue).

The WEC has benefitted from a close coordination of technical activities, preparation of WEC operations in orbit, data treatments and scientific data analysis.

A Technical Co-ordinator has organised the predelivery tests and spacecraft integration and tests on an engineering model and five flight units (four flight and one spare unit). For these activities, he has been supported by a closely knit team of engineers from all participating laboratories. The detailed operations of the WEC are dealt with by an Operations Working Group, which will be very active during the actual mission; as discussed further in Section 4. A Data Working Group has been working for several years to prepare a common WEC data system. Their prime aim is to develop common tools for the WEC, but also to facilitate data exchange with other Cluster experimenters and with other projects; this activity is described in Section 5. The WEC has a Scientific Board, which comprises the five Principal Investigators, and an elected chairman who serves for a limited period. This Board receives reports from the Working Groups and is the forum for all agreements and decisions, both during the development and later during the in- orbit operational phase.

2. Scientific Objectives of the WEC

The WEC experiments on each of the four Cluster spacecraft are capable of measuring the spin-plane component of the electric field, from dc to several hundred kHz, and the magnetic field vector from 0.1 Hz to $\sim$4 kHz. Telemetry permits full electric and magnetic wave form sampling, with 25 samples per second and 450 samples per second in respectively normal and burst modes. On-board analysis of magnetic wave-fields can be carried out up to 4 kHz and up to $\sim$500 kHz for electric wave fields. In addition, active sounding of the plasma can be used to determine plasma densities in the range 0.2–80 cm^{-3}, and the electric field probes can be used as a reference to determine the electric potential of the spacecraft, which is a parameter of interest for low-energy ion and electron measurements. Electron densities and their fluctuations can be obtained from spacecraft potential measurements (sensitive down to lobe densities), and by operating electric field probes in 'Langmuir mode' (1–100 cm^{-3}).

The fluxgate magnetometer on Cluster (FGM) will measure the magnetic field from dc to several tens of Hz and will overlap with the WEC search coil magnetometer in the range 0.1 to $\sim$10 Hz. A close cooperation with FGM is essential for most of the scientific objectives outlined in the following. The Cluster tetrahedron, with typical dimensions from a few hundred to a few thousand km, makes it possible to obtain information about spatial scales in this range. Although small-scale structures have been identified on numerous earlier missions – in the solar wind and in the magnetosphere – only with Cluster will it be possible to obtain three-dimensional plasma physics data and to distinguish temporal from spatial variations.

Regions of particular interest are the solar wind, the bow shock, the magnetopause, the cusp and the magnetospheric boundary layers. The crossing of auroral field lines near perigee (5–6 R_E), at different local times with a more elongated Cluster tetradedron, will permit new field and wave measurements and their comparison with ground observations.

Given these possibilities for new observations, the scientific objectives of the WEC are:

(1) To determine power conversion $\mathbf{J} \cdot \mathbf{E}$ for structures (e.g., magnetopause) where $\mathbf{J}$ has a component perpendicular to $\mathbf{B}$. $\mathbf{J}$ will be calculated from four-point measurements of the fluxgate magnetometer (FGM) complemented by STAFF data.

(2) To obtain three-dimensional information on the Poynting flux.

(3) To calculate shear flows and charge separation from div $\mathbf{E}$.

(4) To unambiguously determine the parameters which characterise plasma turbulences (distribution of $\mathbf{k}$-vectors) and small-scale field-aligned current structures (geometry, current density, etc.). This can be achieved by studying interspacecraft correlations of field fluctuations.

(5) To carry out four-point wave/particle interaction studies, via correlations performed on-board between wave and particle measurements.

(6) To identify source locations for the wave vectors measured at various spacecraft positions.

(7) To establish the role of high-frequency waves and to study their fine structure and its bearing on nonlinear wave/particle interactions. This will be done by using wideband data.

(8) To measure plasma density structures varying in time and space.

(9) To measure the spacecraft potential.

3. Technical Organisation of the WEC

3.1. Interfaces between WEC Instruments and the Spacecraft

Figure 1 is the general block diagram of the WEC showing the interfaces between the five experiments. The five sensors: three magnetic serach coils (STAFF) and two electric double-probe antennas (EFW) can be seen on the left. The electronic units (centre) include:

(i) The magnetic waveform unit, the calibration unit and the spectrum analyser (two electric and three magnetic components) belonging to STAFF.

(ii) The wide-band receiver from WBD, which can process any of four components (two electric and two magnetic).

(iii) The EFW electronic unit is microprocessor-controlled and has the added capacity of being able to store selected events in a separate burst memory.

(iv) The sounder receiver and transmitter.

The DWP experiment, shown on the right of Figure 1 contains a data processing unit which is the main interface with the spacecraft for data transmission. It also operates the experiments' ON/OFF commands and the selection of the modes of operation, either via time-tagged orders received from the ground stations, or via software. Although the WBD high-speed data can be routed via DWP and stored on-board, most of the time WBD does not utilise the DWP; instead, the WBD information is sent, via spacecraft systems, directly to the NASA Deep Space Network. In addition to data formatting and packaging, control of experiments etc., the DPU has specific capabilities, such as compaction and selection of data, and wave/particle correlations. The spacecraft potential measurements are sent to three Cluster experiments from the WEC, as indicated.

3.2. Dynamic Ranges for Electric and Magnetic Measurements

Figure 2(a) shows respectively the noise level and the highest measurable level (dashed lines) for different electric field measurements as a function of frequency, compared with the expected levels for several kinds of waves (solid lines). The noise level is an estimate based on flight data from experiments on GEOS-1/GEOS-2 and ISEE-1 similar to the WEC. The capability of EFW/STAFF to observe large signals in the 1–10 Hz and 0–180 Hz filter bandwidths is also indicated; in reality there will be one or more near monochromatic waves to be observed. The WHISPER dynamic range is close to that of the WBD, but limited to the frequency range 2–80 kHz.

The STAFF search coil noise level (dashed lines) and expected magnetic wave spectral amplitudes (solid lines) are presented in Figure 2(b). Notice that the max-

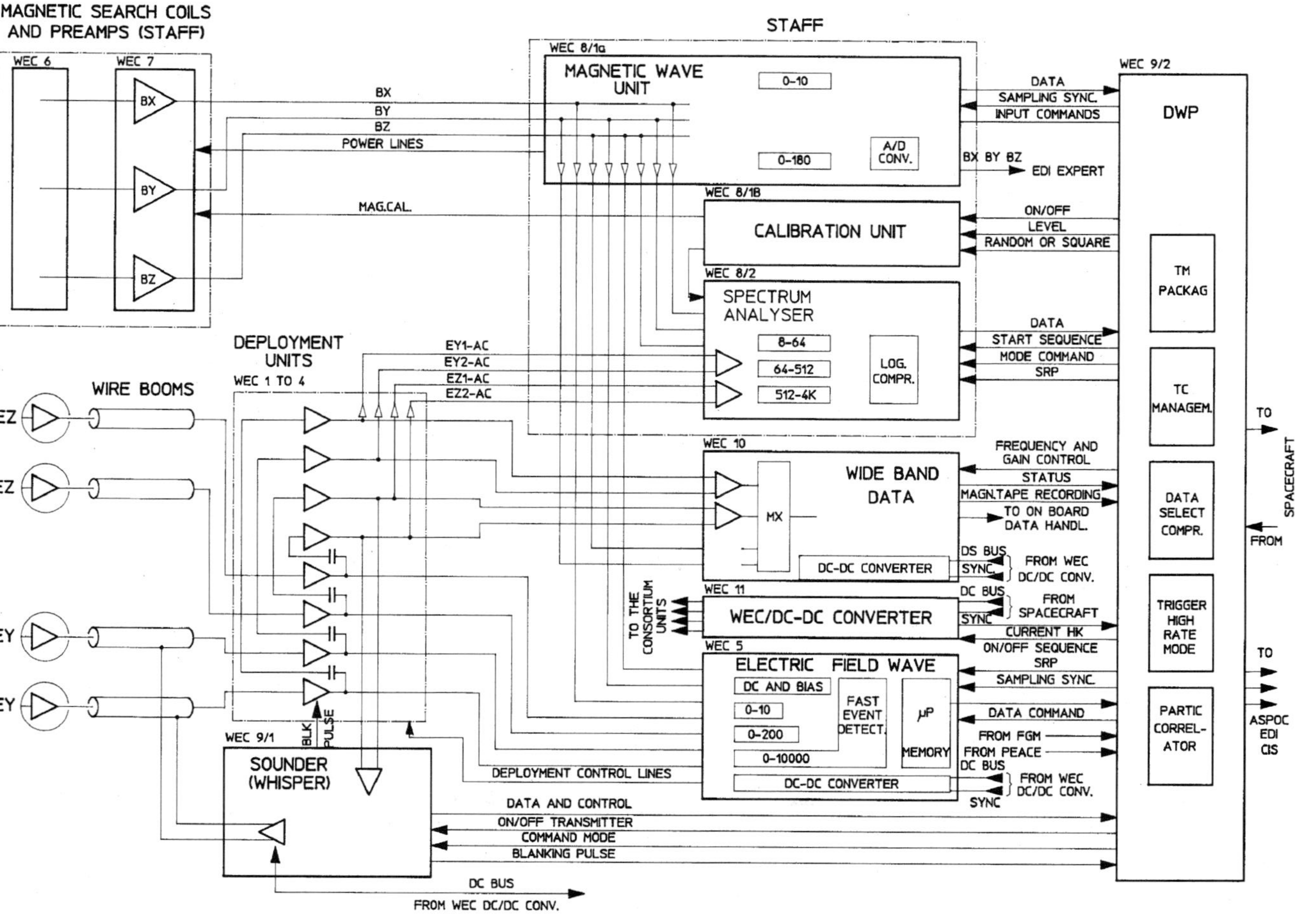

Figure 1. Block diagram of the five experiments which have been combined to form the Wave Experiment Consortium (WEC).

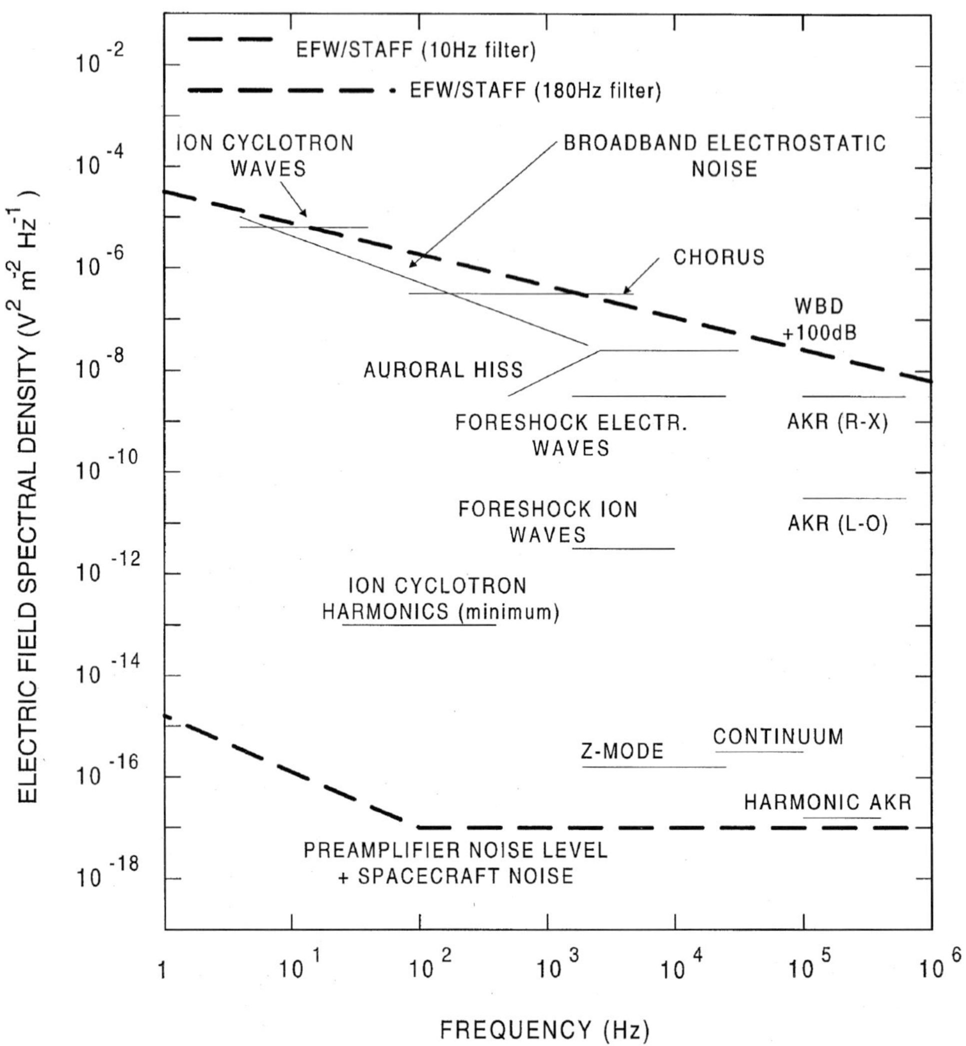

Figure 2a. This figure shows the dynamic range of electric field and wave experiments and the expected levels of signals. The upper three dashed lines show the highest measurable levels for respectively: EFW/STAFF in the 0–10 Hz filter band, EFW/STAFF in the 0–180 Hz band and WBD. The highest measureable level for WHISPER is close to that of WBD, however the upper frequency is 80 kHz. The noise level is well below the expected levels of phenomena to be observed as indicated by solid lines.

imum observable levels are 100 db above the noise, providing a safe coverage of the expected wave phenomena.

The Cluster WEC experiments address particularly well large, low-frequency electric and magnetic wave components, but will also give good coverage of these components up to ~4 kHz. Above this frequency, only the electric component can be measured with good sensitivity.

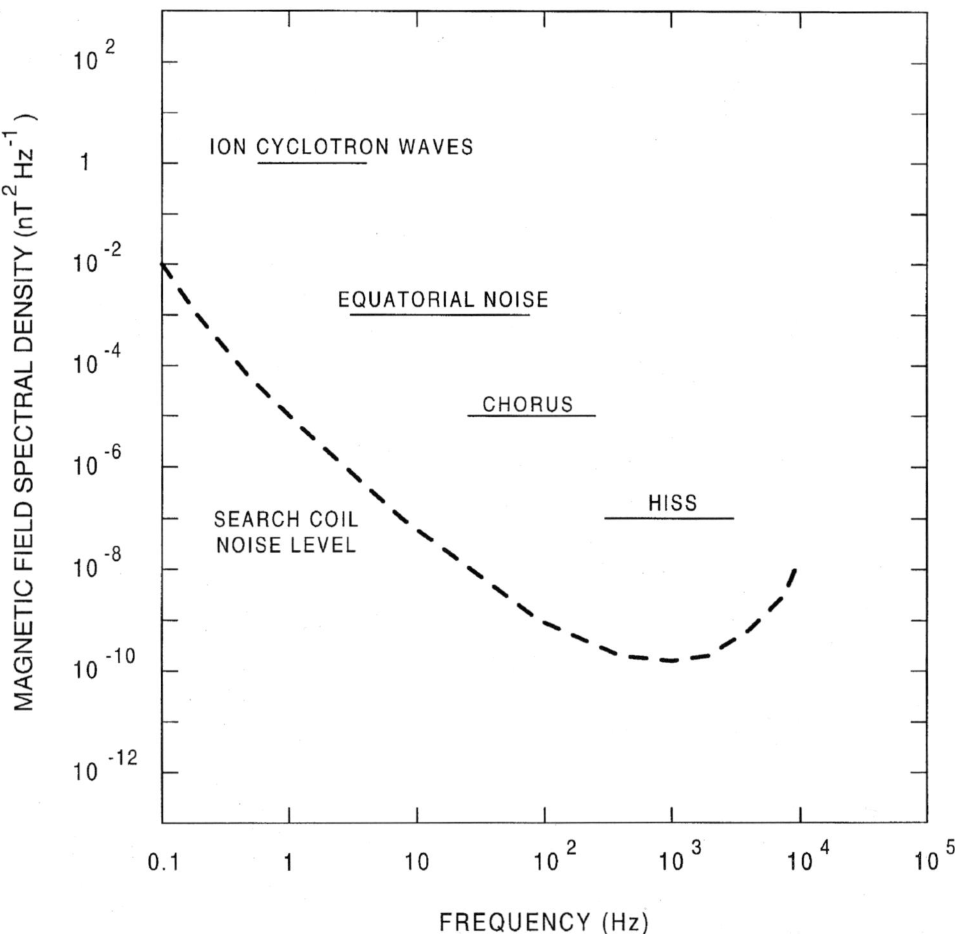

Figure 2b. Magnetic wave measurements have a dynamic range of 100 db above the search coil noise level indicated. This will provide good coverage of the expected signals indicated by solid lines.

3.3. Technical Management of the WEC

To cope with the large number of connections between the various WEC instruments, and with the desire of the Cluster Project to have a single interface, the Wave Experiment Consortium Technical Coordination Office (WECTCO) has been set up. The WECTCO is led by B. de la Porte (CETP), A Bouabdellah (CETP), J.A. Thompson (U. of Sheffield), P.N.H. Davies (U. of Sussex), R. Huff (U. of Iowa) and F.X. Sène (LPCE). The WECTCO is responsible for:

- WEC interface with the spacecraft.
- Interfaces between WEC experiments.
- Management of WEC schedule.
- Integration at WEC level.

Table I
Membership and organisation of the WEC operations group

Subgroup	WHISPER	STAFF search coil	STAFF spectrum analyser	EFW	WBD	DWP
Scientific	PI	PI		PI	PI	PI
Commissioning	J. G. Trotignon	C. De Villedary	C. C. Harvey	G. Holmgren	R. Huff	H. St. C. Alleyne
Technical	Ph. Martin	C. De Villedary	C. C. Harvey	A. Lundgren	R. Huff	K. H. Yearby

The WEC Operations Group has a 'duty coordinator' and is chaired by the DWP PI, assisted by B. de la Porte for WEC hardware and power considerations and A. Buckley for use of the correlator, and by other members of subgroups listed below. H. St. C. Alleyne leads the commissioning subgroup and K. H. Yearby the technical subgroup.

- Participation in assembly, integration and verification.
- Ground operations and EGSE.

4. WEC Operations

The WEC was formed to share resources to maximise the payload scientific return. This is particularly important for telemetry: WEC allocations on each spacecraft are

5217 bits s^{-1} in normal bit rate (NBR) and 43.898 kbits s^{-1} in high bit rate (HBR). In general, WBD will transmit real time data (220 kbits s^{-1}) directly to DSN in parallel with other WEC experiments operated in NBR. This sharing of resources means that scheduling of the components of the WEC cannot be done by individual instrument teams working in isolation; and so, the WEC has established a coordinating operations group, the membership of which is shown in Table I.

The role of this group is to convert the Master Science Plan, which has been agreed by the Cluster Science Working Team, into command sequences which will be merged by the JSOC (Hapgood *et al.*, this issue) with the scientific command requests from the rest of the payload. As such, the operations group performs a task which is rather similar to that of the JSOC for issuing commands. This group is also responsible for the generation of new modes for the WEC and new command sequences for special scientific operations, commissioning plans and for dealing with operations in the event of problems in orbit. The operations group forms the WEC's formal point of contact with the JSOC of the WEC for issuing of commands.

Most normal operations are based on a restricted number of standard WEC modes. This is for the clear reason that it will yield datasets which are readily understood without frequent discontinuities associated with mode changes. However, even within these normal WEC modes there are additional parameters which still need to be defined. These are described more fully in the associated individual WEC instrument papers (Gustafsson *et al.*; Cornilleau-Wehrlin *et al.*;

Table II

Some of the main WEC Operation Modes (simplified). Terms such as 'normal' or 'fast' cover several possibilities and do not have the same meaning in each case. The WBD instrument is capable of many modes but is expected to be operated in NBR-basic or HBR-basic mode much of the time that it is operational.

Name	WHISPER	STAFF	STAFF	EFW	WBD	Correlator
NBR-basic	3 s active, 25 s passive	Normal	Normal	Normal	Normal	On
NBR-low recurrence	4 s active, 100 s passive	Normal	Normal	Normal	Normal	On
NBR-Langmuir	4 s active, 100 s passive	Normal	Normal	Normal (Langmuir)	Normal	On
NBR-spin synchronized	129 spins 516 s active	Normal	Normal	Normal	Normal	Off (on)
NBR-active	104 s active	Normal	Normal	Normal	Normal	On
NBR-WBD	4 s active, 4 × 25 s passive	Normal	Normal	Normal	Many modes	On
HBR-basic	3 s active, 25 s passive	High	Fast	High	Off	Off
HBR-low	4 s active, 4 × 25 s passive	High	Fast	High (Langmuir)	Off	Off
HBR-Langmuir	4 s active, 4 × 25 s passive	High	Fast	High (Langmuir)	Off	Off
HBR-spin synchronized gliding	580 s active	High	Fast	High	Off	Off
HBR-active continuous	185 s active, 13 s passive, 2 s off	High	Fast	High	Off	Off
HBR-EFW	Passive (off)	High (off)	Off	High (many modes)	Off	Off (on)
HBR-WBD	Off	Normal	Normal	Normal	Special mode	Off

Décréau *et al.*; Woolliscroft *et al.*; this issue). Table II gives a summary of the standard WEC modes. The DWP instrument paper (Woolliscroft *et al.*; this issue) gives more details on how these modes correspond to measurement cycles and the use of macro command sequences.

5. WEC Scientific Data Handling

There are essentially three means by which the WEC co-investigators can access the Cluster data: firstly, via the Cluster Science Data System, CSDS; secondly, via

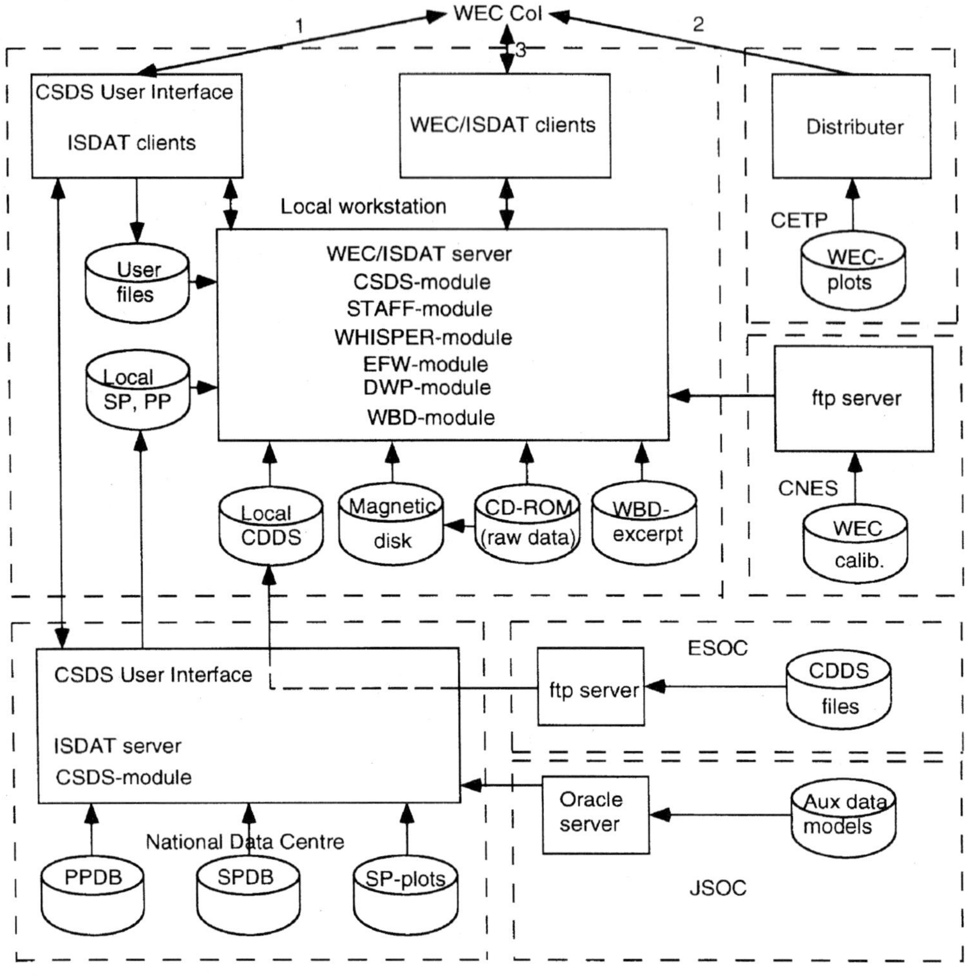

Figure 3. The WEC data handling system.

the WEC Summary Plots; thirdly, via the WEC Interactive Science Data Analysis Tool (WEC/ISDAT). They are all illustrated in Figure 3 and described in more detail below.

5.1. Science data system

CSDS is an undertaking, at Cluster project level, to make Cluster scientific data available to a wide scientific community. The CSDS is described in Schmidt and Escoubet (this issue), here we emphasise the parts that are of particular importance to the WEC data handling, notably the fact that the ISDAT is being used both for the WEC detailed data analysis and as a component of the CSDS.

WEC related CSDS databases are produced at the national data centres in France, U.K., and Sweden and copied to all CSDS national data centres. The CSDS Prime Parameters (PP) and Summary Parameters (SP), residing at the National Data Centres, are accessible via the CSDS User Interface as illustrated by the left path of Figure 3.

The WEC scientists can access the databases via a CSDS User Interface client software package, installed on his local work stations, which is connected to a national data network as shown in Figure 3. By the CSDS User Interface the user can:

- Browse CSDS parameter catalogues.
- Search in scientific data using personal search criteria.
- Analyse, manipulate, display and locally store CSDS parameters.
- Retrieve summary plots.
- Browse catalogues at the Joint Science Operations Centre (JSOC).
- Copy data files to the local work station.
- Analyse and display locally stored CSDS parameters.

Of particular interest in connection with WEC data analysis is that the ISDAT software constitute an integral part of the CSDS user interface. This means that it will be possible to access locally-stored CSDS data files from the WEC software package, and integrate the low time-resolution parameters from all instruments with the WEC full-time resolution data.

5.2. The WEC summary plots

Special WEC summary plots will be prepared by and will be accessible from CETP, Velizy, France, as illustrated on the far right section of Figure 3. They will serve a similar purpose as the CSDS summary plots, but will give more details and be specially designed for the WEC. Dynamic spectra of magnetic and electric field wave components will be produced for the whole frequency range of WEC. The status of WBD data, which is not on the Cluster master CD-ROM, will be provided. The WEC summary plots will also include 'relative wave parameters', or comparisons calculated on the basis of WEC measurements on the four spacecraft.

5.3. The WEC/ISDAT system

5.3.1. *The WEC/ISDAT Software Development*

WEC/ISDAT is designed for the detailed, full-time-resolution, scientific data analysis of the WEC data. The software development has been a cooperative effort within the WEC scientific laboratories, based on the ISDAT system that the WEC EFW team has been developing over the last few years.

The development team realised at an early stage that a working group would never be able to provide a system that would satisfy all involved co-investigators. On the other hand, it was realised that the WEC team would collectively have a large scientific and technical expertise; a leading principle of the work has therefore been that each group or individual would contribute to the system within their respective fields of expertise. This has led to the active involvement of a majority of WEC laboratories. This work has been carried out to make extensive use of network communication programmes, such as: electronic mail, ftp file transfer, remote login, world wide web etc. A limited number of personal progress meetings have been held.

5.3.2. *The WEC/ISDAT Software Design*

The WEC/ISDAT software design is based on a client/server model. The original development is made for UNIX systems in general, but it is also being ported to Open VMS in conjunction with the implementation in the CSDS User Interface. A guiding principle is to 'hide' all instrument-specific technical items, such as de-commutation or calibration in server modules and to use a well defined, general data structure as an interface between the server and the clients. The server is modular in the sense that a local system can be built with an arbitary set of Cluster instrument modules. The modular design can also support the eventual inclusion of non-Cluster instrument modules for inter-project studies. On the client side, a limited number of standard analysis and display clients are provided, but the individual user can also add his personal display and analysis software written either in C or IDL languages. The totality of the Cluster raw data will be distributed to all WEC laboratories. The ISDAT is a flexible system distributed over wide-area networks. It is based on a central workstation with one server at each outside laboratory and clients at co-investigators workstations.

5.3.3. *The WEC/ISDAT Data Flow*

The WEC/ISDAT data flow is illustrated by the middle path in Figure 3. The WEC/ISDAT server will give access to the following databases:

– The locally stored CSDS parameters.

– The locally stored Cluster Data Disposition System (CDDS). These are files which are identical to the final data delivered on CD-ROM, but available immediately after data-taking from ESOC through the national data centres. The amount of data is limited and the CDDS are primarily intended for the PI's to perform instrument health checks or quick science analysis for mode settings.

– Final science raw data distributed to all WEC co-investigator laboratories from ESOC. WEC/ISDAT can use the CD-ROM directly, but it is anticipated that normally data will first be copied to a magnetic disk for improved performance. Depending on the locally available WEC/ISDAT server modules, the WEC co-investigator can access data from either one or several instruments from the CD-ROM.

– Data from the WBD instrument is distributed separately from the general Cluster CD-ROM. If data is locally or remotely available to the WEC co-investigator and if the WEC/ISDAT server has the WBD-module, then the WBD data is accessible via the WEC/ISDAT software.

– Files generated by the user via the CSDS User Interface.

In order to maintain up to date calibrations, all WEC instrument calibrations are available via an ftp server at CNES as illustrated in Figure 3.

6. Conclusion

The Wave Experiment Consortium (WEC) was established, already at the time of the writing of experiment proposals, to coordinate all activities related to five field and wave experiments on Cluster (STAFF, WHISPER, WBD, EFW, and DWP). These experiments are described in detail in this issue. This coordination resulted in technical solutions which ensured the best possible use of resources and uniformity in the ways in which data were sampled. Testing of the WEC experiments as a block, before integration on the four spacecraft, also contributed to a well organised activity during the lengthy test programme.

Working groups for data analysis and operations of the experiments in orbit are part of the WEC set-up, and a large number of technical and scientific staff have been working for several years to prepare the science data-taking in orbit and the subsequent data analysis.

The ultimate aim of WEC is to have a well organised data output from the five WEC experiments and, within the framework of CSDS, to combine WEC data with other Cluster data sets.

The tools and organisation are in place for carrying out new and exciting science with four-point data sets from Cluster.

Acknowledgements

A large team of engineers from WEC groups have participated in the integration and tests of experiments. Some of these engineers have carried out work on one particular instrument and others have carried out overall WEC tasks during integration and tests or by working within the Operations and Data Working Groups. The WEC would like to express appreciation to the following for their dedication, for doing an excellent job and for keeping up a good team spirit.

CETP, Velizy: D. Bagot, V. Bouzid, A. Meyer, J. M. Napa, C. de Villedary.

LPCE/CNRS, Orleans: P. Fergeau, M. Lévêque, Ph. Martin, M. Parrot, J.C. Trotignon.

Obs. de Paris, Meudon: Y. de Conchy, C. C. Harvey, F. Wouters.

IRF-U, Uppsala: L. Åhlen, A. Lundgren.

SSL, Berkeley: P. Berg, P. Harvey, D. Pankow, R. Ulrich.

SSD/ESA, Noordwijk: A. Butler, D. Klinge, M. Thomsen, J. Zender.

University Oulu: K. Lappalainen.

University Oslo: B. Lybekk, T. Sten.

University Sheffield: H. Aleyne, S. Walker, K. H. Yearby.

University Sussex: A. Buckley, P. N. H. Davies, S. J. Davies.

THE CLUSTER SPATIO-TEMPORAL ANALYSIS OF FIELD FLUCTUATIONS (STAFF) EXPERIMENT

N. CORNILLEAU-WEHRLIN, P. CHAUVEAU, S. LOUIS, A. MEYER, J. M. NAPPA, S. PERRAUT, L. REZEAU, P. ROBERT, A. ROUX and C. DE VILLEDARY
CETP/UVSQ, Vélizy, France

Y. DE CONCHY, L. FRIEL, C. C. HARVEY, D. HUBERT, C. LACOMBE, R. MANNING and F. WOUTERS
Observatoire de Paris, Meudon, France

F. LEFEUVRE, M. PARROT, J. L. PINÇON and B. POIRIER
LPCE/CNRS, Orléans, France

W. KOFMAN
CEPHAG, Grenoble, France

PH. LOUARN
Observatoire Midi-Pyrénées, Toulouse, France

AND THE STAFF INVESTIGATOR TEAM

Abstract. The Spatio-Temporal Analysis of Field Fluctuations (STAFF) experiment is one of five experiments which together comprise the Wave Experiment Consortium (WEC). STAFF consists of a three-axis search coil magnetometer to measure magnetic fluctuations at frequencies up to 4 kHz, and a spectrum analyser to calculate in near-real time aboard the spacecraft, the complete auto- and cross-spectral matrices using the three magnetic and two electric components of the electromagnetic field. The magnetic waveform at frequencies below either 10 Hz or 180 Hz is also transmitted. The sensitivity of the search coil is adapted to the phenomena theo be studied: the values 3×10^{-3} nT Hz$^{-1/2}$ and 3×10^{-5} nT Hz$^{-1/2}$ are achieved respectively at 1 Hz and 100 Hz. The dynamic range of the STAFF instruments is about 96 dB in both waveform and spectral power, so as to allow the study of waves near plasma boundaries. Scientific objectives of the STAFF investigations, particularly those requiring four point measurements, are discussed. Methods by which the wave data will be characterised are described with emphasis on those specific to four-point measurements, including the use of the Field Energy Distribution function.

1. Introduction

The Cluster mission has been designed to study the thin layers of the interaction regions between the solar wind and the Earth's magnetosphere. The very existence of these regions, with their different plasma bulk properties, is largely due to wave-particle interactions which, in a collisionless plasma, provide the only means of modifying the bulk properties of plasma crossing the frontier. Within these regions, waves again provide the only effective coupling between particles of the same and of different species, and give rise to anomalous transport effects; the basic physics of these regions requires a comprehension of understanding of the

Space Science Reviews **79:** 107–136, 1997.

wave-particle interactions present. Thus it is important to characterise the waves and turbulence: this is the objective of the Cluster STAFF measurements. Four point measurements will allow, for the first time, a clear separation of spatial and temporal effects. A major consideration for wave observations in a fast-flowing medium is the Doppler effect. Waveform data from four spacecraft in a tetrahedral configuration allow correction for this effect when the wavelength is comparable with the inter-spacecraft separation. On the other hand, when the wavelength is small compared to the inter-spacecraft separation, the determination of the wave normals on the four separate spacecraft may yield information about the source location. To understand turbulence it is important to measure over a frequency range wide enough to determine any cut-off frequency; instrumentation has sometimes been inadequate for this purpose on earlier missions. Earlier missions have been even less adapted to investigate spatial wavenumber spectra. Furthermore, some geophysically important regions have been rather neglected: for example, the cusp has been visited only by the HEOS spacecraft.

In the next section of this paper, the principle scientific objectives are discussed. Methods by which the wave data will be characterised are described in Section 3, with emphasis on those specific to four-point measurements.

This is followed by a technical description of the STAFF experiment, including the various in-flight modes of operation of the instrument. The experiment has identical instruments on each of the four Cluster spacecraft. Each instrument comprises a three-axis search coil magnetometer to measure the magnetic fluctuations up to 4 kHz, and a spectrum analyser to calculate in near-real time the 5×5 cross-spectral matrix formed from the three magnetic and two electric field components, at 27 frequencies, provided by Electric Field and Wave (EFW) experiment, of the electromagnetic field. The vector magnetic field waveform is also transmitted, in a frequency band extending to either 10 Hz or 180 Hz, selectable by telecommand. STAFF is one of the five wave instruments aboard Cluster which form the Wave Experiment consortium (WEC, see Pedersen *et al.*, this issue).

2. Scientific Objectives

In this section we present areas where we anticipate that the STAFF experiment will make a significant contribution to our current understanding of the plasma physics of Earth's environment. Telemetry and ground station limitations do not allow full 24 hr per day data coverage along the whole Cluster orbit. Thus, in what follows, we mainly discuss those regions which are primary objectives of the Cluster mission and for which data acquisition is a priority. The papers cited are mostly recent work, or reviews in which references to earlier pioneer work can be found.

2.1. Solar Wind and Upstream Waves

The frequency range of the STAFF experiment is *a priori* more suitable for studying the proton foreshock than the electron foreshock. However, waves at ~ 1 Hz called 'upstream propagating whistlers' have been detected in front of the ion foreshock, in the electron foreshock (Russell *et al.*, 1971). The origin of these waves is still controversial: they could be anisotropy-driven instabilities amplified locally by electrons (Sentman *et al.*, 1983) or they could be generated at the shock itself by the ions and then propagate upstream (Krauss-Varban *et al.*, 1995). Data from the four STAFF experiments will allow the source to be localised and, together with four point measurements of the particle distribution function, should allow this question to be answered. Also in the electron foreshock, possible non linear mode coupling between Langmuir waves and whistler mode waves can be investigated by combining Whisper and STAFF spectrum analyser measurements.

Entry into the ion foreshock is accompanied by the onset of strong electrostatic noise in the frequency range from about 100 Hz to 10 kHz, often called 'ion-acoustic' noise. This noise is associated with 1–40 keV ions streaming into the solar wind (Gurnett, 1985). The identification of the mode of propagation of this noise requires a better description of the surrounding plasma. Deep inside the ion foreshock, the bulk flow velocity of the solar wind is found to be reduced by 5 to 10% (Zhang *et al.*, 1995); in a collisionless medium, this can only be a consequence of wave-particle interactions. The electric field spectrum from the STAFF spectrum analyser will allow the complete spectrum, from the EFW to Whisper frequencies, to be measured, thus fully describing this noise. It is well known that different kinds of waves (Greenstadt *et al.*, 1995) are related to the shape of the different kinds of upstream ion distribution functions observed in the ion foreshock (Fuselier, 1995). Using the four point measurements, waves can be used to probe the ambient plasma and help localise the sources of reflected and diffuse upstream ions. Numerical simulations demonstrate the close connection between diffuse ions and, upstream waves, and their effect on the solar wind (Scholer, 1995).

A new class of ULF waves (with frequency ~0.3 Hz) upstream of the bowshock has been discovered by Le *et al.* (1992) when the solar wind plasma β is high. Neither the intrinsic wave mode nor the free energy source have yet been determined unambiguously. In regions where β is particularly large these waves appear to be ion cyclotron waves (Le *et al.*, 1992). A careful analysis of waves in the same frequency range, but when the solar wind plasma β is low (<1), has led Blanco-Cano and Schwartz (1996) to conclude that these waves are extended whistler mode trains. The mode of these waves can be unambiguously determined by the use of the waveform data from four spacecraft. Comparison with particle data obtained simultaneously should indicate the source of the free energy responsible for their amplification.

2.2. The Earth's bow shock

Upstream of the quasi-parallel bow shock, ULF waves steepen to form shocklets and Short Large Amplitude Magnetic Structures (SLAMS); and high-frequency whistler wave packets are amplified. Because their phase velocity is less than the solar wind flow speed, the majority of the waves in the foreshock must eventually be convected back into and through the shock itself. This implies that the foreshock and the shock cannot be treated separately (Burgess, 1995). The dominant instabilities and the two-dimensional scales found in numerical simulations (Dubouloz and Scholer, 1995) must be compared with the four-point observations of STAFF. The notion of cyclical shock reformation has been investigated by simulations of high Mach number quasi-parallel shocks (Winske *et al.*, 1990). This work prompts the following questions which should be answered using Cluster wave measurements: What is the role of high-frequency waves in the cyclical reformation? Which kind of turbulence is associated with non-gyrotropic ion distributions observed in space and simulations at shocks?

Similarly, recent simulations of supercritical quasi-perpendicular shocks have shown that shock-reflected ions generate upstream propagating whistler waves at frequencies of the order of a few tens of Hz in the plasma rest frame (Hellinger *et al.*, 1995; Krauss-Varban *et al.*, 1995). These waves propagate obliquely with respect to both the shock normal and the local magnetic field and are most intense in the shock ramp. This suggests that quasi-perpendicular shocks are intrinsically three-dimensional, a conjecture which should be tested byusing four-point STAFF measurements, especially when operating in burst mode which yields the waveform up to 180 Hz. Simulations also suggest that electron heating through quasi-perpendicular shocks is essentially adiabatic except for some slight heating by the upstream whistlers, and eventually by lower hybrid waves located within the main shock transition (Krauss-Varban, 1994; Savoini and Lembège, 1995). While whistler waves have been extensively observed, the identification of waves near the lower hybrid resonance frequency (typically in the range from about 3 to 15 Hz) is almost non-existent, due to the inappropriate frequency ranges of electric and magnetic sensors on previous missions (Scudder *et al.*, 1986). A detailed analysis of the waves detected with Cluster instruments in the shock may give evidence of lower hybrid waves and their relationship to the whistler mode turbulence, and even to the low-frequency electromagnetic noise. In the downstream region, characteristic electron distributions have been observed; they require a more efficient non-adiabatic heating process (Veltri *et al.*, 1990; Savoini and Lembège, 1995). Savoini and Lembège (1994) have reported magnetised simulations that reproduce simultaneously both the observed flat-topped electron distribution functions and the electromagnetic shock structure. Nevertheless, the origin of the non-adiabatic heating is not completely elucidated; it is probably intimately related to the frequently observed shock-associated higher frequency waves. These waves lie in the kHz frequency domain and will be detected easily by STAFF. Multi-spacecraft observations will provide useful information on the relation between local shock structure and wave intensity and, in particular, the possible relationship between the intensity of the high-frequency waves and the thickness of the bow shock.

Finally, one of the main objectives of Cluster is to determine unambiguously the frame-dependent wave properties such as frequency, phase and group velocities in the strongly Doppler-shifted environment of the solar wind. Four satellites will help identify the different kinds of waves which are being amplified or which can propagate to a given measurement point. So far very few determinations of the coherence length of the turbulence have been published (Le *et al.*, 1993); filling this gap in our knowledge is certainly one of the main tasks of Cluster. The variation of the Cluster inter-spacecraft separation during the course of the mission, and the changing geometrical configuration, will enable different wavelength ranges to be studied.

2.3. The Magnetosheath

The full importance of the role played by the magnetosheath in the interface between the free solar wind and the magnetosphere is gradually being recognised. The magnetosheath is a magnetofluid in which the flow and magnetic field patterns change from the bow shock to the magnetopause; a slow mode transition region has been identified and plays a crucial role in these changes (Song, 1994). The properties of the plasma in the outer magnetosheath depend on the bow shock properties, while the properties of the plasma of the inner sheath depend on the shear angle between the magnetosheath magnetic field and the geomagnetic field (Phan *et al.*, 1994).

The dissipation processes are not the same in different regions of the magnetosheath. Using one or two spacecraft, several wave modes have been identified, at scales larger than or equal to the ion Larmor radius (Hubert, 1994; Lacombe and Belmont, 1995). Anticorrelation is observed between the proton temperature anisotropy and the proton β in the sheath depletion layer (close to the magnetopause) when the magnetosheath is strongly compressed; this probably indicates a quasi-linear equilibrium reached through unstable mirror and Alfvén ion cyclotron waves (Anderson *et al.*, 1994). The bow shock, the inner sheath, and the magnetopause can all affect the waves convected from the solar wind to the magnetosphere, through mode coupling, damping, mode conversion and reflection (Hubert, 1994; Krauss-Varban, 1994). For deterministic reasons, information must also propagate from the inner to the outer magnetosheath, but the nature of its transmission is still an open question.

Microscale plasma phenomena are critical in this high-β plasma. In particular, whistler-mode noise is known to play an important part in controlling electron thermal anisotropy.

2.4. The Magnetopause and the Low-Latitude Boundary Layer

There is ample indirect (and some direct) evidence that the magnetopause is a permeable boundary. The investigation of the physical processes by which mass and momentum are transferred through the magnetopause, from the solar wind to the magnetosphere, is one of the prime goals of the mission. Our objective is to assess the role that plasma waves play in affecting and controlling these transfers.

Different models have been proposed, such as the reconnection model or the Kelvin–Helmholtz instability. Also, there is evidence for localised flux tubes, known as Flux Transfer Events (FTEs), connecting the magnetosheath to the magnetosphere, but whether FTEs are the remnants of reconnection events or the nonlinear consequence of tearing or Kelvin–Helmholtz instability is still an opened question. Hence, for all models, the key issue is the identification of the plasma waves that permit the anomalous cross-field diffusion and/or resistivity at a global/local scale. Many experiments aboard single spacecraft have shown that a very high level of fluctuations is observed in all frequency ranges during magnetopause crossings (Labelle and Treumann, 1988; Anderson *et al.*, 1991; Cattell *et al.*, 1995). The estimation of diffusion coefficients indicate that the ULF fluctuations are the more likely to provide the anomalous diffusion (Gendrin, 1983). Nevertheless the question of the origin of these fluctuations has seldom been addressed in the literature. Two different hypotheses can be considered: either they are due to local instabilities of the boundary, or they are generated in the magnetosheath and are amplified at the magnetopause. The second hypothesis has been tested on a simple model by Belmont *et al.* (1995). This study has shown that most of the experimental characteristics of the fluctuations observed in this region (Rezeau *et al.*, 1989) can be explained by the resonant amplification of magnetosheath fluctuations at the magnetopause. These results are encouraging but they should be confirmed by including more realistic characteristics of the magnetopause in the model. The STAFF experiment will permit a detailed comparison of the magnetosheath and magnetopause fluctuations with very high resolution four-point measurements. It will be possible to compare data recorded at the same time in the magnetosheath, the magnetopause and the magnetosphere, and thus hopefully confirm this theoretical interpretation. Moreover, when observing the same wave event aboard the four spacecraft, the mode identification can be done unambiguously, without any *a priori* hypothesis.

2.5. The Cusp

Because of its singular magnetic field configuration, the cusp is thought to be one of the regions where magnetosheath plasma entry occurs. In spite of the key role it may play, the exterior cusp is one of the less explored regions of space that Cluster will visit. From a few crossings by the HEOS spacecraft (Haerendel *et al.*, 1978), it has been inferred that the flow in the cusp region shows very turbulent behaviour. At lower altitudes, but still $R > 2\ R_E$, strong field-aligned currents were observed by the OGO-5 spacecraft. These current sheets or filaments were found to be accompanied by Ion Cyclotron Waves (ICWs) and by higher frequency waves, which are presumably electrostatic (Fredericks *et al.*, 1973). The wave characteristics of the cusp region are relatively unexplored, and the complete frequency coverage of the Wave Experiment Consortium is particularly desirable. Furthermore, the STAFF experiment ability to perform spatial correlation of waveforms at lower frequencies, and comparison of spectra at higher frequencies, will help characterise the plasma waves or turbulence, and hence allow the study of their role in accelerating particles along the magnetic field lines.

2.6. The Plasmasheet

The central part of the magnetotail, the plasmasheet, is a very complex region where the plasma is accelerated up to tens of keV; this acceleration is particularly efficient during substorms. We are still far from a full understanding of the processes that lead to the reconfiguration of the tail magnetic field and to the acceleration and Earthward injection of plasma during substorms. At least, there is as yet no consensus on a scenario explaining this chain of processes. One of the key questions is the relation between dipolarisation/injection occurring in the inner plasmasheet, and the signatures observed further out in the tail-like flux ropes and/or plasmoids. In a plasma where binary collisions are essentially absent, the dramatic topology changes mentioned above are controlled by collisionless processes. There are two types of collisionless processes involving breakdown of adiabatic invariants: (1) waves and turbulence at frequencies of the order of the gyrofrequency or of the order of the bounce frequency, and (2) non-adiabaticity associated with curvature effects within thin current sheets. In fact, these two types of processes are related, as will be seen below. The coordinated measurements to be carried out aboard the four Cluster spacecraft will provide a powerful new tool to study the role played by these two kinds of processes.

It is well known that the first adiabatic invariant may not be conserved in the tail, at least for ions, and possibly for electrons. This loss of adiabaticity makes it more difficult to describe transport in the tail. Büchner and Zelenyi (1988) have proposed to use the parameter κ to classify the various types of orbits; κ is the square root of the ratio between the curvature radius and the ion Larmor radius, and the usual adiabatic invariants are conserved in regions where $\kappa \gg 1$. The transition to a chaotic regime can have important macroscopic consequences; for instance, it has been suggested that it could play a key role as a substorm trigger. Delcourt *et al.* (1995) have investigated the transition from the adiabatic to the chaotic regime. Using a simple analytical model of short-lived centrifugal impulses and simulation of particle motion, they also showed that the $\kappa = 1$ regime is characterised by prominent bunching effects in gyration phase, which can lead to the formation of non gyrotropic distributions. In turn, these distributions can be strongly unstable, thereby leading to the generation of intense plasma waves around the proton gyrofrequency: this hypothesis should be tested with simultaneous Cluster measurements of waves and particles.

Thin Current Sheets (TCS) have been observed in the central part of the plasma sheet (see for instance Mitchell *et al.*, 1990). A direct relation between the thinning of TCS and substorm break up has been evidenced in a number of cases (see, e.g., Lui *et al.*, 1992; Perraut *et al.*, 1995). Other triggers can be involved, however. Lui *et al.* (1992) and Perraut *et al.* (1993) have shown that intense waves are observed just prior to break up or at break up. These waves have frequencies typically in the proton gyrofrequency range. These measurements were carried out relatively close to the Earth. Nevertheless there are observations made at larger distances by *Galileo* and Geotail: *Galileo* measurements (Khurana *et al.*, 1995) confirm that flux ropes and thin current sheets also develop in the far plasma sheet; small scale structures are regularly observed inside the plasmasheet with Geotail (Matsumoto *et al.*, 1994), associated with broad band electrostatic noise (BEN). Therefore turbulence can play a role in triggering substorms. The assessment of this role requires the spatial coverage that will become available with Cluster.

At lower frequencies or larger scales, the existence of small-scale Field-Aligned Current (FAC) structures has been inferred from GEOS measurements (Robert *et al.*, 1984). While interesting, these observations carried out aboard a single spacecraft remain inconclusive because spatial and temporal variations cannot be separated. The electric and magnetic measurements on-board the four Cluster spacecraft will allow the removal of this ambiguity for the first time, and hopefully this will clarify the link between these FACs and nonlinear Kinetic Alfvénic Structures (KAS). During the preparation of the mission, techniques have been developed to fully characterise the small-scale FAC/KAS, based on the measurements that will be carried out by WEC and by other experiments aboard the four satellites. These techniques involve: (1) inter-spacecraft correlations, (2) wavelet analysis, adapted to varying distances between the spacecraft, and/or to spatial scales determined from data obtained from other experiments, both within and outside WEC, (3) singular spectrum analysis (Vautard *et al.*, 1992) to help characterise localised structures, and (4) sophisticated methods to determine the 'k' spectrum from four points. These methods have been developed in close collaboration with (and tested by) the European Network on Numerical Simulation.

2.7. The Auroral Region

The present plan for Cluster data acquisition makes it possible to investigate the auroral zones, with the four satellites crossing this region close to perigee. Cluster will be at a rather high altitude (above 4 R_E) and thus will not cross the most 'active' auroral region, i.e., the region where data from Viking suggest that most of the parallel acceleration and non thermal radiation take place (Louarn *et al.*, 1990). Nevertheless the physical conditions that prevail above, and below, the 'main' acceleration region are very important; they fix the boundary conditions, and therefore control the parallel acceleration. Studies of the region below the main acceleration region, using *Freja* data (Wahlund *et al.*, 1994) have produced evidence for small-scale field-aligned current structures, identified as Solitary Kinetic Alfvén Waves (SKAWs), by Louarn *et al.* (1994). Preliminary results show that the parallel electric fields of the SKAWs are quite large, thus suggesting that they might play a role in the acceleration process. It is interesting to study the field aligned currents and the SKAWs at higher altitudes, to determine their possible role in the direct acceleration of particles, and the determination of boundary conditions. Given the small transverse scale of the SKAWs, and the large Doppler shift associated with the spacecraft motion, this study is quite difficult. More generally small scale field aligned currents, be they stationary or unsteady (Alfvén waves) are important at these altitudes ($\sim$ 4 R_E). Waveform measurements to be carried out aboard the four spacecraft as they pass by, one after the other, will allow an appropriate characterisation of these structures. It is important that the magnetic field, during these periods, will be approximately in the spin plane, which helps the determination of the parallel electric field of the SKAWs. At higher frequencies large amplitude electrostatic waves, presumably ion acoustic waves, should be observed, in close association with the SKAWs. The STAFF Spectrum Analyser is well suited to investigate these waves, which should have frequencies below the ion plasma frequency. Correlations between STAFF and the particle experiments will help in assessing the role of these waves in accelerating particles. In the acceleration region, electromagnetic ion cyclotron waves in the ELF frequency range (some 100 Hz) have been observed (Temerin and Lysak, 1984); they are presumably generated by accelerated electrons beams. Will they be observed at higher altitude? More generally very few, or even no measurement of electromagnetic waves has been performed yet in the high latitude auroral zone, and STAFF measurements will be somewhat exploratory here.

3. Scientific Wave Data Analysis

The innovation of the Cluster project is that, for the first time, a set of four identical spacecraft will produce a powerful tool for disentangling spatial and temporal variations. This has led the STAFF team to prepare specific tools to analyse the wave data. Some are discussed below, together with the method applied to validate the on-board analysis performed by the STAFF spectrum analyser.

The current structures which Cluster will encounter will normally be characterised by the Fluxgate Magnetometer (FGM). Nevertheless, some of these structures are of very small scale, corresponding to temporal signatures of the order of 1 s or less (Rezeau *et al.*, 1993). In such cases, the search coil data are complementary to FGM ones, STAFF being more sensitive for frequencies greater than about 1 Hz. When a small-scale structure, such as a small-scale current tube or a solitary wave, is observed by one spacecraft, the immediate question is whether the other spacecraft observed the same structure? To answer this question we must be able to identify that the different signatures possibly observed aboard two, three or four spacecraft do correspond to the same structure. A two-spacecraft study of magnetic fluctuations at the magnetopause has been performed with ISEE data. It has shown that in some cases the same structure can be identified on ISEE-1 and -2 (Rezeau *et al.*, 1993). Nevertheless, the most simple method of identification, the computation of the correlation function of the two signals, proved to be difficult to use for turbulent signals as observed at the magnetopause. The reason is that the correlation is efficient for identifying the same frequency in the two signals. This has two major consequences: (i) as the spectrum of the fluctuations is a continuous spectrum, a given frequency is always present in the data and the correlation coefficient is never very small, (ii) there is a perspective effect: a given structure will have a typical frequency which is different on two spacecraft, if the distance of closest approach is different (Rezeau *et al.*, 1990, 1993) and thus the two signatures will appear weakly correlated. In conclusion, the correlation will be efficient only in the cases when the inter-spacecraft distance is small and the conditions of observation of a given structure are similar on both spacecraft. But, as in a more general context it does not work, we are now developing a new tool for inter-spacecraft correlation.

Plasma waves will be characterised by complementary means involving ground-based and on-board calculations of the auto- and cross-correlation functions, as well as inter-correlation between waveforms measured at the various spacecraft locations.

From auto-correlations the energy density of electric and magnetic components will be inferred, together with the electrostatic/electromagnetic nature of the observed waves. The full knowledge of the polarisation characteristics of an electromagnetic wave field requires the computation of the cross-power spectra. Assuming knowledge of the dispersion relation, the Wave Distribution Function (WDF) is determined from the auto- and cross-power spectra at a given spacecraft location (Lefeuvre *et al.*, 1981). The WDF approach allows to remove the sign ambiguity in the wave normal direction (parallel or anti-parallel to the Earth magnetic field) when the measurement of at least one electric component is added to the measurements of the three magnetic components of the wave field. Depending on the frequency range of the waves studied, the spectral matrix will be calculated either on the ground from the waveform data, or on-board each spacecraft by the STAFF spectrum analyser (see next section). Using the waveform measurements from the four spacecraft a generalised spectral matrix containing all the available auto-, cross- and inter-spacecraft power spectra can be computed. From it, the Field Energy Distribution (FED) function will be determined which specifies how the field energy is distributed in a four dimensional space, i.e., as a function of the frequency and the three components of the wave vector, without assuming any dispersion relation (Pinçon and Lefeuvre, 1991). This kind of measurement will be achieved for the first time, thanks to the simultaneous measurements at four locations. To compute on the ground the general spectral matrix, we will need both the STAFF search coil data and the electric field waveform from the EFW experiment (Gustafsson *et al.*, this issue). This will yield five components of the electromagnetic waves on each spacecraft. STAFF and EFW hardware have been designed to simplify this correlation: the low pass filters are identical, the sampling frequency is the same, and a synchronisation signal for simultaneous sampling is sent to both experiments by the DWP experiment (Woolliscroft *et al.*, this issue). In addition, it is possible to use electric field measurements from the Electron Drift Instrument (EDI) (Paschmann *et al.*, this issue). The combination of data from the EFW, EDI and the search coil will allow the determination of the six components of the electromagnetic field. The inter-experiment link between STAFF and EDI will allow to synchronise EDI and WEC data.

The WDF strongly depends upon the geometry of the ray from the source to the point of observation. Simultaneous measurements at the various spacecraft locations will allow triangulation via multiple ray-tracing calculations, thus permitting the study of source location. An application of this method has been performed in order to validate the STAFF spectrum analyser calculations (Santolik, 1993; Belkacemi, 1993, 1994). An example, using a known simulated input signal and data output from one satellite, is shown in Figure 1, where the error in the determination of the wave normal direction is studied as a function of the signal-to-noise ratio. The WDF method applied here to recover the input direction of the injected signal has been compared with two other methods (Means, 1972; McPherron *et al.*, 1972) which can be used when the wave is assumed to be planar. Several independent realisations of noisy samples have been used to calculate the spectral matrices. Except at high signal to noise ratios, it is observed that the wave normal direction (θ and ϕ) is well recovered with the three methods.

The FED function calculation method has been applied to simulated data, under different conditions: number of wave components, respective spacecraft location, frequency and signal to noise dependence (Pinçon and Lefeuvre, 1992). This has allowed the determination of the experimental constraints under which the FED function can be determined. In particular, a study by Pinçon *et al.* (1990) has shown the influence of the inter-spacecraft distance on the results (see Figure 2); valid results will be obtained whenever one of the maximum inter-spacecraft distances is less than five times the minimum one.

An extensive characterisation of waves by Cluster needs, as seen above, correlative studies with data from different experiments. It is planned that all WEC Co-Investigators will have access to all WEC high-resolution data, through a common data analysis system, ISDAT (Pedersen *et al.*, this issue). Correlative studies with non-WEC experiment are foreseen, first through the Cluster Science Data System (CSDS) in which STAFF and all the other Cluster teams participate (Schmidt and Escoubet, this issue), second using case-by-case high-resolution data.

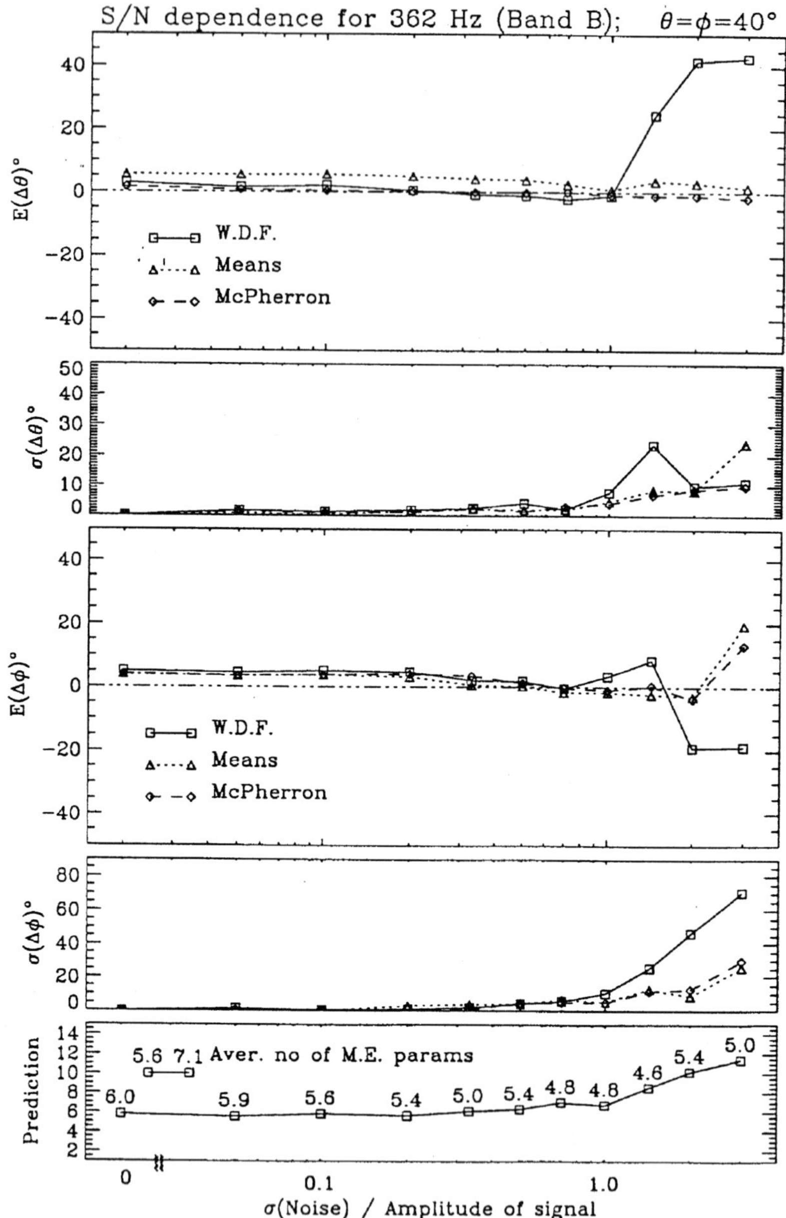

Figure 1. Electromagnetic plasma waves are simulated, using Maxwell equations. For a given plasma (here f_p = 100 kHz and f_{ce} = 6 kHz), the ratio between the variance of the random noise added to the electromagnetic field waveform and the signal is varied. The wave normal direction is fixed and known ($\theta = 40°$, $\phi = 40°$). These two angles are estimated by the spectral matrix calculated with a simulation of the STAFF spectrum analyser. From the top to the bottom, the panels of the figure give the average value of the angle error $E(\Delta\theta)$, the corresponding standard deviation $\sigma(\Delta\theta)$, the average value of the angle error $E(\Delta\phi)$, the corresponding standard deviation $\sigma(\Delta\phi)$, and a prediction parameter used to qualify the WDF solutions (see Lefeuvre *et al.*, 1981), as function of the signal to noise ratio. The squares refer to the WDF method, the triangles to the Means' method, and the diamonds to the McPherron's method. The figure gives the angle errors $\Delta\theta$ and $\Delta\phi$, that is always less than 10°.

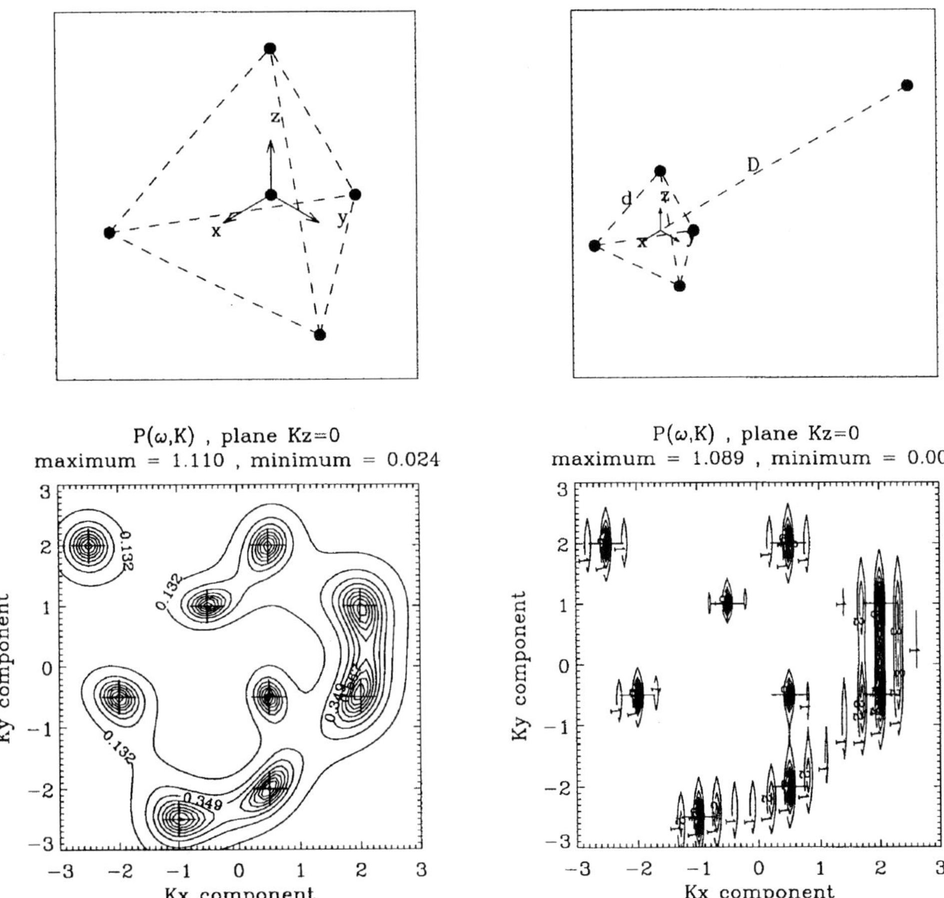

Figure 2. Example of the Field Energy Distribution (from Pinçon *et al.*, 1990). For a given distribution, here nine plane waves, the calculation of the FED is performed for different spacecraft respective position (top panels). Four satellites are located at top of a tetrahedron. On the left, the fifth satellite is added at the centre of the tetrahedron, whereas on the right the fifth satellite is placed at a large distance from the centre of the tetrahedron. The results derived from the 5 satellites are respectively given in the bottom panels as energy contours in the kx, ky plane for $kz = 0$. The initial FED is represented by crosses. The results are good when the fifth satellite is inside the tetrahedron (left case) or at distances D of the same order of magnitude as d. Differences arise when D is large relatively to d (right case, the ratio of distances is 20) (after Pinçon *et al.*, 1990). The results are similar with four spacecraft.

4. Experiment Technical Description

The STAFF experiment comprises a boom-mounted three-axis search coil magnetometer and two complementary data-analysis packages: a digital spectrum analyser, and an on-board signal-processing unit. The latter permits the observation of the three magnetic waveforms up to either 10 Hz or 180 Hz, depending upon mode. The spectrum analyser also receives the signals from the four electric field probes

Table I
Group tasks

CETP	STAFF co-ordination (PI + technical manager) manufacturing and testing of: search coil, magnetic waveform unit, calibration check-out software support to integration and testing data analysis
DESPA- Meudon	manufacturing and testing of the spectrum analyser check-out software support to integration and testing data analysis
LPCE-Orléans	design, calibration and tests of the filters data analysis
SSD-ESTEC	manufacturing of the filters
CEPHAG	theoretical support for data analysis relationship with ground-based measurements
Co-Is from other institutes:	link between STAFF and the other WEC experiments
LPCE, Orléans, France	P. M. E. Décréau
Sussex University, U.K.	M. P. Gough
University of Iowa, U.S.A.	D. A. Gurnett
SISP, Uppsala, Sweden	G. Gustafsson
SSD-ESTEC, The Netherlands	A. Pedersen
Sheffield University, U.K.	H. St. C. Alleyne, L. J. C. Woolliscroft

of the EFW experiment, which are used to form a pair of orthogonal electric field dipole sensors. All five inputs ($2E + 3 \times \mathbf{B}$) are used to compute in real time the 5×5 Hermitian cross-spectral matrix at 27 frequencies distributed logarithmically in the frequency range 8 Hz to 4 kHz.

STAFF is one of the five experiments of the Wave Experiment Consortium (WEC) (see Pedersen *et al.*, this issue). The STAFF team includes scientific and hardware contributions from a number of institutes, as shown in Table I. To optimise coordination within WEC, the STAFF investigator team includes all the WEC Principal Investigators.

4.1. The Search Coil Sensors and the Pre-Amplifier

Three mutually orthogonal sensors are mounted on a rigid boom away from the spacecraft body (see Figure 3). Two sensors, B_y and B_z, lie in the spin plane and are aligned mechanically with the long wire dipole antennas of the EFW experiment; the third is parallel to the spacecraft spin axis. Each sensor consists of a high

Figure 3. One set of the 3 STAFF search coil antennas under their thermal blanket, on the FM1 spacecraft.

permeability core embedded inside two solenoids. The main winding has a very large number of turns mounted in separate sections. The frequency response of the sensor is flattened in the frequency range 40–4000 Hz by a secondary winding used to introduce flux feedback. The secondary winding is also used as a calibration loop on which an external AC signal is applied through a calibration network included in the pre-amplifiers. The search coils are designed so as to minimise their sensitivity to electric fields. The transfer function and the experiment sensitivity are given in Figure 4. The measured sensitivity is 3×10^{-3} nT Hz$^{-1/2}$ at 1 Hz and 3×10^{-5} nT Hz$^{-1/2}$ at 100 Hz. The similarity of the search coils mounted on the four spacecraft is good, as can be seen on lowest panel of Figure 4 which gives the difference between two models of the uncalibrated amplitude and phase responses of one component (B_x): the phase is reproductible within $\pm 1°$, and the amplitude within 0.2 dB above 1 Hz, at the exception of the vicinity of 50 Hz. Much the same result is obtained for any combination of sensors.

In order to compare accurately data from the four satellites, a careful measurement of the angle between each magnetic axis and its corresponding mechanical

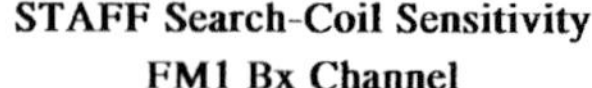

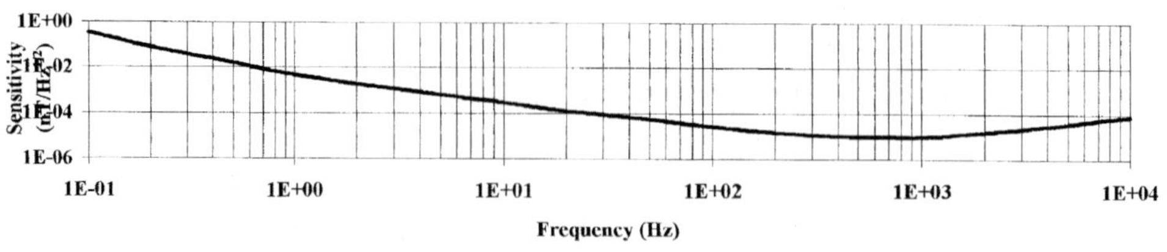

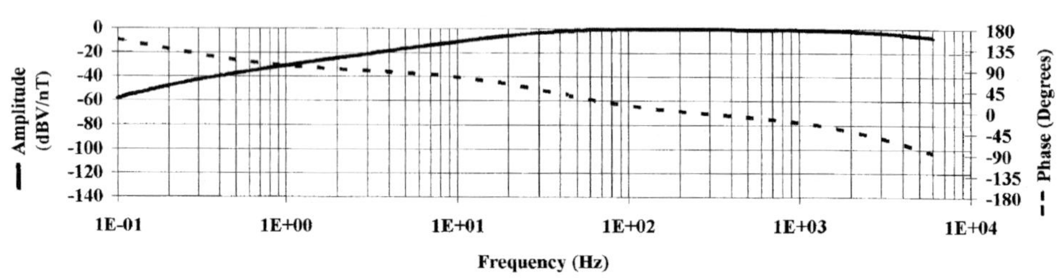

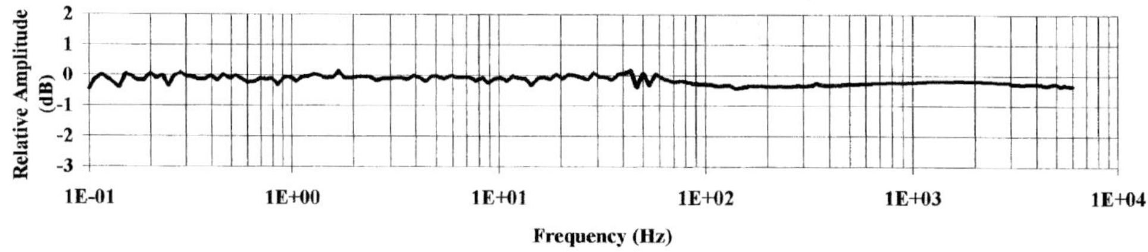

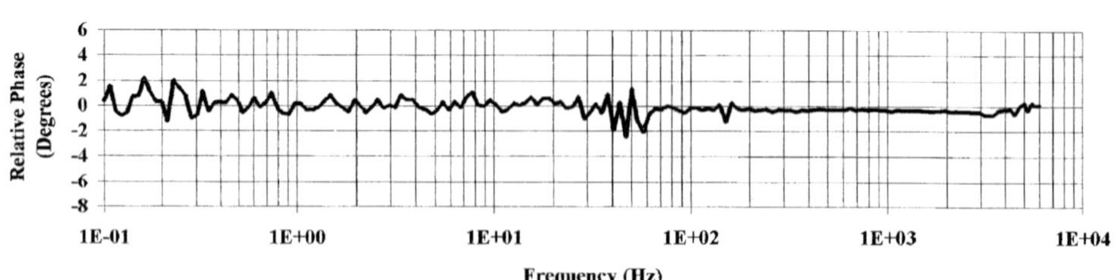

Figure 4. STAFF search coil and pre-amplifier transfer function and sensitivity measured in calibration facility at Chambon la Forêt, France. *Upper panel*: Flight Model 1 sensitivity for the B_x channel. second panel: transfer function for the same antenna in amplitude and phase. *Bottom panels*: example of the similarity of the search coil on-board the four spacecraft, the relative amplitude and phase responses of channel B_x on-board spacecraft 1 and 2.

axis has been made. These angles may be a few degrees, but they are known with a precision of 0.1°. Thus the magnetic field may be accurately transformed into any required reference frame.

Three pre-amplifiers are mounted in an electrical unit, located on the spacecraft deck. The low-power-consumption pre-amplifiers have a low-noise input stage and high-input impedance since they are connected to the magnetic sensors which are characterised by a low DC resistance and a very high impedance in the vicinity of the resonant frequency. The dynamic range of the pre-amplifiers is about 100 dB, which allows weak signals to be measured in the presence of the large voltage signals induced by the rotation of the spacecraft in the DC magnetic field. A new pre-amplifier using hybrid technology has been developed (Youssef *et al.*, 1991) and will be flown for the first time on Cluster. This technique has the advantages of including protection against radiation, the possibility of thermal control of the pre-amplifier, while being lighter than more traditional technology.

The output signals of the magnetic pre-amplifiers are conditioned for use at frequencies below 180 Hz by (i) the magnetic waveform unit, and for use up to 4 kHz by (ii) the spectrum analyser, (iii) the Wide-Band Data unit, (iv) the EFW experiment for the fast event detector, and (v) the EDI experiment.

4.2. The Magnetic Waveform Unit

The magnetic waveform unit (see Figure 5) consists of three sections which assure respectively: waveform digitalisation, data output interface, and on-board calibration. The latter is discussed in Section 4.4.

The three magnetic components B_x, B_y, B_z, at the output of the search coil pre-amplifier are passed through low-pass anti-aliasing filters with -3 dB cut-off at either 10 Hz or 180 Hz. These filters are of seventh order, i.e., they have an attenuation of 42 dB per octave. They are stable to better than 1% in amplitude and 1° in absolute phase; comparison between the different spacecraft show that they maintain this accuracy. The sampling frequency is 2.5 times the filter frequency, 25 or 450 Hz. Thus, the rejection of aliased components is at least 40 dB. Identical filters are used in the EFW experiment so as to optimise the correlation of electric and magnetic waveforms.

The filtered signals are applied to three sample and hold devices synchronised by the DWP experiment, then digitised, and sent to the DWP experiment. The same synchronisation signal is sent to both the STAFF and the EFW experiments. The bandwidth for the waveform measurements is selected by telecommand. The filtered signals are simultaneously sampled in a large dynamic range within a very short sampling time of about 10 μs in order to guarantee a relative error of less than one degree at 180 Hz between the three components. The sampling signal, provided by DWP, is common between the STAFF and EFW experiments in order to ensure the best simultaneous analysis of the five available components of the electromagnetic waves.

Figure 5. The STAFF electronic box on the Cluster platform at Dornier (WEC 8/1, the waveform unit is on top of WEC 8/2, the spectrum analyser) for the flight model number 4.

The samples are digitised in a real 16-bit analogue-to-digital converter and transmitted to the DWP experiment through one parallel interface. The 16-bit digitalisation allows to analyse simultaneously natural waves of a few 10^{-5} nT Hz$^{-1/2}$ and the large signal induced by the rotation of the spacecraft in the environmental DC field, up to some 100 nT at 0.25 Hz. With such a dynamic we expect to have accurate measurements, even at the inversion of the DC magnetic field at the magnetopause.

Owing to telemetry limitations, a reduction of the dynamic data range from 16 to 12 bits is performed inside DWP (Woolliscroft *et al.*, this issue). The principle is to transmit the full 16-bit word at the beginning of each telemetry packet, and later the difference between the successive samples, coded on 12 bits in such a way that the dynamics of the experiment should be preserved even at boundary crossings. Conservative back-up solutions can be selected by telecommand, being either a more crude compression, or no compression at all. The back-up compression will be used for instance at perigee where the spin signal is expected to be above some 200 nT.

4.3. THE SPECTRUM ANALYSER

The spectrum analyser calculates the complete auto- and cross-spectra of three components of the magnetic and two of the electric field, over a frequency range of nine octaves, with commendable time resolution. More precisely, the spectrum analyser determines the complete 5×5 Hermitian cross-spectral matrix of the signals from five input channels, over the frequency range of 8 Hz to 4 kHz, as follows (Figure 6).

The five auto-spectral power estimates are obtained with:

– a dynamic range of approximately 100 dB,

– an average amplitude resolution of 0.38 dB,

– a sensitivity as shown in Figure 4 for the magnetic components and in Gustafsson *et al.* (this issue) for the electric components.

The 10 cross-spectral power estimates are normalised to give the coherence, which is obtained with the following precision:

– the magnitude is sampled into one of 8 bins with upper limits distributed approximately as 2^{-n}, for $n = 0$ to 7;

– the precision of the phase depends upon the magnitude of the coherence: for a signal with magnitude in the highest bin, it is approximately 5° close to 0°, 180°, and ±90°, increasing to about 10° midway between these angles.

The spectral estimates are made at 27 frequencies distributed logarithmically over the range from 8 Hz to 4 kHz. All channels are sampled simultaneously, and the integration time for each channel is the same as the overall instrument time resolution, which can be commanded to values between 125 ms (except at the lowest frequencies) and 4 s. The cross-spectra are generally telemetered 4 times less frequently than the auto-spectra.

The frequency range of 8–4000 Hz is divided into three logarithmically distributed frequency sub-bands, each with a maximum frequency eight times the minimum frequency:

Band A: 8–64 Hz,

Band B: 64–512 Hz,

Band C: 512–4000 Hz.

The 'front end' of the analyser is analogue. For each of the three bands and for each of the five sensors there is a separate automatic gain-controlled (AGC) amplifier and separate band-pass filtering. This pre-conditioning normalises the overall output signal level within each sub-band to an optimum level for digitisation. The subsequent digital filtering performs the fine frequency analysis. The gain of these AGC amplifiers has the role of a multiplying factor in the determination of the absolute measurement. In the case of the spin-plane components (Ey, Ez and By, Bz) the total power from the two sensors is used for the normalisation, to remove the spin modulation. Separate high- and low-pass filters ensure that the gain normalisation is performed only for signal components with frequency within the

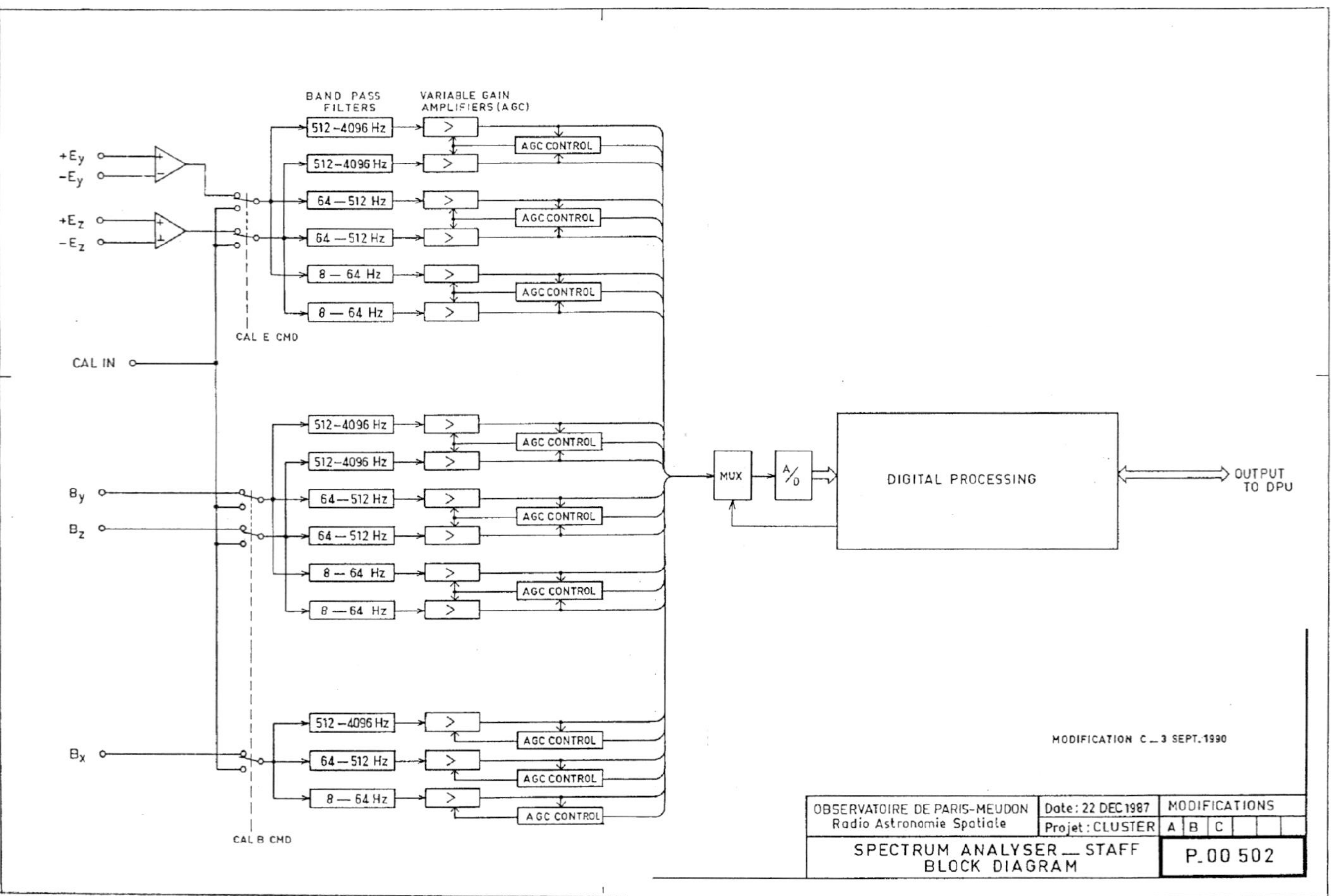

Figure 6. Spectrum analyser block diagram. *From left to right*: interface with EFW and STAFF search coil, analogue filtering in 3 bandwidths, the 9 AGC controls (that couple 2 by 2 the spinning components) and the digital processing to get the spectral matrix coefficients, transmitted to the DWP DPU.

band which will be further analysed digitally and, more importantly, they prevent 'aliasing' by frequencies above the Nyquist frequency.

The outputs from the 15 amplifiers are multiplexed to a single 8-bit 'flash' analogue/digital converter. They are digitised at a rate of 16 kHz, in a rapid-fire mode by groups of 5 or 10, as needed. The 9 AGC gain-control signals are digitised separately for inclusion in the telemetry.

The digital processing (block at the right hand side of Figure 6) of the sampled inputs is performed in three distinct steps:

– de-spin of the spin-plane sensor outputs;
– determination of the complex Fourier coefficients;
– calculation of the correlation matrices.

The de-spinning operation involves taking the signals from the two pairs of sensors, the electric dipoles and the y and z components of the search coil, and combining them so as to create the inputs which would have been received from non-rotating sensors. This transformation is necessary because the instrument measurement time interval is not short compared with the spacecraft spin period. Each pair of samples S_y and S_z are used to calculate

$$S_u = S_y \cos(m) + S_z \sin(m) ,$$

$$S_v = S_z \cos(m) - S_y \sin(m) ,$$

where m is the instantaneous angular position of the spacecraft as derived from the on-board Sun Reference Pulse (SRP), and u and v are the fixed coordinates.

These calculations are time-consuming. Therefore in band C the calculations are not performed on each individual data pair, but rather at the level of successive spectral matrices, which are computed every 16 ms, during which time the spacecraft has rotated very little.

The Fourier coefficients are determined using algorithms which are extensions of the Remez exchange algorithm (Rabiner and Gold, 1975). Each of the three analogue receiver bands is analysed using an algorithm very similar to the wavelet transform. The analysis divides the 3-octave band into 9 logarithmically-spaced channels, each with a relative 3 dB bandwidth of 26% of its central frequency. The time required for this analysis depends on the frequency band, ranging from 0.016 to 1 s.

The auto- and cross-spectra are calculated by multiplication of the complex Fourier coefficients and accumulation of the products; the 27 frequency channels yield a total of 135 auto-spectral coefficients and 270 complex cross-spectral coefficients. In Normal mode the measurement cycle is 4 s, during which time the analyser returns the auto-spectral coefficients accumulated during four consecutive intervals of 1 s, and the cross-spectral coefficients averaged over 4 s. During accumulation, any inbalance between the spin-plane analogue receivers gives rise to an apparent coupling between the y and z sensors; this will be corrected during ground data processing.

Internally the coefficients are accumulated as 40-bit integers representing spectral power. Of these 40 bits, 24 are significant for the final auto-spectrum; they represent a dynamic range of $10\log_{10}(2^{24}) = 72$ dB. To optimise use of the allocated telemetry and to simplify the interface with the DWP the 24-bit amplitudes N are logarithmically compressed into 8-bit telemetry words using 5 bits for the exponent E and 3 bits for the mantissa M, so that

$$N = 2^{(E-3)}(8 + M) .$$

The overall dynamic range for this data representation is 96 dB, while the average resolution is 0.38 dB. The cross-spectral coefficients S_{ij} are normalised on-board by division by the associated auto-spectral coefficients S_{ii} and S_{jj}, to yield the coherence C_{ij},

$$C_{ij} = S_{ij}/\sqrt{S_{ii}S_{jj}} ,$$

which is always less than unity. The real and imaginary parts of C_{ij}, including their sign, are each encoded in 4 bits.

The dynamic range of the front-end AGC analogue amplifiers is 75 dB, while the digital processor has a dynamic range of 45–50 dB. For the magnetic field, the dynamic range of the analogue amplifiers has been adjusted so that in the highest band of frequencies their maximum gain corresponds to the sensitivity (noise level) of the magnetic antennas. The digital noise introduced by the spectrum analyser is thus negligible, while the analogue receivers still cover the entire dynamic range of 70 dB of the magnetic antennas. In the lower two frequency bands the dynamic range is offset by factors of respectively 8 and 64, i.e., 9 and 18 dB; the lower end of the range is extended using the dynamic range of the digital analyser. For the electric field the dynamic range of the analogue amplifiers has similarly been adjusted with respect to the sensor noise level, to yield an effective dynamic range better than 70 dB at any frequency.

4.4. In Flight Calibration

Calibration can be commanded by the DWP to calibrate in flight the STAFF experiment (the magnetic wave sensors, the waveform unit and the spectrum analyser), either at normal bit rate or at high bit rate. Two kinds of calibration signals are generated in the magnetic waveform unit: either two simultaneous sine waves at around 7 Hz and 100 Hz, or a pseudo-random noise covering 4 kHz bandwidth. One output attenuator covers a 80 dB dynamic range in steps of 13 dB. An example of dynamic spectrum obtained from one component of the wave form data, up to 12 Hz, during an internal calibration period is shown in Figure 7. In the first part of the calibration, one can see the white noise, whose amplitude decreases after 64 s by steps of 13 dB every 16 s, and increases again before sending sine waves,

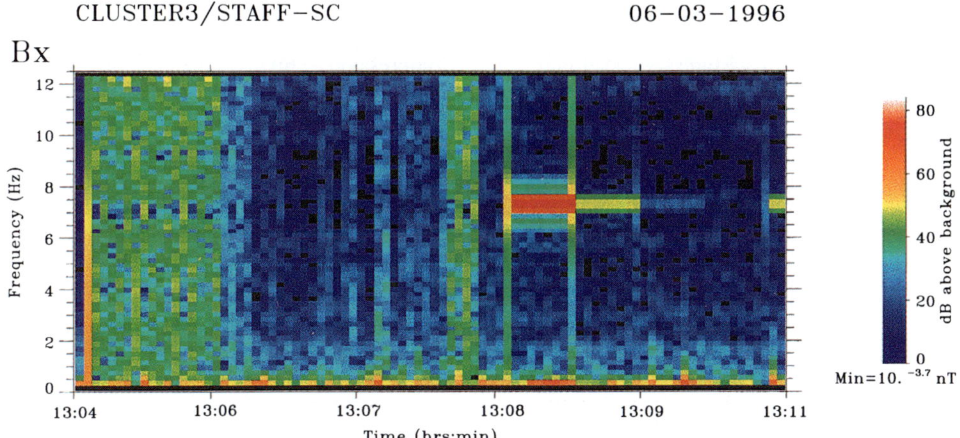

Figure 7. Dynamic spectra of STAFF B_x component from the waveform data obtained during an internal calibration period in normal bit rate. The calibration comprises 23 steps of 16 s each. In the first part of the calibration, a white noise is generated, whose amplitude decreases after 64 s by steps of 13 dB, and increased again before sending sin waves of about 7 Hz, whose amplitude is also varied by steps of 26 dB.

whose amplitudes are also varied, by steps of 26 dB. In the noisy environment of the ground integrations, the calibration dynamic of 80 dB is achieved, that will be improved in flight conditions. The signal of calibration is transmitted in the telemetry packets, together with the signals coming from the output of the STAFF analysers. It is used as a reference signal for the phase measurements between the different channels. The calibration allows the location of a failure, if any, to be identified in the STAFF chain. The transfer function can be re-established on ground. Special care has been taken with this tool, as it is very important to have accurate calibrations to compare measurements performed at the four spacecraft, as for instance in the calculation of inter-spectra in applying the FED method (see Section 2). The different spectrum analyser modes are also tested during this sequence, allowing to identify possible calculations errors. Modifications of the onboard program can be patched by telecommand, if necessary.

4.5. Operational Modes

Different operational modes are foreseen, mainly depending on the bit rate. A short description of these modes, for both the waveform and the spectrum analyser is given below (for details, see Tables II–IV). The STAFF modes are part of the WEC modes that are monitored by DWP (Pedersen *et al.*, this issue). Those WEC modes have been defined in order to get the best scientific return. The different constraints, telemetry, Whisper active, Langmuir mode of EFW, and available power have been

Table II
STAFF waveform data modes (search coil)

Mode	Bit rate	Compression	bit s^{-1}
NM	Normal	Yes (12 bits)	928
BM	High bit rate	Yes (12 bits)	16 480
EM NBR	Normal	No (16 bits)	1 216
EM HBR	High bit rate	No (16 bits)	21 760

Table III
STAFF spectrum analyzer normal bite rate operation modes

Band	Auto		Cross		AGC	
	resolution	bit s^{-1}	resolution	bit s^{-1}	resolution	bit s^{-1}
Normal mode 1: NM1 ($3 \times B + 2 \times E$) 1696 bit s^{-1} (including status)						
A: 8–64 Hz	1 s	360	4 s	180	1 s	24
B: 64–512 Hz	1 s	360	4 s	180	1 s	24
C: 512–4096 Hz	1 s	360	4 s	180	1 s	24
Normal modes 1′: NM1′b ($3 \times B$) or NM1′e ($2 \times E + B_x$) 864 bit s^{-1} (including status)						
A: 8–64 Hz	1 s	216	4 s	54	1 s	16
B: 64–512 Hz	1 s	216	4 s	54	1 s	16
C: 512–4096 Hz	1 s	216	4 s	54	1 s	16
Normal modes 2: NM2b ($3 \times B$) or NM2e ($2 \times E + B_x$) 1840 bit s^{-1} (including status)						
A: 8–64 Hz	1 s	216	1 s	216	1 s 16	
B: 64–512 Hz	0.5 s	432	1 s	216	0.5 s	32
C: 512–4096 Hz	0.5 s	432	1 s	216	0.5 s	32
Special mode: SM ($3 \times B + 2 \times E$) 3032 bit s^{-1} (including status)						
A: 8–64 Hz	1 s	360	2 s	360	1 s	24
B: 64–512 Hz	0.5 s	720	2 s	360	0.5 s	48
C: 512–4096 Hz	0.5 s	720	2 s	360	0.5 s	48
Emergency mode: EM ($3 \times B + 2 \times E$) 1120 bit s^{-1} (including status)						
A: 8–64 Hz	2 s	180	4 s	180	2 s	12
B: 64–512 Hz	2 s	180	4 s	180	2 s	12
C: 512–4096 Hz	2 s	180	4 s	180	2 s	12

considered in the choice of the modes (Pedersen *et al.*, this issue). The WEC modes can be modified if necessary.

The principle is to cover the full STAFF frequency range in all modes, but the methods are different depending on the bit rate. In normal bit rate, the waveform data covers the 0.1–10 Hz frequency range, whereas the spectrum analyser covers

Table IV
STAFF spectrum analyser high bit rate operation modes

Band	Auto		Cross		AGC	
	resolution	bit s^{-1}	resolution	bit s^{-1}	resolution	bit s^{-1}
Fast mode 1: FM1 ($3 \times B + 2 \times E$) 7600 bit s^{-1} (including status)						
B: 64–512 Hz	0.125 s	2880	1 s	720	0.125 s	192
C: 512–4096 Hz	0.125 s	2880	1 s	720	0.125 s	192
Fast mode 2: FM2 ($3 \times B + 2 \times E$) 4528 bit s^{-1} (including status)						
B: 64–512 Hz	0.25 s	1440	1 s	720	0.25 s	96
C: 512–4096 Hz	0.25 s	1440	1 s	720	0.25 s	96
Fast mode 3: FM3b ($3 \times B$) or FM3e ($2 \times E + B_x$) 4160 bit s^{-1} (including status)						
B: 64-512 Hz	0.125 s	1728	1 s	216	0.125 s	128
C: 512–4096 Hz	0.125 s	1728	1 s	216	0.125 s	128

the frequency range 8 Hz–4 kHz, working in its three frequency bands. In high bit rate (also called burst mode), the waveform data covers the 0.1–180 Hz frequency range, and in order to save telemetry, the spectrum analyser only operates in its two upper frequency bands, from 64 Hz to 4 kHz.

For the waveform data, two combinations of commands can be sent, one is the sampling frequency rate, the other is whether a data compression is applied or not (see Section 4.2). In normal bit rate, the sampling frequency is 25 Hz, associated with the 10 Hz low pass filter. In high bit rate, the sampling frequency is 450 Hz, with the 180 Hz filter. STAFF and EFW use the same sampling frequency, synchronised by DWP. Thus there are four STAFF waveform data modes (see Table II). The NM mode, in normal telemetry rate with compression, giving a bit rate of 928 bit s^{-1}. In emergency mode the 16-bit words are telemetered (EM NBR mode). The principle is the same in high bit rate; with data compression, the needed telemetry is 16 480 bit s^{-1} (BM mode).

For the spectrum analyser, the different modes have been defined by combining three parameters: the time resolution, the number of frequencies computed and the number of wave components considered. The modes are defined to fulfil different scientific objectives, in the framework of three constraints, first the telemetry limitation, second the total WEC power limitation, then the operations mode of the other WEC experiments, as discussed in Pedersen *et al.* (this issue).

Normal Mode 1 (NM1) is the basic mode in normal bit rate. The auto-spectra are averaged over 1s, and the complete matrix over 4s for five components (25 coefficients). The other modes are variations of this.

In Normal Mode 1′ (NM1′), the calculation is performed for only three components, either $3 \times \mathbf{B}$ (NM1′b) or B_x plus $2 \times E$ (NM1′e). Only nine elements of the spectral matrix are computed (out of 25). This mode is used in time-sharing with NM1, during periods when Whisper is active.

In Normal Mode 2 (NM2), three of the five wave components are selected as in NM1′. The time resolution is 0.5 or 1 s for the auto-spectra and 1 s for the cross-spectra. This mode is a 'low power' mode, and the input of the components that are not used are powered off. When the modes NM2b and NM2e are used in time-sharing, the AGC needs some time to recover. The mode NM2b can also be used when EFW is in Langmuir mode and not in electric field measurement mode on all four booms, as the calculation of the de-spun electric components needs the two spinning components.

In Special Mode the time resolution is improved. This mode can be used if for some reason, more telemetry is available to STAFF inside WEC in normal bit rate.

In Emergency Mode, the five components are used, with a lower time resolution of 2 and 4 s for the auto- and cross-spectra, respectively. This reduction in time resolution, and thus in telemetry is intended to compensate for the telemetry increase due to the non-compression of the waveform data, in case DWP cannot perform the waveform compression.

In high bit rate (also called burst mode), the spectral matrix coefficients are calculated only in the two highest frequency bands. In the Fast Modes the time resolution is 1 s for the cross-spectra and either 0.125 s or 0.25 s for the auto-spectra. Here again five or three components can be considered (see Table IV).

The calibration mode calibrates both parts of the experiment, the magnetic wave form and the spectrum analyser (see Section 4.4). It is foreseen to operate the calibration program no more than once a day, or rather once per orbit, and preferably at the beginning of a data acquisition sequence. The duration of the calibration sequence is about 6 min at normal bit rate and 2 min in burst mode.

4.6. GSE AND INTEGRATION SOFTWARE

Specific STAFF software to test the instrument capabilities has been written by the STAFF team (Table I), whereas the overall WEC testing is under responsibility of DWP (Pedersen *et al.*, this issue). The decommutation of the WEC packets into STAFF data packets is done by DWP, as for other WEC instruments. The GSE software has been used as the starting point of the ground segment software. The ground software includes the instrument monitoring, the preparation of STAFF parameters for the Cluster Science Data System (CSDS) (Schmidt and Escoubet, this issue), and of course scientific data analysis. The role of DWP in this facilitates the use of a common software to analyse simultaneously data coming from different WEC instruments. The definition and the implementation of this common software are coordinated by the WEC Data Working Group (Pedersen *et al.*, this issue).

5. Conclusion

The STAFF experiment consists of a set of four state-of-the-art instruments to measure and analyse the vector magnetic field fluctuations and, in the appropriate frequency range, correlate them with the two components of the electric field measured by the EFW experiment. STAFF will thus optimise the global Cluster scientific return, particularly in those regions of space that are primary scientific targets for the mission. The instruments on each individual spacecraft will produce significant new data and, furthermore, the four instruments are identical, and have sufficient resolution, to allow good separation of spatial and temporal variations of the measured parameters. The experiment is well characterised by its calibration model; the parameters of this model will be checked regularly and updated if necessary, using data acquired during the calibration cycle which is executed routinely throughout the mission.

Thus STAFF will make accurate estimates of important wave properties in the frame of reference of each spacecraft. Using data from the four instruments, alone or in combination with other WEC data such as the EFW convection (DC) and waveform data, it will generally be possible to determine wave properties in the physically more important rest frame of the plasma. Comparison with simultaneous particle measurements will be essential to study the physical processes. To this end, STAFF is participating actively in the implementation of the Cluster Science Data System, which is the essential first step towards the coordinated studies of high-resolution data.

Acknowledgements

The realisation, testing, integration of the STAFF experiment, as well as the preparation of the ground software for instrument commanding, health verification and science analysis have benefited of the help of many people, in many places. We would like to thank: M. Belkacemi, D. Carrière, A. C. Gueriau, J. P. Mengué, A. Rapin, J. P. Rivet, L. Sitruk (DESPA/Meudon Observatory), D. Bagot, V. Bouzid, N. Denis, E. François, A. Mayaki (CETP), A. Butler, D. Klinge (SSD/ESTEC). The WEC, ESA, and Dornier teams are thanked for the co-operative help in the design, testing and integration of the experiment. H. Poussin, M. Nonon, J. Y. Prado, and J. P. Thouvenin at CNES are thanked for their continuous assistance and co-operation in the STAFF CSDS data handling preparation, as are the JSOC and ESOC teams in the preparation of the operations. The STAFF experiment and software are realised thanks to CNES support, through contract with CETP, DESPA, and LPCE. The University of Versailles Saint-Quentin is thanked for its support to CETP Cluster team. The referees are thanked for their fruitful comments.

References

Anderson, B. J., Fuselier, S. A., and Murr, D.: 1991, 'Electromagnetic Ion Cyclotron Waves Observed in the Plasma Depletion Layer', *Geophys. Res. Letters* **18**, 1955-1958.

Anderson, B. J., Fuselier, S. A., Gary, S. P., and Denton, R. E.: 1994, 'Magnetic Spectral Signatures in the Earth's Magnetosheath and Plasma Depletion Layer', *J. Geophys. Res.*. **99**, 5877.

Belkacemi, M.: 1994, 'Modélisation et Qualification de l'Analyseur Multicanal Numérique Temps Réel à Corrélation STAFF SA pour l'Expérience Spatiale Européenne Cluster', PhD Thesis, Paris 6 University.

Belkacemi, M., Bougeret, J.-L., Cornilleau-Wehrlin, N., Friel, L, Harvey, C. C., Manning, R., and Parrot, M.: 1993, 'Determination Characteristics of the STAFF Spectrum Analyser using Simulated Data', *Proc. of ESA conference on Spatio-Temporal Analysis for Resolving Plasma Turbulence*, Aussois, France, 31 January–5 February, 1993, ESA WPP-047.

Belmont, G., Reberac, F., and Rezeau, L.: 1995, 'Resonant Amplification of Magnetosheath MHD Fluctuations at the Magnetopaus'e, *Geophys. Res. Letters* **22**, 295.

Blanco-Cano, X. and Schwartz, S. J.: 1996, 'AMPTE-UKS Observations of ULF Waves in the Quasi-Parallel Ion Foreshock', *J. Geophys. Res.*, in press.

Büchner, J. M. and Zelenyi, L. M.: 1988, *Adiabatic Chaotic and Quasi-adiabatic Charged Particle Motion in Two-dimensional Magnetic Field Reversals*, ESA SP-285, 1, 219.

Burgess, D.: 1995, 'Foreshock-Shock Interaction at Collisionless Quasi-parallel Shocks', *Adv. Space Res.* **15** (8/9), 159.

Cattell, C., Wygant, J., Mozer, F. S., Okada, T., Tsuruda, K., Kokubun, S., and Yamamoto, T.: 1995, 'ISEE 1 and Geotail observations of Low-Frequency Waves at the Magnetopause', *J. Geophys. Res.* **100**, 11823.

Delcourt, D. C., Martin, R. F., Jr., Sauvaud, J.-A., and Moore, T. E.: 1995, 'The Centrifugal Trapping in the Magnetotail', *Ann. Geophys.* **13**, 242.

Dubouloz, N. and Scholer, M.: 1995, 'Two-Dimensional Simulations of Magnetic Pulsations Upstream of the Earth's Bow Shock', *J. Geophys. Res.* **100**, 9461.

Fredericks, R. W., Scarf, F. L., and Russell, C. T.: 1973, 'Field-Aligned Currents, Plasma Waves, and Anomalous Resistivity in the Disturbed Polar Cap', *J. Geophys. Res.* **78**, 2133.

Fuselier, S. A.: 1995, 'Ion Distributions in the Earth's Foreshock Upstream from the Bow Shock', *Adv. Space Res.* **15** (8/9), 43.

Gendrin, R.: 1983, 'Magnetic Turbulence and Diffusion Processes in the Magnetopause Boundary Layer', *Geophys. Res. Letters* **10**, 769.

Greenstadt, E. W., Le, G., and Strangeway, R. J.: 1995, 'ULF Waves in the Foreshock', *Adv. Space Res.* **15** (8/9), 71.

Gurnett, D. A.: 1985, in B. T. Tsurutani and R. G. Stone (eds.), 'Plasma Waves Instabilities, Collisionnless Shocks in the Heliosphere: Reviews of Current Research', *Geophys. Monogr. Ser.*, Vol. 35, AGU, Washington, D.C., p. 207.

Gustafsson, G. *et al.*: 1996, this issue.

Haerendel, G., Paschman, G., Sckopke, N., Rosenbauer, H., and Hedgecock, P. C.: 1978, 'The Frontside Boundary Layer of the Magnetosphere and the Problem of Reconnection', *J. Geophys. Res.* **83**, 3195.

Hellinger, P., Mangeney, A., and Matthews, A.: 1995, 'Whistler Waves in 3D Hybrid Simulations of Quasi-Perpendicular Shocks', *Geophys. Res. Letters* **22**, 2091.

Hubert, D.: 1994, 'Nature and Origin of Wave Modes in the Dayside Earth Magnetosheath', *Adv. Space Res.* **14** (7), 55.

Krauss-Varban, D.: 1994, 'Bow Shock and Magnetosheath Simulations: Wave Transport and Kinetic Properties, Solar Wind Sources of Magnetospheric Ultra-Low-Frequency Waves', *Geophysical Monograph* **81**, 121.

Krauss-Varban, D., Pantellini, F. E. G., and Burgess, D.: 1995, 'Electron Dynamics and Whistler Waves at Quasi-Perpendicular Shocks', *Geophys. Res. Letters*, in press.

Khurana, K. K., Kivelson, M. G., Frank, L. A., and Paterson, W. R.: 1995, 'Observations of Magnetic Flux Ropes and Associated Currents in Earth's Magnetotail with the Galileo Spacecraft', *Geophys. Res. Letters* **22**, 2087.

Labelle, J. and Treumann, R. A.: 1988, 'Plasma Waves at the Dayside Magnetopause', *Space Sci. Rev.* **47**, 175.

Lacombe, C. and Belmont, G.: 1995, 'Waves in the Earth's Magnetosheath: Observations and Interpretations', *Adv. Space Res.* **15** (8/9), 329.

Le, G., Russell, C. T., Thomsen, M. F., and Gosling, J. T.: 1992, 'Observations of a New Class of Upstream Waves with Periods near 3 Seconds', *J. Geophys. Res.* **97**, 2917.

Le, G., Russell, C. T., and Orlowski, D. S.: 1993, 'Coherence Length of Upstream ULF Waves: Dual ISEE Observations', *Geophys. Res. Letters* **20**, 1755.

Lefeuvre, F., Parrot, M., and Delannoy, C.: 1981, 'Wave Distribution Functions Estimation of VLF Electromagnetic Waves Observed On-Board GEOS-1', *J. Geophys. Res.* **86**, 2359.

Louarn, P., Roux, A., de Féraudy, H., and LeQuéau, D.: 1990, 'Trapped Electrons as a Free Energy Source for the Auroral Kilometric Radiation', *J. Geophys. Res.* **95**, 5983.

Louarn, P., Wahlung, J. E., Chust, T., de Feraudy, H., Roux, A., Holback, B., Dovner, P. O., Eriksson, A., and Holmgren, G.: 1994, 'Observation of Kinetic Alfvén Waves by the Freja Spacecraft', *J. Geophys. Res.* **99**, 23 623.

Lui, A. T. Y., Lopez, R. E., Anderson, B. J., Takahashi, K., Zanetti, L. J., McEntire, R. W., Potemra, T. A., Klumpar, D. M., Greene, E. M., and Strangeway, R.: 1992, 'Current Disruption in the Near-Earth Neutral Sheet Region', *J. Geophys. Res.* **97** 1461.

Matsumoto, H., Kojima, H., Miyatake, T., Omura, Y., Okada, M., Nagano, I., and Tsutsui, M.: 1994, 'Electrostatic Solitary Waves (ESW) in the Magnetotail: BEN Waveforms Observed by Geotail', *Geophys. Res. Letters* **25**, 2915.

McPherron, R. L., Russel, C. T., and Coleman, J. R., Jr.: 1972, 'Fluctuating Magnetic Waves in the Magnetosphere, 2, ULF Waves', *Space Sci. Rev.* **13**, 411.

Means, J. D.: 1972, 'Use of the Three-Dimensional Covariance Matrix in Analysing the Polarisation Properties of Plane Waves', *J. Geophys. Res.* **77**, 5551.

Mitchell, D. G., Williams, D. J., Huang, C. Y., Frank, L. A., and Russel, C. T.: 1990, 'Current Carriers in the Near-Earth Cross-Tail Current Sheet During Substorm Growth Phase', *Geophys. Res. Letters* **17**, 3131.

Paschman, G. *et al.*: 1996, this issue.

Pedersen, A. *et al.*: 1996, this issue.

Perraut, S., Morane, A., Roux, A., Pedersen, A., Schmidt, R., Korth, A., Kremser, G., Apariciao, B., and Pellinen, R. J.: 1993, 'Characterisation of Small Scale Turbulence Observed at Substorm Onset: Relationships with Parallel Acceleration of Particles', *Adv. Space. Res.* **13**, 217.

Perraut, S., Roux, A., Holter, O., Korth, A., Kremser, G., and Pedersen, A.: 1995, *Current Sheet Structure and Relation to Breakup*, ESA SP-371, p. 239.

Phan, T.-D., Paschmann, G., Baumjohann, W., Sckopke, N., and Lühr, H.: 1994, 'The Magnetosheath Region Adjacent to the Dayside Magnetopause: AMPTE/IRM Observations', *J. Geophys. Res.* **99**, 121.

Pinçon, J. L. and Lefeuvre, F.: 1991, 'Local Characterization of Homogeneous Turbulence in a Space Plasma from Simultaneous Measurements of Field Components at Several Points in Space', *J. Geophys. Res.* **96**, 1789.

Pinçon, J. L. and Lefeuvre, F.: 1992, 'The Application of the Generalized Capon Method to the Analysis of a Turbulent Field in Space Plasma: Experimental Constraints', *J. Atmospheric Terrest. Phys.* **54**, 1237.

Pinçon, J. L., Lefeuvre, F., and Parrot, M.: 1990, 'Experimental Constraints in the Characterisation of a Turbulent Field from Several Satellites', *Proc. Int. Workshop on Space Plasma Physics Investigations by Cluster and Regatta*, Graz, ESA SP-306, p. 57.

Rabiner and Gold: 1975, *Theory and Application of Digital Signal Processing*, Prentice-Hall, New York.

Rezeau, L., Morane, A., Perraut, S., Roux, A., and Schmidt, R.: 1989, 'Characterization of Alfvénic Fluctuations in the Magnetopause Boundary Layer', *J. Geophys. Res.* **94**, 101.

Rezeau, L., Roux, a., and Cornilleau-Wehrlin, N.: 1990, 'Multipoint Study of Small Scale Structures at the Magnetopause', *Proc. Int. Workshop on Space Plasma Physics Investigations by Cluster and Regatta*, Graz, February 1990, ESA SP-306, p. 103.

Rezeau, L., Roux, A., and Russell, C. T.: 1993, 'Characterization of Small Scale Structures at the Magnetopause from ISEE Measurements', *J. Geophys. Res.* **98**, 179.

Robert, P., Gendrin, R., Perraut, S., Roux, A., and Pedersen, A.: 1984, 'GEOS-2 Identification of Rapidly Moving Current Structures in the Equatorial Outer Magnetosphere During Substorms', *J. Geophys. Res.* **89**, 819.

Russell, C. T., Childers, D. D., and Coleman, P. J., Jr.: 1971, 'OGO-5 Observations of Upstream Waves in the Interplanetary Medium: Discrete Wave Packets', *J. Geophys. Res.* **76**, 845.

Santolik, O.: 1993, 'Etude sur les Directions Normales des Ondes Electromagnétiques dans un Plasma', LPCE/NTS/018.A.

Savoini, P. and Lembège, B.: 1994, 'Electron Dynamics in Two and One Dimensional Oblique Supercritical Collisionless Magnetosonic Shocks', *J. Geophys. Res.* **99**, 6609.

Savoini, P. and Lembège, B.: 1995, 'Heating and Acceleration of Electrons through the Whistler Precursor in 1D and 2D Oblique Shocks', *Adv. Space Res.* **15** (8/9), 235.

Schmidt, R. and Escoubet, C. P.: 1996, 'The Cluster Science Data System', this issue.

Scholer, M.: 1995, 'Interaction of Upstream Diffuse Ions with the Solar Wind', *Adv. Space Res.* **15** (8/9), 125.

Scudder, J. D., Mangeney, A., Lacombe, C., Harvey, C. C., Wu, C. S., and Anderson, R. R.: 1986, 'The Resolved Layer of a Collisionless, High b, Supercritical, Quasi-perpendicular Shock Wave: 3. Vlasov Electrodynamics', *J. Geophys. Res.* **91**, 11 074.

Sentman, D. D., Thomsen, M. F., Gary, S. P., Feldman, W. C., and Hoppe, M. M.: 1983, 'The Oblique Whistler Instability in the Earth's Foreshock', *J. Geophys. Res.* **88**, 2048.

Song, P.: 1994, 'ISEE Observations of the Dayside Magnetosheath', *Adv. Space Res.* **14**, 71.

Temerin, M. and Lysal, R. L.: 1984, 'Electromagnetic Ion Cyclotron Mode (ELF) Waves Generated by Auroral Electron Precipitation', *J. Geophys. Res.* **89**, 2849.

Vautard, R., Yiou, P., and Ghil, M.: 1992, 'Singular-Spectrul Analysis, A Toolkit for Short, Noisy Chaotic Signals', *Physica* **D58**, 95.

Veltri, P., Mangeney, A., and Scudder, J. D.: 1990, 'Electron Heating in Quasi-Perpendicular Shocks: a Monte Carlo Simulation', *J. Geophys. Res.* **95**, 14939.

Wahlund, J. E., Louarn, P., Chust, T., de Feraudy, H., Roux, A., Holback, B., Cabrit, B., Eriksson, A. I., Kintner, P. M., Kelley, M. C., Bonnel, J., and Chesney, S.: 1994, 'Observations of Ion Acoustic Fluctuations in the Auroral Topside Ionosphere by Freja S/C', *Geophys. Res. Letters* **21**, 1835.

Winske, D., Omidi, N., Quest, K. B., and Thomas, V. A.: 1990, 'Reforming Supercritical Quasi-Parallel Shocks, 2., Mechanisms for Wave Generation and Front Re-formation', *J. Geophys. Res.* **95**, 18821.

Woolliscroft, L. J. C. *et al.*: 1996, this issue.

Youssef, A., Meyer, A., Ducrocq, J. B., and Roux, A.: 1991, 'New Technologies for Integrating Thermal Control and Radiation Protection in Hybrid Technology', *Proc. ESA Electronic Components Conference*, ESTEC, Noordwijk, ESA SP-313.

Zhang, T.-L., Schwingenschuh, K., and Russell, C. T.: 1995, 'A Study of the Solar Wind Deceleration in the Earth's Foreshock Region', *Adv. Space Res.* **15** (8/9), 137.

THE ELECTRIC FIELD AND WAVE EXPERIMENT FOR THE CLUSTER MISSION

G. GUSTAFSSON, R. BOSTRÖM, B. HOLBACK, G. HOLMGREN, A. LUNDGREN, K. STASIEWICZ and L. ÅHLÉN
Swedish Institute of Space Physics, Uppsala Division, S-755 91 Uppsala, Sweden

F. S. MOZER, D. PANKOW, P. HARVEY, P. BERG and R. ULRICH
University of California, Berkeley, U.S.A.

A. PEDERSEN, R. SCHMIDT, A. BUTLER, A. W. C. FRANSEN, D. KLINGE and M. THOMSEN
SSD/ESTEC, Noordwijk, The Netherlands

C.-G. FÄLTHAMMAR, P.-A. LINDQVIST and S. CHRISTENSON
Royal Institute of Technology, Stockholm, Sweden

J. HOLTET, B. LYBEKK and T. A. STEN
University of Oslo, Norway

P. TANSKANEN and K. LAPPALAINEN
University of Oulu, Finland

J. WYGANT
School of Physics and Astronomy, Minneapolis, U.S.A.

Abstract. The electric-field and wave experiment (EFW) on Cluster is designed to measure the electric-field and density fluctuations with sampling rates up to 36 000 samples s^{-1}. Langmuir probe sweeps can also be made to determine the electron density and temperature. The instrument has several important capabilities. These include (1) measurements of quasi-static electric fields of amplitudes up to 700 mV m^{-1} with high amplitude and time resolution, (2) measurements over short periods of time of up to five simualtaneous waveforms (two electric signals and three magnetic signals from the seach coil magnetometer sensors) of a bandwidth of 4 kHz with high time resolution, (3) measurements of density fluctuations in four points with high time resolution. Among the more interesting scientific objectives of the experiment are studies of nonlinear wave phenomena that result in acceleration of plasma as well as large- and small-scale interferometric measurements. By using four spacecraft for large-scale differential measurements and several Langmuir probes on one spacecraft for small-scale interferometry, it will be possible to study motion and shape of plasma structures on a wide range of spatial and temporal scales. This paper describes the primary scientific objectives of the EFW experiment and the technical capabilities of the instrument.

List of EFW PI and Co-Investigators (alphabetical order)

L. Blomberg[2], R. Boström[1], C. Cattell[3], N. Cornilleau-Wehrlin[4], P. Decreau[5], A. Egeland[6], C.-G. Fälthammar[2], R. Grard[7], D. Gurnett[8], G. Gustafsson[1] (PI), C. Harvey[9], B. Holback[1], G. Holmgren[1], Ø. Holter[6], J. Holtet[6], P. Kellogg[3], P. Kintner[10], S. Klimov[11], J.-P. Lebreton[7], P.-A. Lindqvist[2], R. Manning[9], G. Marklund[2], N. Maynard[12], F. Mozer[13], K. Mursula[14], H. Opgenoorth[1], A. Pedersen[7], H. Pecseli[6], R. Pfaff[15], I. Roth [13], A. Roux [4], J. Singer[16], R. Schmidt[7], K. Stasiewicz[1], P. Tanskanen[14], M. Temerin[13], E. Thrane[17], L. Woolliscroft[18], J. Wygant[3].

Space Science Reviews **79:** 137–156, 1997.

Institutions

[1] Swedish Institute of Space Physics, Uppsala, Sweden.
[2] Division of Plasma Physics, Alfvén Laboratory, Stockholm, Sweden.
[3] School of Physics and Astronomy, University of Minnesota, U.S.A.
[4] CETP/USQV, Velizy, France
[5] LPCE/CNRS, Orleans, France
[6] University of Oslo, Oslo, Norway
[7] Space Science Department/ESTEC, Noordwijk, The Netherlands
[8] University of Iowa, Iowa City, Iowa, U.S.A.
[9] DESPA, Meudon, France
[10] School of Electrical Engineering Cornell University, Ithaca, New York, U.S.A.
[11] Space Research Institute Moscow, Russia
[12] Mission Research Corporation, Nashua, New Hampshire, U.S.A.
[13] University of California, Berkeley, California, U.S.A.
[14] University of Oulu, Oulu, Finland
[15] Goddard Space Flight Center, Greenbelt, Maryland, U.S.A.
[16] NOAA, Boulder, Colorado, U.S.A.
[17] Norwegian Defence Research Establishment, Kjeller, Norway
[18] University of Sheffield, Sheffield, England

1. Introduction

The Cluster mission of four identical spacecraft in nearby, non-coplanar orbits will for the first time give us measurements of magnetospheric structures in three dimensions and will improve the possibilities to separate between time and space variations.

The electric-field and plasma densities are esssential quantities to measure in order to understand magnetospheric processes. The detailed high resolution measurements at each spacecraft combined with measurements of the large-scale gradient from the four spacecraft will be an important contribution to the mission.

Spherical double probes for measurements of electric-fields in magnetospheric plasma were first launched on the S3-3 spacecraft in 1976 (Mozer *et al.*, 1979) and has since then been used on a number of satelllite missions. A review of the performance of double probes was given by Pedersen *et al.* (1984).

The four Cluster spacecraft will pass through numerous plasma regimes separated by different boundaries and discontinuities. This implies that the spacecraft will encounter a large variety of plasma phenomena at different scales both in time and space domain, as well as different field intensities. The EFW experiment has four spherical sensors located at the ends of 50 m long booms in the spin plane of the satellite. The electric-field component along the spin axis can only be constructed with the assumption that $\mathbf{E} \cdot \mathbf{B} = 0$. The experiment is well suited to study both the large-scale structures, with four-spacecraft interferometry, as well as small (micro)-scale structures, with two (or more) probes operated in Langmuir mode on one spacecraft. Two analogue-to-digital converters can sample waveforms up to 36 ksamples s^{-1} and can redirect the output to an internal (burst) memory of one Megabyte allowing sampling frequencies much higher than those permitted by the spacecraft telemetry system (i.e., 450 samples s^{-1}). To provide an adequate data base for studying physical phenomena in various magnetospheric regions, the electric-field and plasma fluctuation experiment is able to measure wave phenomena with temporal resolution of 110 μs. This will cover wave frequencies from DC up to those of lower hybrid frequency and will make it possible to resolve small-scale electromagnetic structures. EFW will concentrate on taking full waveforms up to 180 Hz. Complementary electric and magnetic field data will be obtained by other experiments within the Wave Consortium (WEC). With interferometric (differential) timing between opposite probes the instrument will make it possible to measure small-scale plasma structures (viz., cavitons) and distinguish between spatial and temporal variations. This paper describes the primary scientific objectives of the experiment, its technical capabilities, and planned data-processing systems.

2. Scientific Objectives

2.1. Electric Fields at Small Scales

Some of the most interesting magnetospheric phenomena are associated with non-linear processes that result in the acceleration of plasma. Electric-field measurements on several satellites: S3-3, Viking, ISEE, and Freja have shown that particle acceleration is associated with large amplitude, short duration electric-field signatures. An example of a short duration (spiky) electric-field observed at bow shock is shown in Figure 2.1.

ISEE observations in the bow shock revealed intense, often bipolar, electric-field spikes with amplitudes of 100 mV m^{-1} and durations of 50 ms (Figure 1). It was not possible to fully determine the properties of these waves with the ISEE experiment but will be possible with Cluster. The phase velocities, scale size, polarisations relative to the ambient magnetic field, three-dimensional structures, coherence lengths, relation to comparable temporal scale density and magnetic-field fluctuations can be determined with the Cluster electric-field instrument. These quantities are needed to evaluate the role of the spikes in providing anomalous resistivity, thermalised ion and electron distributions, and in the production of high-energy electron beams commonly seen at the bow shock. Small localised plasma structures are inherently nonlinear and represent interesting objects for study both for fundamental plasma physics and for applied magnetospheric research. High-rate data sampling usually requires internal burst memory because of limits imposed by standard spacecraft telemetry. Whenever higher time resolution data have become

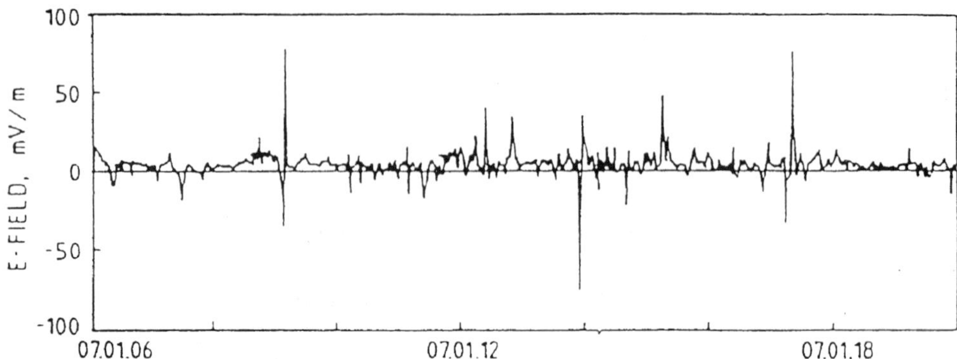

Figure 1. Measurements on ISEE-1 at 32 samples s^{-1} of spiky electric-field of 100 mV m^{-1} and a duration of 50 ms during bow shock crossing. This is a single-axis measurement, so the direction relative to the magnetic field is not known. On Cluster, the direction and phase velocity can be determined.

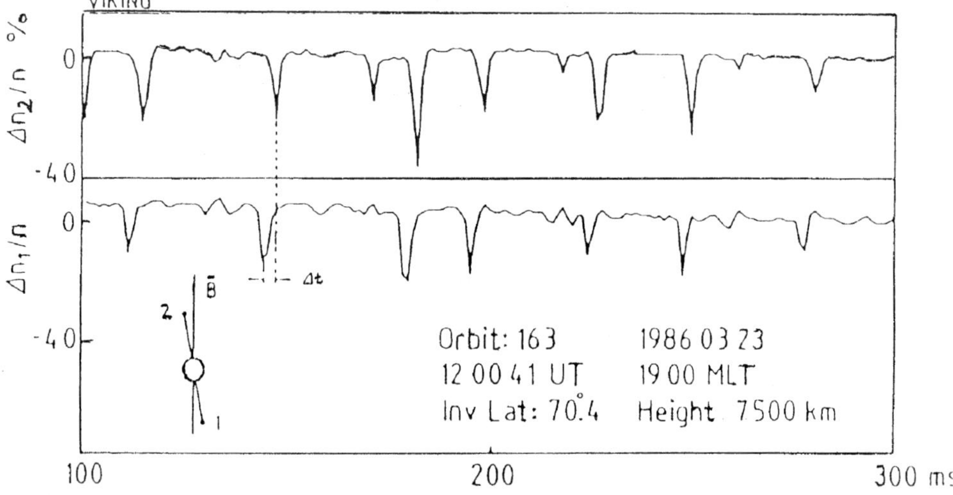

Figure 2. Weak double layers observed on *Viking*. These measurements show high time resolution recordings of density fluctuations on two Langmuir probes. From these data one can determine propagation velocity, scale length and net electric field (see Boström, 1988). Similar phenomena will be studied with Cluster in different regions of the magnetosphere.

available, new phenomena have been discovered. With the one Megabyte internal burst memory of EFW, it is expected that new phenomena will be observed requiring observations with high time resolution. An example is the observations of weak double layers on Viking shown in Figure 2.

The electric-field experiment and wave experiment on Freja (Marklund *et al.*, 1994; Holback *et al.*, 1994) provided many interesting observations on large-amplitude electric fields, plasma cavities associated with lower hybrid waves and large-amplitude modulated Langmuir waves in the auroral ionosphere (see special issue *Geophys. Res. Letters* **21**, No. 17, 1994). We anticipate the discovery of similar and entirely new features in the high time resolution data to be obtained with the Cluster satellites at higher altitudes.

2.2. Electric Fields at Intermediate Scales

Many of the central scientific goals of the Cluster mission relate to electric fields at intermediate scales (hundreds of km to a few R_E). These include MHD turbulence in the solar wind, magneto-sheath and cusp, instabilities driven by velocity shears, waves associated with quasi-parallel shocks, flux transfer events, impulsive penetration of plasma into the magnetosphere, and slow mode shocks in the near tail in association with reconnection. Electric-field data will provide information which is vital for determining wave modes, wave vectors, phase velocities and energy flow. The spatial and temporal variation in the auroral zone electric-field is another interesting topic addressed by measurements on these spatial scales. Observations on several low-orbiting satellites have shown that discrete auroral arcs are associated with solitary kinetic Alfvén waves. The Cluster spacecraft will make it possible to search for and study possible magnetospheric footprints of solitary alfvénic waves. The average behaviour of the auroral zone electric-field and related electrodynamical parameters for various geophysical conditions is relatively well documented. This is, however, not the case for the 'instantaneous' auroral electrodynamics on which very few studies have been conducted (Heelis *et al.*, 1983; Marklund *et al.*, 1986). A new technique to obtain global realistic and self-consistent distributions of auroral electrodynamical parameters has recently been developed (Marklund *et al.*, 1986). Simultaneous observations on the different spacecraft involved are used both for calibration of the model input data (field-aligned currents and conductivities) and for tests of the results (equipotential pattern). For these kinds of studies the Cluster mission with simultaneous four-point measurements will be ideally suited. The electric-field is an important parameter in studies of magnetohydrodynamic (MHD) turbulence. MHD turbulence is expected in all the important regions to be investigated by Cluster. The electric-field, together with the magnetic field, determines the Poynting flux and the propagation direction of the waves and plasma flows in these regions. The scale size and the magnitude of the electric fields involved have a wide range of values in these various regions. In the solar wind, scale sizes are several hundred kilometres to many thousands of kilometres and electric-field magnitudes are of the order of 1 mV m^{-1}. On auroral field lines, the electric-field is typically much larger. Magnitudes of over 100 mV m^{-1} are common and scale sizes from less than a kilometre to tens of kilometres. (Block *et al.*, 1987; Fälthammar *et al.*, 1987; Marklund *et al.*, 1987). The quasi-parallel shock structure provides a strong challenge to space physicists attempting to understand the role of different wave modes and scale sizes in the deceleration and thermalisation of upstream plasma. The four Cluster spacecraft can determine the phase velocity and Poynting flux of waves and 'shocklets' upstream and downstream of the shock. In the absence of such measurements it has been difficult to distinguish between waves standing in the shock frame of reference, waves created in the upstream region and convected and amplified across the shock transition region, and waves created in the transition region.

2.3. Electric Fields at Large Scales

At the longest scale lengths of interest for the Cluster study are the electric fields associated with processes such as polar-cap-potential drop, convection in the 'quiet' magnetotail, and the formation of a near-Earth neutral line. Steady-state convection in the magnetotail has been examined using analytic and numerical methods as well as simulations (Erickson and Wolf, 1980). Their models will be tested for the first time using the Cluster electric-field data. Another aspect of large scale electric fields is the distinction between potential and inductive fields. ISEE-1 electric-field data have led to the suggestion that the electric fields associated with the hypothesised near-earth neutral line are confined to $\approx$ 10–15 R_E in the dawn-dusk direction (Cattell *et al.*, 1986; Pedersen *et al.*, 1984). Comparisons of observed electric fields in the tail (20–40 mV m^{-1} during candidate neutral line events) with cross-polar-cap potentials imply that the field is primarily inductive or associated with structures of limited size. The Cluster satellite electric and magnetic-field data will allow direct determination of associated electric fields, the relative importance of inductive and potential fields, as well as the relationship of the neutral-line-propagation speed to the electric field and where reconnection is initiated in the plasma sheet. The four satellites will also be able to determine the extent of substorm electric fields which may not be directly related to the neutral line (Pedersen *et al.*, 1985), in particular how these fields are related to substorm injection. Cluster electric-field data can provide additional, more detailed information for understanding plasma sheet motion, in particular, variations in the dawn-dusk and earthward-tailward directions. Multipoint measurements on Cluster will be complemented by ground-based observations of the ionospheric convection using the EISCAT/ESR facility on Svalbard, SuperDarn and other available ground facilities. These data will provide important information on the quality of magnetospheric models and equipotentiality of magnetic-field lines. We plan extensive use of software tools, developed by Stasiewicz (1994), for field line mapping of magnetospheric/ionospheric structures and visualisation of satellite orbits. The four-spacecraft configuration will make possible comprehensive investigations of the energy transfer by means of Poynting flux. The amount of electromagnetic energy that is converted to particle energy in a region is given by the integral of the flux over a closed surface bounding that region. It is important to note that it is not the Poynting flux but, rather, its integral over a closed surface, that is the physically meaningful quantity. Four-point simultaneous measurements of the Poynting flux on the four Cluster spacecraft can provide sufficient information to estimate the surface integral. When this surface encompasses a physically interesting region such as the polar cusp, current sheet, magnetopause, etc., it should be possible to reach conclusions on whether acceleration processes or plasma heating are occurring at these locations.

3. Instrument Description

3.1. General

To meet the scientific objectives the electric-field instrument is capable of measuring, in various modes:

– Instantaneous spin plane components of the quasi-static electric-field vector, over a dynamic range of 0.3 to 700 mV m^{-1}, and with variable time resolution down to 0.1 ms.

– Oscillating electric-field in the range 50–8000 Hz and amplitude range 10 mV m^{-1} to 1μV m^{-1} .

– The thermal plasma density, over a dynamic range of 1 to 100 cm^{-3}.

– Plasma density fluctuations over a dynamic range of 1 to 50% with a time resolution of 0.1 ms.

– Time delays between signals from up to four different antenna elements on the same spacecraft, with a time resolution of 110 μs.

– The spacecraft potential which can give information about electron density in the range 10^{-2}–10 cm^{-3} with time resolution down to 0.2 s.

The sensor system of the instrument consists of four orthogonal cable booms carrying spherical sensors and deployed to 50 m in the spin plane of the spacecraft. Most of the electronics are contained in one electronics unit as shown in the block diagram in Figures 3 and 4.

The instrument has several important features. The potential difference between two opposite spherical sensors on orthogonal booms provides the average electric fields in two directions. The instrument can also operate the probes in Langmuir mode with voltage bias. By stepping the voltage bias the current-voltage characteristics of the plasma can be achieved to calculate the electron temperature and the plasma density. The output signals from the spherical sensor pre-amplifiers are made available to the other wave instruments for high-frequency wave measurements. The experiment has two fast Analogue-to-Digital, (A/D), conversion circuits for simultaneous sampling of two waveform signals at a rate of 36 000 samples s^{-1}. At data rates above 450 samples s^{-1}, the data are buffered in a one Megabyte burst memory and played back through the ordinary telemetry stream in the playback mode. On-board least-square fit calculations of two components of the electric-field, averaged over one spacecraft spin period (4 s), will always be present in the telemetry stream, when two or more probes are operated in voltage mode. The measured signals are continuously monitored on board in order to detect conditions that are appropriate for triggering a burst collection of data.

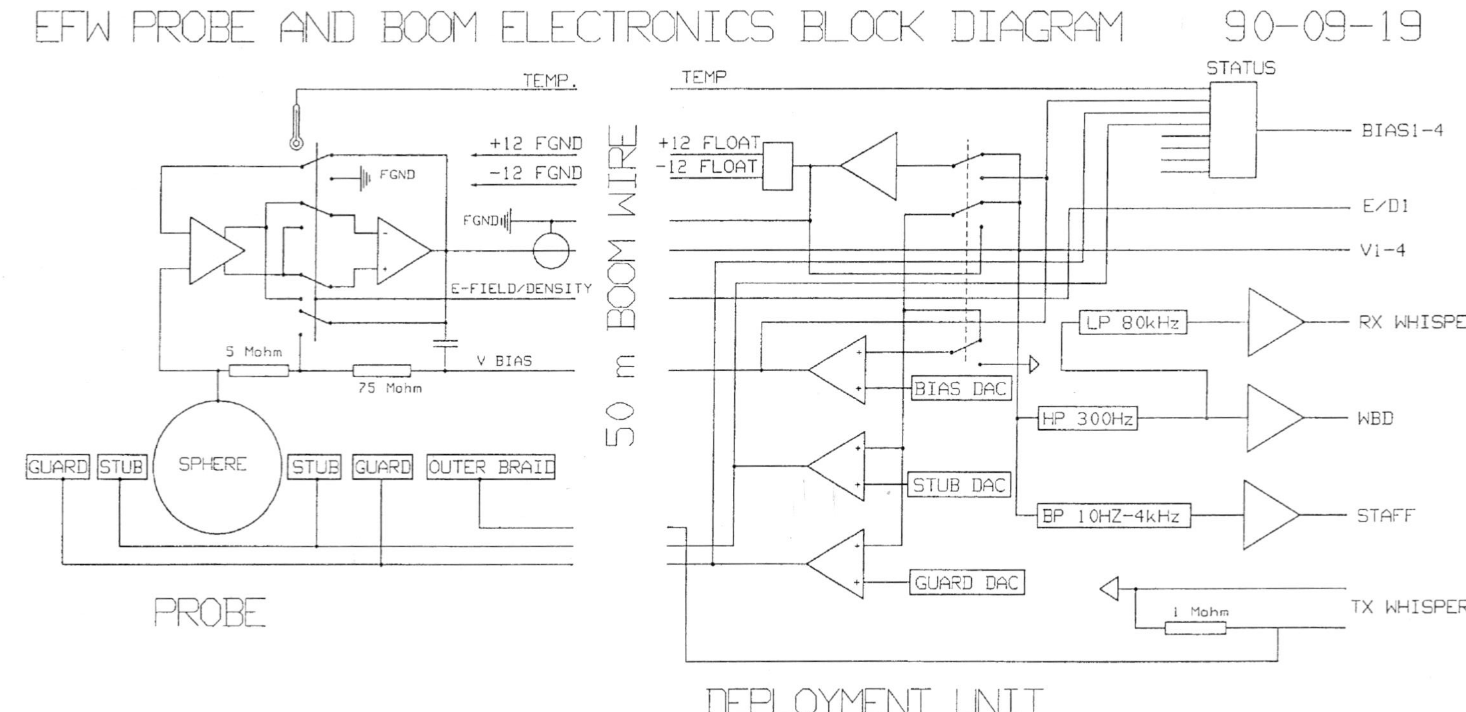

Figure 3. Block diagram of EFW.

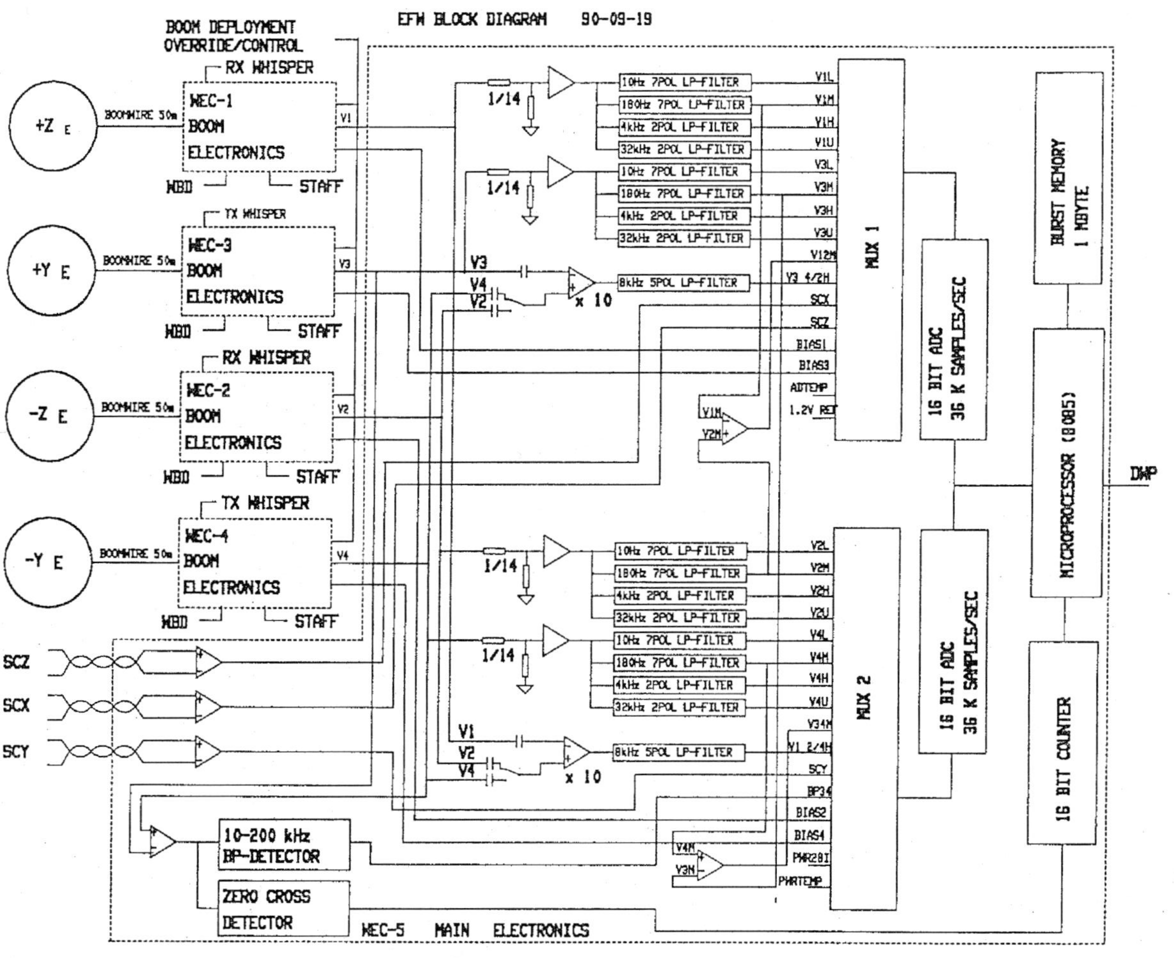

Figure 4. Block diagram of EFW.

Table I
Measured quantities

Quantity	Frequency	Range	Ampl. res.	Time res.
DC(E_y, E_z)	0–10 Hz	0.3 mV m^{-1}–700 mV m^{-1}	22 μV m^{-1}	1/25 s
	0–180 Hz		22 μV m^{-1}	1/450 s
	0–4000 Hz		22 μV m^{-1}	1/9000 s
	0–32 000 Hz		22 μV m^{-1}	1/18 000
AC(E_y, E_z)	50–8000 Hz	1 μV m^{-1} – 10 mV m^{-1}	0.15 μV m^{-1}	1/18 000s
$n, \delta n$	0–10 Hz	1–100 cm^{-3}		1/25 s
	0–180 Hz			1/450 s
	0–4000 Hz			1/9000 s
n, T_e	DC	1–100 cm^{-3}, eV range		
Freq. counter	10–200 kHz			

Table II
Data rates

Nominal mode data rate	1 440 bit s^{-1}
Burst mode rates	15 040 bit s^{-1}
	22 240 bit s^{-1}
	29 440 bit s^{-1}
Internal burst memory loading data rate	$\leq$ 1152 kbit s^{-1}

Characteristics of the instrument are shown in Tables I, II, and III. The search coil magnetometer data (not included in Table I) can also be routed through the 4 kHz channel of the EFW instrument.

3.2. Wire Booms and Probes

Each of the four EFW boom units is a small, self-contained package containing two major components: a deployment mechanism and the multi-conductor cable with the spherical sensor. The mechanism design has evolved from a series of successful satellite experiments including S3-2 and S3-3, ISEE, Viking, Freja, CRRES and Polar. The deployment unit contains a rotating reel for the cable, a brush

Table III
Physical data

Item	Mass (kg)	Power (W)
Main electronics box	1.8	3.7
Wire booms 4 units	13.6	

Figure 5. A picture showing the EFW instrument.

DC gear motor, an over-tension and end-of-cable indicator, an analogue cable-length indicator, a pyrotechnic-released spherical sensor housing, and a cable-oscillation Coulomb damper through which the cable exits the mechanism. In orbit, opposite booms will be symmetrically deployed to a probe-to-probe distance of approximately 100 m. However, the z-boom pair will be deployed to 70 m and kept at that length for about a week before being deployed to full length (100 m). A possibility for the satellite control to deploy the booms is also built into the system. The hardware elements of the experiment are shown in Figure 5.

The boom cable consists of eight wires and one coaxial cable, a kevlar braid that takes care of the mechanical stress due to the centrifugal force and, on top of it all, there is an outer braid made of conducting wire. The eight wires in the cable are used for power supply to the pre-amplifier housed in the sphere, for relay switching, for biasing voltages to the stub and guard, for probe temperature sensing and for biasing of the probe. The outer braid of one pair of booms is connected to the transmitter of the sounder experiment.

In order to limit (and control) the flux of photoelectrons from the boom wires to the spherical sensors and thereby minimise error sources, there is a special arrangement with the boom cable of symmetrical stubs and guards close to the probe. The outer braid is sectioned into stubs of 141 cm (from the probe surface) on both sides of the probe and the stubs are connected to guards of 10 cm length. Corresponding Stubs and a Guard are attached to the outer side of the probe for symmetry reasons. The outer stub and guard wire is made similar to the boom wire and is wound on a reel in stowed configuration with a cover called Top Hat. When the centrifugal force exceeds a certain value during deployment of the boom, the Top Hat releases and the outer stub and guard wire stretches out. The outer stub and guard wire will cause a shadow effect on its 'own' probe that is similar to the shadow made by the boom wire on the opposite probe. This evens out asymmetries that otherwise would cause errors in the probe potentials.

The inner and outer stubs, hereafter only called the stub, are electrically connected to each other. Also the inner and outer guards, hereafter called the guard, are electrically connected. The stub and guard can be given different bias voltages in order to affect the flow of photoelectrons to and from the probe. The bias voltages are referred to the probe voltage and adjusted individually with an offset, one for each of the stubs and guards. The bias voltages are generated by 8 bit Digital-to-Analogue-Converters (DAC) located in the boom units. The stub voltage follows the sphere voltage with an offset which can be varied between ± 1.44 V in 256 steps. The guard voltage follows the sphere potential with an offset between ± 35.6 V in 256 steps.

The spheres are 8 cm in diameter and are made of aluminium. They are coated by electrically conducting paint containing graphite (DAG 213) which also provides a suitable temperature balance on the probes.

3.3. Analogue Electronics

3.3.1. *Probe Modes*

There are two main modes of operation of the probes:

– The electric-field or voltage mode for potential measurement.

– The Langmuir or current mode for plasma density and temperature measurements.

In electric-field mode the pre-amplifier is set to high impedance input to allow the probe to adopt the potential of the surrounding plasma. In Langmuir mode the input impedance is low compared to the probe-to-plasma impedance so the probe can be given a bias potential and measure the resulting current. Switching between the two modes is achieved via a microprocessor-controlled relay in the probe and in the main electronics.

3.3.2. *Electric-field mode*

For the electric-field mode, the probe should be given a constant bias current. This is achieved by a voltage that is constant with respect to the probe potential over bias resistors. Thus, an offset voltage is added to the probe voltage and the resulting bias current can vary from -0.47 to $+0.47$ μA in 256 steps. Within these limits the bias current can be large enough to balance the maximum possible photoelectron flux to/from the spheres.

Each sphere is connected to the experiment electronics via a high impedance input pre-amplifier housed inside the sphere allowing the probe to adopt the potential of the surrounding plasma. Since the probe sheath has high resistance (10^7–10^{10} Ohms) and a capacitance of about 5 pF, the pre-amplifier must have high input impedance, low leakage current ($<$10 pA) and low input capacitance ($<$1 pF) to avoid attenuation of input signals. Since the potential of a biased sphere can differ from that of the spacecraft by 5–50 V in the absence of an electric-field, and fields as large as 500 mV m^{-1} have been observed, the dynamic range of the pre-amplifier and associated sensor electronics is $\pm$60 V from DC to 300 Hz. The small signal response exists to 600 kHz for use by other instruments on Cluster.

The sphere potential is determined by the balance of plasma electron current, photoemission current, and a bias current to the sphere. The magnitude of the bias current controls the 'working point' for the probe on the voltage-current characteristic curve and the aim is to find the working point that gives the lowest probe-to-plasma impedance.

3.3.3. *Langmuir Mode*

The spherical sensors can be operated as current-collecting Langmuir probes to provide information on the plasma density and electron temperature.

For this mode, the pre-amplifier in the probe is switched to low impedance input and the probe is given a bias voltage rather than a bias current as for the electric-field mode. The bias voltage is referred to satellite ground. Ideally, a positively biased probe collects an electron current that is proportional to the plasma density, provided the electron temperature is constant. Thus, variations in density will result in corresponding variations in the probe current. Also in this mode it is important to control the photoelectrons in order to minimise the errors that they cause.

3.4. Digital Electronics

The digital electronics contains two fast, 16-bit, analogue-to-digital systems, a set of digital-to-analogue converters for biasing, a single 8-bit radiation-hardened microprocessor, and a one Megabyte burst memory. Extensive software functions increase the instrument's capabilities and data coverage. The strategy for measuring the electric-field over a range of plus or minus 700 mV m^{-1} to a high relative accuracy can be achieved in a number of ways. With a 16-bit converter, the single-ended measurement of the sphere voltage (V1 through V4) will be possible from 22 μV m^{-1} to 700 mV m^{-1}. That gives a resolution of 22 μV m^{-1} bit^{-1} over the whole range, but the noise level for DC measurements is expected to be about 0.1 to 0.3 mV m^{-1} due to photoelectrons from experience with previous missions. For AC measurements an additional gain of 148 times gives a range of 10 mV m^{-1} and a one-bit resolution of 0.15 μV m^{-1}. The noise in this case corresponds to about 1 μV m^{-1}.

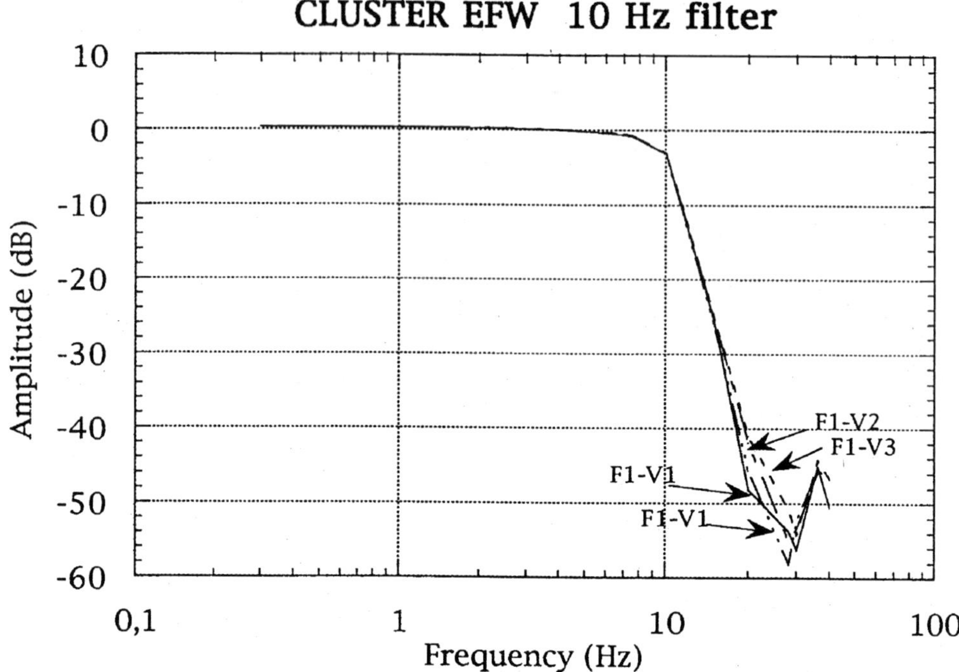

Figure 6. Instrument calibrations: frequency response of the 10 Hz filter.

3.5. Instrument Calibration

An extensive series of amplitude, frequency response and phase calibrations were carried out at various stages of the integration of each of the instruments for the four satellites. The frequency response calibration was made for a large number of input amplitudes, DC offsets for each of the EFW filters. As an example we show the calibration of a 10 Hz and a 180 Hz filter, see Figures 6 and 7.

An example of a phase calibration is shown in Figure 8. A design goal has been to manufacture all the instruments as identically as possible. The calibrations made confirm that the instruments give similar results, so the calibration curves shown here are representative for all the instruments. Several calibrations will also be made after launch. Prior to boom deployment, small plasma simulators, included in the instruments, will be used to generate calibration signals. After deployment, the oscillating $\mathbf{V} \times \mathbf{B}$ electric-field signals generated by the spinning system will be used to compare the outputs from the four probes on each satellite. This is a high accuracy test if made in areas of space with constant plasma parameters for the time duration of one spin. Finally, in regions of space where plasma parameters can be assumed to be constant over large areas of space, a comparison can be made between instruments on the four spacecraft. Other methods that can be used for in-flight calibrations are statistical studies and measurements from other instruments.

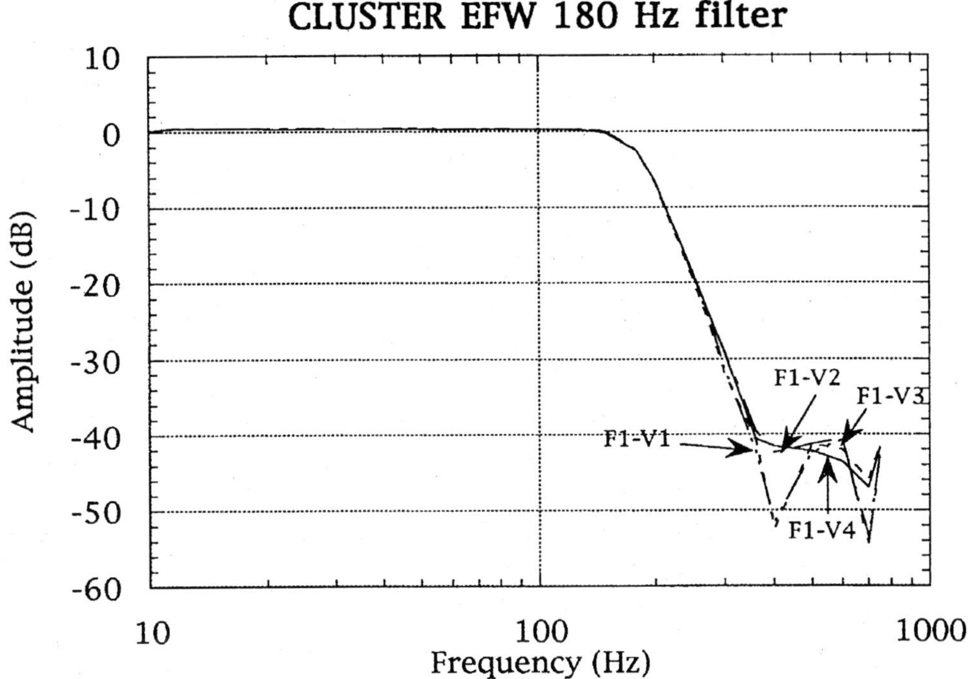

Figure 7. Instrument calibrations: frequency response of the 180 Hz filter.

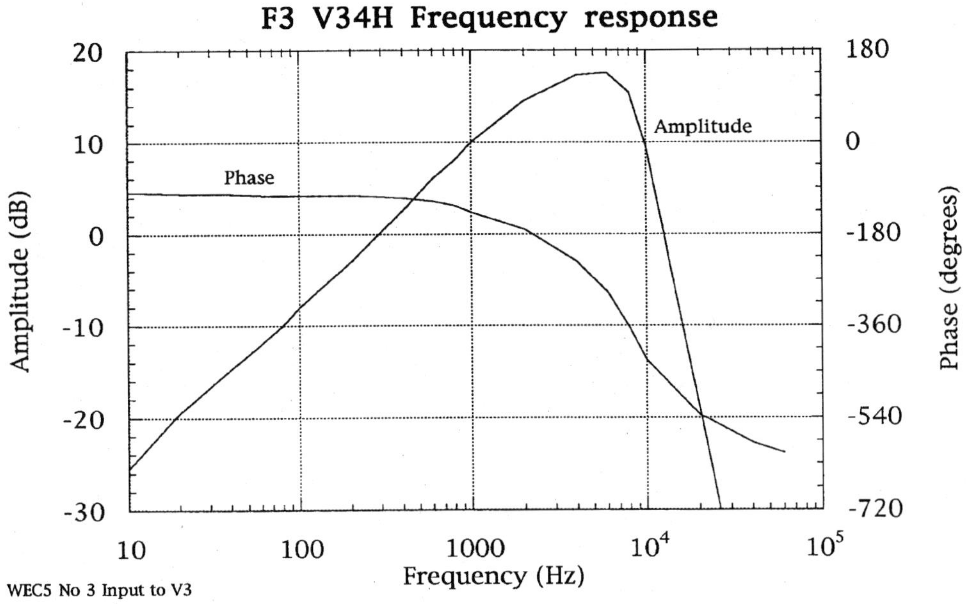

Figure 8. Instrument calibrations: an example of phase and amplitude response of the instrument.

3.6. EFW OPERATIONAL MODES

The EFW instrument has a large number of possible sampling modes. The main parameters to be selected in each case are: probe bias, probe mode, filter frequency, data rate and internal memory use. Several of the parameters are interrelated which limits the possible number of combinations, the telemetry rate is also dependent on other instruments in the WEC. The main combinations of probes and data rates are shown schematically in Figure 9.

The four probes can all be operated individually in current mode (to measure density/temperature), or voltage mode (to measure electric field) (see Figure 9). In some combinations voltages are measured from probe to satellite and in some cases differentially between two probes. The three components from the search coil instrument are also available in EFW with a bandwidth of 4 kHz. A diagnostic sweep, of 1–2 s duration, is performed approximately every 30 min. A bit in the EFW telemetry will indicate the occurrence of the sweep. Each probe signal is transferred through an anti-aliasing filter and sampled by one of the two A/D converters. Low pass filters at 10 Hz, 180 Hz, 4 kHz, or about 32 kHz, or a bandpass filter for 50 Hz to 8 kHz are available. The low telemetry rate (normal mode) limits the direct transmission to 10 Hz (25 samples s^{-1}) but for the high telemetry rate (burst mode) 180 Hz (450 samples s^{-1}) can be obtained. Frequencies above that may be recorded with the internal burst memory. There is also a simple frequency counter and an rms detector for 10–200 kHz available to detect the plasma frequency. The data transfer from EFW to telemetry goes via the Digital wave processor instrument (DWP)and can be made at four different rates: 1440, 15 040, 22 240, or 29 440 bit s^{-1}. All data rates contain 640 bits of housekeeping data (except for a few very exceptional cases) and in addition 2×25 samples s^{-1} of data, e.g., voltage differences between between two pairs of probes for the 1440 bit s^{-1} case; 2×450 samples s^{-1} of data, e.g., voltage difference between two pairs of probes for 15 040 bit s^{-1}; 3×450 samples s^{-1} of data, e.g., voltage difference between one pair of probes and currents from two probes for 22 240 bit s^{-1}; and finally 4×450 samples s^{-1} of data, e.g., currents from four probes for the data rate of 29 440 bit s^{-1}. These and more combinations for each of the EFW data rates are shown in Figure 9. Each square of the figure shows one combination of probes for a particular data transmission rate, shown as one row. As an example, mode EFW1a corresponds to two perpendicular vector measurements of the electric-field, in the spin plane, giving a total data rate of 1440 bit s^{-1}. After gaining experience with the instrument in orbit the number of combinations may be reduced. There is also a possibility, not shown in Figure 9, to measure two parallel electric-field vectors separated in space in, e.g., mode EFW2a. EFW carries out on-board estimates of the spacecraft potential that are sent to DWP for distribution to other instruments. For this calculation at least two probes operated in the voltage mode are required, which is not the case when three or four probes are operated in current mode. EFW also has the capability to make on board preliminary estimates of the DC electric-

EFW PROBE CONFIGURATION

TM MODE EFW 1 1440 bits/s

Signals sampled at 25 samples/s

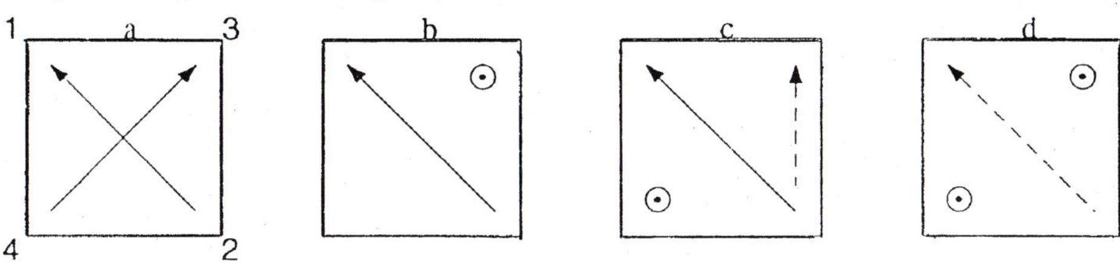

TM MODE EFW 2 15040 bits/s

Signals sampled at 450 samples/s unless otherwise noted

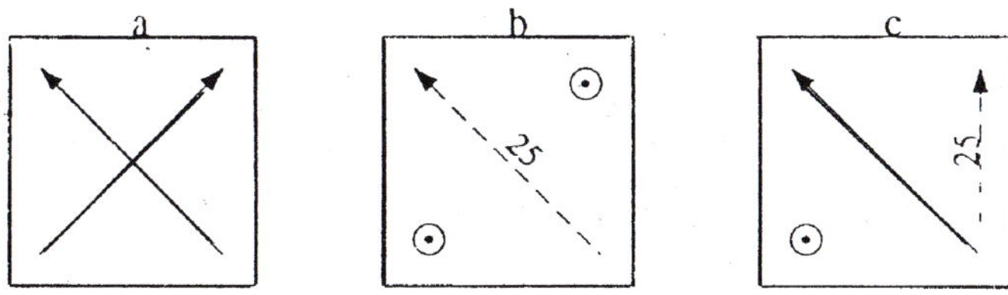

TM MODE EFW 3 22240 bits/s

Signals sampled at 450 samples/s

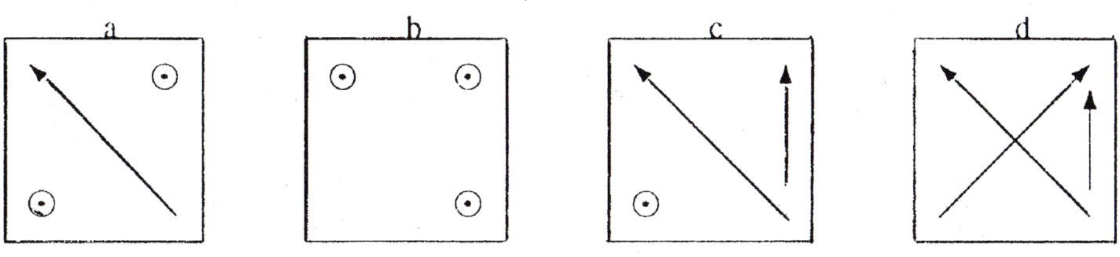

TM MODE EFW 4 29440 bits/s

Signals sampled at 450 samples/s

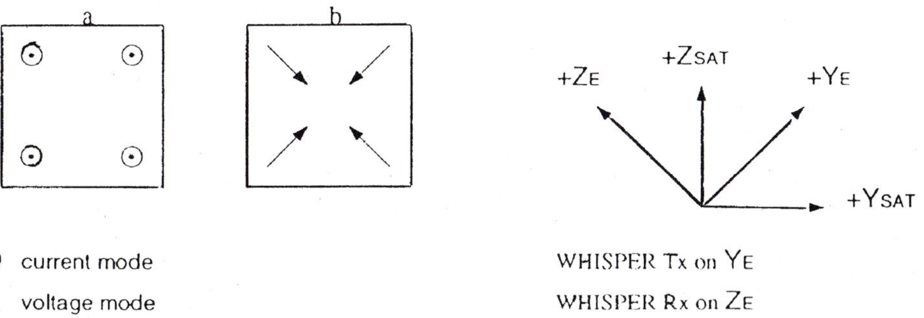

Figure 9. The software and hardware has been made to accommodate the probe combinations shown in each box in the figure. The broken lines indicate that these modes imply some compromise regarding the content of the telemetry format to get sufficient space.

field by means of spin fit calculations. The EFW internal memory storage will be used to register signals from the filters with frequencies 0–4 kHz, 50 Hz to 8 kHz and 0–32 kHz, as data from these frequencies cannot be transmitted in real time due to telemetry limitations. The 4, 8, and 32 kHz signals are sampled at 9, 18 and 36 ksamples s^{-1}, respectively, and stored in the internal memory. The triggering of the internal memory can be made by ground commands, or by an internal algorithm based on EFW data, search coil magnetometer data or on fluxgate magnetometer data via DWP or an event flag from the fluxgate magnetometer.

3.7. Data handling and telemetry

3.7.1. *Telemetry description*

The EFW instrument sends scientific and housekeeping data to DWP for packaging and telemetry of the data. The telemetry contains 16-bit measurements of currents or voltages. The housekeeping data contains; on-board calculation of the electric-field, where the electric-field, in the spin plane, is measured at 32 equidistant angles and a least squares sine fit is made to obtain the field, housekeeping parameters about the instrument every 0.2 s and finally low rate data of voltages sampled at 5 samples s^{-1}. A special telemetry mode (called BM3) is available where the internal burst memory can be emptied in 6 min. Dumps are also possible in the modes called BM2 and BM3. Parts of the memory may also be emptied at a very low rate through the regular telemetry modes.

3.7.2. *On-Board Data Handling*

The Cluster EFW instrument communicates with DWP by a single 38.4 kbit s^{-1} serial link. EFW accepts commands, telemetry codes and fluxgate magnetometer data (8 vectors s^{-1}) from the DWP unit, and transmits telemetry data to DWP. The EFW software coordinates real-time sampling along with Internal Burst (high sampling rate) data collections. The total A/D rate is 36 000 sample-pairs per second. EFW software must allocate samples out of this total collection rate between real-time samples, burst samples, boom monitoring, sweep collection points, etc. The user controls how many of the samples may be used for real-time and non-burst versus how many can be used in bursts.

3.7.3. *Scientific Data Handling*

A major effort has been spent on software development in preparation for the EFW scientific data analysis. Two major software packages have been developed: the Orbit Visualisation Tool (OVT) and the Interactive Science data Analysis Tool (ISDAT). They are both general, not project specific, tools, implying that they can be utilised for any similar project and facilitate an inter-project and inter-instrument analysis of the data. In addition, the EFW team has actively contributed to the Cluster Science Data System (CSDS).

Table IV
EFW CSDS parameters

Parameter	Units
Duskward electric-field	$mV\ m^{-1}$
Standard deviation of E-dusk	$mV\ m^{-1}$
Electric-field power 1–10 Hz	$(mV\ m^{-1})^2\ Hz^{-1}$
Electric-field power 10–180 Hz	$(mV\ m^{-1})^2\ Hz^{-1}$
Probe current	A
Probe-spacecraft voltage	V

The Orbit Visualisation Tool (OVT)

The OVT is a versatile interactive graphical tool by which the user can:

- Visualise single or multi-spacecraft positions and orbit predictions.
- Relate the spacecraft position to magnetic-field models.
- Trace magnetic-field lines.
- Visualise magnetic iso-surfaces and borders.
- Visualise experimental data in magnetospheric model frames.

The Interactive Science Data Analysis Tool (ISDAT)

The ISDAT is a general-purpose tool that facilitates science data access in a uniform way locally or over networks. It is based on a client server architecture with a well-defined general data structure for the communications between the server and the clients. The system provides a limited number of data analysis and visualisation tools but, more importantly, allows the user to easily add on his own data analysis and display clients. The ISDAT has been adopted as the common tool for WEC scientific data analysis and it also constitutes an integral part of the CSDS User Interface. Hereby, for the EFW and WEC, full resolution data analysis can be combined with access to lower resolution data from other instruments provided by the CSDS (see also Pedersen *et al.*, this issue, for a more detailed description of the WEC data analysis plans).

The Cluster Science Data System (CSDS)

The CSDS is a Cluster project-wide effort to provide Cluster scientific data to a wide scientific community (see Schmidt *et al.*, this issue). The EFW CSDS parameters are produced at the CSDS Scandinavian Data Centre, Alfvén Laboratory, Royal Institute of Technology, Stockholm. The EFW parameters provided by CSDS are listed in Table IV.

Acknowledgements

The funding of the instruments was provided in Sweden by Swedish National Space Board; in U.S.A. by NASA; at ESTEC from ESA; in Norway by the University of Oslo; in Finland by University of Oulu and the Academy of Finland.

The design and fabrication of the instrument was a cooperative effort by several laboratories under management by the Swedish Institute of Space Physics in Uppsala. The following laboratories have contributed to hardware, software and/or ground support instruments: Space Science Laboratory at Berkeley, Space Science Department of ESTEC, Alfvén Laboratory in Stockholm, University of Oslo and University of Oulu. We would like to thank Anita Rogelius and Gunny Janzon at IRF-Uppsala for assistance with administration of the project.

References

Block, L. P., Fälthammar, C. G., Lindqvist, P.-A., Marklund, G., Mozer, F. S., and Pedersen, A.: 1987, *Geophys. Res. Letters* **14**, 435.

Boström, R., Gustafsson, G., Holback, B., Holmgren, G., Koskinen, H., and Kintner, P.: 1988, *Phys. Rev. Letters* **61**, 82.

Cattell, C. A., Mozer, F. S., Jr., Hones, E. W., Anderson, R. R., and Sharp, R. D.: 1986, *J. Geophys. Res.* **91**, 5663.

Erickson, G. M. and Wolf, R. A.: 1980, *Geophys. Res. Letters* **7**, 900.

Eriksson, A. and Boström, R.: 1995, Irf Scientific Report 220. Technical Report, Swedish Inst. of Space Phys., Uppsala.

Fälthammar, C. G., Block, L. P., Lindqvist, P.-A., Marklund, G., Pedersen, A., and Mozer, F. S.: 1987, *Ann. Geophys.* **5**, 4.

Gustafsson, G. *et al.*: 1993, in R. Schmidt (ed.), *Cluster: Mission, Payload and Supporting Activities*, ESA SP-1159, p. 17.

Heelis, R. A., Foster, J. C., de la Beaujardière, O., and Holt, J.: 1983, *J. Geophys. Res.* **88**, 111.

Hilgers, A., Holback, B., Holmgren, G., and Boström, R.: 1992, *J. Geophys. Res.* **97**, 8631.

Holback, B., Jansson, S.-E., Åhlén, L., Lundgren, G., Lyndgal, L., and Powell, S.: 1994, *Space Sci. Rev.* **70**, 577.

Marklund, G. T., Heelis, R. A., and Winningham, J. D.: 1986, *Can. J. Phys.* **64**, 1417.

Marklund, G. T., Blomberg, L. G., Potemra, T. A., Murphree, J. S., Rich, F. J., and Stasiewicz, K.: 1987, *Geophys. Res. Letters* **14**, 329.

Marklund, G., Blomberg, L., Lindqvist, P. A., Fälthammar, C.-G., Haerendel, G., Mozer, F., Pedersen, A., and Tanskanen, P.: 1994, *Space Sci. Rev.* **70**, 483.

Mozer, F. S., Cattell, C. A., Temerin, M., Torbert, R. B., von Glinski, S., Woldorff, M., and Wygant, J.: 1979, *J. Geophys. Res.*, **84**, 5875.

Pedersen, A., Cattell, C. A., Fälthammar, C. G., Formisano, V., Lindqvist, P.-A., Mozer, F. S., and Torbert, R. B.: 1984, *Space Sci. Rev.* **37**, 269.

Pedersen, P., Cattell, C. A., Fälthammar, C.-G., Knott, K., Lindqvist, P. A., Manka, R. H., and Mozer, F. S.: 1985, *J. Geophys. Res.* **90**, 1231.

Schindler, R. and Birn, J.: 1982, *J. Geophys. Res.* **87**, 2263.

Stasiewicz, K.: 1994, in A. Egeland, J. Holtet, and P. E. Sandholt (eds.), *Physical Signatures of Magnetospheric Boundary Layers*, Kluwer Academic Publishers, Dordrecht, Holland, p. 433.

WHISPER, A RESONANCE SOUNDER AND WAVE ANALYSER: PERFORMANCES AND PERSPECTIVES FOR THE CLUSTER MISSION

P. M. E. DÉCRÉAU
LPCE et Université d'Orléans, Orléans, France

P. FERGEAU, V. KRANNOSELS'KIKH, M. LÉVÊQUE and PH. MARTIN
LPCE/CNRS, Orléans, France

O. RANDRIAMBOARISON
LPCE et Université d'Orléans, Orléans, France

F. X. SENÉ and J. G. TROTIGNON
LPCE/CNRS, Orléans, France

P. CANU
CETP/VSQP, Vélizy, France

P. B. MÖGENSEN
DSRI, Copenhagen, Denmark

AND WHISPER INVESTIGATORS

Abstract. The WHISPER sounder on the Cluster spacecraft is primarily designed to provide an absolute measurement of the total plasma density within the range 0.2–80 cm^{-3}. This is achieved by means of a resonance sounding technique which has already proved successful in the regions to be explored. The wave analysis function of the instrument is provided by FFT calculation. Compared with the swept frequency wave analysis of previous sounders, this technique has several new capabilities. In particular, when used for natural wave measurements (which cover here the 2–80 kHz range), it offers a flexible trade-off between time and frequency resolutions. In the basic nominal operational mode, the density is measured every 28 s, the frequency and time resolution for the wave measurements are about 600 Hz and 2.2 s, respectively. Better resolutions can be obtained, especially when the spacecraft telemetry is in burst mode. Special attention has been paid to the coordination of WHISPER operations with the wave instruments, as well as with the low-energy particle counters. When operated from the multi-spacecraft Cluster, the WHISPER instrument is expected to contribute in particular to the study of plasma waves in the electron foreshock and solar wind, to investigations about small-scale structures via density and high-frequency emission signatures, and to the analysis of the non-thermal continuum in the magnetosphere.

1. Introduction

The Waves of High frequency and Sounder for Probing of Electron density by Relaxation (WHISPER) experiment, an element of the Wave Experiment Consortium (WEC) presented in Pedersen *et al.* (this issue), is to be flown as four identical instruments on the four spacecraft of the Cluster mission. The WHISPER investigator team is shown in Table I. In addition, several members of the WEC have

Space Science Reviews **79:** 157–193, 1997.

Table I
WHISPER team organisation

Institution	Main role		
	PI and Co-Is	Hardware, Data analysis Onboard software	
LPCE, Orléans	P. M. E. Décréau V. Krasnosels'kikh O. Randriamboarison J. G. Trotignon	F.X. Sené Ph. Martin P. Fergau M. Lévêque C. Vasiljevic	E. Guyot L. Launay
CETP, Vélizy	P. Canu N. Cornilleau H. de Féraudy		
DSRI, Lyngby		P. B. Mögensen	
SPW, Lyngby	I. Iversen		
SZSP, Uppsala	G. Gustafsson		
University of Iowa	D. Gurnett		
University of Sheffield	L. Woolliscroft		

contributed directly to the WHISPER experiment as a whole: the sensors, in particular, are part of the Electric Field and Wave (EFW) instrument (Gustafsson *et al.*, this issue), and the interfaces to the satellite are part of the Digital Wave-Processing (DWP) instrument, (Woolliscroft *et al.*, this issue).

The WHISPER experiment fulfils two functions:

(1) as a resonance sounder, it measures the total electron density via active stimulation and subsequent detection of the resonances of the local plasma,

(2) in passive operations, it provides a survey of natural emissions in the 2–80 kHz range.

Both functions directly serve the main objectives of the Cluster mission, namely the study of small-scale structures in the magnetosphere and its boundaries. On one hand, electron density is a key parameter of the medium; access to its absolute value is especially important on Cluster, as density gradients have to be measured. The role of the sounder, unaffected by the plasma perturbations due to the presence of the spacecraft, is to provide reliable density values, which can also be used as reference values by other instruments. On the other hand, plasma wave characterisation is a fundamental key to the understanding of plasma structure formation and dynamics. WHISPER will analyse continuously the electric waves in the upper part of the WEC frequency range. It will return frequency/time spectrograms and a high-resolution time series of the electric power in the total frequency range. The Wide-Band Data (WBD) instrument (Gurnett *et al.*, this issue) will complement this information a small part of the time, in returning wideband waveform data.

Similar resonance sounders have been operated successfully in the last decades: in the equatorial magnetosphere, explored by the GEOS-1 and-2 satellites (Etcheto *et al.*, 1983); crossing magnetospheric boundaries with ISEE-1 (Trotignon *et al.*, 1986); on a polar orbit visiting the cusp and polar cap regions at middle altitudes, with the Swedish *Viking* satellite (Perraut *et al.*, 1990); and more recently in interplanetary space with *Ulysses* (Stone *et al.*, 1992a, b). All these instruments have basically included a receiver and wave analyser, associated with a transmitter which is activated whenever a density diagnosis is desired. Experience acquired with these projects has guided the design of the WHISPER instrument. However, the specific constraints of the Cluster project, the use of a different technology for the wave analysis (on-board FFT instead of a Swept Frequency Analyser), result in performances that are unique to WHISPER in a number of respects.

The aim of this paper is to present the intrinsic capabilities of the WHISPER instrument, and, from a pre-launch viewpoint, which scientific objectives they serve and how. This information also forms a baseline for the choices in terms of operational strategy that have to be made at several stages during the mission. Indeed, the operational strategy will evolve as the analysis of the four-point measurement data progresses, in line with the priorities set by the Cluster scientific team.

In Section 2 we present the scientific objectives addressed by the WHISPER experiment, and the target specifications. A technical description of the instrument is presented in Section 3. Its performance as measured after integration with the spacecraft are described in Section 4. Because the instrument has two partially mutually exclusive facets, namely an active sounder and a sensitive recorder of natural emissions, choices have to be made during operations. There is wide range of possibilities under the telemetry constraints via suitable WHISPER and WEC commanding. The general operational strategy at WEC-level is presented in Pedersen *et al.*, this issue. In this paper, we describe the functioning of WHISPER in the two basic WEC modes that will be operated a major part of the time (Section 5). We give examples of variations to these basic modes, required when specific objectives are being addressed. In the same section, we describe briefly the commissioning activity, a necessary step towards the validation and optimisation of operational modes. An important element of the Cluster project is the Cluster Science Data System (CSDS) concept (Schmidt and Escoubet, this issue). Section 6 describes the WHISPER data handling in the CSDS context and as foreseen for the scientific analysis. A summary of WHISPER performances (Section 7) concludes the paper.

2. Scientific Objectives

The primary objective of the Cluster mission is the study in three dimensions of small-scale plasma structures in key regions of the Earth's environment. The specific contribution of WHISPER is in the general context of wave and turbulence studies, to be carried out by the Wave Experiment Consortium (Woolliscroft *et al.*, 1995; Pedersen *et al.*, this issue).

A description of the basic investigations that will be performed, with mostly active or mostly passive WHISPER measurements respectively, is given in Décréau *et al.* (1993). Indeed, natural emissions are better characterised when the transmitter is inactive which, as a general rule, results in alternate active and passive operation of sounders, as is the case for Cluster. Continuous active operation of the sounder will hereafter be referred to as (S), for Sounding modes, and continuous passive operation as (N), for Natural wave measurement modes. It is worth emphasising that less information about natural emissions is obtained during (S) mode operation, as described in Section 3, but more about local plasma properties is gained.

Measurements in (S) mode will address the following objectives:

(a) To identify regions in space and mass transport processes, via measurements of the absolute density.

(b) To study the spatial extension and drift speed of key structures.

(c) To observe density fluctuations for value-added analysis of low-frequency waves.

(d) To identify high-frequency wave modes, via precise measurement of characteristic electron frequencies.

(e) To estimate the cold-to total-electron density ratio.

In (N) mode, the main objectives are:

(a) To study energy and mass transfers, using electrostatic waves as local tracers of small structures, and more generally investigating the role of turbulence.

(b) To identify the mode and source mechanism of high-frequency emissions, electrostatic or electromagnetic.

We will now concentrate on some specific topics where new findings are anticipated, due to Cluster's multi-point observation aspect, combined with the high time-resolution possibilities offered by the FFT wave analysis technique. These examples serve to illustrate the desired performances.

2.1. Plasma Waves in the Electron Foreshock

The study of the region upstream of the Earth's bow shock, the foreshock, has received considerable interest in the past 15 years. Due to the simple geometry of the foreshock, to the various angle values between the solar wind magnetic field and the shock normal, this region forms a natural laboratory for the study of basic plasma phenomena. Among them are electron beam relaxation processes and Langmuir wave turbulence (see Figure 1). The main role of WHISPER in this region

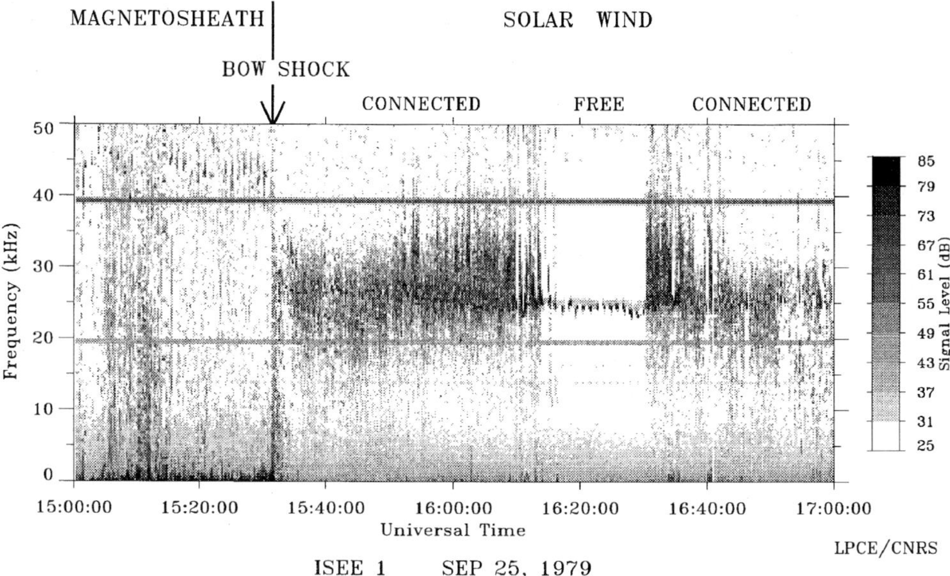

Figure 1. Sounder data from ISEE 1. The plasma resonance, triggered by the active sounding, appears clearly, above 40 kHz in the magnetosheath region and around 25 kHz in the solar wind. Natural noise is observed, essentially around the plasma resonance in the foreshock region ('CONNECTED' solar wind) in the form of broad-band sporadic emissions and at low frequency in the magnetosheath and bow shock.

will be to observe the high-frequency wave emissions from four points. Despite the studies undertaken with spacecraft data, numerical simulation and theories, many phenomena are still not totally understood. For example the plasma waves observed in the electron foreshock at a frequency close to the local electron plasma frequency are triggered by electrons streaming back from the shock time-of-flight distribution function. Despite being one of the most simple interactions in plasma physics, the questions addressing the precise origin of the observed emissions have not yet fully been answered. How does the instability saturate? Why are the non-Maxwellian features of the electron distribution function still present at a great distance from the shock, despite a large instability growth rate? Is strong or weak turbulence involved? Are the waves locally generated or convected from the foreshock boundary?

Owing to the three-dimensional capabilities allowed by the four Cluster spacecraft, a number of topics can be explored, including:

(a) Location of the plasma waves in the foreshock. One of the key parameters for describing the observed waves is the depth of penetration inside the foreshock with respect to the foreshock boundary (Etcheto and Faucheux, 1984; Filbert and Kellogg, 1979; Hospodarsky *et al.*, 1994). This requires a knowledge of the shape and location of the shock and a steady solar wind, conditions which are seldom met. Furthermore, the computation of the depth of penetration and the location of the shock boundary is based on a textbook paraboloid, probably far different from an actual wavy shock, which will generate multiple foreshock boundaries. The use of four spacecraft will help in the determination of the foreshock boundary and, conversely, could provide some remote information on the shape of the shock.

(b) Wave characterisation. The wave spectra often consist of short duration wave packets, typically a few tens of ms (Etcheto and Faucheux, 1984; Hospodarsky *et al.*, 1994; Lacombe *et al.*, 1985). The burst level observed on ISEE-1 is usually around 0.1 mV m^{-1}, but can go above 0.1 V m^{-1} (saturation level), the observed bandwidth usually stays between 2 and 10 kHz. The comparison between the characteristics of the wave spectra (occurrence, power, multiple peaks, frequency bandwidth, downshift or upward shift of the peak frequency) at high time resolution from four spacecraft will provide key information like duration and spatial extension, on these wave packets. Another possible approach to characterise the nature of the Langmuir turbulence is to study the correlation between the electric-field power and the density fluctuations. Indeed, parametric processes like decay or modulation instability are characterised by specific correlation functions between these two parameters. Measuring rapid density fluctuations is outside WHISPER's capabilities as a sounder. However, in its Langmuir mode, the EFW instrument has access to the density fluctuations, which can be correlated with WHISPER data on electric-field power fluctuations in the Langmuir wave range. When combined with the detailed electron distributions gathered by the PEACE detectors (Johnstone *et al.*, this issue) and by the higher time/frequency resolution spectra of the WBD instrument (when available), all these observations will provide key inputs for modelling the generation mechanisms of the waves.

2.2. Low-Frequency (MHD-type) Electromagnetic Turbulence

In another frequency domain, the study of magneto-hydrodynamic (MHD) turbulence, the density data provided by the WHISPER sounder can be of use as a value-adding parameter, if they are taken with a sufficient time resolution (of the order of one second). The study of general properties (dimension, velocity) of large structures present in nonlinear turbulent regions (like the downstream region of quasi-perpendicular shock, the quasi-parallel shock region, the solar wind and magnetospheric tail), will benefit from absolute density measurements in four points. An important step for the study of turbulence is the statistical analysis of fluctuations, beginning with acquiring the dispersion curves in linear and nonlinear regimes. Such an analysis has already been carried out, making use of two-satellite (AMPTE UKS and AMPTE IRM) magnetic-field data (Dudok de Wit *et al.*, 1995). The use of four points will give access to multi-scale comparative correlations. The analysis of phase information between magnetic, electric and density variations will provide new information on the general properties of turbulence. In particular, the behaviour of density fluctuations is crucial in order to distinguish between magnetosonic and Alfvén waves.

2.3. Boundary crossings

2.3.1. *Magnetopause Structures*

The study of the magnetopause remains a challenging and intriguing problem that is equally important for general plasma physics and for an understanding of solar-terrestrial relations. *In-situ* observations in the early 1960s suffered from the intrinsic inadequacy of single-satellite measurements, aggravated by the spatial mobility of the boundary. Significant progress has been achieved with dual-satellite observations (ISEE, AMPTE). Although much was learned, many questions about the magnetopause structure and dynamics remain (Elphic, 1988). What controls the thickness which has been observed to be much larger – 500–1500 km (Russell and Elphic, 1978) – than the expected scale of one ion gyroradius? Is the magnetopause a thin boundary which slowly crosses the spacecraft's path, or a thick one crossing rapidly? The origin of small-scale (hundreds of km) and time-varying currents within the magnetopause current sheet are still unexplained. The Cluster mission promises to go a step further toward answering these questions. In particular, the tetrahedron is capable of resolving the topology and velocity of a rigid moving surface (see for instance Chapman and Dunlop, 1993; Motez and Chanteur, 1994). WHISPER is expected to contribute to the identification of boundaries, via density and natural wave signatures. The resolution to be achieved is a few seconds or better, i.e., at least comparable to the best resolution achieved on ISEE-1 (see Figure 2).

The magnetopause boundary is also of crucial importance in relation to the transfer of solar wind mass and momentum into the Earth's magnetosphere. Magnetic reconnection via Flux Transfer Events (FTEs) (Russell and Elphic, 1978) is the dominant process when the interplanetary magnetic field has a southward directed component. However, other processes are under consideration. Lemaire proposed a mechanism for the impulsive penetration of solar wind plasma through the magnetopause, resulting from an impact upon the current layer of plasmoid having excess momentum (Lemaire, 1977, Lemaire and Roth, 1992). Related, but distinct, ideas were presented by Heikkila (1982) (see also Heikkila, 1990). The mechanism has been widely quoted in relation to possible sources of low-latitude boundary-layer plasma, and some ionospheric signatures near the polar cusp. Moreover, events which are commonly addressed as FTEs could actually be caused by boundary waves due to the impact of a high-pressure region on the current sheet (Sibeck, 1992). Simulation has shown that signatures of reconnection and boundary motion can be surprisingly similar (Otto *et al.*, 1995). Cluster will help to test these scenarios. WHISPER's contribution will again be its capability to measure the absolute value of the total density in four points. To test the Lemaire scenario in particular, WHISPER will provide more precise information about the structure and velocity of the plasmoids in the solar wind and magnetosheath region as well as during some low-latitude boundary-layer crossings planned at the

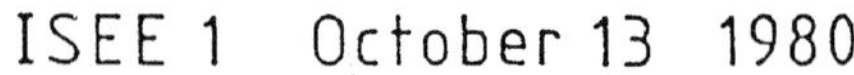

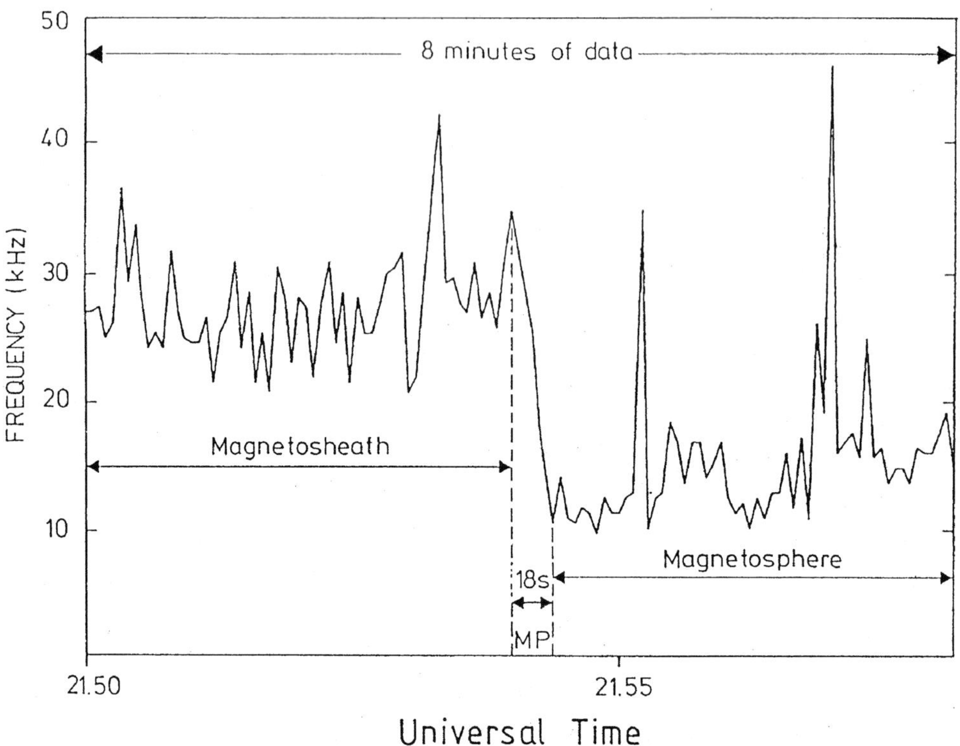

Figure 2. Plasma frequency profile obtained from the ISEE-1 relaxation sounder data as the satellite was travelling inbound from the magnetosheath to the magnetosphere. The time resolution was 4 s. The magnetopause was crossed several times between 21:54 UT and 21:58 UT. The plasma frequency drops from 30 kHz (10 cm^{-3}) in the magnetosheath to 10 kHz (1 cm^{-3}) just before 21:55 UT.

flank of the magnetosphere. In addition, correlated multi-satellite and ground-based observations will be needed.

2.3.2. *Plasma-Sheath Boundary Layer*

Observations performed during the past decade have established that the plasma-sheet boundary includes a boundary layer that supports a considerable amount of field and particle activity (Eastman *et al.*, 1984; Parks *et al.*, 1984). Understanding the structure and dynamics of this region is a key element in the understanding of mass, momentum and energy transport across the magnetospheric boundaries, which is one of the main goals of Cluster. The ISEE relaxation sounder has revealed the existence of a cold electron layer outside the plasma-sheet boundary (Etcheto and Saint-Marc, 1985). The electron density profile found outside the plasma-sheet boundary by this instrument (Figure 3) peaked at 5 cm^{-3} on 22 February, 1982, which is rather high in this region. Although some recent works have provided new information, like the existence of counter-streaming low-energy electron and ions

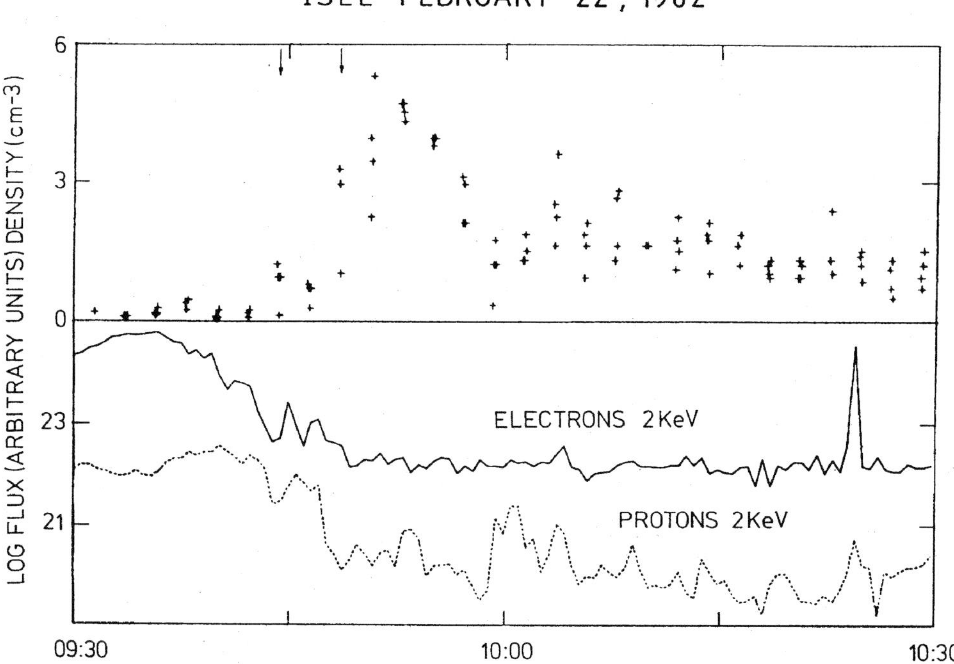

Figure 3. High-density plasma region in the plasmasheet boundary layer as observed by the ISEE-1 relaxation sounder. Spacecraft location was $X = 13.6$, $Y = 15.5$, $Z = -8.6\ R_E$ in the GSM coordinate system. *From top to bottom*: electron plasma density; electron and proton fluxes. The large density increase just after the spacecraft leaves the central plasmasheet at 09:45 UT is clear.

beam, this new region, termed the Low-Energy Layer (LEL) by Parks *et al.* (1992), is still little known, and the origin of these low-energy particles still debated (Feldstein and Galperin, 1994). New and better observations are essential to determine the characteristics, extent and dynamics of the LEL and its role in the overall magnetotail dynamics (Parks *et al.*, 1994). WHISPER, as already demonstrated by the ISEE sounder, will be a key instrument in providing this information. Firstly, it can provide the absolute density of the low-energy electron populations. The four-spacecraft configuration will allow the 3D geometry, gradients and time evolution of the LEL to be characterised. Secondly, the study of natural wave spectra (broadband electrostatic noise, Langmuir waves) routinely surveyed by WHISPER will help to characterise the fast varying source populations such as beams (Parks *et al.*, 1984; Onsager *et al.*, 1991), and their 3D spatial extension.

2.4. Electromagnetic continuum

The pioneering studies on the terrestrial non-thermal continuum radiation (see Gurnett, 1975) have been pursued with the GEOS and ISEE magnetospheric satellites (Christiansen *et al.*, 1984; Etcheto *et al.*, 1982; Gough, 1982). The detailed frequency structuring of the emissions, their directivity properties, combined with triangulation from GEOS and ISEE, have led to interesting findings about the location and dimensions of source regions, as well as about the generation mechanisms. However, questions remain open, such as the possible role of magnetopause sources and the exact nature of the conversion mechanism creating the continuum.

Some parts of the Cluster orbit will offer a viewpoint for the remote observation of plasmapause source regions. The presence of a multi-spacecraft measurement base gives a possibility for localisation by triangulation, if the individual emission regions are indeed small, as GEOS and ISEE observations have indicated (Etcheto *et al.*, 1982). In addition, dynamic studies can be made possible if source directions are measured with a good time resolution.

In the close vicinity of sources, simultaneous observations from, on one hand, the strong local electrostatic emissions (levels of a few mV m^{-1}) thought to be the primary wave source of continuum, on the other hand the radiated narrow band electromagnetic radiations (levels of a few (V m^{-1}) resulting from the conversion process, are made possible by the multi-spacecraft Cluster. This should provide new findings on the generation mechanism under study. For this purpose, the experiment should be able to resolve the frequency separation of banded emissions, of the order of 1 to a few kHz. The dawn and nightside plasmapause sources are likely to be approached and crossed on few occasions (near perigee, under favourable magnetic activity conditions). The magnetopause crossings will be more frequent. There, WHISPER observations will provide information about the possible existence of continuum sources on the magnetopause surface explored by Cluster: at medium latitude on the dayside and at low latitude on the flanks.

Such studies require a large dynamic range for the receiver, high precision on signal level (relative levels could be used to provide information about source-to-satellite distances), a measurement of directivity properties, if possible for each individual spectral component, thus a good frequency resolution and, finally, a reasonable time resolution for the complete diagnosis (about one minute is aimed for). In addition, large separation distances in local time are required for remote-sensing.

2.5. Detailed plasma diagnosis

The measurement of the plasma distribution function is a complex task, especially in the domain of low-energy plasmas, commonly affected by the presence of sheaths surrounding the spacecraft (Décréau *et al.*, 1978; Olsen, 1982). One role of WHISPER is to provide a reference value for the total density in the plasma.

The presence of cold plasmas, even in small quantities, is important for two main reasons: (a) it provides a clue for studying global transport processes in the magnetosphere, (b) this component can change the conditions of plasma stability, and be a driving factor for wave generation and energy transfer studies. In order to achieve a good plasma diagnosis, it is of the utmost importance that all data contributing to complementary information about the plasma distribution function (particle counts, spacecraft potential, Langmuir currents), can be obtained simultaneously. Particular effort has been made to achieve this: on one hand, WHISPER has been designed toward low-level perturbations, on the other hand dedicated common modes have been arranged to minimise the effect of possible perturbations on particle counting.

3. Instrument Description

3.1. Principle of the Measurement

The principle of a relaxation sounder, illustrated in Figure 4, is similar to that of a classical radar flown in a plasma. Inside the active period of a frequency step of total duration δt, a radio wave transmitter sends a wave train over a limited time period T (1 ms in the case shown), at a fixed frequency f. This corresponds, in the frequency domain, to a distribution of the signal centred at f, with a full width at half maximum of $1/T$ (1 kHz in case of the Figure 4). Such a burst will excite natural resonances of the plasma in the frequency range it covers. Shortly after this active period, a radio receiver is connected to an electric sensor (a double-sphere dipole in the case of WHISPER) and listens to the signal around f until the end of the time δt. Then, the frequency is translated and the process is repeated at a new frequency step. A succession of such steps, constituting a sweep, allows the properties of the neighbouring plasma to be explored throughout the range of interest.

This technique is suitable for observing with a high spectral resolution the slow waves (mainly electrostatic) which are able to propagate near the characteristic frequencies of the plasma. These low-group-velocity waves, after excitation by the transmitted pulse, travel either together with the satellite (accompanying waves), or away and back to it because of the curvature of the ray path by the density and/or magnetic field gradients (oblique echoes) (see Bitoun *et al.*, 1975; Etcheto *et al.*, 1981, for reviews). The very-narrow-bandwidth (almost monochromatic) plasma-resonance signals received provide accurate density measurements thanks to the direct or indirect (via other resonances) identification of the electron plasma frequency F_{pe}.

A resonance is identified by two main features:

(1) the presence of a decay in the received signal following the active period,

(2) a signal level higher than the natural emissions at the same frequency.

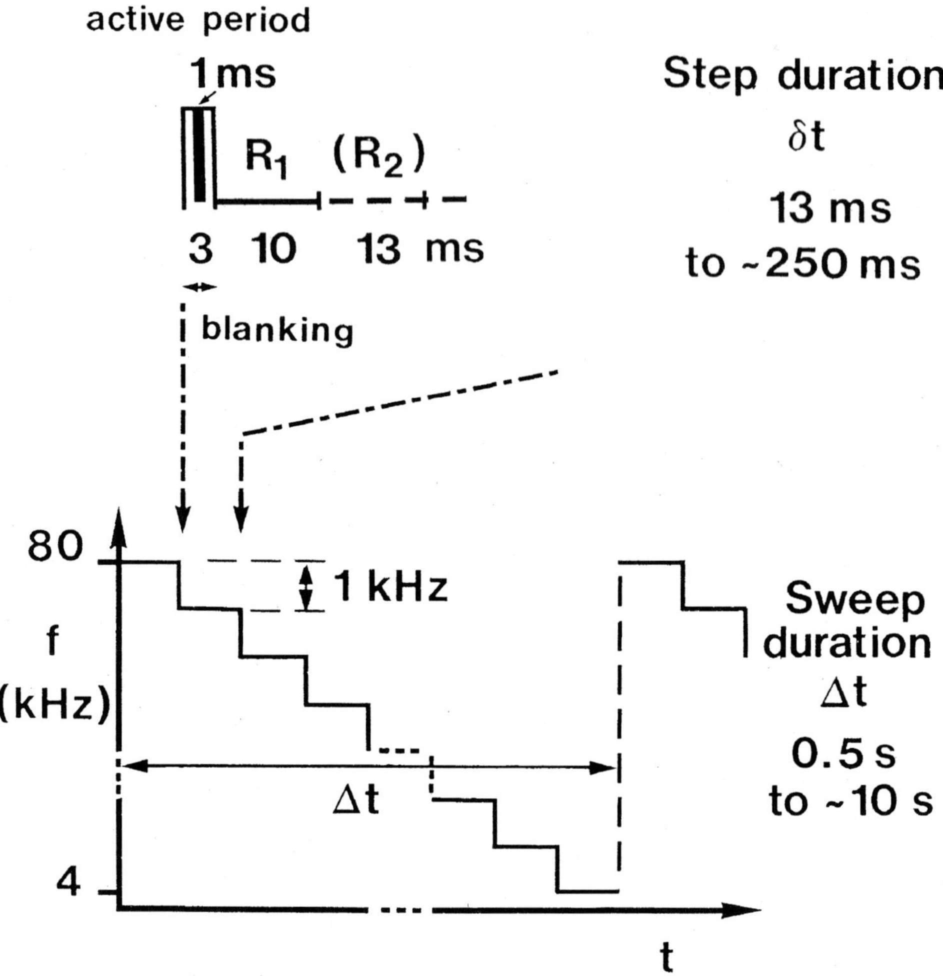

Figure 4. Configuration of a frequency sweep, illustrating the principle of the relaxation sounder.

The comparison of the 'active' echoes (received immediately after a transmitted burst) with 'passive' emissions (received a long time after an eventual transmission at the same frequency) is a mandatory step in conducting a sounding experiment.

When compared to other means of providing the plasma density, a sounder has distinct advantages:

(a) It can provide the total local electron density, including the cold component, which is difficult to measure from particle experiments.

(b) The Langmuir probe technique, which is well suited to the detection of rapid variations, needs to be regularly calibrated in flight: the current collected depends not only on the density of the medium but also on the electron energy spectrum, both quantities varying across plasma boundaries.

(c) Over a passive technique, where natural emissions are analysed for the diagnosis of plasma density, three points favour the relaxation sounder: (1) it triggers plasma resonances when the medium does not show them naturally, (2) those resonances are identified in a reliable way, which is not always the case with natural emissions, (3) lastly, it guarantees an *in situ* measurement. Indeed, natural waves may be generated at a considerable distance from the satellite, without the experimenter being aware of this situation.

(d) It is a reliable instrument, not subject to ageing phenomena. A sounder provides an absolute value of the density and needs no real calibration during the whole mission, only integrity checks.

3.2. Characteristics of the Instrument Design

From the scientific objectives described above (Section 2), one can derive the basic constraints that had to be satisfied by the Cluster sounder design:

(1) trigger the plasma resonances in an efficient way in all the regions studied,

(2) analyse them with the frequency resolution needed to measure the plasma frequency with accuracy (about 1%),

(3) measure in parallel the natural emissions, in order to identify their propagation modes by comparison with the active resonances,

(4) track the presence of wave bursts of short (a few 10 ms) duration,

(5) measure the directivity pattern of natural emissions,

(6) complete the density and natural noise measurements with the required time resolution (a few seconds or better),

(7) provide a possibility of sounding without perturbing simultaneous operation of particle counters (in general, special care about the design and interfaces with other instruments must be taken, in order to avoid possible cross-talk effects during burst transmissions),

(8) lastly, operate the instrument with the resources available on Cluster, in particular a limited telemetry rate and low weight and power.

3.2.1. *The Sounder*

In (S) mode operations, the principal goal is to operate the transmitter so that detectable echoes will be received in all regions. Two extreme cases expected are:

(a) the solar wind and the magnetosheath, where the resonances are intense, of long duration (about 100 ms), and located at high frequencies (a few 10 kHz),

(b) the magnetotail, where the observed resonances are weak, short (about 20 ms and less) and at a lower frequency (typically below 20 kHz).

The level of transmission will be adapted to the situation encountered: mainly a low value on the front side of the magnetosphere, with the highest value being chosen for the tail and other regions of warm plasma. Adjustment of the receiver gain and of the transmission level will permit the measurement of passive and active emissions within the dynamic range window available at the time.

The transmission is organised in steps of frequency, which are organised in sequences or sweeps to cover the desired frequency range, as shown in Figure 4. The signal received is analysed by FFT, i.e., a whole spectrum is available for each receiving period R_n. The basic characteristics of the sweep have been defined according to the following logic:

– The choice of the useful frequency range is based on results from the ISEE-1 mission which explored the main regions to be crossed by the Cluster orbit. The ISEE-1 sounder operated in the 0–50 kHz range; it missed the highest magnetosheath densities, so this range was increased to about 0–80 kHz.

– The pulse duration, (1ms in standard mode, is similar to the value found efficient on other missions; it fixes the frequency step separation (980 Hz) and the total step number (83).

– The frequency resolution is linked to the FFT bin number. In order to achieve a precision of the order of 1% for the average frequency value, we have chosen to cover the 980 Hz frequency slice by 6 bins, which means a 162 Hz bin separation. This is achieved with a FFT size of 512 bins over the total frequency range analysed (precisely 0–83 kHz). The frequency range fixes the sampling frequency (166 kHz) and the duration of the acquisition phase (6.2 ms) corresponding to the 1024 samples required by the FFT.

– The acquisition samples are located close in time to the active period of the step, in order to catch echoes of short duration which are expected in some regions.

– The minimal step duration must include the transmission and acquisition phases. It must also be synchronised with the EFW instrument sampling period (40 ms for the low frequency wave form). It has thus been fixed to 13.3 ms.

The receiving part of the earlier generation of relaxation sounders used the Swept Frequency Analyser (SFA) technique. In the case of GEOS, the waveform received on individual steps was transmitted on ground, and selected events could be further analysed, while in the case of ISEE-1 a limited amount of waveform data was routinely transmitted and analysed on ground. In both cases the spectral analysis showed very-narrow-bandwidth resonances (Etcheto *et al.*, 1981). On the *Ulysses* URAP sounder, the frequency range of 1.25 kHz–50 kHz is swept through 64 contiguous steps of 750 Hz bandwidth. At each step, the signal is analysed on board by the URAP DPU. A modified Walsh transform is applied on the waveform to improve the 750 Hz frequency resolution of the step down to 64 Hz, thus providing a better signal-to-noise ratio and a frequency resolution close to the actual resonance bandwidth. The FFT technique chosen on WHISPER plays the role of a narrow band receiver, albeit on the complete frequency range. The useful part of the spectrum, i.e., the few frequency lines which are inside the stimulated frequency slice, is selected by a straightforward on-board data compression (Woolliscroft *et al.*, this issue). In addition, this software treats the previous spectrum. It retains the natural level information in the same frequency slice (about to be stimulated) within a few milliseconds of the actual stimulation, whereas with the previous SFA technique, the corresponding delay was the sweep duration, i.e., from a few seconds to several tens of seconds. This important new capability so improves the 'active to passive level' clue for resonance recognition that the study of the signal decay, telemetry consuming, has been abandoned.

3.2.2. *The Natural Wave Analyser*

In (N) mode operations, the goal is to measure the electric component of wave emissions with a sufficient signal level precision and a good resolution both in time/directivity and in frequency, this within limited telemetry resources.

This challenge is classically solved by the use of two different devices: Filterbanks (FTB), which provide a high time and directivity resolution with poor frequency resolution, and SFAs, providing good frequency resolution with relatively poor time resolution. In the case of WHISPER, the FFT technique potentially offers both a good time and frequency resolution, as a complete spectrum is available every 13.3 ms. In practice, the TM rate severely limits those performances, and a compromise between time and frequency resolution has to be achieved. Four FFT dimensions are available on WHISPER: 512, 256, 128, and 64 bins. The associated bin separations in frequency, Δb, are respectively 162, 324, 648, and 1296 Hz. They give access to different typical time resolutions, according to the TM rate available, about 1 kbit s^{-1} in normal mode (NM), 5 kbit s^{-1} in burst mode (BM). Respective figures of $\Delta t \times \Delta b \approx 850$ in NM, $\Delta t \times \Delta b \approx 100$ in BM are achieved. A better time resolution is attached to the E_{pow} parameter: the electric field power calculated on-board every 13.3 ms from the accumulated squared time samples of the acquisition phase. The E_{pow} time series provides a meaningful global survey of fluctuations in the 2–80 kHz range, for phenomena of typical duration larger than 13 ms (stationarity has to be assumed in between two acquisitions). Only averaged values of this parameter can be transmitted on-ground in nominal telemetry rate. Lastly, with a spin rate of 4 s, the precision in directivity measurement is:

$$\Delta\theta \text{ (deg)} = 90\delta t \text{ (s)} ,$$

where $\Delta\theta$ is the spin angle covered by the antenna rotation during a measurement of duration δt. This gives about one degree in the case of a single acquisition, i.e., for E_{pow} in burst TM rate. When a few acquisitions are treated to provide an averaged spectrum, which is the case in (N) mode, this precision is degraded. A compromise has to be achieved between a good signal-to-noise ratio, gained from the bin-to-bin averaging of individual spectra, and a good directivity measurement, thus the averaging rate must be programmable.

3.3. TECHNICAL REALISATION

The relaxation sounder instrument on Cluster consists of three main parts (Figure 5): (a) a sensitive double-sphere antenna, part of the EFW experiment, (b) a transmitter and a receiver including a spectrum analyser, which forms the WHISPER experiment proper, (c) a data-acquisition and a data-processing system, which

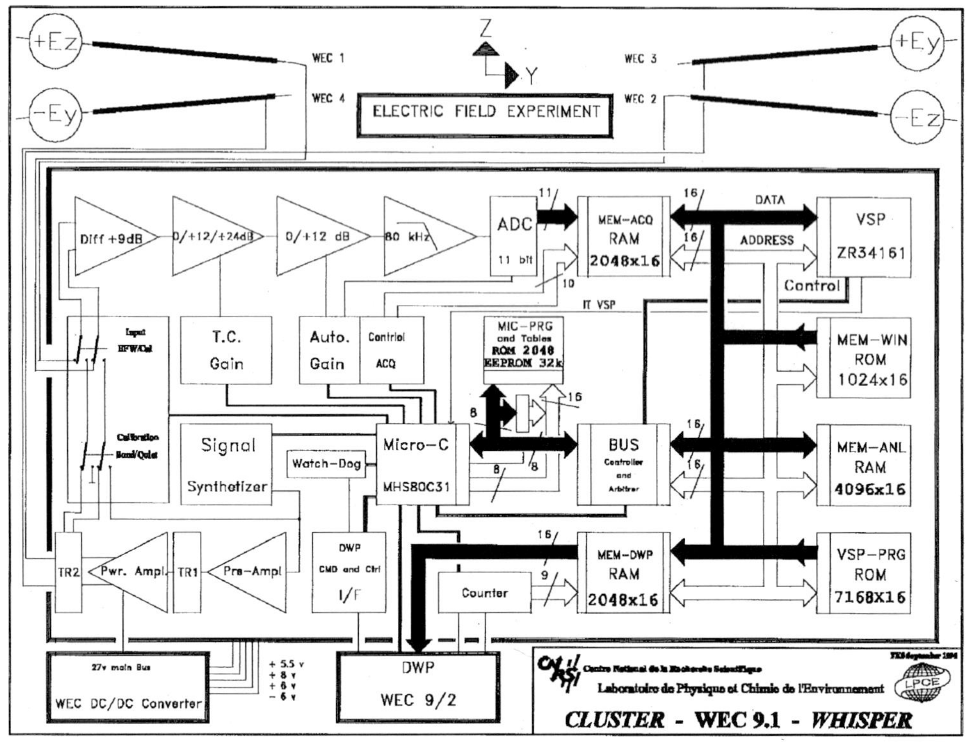

Figure 5. Block diagram of the WHISPER unit.

are part of the DWP experiment. The latter device also ensures the control of telecommand and telemetry for the entire WEC. The WHISPER power supply is part of the complete WEC setting, described in Pedersen *et al.* (this issue).

3.3.1. *Transmitter*

The transmitter of WHISPER is connected to the shields of one electric wire boom pair, Ey, through the EFW experiment module. In (N) mode operations, it remains inactive. In (S) modes, a signal synthesiser delivers a pulse of sine waves, of either 1 ms or 0.5 ms duration (always an entire number of periods, to minimise perturbation by transitory phenomena), at a frequency set by a microcontroller according to a command word transmitted via DWP to the microcontroller. Different levels can be selected (see Table II for the options offered by telecommand). WHISPER transmissions are triggered by the blanking signal transmitted to WHISPER by DWP. The time lag between two pulse occurrences (i.e., the step duration) is variable. The successive frequency values in a sweep follow a table chosen by telecommand among about 256 options (Sené, 1994). The selected table, adapted to the pulse duration, can cover various frequency ranges and various patterns in the succession of triggered frequencies. In planned nominal operations, the sweep is not explored uniformly, as shown in Figure 4, but according to a pattern which ensures that two frequency steps triggered successively in time are sufficiently apart from each other in frequency. Indeed, the frequency slice stimulated by a transmission is not strictly limited to 1 or 2 kHz, hence the adjoining frequency slices are prompted to be stimulated too, at a lower level. The frequency pattern is designed to prevent the 'passive' information recorded before a pulse transmission being corrupted by the preceding sounding. In addition, the frequency range can be swept in its upper part with pulses 0.5 ms wide, whereas 1 ms pulses are chosen in the lower part, in order to optimise the sounding efficiency within a fixed sweep duration. The total power consumed for the transmission is about 250 mW, averaged over a step. This can be increased in specific operations.

3.3.2. *Receiver and Analyser*

The receiver is connected to the double sphere dipole probe Ez, placed at right angles to the Ey boom pair, through the electric field module. It is operated in each 13.3 ms listening step, after the eventual transmission of the pulse (the receiver is inhibited during the pulse), during the acquisition slot. The signal is amplified by a gain amplifier including three stages. The total gain can take four different values (about 9, 21, 33, and 45 dB), with an exact ratio of four in between any two consecutive ones. Different options can be selected by telecommand. The gain is either fixed or automatically controlled. The automatic control selects one of two consecutive stepped gains. It responds to the characteristics (number of overflows) of a short sampling (64 samples) taken before the real acquisition, placed in the second part of the 13.3 ms basic time structure. The number of overflows occurring during the actual acquisition is recorded and transmitted as a status parameter.

After an 11-bit analogue-to-digital (A/D) conversion, corresponding to a 66 dB dynamic range (ratio between the saturation level and the quantification level), each signal sample is placed in the Most Significant Bit (MSB) part of a 16-bit word when the gain receiver value has a fixed value. When the automatic gain control option is chosen, the 11 bit issued from the converter are placed according to the actual gain value, either in the MSB part of the 16-bit word of the acquisition memory, or 2 bit below to take account of the factor of 4 difference in gains. The next operation is the frequency analysis. The device used is a Zoran Vector Signal Processor (VSP) (Vasiljevic, 1990) which performs an FFT, and delivers 64 to 512 useful bins in the range 0–83 kHz. Each bin is a complex value. In the WHISPER application, the VSP calculates the module and the result is placed in a 16-bit word. The 512 bins option is always chosen in (S) modes, which require a good frequency resolution. The A/D converter and VSP combination yields to a dynamic range for the bin values larger than the 66 dB available for the time samples. In order to control the calculation noise over the complete frequency range, windowing of the time signal is necessary. The chosen window is the three terms Blackmann–Harris window (Harris, 1978).

Table II
Main options offered by the WHISPER design

GENERAL OPTIONS	
Receiver gain	
Fixed values (dB):	45, 33, 21, 9
Auto. controlled values (dB):	45/33, 33/21, 21/9
Switch threshold (overflow number):	2, 4, 8, 16, 30
DWP processing	
Word size after compression (bit):	6, 8

(N) MODE SPECIFIC OPTIONS	
FFT size (bin):	512, 256, 128, 64
Number of averaged spectra:	1, 2, 4, 8, 16, 32, 64
DWP processing	
Spectrum selection rate:	All, 2/3, 1/2, 1/3, 1/4, 1/6, 1/8, 1/10
Averaging of E_{pow} data:	ON or OFF

(S) MODE SPECIFIC OPTIONS				
Step duration (ms):			13.3, 26.6, 40, 66.6, 106.6, 125 or 250 (spin synchronised)	
Transmission levels (Vpp):			50, 100, 200	
Frequency table (Examples)	Pulse duration (ms)	Number of steps	Range (kHz)	Comment
(a)	1	80	3.6-81.5	Complete table
(b)	1	6	21.2-26.8	Shortest table
(c)	0.5/1	52	3.6-81.0	Standard (dayside)
(d)	0.5/1	32	3.6-45.9	Standard (nightside)
(e)	0.5	12	10.9-34.2	Specific Solar Wind
(f)	0.5	32	4.1-66.6	Gliding mode
(g)	0.5	36	4.1-74.2	Synchronised mode
DWP processing				
Strategy A: selection of ½ or ¼ active bins, Active/Passive ratios				
Strategy B: all active bins, Active/Passive ratios				
Strategy C: all active and passive bins				
Strategy D: all active bins				
Reduced passive spectrum information: ON or OFF				

The total duration of the basic operations cover three time structures of 13.3 ms: one for the acquisition, one for the FFT calculation, one for the transfer of FFT frames to DWP. This requires several RAM memories, referred to in Figure 5 as MEM-ACQ for the acquired time samples, MEM-ANL for the FFT analysis, MEM-DWP for buffering the results to be transferred to DWP. The details of operations are managed by the microcontroller. In sounding operations, when the medium reacts dynamically, no accumulation of spectra is foreseen. In (N) modes, on the contrary, a number of successive individual spectra are accumulated by the DWP software which treats WHISPER data. This number is selected by telecommand. The microcontroller marks each individual spectrum placed in MEM-DWP by its rank in the series to be accumulated.

3.3.3. *On-Board Control and Data Processing*

The WHISPER instrument includes two sets of on-board software. One of them is used by the VSP for performing the calculation of the average power E_{pow}, the various FFTs, the module of the complex bin data delivered by the FFT, the bin-to-bin accumulation of spectra when a WHISPER internal data compression is requested and the instrument is in (N) mode.

The other software is used by the microcontroller. Its main functions are: (1) to set-up the experiment according to the WHISPER command word, (2) to load the needed FFT program into the VSP internal RAM, (3) to generate and control the transmitted wave, (4) to run a 'watchdog' program against the radiation effects, (5) to reshape the data output from the analyser before transferring them in MEM-DWP by adding status, sequencing, and E_{pow} information, (6) to run the calibration modes described in Section 5.

The onboard data compression implemented by DWP is described in Woolliscroft *et al.* (this issue). Each bin level information extracted from the MEM-DWP buffer is equivalent to a 24-bit word. This size reduced to 6 or 8 bits per word via a quasi-logarithmic compression. In (S) modes, two spectra are built after a sweep: the 'active' spectrum, i.e., bin levels calculated from a post-pulse acquisition, and the 'passive' one, i.e., bin levels calculated from a pre-pulse acquisition. All or part of this information is transmitted to the ground, according to a chosen strategy. In (N) modes, bin-to-bin accumulated spectra are built. All of them, or only a sample of them, are transmitted to the ground, according to one of the selection rate possibilities (Table II). The spectrum selection rate, combined with the number of accumulations in each spectrum, leads to the overall time resolution, driven by telemetry constraints. The internal WHISPER data compression, less flexible, is available as an option, in case a back-up is needed. It reconstructs the active spectrum in (S) mode, accumulates successive spectra bin -to-bin in the (N) mode.

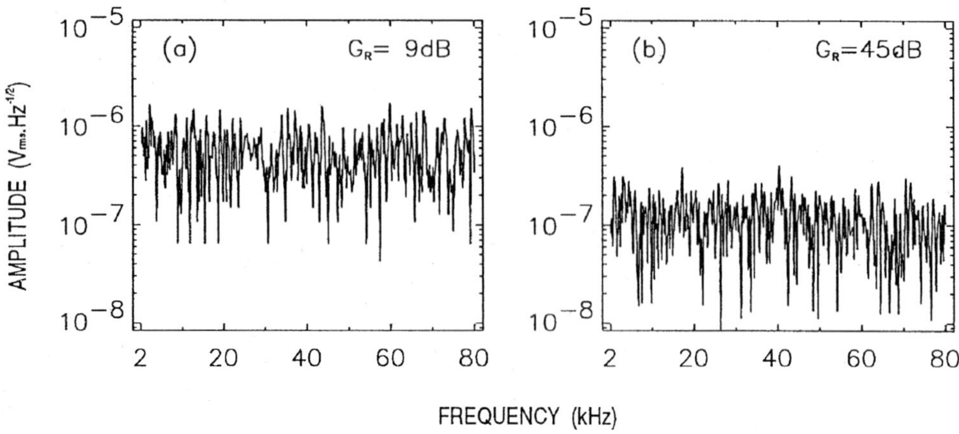

Figure 6. Noise levels recorded by WHISPER during special WEC tests on the F1 satellite.

4. Technical Performances

4.1. NATURAL WAVE MODE

4.1.1. *Sensitivity and Dynamic Range*

The absolute sensitivity of the WHISPER experiment (including the EFW sensors) has been measured during a special WEC test carried on with the F1 satellite model in a low-level noise environment (Martin, 1994). Figure 6 displays the levels recorded for an individual spectrum under the conditions leading to the best achievable sensitivity: FFT size at 512 bins, for two extreme values of the receiver gain, $G_R = 9$ dB and $G_R = 45$ dB. Remarkably, there is no single spurious line over the whole frequency range, and this has been confirmed by other records. For the other satellites, the noise measurement tests have been conducted in classical integration rooms. They show in consequence a few spurious lines due to the ground equipment. However, the noise levels are reliably measured. They are comparable (within at most 2 dB) for the four different models. Two figures can be noted from the measurements shown in Figure 6: the average noise level and the minimum signal detectable, respectively, $V_{\text{noi}} \approx 1.4 \times 10^{-7}\ V_{\text{r.m.s.}}\ \text{Hz}^{-1/2}$ and $V_{\text{min}} \approx 10^{-5}\ V_p$ for $G_R = 45$ dB. The latter value is estimated from the highest level observed on the spectrum, $S_{\text{max}} \approx 4 \times 10^{-7}\ V_{\text{r.m.s.}}\ \text{Hz}^{-1/2}$, translated as follows:

$$V_{\text{min}} = S_{\text{max}} \sqrt{2 w_f \Delta b}\,,$$

where $w_f = 1.71$ is a factor corresponding to the chosen window (Harris, 1978), and Δb the bin separation. The V_{noi} and V_{min} figures recorded for a receiver gain $G_R = 9$ dB are only 12 dB above those obtained for $G_R = 45$ dB. This indicates that, when moving from a 9 dB to a 45 dB gain, the benefit of a lower (by 36 dB) quantification level is partly lost by the presence of the electronic noise. More precisely, going from $G_R = 9$ dB to $G_R = 21$ dB improves the floor level of 9 dB, the improvement is of 2 dB when going from $G_R = 21$ dB to $G_R = 33$ dB, and only of 1 dB when going from 33 dB to 45 dB. In practice, the electronic noise contribution is quite dominant in the $G_R = 45$ dB option, which reduces the dynamic range available when using the spectral analysis. This option can still be of interest for the E_{pow} measurement, which is never affected by the electronic noise level, due to a dynamic range limited as explained below.

The maximal sine wave measurable corresponds to the upper level of the A/D converter (2.5 Vp). This value, converted at WHISPER input, varies according to the receiver gain. It has to be taken into account both for the measurement of spectral lines and for the E_{pow} parameter measurement. In the latter case, the minimum amplitude detectable is the effective detection threshold, which corresponds to word bit size constraints in the calculation performed by the VSP. In practice, only the 8 MSB part of the word recorded in the acquisition memory is used. This leads to a 48 dB dynamic range, not improved in the automatic gain control option.

Table III summarises the sensitivity performances of WHISPER. The sensitivity level indicated for E_{pow} is simply a translation of the minimum signal measurable (the quantification threshold value), taking account of the 80 kHz frequency width of the WHISPER receiver. The performances achieved with the spectral analysis, in particular the dynamic range, vary in a more complicated way. The use of an FFT size lower than 512 bins does not degrade the sensitivity level expressed in $V_{\text{r.m.s.}}$ $\text{Hz}^{-1/2}$, however it does modify the minimum sine wave amplitude detectable, according to the size of the narrow band analysis; thus the various figures presented in Table III. In practice, we foresee to telecommand in a standard way the 33/21 dB or the 21/9 dB automatic control gain, as the resulting dynamic ranges cover the main emissions expected in the WHISPER frequency domain (see Pedersen *et al.*, this issue). The lowest levels quoted in section 2 are those of the weak electromagnetic radiation of the non-thermal continuum (a few μV m^{-1}). They will be detected by the FFT analyser in the standard gain position, but hidden to the E_{pow} series. The usual levels of the foreshock waves (10^{-2} Vp) will be present in both measurements. At the other extremity of the range, the WHISPER saturation level in the standard gain 21/9 dB is about a factor of two higher than the ISEE-1 value. It should be adapted to most Cluster situations.

In order to complete this presentation, one should note two effects which can modify the performances presented above. Firstly, the detection of low-level signals embedded in noise can be improved via the bin-to-bin averaging of a number of spectra, as illustrated in Figure 7. In a test run during WEC integration of the F1 instrument models, two signals were injected in the measurement chain: a sine wave of 10 mVp, and a 'white' noise of 30 $\mu V_{\text{r.m.s.}}$ $\text{Hz}^{-1/2}$. The FFT size is 512 bins. Figure 7 displays two averaged spectra, for a number of accumulations of respectively 8 and 64. The variance of the noise is diminished as expected (a factor 4) for the high accumulation value, with respect to the low one. In practice, a factor

Table III
Summary of WHISPER sensitivity and dynamic range performances

Receiver Gain (dB)	Parameter	Sensitivity $V_{rms}/Hz^{-1/2}$	Max. signal Vp (V)	Min. signal Vp (V)	Dynamic range (dB)
45	Spectrum	$1.4x10^{-7}$	$1.4x10^{-2}$	FFT 512/256/128/64: $0.95/1.3/1.9/2.6x10^{-5}$	FFT 512/256/128/64: 63/60/57/54
	Epow	$1.4x10^{-7}$	$1.4x10^{-2}$	$5.6x10^{-5}$	48
33	Spectrum	$1.6x10^{-7}$	$5.6x10^{-2}$	FFT 512/256/128/64: $1.1/1.5/2.2/2.9x10^{-5}$	FFT 512/256/128/64: 74/71/68/65
	Epow	$5.6x10^{-7}$	$5.6x10^{-2}$	$2.3x10^{-4}$	48
21	Spectrum	$2.0x10^{-7}$	$2.2x10^{-1}$	FFT 512/256/128/64: $1.4/1.8/2.7/3.7x10^{-5}$	FFT 512/256/128/64: 84/81/78/75
	Epow	$2.3x10^{-6}$	$2.2x10^{-1}$	$0.9x10^{-3}$	48
9	Spectrum	$5.6x10^{-7}$	$9.0x10^{-1}$	FFT 512/256/128/64: $3.8/5.2/7.6/10x10^{-5}$	FFT 512/256/128/64: 87/84/81/78
	Epow	$9.0x10^{-6}$	$9.0x10^{-1}$	$3.6x10^{-3}$	48
45/33	Spectrum	$1.4x10^{-7}$	$5.6x10^{-2}$	FFT 512/256/128/64: $0.95/1.3/1.9/2.6x10^{-5}$	FFT 512/256/128/64: 75/72/69/66
	Epow	$5.6x10^{-7}$	$5.6x10^{-2}$	$2.3x10^{-4}$	48
33/21	Spectrum	$1.6x10^{-7}$	$2.2x10^{-1}$	FFT 512/256/128/64: $1.1/1.5/2.2/2.9x10^{-5}$	FFT 512/256/128/64: 86/83/80/77
	Epow	$2.3x10^{-6}$	$2.2x10^{-1}$	$0.9x10^{-3}$	48
21/9	Spectrum	$2.0x10^{-7}$	$9.0x10^{-1}$	FFT 512/256/128/64 $1.4/1.8/2.7/3.7x10^{-5}$	FFT 512/256/128/64: 96/93/90/87
	Epow	$9.0x10^{-6}$	$9.0x10^{-1}$	$3.6x10^{-3}$	48

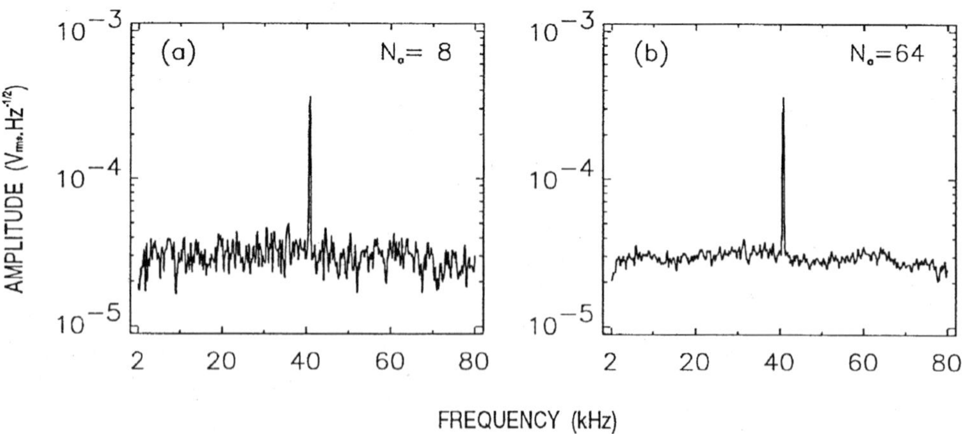

Figure 7. Signal over Noise behaviour resulting from two different options of the bin to bin averaging of spectrum.

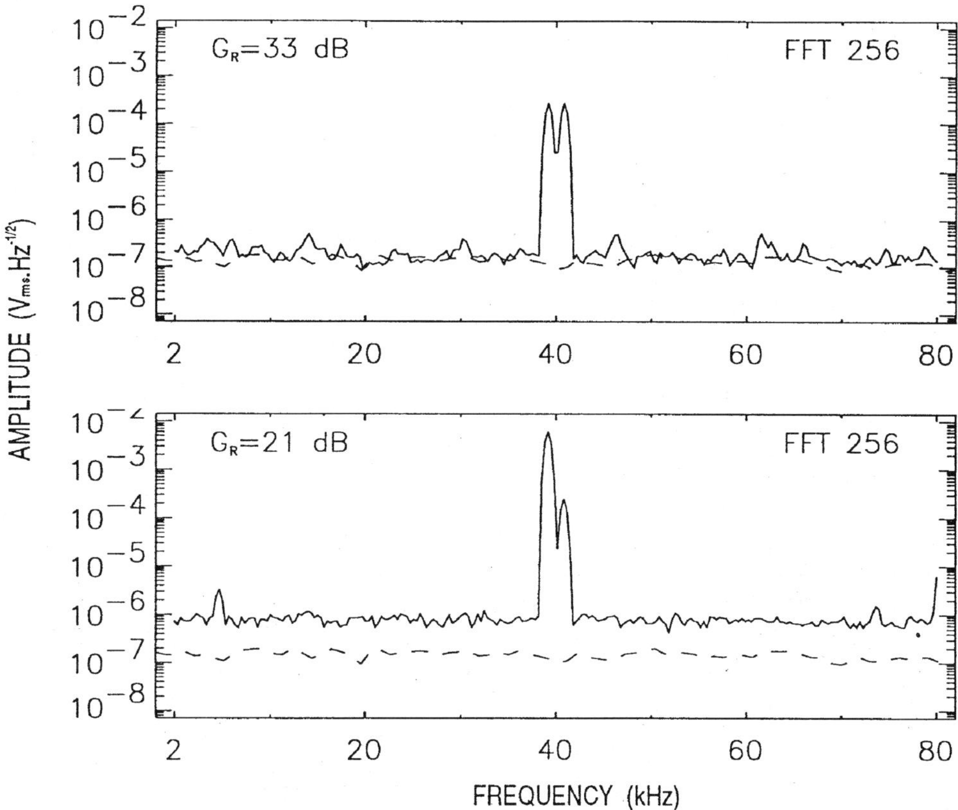

Figure 8. Calculation noise due to the Blackman–Harris windowing. Placed at about 75 dB of the strongest analysed signal, it is hidden by the electronic noise level in the upper panel, while it appears clearly in the bottom panel. The dashed lines are the output obtained in the absence of stimulus.

of about 2.5 dB can be gained on the minimum signal detectable by averaging 16 individual spectra, as used in the basic WHISPER functioning at NM telemetry rate.

Second, it should be noted that, as a consequence of the windowing process, the FFT calculation noise can exceed the electronic noise level, as shown in Figure 8. The two spectra displayed are obtained under the same functioning mode: FFT size of 256 bins, automatic gain change 21/33 dB, averaging of 8 spectra. The upper panel spectrum corresponds to an input signal of two sine waves of the same amplitude, 10 mVp, close in frequency (placed 5 bins apart). The floor level recorded is similar to the electronic noise level recorded in the absence of input, shown in the same panel. The lower panel spectrum has been obtained after increasing one sine wave amplitude from 10 mVp to 250 mVp, i.e., at the saturation level. The receiver gain is automatically switched from 33 to 21 dB, which increases slightly (2 dB) the noise level reference in the absence of signal. The floor level in the presence of the 250 mVp signal is significantly higher than the reference level.

This is due to the Blackmann-Harris window noise, placed at about 75 dB below the spectral component of highest level (Harris, 1978). Consequently, one has to distinguish two dynamic range values: (1) the range over which a single sine wave is measurable, (84 dB in the case of Figure 8, (2) the ratio between the largest and the lowest signal measurable simultaneously ($\approx$ 75 dB at best).

4.1.2. *Frequency resolution*

The frequency resolution is a direct function of the FFT size selected. To illustrate the behaviour of the WHISPER analyser in this respect, we show in Figure 9 the spectral features obtained for an input signal composed of two sine waves of same amplitude (10 mVp) and neighbouring frequencies (40.04 and 41.61 kHz). The FFT size varies from 512 bins (top panel) to 128. Two different questions are posed: (1) what is the precision of the measured frequency position when a single sine wave is present? (2) what is the minimal frequency separation of two sine waves for them to be individually characterised? WHISPER capabilities are degraded as follows when the FFT size diminishes:

– In the first case (FFT 512, 10 bins frequency separation), the spectral lines are sufficiently far apart to allow the precise measurement of the energy attached to each of them. The precision on the frequency localisation of a line is the bin separation.

– In the second case (FFT 256, 5 bins separation), the spectral lines are well separated, and this is still true for a large amplitude difference of the sine waves (see bottom panel of Figure 8), but sorting respective energy contributions is less straightforward.

– In the last case (FFT 128, 2.5 bin separation), the frequency separation is marginal, it can even fall down if the data compression rounding is unfavourable (see Figure 10, illustrating the effect of a 6-bit word size compression). In this case, one can also observe how the detailed shape of the line evolves according to the position of the sine wave frequency relative to an exact bin location: the lowest frequency is placed in between two bins, the upper frequency at a bin.

In conclusion, we will consider that the actual frequency resolution attached to a spectrum calculated with a bin separation Δb is $\Delta f = 2.5\Delta b$, i.e., 400 Hz in the best case, and that the precision of the frequency measured for a single line is Δb, i.e., 162 Hz at best. Under the best case, therefore, gyroharmonics can be separated for magnetic field values down to about 15γ. This includes a large part of the magnetosphere, but not the high altitude portions if the orbit in the nightside, nor the regions above the magnetopause. It is worth noting that the measured frequencies are absolute values, a consequence of the high precision and stability of on-board frequencies.

4.1.3. *Signal Level Precision*

The actual precision on the effective value of the signal, as calculated at the EFW sensors' input, results from three main factors: the precision in the knowledge of

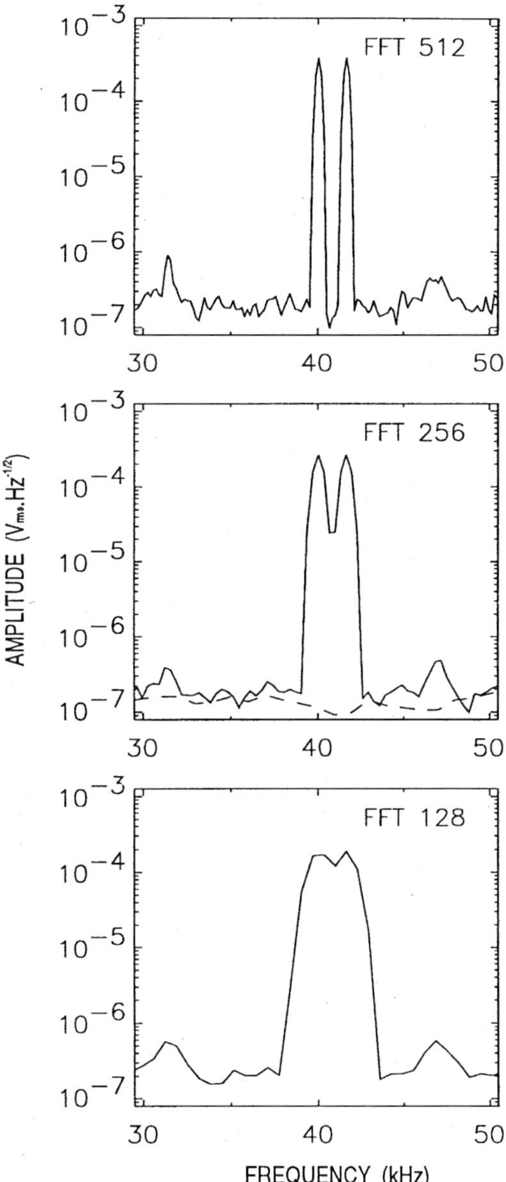

Figure 9. Frequency separation capabilities as seen by the WHISPER analyser for different FFT sizes.

the transfer function of the analogue part of the measurement chain, the precision of the FFT calculation, the uncertainty added by the on-board data compression.

The uncertainty attached to the FFT calculation is negligible. The main contribution corresponds to the algorithm calculating the modules of the bin values, from their real and imaginary parts. It always stays below 0.17 dB. The uncertainty

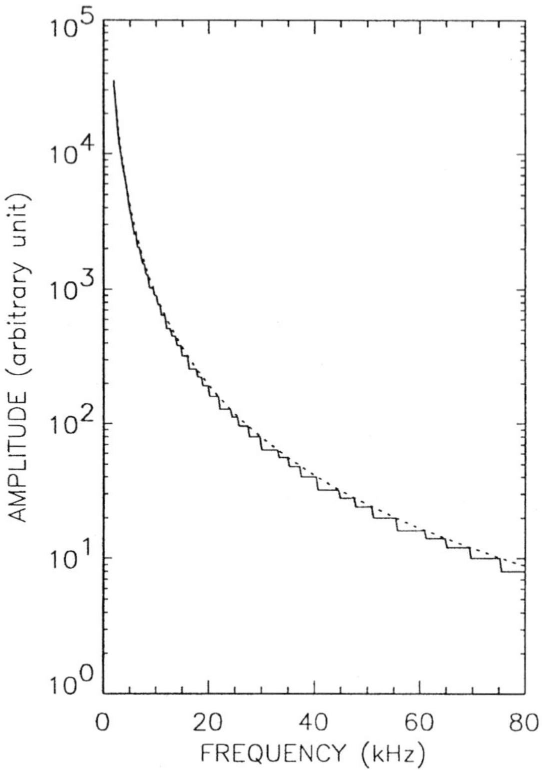

Figure 10. Role of the quasi-logarithmic word compression on the signal amplitude resolution. The dotted line shows an ideal spectrum before compression. The solid line shows the decompressed amplitudes when the input data are compressed in 6 bit words, adjusted to the total dynamic range of the spectrum.

attached to the on-board word compression is ±1 dB in the worst case (Figure 10), which happens in a spectrum whose amplitudes cover the largest range offered by the windowing, and when the word dimension is limited to 6 bits. It is at most ±0.2 dB for a word dimension of 8 bits.

The calibrations of the five flight model units of WHISPER have been performed in the course of the several integration campaigns and environment tests made at WHISPER, WEC and Cluster levels (Martin *et al.*, 1994). Figure 11 presents the transfer functions of the complete EFW and WHISPER chain, for the four values of the WHISPER receiver gain, and for the flight models F1 to F4, at ambient temperature. Two features can be noted: (1) except in the lower part of the range (below 2 kHz), which is actually outside WHISPER's domain, all models behave in the same way, to within less than 1 dB; (2) for a same model, on the other hand, the transfer functions for two adjoining gains are in a ratio of 4, to within less than 1 dB. This property is a constraint of the design in the automatic gain control option, as has been explained above (Section 3). The variations of

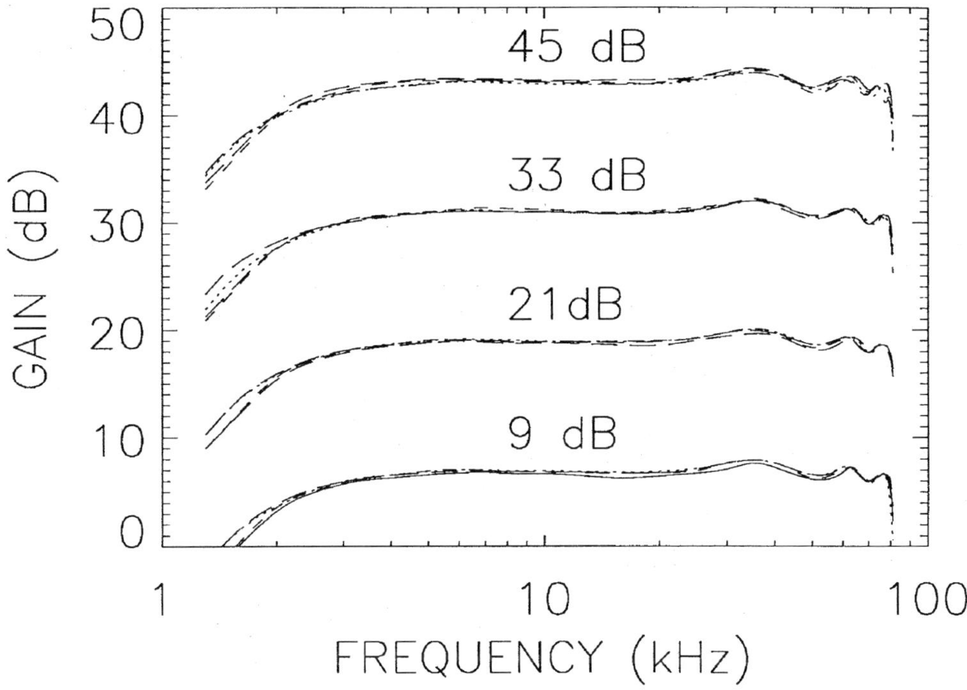

Figure 11. Transfer functions of the WHISPER experiment for the F1 (solid line), F2 (dots), F3 (dashes) and F4 (long dashes) models, and for the four WHISPER gain values.

the transfer functions with temperature are not high ((2 dB at most). They are, however, taken into account in the data-handling chain. One advantage of a wave instrument is that transfer functions are known to be very stable, hence the on-board calibration activity is reduced to verification of the stability of both the receiver and the transmitter characteristics. Two calibration modes, referred to as 'Quiet' and 'Sound' (see Figure 5) are implemented in WHISPER. In the first one the inputs of the receiver are connected to the internal synthesiser. In the second, the transmitter is running while connected to the inputs of the receiver. In both calibrations, 10 frequency values, all the gain steps and all the output levels are swept. According to a specific on-board processing, the results are compared to an internal table. A single frame including an error level table (4 bytes in the housekeeping telemetry) is transmitted. This mode has been used as a WHISPER go-nogo test mode along the integration phases. It will also be used to provide quick health tests during the mission (see Section 6). The complete calibration will be run once per orbit and the results analysed on ground. If necessary, the calibration tables implemented in the data-handling chain will be revised.

In summary, the overall uncertainty on the amplitude of a spectral line is ± 3 dB at most. The $E_{\rm pow}$ parameter is transmitted in an 8-bit word. It is measured with a comparable precision.

4.2. Sounding Mode

The basic performances presented above are valid in sounding modes. However, a few specific features are worth noting:

(a) The E_{pow} parameter, not relevant, is not transmitted.

(b) The active spectrum and the passive spectrum obtained after a sweep are reconstructed from a number of individual spectra, obtained under different stimulation. Due to the resonance sounding principle, an active sweep is made of both large and small amplitude lines. This feature is not attenuated by the windowing dynamic range. Consequently, the resultant dynamic range of active spectra is expected to be more important that the instantaneous range of (N) spectra: up to 96 dB. This is the reason why the word size attached to the spectral lines is nominally 8-bit, whereas it is generally 6-bit in (N) modes.

(c) No averaging is made for the reconstructed passive spectrum. In addition, if the receiver has been saturated by a very strong resonance, the following measurements are likely to be affected. Hence, the information given by the passive spectrum will have to be taken with caution.

(d) The FFT size is always 512 bins, which ensures that single resonances will be localised with a 162 Hz precision. This results in a precision of 8% on the lowest measurable density (0.1 cm^{-3}), of 0.5% on the highest one (80 cm^{-3}). On the other hand, the instrument can separate frequencies at about 400 Hz from each other. Detailed studies implying the characterisation of Bernstein modes (Etcheto *et al.*, 1983) will have to be carried out close enough to the Earth for the interesting emissions to appear well separated.

(e) There are cases in NM telemetry where only 1 in 2 (or 1 in 4) spectral lines are transmitted to the ground. Studies based on the measure of the total power carried by a resonance are not possible in this case.

5. Operations

5.1. Basic Modes

The sharing of on-board resources demands precise coordination of the commanding of the wave instruments. On one hand, a number of functions cannot be satisfied in parallel, thus time sharing has to be organised, on the other hand there is an adaptive allocation of telemetry resources inside the WEC, needed for specific studies. The operational constraints for WHISPER are: (1) the instrument's output has no scientific interest for specific modes of EFW (Langmuir mode on all sensors, then kept at constant voltage), (2) it will be set OFF when WBD uses the payload telemetry resources. Both circumstances are expected to happen rarely (less than once a month). Another constraint is that active operations must be optimised. Both the occurrence and the level of sounding have to be adapted to the regions under study, and to other instruments' environmental constraints. Moreover, during active

WHISPER operations, the other WEC instruments wish to run specific modes. This is the reason why most WEC modes are organised according to the recurrence of sounding operations (Woolliscroft *et al.*, 1995; Pedersen *et al.*, this issue).

Stable operations of the whole payload are desirable for most of the mission life. Two basic modes, one for each Cluster telemetry rate (NM or BM) have been designed by the WEC investigators for an extensive utilisation. They are referred to respectively as NBR (Normal Bit Rate) Basic and HBR (High Bit Rate) Basic modes. The recurrence period of sounding is 28 s in both basic modes. Figure 12 shows the organisation of WHISPER during those 28 s, for a nominal spin rate (4 s). The first 3 s are devoted to sounding. Two sweeps are performed successively, for reliability of measurement. Actually, the power measured at a plasma resonance is expected to be modulated with the antenna rotation, according to the wave vector direction in the spin plane (see example in Etcheto *et al.*, 1981). During a spin, the signal may thus be almost extinguished within a few degrees of two spin angle values. The sweep duration chosen ensures that, during the second sweep, the antenna listens to a given resonance with an orientation at about 45° from the orientation it had at that resonance during the first sweep. A resonance hidden in the first sweep should then be visible in the second one. The sounding operations are the same in NM or BM telemetry rates: frequency tables (c) or (d) (see Table III), step durations 26.6 or 40 ms, transmission level and gain to be adjusted, but the data return is complete in BM (strategy C for DWP processing), whereas it is reduced in NM rate (strategy A and reduced passive information). The last 25 s are devoted to (N) mode measurements. The FFT size (256 bins) and word size (6 bit) are identical for NM and BM rates, the spectra are averaged by 16 in NM, by 8 in BM. The time resolutions achieved are 2.15 s and 0.32 s, respectively for the spectra, 213 ms and 13 ms, respectively, for the E_{pow} parameters. The Figure shows how the spin angles are sampled for each measurement, at each TM rate. The resolution in directivity goes from excellent for E_{pow} in BM (one measure every 1.2°) to poor for spectral data in NM (each measure covers 20°, and the antenna pattern is not completely explored).

5.2. Specific Modes

Several WEC modes have been designed to provide the time resolution of the measured density close to the needs described in Section 2. They are presented in the upper panel of Table IV. Two of them (NBR and HBR active continuous) are designed to optimise the time resolution while satisfying power and TM constraints. The resolution can be increased by reducing the frequency range explored (see alternate option in HBR active continuous, designed in particular for the solar wind region). The two other modes are spin-synchronised, in an attempt to minimise the interactions from the sounder with the particles measured by the low-energy analysers, PEACE (Johnstone *et al.*, this issue) and CIS (Rème *et al.*, this issue). In NBR spin-synchronised mode, WHISPER transmissions are synchronised to the

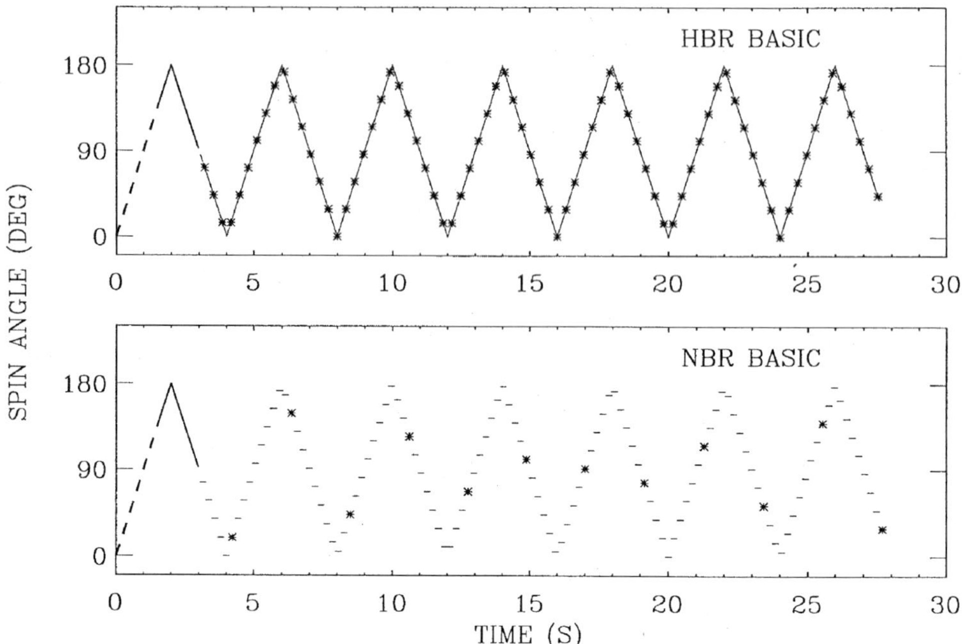

Figure 12. Structure of WHISPER operations in the WEC basic modes, indicated as a function of time and spin angle variations. Each symbol marks a specific measurement: dashed and solid lines in the first 3 s represent each a sounding sweep, asterisks refer to spectral measurements, and minus symbols (too close to be distinguished from each other in the upper panel) to E_{pow} measurements.

voltage flyback period of the electron detection plates. In HBR gliding mode, a short active sweep is run in a time slot shorter than one eighth of the spin (0.5 s), the transmitter is then quiet for a whole spin, natural emissions being measured during that second phase. This basic structure (1.25 spin) is repeated according to a repetition factor. One active sweep and four accumulated 'natural' spectra are transmitted to the TM file in each 1.25 spin. This mode provides both the density and wave measurements with a reasonable time resolution. It is a good candidate for the study of boundaries.

Wave measurement options are presented in the lower panel of Table IV. They alternate (S) and (N) operations. Various frequency and time resolutions are available. One can note different TM consumptions in a same TM mode (NBR or HBR, respectively). This reflects the use of priorities, either given to WHISPER (for example in NBR Langmuir a) or to STAFF (NBR Low Rec.). The HBR Special WHI mode is designed for foreshock studies, needing a high time resolution for burst wave packets. Here, the time resolution has been favoured with respect to the frequency resolution. The latter one is still sufficient to resolve the wide band noise (typically a few kHz wide) quoted in Section 2. The HBR Angle mode is designed for studies of detailed directivity patterns. The actual angle resolution will depend on the amplitude of the modulation, and on the cleanliness of the signatures.

Table IV
WHISPER characteristics for a selection of WEC modes

DENSITY MEASUREMENTS					
WEC NAME	SEQUENCING	Δt (Ne) (s)	Frequency range (kHz)	DWP PROCESSING	TM rate (bps)
NBR Active Continuous	Continuous Sounding 40ms pulse recurrence	3.25	4-80	D: Active spectrum, Passive reduced	1410
NBR Spin Sync.	Continuous Sounding 1/32 spin pulse rec.	4.5	4-74	D: Active spectrum, Passive reduced	940
HBR Active Continuous	Continuous Sounding and short (N) periods 27ms pulse recurrence/ (40ms pulse recurrence)	1.44/ (0.5)	4-80/ (11-34)	C: Active spectrum, Passive complete	5400/ (5000)
HBR Gliding	Spin Synchronized sweeps (13ms pulse recurrence): 0.125 spin (S)/1 spin (N)	4.5	4-67	B:Active spectrum, A/P ratios (S)/ 4 passive spect.(N)	4500

WAVE MEASUREMENTS						
WEC NAME	SEQUENCING	Δt (spectra) (s)	Δt (Epow) (s)	Δb (Hz)	Δθ (spectra) (deg)	TM rate (bps)
NBR Basic	3s (S) / 25s (N)	2.15	0.213	324	19	970
NBR Low Rec.	4s (S) / 100s (N)	2.6	0.85	324	77	850
NBR Langmuir a	4s (S) / 100s (N)	2.6	0.85	162	77	1590
HBR Basic	3s (S) / 25s (N)	0.32	0.013	324	9.6	5670
HBR Low Rec. a	4s (S) / 100s (N)	0.65	0.013	324	19	3350
HBR Langmuir	4s (S) / 100s (N)	0.65	0.213	162	19	6160
HBR Special WHI	4s (S) / 100s (N)	0.107	0.013	648	9.6	8560
HBR Angle a	4s (S) / 104s (N)	0.32	0.013	324	4.8	6280

5.3. Commissioning

All the options presented above, and many others available according to the programmable options (Table II), will be tested in space during the commissioning phase of the mission. In (N) modes, we will mainly check the instrument's behaviour with respect to the several gain options. The sounder's behaviour will be studied with special care. Indeed, there is a need to validate in space a design with a number of new functioning features. Three main points will be examined:

(1) The efficiency of sounding with respect to different transmission levels, pulse durations, and step durations. The duration of a sweep can be adjusted from below 0.5 s (highest pulse recurrence, 13.3 ms step duration), to a few seconds.

(2) The quality of the 'passive' measurements recorded during soundings.

(3) The effects of soundings on the spacecraft environment, especially on the particle populations measured by the analysers. A specific WHISPER interference

campaign is organised to this end. The active participation of several instruments is required: EFW and ASPOC (Riedler *et al.*, this issue) in relation to the spacecraft potential control, PEACE, CIS and the DWP correlator for the analysis of possible count fluctuations. In particular, the advantage of spin synchronisation will be questioned.

We wish to validate one or several of the detailed sounding mode options, in order to run only a small selection of them during the mission. The exact duration of a density measurement will thus be known in flight, and the WEC mode strategy may have to be re-examined, according to the scientific objectives which can realistically be met. For example, we may, or may not, push forward studies implying a good density time resolution (like the study of MHD turbulence described in Section 2). Lastly, it is worth saying that the WHISPER (S) and (N) modes will be fine-tuned not only during the commissioning phase of the mission, but also later, with the appearance of regions as yet unexplored by a sounder, of different separations, and possibly also of unexpected operational constraints.

6. WHISPER Data Processing

Several sets of data reduction software have been developed to program and monitor the WHISPER instrument, process the data delivered by WHISPER and analyse the result of the mission for scientific purposes in cooperation with other investigators. Most of the time these software have been implemented in different institutions: Cluster French Centre (CFC) at CNES, Joint Science Operation Centre (JSOC) at RAL, Centre de Surveillance et de Programmation de l'Instrument (CSPI) WHISPER at LPCE. Systems and computers used at those several places are compatible, but often different.

The WHISPER software is relatively complicated. It is mainly to open the science data packets, recognise the operating mode and transfer raw data into physical values by applying the WHISPER transfer functions that are available on the CNES server. Then the Prime Parameter Data (PPD) and Summary Parameter Data (SPD) have to be computed and delivered to a wide scientific community in order to fulfil the Cluster Science Data System requirements.

One of the parameters that has to be determined, namely the total plasma density, requires a specific program, because WHISPER does not measure it directly. As explained above, WHISPER is primarily an active experiment, aimed at exciting radio waves at the characteristic frequencies of the plasma. The number and type of resonances which may be stimulated depend upon the magnetospheric region that is probed. So, the program needs to know the region crossed by the satellite and, more generally, the satellite position, to improve the quality of the density determination process. As some resonances are observed as harmonics of the electron cyclotron frequency, the magnetic field data provided by the FGM instrument (Balogh *et al.*, this issue) are also necessary.

The main objectives and some technical characteristics of the WHISPER software that will run at JSOC, CFC, CSPI, and inside the Swedish Interactive Science Data Analysis Tool (ISDAT) system are recalled in the following subsections.

6.1. WHISPER DATA-PROCESSING AT JSOC

Two software packages have been delivered to JSOC, a calibration check software (CALCHK) also known as 'health and safety' and an Inter-Experiment Calibration/Science Performance Monitoring (IEC/SPM) software. The CALCHK program analyses the result of the error level table (see section above) and fills in a return code, zero if the test is OK, one if it is not, and greater than one if an error occurred during computation. When the return code is greater than or equal to one, a text file will be produced and sent by mail to the WHISPER team.

The IEC/SPM software uses the science data packets and the FGM magnetic field modules and spacecraft position files to compute the WHISPER PPD plus data integrity information such as data gap percentage, numbers of valid and non-valid blocks and, for passive modes, averaged E-field overflows and electric power in the frequency range 2–80 kHz. The total plasma density determination is included in this software package, although the way to compute this fundamental parameter has been simplified due to technical constraints.

6.2. WHISPER DATA-PROCESSING AT CFC

At CFC, three software components have been delivered. The first one, called WHISPER PHYSICAL DATA, decommutates the WHISPER data from each satellite and calibrates them. The data are checked to verify their integrity. The second, WHISPER PPD SPD, creates the PPD and SPD which are part of the CSDS products. A library written by J. C. Kosik from CNES is used to determine the magnetospheric regions crossed by Cluster. The PPD and SPD (Schmidt and Escoubet, this issue) include the electric power in the frequency range 2–80 kHz, the variance of this power, and the total plasma density. The third one, called WHISPER 4SAT SPD VIEWS, creates a plot which is similar to the CSDS SPLOTS, but includes all four satellites and is specific to WHISPER.

6.3. WHISPER DATA-PROCESSING AT CSPI

For maintenance, testing, and validation purposes, all the software delivered to institutes other than LPCE (the PI laboratory) will be implemented at LPCE where the 'Centre de Surveillance et de Programmation de l'Instrument WHISPER' is located. One important output from the main WHISPER software package is a dynamic spectrogram, displaying the E-field power as a function of time and frequency (Figure 13). Specific software or more refined versions of the existing software have also been developed. One of these programs concerns the total plasma density determination which requires more detailed analysis than that used

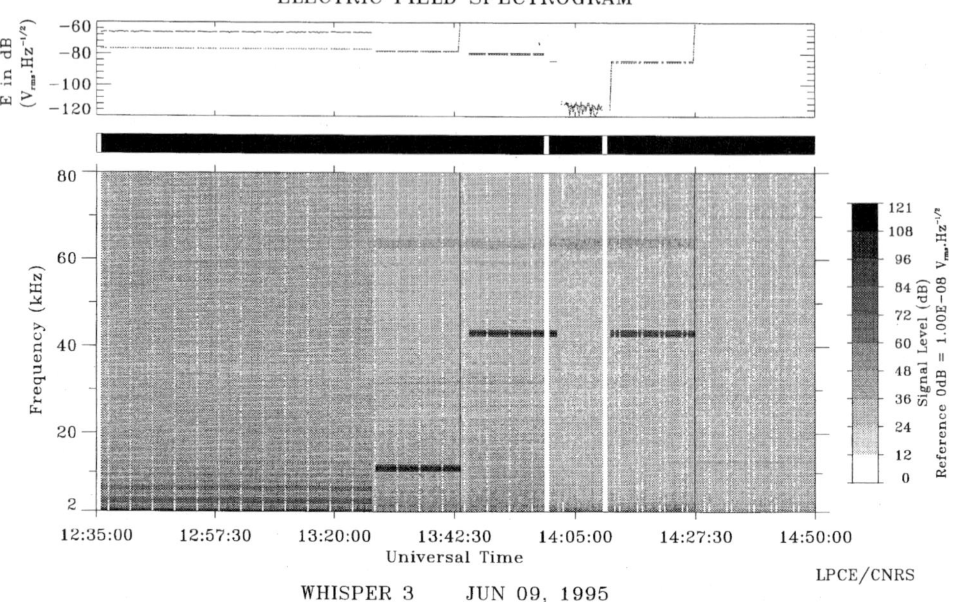

Figure 13. Frequency/time spectrogram (bottom panel) obtained during mission commanding tests, where sine waves are injected on the electric field sensors. The central panel indicates (in white) periods when, as required by telecommand, WHISPER is in standby. The upper panel shows the electric field power (dotted line), calculated from the frequency bins amplitudes integrated over the displayed spectrum, compared to the E_{pow} variations (solid line) calculated from the time samples at the input of the spectral analyser. The electric field power E_{pow} may exceed the power contained in the displayed spectrum, strictly restricted to the 2–80 kHz range. This is the case from about 12:35 to 13:30 UT, where a sine wave of large amplitude is injected at 1.45 kHz.

at CFC or JSOC. In addition, some facilities such as Graphical User Interfaces or tools to facilitate communications with ESOC, JSOC, CFC, and the other CSDS data centres have been expanded.

6.4. WHISPER DATA-PROCESSING USING ISDAT

ISDAT is part of the Wave Experiment Consortium package. The WHISPER part for the server has been developed. This system is described in Pedersen *et al.*, this issue. The main part comes from the decommutation kernel, which has been adapted to ISDAT.

7. Summary of Performances

The total measured mass of the WHISPER module, including the WHISPER-DWP interconnection cable and mounting bolts is 2.005 kg. The power consumption is 2.7 W in natural mode, 3.0 W in sounding mode. Those figures are primary power including the efficiency of the WEC power supply. The TM rates allocated to WHISPER are variable. In the standard WEC modes, it is about 980 bit s^{-1} in Nominal (NM) TM rate, 5800 bit s^{-1} in Burst (BM) TM rate. The operational performances of the instrument correspond to the following characteristics:

– Plasma densities from 0.2 to 80 cm^{-3} are measured with an accuracy of order 1% and a time resolution of 1.5 s at best (28 s in basic nominal operations).

– Natural wave electric fields are measured between 2 and 80 kHz, with a sensitivity of 2×10^{-9} V m^{-1} $Hz^{-1/2}$ and variable time and frequency separation properties (for basic operations, respectively, 320 ms and 800 Hz in BM telemetry, 2.15 s and 800 Hz in NM telemetry).

– The dynamics of measurable signals for a given mode covers about 80 dB. A larger absolute range (100 dB) can be reached by programmable gain.

– The high stability of on-board frequencies leads to a negligible uncertainty on the frequency determination. The stability of the gain of the amplifiers, and the precision of their determination derived from ground and on-board calibrations, guarantees an absolute precision of about ± 3 dB in signal amplitude measurements. The on-board data compression algorithms have been designed to keep that level of precision. The matching of the transfer function between the four Cluster spacecraft will be performed with the same precision.

Acknowledgements

The authors wish to acknowledge the crucial efforts of a number of people who, at one time or another, worked hard to make WHISPER what it is: A. Bahnsen, from DSRI, studied early versions of the sounder; C. Vasiljevic (LPCE) qualified the ZORAN VSP for the WHISPER application; J. P. Villain and D. Lagoutte (LPCE) contributed to the first VSP software; M. Steinbeck, T. Lemaire, and Y. Tessier (LPCE) participated to the hardware and software conception; J. P. Labat, R. Kirche, J. P. Olivo, D. Sautarel, F. David, and P. Ferrerro from AETA, were responsible for the fabrication of the WHISPER flight models and of the WEC power supply; A. Sumner (University Sheffield) developed the DWP application for WHISPER; E. Guyot and L. Launay (LPCE) developed the data handling software. The WEC, ESA, Dornier, RAL, and IABG teams are thanked for their participation and help in the testing and integration of the experiment. The authors extend their thanks to those who helped the WHISPER investigators in their contribution to the CSDS, in particular J. P. Thouvenin, H. Poussin, M. Nonon, J. Y. Prado, and J. C. Kosik at CNES, for the preparation of CSDS data products, the JSOC and ESOC teams for the preparation of the operations. The WHISPER experiment and software are realised thanks to a CNES contract. The transmitter hardware was designed and realised at DSRI.

References

Balogh, A. *et al.*: 1996, this issue.
Bitoun, J., Fleury, L., and Higel, B.: 1975, *Radio Sci.* **10**, 875.
Chapman, S. C. and Dunlop, M. W.: 1993, *Geophys. Res. Letters* **20**, 2023.
Christiansen, P. J., Gough, P., Etcheto, J., Trotignon, J. G., Rönmark, K., and Stenflo, L.: 1984, *Proc. Conf. Achievements of the IMS*, ESA SP-217, p. 517.
Décréau, P. M. E., Etcheto, J., Knott, K., Pedersen, A., Wrenn, G. L., and Young, D. T.: 1978, *Space Sci. Rev.* **22**, 633.
Décréau, P. M. E., Martin, Ph., Sené, F. X. *et al.*: 1993, *ESA SP-1159*, p. 51.
Dudok de Wit, T., Krasnosel'skikh, V. V., Bale, S. D. *et al.*: 1995, *Geophys. Res. Letters* **22**, 2653.
Eastman, T., Frank, L., Peterson, W., and Lennartson, W.: 1984, *J. Geophys. Res.* **89**, 1553.
Elphic, R. C.: 1988, *Adv. Space Res.* **8**, 238.
Etcheto, J., de Féraudy, H., and Trotignon, J. G.: 1981, *Adv. Space Res.* **1**, 183.
Etcheto, J., Christiansen, P. J., Gough, M. P., and Trotignon, J. G.: 1982, *Geophys. Res. Letters* **9**, 1239.
Etcheto, J., Belmont, G., Canu, P., and Trotignon, J. G.: 1983, *ESA SP-195* **39**.
Etcheto J. and Faucheux, M.: 1984, *J. Geophys. Res.* **89**, 6631.
Etcheto, J. and Saint-Marc, a.: 1985, *J. Geophys. Res.* **90**, 5338.
Feldstein, Ya. I. and Galperin, Yu. I.: 1994, *J. Geophys. Res.* **99**, 13537.
Filbert, P. C. and Kellogg, P. J.: 1979, *J. Geophys. Res.* **84**, 1369.
Gough, M. P.: 1982, *Planetary Space Sci.* **30**, 657.
Gurnett, D. A.: 1975, *J. Geophys. Res.* **80**, 2751.
Harris, F. J.: 1978, *Proc. of the IEEE* **66**, 51.
Harvey, C. C., Etcheto, J., and Mangeney, A.: 1982, *Space. Sci. Rev.* **39**, 1979.
Heikkila, W. J.: 1982, *Geophys. Res. Letters* **9**, 159.
Heikkila, W. J.: 1990, *Space Sci. Rev.* **53**, 1.
Hospodarsky G. B., Gurnett, D. A., Kurth, W. S., Kivelson, M. K., Strangeway, R. J., and Belton, S. J.: 1994, *J. Geophys. Res.* **99**, 13363.
Johnstone, A. *et al.*: 1996, this issue.
Lacombe, C., Mangeney, A., Harvey, C. C., and Scudder, J. D.: 1985, *J. Geophys. Res.* **90**, 73.
Lemaire, J.: 1977, *Planetary Space Sci.* **25**, 887.
Lemaire, J. and Roth, M.: 1992, *Planetary Space Sci.* **40**, 193.
Martin, Ph.: 1994, Integration Report, LPCE Doc.
Martin, Ph., Décréau, P. M. E., and Sené, F. X.: 1995, *LPCE NTS*, **036**.
Motez, F. and Chanteur, G.: 1994, *J. Geophys. Res.* **99**, 13 499.
Olsen, R. C.: 1982, *J. Geophys. Res.* **87**, 3481.
Onsager, T. G., Thomsen, M. F., Elphic, R. C., and Gosling, J. T.: 1991, *J. Geophys. Res.* **96**, 20999.
Otto, A., Lee, L. C., and Ma, Z. W.: 1995, *J. Geophys. Res.* **100**, 14 895.
Parks, G. K., McCarthy, M., Fitzenreiter, R., Etcheto, J., Anderson, K. A., Anderson, R. A. *et al.*: 1984, *J. Geophys. Res.* **89**, 8885.
Parks, G. K., Fitzenreiter, R., Ogilvie, K. W., Huang, C., Anderson, K. A. *et al.*: 1992, *J. Geophys. Res.* **97**, 2943.
Parks, G. K., Fitzenreiter, R., Ogilvie, K. W., Huang, C., Anderson, K. A. *et al.*: 1994, *J. Geophys. Res.* **99**, 13541.
Pedersen, A. *et al.*: 1966, this issue.
Perraut, S. *et al*: 1990, *J. Geophys. Res.* **95**, 5997.
Rème, H. *et al.*: 1996, this issue.
Riedler, W. *et al.*: 1996, this issue.
Russell, C. T. and Elphic, R. C.: 1978, *Space Sci. Rev.* **22**, 681.
Schmidt, R. and Escoubet, Ph.: 1996, this issue.
Sené, F. X.: 1994, WHISPER Internal EID, LPCE Doc.
Sibeck, D. G.: 1992, *J. Geophys. Res.* **97**, 4009.
Stone, R. G., Bougeret, J. L., Caldwell, J. *et al.*: 1992a, *Astron. Astrophys. Suppl. Ser.* **92**, 291.
Stone, R. G., Pedersen, B. M., Harvey, C. C. *et al.*: 1992b, *Sciences* **257**, 1524.

Trotignon, J. G., Etcheto, J., and Thouvenin, J. P.: 1986, *J. Geophys. Res.* **91**, 4302.
Vasiljevic, C.: 1990, Mémoire de Thèse, Université d'Orsay.
Woolliscroft, L. J. C., Cornilleau-Wehrlin, N., Décréau, P. M. E. *et al.*: 1995, *Radio Sci. Bull.* **273**, 36.

THE WIDE-BAND PLASMA WAVE INVESTIGATION

D. A. GURNETT, R. L. HUFF and D. L. KIRCHNER
Department of Physics and Astronomy, The University of Iowa, Iowa City, IA 52242, U.S.A.

Abstract. As part of the Cluster Wave Experiment Consortium (WEC), the Wide-Band (WBD) Plasma Wave investigation is designed to provide high-resolution measurements of both electric and magnetic fields in selected frequency bands from 25 Hz to 577 kHz. Continuous waveforms are digitised and transmitted in either a 220 kbit s^{-1} real-time mode or a 73 kbit s^{-1} recorded mode. The real-time data are received directly by a NASA Deep-Space Network (DSN) receiving station, and the recorded data are stored in the spacecraft solid-state recorder for later playback. In both cases the waveforms are Fourier transformed on the ground to provide high-resolution frequency-time spectrograms. The WBD measurements complement those of the other WEC instruments and also provide a unique new capability for performing very-long-baseline interferometry (VLBI) measurements.

1. Introduction

The Cluster Wide-Band (WBD) Plasma Wave investigation is designed to provide very high-resolution frequency-time measurements of plasma waves in the Earth's magnetosphere. The WBD instrumentation is part of the Wave Experiment Consortium (WEC) and consists of a digital wide-band receiver that can provide electric- or magnetic-field waveforms over a wide range of frequencies. Wide-band receivers have played an important role in many previous spacecraft missions, including GEOS, ISEE, Voyager, DE, and Galileo. The WBD instruments on the four Cluster spacecraft are similar in design to those being flown on the Polar and Cassini spacecraft. The investigators responsible for the WBD investigation on Cluster are listed in Table I.

The wide-band technique involves transmitting band-limited waveforms directly to the ground using a high-rate data link. The primary advantage of this approach is that continuous waveforms are available for detailed high-resolution frequency-time analyses. The frequency-time resolution is limited only by the uncertainty principle, $\Delta\omega\Delta t \sim 1$. Since the frequency resolution ($\Delta\omega$) and time resolution (Δt) can be selected on the ground, the wide-band technique has the advantage that the resolution can be adjusted to provide optimum analysis of the phenomena of interest.

The high-resolution nature of wide-band measurements is particularly important for the study of plasma emissions that have very complex frequency-time characteristics. To illustrate the flexibility and high resolution available from wide-band data, Figure 1 shows representative spectrograms of several types of plasma wave emissions that are commonly observed in the Earth's magnetosphere. The spectrogram in panel (a), for example, has been processed using a very expanded time

Space Science Reviews **79:** 195–208, 1997.

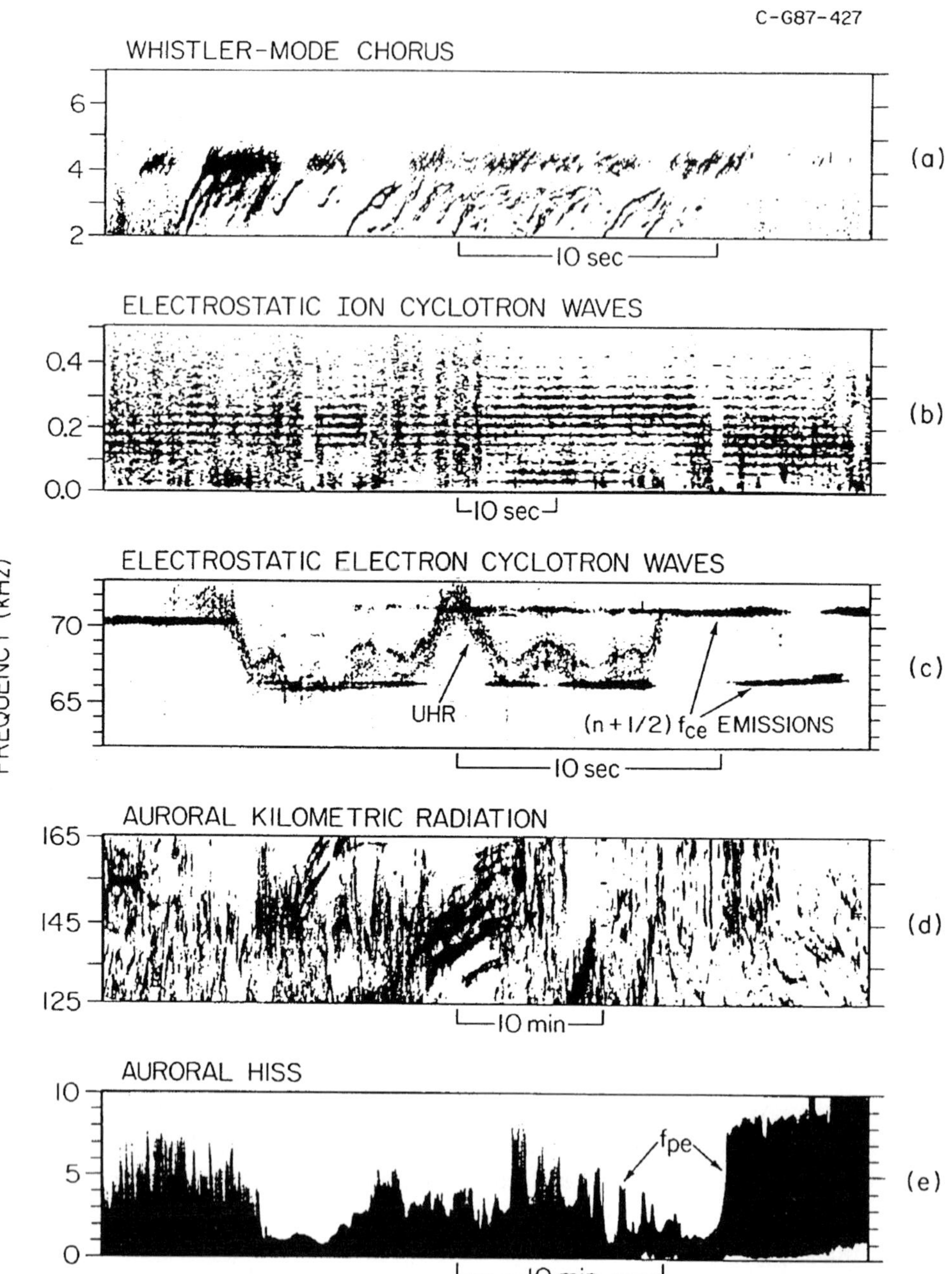

Figure 1. Frequency-time spectrograms of plasma emissions observed in the Earth's magnetosphere using wide-band instrumentation.

Table I
Cluster WBD investigators

Principal investigator	Affiliation
Donald A. Gurnett	The University of Iowa
Co-investigators	
Jean Louis H. Bougeret	Observatoire de Paris (Meudon)
Patrick Canu	Centre d'études des Environnements Terrestre et Planetaires/UVSQ
Georg Gustafsson	Swedish Institute of Space Physics
Gerhard Haerendel	Max-Planck-Institut für extraterrestrische Physik
Robert A. Helliwell	Stanford University
Umran S. Inan	Stanford University
Warren L. Martin	Jet Propulsion Laboratory
Robert L. Mutel	The University of Iowa
Bent Pedersen	Observatoire de Paris (Meudon)
Alain Roux	Centre d'études des Environnements Terrestre et Planetaires/UVSQ
Steven R. Spangler	The University of Iowa
Eigil Ungstrup	Geophysical Institute (Copenhagen)
Les Woolliscroft[a]	The University of Sheffield

[a] Deceased.

scale to resolve the fine structure of whistler-mode chorus emissions. Chorus is an electromagnetic emission that propagates in the whistler mode at frequencies below the electron cyclotron frequency (Helliwell, 1965). These emissions are believed to be produced by energetic electrons trapped in the terrestrial radiation belts (Kennel and Petschek, 1966). The spectrogram in panel (b) has been processed with greatly expanded frequency resolution at low frequencies to resolve the narrow-band structure of electrostatic ion cyclotron waves. Electrostatic ion cyclotron waves are electrostatic waves that propagate near harmonics of the ion cyclotron frequency (Kintner *et al.*, 1978). These waves are believed to be excited by field-aligned currents, and by highly anisotropic low-energy ion distributions. The spectrograms in panels (c) and (d) were obtained in a frequency conversion mode of operation and show the extremely complicated fine structure of $(n + \frac{1}{2})f_c$ electron cyclotron waves near the upper hybrid resonance (UHR) and the complex narrow-band structure of auroral kilometric radiation. The $(n+\frac{1}{2})f_c$ electron cyclotron waves are electrostatic waves that occur between harmonics of the electron cyclotron frequencies. These waves are believed to be excited by highly anisotropic low-energy electron distributions (Ashour-Abdalla and Kennel, 1978). Auroral kilometric radiation is an intense electromagnetic emission generated in the Earth's auroral regions (Gurnett, 1974). These emissions often have an extremely complex fine structure (Gurnett *et al.*, 1979). Finally, to illustrate the use of wide-band data as a means for obtaining very high-resolution measurements of basic plasma parameters, panel (e) shows a spectrum of auroral hiss that has a sharp upper frequency cutoff at the electron plasma frequency, f_p. The cutoff at f_p provides very accurate (few percent), high resolution (0.1 s) measurements of electron densities in the auroral zone and over the polar cap (Persoon *et al.*, 1983).

2. Scientific Objectives

The eccentric, highly inclined orbit of the Cluster spacecraft carries the spacecraft through most of the important regions of the magnetosphere, including the polar cusp, the plasma mantle, the magnetopause boundary layer, the polar cap, the auroral zones, the plasma sheet, and the plasma sheet boundary layer. Many different types of plasma waves occur in these regions. Some of the more important include electrostatic ion cyclotron waves, whistler-mode chorus, auroral hiss, auroral kilometric radiation, continuum radiation, electrostatic electron cyclotron waves, upper hybrid waves, and lower hybrid waves.

Within the Wave Experiment Consortium, each experiment has its own areas of specialisation as described in the WEC overview (see Pedersen *et al.*, this issue) and the WEC companion papers. The primary purpose of the WBD investigation is to support WEC science objectives by providing high-resolution spectral analysis. At boundaries and other regions with steep spatial gradients, WBD provides high-time resolution single-spacecraft measurements for comparison with data from other instruments, such as the magnetometer and plasma instruments. From these data, waves produced by current-driven instabilities and other mechanisms involving spatial inhomogeneities can be clearly identified. In addition to single-point measurements, wide-band data from two or more spacecraft can also be used to resolve space–time ambiguities as the spacecraft pass through complex spatial structures. The study of such structures is one of the primary objectives of the Cluster mission. In cases where the upper hybrid frequency or electron plasma frequency can be identified, WBD can also provide very high resolution measurements of the electron density. Also, multiple-point comparisons of electron densities can be used to analyse the motion and evolution of plasma structures in the auroral zone and polar cap.

In addition to providing measurements that support the overall objectives of the WEC, the WBD instrument can be used to provide multi-spacecraft long-baseline radio interferometry measurements. The basic principles of a radio interferometer are illustrated in Figure 2 which shows an electromagnetic wave of wavelength λ incident on two antennas separated by a baseline distance L. The signals from the two antennas, E_1 and E_2, are multiplied and averaged to produce a cross-correlation $\langle E_1 E_2 \rangle$. The sign and amplitude of the cross-correlation is determined by the angle of arrival θ relative to the symmetry axis of the interferometer. If the wavelength of the radiation is less than the separation distance between the antennas, then the cross-correlation varies as the cosine of the angle of arrival. The antenna pattern

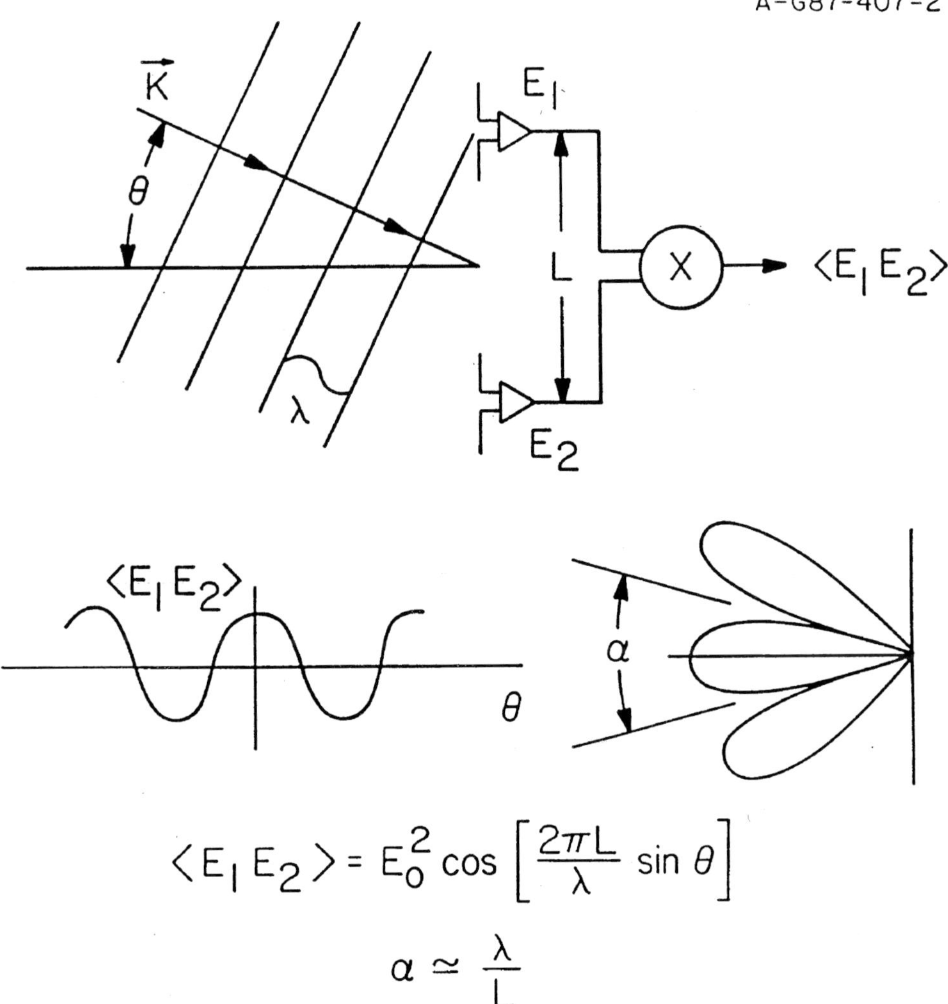

Figure 2. The angular response of a two-point radio interferometer. The response consists of a series of lobes, each with a beamwidth, $\alpha = \lambda/L$.

of the interferometer then has a series of lobes, each with a beamwidth α. If a source moves through this fan-shaped antenna pattern, a sinusoidal 'fringe' pattern is produced in the cross-correlation.

Two-point interferometry measurements of the type described above have been performed by the ISEE spacecraft (Baumback *et al.*, 1986). Figure 3 shows simultaneous observations of auroral kilometric radiation from the ISEE-1 and -2 spacecraft at a geocentric radial distance of about 14 R_E, and a baseline separation distance of 3260 km. The top two panels show the auroral kilometric radiation spectrum received at the two spacecraft, and the bottom panel shows a plot of the

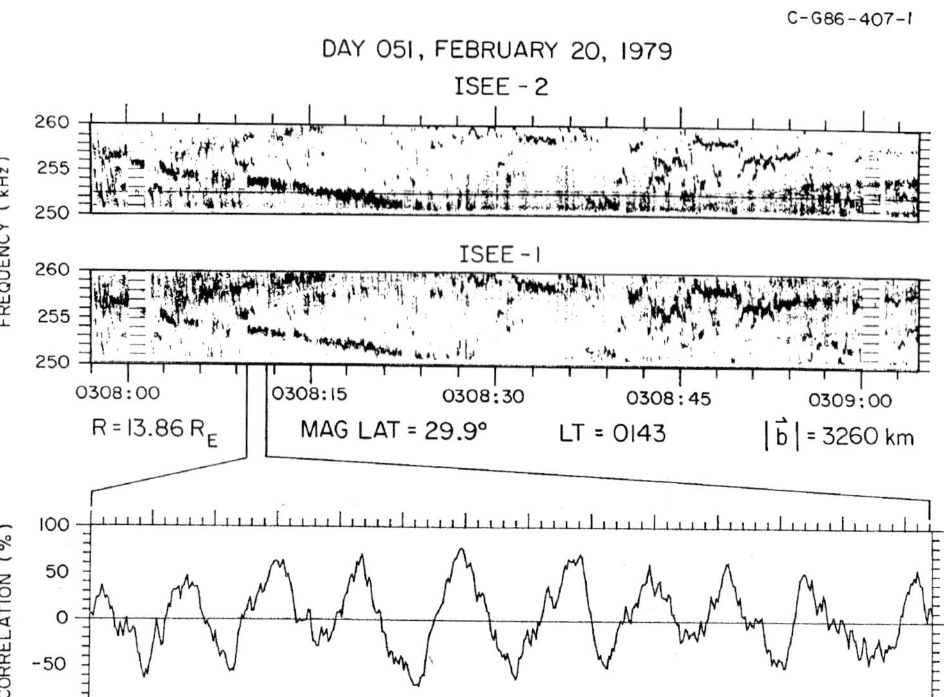

Figure 3. Simultaneous frequency-time spectrograms of an auroral kilometric radiation event detected by ISEE-1 and -2. The bottom panel shows an interference fringe pattern in the cross-correlation between the two signals.

cross-correlation on an expanded time scale that covers a time interval of about 1.3 s. The sinusoidal waveform in the cross-correlation is the characteristic 'fringe pattern' produced by interference between the two received signals. By analysing the fringe amplitude, the angular size of the source can be determined. An important parameter in any interferometer experiment is the number of baseline separations involved between the various antennas. With ISEE-1 and -2 only one baseline was available, which meant that only one component of the source brightness distribution could be studied. With Cluster a total of six baselines can be achieved between the four spacecraft. Up to ten baselines can be obtained if simultaneous measurements are obtained between Cluster and the Polar spacecraft (Gurnett *et al.*, 1995), which has a similar wide-band receiver. Multi-spacecraft interferometer measurements have the potential of giving important new information on the angular motion and size of various terrestrial and astronomical radio sources.

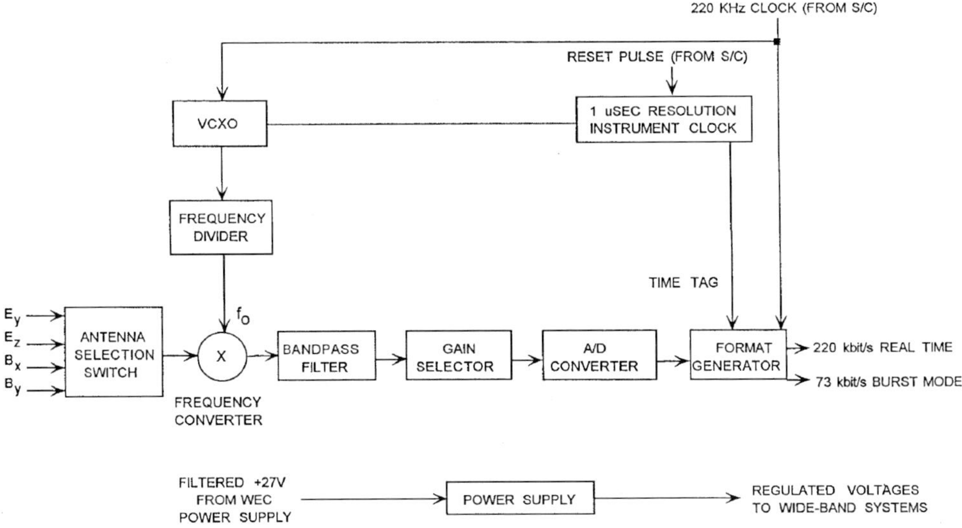

Figure 4. Block diagram of the Cluster wide-band receiver.

3. Description of the Wide-Band Receiver System

A simplified block diagram of the WBD instrument is shown in Figure 4. The instrument processes signals from four antennas, any one of which can be chosen via the antenna selection switch located at the input to the receiver. The four selectable inputs consist of two electric-field signals (E_y and E_z), and two magnetic-field signals (B_x and B_y). These inputs are provided by the electric-field (EFW) and magnetic-field (STAFF) experiments. Unregulated power is supplied to WBD by the WEC power supply. Power switching and all commands to the WBD are provided by the WEC data-processing unit (DWP). The WBD electronics are enclosed in a gold-plated magnesium housing. A photograph of WBD Flight Model 3 is shown in Figure 5. A summary of WBD instrument parameters is given in Table II. Individual aspects of the wide-band receiver system are discussed in detail in the following sections.

3.1. Sensors and Sensor Interfaces

The Cluster plasma-wave sensors consist of two orthogonal spherical electric antennas located in the spin plane of the spacecraft, and a triaxial search coil magnetometer oriented with two axes in the spin plane and the third axis parallel to the spacecraft spin axis. The electric antennas, designated E_y and E_z, are provided by the EFW investigation and have sphere-to-sphere separations of 100 m when fully deployed. The spheres each contain a high-impedance preamplifier that provides signals to the EFW main electronics, to the WBD, and to the other wave instruments via buffer amplifiers. The EFW/WBD buffer amplifier is a

Table II
WBD instrument parameters

Sensors	Two electric-field components (E_y, E_z); two magnetic-field components (B_x, B_y)
Conversion frequencies	0.125 kHz, 250 kHz, 500 kHz
Band pass filter ranges	1 kHz to 77 kHz
	50 Hz to 19 kHz
	25 Hz to 9.5 kHz
Frequency resolution	Determined by FFT
Time resolution	10–20 ms (per FFT spectrum)
Gain select	5 dB steps, 16 levels, dynamic range 75 dB, automatic rnaging or set by command
A/D converter	1-bit, 4-bit, or 8-bit resolution for a selection of samples rates
Mass (flight models, measured)	1.67 kg
Power (flight models, measured)	1.57 W

Figure 5. WBD Flight Model 3.

low-noise, low-power design that maintains a flat response up to about 600 kHz. The boom-mounted three- axis search coil magnetometer (B_x, B_y, and B_z) is part of the STAFF instrumentation, and provides magnetic-field measurements up to 4 kHz. The WBD instrument has the capability of processing signals from only one of the four sensors. The sensor selection is controlled by spacecraft command via the DWP.

3.2. Frequency Bands

The input frequency range of the wide-band receiver can be shifted by a frequency converter to any one of four frequency ranges, where the conversion frequency, f, determines the lower edge of the frequency range selected. The conversion frequency is obtained by dividing a 14-MHz reference oscillator that is located within the WBD instrument. To maintain phase stability in the entire system, this oscillator is synchronised to the spacecraft 220.752-kHz high-frequency clock. The high-frequency clock is obtained by dividing the spacecraft Ultra-Stable Oscillator (USO). The USO operates at a frequency of 2^{23} Hz with a stability ($\Delta f/f$) of 6.5×10^{-9} over a 12-hour period.

A spacecraft command to select a particular frequency band causes DWP to switch the wide-band receiver to an appropriate input bandpass filter and to a conversion frequency of either 0, 125, 250, or 500 kHz. If baseband ($f = 0$) is selected, the mixing stage is bypassed so that the signal is routed directly to the output with no frequency conversion.

The bandwidth of the WBD output waveform is determined by one of three bandpass filters selected in combination with a given output mode. The conversion frequencies and bandpass filter ranges are summarised in Table II.

3.3. Gain Control

The gain of the wide-band receiver is determined by a set of four dual-gain amplifiers that may be selected to provide various gains in increments of 5 dB. The dual-gain amplifiers have gains of 0 or 5 dB, 0 or 10 dB, 0 or 20 dB, and 0 or 40 dB. The gain control has two modes of operation: fixed and auto-ranging. The gain control modes are controlled by spacecraft command.

In the fixed-gain mode, the receiver gain can be set to any one of the sixteen levels (from 0 dB to 75 dB). In the auto-ranging mode, the output from the programmable amplifier is compared to a pair of reference amplitudes. If the criteria for changing the gain are met, the gain is either increased by one step (5 dB) or decreased by one step. In order to avoid excessive toggling between gain steps, a commandable threshold must be exceeded, thereby introducing hysteresis in the Automatic Gain Control (AGC) loop. The gain is updated at a rate determined by the gain update clock, which is a DWP function selected by spacecraft command. The period of the clock is programmable from 0.1 to 27 s in increments of 0.1 s. The actual gain change takes place at the beginning of the next WBD major frame.

Table III
WBD output modes

Mode	Bandwidth	Sample rate	Bits/sample	Duty cycle	Comments
0	25 Hz–9.5 kHz	27.4 kHz	8	100%	Default mode
1	25 Hz–9.5 kHz	27.4 kHz	8	100%	
2	50 Hz–19 kHz	54.9 kHz	4	100%	
3	50 Hz–19 kHz	54.9 kHz	8	50%	
4	1 kHz–77 kHz	219.5 kHz	8	12.5%	
5	1 kHz–77 kHz	219.5 kHz	1	100%	
6	1 kHz–77 kHz	219.5 kHz	4	25%	
7	1 kHz–77 kHz	219.5 kHz	8	12.5%	[a]

[a] This mode is a repeat of output Mode 4, but also toggles the primary and redundant OBDH interfaces.

3.4. A/D CONVERTER AND FORMAT GENERATOR

The output analogue waveform is sampled by an 8-bit analogue-to-digital converter that provides the sampling resolution and data output rates listed in Table III. For sample rates where the bit rate exceeds the spacecraft telemetry rate (220 kbit s^{-1}), the digitised wide-band data is duty-cycled by a format generator that reduces the average bit rate to 220 kbit s^{-1}. The format generator organises the digitised waveform data into a 1096-byte output frame, which includes appropriate timing and status information.

3.5. WBD DATA INTERFACE

The WBD instrument utilises two separate paths for transferring frames of digitised waveform data to the spacecraft data-handling system. The primary path supports real-time acquisition of WBD data by the NASA DSN. The secondary path supports non-real-time data acquisition via the spacecraft solid-state recorder.

3.5.1. *Real-Time Data Acquisition*

A special serial data interface between the WEC and the spacecraft Central Data Management Unit (CDMU) supplies the primary path for data from the WBD instrument. The interface functions consist of a 220 kHz sampling clock, a timing pulse, and the data output. These interfaces are all redundant. Data are present at this interface whenever WBD is powered. During real-time data acquisition (TDA Mode 8), the WBD data appears on a dedicated virtual telemetry channel (VC5), embedded in the 1096-byte data field of the standard 1279-byte transfer frame. The WBD transfer frames are acquired by a DSN receiving station.

3.5.2. *Non-Real-Time Data Acquisition*

The secondary path for the WBD data is provided by a serial interface between WBD and DWP. This interface supports a non-real-time time mode of data collection (TDA Mode 5.2, or BM2) dedicated mainly to WBD. When the BM2 operational mode is enabled, WBD data are transferred to DWP at 220 kbit s^{-1}, and the DWP, in turn, reduces the wide-band data by a factor of three either by digital filtering or duty cycling (accepting only one out of three frames). At the new average bit rate of 73 kbit s^{-1}, the WBD data are transferred to the On-Board Data-Handling (OBDH) system for recording and subsequent playback using the Solid-State Recorder (SSR).

4. Hardware Calibration and Performance

A detailed baseline calibration was performed for each WBD model (including the Flight Spare) at the unit level. These unit-level measurements, designed to establish the absolute calibration of the instrument, included amplitude/frequency responses for all filters and conversion modes, noise levels for all bands, characterisation of non-linear effects, and response over temperature. Each of these calibrations was carried out after final assembly of a given flight model in accordance with a fixed procedure so that any differences in performance from unit to unit could be identified. Since WBD utilises signals from sensors provided by other experiments (EFW and STAFF), each WBD unit-level calibration was augmented by special WEC-level testing carried out during the WEC flight model integrations performed in Velizy, France. Finally, in order to provide true end-to-end performance verification of the WBD flight units, additional calibration data were obtained during system level integration/test for comparison with measurements taken earlier.

Various WBD performance criteria were established at the beginning of the hardware design phase, based mainly on the characteristics of signals expected in the regions of the magnetosphere traversed by the Cluster orbit (see Pedersen *et al.*, this issue). In particular, the expected signal strengths as a function of frequency established requirements for receiver sensitivity and dynamic range.

Based on results from calibration and other testing, the measured performance for the WBD flight models is considered to be excellent, consistent with design goals and predicted capability. In the baseband modes (no frequency conversion) signals of 5 μV (4×10^{-18} V^2 m^{-2} Hz^{-1}, assuming 100 m sphere-to-sphere separation and a bandwidth of 10 kHz) can be detected. In the frequency conversion modes (125 kHz, 250 kHz, 500 kHz) signals of 10 μV (1.5×10^{-17} V^2 m^{-2} Hz^{-1}) can be detected. With the combined receiver and AGC design, signals of 1.5 V can be measured without saturation effects, giving a dynamic range of over 100 dB. Considerable effort went into making the WBD flight units as identical to one another as possible. The result is that frequency/amplitude response characteristics of the WBD flight models are the same to within about 1 dB of accuracy. End-to-end (system level) measurements of the background noise floor are the same to within 2–3 dB. Variations in the background noise floor from unit to unit are attributed to small differences in the EMC environment associated with each individual spacecraft.

5. WBD Operations

Although WBD can be commanded to a large number of internal configurations (antenna, bandwidth, gain select, etc.), the most significant WBD operational aspect relates to the manner of telemetry acquisition. Because the wide-band approach carries a basic requirement for high data rates, WBD data (other than housekeeping parameters) do not appear in the low rate telemetry. Rather, two special high-data-rate telemetry acquisition modes were implemented to support WBD operations. As discussed earlier, these are acquisition mode TDA 8, which involves real-time data reception by a DSN ground station, and acquisition mode TDA 5.2, which is a special record mode (BM2) dedicated to WBD. The use of these modes is constrained by various operational considerations, including the availability of appropriate ground-based receiving antennas.

During the Cluster mission, the primary mode for acquiring WBD data will be the single-spacecraft TDA 8 mode of operation using the DSN. By agreement, DSN will provide a minimum of 2 hours of coverage per spacecraft per 57-hour orbit. Also, to avoid interference with ESOC operations, the Madrid DSN station will not be used on a regular basis, since it resides at roughly the same latitude as the ESOC Redu and Odenwald stations. Hence, the bulk of the WBD coverage will be provided by the Canberra and Goldstone DSN stations. Since the TDA 8 mode utilises the entire bandwidth of the downlink, data from the other Cluster experiments must be recorded on the solid-state recorder during the DSN acquisition periods. Multi-spacecraft DSN operations will be performed in support of the interferometry objective discussed earlier. Given the logistical difficulty of the associated scheduling task, these specialised interferometer operations will be carried out relatively infrequently (on the order of 1 hour per month). In the event that simultaneous DSN acquisition using four spacecraft cannot be accomplished due to scheduling or other constraints, the dedicated BM2 record mode will be used. However, since BM2 usage results in the acquisition of only small amounts of data from experiments other than WBD, by agreement BM2 will be used only a few times during the two-year Cluster mission.

Since WBD is part of the WEC, WBD mission operations must take place within the context of WEC operations. As is the case for all aspects of WEC operations, the Joint Science Operations Centre (JSOC) will provide the coordinating interface for DSN scheduling.

6. WBD Data-Processing

During the Cluster mission, WBD data are acquired by JPL/DSN at an average rate of about 3 Gbytes per week. These data are placed on 8 mm tape cartridges and mailed to Iowa on a timely schedule. A pre-processing step then frame-synchronises the data, resulting in a permanent WBD data archive of original waveforms residing on CD-ROM at Iowa. By agreement, duplicate CD-ROMs are prepared and distributed to the French, British, and Scandinavian Cluster data centres on a regular basis. WBD BM2 data and WBD housekeeping data are contained in the raw data medium CD-ROMs distributed by the Cluster Data Disposition System (DDS).

As discussed in Pedersen *et al.* (this issue), the Interactive Science Data Analysis Tool (ISDAT) is the primary software package used by the WEC for detailed data analysis. For WBD, this requires the development of database handler front-end software so that ISDAT can read the TDA mode 8 and BM2 data. Likewise, ISDAT client application routines that process and display WBD-specific data products are to be completed prior to launch.

7. Summary

WBD instruments are included on all four of the Cluster spacecraft. Performance of the wide-band receivers has been determined by test to be very good in terms of both sensitivity and dynamic range, easily meeting design expectations. Although WBD has a relatively low operational duty cycle ($\sim$ 4% of orbit coverage per spacecraft), the WBD waveform measurements are expected to contribute significantly to WEC science, and thereby to the attainment of the Cluster Mission science objectives.

Acknowledgements

The authors wish to thank the engineers and technicians at The University of Iowa who participated in the Cluster WBD hardware effort, particularly Mr Michael Mitchell and Ms Bao-Tram Pham Chamberlain. The authors also wish to acknowledge the important contributions of their close colleague and friend, Mr Les Woolliscroft, who died in March of 1996. This work was carried out under NASA contracts NAGS5-30365 and NAS5-30730.

References

Ashour-Abdalla, M. and Kennel, C. F.: 1978, 'Nonconvective and Convective Electron Cyclotron Harmonic Instabilities', *J. Geophys. Res.* **83**, 1531.

Baumback, M. M., Gurnett, D. A., Calvert, W., and Shawhan, S. D.: 1986, 'Satellite Interferometric Measurements of Auroral Kilometric Radiation', *Geophys. Res. Letters* **13**, 1105.

Gurnett, D. A., Anderson, R. R., Scarf, F. L., Fredricks, R. W. and Smith, E. J.: 1979, 'Initial Results from the ISEE 1 and 2 Plasma Wave Investigation', *Space Sci. Rev.* **23**, 103.
Gurnett, D. A., Persoon, A. M., Randall, R. F., Odem, D. L., Remington, S. L., Averkamp, T. F., DeBower, M. M., Hospodarsky, G. B., Huff, R. L., Kirchner, D. L., Mitchell, M. A., Pham, B. T., Phillips, J. R., Schintler, W. J., Sheyko, P., and Tomash, D. R.: 1995, 'The Polar Plasma Wave Instrument', *Space Sci. Rev.* **71**, 597.
Helliwell, R. A.: 1965, *Whistlers amd Related Ionospheric Phenomena*, Stanford University Press, Stanford, CA.
Kennel, C. F. and Petschek, H. E.: 1966, 'Limit on Stably Trapped Particle Fluxes', *J. Geophys. Res.* **71**, 1.
Kintner, P. M., Kelley, M. C., and Mozer, F. S.: 1978, 'Electrostatic Hydrogen Cyclotron Waves Near one Earth Radius Altitude in the Polar Magnetosphere', *Geophys. Res. Letters* **5**, 139.
Pedersen, A. *et al.*: 1996, this issue.
Persoon, A. M., Gurnett, D. A., and Shawhan, S. D.: 1983, 'Polar Cap Electron Densities from DE-1 Plasma Wave Observations', *J. Geophys. Res.* **88**, 10 123.

THE DIGITAL WAVE-PROCESSING EXPERIMENT ON CLUSTER

L. J. C. WOOLLISCROFT, H. ST. C. ALLEYNE, C. M. DUNFORD, A. SUMNER, J. A. THOMPSON, S. N. WALKER and K. H. YEARBY
The University of Sheffield, U.K.

A. BUCKLEY and S. CHAPMAN
University of Warwick, Coventry, U.K.

M. P. GOUGH
University of Sussex, Sussex, U.K.

THE DWP CO-INVESTIGATORS *

Abstract. The wide variety of geophysical plasmas that will be investigated by the Cluster mission contain waves with a frequency range from DC to over 100 kHz with both magnetic and electric components. The characteristic duration of these waves extends from a few milliseconds to minutes and a dynamic range of over 90 dB is desired. All of these factors make it essential that the on-board control system for the Wave-Experiment Consortium (WEC) instruments be flexible so as to make effective use of the limited spacecraft resources of power and telemetry-information bandwidth. The Digital Wave Processing Experiment, (DWP), will be flown on Cluster satellites as a component of the WEC. DWP will coordinate WEC measurements as well as perform particle correlations in order to permit the direct study of wave/particle interactions.

The DWP instrument employs a novel architecture based on the use of transputers with parallel processing and re-allocatable tasks to provide a high-reliability system.

Members of the DWP team are also providing sophisticated electrical ground support equipment, for use during development and testing by the WEC. This is described further in Pedersen *et al.* (this issue).

1. Introduction

The Cluster mission employs four satellites to study small-scale structures (of the order of a few ion Larmor radii) in three dimensions in the Earth's plasma environment (Schmidt *et al.,* 1988; Schmidt and Goldstein, 1988). It is the first of the 'cornerstone' missions in ESA's 'Horizon 2000' programme and includes the experiments of the WEC (Roux and de la Porte, 1988). The other individual component instruments of the WEC, as shown in Figure 1, are described in companion papers in this volume (STAFF, Cornilleau-Wehrlin *et al.*, EFW, Gustafsson *et al.,* WHISPER, Decreau *et al.,* WBD, Gurnett *et al.,)* and in previous papers (Cornilleau-Wehrlin *et al.,* 1988, Gustafsson *et al.,* 1988, Decreau *et al.,* 1988, Gurnett *et al.,* 1988, Woolliscroft *et al.,* 1988, Woolliscroft *et al.,* 1992). This paper first discusses briefly the specific scientific objectives which served as design

* The DWP team deeply regrets the untimely loss of three colleagues, Les Woolliscroft, Dyfrig Jones, and Peter Christiansen.

Space Science Reviews **79:** 209–231, 1997.

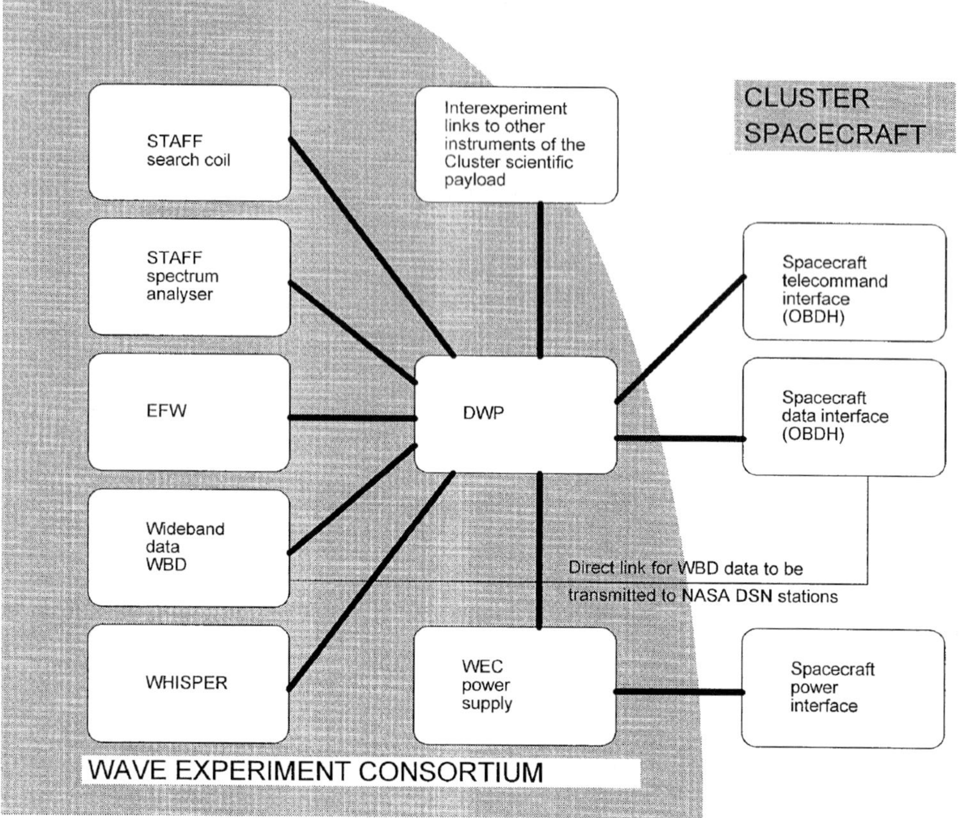

Figure 1. A simplified block diagram of the Wave Experiment Consortium (WEC).

objectives for the DWP instrument. The rest of the paper is largely devoted to a technical description of the instrument.

The DWP hardware was built at the University of Sheffield and most of the embedded software has been provided by the University of Sheffield team. Other co-investigators (see Table I) have provided design advice, software specifications, responsibility for other parts of the WEC or theoretical help.

2. Scientific Objectives which Served as Design Constraints for DWP

2.1. DWP OVERVIEW

DWP is responsible for coordinating WEC operations at several levels. At the lowest level, DWP provides electrical signals to synchronise instrument sampling. At higher levels, DWP time tags data in a consistent manner and provides a facility for constructing more complex WEC modes by means of macros. Several

Table I
Current DWP investigators

H. St. C. Alleyne	The University of Sheffield, UK
M. A. Balikhin	
K. H. Yearby	
D. Burgess	QMWC, UK
S. Chapman	University of Warwick, UK
N. Cornilleau-Wehrlin	CETP-Velizy, France
A. Roux	
P. M. E. Decreau	CNRS-LPCE, France
V. V. Krasnosel'skikh	
F. Lefeuvre	
M. Parrot	
A. Egeland	University of Oslo, Norway
M. Gedalin	Ben Gurion University, Israel
M. P. Gough	University of Sussex, U.K.
D. Gurnett	University of Iowa, U.S.A
G. Gustafsson	Uppsala, Sweden
G. Holmgren	
C. C. Harvey	Meudon, France
R. Horne	British Antarctic Survey, UK
I. B. Iversen	SPW, Birkerod, Denmark
W. Kofman	CEPH AG, France
H. C. Koons	Aerospace Corporation, U.S.A.
J. A. LaBelle	Dartmouth College, U.S.A.
F. Mozer	University of California, U.S.A.
A. Reznikov	Izmiran, Russia

The DWP team deeply regrets the untimely loss of three colleagues, Les Woolliscroft PI until his death on 12 March 1996, Dyfrig Jones and Peter Christiansen.

particular aspects of DWP, namely, fault tolerance, flexible data-handling including data compression and the implementation of a particle correlator, timing and mode control are described.

2.2. Time Resolution and Timing

The transducers, i.e., antennas, are capable of measuring the parameters of the wave field over all frequencies of interest to space plasma physicists. The study of polarizations and phases of the waves requires accurate timing of the data. The acquisition of Cluster data by the on-board data-handling system are normally timed with an accuracy of ± 2 ms with respect to UTC for each spacecraft. Data from two spacecraft can then be compared with an accuracy of ± 4 ms. Within DWP the timing is more accurate and certain time differences may be measured to a few tens of microseconds by comparing the time of the active edge of the DWP

Table II
DWP data compression in high bit rate

Instrument	Raw bit rate	Compressed bit rate
Wideband	220 kbit s^{-1}	74 kbit s^{-1}
STAFF SC	22 kbit s^{-1}	16 kbit s^{-1}
WHISPER	$\approx$ 600 kbit s^{-1}	3 kbit s^{-1}

900 Hz clock with the spacecraft reset pulse (Woolliscroft *et al.*, 1992). This time is measured using a transputer timer to a precision of 1 μs, but the accuracy of each individual measurement is limited by the interrupt latency of the transputer. The question of time resolution led to the need for data compression.

2.3. Data compression

A problem common to most space instruments is that of restrictions on the available telemetry bandwidth. This can be particularly severe for wave experiments, where the mismatch between information and telemetry bandwidths can be several orders of magnitude. Data-compression techniques are employed in the DWP instrument to optimise the use of available telemetry.

Data compression permits more useful information to be transmitted over a given telemetry system than would otherwise be possible. This is achieved by removing redundant information from the data. Various data compression methods are currently being implemented within the WEC and DWP. To simplify allocation of telemetry bandwidth, these are restricted to methods providing a fixed degree of compression independent of the variability of the data. The methods used for the WEC are:

(a) Wideband instrument – digital filtering with resampling,
(b) STAFF Search Coil (MWF) – differential encoding,
(c) WHISPER – data selection followed by pseudo-logarithmic compression,
(d) Correlator – averaging.

The compressions obtained (in high-bit-rate mode) are listed in Table II.

2.4. Particle correlation

This novel diagnostic technique is based on forming auto-correlation functions of the time series of particle detector counts as a function of energy and pitch angle. The basic operations will be carried out in DWP resident software using algorithms based on those developed for the previous AMPTE and CRRES missions, rocket experiments and also for computer simulations.

The particle correlator technique permits:

(a) the detection of particle flux bursts on time scales short compared with the energy and dwell time, and

(b) indication of regions of velocity space in which wave/particle interactions are occurring.

2.5. Mode control for WEC

The various WEC instruments are capable of very many modes. These have been designed to be flexible and to allow for good science to be conducted in different regions of the space plasma.

The detailed choice of mode for each of the WEC instruments is quite complicated and will depend on, for example, the availability of NASA's Deep-Space Network (DSN) and whether the spacecraft data-handling system is being operated in a high data rate (often called HBR or BM1). The choice of data rate is selected (within many constraints) by the Science Working Team so as to make the most effective use of the on-board data recorders. The data system has limited capacity and may handle only about half of the data per orbit which can be collected at normal data rate or one twelfth of the data if the high data rate is used throughout.

DWP provides the control for the WEC. The normal way in which the WEC will operate is through macro commands which are issued by DWP to the other WEC instruments. The macro command sequences correspond to the WEC modes and are described more fully in Woolliscroft *et al.* (1992). In order to have a reasonably consistent set of data throughout the mission the number of WEC modes has been restricted. Table II in Pedersen et al. (this issue) gives a summary of currently defined WEC modes. Other modes can be created by uplinking new macros to the spacecraft. Most of the macros defined for the implementation of WEC modes will loop indefinitely or until a stop command is received.

3. DWP Hardware Design

The DWP instrument is based on three T222 transputers in a distributed fault-tolerant architecture. This architecture might be termed a federated architecture and is illustrated in Figure 2. It is based on an instrument bus to which each processor module is connected.

The interfaces to other parts of the WEC and the spacecraft are attached to this bus and hence any processor can handle any task. Redundant interfaces are provided to the telemetry and command systems of the spacecraft and the Inter-Experiment Link (IEL). Redundant power supplies within the WEC power supply are provided exclusively for DWP. Resource limitations (both for DWP and the other WEC instruments) have forced non-redundant interfaces within the WEC. The electronic circuits are mounted on three circuit boards and, together with housing and connectors, have a mass of approximately 2 kg.

The three processor modules are interconnected by links to provide inter-process communications.

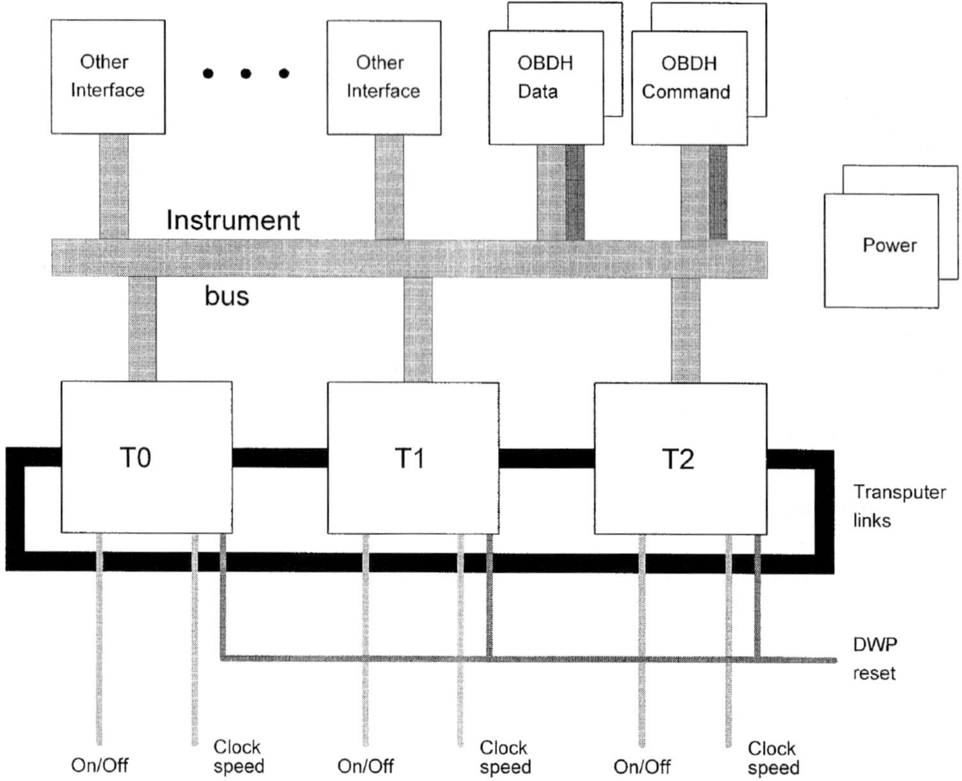

Figure 2. A block diagram of DWP where T0, T1, and T2 are the transputer processor modules. The 'Other Interfaces' are to the WEC instruments and the interexperiment link as indicated in Figure 1.

The Engineering Model is shown in a semi-assembled form in Figure 3. In this photograph the printed circuit boards are attached to a backplane to permit testing out of their mechanical housing. Table III gives an approximate mass breakdown and outline specification of the DWP experiment.

3.1. The Processor

A T222 transputer with 32 Kbytes of external RAM (internal memory is disabled to provide increased radiation tolerance) and 32 Kbytes of PROM forms the core of each module. Each PROM contains the control program. In addition, each module has an HS82C37A DMA controller, event multiplexer and instrument bus interface. A 16-bit transputer is the microprocessor chosen in preference to a 32-bit device. The main reasons for this are that fewer RAM and ROM chips are accessed in one cycle, thus reducing power consumption, and most instruments use a 16-bit data length. A programme of test irradiation of the T222 transputer was carried out in the test facilities at ESTEC. The results established its radiation hardness to

Figure 3. In this photograph of the engineering model, EM, of DWP, the three printed-circuit boards are attached to a temporary and non flight backplane to permit testing when not in the mechanical housing.

Table III
The main characteristics of DWP flight models

	Mass *
Mechanical structure etc.	550 gm
Electronics	1500 gm
Total	2050 gm
	Power
1 processor (half speed)	620 mW
3 processors (half speed)	1410 mW
	Computing Power
3 processors (average)	30 Mips
3 processors (peak)	60 Mips

* Figures given for EM.

>30 krad and thus its viability for the Cluster mission where the DWP electronics is expected to experience <15 krad.

Power-down circuitry enables processor modules to be switched off to conserve power. Series resistors or tri-state buffers are employed to ensure that the inputs of powered-down circuits are not driven. Processor modules are powered down by a hardware-decoded command from the On-Board Data Handling (OBDH) interface.

The design of the DWP permits the transputers to be operated at input clock frequencies of 2.5 or 5 MHz, the lower rate requiring less power. The clock frequency select signal is applied to the transputer link speed input (as well as to the clock divider) in such a way that the links always operate at the standard speed of 10 Mbit s^{-1}. This allows processor modules operating at different speeds to communicate. Speed selection is made by hardware command from the OBDH interface.

3.2. The Instrument Bus and Interfaces

The instrument bus permits 16-bit parallel communication between the processor modules and the instrument interfaces. Events and Direct Memory Access, DMA, transfers are also requested over this bus. Only one transputer (chosen at boot time) is allowed access to the instrument bus (Woolliscroft *et al.*, 1995). This removes the requirement for any complex instrument bus arbitration logic. However, simple logic is provided to prevent damage to DWP if, owing to some fault, more than one processor does try to drive the instrument bus simultaneously.

A variety of interfaces are used to control and obtain data from the instruments of the WEC. Interfaces with a relatively low data rate employ registers which are allocated one of the 16 instrument bus addresses and in some cases an event request line. High-data-rate interfaces use the DMA channels.

3.3. The spacecraft OBDH interface

The OBDH interface contains telemetry, telecommand, and service signal circuitry, duplicated for telemetry and telecommand to provide redundancy in case of failure. The telemetry interface is the route for all data leaving the DWP. A DMA channel is used to provide high data throughput. Data may be output on either of the redundant interfaces, as requested by the spacecraft OBDH.

The telecommand interface enables the DWP software to receive commands from the spacecraft OBDH. Separate command-registered and interrupt request lines are assigned to the two redundant parts of this interface. An extension of this interface (hardware commands) permits the OBDH to command parts of the DWP hardware directly without software intervention (processor speed and power on/off).

3.4. Watchdog timer

Watchdog timer hardware is interfaced to the instrument bus. The kernel process should signal the watchdog at regular intervals; if it fails to do so for any reason, the watchdog will reset all three processors. The reset inputs to the processor modules are edge-sensitive to ensure that DWP could still function if the watchdog failed with its output in the active state.

4. Software Design and WEC Functions

4.1. Operation of the WEC instruments

DWP is responsible for coordinating WEC operations at several levels. At the lowest level, DWP provides electrical signals to synchronise instrument sampling. At higher levels, DWP time tags data in a consistent manner and provides a facility for constructing more complex WEC modes by means of macros. These facilities are briefly described in the following sections.

4.2. EFW and STAFF search coil (MWF)

DWP has an internal clock running at a fixed frequency of 900 Hz. Pulses of this clock are counted by software which derives either a 25 Hz or a 450 Hz signal known as the WEC Sample Sync (WECSS). This controls STAFF Magnetic Waveform Analyser (MWF) and EFW electric field sampling and ensures that sample taking is synchronised to this clock. The WECSS clock is not synchronous with the spacecraft's spin rate.

DWP has direct digital control of STAFF MWF by means of a number of control signals that enable DWP to select the $\mathbf{B}_x$, $\mathbf{B}_y$, or $\mathbf{B}_z$ field components and to control the analogue to digital converter, ADC. DWP reads $\mathbf{B}_x$, $\mathbf{B}_y$, and $\mathbf{B}_z$

values to 16-bit resolution at the frequency of WECSS and assembles the values in a buffer within DWP. By contrast, EFW has its own microprocessor that fills a sampling buffer within EFW and, one per second, outputs a data packet to DWP through a serial interface (uart). To facilitate time correlation of EFW and STAFF MWF data, DWP samples one second's worth of MWF data before outputting it to the spacecraft. Furthermore, DWP controls the start of EFW sampling (by means of the transmission of a synchronisation character) so that EFW and MWF sampling start at the same time and both generate a complete packet of data one second later. DWP ensures that both of these packets carry the same time tag so as to aid subsequent correlative studies.

It can be seen that for EFW and STAFF operations there is a one-second period where sample buffers are being filled before they are output. DWP delays any telecommanded mode changes for EFW and MWF until the sample buffers have been flushed of the samples taken during the last second. This point is termed the One Second Timing Boundary (OSTB), and it will be shown later that all WEC instrument mode changes are synchronised to OSTBs to ensure continuous time coverage in known instrument modes.

DWP compresses STAFF MWF data using a bit saving differential algorithm.

4.3. STAFF SPECTRUM ANALYSER

The STAFF Spectrum Analyser (SA) instrument contains a Digital Signal Processing (DSP) chip and has its own analogue interfaces to the STAFF search coils and EFW. DWP can be programmed to start an SA analysis at set intervals. These analyses usually take one second or multiples thereof. DWP can then read the results of the analysis whilst SA does analysis on the next interval of data.

DWP transmits Start Analysis commands to SA synchronous with the OSTBs. Thus synchronisation with STAFF MFW and EFW is guaranteed.

4.4. WHISPER

Sampling by the WHISPER instrument is controlled by the WHISPER Sample Synchronisation (WHSS) signal which is generated by DWP and is a 75 Hz signal synchronous with the WECSS. On every pulse of this signal, WHISPER may take up to 1024 samples and uses a dedicated DSP to perform FFTs over these samples and to compute the modulus of the FFTs. This can result in a maximum data rate to DWP of 75 frames of 512 16-bit values per second, which is equivalent to 615 kbit s^{-1}.

In nominal functioning, the WHISPER output frame is read and compressed by the DWP according to the processing strategy defined in the WHISPER mode selected by telecommand. The DWP compression treats each part of the WHISPER data frame according to a fully determined process, so that the volume output is known for each processing option. This method permits the discarding of a large amount of data, in order to achieve the severe compression required: the WHISPER FFT frames, available every 13.33 ms, correspond to a flow of 615 kbits s^{-1} in the 512-bin FFT option, whereas WHISPER is routinely allocated less than 1kbit s^{-1}.

4.4.1. *Natural Modes*

The high compression needed is obtained in three main phases. First, the useful parts of successive spectra (480 bins in the 512-bin FFT option) are accumulated bin to bin to form a single 'accumulated' spectrum. The energy information may optionally be treated in the same way. When it is not accumulated, this data provides information about the variability of high frequency emissions, at a time resolution of 13.33ms (up to 30 Hz in the frequency domain). In a second phase, each accumulated bin , which is 32-bit data (24 bits useful), is compressed in a quasi logarithmic way, within a word of 6 or 8 bits. The compression may be fixed or dynamically adjusted. In the fixed compression case, the number of bits for coding respectively an exponent part and a fraction part of a word are fixed (in the 8-bit word case, the exponent part of the word is coded on 5 bits, the fraction part on 3). In the dynamic compression case, the exponent part is adjusted to the size of the word. The energy information, a 32-bit data word, is compressed to 8 bits according to a similar algorithm. The third phase consists simply in discarding accumulated spectra, according to a selection criterion. The status (command words, number of overflows, gain changes...), treated separately, are part of the WHISPER science data flow.

The internal WHISPER processing in natural mode reflects the first phase only: it is the bin to bin averaging of 8 (256-bin FFT option) or 16 (128-bin FFT option) spectra.

4.4.2. *Sounding Modes*

The first phase of the processing is to sort out the 'active' FFTs and 'passive' FFTs (as above). A large part of the FFT WHISPER frames is thus thrown out. The TM rate is so severely limited that we abandoned the idea of studying the decay properties of the resonances. The active to passive ratio taken within a short time slot is thought to be a sufficient clue for resonance identification. A second phase of the processing is optional (strategy A). It consists in grouping the bins in packets of 2 or 4 consecutive in frequency. One bin (the largest one) is selected in each group. If the strategy A is chosen, the position of the bin in the group, and also the active-to-passive ratio, coded on 4 bits, are included in the data flow. In a strategy B, the bin selection is not implemented, but the active-to-passive ratios are coded on 4 bits, which is not done in strategies C (where the complete passive and active information is kept) and D (where all the active information is kept). The third phase is similar to natural mode. The active bins levels (after the eventual selection) are compressed from a 32-bit data word to a 6- or 8-bit data word. A fourth phase is optional. It consists of the reduction of the passive information: 6 bins consecutive in frequency are accumulated. A last phase takes place when the passive information is transmitted (i.e., when the passive reduction is asked for and in strategy C). It consists in the quasi-logarithmic compression of passive levels in the same way as active levels. The resulting data are called the 'passive reduced' information, or simply the passive information (strategy C). The status (command words, number of overflows, gain changes...) are treated separately. No energy information is transmitted.

The WHISPER internal processing is available only for one specific sounding mode (fixed frequency table and step duration). It performs the first phase of the compression, only for the active bins.

When WHISPER is in a sounding mode, DWP controls the timing of the resonance sounder transmission by means of a Blanking Pulse signal synchronous with the WHSS. This signal is also transmitted across the IEL interface and to other WEC instruments to warn them that WHISPER sounding is taking place. This may be useful in the event that electromagnetic interference from the sounder transmitter is suspected. DWP can also synchronise WHISPER sounding with the spacecraft spin if required. The significance of spin synchronisation is that it may be used to minimise any possible interference. However, this imposes limitations on WHISPER operations and is not the nominal mode. WHISPER mode changes (e.g., passive to sounding) are synchronised with OSTBs to ensure synchronisation with EFW and STAFF.

4.5. WBD

Wideband Data (WBD) has direct connections to the spacecraft data handling system and this is the path that is usually used for data transmission. Sampling and output are not controlled by DWP. WBD data may also be routed to one of the two spacecraft tape recorders through the DWP interface to the spacecraft. To reduce the bit-rate to fit within the WEC science telemetry allocation, DWP reduces the data rate by a factor of three by applying a digital filtering algorithm to this data stream and then resampling it.

4.6. Particle correlator

The DWP particle correlator takes raw electron detection pulses from the PEACE instrument (Johnstone *et al.,* 1993) and performs software auto-correlation functions (ACF) that are sorted and summed according to instantaneous PEACE selected electron energy.

The energy level range is partitioned into 15 contiguous energy bins irrespective of the PEACE energy sweeping rate. ACFs have 8 lags including zero lag. As the lag time is 45 μs this corresponds to modulation frequencies up to 11.1 kHz with only low, 1.4 kHz resolution.

The basic summation period is one spin at normal rate, and 1/32 spin at high data rate. The data rate is limited to transmitting the summed ACF at two selected energies per summation period, plus a third ACF that is stepped in energy once per summation period, through the other 13 levels. Thus in normal mode two selected energies are covered with a time resolution of 4 s, while other energies have a time resolution of 52 s. In high data rate these periods are 0.125 s and 1.625 s, respectively. Note that in normal mode the data corresponds to a range of electron pitch angles, being summed over spins. Assuming that the analysed zone includes the direction parallel to the magnetic field, **B**, then this range in pitch angle is in general, 0 degrees to an angle twice that between the spin axis and the Earth's magnetic field direction. Thus operation in high data rate will be needed to provide the pitch angle dependence of particle modulations. The low data rate mode will obviously operate best when the spin axis is close to **B**.

4.7. Time Tagging of WEC Data

DWP time-stamps the WEC data with a time tag relative to the spacecraft on-board time, which is approximately as accurate as the timing given by the spacecraft systems. As is shown in (Woolliscroft *et al.,* 1992), higher accuracy for the data is available from DWP in certain circumstances.

STAFF and EFW data are stamped with the time of the OSTB at the start of the sampling period. This facilitates time correlation between data from these instruments. WBD and WHISPER data are stamped with the time at which the data packet was received by DWP. From this, and from an understanding of the internal timing within the instrument, it is possible to determine when the samples were actually taken.

4.8. WEC Macros

An area of DWP memory is allocated as a buffer for WEC macros. The WEC macro buffer can hold a number of command sequences that can be invoked with a single telecommand from the spacecraft. The WEC macros exist for two main reasons:

(i) The number of telecommands that can be uplinked to the spacecraft in any orbit is limited. By storing the most commonly used sequences of commands in the macro-buffer, efficient use can be made of the spacecraft telecommanding allocation.

(ii) Commands can be inserted into the macro buffer to enable features such as looping and timed delays to implement automatic WEC instrument mode switching. These permit the construction of more scientifically interesting WEC modes.

DWP holds 256 instrument commands in a part of the program PROM. Table IV shows the initial content of the FM ROMs showing which WEC modes are located in ROM. When DWP is powered on in flight, these are copied into the DWP volatile RAM. These sequences may be modified at any time by telecommand. When DWP is powered off, any modified macro sequences in volatile memory are lost. This

Table IV
Default WEC macros

Slot	General use	ROM definition
31	All off subroutine	All off subroutine
30	Leave STANDBY subroutine	EFW, SA, SC, WH, WH TX on subroutine
29	EFW, SA, SC on subroutine	EFW, SA, SC on subroutine
28	Enter STANDBY subroutine	Spare/scratch space
27	Spare/scratch space	Spare/scratch space
26	Spare/scratch space	Spare/scratch space
25	General	
24	NBR	Gliding/
23		NBR Spin Sync (a)
22	General	
21	HBR	
20	HBR EFW (a)	HBR EFW (a)
19		
18		
17	NBR Langmuir	NBR Langmuir (a)
16		
15		
14	NBR Continuous	NBR Continuous (a)
13		
12		
11	HBR basic	HBR basic
10		
9		
8	HBR low recurrence	HBR low recurrence (a)
7		
6		
5	NBR basic	NBR basic
4		
3		
2	NBR low recurrence	NBR low recurrence (c)
1		
0	Default emergency mode	Default emergency mode

buffer space is currently divided into 32 slots of 8 commands. A macro command sequence must start at the beginning of a slot, but may overrun into following slots. Macro primitives are available as listed in Table V. With these primitives, it is possible to build nested loops and timed delays.

Table V
DWP macro primitives

Commands available	Parameter
Call macro	slot number (0–31)
Terminate macro	None
Load macro slot	Slot number (0–31)
Edit macro telecommand	Command number (0–255)
Short wait	Delay in seconds (0–63)
Long wait	Delay in minutes (0–63)
Unconditional jump	Backwards offset (1–64 commands)
Set counter-n	Counter select (0–3) and value (1–64)
Conditional jump	Jump back 1–64 instructions if counter n is not zero

4.9. Example of a WEC macro command sequence

To give an example of the application of WEC macros, the implementation of a proposed nominal WEC mode when a high telemetry bit rate to the spacecraft is available is described below. For the sake of clarity, the sequence of commands is simplified slightly.

This mode has a cycle of 28 s. For the first 2.85 s of this cycle, WHISPER performs a sounding sweep measuring induced plasma resonances. For the last 25 s, WHISPER is in passive mode in which it acts as a receiver measuring the natural emissions. Whilst WHISPER is sounding, SA performs a single spectrum analysis (duration 4 s) with three magnetic field components. Whilst WHISPER is passive, SA performs a five-component (two electric and three magnetic) spectrum analysis. As can be seen from this example, which is detailed in Table VI, the macro will run indefinitely or until the Terminate Macro command is received. An important point to note is that any time whilst a macro command is being executed, additional telecommands may be transmitted to DWP from the spacecraft. It is assumed that the command sequence will be vetted before transmission, to ensure that it will not conflict with the execution of the macro (e.g., turning off one of the instruments in use). If this were to occur, DWP would flag warning messages in the WEC housekeeping data. The sorts of telecommand that are likely to be useful are those which adjust operational parameters such as bias current or gain settings (these are referred to sometimes as switches).

To summarise, the WEC macros provide a skeleton for each WEC mode which can be adjusted by additional telecommands transmitted from the spacecraft as may be required for any particular campaign.

Table VI
An example of a macro command sequence

Sequence	Command
1	Set EFW Fast Mode
2	Set WECSS frequency to 450 Hz
3	Leave STANDBY
4	Wait for next OSTB
5	Select WHISPER application processing (soundings)
6	Set WHISPER to sounding mode (2 commands)
7	Transmit 'Start 4-second analysis' command to SA
8	Wait for 3 OSTBs
9	Select WHISPER application processing (passive)
10	Set WHISPER to Passive mode
11	Set Counter-0 to 25
12	Wait for 1 OSTB
13	Transmit 'Start 1-second analysis' command to SA
14	Decrement Counter-0, if not zero jump to Sequence 12
15	Jump to Sequence 5

4.10. Power-on Default

A power-on default/emergency mode is defined. This mode will not be the normal scientific mode, but will yield a reasonable set of scientific data in the event of any of four possible worst case or emergency situations:

(a) low spacecraft bit-rate (a safe assumption)
(b) spacecraft power limited (again a safe assumption)
(c) no command link into DWP (perhaps due to a failure)
(d) DWP capabilities limited (perhaps due to radiation damage)

If these emergency conditions do not apply, it should be possible to command the WEC experiments into other modes which take the actual circumstances into account.

4.11. WEC Modes

The main WEC modes are described in Table II of Pedersen *et al.* (this issue). Table VII takes three of these modes and shows how they are 'expanded' to specify the operations of the various parts of the WEC. It does not, at this level, specify the switches and parameters which still need to be set for scientific operations. These are beyond the scope of this paper and are the work of the WEC Operations Group (Pedersen *et al.*, this issue).

Table VII

The cycle of measurements within the WEC for some of the main WEC Operational Modes as listed in Table II of [1]. Note that the detailed measurements in each part of these cycles must be interpreted by reference to the papers for the individual WEC instruments in the ESA SP1159 references at the end of this paper and accompanying papers in this volume.

Normal Bit Rate Basic Mode (cycle time 28 seconds)			
	First 3 seconds of cycle	or first 4 seconds of cycle	rest of cycle
WHISPER	4 - 80 kHz, 1.5 s resolution, active	4 - 80 kHz, 2.15 s resolution, passive (natural)	
STAFF spectrum analyser	3 B axes, autospectrum (1 s), cross spectra (4 s) resolution, in bands 8 - 64, 64 - 512 and 512 -4096 Hz		3 B + 2 E axes, autospectrum (1 s), cross spectra (4 s) resolution, in bands 8 - 64, 64 - 512 and 512 -4096 Hz
STAFF search coil	0.1 - 10 Hz, 3 B axes		
EFW	0 - 10 Hz, 2 E axes, with antennas in voltage mode		
Wideband	If on (depending on DSN coverage) probable mode covers 25 Hz - 9.5 kHz but others are possible		
DWP Correlator	On, details of operation depend on mode of operation of PEACE		
High Bit Rate Basic Mode (cycle time 28 seconds)			
	First 3 seconds of cycle	or first 4 seconds of cycle	rest of cycle
WHISPER	4 - 80 kHz, 1.5 s resolution, active	4 - 80 kHz, 0.32 s resolution, passive (natural)	
STAFF spectrum analyser	3 B axes, autospectrum (0.125 s), cross spectra (1 s resolution) in bands 64 - 512 and 512 -4096 Hz		3 B + 2 E axes, autospectrum (0.25 s), cross spectra (1 s resolution), in bands 64 - 512 and 512 -4096 Hz
STAFF search coil	0.1 - 180 Hz, 3 B axes		
EFW	0 - 180 Hz, 2 E axes, with antennas in voltage mode		
Wideband	If on (depending on DSN coverage) probable mode covers 25 Hz - 9.5 kHz but others are possible		
DWP Correlator	Off		
Normal Bit Rate Langmuir Mode (cycle time 104 seconds)			
	First 4 seconds of cycle	rest of cycle	
WHISPER	4 - 70 kHz, 2 s resolution, active	4 - 80 kHz, 2.6 s resolution, passive (natural)	
STAFF spectrum analyser	3 B axes, autospectrum (1 s), cross spectra (4 s resolution), in bands 8 - 64, 64 - 512 and 512 -4096 Hz, or autospectrum (1, 0.5, 0.5 s) and cross spectra (1 s resolution), in same frequency bands		
STAFF search coil	0.1 - 10 Hz, 3 B axes		
EFW	0 - 10 Hz, 1 E axis plus 2 probes in current mode sampled at 25 samples/s		
Wideband	If on (depending on DSN coverage) probable mode covers 25 Hz - 9.5 kHz but others are possible		
DWP Correlator	On, details of operation depend on mode of operation of PEACE		
Normal Bit Rate Spin Synchronized Mode (cycle time 129 spacecraft rotations)			
	0 - 129 spins		
WHISPER	4 - 68 kHz, active with sounder synchronized to spin over a 516 s cycle		
STAFF spectrum analyser	3 B axes, autospectrum (1 s), cross spectra (4 s resolution), in bands 8 - 64, 64 - 512 and 512 -4096 Hz, or autospectrum (1, 0.5, 0.5 s), cross spectra (1 s resolution), in same frequency bands		
STAFF search coil	0.1 - 10 Hz, 3 B axes		
EFW	0 10 Hz, 2 E axes or 0 - 10 Hz, 1 E axis plus 2 probes in current mode sampled at 25 samples/s		
Wideband	If on (depending on DSN coverage) probable mode covers 25 Hz - 9.5 kHz but others are possible		
DWP Correlator	Normally off		

5. Fault Tolerance

During operational life there are two main sources of problems: the first is collisions with micrometeoroids and space debris (which, for the orbit of Cluster, is a small problem); the second is radiation damage. The latter can have three consequences: latch-up followed rapidly by thermal run-away and destruction of a device; a Single Event Upset (SEU) which causes the value of a register to be reversed (this is a non-permanent effect); and premature ageing. Of these three, SEUs are the hardest to combat by circuit design and component selection, especially in the case of volatile memory devices. Since the executing program and data are stored in such devices, this constitutes a potentially serious problem. In the next section, we explore the techniques used to deal with SEUs (or transient errors), and premature ageing which leads to a permanent failure of a device.

5.1. Designing for Reliability

Conventionally, the techniques for reliability design consider hardware and software as distinct topics. Hardware reliability design uses techniques such as Failure Mode Event Criticality Analysis to identify critical areas (often these are single-point failures). Software reliability frequently uses exception handling and related concepts. Of course both concentrate on good design and implementation practice as well (for example, derating components, high-quality manufacturing and modular software). We have attempted to consider the system reliability and functionality independently of the hardware and software in which it is implemented, although the traditional approaches have been used as well.

Possible faults, whatever their cause, may conveniently be classified according to the required recovery procedure. Five classes have been identified as in Table VIII and are discussed below.

5.2. Faults that May Be Ignored

Faults that fall into this class may be caused by SEUs or some software bugs. The corruption, by an SEU or bug, of a memory word that contains only sampled science data will cause no observable upset in operation and will have no serious consequences if only a single word is corrupted.

5.3. Software Protectable Faults

Faults that fall into this class may again be caused by SEUs, some software bugs and faults in other WEC instruments. The corruption, by an SEU or bug, of a memory word that contains certain kinds of program variable can be protected against by a defensive style of programming (e.g., always checking variable for valid contents). The operation of this protection may, at worst, cause a hiccup in data output.

Table VIII

An hierarchy of faults which can be caused by radiation (or other) damage or as a result of software faults

Class	Type of fault	Typical cause	Solution employed in DWP
1	May be ignored	Software bug or some single event upsets (cosmic-ray bit flip)	Ignore or fix on the ground in the data handling system
2	Faults which can be protected by software	SEUs, software bugs or faults in other WEC instruments	Automatic checks which can power off other WEC instruments or take other corrective action
3	Restart required without loss of mode or command history	SEUs or software bugs	Watchdog reset
4	Complete restart	DWP fault requiring a reconfiguration	Restart is in default mode
5	Irrecoverable faults	Major hardware failures	Limited redundancy. Commands can be used to reconfigure DWP and allocate tasks

DWP has some automatic checks built into its software that monitor performance of WEC instruments interfaced to it. These automatic checks can be disabled from the ground if, for example, it is found necessary to cope with voltage sensor failures that do not generally affect instrument operation. Typically, if an instrument fails a check, it is automatically powered off and then powered on again a short time later. This sequence is repeated (a maximum of three times, if necessary) before DWP decides that the instrument has terminally failed and leaves it powered off. The instrument can still be powered back on by a telecommand. In the event of a catastrophic WEC instrument failure causing the operation of the instrument to endanger the operation of the other WEC instruments, a telecommand may be transmitted to inform DWP that the interface or instrument has failed. DWP will then ignore all further commands for that instrument without flagging errors in the telemetry.

5.4. Faults Requiring a Restart without Loss of Mode

Most of the remaining faults caused by corruption of memory by an SEU or software bug fall into this class. If the word holds a transputer instruction in the currently executing path, then the software will probably crash owing to execution of an illegal transputer instruction or arithmetic overflow. This will cause the watch-dog timer to trip a short time later, which will reset the DWP instrument, and restart software execution.

In general, a reset caused by a failure within DWP should not cause the mode of an external WEC instrument to change, although some telemetry data loss is allowable. The reason for this is that many commands may have been uploaded to an instrument to place it in a specific mode. In many cases, if DWP were to reset the instrument, this information could not be repeated, owing to the complications of storing large command sequences for each instrument. These command sequences might also have included code patches for those instruments that are microprocessor based. In the case of one instrument that had a moderately complex interface protocol, the software had to be carefully designed to take account of the difficulty in recovering from errors at any point in the exchange. Other instruments generate data almost independently of the DWP input and recovery is accomplished without difficulty.

When DWP software starts execution after a reset (either watchdog or power on), it examines the state of the watchdog flag. This is set if the reset was caused by a watchdog time-out. DWP then attempts to restore itself to the state it was in before the reset, without disturbing the state of the connected WEC instruments. This is done by storing vital system parameters such as mode information and recently received commands in an area of memory known as checksummed memory. This is an area of conventional volatile RAM which regularly has a checksum computed from its contents.

After a watchdog reset, DWP attempts to restore its old state by using the contents of checksummed memory. Before it can do this, it has to verify that the contents have not been disturbed by the fault. It does this by computing the checksum over the checksummed memory, and checking it against the stored checksum. In the event of corruption, DWP accesses a redundant copy of checksum memory that is kept up to date at all times.

The concept of a complete redundant copy of checksum memory is feasible, as its size is very small (about 100 words). Furthermore, most words stored in checksummed memory are only updated at a low rate (less than once per second) so the overhead of computing a checksum and maintaining a fully redundant copy is not great.

5.5. Faults Requiring a Complete Restart

The only cause of this class of fault is a DWP hardware fault that requires reconfiguration (e.g., a processor failure or latch-up). Such a fault will cause DWP to restart (in some cases a power cycling command may be required) in its default mode. All instruments are reset and information about the mode prior to the fault is lost.

5.6. Irrecoverable Faults

Irrecoverable faults are caused by major hardware failures (possibly due to radiation damage or impact) and redundancy is often used to reduce their probability by eliminating single points of failure. As was stated in Section 2, resources for full redundancy are not available to DWP. However, certain major single-point failures have been identified and redundancy used for them. They are mainly in the interfaces between DWP and the spacecraft systems. Main and redundant command and data interfaces are provided and dual power supplies are used with blocking diodes. The most vulnerable hardware is the instrument bus, which is non-redundant. The interfaces to the WEC instruments are not redundant, so a failure in one of them would be serious (but not an overall catastrophe). The three processor modules, each with a transputer, program ROM, RAM and associated circuitry, form a fault-tolerant system. Any one processor (at half speed) can run the kernel programme which provides the essential communications and system control. The other 'applications' tasks are then run in one or both of the other transputer modules.

6. Discussion and Conclusions

DWP with the Cluster WEC experiments will offer many opportunities for space plasma physics and the study of solar-terrestrial processes. The wide range of frequencies to be expected may be illustrated by Figure 4. Among the WEC capabilities are:

– The determination of those parameters which characterise plasma turbulence (distribution in the **k**-vectors) and small-scale field-aligned current structures (such as geometry, current density) from inter-spacecraft correlations of field fluctuations.

– Evaluation of magnetic vorticity, charge separation, etc.

– Wave/particle interactions, via correlations performed on-board using particle measurements from the PEACE instrument and analysis with wave data.

– Measurements of plasma density and assessment of its spatial variations.

– The evaluation of spacecraft potential both as a scientific diagnostic and an engineering parameter.

The applicability of these different measurement and analysis techniques obviously varies with the kind of data that will be considered and with the time resolution of the different instruments. The wideband data of WBD are particularly appropriate for studying high frequency wave phenomena, such as nonlinear interactions between Langmuir waves. For some detailed studies of the frequencies observed the high frequency resolution of WHISPER over a wide frequency range will be of value. The discontinous operation mode of WBD, however, prevents large ranges of time-scales from being investigated. For that reason, techniques such as the structure function are not fully applicable.

The time-series of the STAFF and EFW instruments are more relevant for the analysis of low-frequency phenomena, such as nonlinear magnetic structures, whistler waves etc. These have synchronized (on one spacecraft) measurements of the wavefield at frequencies up to 180 Hz in high bit rate and 10 Hz in normal bit

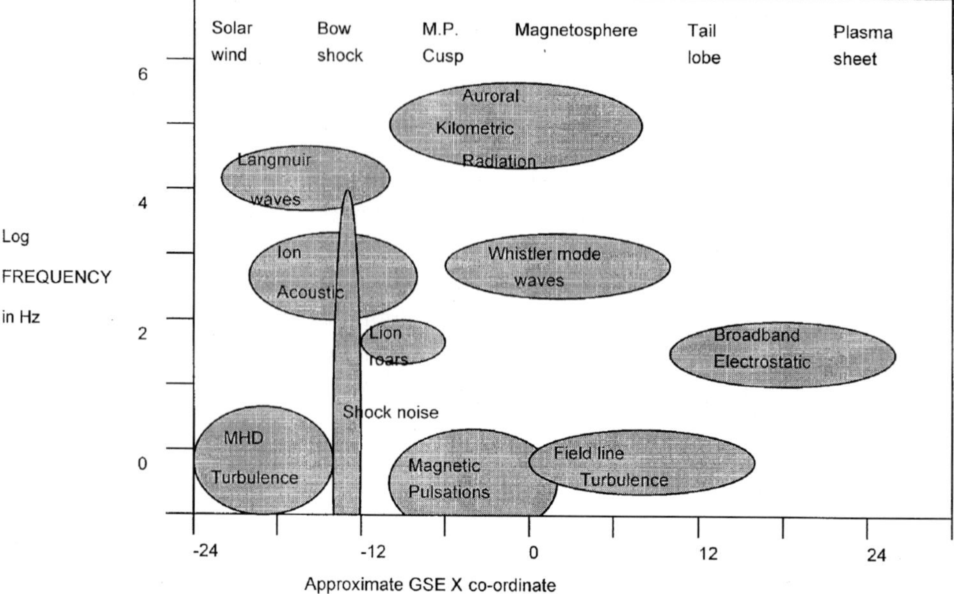

Figure 4. The main frequencies of natural waves which may be expected along a line crossing the polar region with GSE coordinates approximately (x, 0, +8). This line starts in the solar wind and goes through to the tail of the magnetosphere.

rate modes. The data from the magnetometer, FGM, overlap with STAFF at low frequencies. Techniques such as that discussed by Pincon and Lefeuvre (1991) in principle allow the spatio-temporal properties of the wavefield to be inferred from 3 components of the magnetic field and the 2 components of the electric field.

A more complete analysis of the wavefield properties (such as the separation of different types of waves) will eventually require electron density data from WHISPER or other instruments on the spacecraft.

With Cluster we will have data which are good, even with a single spacecraft. With four spacecraft we get improvements in solving spatio-temporal ambiguities. But the analysis is still quite difficult and the goal of being able to avoid model dependent analysis is not possible.

Acknowledgements

The DWP team acknowledges financial support from the PPARC and its predecessor, SERC, and technical support from Rutherford Appleton Laboratory. The present and past WEC Chairs (A. Pedersen, F. Lefeuvre, and A. Roux in reverse chronology) have guided the WEC team from the initial stages of proposal preparation through to the current activity of detailed operational planning. WEC colleagues must be thanked for their interesting discussions and the Cluster Project

Team at ESA for their support and encouragement. The ESA and industry (Dornier) spacecraft teams have been both professional and helpful during the integration and testing phases.

References

Cornilleau-Wehrlin, N. *et al.*: 1997, this volume.

Cornilleau-Wehrlin, N. *et al.*: 1988, *STAFF – A Spatio-temporal Analysis of Field Fluctuations Experiment. The Cluster Mission: Scientific and Technical Aspects of the Instruments*, ESA SP-1103, p. 25.

Decreau, P. M. E. *et al.*: 1997, this volume.

Decreau, P. M. E. *et al.*: 1988, *WHISPER – A Sounder and High Frequency Wave Analyser Experiment. The Cluster Mission: Scientific and Technical Aspects of the Instruments*, ESA SP-1103, p. 37.

Gurnett, D. A. *et al.*: 1997, this volume.

Gurnett, D. A. *et al.*: 1988, *The Wideband Plasma Wave Investigation. The Cluster Mission: Scientific and Technical Aspects of the Instruments*, ESA SP-1193, p. 45.

Gustafsson, G. *et al.*: 1997, this volume.

Gustafsson, G. *et al.*: 1988, *The Spherical Probe Electric Field and Wave Experiment. The Cluster Mission: Scientific and Technical Aspects of the Instruments*, ESA SP-1103, p. 31.

Johnstone, A. D. *et al.*: 1993, *PEACE: a Plasma Electron and Current Experiment*, ESA SP-1159, p. 163.

Pedersen A. *et al.*: 1997, this volume.

Pincon, J. L. and Lefeuvre, F.: 1991, 'Local Characterization of Homogeneous Turbulence in a Space Plasma from Simultaneous Measurements of Field Components at Several Points in Space', *J. Geophys. Res.* **96**, 1789.

Roux, A. and de la Porte, B.: 1988, *The Wave Experiment Consortium. The Cluster Mission: Scientific and Technical Aspects of the Instruments*, ESA SP-1103, p. 21.

Schmidt, R. *et al.*: 1988, 'Cluster and SOHO: A Joint Endeavor by ESA and NASA to Address Problems in Solar, Heliospheric and Space Plasma Physics', *EOS* **69**, 177, 179.

Schmidt, R. and Goldstein, M. L.: 1988, *Cluster – A Fleet of Four Spacecraft to Study Plasma Structures in Three Dimensions. The Cluster Mission: Scientific and Technical Aspects of the Instruments*, ESA SP-1103, p. 7.

Woolliscroft, L. J. C. *et al.*: 1988, *The Digital Wave Processing Experiment. The Cluster Mission: Scientific and Technical Aspects of the Instruments*, ESA SP-1103, p. 49.

Woolliscroft, L. J. C. *et al.*: 1992, *The Implementation of the Digital Wave Processor for the ESA/NASA Cluster Mission. Proc. of 26th ESLAB Symposium - Study of the Solar-Terrestrial System*, Killarny, Ireland, 16–19 June, 1992, ESA SP-346, p. 243.

Woolliscroft, L. J. C. *et al.*: 1995, *On the Design and Implementation of a Dependable Space Instrument, Proc. Euromicro Parallel and Distributed Processing*, San Remo, Italy, IEEE, p. 366.

THE ELECTRON DRIFT INSTRUMENT FOR CLUSTER

G. PASCHMANN, F. MELZNER, R. FRENZEL, H. VAITH, P. PARIGGER, U. PAGEL, O. H. BAUER, G. HAERENDEL, W. BAUMJOHANN and N. SCOPKE
Max-Planck-Institut für extraterrestrische Physik, 85740 Garching, Germany

R. B. TORBERT, B. BRIGGS, J. CHAN, K. LYNCH and K. MOREY
University of New Hampshire, Durham, NH 03824, U.S.A.

J. M. QUINN, D. SIMPSON and C. YOUNG
Lockheed Space Science Laboratory, Palo Alto, CA 94304, U.S.A.

C. E. McILWAIN, W. FILLIUS, S. S. KERR and R. MAHIEU
University of California at San Diego, La Jolla, CA 92093, U.S.A.

E. C. WHIPPLE
University of Washington, Seattle, WA 98195-1650, USA

(Received 2 May, 1996)

Abstract. The Electron Drift Instrument (EDI) measures the drift of a weak beam of test electrons that, when emitted in certain directions, return to the spacecraft after one or more gyrations. This drift is related to the electric field and the gradient in the magnetic field, and these quantities can, by use of different electron energies, be determined separately. As a by-product, the magnetic field strength is also measured. The present paper describes the scientific objectives, the experimental method, and the technical realization of the various elements of the instrument.

1. Introduction

To achieve the objectives of the Cluster program, it will be necessary to make sensitive and accurate measurements of the relevant electrodynamical parameters. The electric field is one of the essential quantities, yet it is one of the most difficult to measure. This is because in many important circumstances the electric fields are very small (less than 1 mV m^{-1}) and the plasma is very dilute, with densities less than 10 cm^{-3}. Under such circumstances, it is often difficult for the conventional double-probe technique to distinguish natural fields from those induced by spacecraft wakes, photoelectrons, and sheaths.

The instrument described in this paper is based upon the electron drift technique. This method involves sensing the drift of a weak beam of test electrons emitted from small guns mounted on the spacecraft. When emitted in certain directions, the electron beam returns to dedicated detectors on the spacecraft after one or more gyrations. During these gyrations, the beam probes the ambient electric field at a distance of some kilometers from the spacecraft, and therefore essentially outside the latter's influence. The operational principle was originally proposed by F. Melzner and was proven on ESA's GEOS spacecraft (Melzner *et al.*, 1978).

In the GEOS application the electron drift was measured only once per spacecraft revolution, and only for a restricted range of directions. These restrictions are

Space Science Reviews **79:** 233–269, 1997.

removed in the instrument described here. Two electron guns are used, each of which can be aimed electronically in any direction over more than a hemisphere. A servo loop continuously re-aims the electron guns so that the beams return to the detectors. The electron drift can be calculated by triangulation of the two emission directions. This method was developed by McIlwain and Quinn at UCSD and proposed for NASA's Equator mission. Comparing the drifts of electrons emitted at different energies enables electric fields and magnetic field gradients to be determined separately.

For small magnetic fields, the triangulation method becomes inaccurate, and the drift will instead be calculated from the differences in the time of flight of the electrons in the two nearly oppositely directed beams. This technique was developed by Tsuruda at ISAS and is being employed on the NASA/ISAS Geotail mission (Tsuruda *et al.*, 1985). The time-of-flight measurements also yield an accurate determination of the magnetic field strength.

The electron drift technique has a number of limitations. First, the measurements will be interrupted whenever electrons are strongly scattered by instabilities or interactions with ambient fluctuations. Second, beam tracking will be disrupted by very rapid changes in either the magnetic or the electric field. Third, accurate separation of the electric and magnetic components of the drift may not always be possible with only a limited range of electron energies.

From the respective strengths and weaknesses of the electron drift and double-probe techniques, it is quite obvious that they complement, rather than replace, each other.

The present paper is an expanded version of that published earlier (Paschmann *et al.*, 1993).

2. Scientific Objectives

The ability of the EDI instrument to make accurate and highly sensitive measurements of the electric field and of the perpendicular gradient of the magnetic field makes possible a variety of studies that comprise the essence of the Cluster mission.

Cluster has been designed primarily to study small-scale structures in three dimensions in the Earth's plasma environment. Although they are of relatively small scale, the processes leading to the formation of such structures are believed to be fundamental to the key processes of interaction between the solar wind and the magnetospheric plasmas. We refer the reader to the companion papers for an account of the Cluster objectives (Escoubet *et al.*, 1996, this issue, for example).

Table I lists the quantities that can be obtained from the measured electric fields and magnetic field gradients, and the information that can be derived. One of the prime objectives of the Cluster mission is to obtain differential quantities by measurements of particle and field properties at the four spacecraft locations. These differences can be used to form quantities such as the gradient, curl, and

Table I
Derivable information

Tools	Information
$\mathbf{E}^i + (\mathbf{v}^i \times \mathbf{B}^i)$	Resistivity; deviation from frozen flux condition
$\mathbf{E}^i \cdot \mathbf{I}$	Conversion of electromagnetic energy
Variance analysis of $\mathbf{E}^i(t)$	Attitude and motion of boundaries; normal and tangential fields
Least-squares fit of $\mathbf{E}^i(t)$	deHoffmann–Teller frame; intrinsic electric fields
$\int_{t_1}^{t_2} \mathbf{E}^i(t) \cdot \mathbf{u}\,dt$	Potential difference across layer or discontinuity
$\int_{t_1}^{t_2} \nabla_\perp B^i(t) \cdot \mathbf{u}\,dt - [B^i(t_2) - B^i(t_1)]$	Measure of stationarity of magnetic field profiles
$\nabla_\perp B^i(t) \cdot \mathbf{s}^{ij} - [B^j(t) - B^i(t)]$	Comparison of small and large-scale magnetic gradients
$\langle E_x^i(t) E_y^j(t+\tau)\rangle$	From this and other correlations: characterization of turbulence
$\nabla \cdot \mathbf{E}$	Shear flows
$\nabla \times \mathbf{E}$	$\partial \mathbf{B}/\partial t$; induction electric fields

Superscripts i, j indicate the four spacecraft locations.
$\mathbf{u}$ denotes boundary velocity, $\mathbf{s}^{ij}$ denotes inter-spacecraft separation.
Vector derivatives are approximated by finite differences between quantities measured at the different spacecraft locations.

divergence of the fields, and of the plasma moments such as velocity and pressure. These differentials will yield other physical properties such as current densities from $\nabla \times \mathbf{B}$, vorticity flow from $\nabla \times \mathbf{v}$, shear flows from $\nabla \cdot \mathbf{E}$, induction electric fields from $\nabla \times \mathbf{E}$, and momentum balance from the divergence of the pressure and magnetic stress tensors.

EDI can be used to obtain the magnetic field gradient in the plane perpendicular to the magnetic field, $\nabla_\perp B$. This will be a useful supplement to the differentials obtained from measurements made at the four spacecraft. The spacecraft separation will in general be much larger (hundreds of km) than the electron gyroradius (a few to tens of km) which is the scale on which EDI will measure $\nabla_\perp B$. Comparing the two differentials will allow a test of the consistency of the differentials over the two scales.

2.1. Bow Shock

The electric field plays a very important role in the physics of collisionless shocks. In a laminar shock, the electrons are magnetized and follow equipotentials, while the ions are unmagnetized and are decoupled from the electrons because of the inertia. Charge separation occurs, which causes an electric field along the shock normal, which in turn slows down the ion population. How this electric field is distributed in the shock layer is almost completely unknown.

Combining our measurements of $\nabla_\perp B$ with the larger-scale field gradients obtained from the magnetometer records on the four spacecraft will help to assess the stationarity of the magnetic field profiles across the shocks. Our $\nabla_{\perp}B$ measurements are also important in determining the occurrence of the so-called isomagnetic jumps in electric potential across the shock layer.

An important question about the Earth's bow shock is what mechanisms provide the required dissipation in the absence of particle collisions. It is known that some dissipation is provided by the coherent reflection of a fraction of the incident ion population, especially for quasi-perpendicular shocks. The ion distributions are also subject to instabilities leading to waves which scatter particles. Under quasi-parallel geometries there is coupling between reflected and incident particles. This coupling can lead to large-amplitude turbulence. Fast electric and magnetic field measurements at the four Cluster locations will permit a description of the low-frequency turbulence that is an important means of dissipating the solar-wind energy at the shock. The measurement of $\mathbf{k}$ vectors together with fast plasma measurements will make it possible to determine the wave modes and also the wave-particle interaction mechanisms. Shock surface waves can also be studied in this way.

2.2. Magnetopause, Boundary Layer, and Polar Cusp

The magnetopause is an example of a current sheet formed when two magnetized plasmas interact with each other. In the simplest physical picture, in which the magnetic fields are frozen into the plasma, the two interacting plasmas remain separate. Therefore the major interest is in those processes that violate the frozen flux condition and then lead to transfer of mass, momentum and energy across the current sheet.

Violation of the frozen flux theorem implies that the measured electric field, $\mathbf{E}$, differs from the convection electric field, $\mathbf{E}_c = -(\mathbf{v}\times\mathbf{B})$. Such differences will reveal contributions from the resistive term and the Hall current term in the generalized Ohm's law. The search for cases with $\mathbf{E}_c \neq \mathbf{E}$ will therefore be one of the prime objectives of this investigation.

Comparison can also be made with the fields computed from the deHoffmann–Teller transformation velocity. The electric field in the spacecraft frame may be approximated by $\mathbf{E}_{\mathrm{HT}} = -(\mathbf{v}_{\mathrm{HT}}\times\mathbf{B})$ where $\mathbf{v}_{HT}$ is the velocity of the deHoffmann–Teller frame in which the electric field vanishes. Systematic differences between $\mathbf{E}_c$ and $\mathbf{E}_{\mathrm{HT}}$ may reveal details concerning the magnetopause structure, such as the existence of an intrinsic electric field component normal to the layer.

Processes that lead to such deviations are expected to operate only on small spatial scales. In magnetic reconnection, for example, there is the diffusion region around the X-line which separates the regions of different magnetic field topology. But reconnection also implies the presence of non-zero electric fields, E_t, tangential to the magnetopause over much larger scales. Because E_t is necessarily rather small (of the order of 1 mV m^{-1}), it has not been measured in the past except in a few cases. Most of the previous in-situ evidence for reconnection at the magnetopause has come from measurement of high-speed plasma flows, which give no information on the reconnection rate. Thus the systematic measurement of E_t and its spatial scale remains one of the outstanding tasks of the Cluster mission. In addition to such rather laminar transport processes, macroscopic and/or microscopic turbulence is expected to play an important role at the magnetopause. It therefore will be of prime importance to study the fluctuations in the electric field and their four-point correlations.

As a consequence of the transfer processes, a boundary layer of solar wind plasma exists inside the magnetopause. The significance of the various portions of the boundary layer for the transport of magnetic flux, and thus for the cross-magnetotail potential, can be assessed from measurement of the electric potential across the layer. Previous estimates relied on single-spacecraft measurements which become highly suspect in the presence of boundary motions or non-stationary conditions. The availability of measurements on the four Cluster spacecraft will go a long way towards improving the accuracy of the potential measurement.

Not only do the four-spacecraft measurements provide the means to identify the spatial scales for the transport processes, they also allow for approximate determinations of quantities such as $\nabla \times \mathbf{E}$ or $\nabla \cdot \mathbf{E}$. $\nabla \times \mathbf{E}$ is a measure of $\partial \mathbf{B}/\partial t$ and thus helps to assess temporal changes in the magnetic field configuration; $\nabla \cdot \mathbf{E}$, on the other hand, is related to shear flows and therefore complements the direct plasma-flow measurements.

For the polar-cusp region, a major objective will be the study of plasma turbulence, because eddy diffusion or turbulent convection has been invoked as the dominant plasma transport mechanism in that region. Correlations between the four spacecraft will help to confirm or deny this type of transport.

2.3. MAGNETOTAIL

The electron drift instrument will provide reliable surveys of the convection electric field in the tail, not only in the equatorial plane but also along the north-south direction where strong gradients seem to exist near the plasma sheet boundary layer. These surveys should lead to a better understanding of the entry of solar wind/lobe plasma into the central plasma sheet and the circulation of this plasma to the frontside magnetosphere.

Another important topic where electric field measurements can contribute to our understanding is that of current sheets. Current sheets in the magnetosphere, such as in the magnetotail, are critical regions in that they are the most important sites of particle energization. In these current sheets, the magnetic field is small, the gyroradius can be large compared to the scale size, and consequently the electric field can play a dominant role in the particle motion. Attention is being focussed on these regions as sites for magnetic field reconnection, where magnetic field energy can be transformed into particle kinetic energy.

Major questions about current sheets are the mechanisms for their formation and their structure, the mechanisms for dissipation and diffusion, and the mechanisms for disruption and collapse.

In self-consistent studies of the tail current sheet it has been shown that it is possible to have particles trapped in the current sheet because of the mirror geometry that arises from the existence of the minimum of the magnetic field strength at the center of the sheet (e.g., Cowley, 1978). The normal component of **E** is an important factor in this trapping since it reflects particles with the proper sign and energy and also assists in the maintenance of quasi neutrality. It is likely that these trapped populations in the central plasma sheet become accelerated to high energies by the strong inductive electric fields that are present during magnetic substorms, and it is quite probable that they eventually turn into the plasmoids that have been inferred to move with high velocities along the Earth's magnetotail during substorms (Hones, 1979). EDI will measure both the normal and tangential components of **E** in the tail sheet during quiet times in order to help to identify the trapped population there. In addition it will measure the inductive electric fields during substorms which are an important element of the acceleration processes during these events.

Since the EDI instrument will measure the electric field at up to tens of samples per second, the combined electric- and magnetic-field data is well suited for studying ultra-low frequency (ULF) waves in the range of about 10 Hz to several hundred mHz. This range covers many kinds of magnetohydrodynamic (MHD) waves as well as ion plasma waves in regions of low magnetic fields. While most earlier work was based on magnetic field data alone, the combined electric and magnetic data allow the determination of the wave's **k** vectors and Poynting fluxes. Measuring the Poynting flux simultaneously with all four spacecraft also provides a good determination of the resonance regions where wave energy is trapped on a field line in the form of standing oscillations. The ability of EDI to obtain magnetic field gradients becomes important when studying so-called drift-mirror waves which are excited due to ∇B-drifting energetic protons with large perpendicular temperatures (Baumjohann *et al.*, 1987).

The ability to infer an estimate of $\nabla \times \mathbf{E}$ from the four-spacecraft measurements will allow an assessment of temporal changes in the global magnetotail configuration during substorms.

2.4. Inner Magnetosphere

Prime objectives in the inner magnetosphere include studies of the electric fields associated with convection, ULF waves, and particle injections.

The concept of plasma convection in the magnetosphere has unified a number of high-latitude geophysical phenomena. However, there has been a paucity of direct measurements of the convection electric field in the inner (4–12 R_E) equatorial magnetosphere. Early inferences of the electric field were from ground-based observations of plasma drifts – in the ionosphere for example – which were interpreted in terms of $\mathbf{E} \times \mathbf{B}$ drifts. These ionospheric electric fields were then mapped upwards along magnetic field lines into the magnetosphere to obtain estimates of magnetospheric electric fields.

A few electric field measurements have been reported within the plasmasphere from GEOS 1 and ISEE 1 (Pedersen *et al.*, 1978), but most measurements further out were made by the double-probe technique during substorm events when the fields were large (Aggson and Heppner, 1977; Pedersen *et al.*, 1984). Particle measurements in the equatorial region have been used to infer electric fields outside the plasmasphere (McIlwain, 1972; McIlwain, 1981), but it was not until the electron beam technique on GEOS became available that direct measurements in the outer equatorial magnetosphere during quiet times were reported (Baumjohann and Haerendel, 1985; Baumjohann *et al.*, 1985).

Plasma injections are the sudden appearances of energetic plasma at all energies and directions in the equatorial magnetosphere during magnetospheric substorms, frequently within a few tens of seconds of substorm onset (DeForest and McIlwain, 1971). The injected plasma appears to come from a well-defined injection boundary that maps down along the Earth's magnetic field lines to the equatorward edge of the auroral oval. There are probably strong electric fields associated with these plasma injections which may be transient and/or localized at the injection boundary. EDI will measure these fields, including inductive fields, with a time resolution of up to several tens of Hz at all magnetospheric activity levels; these measurements should allow analysis of the plasma motions during these events to yield a better understanding of the injection process.

There has been recent renewed interest in the convection electric field because it is now realized to be central to many magnetospheric processes, including the global MHD equilibrium, reconnection rates, Region-2 Birkeland currents, magnetosphere-ionosphere coupling, ring current and radiation belt transport, substorm injections, and several acceleration mechanisms. New algorithms have been developed to extract electric fields from particle data (Sheldon and Gaffey, 1993; Sheldon and Hamilton, 1994). It is essential, however, that these indirect techniques be supplemented by accurate, high resolution, direct measurements of the electric field of the kind that will be obtained by EDI.

2.5. Small-Scale Structures

Dilute plasmas appear to have a strong tendency to create fine structure. The reasons for the formation of such small-scale structures are manifold: gradient instabilities, current bunching, heat-flux instabilities, cascading from longer wavelength turbulence, beam-plasma interactions, kink-, firehose- or flute-type instabilities. Fine structure may in fact be nature's preferred way to create dissipation. The thinness of auroral arcs testifies that the hard-to-sustain parallel electric fields are a consequence of the fine-structuring and accompanying enhancement of electric currents.

Little is known about fine structure in such situations: their morphology (spatial scales), time scales, and amplitudes. Organization along the magnetic field direction is most frequent, but the coherence lengths (parallel wave numbers) can only be guessed. Cluster will be the first attempt to unravel the origin, dynamics and macroscopic consequences of fine structure in a cosmic plasma with a suitable tool.

The measurement of electric fields deserves special attention in such small-scale structures. Whereas **E** appears as a secondary quantity in magnetohydrodynamics, this is certainly not the case for scale lengths comparable to or smaller than the ion gyroradius. In many situations, the ions can even be regarded as unmagnetized, i.e., the electric force may dominate their dynamics. The electrons, on the other hand, perform an $\mathbf{E} \times \mathbf{B}$-drift that may contribute strongly to the electric current, in contrast to the regular current-free plasma convection on larger scales. Electric fields, electron pressure gradients and magnetic stresses are intimately related under these circumstances:

$$\mathbf{E} = \frac{1}{en}\left[-\left(\nabla p_e + \nabla \frac{B^2}{2\mu_0}\right) + \frac{1}{2\mu_0}(\mathbf{B}\cdot\nabla)\mathbf{B}\right] . \tag{1}$$

The proposed electron beam technique allows a simultaneous measurement of **E** and $\nabla(B^2/2\mu_0)$, two essential quantities in the force balance. Together with measurements of the pressure tensor and of the magnetic tension with four spacecraft, one has a powerful tool for studies of the dynamics of small-scale structures.

EDI should bring a significant enhancement to the study of current sheets and filaments since it will simultaneously measure the electric and magnetic fields, and in particular the perpendicular components of the magnetic field gradient associated with these currents. For example, previous inferences of field-aligned current structures have come from magnetometer data where only the gradient of **B** along the spacecraft path was obtained. EDI will often provide gradients in the plane transverse to **B** as well as along the trajectory. This, together with electric field data, should allow a much better determination of the structure of these current systems.

3. Principle of Operation

3.1. Drift Velocity from Beam Direction Measurements

The basis of the electron drift technique is the injection of test electrons and the registration of their gyrocenter displacements after one or more gyrations in the magnetic field, **B**. The displacement, **d**, is related to the drift velocity, $\mathbf{v}_D$, by:

Table II
Characteristic quantities for key regions

Parameter	Solar Wind	Magneto-sheath	Cusp	Tail lobe	Plasma-sheet	Ring current
Magnetic field, B, nT	8	40	40	30	20	300
Electric field, E, mV m^{-1}	3.6	8.0	4.0	0.5	1.0	1.0
Electron gyroradius, R_g, km	13	3	3	4	5	0.4
Electron gyrotime, T_g, ms	4.5	0.9	0.9	1.2	1.8	0.1
Drift step, d, m	2000	179	89	20	89	0.4
ToF difference, ΔT, μs	214	19	9	2	9	0.04
Angle change, δ, deg	8.7	3.9	1.9	0.3	1.0	0.1
Ambient diff. E-flux	1.0×10^5	1.0×10^6	1.0×10^7	1.0×10^5	2.0×10^7	1.0×10^8
Beam current, nA	1000	200	300	30	400	1
Optics state	6	3	7	3	7	4
Ambient count rate, s^{-1}	2.1×10^4	1.6×10^4	2.2×10^5	1.6×10^3	4.4×10^5	1.4×10^5
Beam count rate, s^{-1}	7.2×10^4	3.8×10^5	3.8×10^5	5.4×10^5	3.6×10^5	2.3×10^5
Contrast	3.4	23.5	2.5	237	0.8	1.7
Signal-to-noise ratio	15.6	93.8	36.5	300	17.2	19.7

Table values are for 1 keV electrons.
Beam currents, beam count rates, contrast, and signal-to-noise ratios are for times the beams are gated on (total of 1 ms per 2 ms sample). Beam and background count rates are computed for the collection areas A and geometric factors H associated with the chosen Optics State (see Table III).

$$\mathbf{d} = \mathbf{v}_D \cdot N \cdot T_g \,, \tag{2}$$

where T_g is the gyroperiod and N denotes the number of such periods after which the electrons are captured. If the drift is solely due to an electric field, $\mathbf{E}_\perp$, transverse to $\mathbf{B}$, then (using MKSA units)

$$\mathbf{d} = \frac{\mathbf{E}\times\mathbf{B}}{B^2} \cdot N \cdot T_g \,. \tag{3}$$

Or, numerically, for $N = 1$

$$d\,(\mathrm{m}) = 3.57\times 10^4\, \frac{E_\perp\,(\mathrm{mV\ m^{-1}})}{B^2\,(\mathrm{nT})} \,. \tag{4}$$

Typical values for B, $E_\perp$, the electron gyroradius and gyrotime, and the drift step d (as well as other quantities referred to later) are listed in Table II for various regions of interest. For any other choice of magnetic and electric fields, the reader is referred to Figure 1.

It is important to realize that after one gyration, all electrons emitted from a common source S in a plane normal to $\mathbf{B}$, are focussed onto a single point that is displaced from S by the drift step, $\mathbf{d}$ (Figure 2). A detector, D, placed at the focus would detect these electrons. As $\mathbf{d}$ is the quantity to be measured, it is not possible

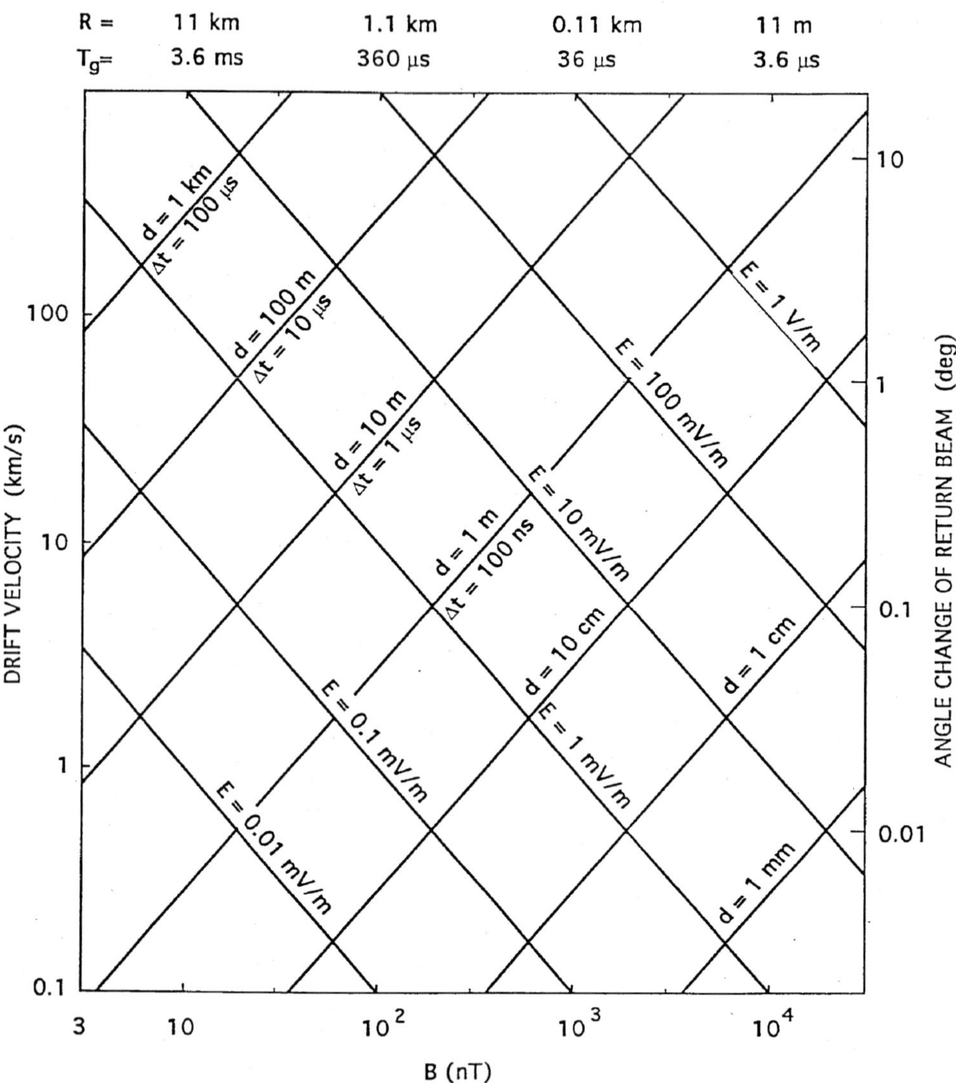

Figure 1. Drift parameters for 1 keV electrons

to put an electron source at S. Fortunately this is not necessary. A beam from an arbitrarily located electron gun will also hit the detector at D, provided the beam is directed towards S. In this case the gun can be thought of as supplying electrons to the source at S, from where they proceed to D, as described. The beam may also be directed away from S, in which case it assumes the role of a beam emanating from the source, as illustrated in Figure 2. If two guns are used, as shown in the figure, measurement of the two emission directions that return a beam onto the detector

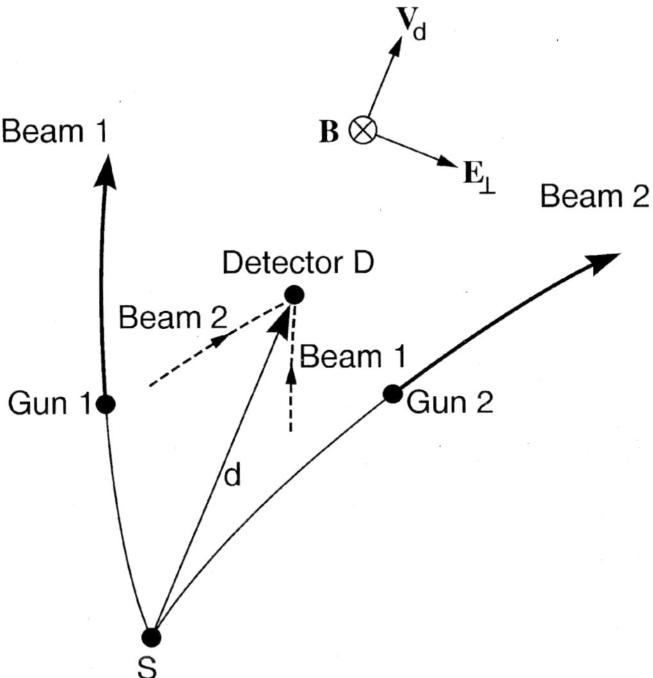

Figure 2. Principle of drift step triangulation. If as a result of the drift all electrons emitted from S reach the detector location D after one gyration, then the beams emitted from two arbitrarily placed guns will also strike the detector if they are directed along lines through S. The drift step **d** is therefore the vector from the intersection point of the two beam directions to the detector D, once the beams are steered such that an 'echo' is received by the detector.

yields the displacement, **d**, and thus the drift velocity, $\mathbf{v}_D$. This is a straightforward triangulation problem that in principle can be done continuously and with high time resolution.

Note that with guns placed at locations other than S, D is no longer a focal point of the beams, nor do the travel times precisely equal the gyrotime, T_g. If the beam is directed towards (away from) S, the travel time will be longer (shorter) than T_g. In subsequent discussions we often refer to S as the target. The angle (in radians) between outgoing and returning beam is given by

$$\delta \approx 2\pi \frac{v_D}{v} , \tag{5}$$

where v is the electron speed. For drift speeds of 100 km s^{-1} (and 1 keV electron energy), the angle δ is 1.9°, increasing to almost 10° at 500 km s^{-1}. As we will see later, a large δ complicates operation.

A solution with a single detector, as illustrated in Figure 2, is not practical, as the detector would have to detect beams from two different directions at the same time. Once two separate detectors are employed, the scheme changes from that

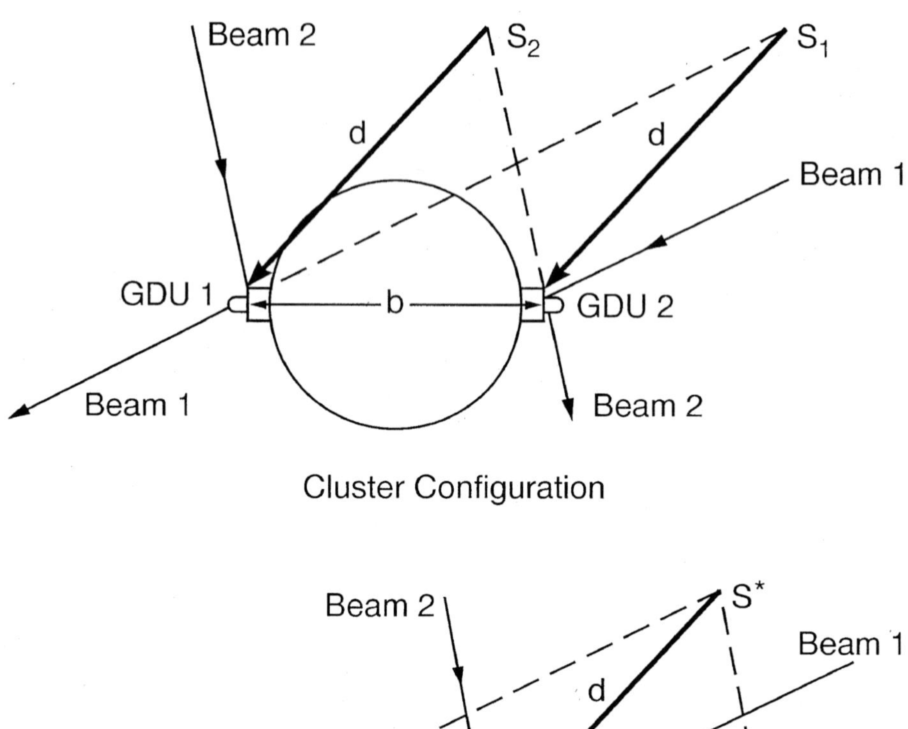

Figure 3. Triangulation scheme for two gun/detector units placed on opposite sides of the spacecraft, at a distance b (top). S_1 and S_2 are the virtual source points for the two detectors. The problem is equivalent to one with two guns spaced $2b$ apart, a single detector, and a single source point S^* (bottom).

in Figure 2 to that shown in Figure 3. Here two gun-detector units (GDU's) are placed on the spacecraft at a distance b, as shown at the top of Figure 3, as required by technical constraints on Cluster (see Section 3.6). In this case one has separate source points, S_1 and S_2, one for each detector. As far as triangulation is concerned, this configuration is equivalent to one where the two guns are spaced $2b$ apart and a single detector is placed half-way inbetween, as shown in the bottom part of Figure 3. Thus one has effectively gained a factor of 2 in triangulation baseline.

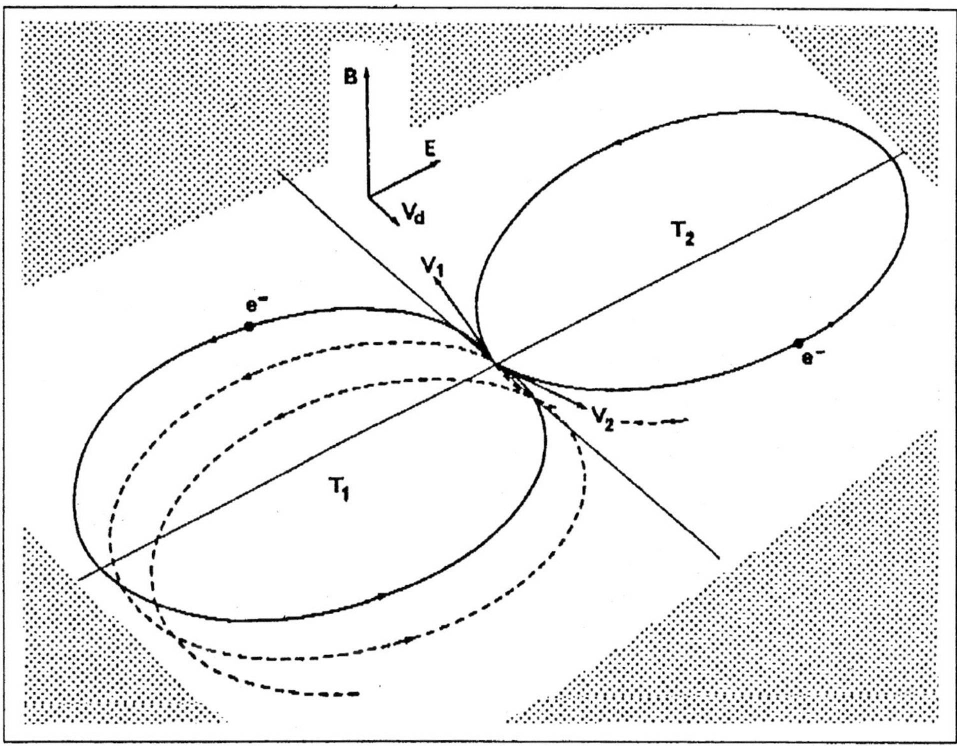

Figure 4. Principle of drift determination via time-of-flight measurements. Electrons emitted in a direction, V_1, opposing the drift, V_d, travel a longer path, and thus have a longer gyrotime, T_1, than electrons emitted along the drift, which take T_2.

As is true for any triangulation problem, the length of the baseline, b, naturally determines the precision with which the displacement, $\mathbf{d}$, can be measured. The baseline is defined as the distance transverse to $\mathbf{v}_D$ between a gun and its associated detector when projected into the plane perpendicular to $\mathbf{B}$. For the configuration with two gun-detector pairs depicted in Figure 3, the maximum effective baseline is twice the actual physical distance between the units, i.e., 6 m. Assuming 1° knowledge of the beam firing direction, one can then determine displacements up to 60 m to better than 20%.

Consulting Table II, one sees that much larger drift steps can occur. For such drift steps the triangulation technique still determines the direction of the drift with high accuracy, but not its magnitude. Under such conditions, electron time-of-flight measurements will be used to determine the magnitude of the drift step. Similarly for very small drift steps, the direction becomes uncertain.

3.2. Drift Velocities from Time-of-Flight Measurements

Appropriate pulse-coding of the beams makes it possible to measure the time of flight of the electrons with a resolution better than 1 μs, if the magnetic field is sufficiently stable over the electron gyroperiod. As illustrated in Figure 4, electrons in the two beams returning to the detectors travel different path lengths. As a result their flight times differ by an amount given by

$$\Delta T = T_{\mathrm{to}} - T_{\mathrm{aw}} = 2T_g \frac{v_D}{v} \propto \frac{d}{v} ,$$

where T_{to} and T_{aw} are the flight times for the beam electrons aimed towards and away from the target, respectively. In the limit of very large drift steps, as depicted in Figure 4, the towards (away) beams are directed essentially anti-parallel (parallel) to the drift velocity. For simplicity we have assumed here that guns and detectors are collocated, or in other words, that the drift step is large compared to gun-detector separations.

Measuring T_{to} and T_{aw} permits determination of $\mathbf{v}_D$. ΔT scales directly as d/v. Hence, while the triangulation becomes increasingly less accurate, the time-of-flight method becomes more accurate with increasing drift step d, limited only by signal-to-noise effects. As shown in Table II, ΔT is 9 μs or larger (and thus accurately measurable) for those regions where the triangulation method starts to fail.

3.3. Measurement of B

The gyroperiod itself is obtained from the mean of the travel times:

$$T_g = \frac{T_{\mathrm{to}} + T_{\mathrm{aw}}}{2} . \tag{7}$$

From T_g the magnetic field strength, B, is obtained via

$$T_g = \frac{2\pi m}{eB} . \tag{8}$$

Values of T_g range from about 0.1 to 10 ms (see Table II). When times of flight are measured, the magnetic field strength is determined with very high accuracy, e.g., to within 0.1% for a 30 nT field. This feature can be used for an accurate in-flight determination of the fluxgate magnetometer offsets.

3.4. Separation of Electric and Magnetic Gradient Drift

The beam electrons are subject not only to electric field drifts, but also to drifts caused by magnetic field gradients, $\nabla_{\perp} B$, directed perpendicular to the magnetic field. When the scale-length, ℓ, of such inhomogeneities becomes small (as it does at the bow shock, the magnetopause, or at the edges or the center of the plasma sheet), the gradient drift will make a significant contribution to the test electrons' displacement. The ratio of this drift, v_B, to that caused by the transverse electric field in the spacecraft frame of reference, v_E, is

$$\frac{v_B}{v_E} = 10^3 \frac{W_e\,(\mathrm{keV})}{E_\perp\,(\mathrm{mV\ m^{-1}})}\,\ell^{-1}\,(\mathrm{km})\,, \tag{9}$$

where W_e is the energy of the electrons. For 1 keV test electrons and a field of 1 mV m^{-1}, v_B/v_E reaches unity if ℓ approaches 1000 km. For this reason we foresee the use of electrons at different energies, typically 0.5 and 1.0 keV. A wider range of energies would help to separate the drifts but is beyond the capabilities of EDI, primarily because of limitations in gun and detector voltage supplies.

When the total drift is measured at two energies, W_1 and W_2, with $r = W_2/W_1$, the electric and magnetic drifts are obtained from the following expressions:

$$\mathbf{v}_E = \frac{(r\mathbf{v}_1 - \mathbf{v}_2)}{(r-1)}\,, \tag{10}$$

$$\mathbf{v}_B(W1) = \frac{(\mathbf{v}_2 - \mathbf{v}_1)}{(r-1)}\,, \tag{11}$$

where $\mathbf{v}_1$ and $\mathbf{v}_2$ are the (total) drifts that are inferred from the triangulation analysis applied to the two measurements at energies W_1 and W_2, respectively.

When the time-of-flight measurement technique is used in the presence of a significant gradient in $\mathbf{B}$, the analysis is more complicated. Since the gyrotime is defined in terms of the magnetic field at the center of the gyro circle, the two beams fired parallel and anti-parallel to the drift direction (Figure 4) will have different gyro times. If there were only the $\nabla_\perp B$ drift (i.e., no electric field) then the gyro times are given by

$$T_g = T_0 \left(1 + \frac{R_g \sin\Phi_0}{\ell}\right)\,, \tag{12}$$

For the anti-parallel beam, $\Phi_0 = \pi/2$ and for the parallel beam $\Phi_0 = -\pi/2$, if one assumes that the gradient in $\mathbf{B}$ is in the same direction as $\mathbf{E}$ in Figure 4. Here T_0 is the gyrotime as given by the magnetic field at the spacecraft and R_g the corresponding gyroradius. The drift velocity and drift step are:

$$v_B = v_{B0} \left(1 + \frac{2R_g \sin\Phi_0}{\ell}\right)\,, \tag{13}$$

$$d_B = d_{B0} \left(1 + \frac{3R_g \sin\Phi_0}{\ell}\right)\,, \tag{14}$$

where again v_{B0} and d_{B0} are the values in terms of the magnetic field at the spacecraft. Use of the measured times now results in

$$T_{\mathrm{to}} + T_{\mathrm{aw}} = 2T_0 \tag{15}$$

as before, but

$$\Delta T = T_{\mathrm{to}} - T_{\mathrm{aw}} = 6T_0 \frac{v_{B0}}{v_0} \; . \tag{16}$$

When both an electric field and a gradient in the magnetic field are present it can be shown that the difference in measured times for two beams fired anti-parallel and parallel to the net drift direction is given to leading order by

$$\Delta T = T_0 \left(4 \frac{v_{B0}}{v_0} \sin \delta\Phi + \frac{2}{v_0} |\mathbf{v}_E + \mathbf{v}_{B0}| \right) \; . \tag{17}$$

Here, $\delta\Phi$ is the starting angle of the anti-parallel beam with respect to the direction of the magnetic field gradient.

In general there are four unknown quantities: the magnitude and direction (or equivalently, the x and y components) of both $\mathbf{v}_B$ and $\mathbf{v}_E$ in the plane perpendicular to $\mathbf{B}$. When two energies are used there will be six measured quantities: two net drift directions (one at each energy) and two pairs of flight times for the parallel and anti-parallel beams. However, the flight times are not all independent since the sum of the pair for each energy is $2T_0$. Nevertheless the net drift directions and the differences in measured flight times for the two energies provide four relations that make it possible to obtain the four unknown drift components of $\mathbf{v}_B$ and $\mathbf{v}_E$.

3.5. Return Beam Intensities

The flux of returning beam electrons incident on the detector depends upon many factors, including the angular current distribution of the outgoing beam, the beam gyroradius, possible beam modification by electrostatic or wave-particle forces, and the geometrical arrangement of the gun and detector with respect to the drift step vector. The outgoing beam has an opening angle, α, of approximately 1°. Thus the beam diverges along the magnetic field direction by a distance $s_{\|} = 2\pi R_g \alpha / 57.3$, where R_g is the gyroradius, but is focussed in the plane perpendicular to $\mathbf{B}$ after one gyro orbit. By definition, this focus is located one 'drift step' from the gun. In general, those beam electrons with the proper firing direction encounter the detector either somewhat before, or somewhat after, this focus point. Because of the angular divergence both along and perpendicular to $\mathbf{B}$, the detector intercepts only a very small part of the emitted beam.

Equation (18) gives the beam flux, F, in $\mathrm{cm}^{-2}\mathrm{s}^{-1}$, at the detector for an emitted beam with a flat current distribution over a square angular cross-section, where I is the gun current in nA, B the magnetic field strength in nT, W the beam energy in keV, x the distance from the gun's gyrofocus to the detector in m, α the beam full width parallel and perpendicular to $\mathbf{B}$ in degrees, and R_g the gyroradius in m.

$$F = 3.1 \times 10^3 \frac{IB}{x\alpha^2 W^{1/2} \left(1 \pm x/2\pi R_g\right)} . \tag{18}$$

The $\pm$-term in the denominator accounts for divergence parallel to $\mathbf{B}$ between the gun's gyrofocus and the detector, depending upon whether the detector intercepts the beam before $(-)$ or after $(+)$ the gyrofocus. Of course a square, uniform cross-section is not a realistic representation of the actual beam. However for that portion of the real beam that has the same angular current density I/α^2 as the uniform beam, the return flux at the detector would be the same.

The beam divergence leads to a large variation in the return beam flux with magnetic field intensity and drift step. In order to compensate partially for this variation, both the beam current and the detector optics are adjusted by EDI's controller unit. The optics may be commanded into a number of different 'states' (see Section 4.2). These states allow a good deal of flexibility in the choice of the detector's effective area (A) to the return beam and its geometric factor ($H = G\Delta E/E$) to ambient electrons.

Table II illustrates sample values of return-beam count rates and signal-to-noise ratios for several regions of interest, using appropriately chosen optics states. The values in Table II are for a 1 keV beam with a 1° width. The beam current has been limited to keep the instantaneous count rates (per anode) below approximately 10^5 counts s^{-1}. The signal-to-noise ratios are based upon counts when the beam is gated on and accumulated over a period of 2 ms. With the 50% duty cycle of the beam, average signal-to-noise ratios are a factor of $\sqrt{2}$ lower. Because x, the distance between the beam focus and the detector, depends upon the relative geometry of the guns, detectors, and drift step, we have taken x to be equal to the larger of either the drift step or 2 m. We have also ignored the beam spreading parallel to the magnetic field that occurs between the focus and the detector, since the sign of this extra term depends upon the specific geometry.

3.6. Requirements for Gun/Detector Configuration

In order to accommodate the time-of-flight measurements, there must be two guns, each steerable over a solid angle of 2π steradian, but facing opposite hemispheres. As the detectors require active steering into the appropriate directions, two such detectors are needed, each able to cover 2π sr. The time-of-flight technique puts no restriction on the relative location of guns and detectors (other than those imposed by field-of-view considerations).

The triangulation technique, on the other hand, requires that guns and detectors are well separated in order to provide adequate baselines. Ideally, they should not be coplanar, but rather form a tetrahedron. Otherwise there will be situations where the baseline vanishes, i.e., when $\mathbf{B}$ and $\mathbf{v}_D$ are in the gun/detector plane. As technical constraints rule out such a tetrahedron solution on Cluster, one gun and one detector are combined into a single package, and two such packages are mounted on opposite sides of the spacecraft (see Figure 3). So they are not only coplanar, but even colinear. As a consequence, the triangulation baseline will vanish each time the projection of the two packages in the plane perpendicular to $\mathbf{B}$ is aligned with $\mathbf{v}_D$. Even though this will cause a spin-modulation of the accuracy with which the drift step is triangulated, the electron guns will stay on track. Furthermore, the time-of-flight technique, which will always be executed simultaneously with the triangulations, will not be affected at all. (Note that in the worst case of a spin axis perpendicular to both $\mathbf{B}$ and $\mathbf{v}_D$, the baseline is always zero.)

3.7. Beam Recognition, Tracking, and Coding

The electron drift technique described in the previous sections requires first a scheme capable of initially finding the beam for arbitrary directions of magnetic and electric fields; secondly, a scheme to keep the beam on target, and, finally, a scheme which determines the time-of-flight of the electrons for each beam.

As described in more detail in Section 6.1, several different schemes are implemented. The simplest of those just sweeps the beam in the plane perpendicular to $\mathbf{B}$, where the latter condition is taken from the magnetometer data received in real-time. From the continuously recorded (and transmitted) counts one can then derive the directions to the target and infer the drift velocity.

As our main operating mode, we have implemented a tracking mode where the beam is rapidly swept back and forth across S. We use correlators to distinguish beam from background electrons, and thus to recognize beam passage over the detector. We will again utilize the on-board magnetometer data to define the scan plane, i.e., the plane perpendicular to the magnetic field.

To obtain the time of flight of the electrons, the beams will be modulated with a coded waveform. By correlating the received signal with the original code, the time delay between emission and reception is measured.

3.8. Capabilities and Limitations of the Technique

The electron drift technique is capable of providing several unique measurements. First, it provides the electric field perpendicular to the magnetic field $\mathbf{E}_\perp$, including its component along the spacecraft spin axis. By contrast, the double-probe technique measures $\mathbf{E}_\perp$ in the plane of the wire booms only. Second, the electron drift technique provides the unique capability of measuring local magnetic field gradients, $\nabla_\perp B$. Third, through its time-of-flight measurements the technique also yields accurate measurements of the magnetic field strength, B. Finally, the measurements are essentially unaffected by the presence of the spacecraft. A time resolution of between 10 and 100 measurements per second is possible depending on the detector signal-to-noise ratio.

On the other hand, the electron drift technique is adversely affected by intrinsic beam instabilities, strong scattering of the beam by ambient fluctuations, large-amplitude 'spikes' in the electric field, and very rapid magnetic field variations. All these effects can cause a loss of beam track and thus a momentary loss of data.

Furthermore, there can be signal-to-noise problems as a result of insufficient beam current and/or excessive fluxes of ambient electrons. Finally, accurate separation between electric and magnetic drifts will not always be possible when the range of beam energies is restricted to between 0.5 and 1.0 keV.

3.9. Spacecraft Potential

The spacecraft is normally at a potential, Φ, that differs slightly from the ambient plasma potential. As the test electrons traverse the sheath surrounding the spacecraft, they are deflected and consequently enter the region of undisturbed ambient electric field with perturbed initial conditions. Normally, this will lead to an additional displacement of the returning beam. Since the proposed measurement of $\mathbf{E}_\perp$ is based on measurements of the direction of the outgoing beam, we best express the perturbation caused by spacecraft fields in terms of the angular deflection of the outgoing beam, β. Upon return the beam may suffer a similar deflection. It is easy to estimate an upper limit of β, not taking into account the peculiarities of the field geometry:

$$\beta < e\Phi/4W \; . \tag{19}$$

In sunlight Φ is of the order of a few tens of volts. With $W = 1$ keV for the electron beam, the error introduced by this effect is comparable to the pointing accuracy of the beam.

The ASPOC instrument on Cluster is designed to keep the spacecraft potential at low values in the outer regions of the magnetosphere and in the solar wind where normal spacecraft potentials may be several volts positive. This system is based on the emission of indium ions at several keV energy and with a current of 1–10 μA. The spacecraft potential will be kept at a low positive potential relative to the ambient plasma. When the potential control system is operating, EDI can use currents up to 10^{-7} A (or possibly more) without influencing the spacecraft potential in any significant way.

Riedler *et al.* (1996, this issue) have estimated the Cluster spacecraft potential both with and without the operation of the Active Spacecraft Potential Control (ASPOC) ion emitter. They show that it takes as least 10 μA emitted ion current to reduce the spacecraft potential to under 10 V over most of the range of expected environmental plasma conditions, as characterized by $N_e\sqrt{kT_e}$. Since the maximum EDI electron current is expected to be 1 μA and the typical current to be on the order of or less than 100 nA, it can be seen that the effect on the spacecraft potential will be essentially negligible when ASPOC is operating with an ion current equal to or greater than 10 μA. We note that there is considerable uncertainty in the ASPOC calculations because of the unknown effective collection area for plasma electrons and the unknown projected area for photoemission. There is also uncertainty in the photoelectron spectrum as the ASPOC authors have noted.

There could be a significant effect of the EDI electron beam current on the spacecraft potential when ASPOC is not operating. The electron beam can be considered to be another 'photoemission' component at an energy of 1 keV. It can be seen from the three curves in Figure 1 of Riedler *et al.* (1996); that it only takes from 10 to 30 V of spacecraft potential to reduce their assumed photoemission current to about 1 μA. At higher (positive) spacecraft potentials, an EDI beam current of 1 μA could dominate over the photoemission current and drive the spacecraft potential even more positive. It is expected in such situations that a much smaller EDI current would be used.

4. Technical Description

The essential elements of the instrument are two electron guns, two detectors with their associated analog electronics, high-voltage supplies, digital controls, and correlators; and a controller unit which includes the interfaces with the spacecraft and with other instruments (cf., Figure 5). As illustrated in Figure 6, guns and detectors are combined in pairs into a single unit, referred to as the gun/detector unit (GDU). The two GDUs are mounted on opposite sides of the spacecraft. For a detailed block diagram of the GDUs, see Figure 7.

4.1. Electron Guns

In order to be able to aim the beam at the target for arbitrary magnetic and electric field directions, the electron guns must be capable of providing a beam that can be steered rapidly into any direction within more than a hemisphere. Electron energies must be variable in order to separate $\mathbf{E} \times \mathbf{B}$ and $\nabla_{\perp} B$ drifts. At the same time the energy dispersion must be small to restrict beam spreading in space and time. Beam currents must be kept sufficiently low to avoid instabilities and/or interference with other experiments on the spacecraft. To maximize the return signal in the detectors, the angular width of the beam must be kept small, but still large enough to account for uncertainties in pointing direction. Electron time-of-flight measurements require that the beam be modulated with frequencies up to 4 MHz.

A design meeting these requirements is illustrated at the top of Figure 6. To our knowledge this is the first electron gun capable of providing narrow beams in any direction within more than a hemisphere. A conventional electron source, consisting of a tungsten cathode and several electrodes (Wehnelt, Focus, and Anode) is used to produce a narrow beam at 2.7 times the required energy. The electron energy is set by the cathode potential and can be varied between 0.5 and 1.0 keV. The spread in energy is determined by the thermal spread (≈ 0.2 eV) and the variation of the potential over the emitting part of the cathode which is ≈ 0.5 eV. The beam

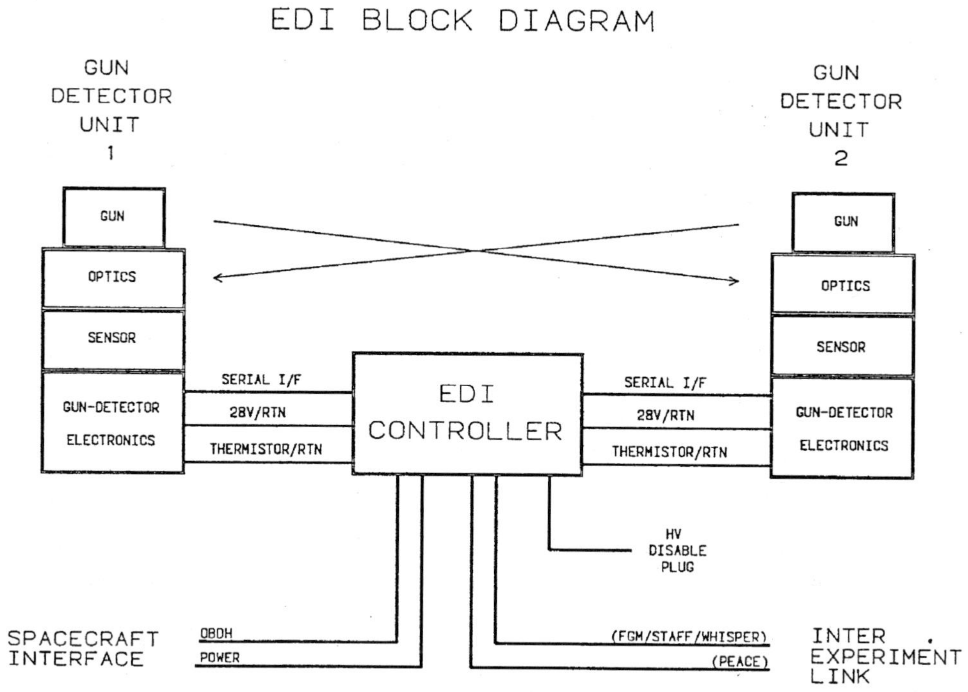

Figure 5. Overall block diagram showing the elements of the EDI instrumentation and their inter-relation.

current is controlled via the current applied to the tungsten filament and can be varied between 0.1 and 2000 nA. The beam is intensity-modulated by superposing the code-signal from the correlator onto the static Wehnelt voltage via a fibre-optic cable.

After exiting from the anode, the beam is deflected into the desired azimuth direction by an octopole arrangement of electrostatic deflectors. The electrons then approach a high-transmission grid at ground potential. This retarding potential decelerates the electrons to their final energy. Since it is mainly the energy along the symmetry axis that is removed when the electrons approach the grid, the deflection angle is amplified. Figure 8 shows the voltages, equipotentials and the computed electron trajectory for the case of 90° deflection. A maximum deflection angle of more than 100° has actually been achieved. The beam width ranges from typically 1° circular for small deflection angles to 2–4° elliptical at large polar angles.

The deflection grid is made of fine copper-beryllium wires woven into a mesh that is then formed into a basket shape, welded to a steel flange, and finally gold-plated. The smaller radius of curvature used for the outer section serves to reduce shadowing by the wires by increasing the angle between beam and grid at large deflection angles. Nevertheless, the mesh introduces a 50% drop-off in beam

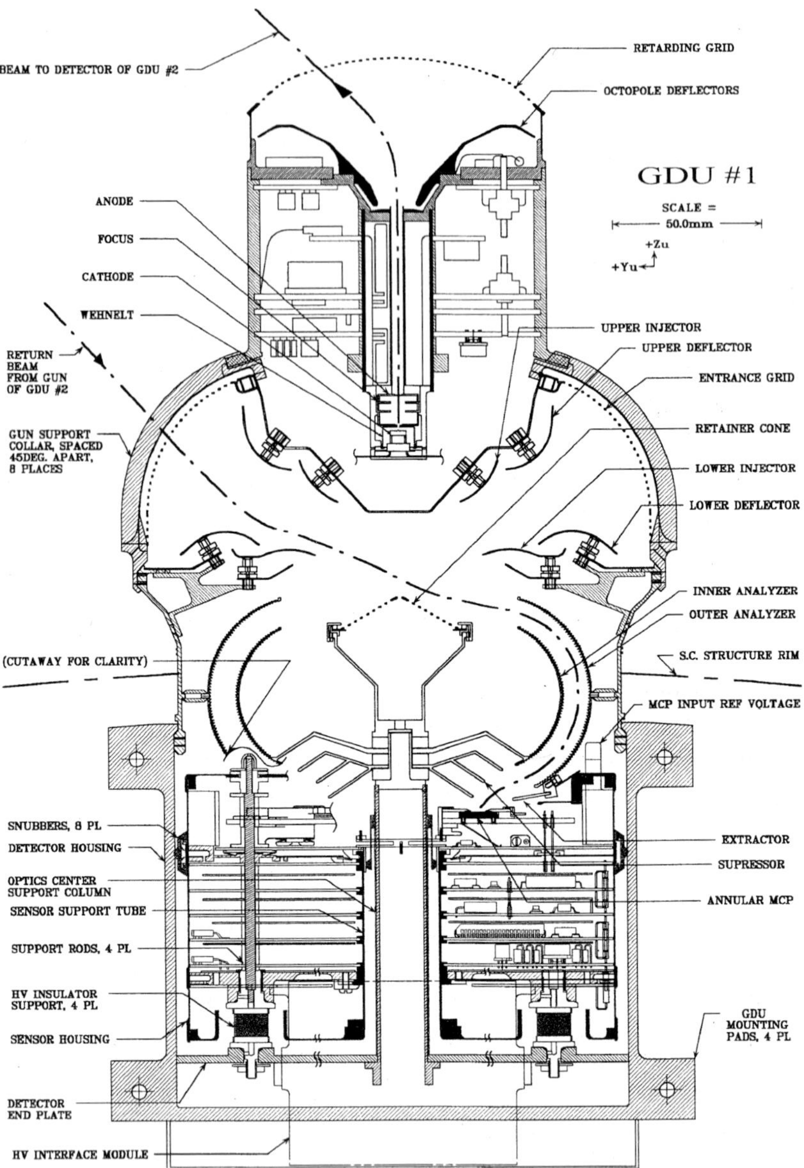

Figure 6. Cross-sectional view of a gun/detector unit (GDU). The cylindrical section at the top is the electron gun, with its filament and control electrodes at the bottom of a long drift tube from where the electrons enter the octopole deflector before they are slowed down and further deflected by the curved retarding grid. The gun is supported by a collar that bridges the detector aperture. Independently selectable voltages on 9 electrodes determine the optical properties of the detector: the polar angle of its look direction, the sensitive area for parallel beams, and the energy-geometric factor for ambient particles. An annular MCP followed by a ring of 128 discrete anodes detects the electrons at their azimuth angle of arrival. The entire sensor section rests on 4 insulator supports and floats at between +2 and +4 kV to pre-accelerate the electrons and bias the MCP. Signals to and from the sensor are routed via optocouplers. Except for 8 narrow struts in the gun-support collar, there is no obstruction of the electron trajectories. For some of the lower optics electrodes that design goal required a support column through the center of the sensor. Note that the GDE section of the unit has been omitted for clarity.

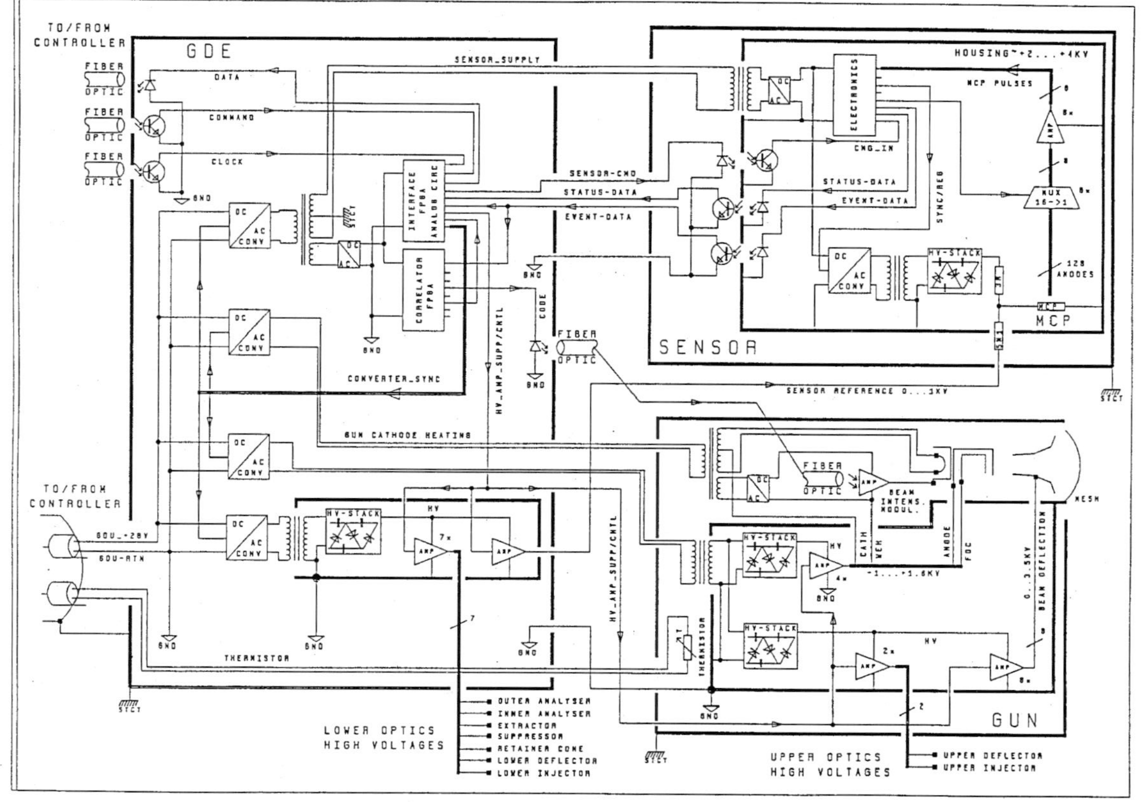

Figure 7. GDU electronics block diagram. The part on the left, designated GDE, provides the power supply for the entire GDU; the D/A and A/D conversions for the signals exchanged via a serial interface over fibre-optics cables with the controller unit; the correlators for the time-of-flight determination; and the high-voltage stack and voltage amplifiers for the lower-optics electrodes. The upper right-hand part shows the sensor section with its programmable MCP high-voltage supply, anode selection logic and the 8 pulse amplifiers/discriminators. The section at the lower right shows the electronics in the gun, with its filament current supply, and its two high-voltage stacks and associated HV amplifiers for the gun and upper optics electrodes, respectively.

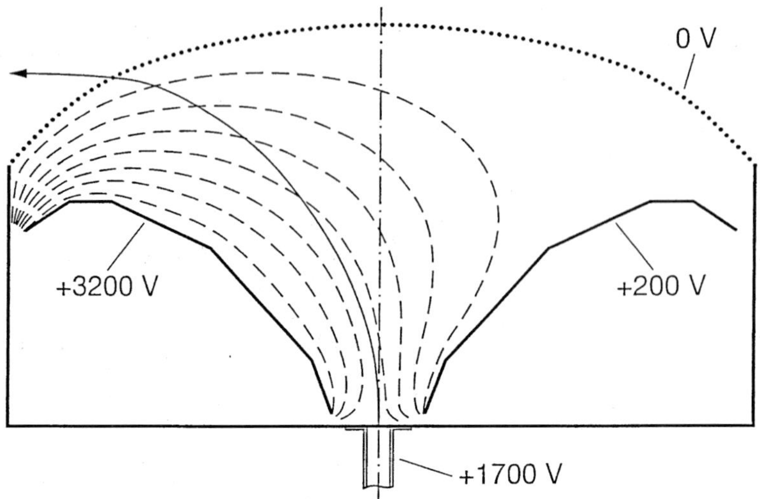

Figure 8. Illustration of the novel beam deflection scheme developed for the EDI electron guns. After exiting the anode that accelerates the 1000 eV electrons emitted by the cathode to 2700 eV, the beam is deflected in the electrostatic potential pattern (dashed lines) produced by two (of eight) deflectors and the outer curved grid (dotted line) that is at ground potential. For the 90° deflection shown the deflectors are at 200 V and 3200 V, respectively. The retarding potential on the grid slows the electrons to their final energy of 1000 eV and bends the trajectories to the desired deflection angle. By changing the voltages on the deflectors, any deflection angle between 0 and 100° can be produced. There are 8 deflectors that are arranged as an octopole to allow deflection of the electron beam at arbitrary azimuth angles. This way the beam can be steered into any direction over more than a hemisphere.

intensity at large deflection angles, as well as some structure in the beam intensity profile.

The gun voltages are generated by individual high-voltage amplifiers that use the light from LEDs to control the leakage current of two high-voltage (HV) diodes connected in a push-pull configuration between the plus- and minus-sides of a HV stack. The control voltages for the eight HV amplifiers of the octopole deflector are derived from two reference voltages, DX, DY, provided by the controller. Deflector voltages are derived from a separate stack and range up to 3.4 kV.

4.2. Detectors

The demands of the Electron Drift Instrument require a detector design that is different from anything flown before. It must be able to look in any direction within a region greater than a 2π steradian hemisphere. To compensate for the low returning beam fluxes, the detectors must have a large effective area, as much as two or three square centimeters. Unlike the natural plasma, the returning beam is monoenergetic and unidirectional. Therefore, the signal-to-noise ratio may be improved by designing the detector to be selective in velocity space. To make continuous electric-field measurements while either magnetic or electric fields vary rapidly, the detectors must be capable of changing their look directions in less than a millisecond. As they cannot be shrouded in any direction, they must have good internal light rejection.

The detector system that we designed to meet these requirements is illustrated in Figure 6. It consists of an optics section, a programmable sensor, associated electronics, and voltage generators. Adjustment of the look direction is achieved in elevation by deflecting the incoming beam, and in azimuth by selecting a contiguous set of sectors of the annular image-region on the sensor micro-channel plate (MCP). The incoming beam of electrons is monoenergetic and monodirectional, and illuminates the entire detector. The large effective area is achieved with double focussing, as explained in Section 4.2.1. Here, 'double focussing' refers to the simultaneous concentration of the beam in two angular planes, rather than in the more commonly understood sense of energy and angle.

To cover all beam directions, two identical detectors are mounted such that they view opposite hemispheres. We have called this detector system 'Janus', after the Roman god with the two back-to-back faces.

4.2.1. *Optics*

After passing through the optics aperture screen, the beam electrons encounter a large transverse electric field generated by the two deflector electrodes on each detector. The injectors also contribute an electric field with components transverse to the particle trajectory and in the radial direction of the detector's cylindrical coordinates. As these electrodes are exposed to sunlight, they cannot be biased negatively or they would expel photoelectrons that would interfere with other spacecraft experiments. The retainer cone provides additional control of transverse and radial electric fields in the central region of the optics. Because the retainer cone at times may be biased negatively, it is constructed from a wire mesh to minimize the surface area from which photoelectrons emanate.

As illustrated in Figure 9, the electron beam is focussed in cross sections parallel to the z axis ('polar' focussing) near the entrance of the electrostatic analyzer, and again near the exit of the analyzer. In the projection perpendicular to the z axis ('azimuthal' focussing), the beam is partially focussed and then diverged before it enters the analyzer, in such a way as to exploit more effectively the final azimuthal focussing that occurs within the analyzer. Together, the deflectors, injectors and retainer cone constitute an 'immersion lens' that projects the beam past the central region into the entrance of the electrostatic analyzer, thereby increasing the effective area of the aperture. The azimuthal focussing also extends the width of the effective aperture area and thus reduces the effects of shadowing caused by the gun-support struts shown in Figure 6.

The deflector potentials play the most important role in determining the polar look direction; the injector potentials are varied primarily to maintain azimuthal beam spread at large deflection angles, but they also influence the look direction;

and the retainer-cone potential provides additional control over the azimuthal focus and polar look direction.

The inner- and outer-analyzer electrodes select electrons in the desired energy range, and their toroidal shapes also contribute to azimuthal focus, as seen in Figure 9. Emerging from the electrostatic analyzer, the electrons pass between two additional electrodes: the extractor and the suppressor. These electrodes adjust the radial position of the MCP image and direct the electrons to strike the MCP with impact angles closer to the surface normal.

Nine independently programmable high-voltage supplies are needed to operate the optics subsystem. A tenth programmable supply sets the sensor-reference voltage. For electron drift measurements, all voltage except the sensor reference are scaled with energy; the voltage on the latter is kept fixed because its capacitative time constant does not permit rapid changes. Although all ten electrode voltages can depend on polar angle, in order to conserve controller resources we vary only the five that depend most strongly on polar angle (deflectors, injectors and retainer cone). All voltages are generated by the same type of HV supplies already referred to in the Gun section.

For a given beam energy and polar angle, different combinations of these voltages can be chosen to obtain different collection areas, A, for the beam and different geometrical factors, H, for the ambient ('background') electrons (cf., Table II). Therefore, with appropriate combinations of voltages we can optimize beam-signal levels or signal-to-noise (SNR) levels, depending on the circumstances. Also, combinations of these ten voltages can be chosen to achieve other special optical characteristics, such as wide or narrow values of polar-angle acceptance or energy bandpass. For example, the detector's energy bandpass can be adjusted for a given incident particle energy by setting independently the voltage difference between the plates (to control the actual bandpass width) and the average voltage on the analyzer plates (to shift the energies of the particles of interest to lie inside the passband as they enter the analyzer). In addition, combinations of voltages can be chosen that achieve almost all of the above characteristics, but measure ions instead of electrons. Figure 10 illustrates sample ion trajectories at a specific energy and initial direction.

To simplify matters, we use a finite number of such voltage combinations, called optics 'states'. Table III lists the beam collection areas A, geometric factors H and acceptance angles for the 'states' presently intended for the electron drift measurements. For each state there is a look-up table from which the voltages that vary as a function of the polar angle are obtained. State 6 has the largest sensitivity A for beams, but also a large H-factor for ambient electrons, and thus not a good SNR. State 3 has the highest SNR, but less than half the area of State 6. States 2 and 7 have only modest SNR, but large acceptance angles, and thus are suited for cases where there is uncertainty in δ, the angle change of the returning beam. State 4, finally, is a 'shut-down' state, to be used when fluxes are very high.

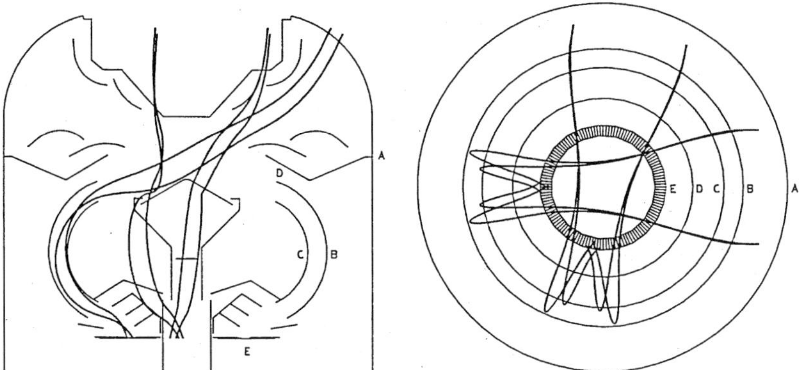

Figure 9. Computer-generated sample trajectories of 1 keV monodirectional electrons at a polar angle of 27°, projected in the (x, z) and (x, y) planes respectively. Two families of four trajectories are shown; these families are identical except for an azimuthal rotation of 80°. Note the two focal points in the (x, z) projection, near the entrance and exit of the analyzer respectively. In the (x, y) projection, note the convergence and divergence of the beam as it traverses the central region of the optics prior to entering the analyzer, followed by the strong azimuthal focussing within the analyzer. The (x, y) projection displays the following selected parts and boundaries of the detector: (A) outer radius of aperture grid; (B) outermost radial extent of outer analyzer plate; (C) outermost radial extent of inner analyzer plate; (D) radial edge of outer analyzer at entrance; (E) sensor annulus.

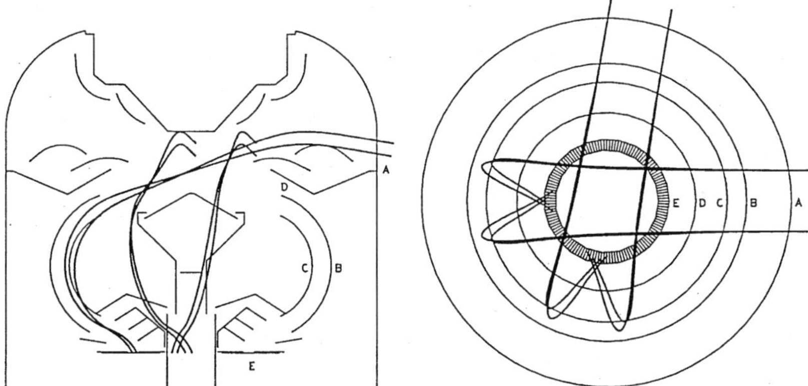

Figure 10. Sample trajectories of 1 keV monodirectional protons at a polar angle of 100°, projected in the (x, z) and (x, y) planes respectively. Two families of four trajectories are shown; these families are identical up to an azimuthal rotation of 80°. Note that focussing in the (x, z) projection occurs twice, as it does for electrons, although not in the same locations as for electrons. In the (x, z) projection, the azimuthal focussing in the central region is more subtle than for electron cases, but the contribution of the analyzers to azimuthal focussing is again clearly evident. The (x, y) projection displays selected parts and boundaries of the detector (see caption to Figure 9).

4.2.2. *Sensor*

Electrons are imaged by the detector optics within the selected polar angle acceptance cone, onto an annular microchannel plate (MCP) stack that is the input to the sensor. Except when the direction of the returning beam is close to the z axis,

Table III
Optics states

State	A	H	Angle
2	2.2	0.110	20
3	2.7	0.016	7
4	0.5	0.001	4
6	5.8	0.210	10
7	1.3	0.022	14

A is the collection area (cm^2) for beams.
$H = G(\Delta E/E)$ is the geometric factor (cm^2 sr eV/eV) for ambient electrons.
Angle is the FWHM acceptance angle (deg).

only a limited azimuthal segment of the image annulus is illuminated by beam electrons, whereas the entire annulus collects ambient electrons that are incident within the optics acceptance cone and energy passband. Under command of the controller, the sensor responds only to events within a selected azimuthal range, thus complementing the polar angle selection that is performed by the optics.

The sensor collects the amplified electron events, which are produced by the MCP stack operated in the pulse-counting mode, on an annular array of 128 discrete anodes. Signals from each anode are routed to one of eight custom hybrid multiplexers, each with 16 single-anode inputs, and whose outputs are routed, in turn, to eight high speed programmable-threshold preamplifier/discriminator hybrids. The sensor control structure is designed so that any eight contiguous anodes in the annular array (22.5° of azimuth) may be selected as the instantaneous field of view. A second-level multiplexer, implemented within an ACTEL programmable gate array, further selects the discriminator outputs to provide two digital output channels. The anodes routed to these two outputs are selected by two arbitrary and independent 'pointer field' bit masks within the eight-anode field of view. Typically, one of the output channels is routed to the instrument correlator for recognition of the electron beam, while the other is available for accumulating total counts or monitoring background.

The electrical potential of the sensor input MCP is established by the detector optical requirements. The sensor internal electronics and housing float at typically +2 kV above this input potential due to the internally generated HV bias voltage across the MCP stack. Sensor power, signal pulses, commands and status are coupled to the rest of the GDU electronics via a high voltage isolation module containing an isolation transformer and 9 fiber optics signal links.

4.3. Correlators

To detect the beam electrons in the presence of background counts from ambient electrons and to measure their flight time, the electron beam is intensity-modulated with a pseudo-noise code (PNC). The modulation is achieved by changing the Wehnelt control voltage such that the beam is successively turned on and off. In order to reject detector signals from the gun in the same unit, the codes for the two gun-detector pairs are inverted relative to each other. The modulation frequency can be chosen between 8 kHz and 4 MHz in order to cover the range of expected time delays and to obtain adequate delay-time accuracy.

The stream of electron event pulses received by the detectors are fed in parallel into an array of counters, each one gated with its individual copy of the PNC, shifted by one chip from counter to counter, and delayed as a whole (by a variable amount) against the PNC used to modulate the outgoing beam. The counter that is gated with a PNC matching the flight time will receive all the beam event ('signal') plus half the background events, while all others receive only half the signal plus half the background. A drift and tracking control loop ('auto-track') varies the delay of the correlator codes relative to the gun code such that the signal is kept in one dedicated counter while the flight time changes as a result of changing magnetic and electric fields.

We first experimented with a long (4095-chip) code. Codes whose lengths exceeds the electron flight time have the advantage that the times can be determined without ambiguity. But since one can only realize a finite number of correlator channels, one must have a very good estimate of the time of flight to properly delay these channels (e.g., to within 0.3% for 15 channels and 4095 chips). Even if the gyro time were known precisely from the on-board magnetometer data, the unknown electric field can cause variations of up to 10%. Thus one would have to vary the code-delay until the time of flight is within the range covered by the 15 correlators.

Because of these problems with long codes, we have finally implemented a short (15-chip) code and 15 correlator channels. Such a solution has the advantage that beam electrons are always counted in one of the channels, regardless of flight time. On the other hand there is the disadvantage that the flight time is determined only modulo the code duration. This ambiguity can usually be removed by starting out with a sufficiently low code-clock frequency such that the entire range of expected flight times fits within one code-length. This initial choice of frequency is based on the gyrofrequency computed from the magnetometer data and an assumed 10% variation in flight-time to account for large electric fields.

The correlator electronics resides in a RAM-based Field-Programmable Gate Array (FPGA) of the XILINX type. Configuration of the FPGA is part of the start-up procedure of the instrument. As the configuration file is held in EEPROM (electrically erasable PROM), other correlator schemes could be uploaded in flight if necessary.

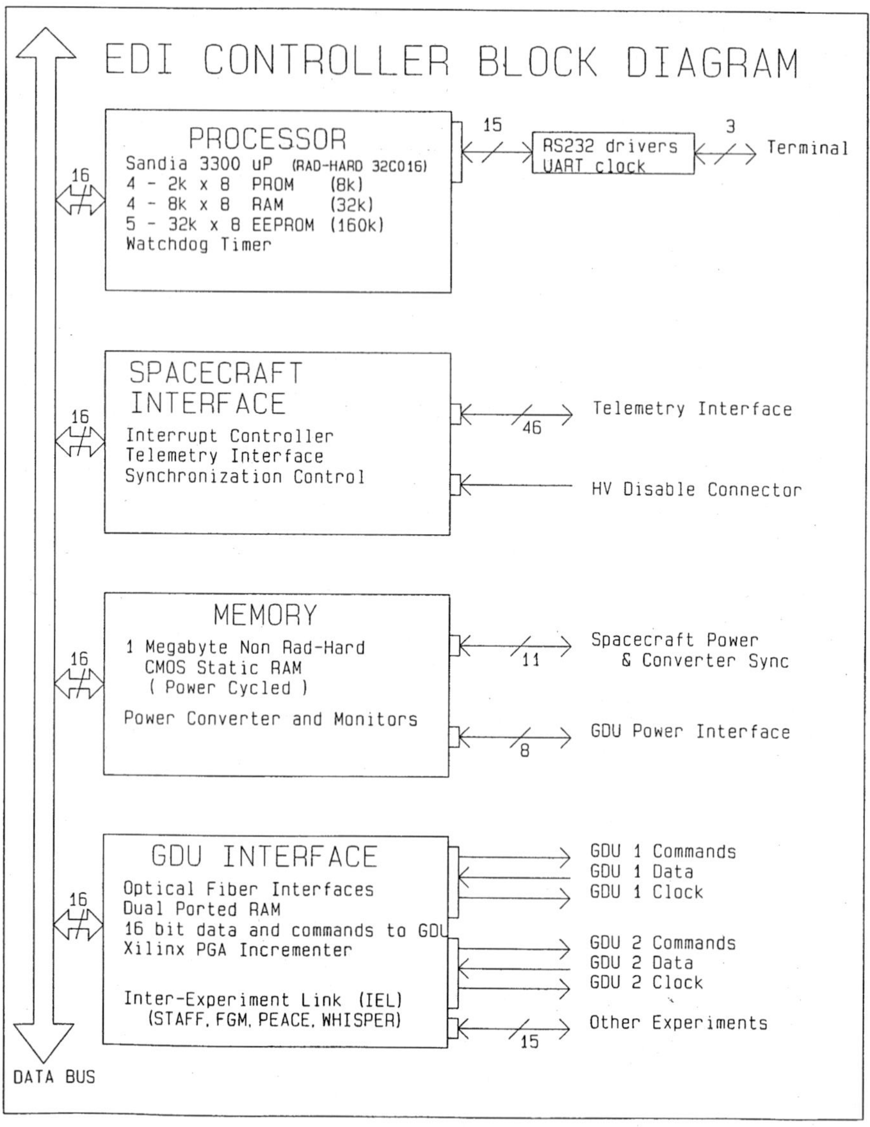

Figure 11. Controller functions.

4.4. Controller

Based upon information from the detectors and the magnetometers, the controller programs the guns and detectors. It establishes beam coding and tracking patterns and handles all interfaces with the spacecraft and with other instruments. This central processor also transfers the programmed gun firing directions (for beams returning to the detectors), plus the timing information from the correlators, to the necessary telemetry data so that the resultant drift-step direction and magnitude can be determined on the ground.

Figure 11 is a sketch of the configuration of the controller. We use a Sandia 3300, a radiation-hardened version of the NSC 32C016, as our central processor. The associated memory has 8 kB of fuse-linked PROM, 160 kB of EEPROM, and 32 kB of static RAM. The fuse-linked PROM contains the software for system initialization, housekeeping, and basic telemetry routines. The EEPROM, which is reprogrammable via spacecraft commands, contains three types of information. Science data algorithms and higher-level telemetry handling make up 48 of the 160 kB of EEPROM. 80 kB are reserved for gun and optics voltage tables. 32 kB are reserved for FPGA configuration data. The 32 kB of CMOS static RAM is available for local storage of various parameters and variables, and also serves as telemetry-data buffer. The controller includes 1 MB of non-hardened memory for storage of burst mode or diagnostic data. This memory can store approximately seven seconds of data recorded at the highest possible time resolution, or approximately one minute of data recorded at debugging speeds.

The controller communicates with the spacecraft, with other instruments, and with the gun/detector electronics. The spacecraft interface handles all science and housekeeping telemetry as well as inputs such as time-tagging and commanding. Inter-experiment links include magnetic field information from FGM and STAFF, a blanking pulse from WHISPER to warn of possible interference from that active experiment, and a blanking pulse sent from EDI to PEACE when the EDI electron beam could interfere with the PEACE electron measurement. This interface with the GDUs is controlled by an FPGA contained in the controller and sends commands from the controller to the GDUs that change the gun and detector hardware parameters, and receives data back from the GDUs, including detector count rates, correlator information, and housekeeping data. The physical interface consists of three optical-fiber cables for each GDU, carrying serial information.

The fundamental functional time step of the EDI instrument is the controller's basis cycle interrupt (BCI), which is nominally 2 or 4 ms. Every BCI, the controller will use the information it receives from the various interfaces to calculate new parameters for directing the beams and the detectors. In the tracking modes of operation, the time step is 4 ms. Tracking tasks that operate on longer time scales, such as slowly changing the beam current or modifying the detector's basic optics state, are driven by a priority-queue that is controlled by a task manager. A 10 ms interrupt or an asynchronous service call initiates the task manager.

4.5. Resources

The GDU mass is 4550 g each and that of the Controller 1610 g. The total power consumption is 9.5 W. The science data rates allocated to EDI are 1520 bit s^{-1} in normal-mode telemetry, and 10780 bit s^{-1} in burst-mode telemetry. An additional 140 bit s^{-1} are used for housekeeping data.

5. Calibrations and Simulations

The electron guns were calibrated at MPE in a dedicated facility at electron energies of 0.5 and 1.0 keV up to the maximum deflection angle of 104° and over the entire 360° in azimuth, with a grid size of 4° and an accuracy of approximately 0.5°. Information on beam profiles and beam intensity control were also gathered during this process.

The EDI detector is calibrated with both a laboratory system and a computer model. The laboratory system simulates the diverged returning EDI beam with an electron gun and two parallel-plate beam mirrors; by manipulating the voltages within these mirrors, the beam can be made to 'raster' across the detector aperture as counts are accumulated on the detector's image plane. Thus a parallel family of narrow beams is used to simulate a broad monodirectional beam. Background is simulated by scattering a second beam from a target. By measuring 'foreground' sensitivity of the optics to the rastered beam and 'background' sensitivity of the optics to the scattered beam, and supplying these quantities to an optimization algorithm that controls the optics voltages, we can determine voltage combinations that achieve desirable optical properties such as high sensitivity and high signal-to-noise ratio, at selected polar angles. A set of such voltage combinations covering all polar angles of interest comprises a 'state'. Adequate rotational symmetry of the optics was verified with a special test setup during EDI acceptance testing.

The computer model of the optics can simulate and display particle trajectories through the detector optics, and accumulate macroparticle 'counts' into bins on the image plane. The user may select sequences of initial conditions for these trajectories so as to cover the regions of phase space to which the detector is sensitive, and thereby determine effective aperture areas, geometrical factors, sensitivity profiles as a function of polar angle, and related quantities. Also, by following a cycle of changing the voltages on the optics electrodes and observing the changes in the displayed particle trajectories, one may evaluate voltage combinations as candidates for inclusion in an optics state.

Because of the large scale of the electron gyroradius and gyroperiods for which the EDI is designed, it is not possible to perform ground tests that fully exercise the instrument. In order to validate its closed loop operation and demonstrate the function of the control algorithms, we have built a tracking simulator that closes the loop between gun and detector even though no beam is actually generated and no electrons detected. The simulator contains a set of tabulated values that are obtained by calculating the proper beam-firing directions and detector-look directions for simulated magnetic and electric fields that are based on actual data measured with instruments on AMPTE-IRM. It feeds these same magnetometer data to the controller via the interface that is used in flight to receive data from the FGM instrument, and taps into the serial interface between controller and GDUs. It compares, against the tabulated values, the instructions which the controller issues to the gun and detector. From the match or mismatch of beam firing directions (the controller-generated instructions vs. the tabulated proper directions), the tracking simulator computes how much, if any, flux arrives at the detector. It adds simulated fluxes of ambient electrons that are also based on AMPTE-IRM measurements. From the detector settings chosen by the controller, the tracking simulator computes the count rates which would result in the sensor, separately for beam and ambient electron fluxes.

Two modes of operation are then possible. Either the count rate information is directly provided to the controller via the serial interface, in which case the correlator characteristics have to be simulated as well; or the simulator directs dedicated hardware pulsers to generate streams of beam and background pulses. The beam pulse stream is modulated with a properly delayed copy of the pseudo-noise code of the outgoing beam in order to reflect the electron time of flight. This way the hardware implementation of the correlators is truly tested instead of being simulated.

The controller tracking software uses these count-rate results, as well as the magnetic-field data it is receiving, as the basis for deciding on the instructions it will send out to the GDUs in the next time step. Running through different simulated data files, we can test and optimize the software and its underlying decision algorithms for their ability to cope with realistic magnetic and electric fields, as well as the implied flux levels and signal-to-noise ratios.

6. Science Modes and Telemetry Data

6.1. Electron Drift Measurements

The simplest operational mode is modelled after the original GEOS application. In this mode, the two beams are steered into directions which are anti-parallel to each other and transverse to the magnetic field. The latter condition will be computed from the magnetometer data received in real-time. Rotation of the satellite will first cause beam 1 and then beam 2 to hit the target S and strike the detectors. Telemetered data consist of a time-series of detector counts from which two drift velocities per satellite rotation can be reconstructed on the ground.

A simple variant of the GEOS Mode, referred to as Rapid Spin (RS), sweeps the beam in the plane perpendicular to **B** at a selectable rate up to 1° per ms. This is a considerable improvement in speed over the GEOS mode where the angular sweep rate is fixed by the satellite rotation, i.e., 0.09° per ms.

Neither GEOS nor RS modes track the target. Our basic tracking mode, referred to as Windshield-Wiper (WW) Mode, uses the RS mode until each beam has hit its target, $S1$ and $S2$. Once this has occurred, we will continuously track $S1$ and $S2$ independently, by sweeping the beams rapidly back and forth across them, in the plane perpendicular to **B**. This is a great improvement over the GEOS and RS schemes, in which the beam is fired at S only when either the spacecraft rotation or the passive angular sweep is at the correct phase. In the 'windshield-wiper' scheme, which is illustrated in Figure 12, the firing direction is controlled actively by the instrument so that the beam hits (or is very near) S all the time.

To recognize that the beam is returning to the detector, the controller constantly (once every 2 ms) computes the signal-to-noise ratio from the maximum and minimum counts in the set of correlator channels. If that ratio exceeds some (selectable) limit, it is assumed that the beam is striking the detector. Initially we will steer each detector to look at a direction anti-parallel to the emission direction of the associated beam. This works as long as the aberration angle δ introduced above is less than the acceptance angle of the detector. For larger aberration angles the detector look-direction must be offset appropriately. This complicates target acquisition.

The 'windshield-wiper' software, which keeps the target direction tracked in angle space, is interwoven intricately with the time-tracking software, which keeps the time-of-flight information available. As noted earlier, the beams are modulated with a coded waveform. Correlating the received signal, after a delay corresponding to the time of flight, with the original coded signal, has two important functions. First, correlating the return signal with the fired signal allows a significant increase in signal-to-noise ratio. Second, the delay that results in the largest correlated signal corresponds to the time of flight of the electron beam.

In the WW-mode, telemetered data consist of the firing angles when the two beams were on target, and the two associated electron times-of-flight, plus timing and quality information. With the allocated telemetry rate it is possible to transmit this set of measurements every 64 ms in nominal-mode (NM) telemetry and every 16 ms in burst-mode (BM1) telemetry. From the telemetered data one can then derive the drift velocities and the magnetic field strength, in BM1 telemetry and under ideal tracking conditions as often as every 16 ms.

We are also working on a '2D-Tracking' Mode in which, after the target has been acquired, the beam is steered according to target-direction estimates based upon Kalman filtering of the observed signals.

6.2. Ambient Particle Measurements

The EDI detectors are capable of measuring both electron and ion particle distributions. Each of the two detectors can be commanded to look in any direction over greater than a 2π steradian hemisphere, and since the two detectors are mounted on opposite sides of the spacecraft with their symmetry axes pointing in opposite directions, full-sky surveys can be achieved without relying on spacecraft spin to complete the coverage of phase space. In addition, since the $\mathbf{B}$ field is known to the EDI controller via the on-board magnetometer data, specialized surveys can be performed in coordinates fixed with $\mathbf{B}$, and they can be performed continuously.

The ambient mode that exploits this ability to scan selectively in coordinates fixed with $\mathbf{B}$ is the Pitch Angle Surveys and Ion Flows (PASIF) mode. This mode contains three sub-modes: (1) pitch-angle surveys, (2) perpendicular ion flows, and

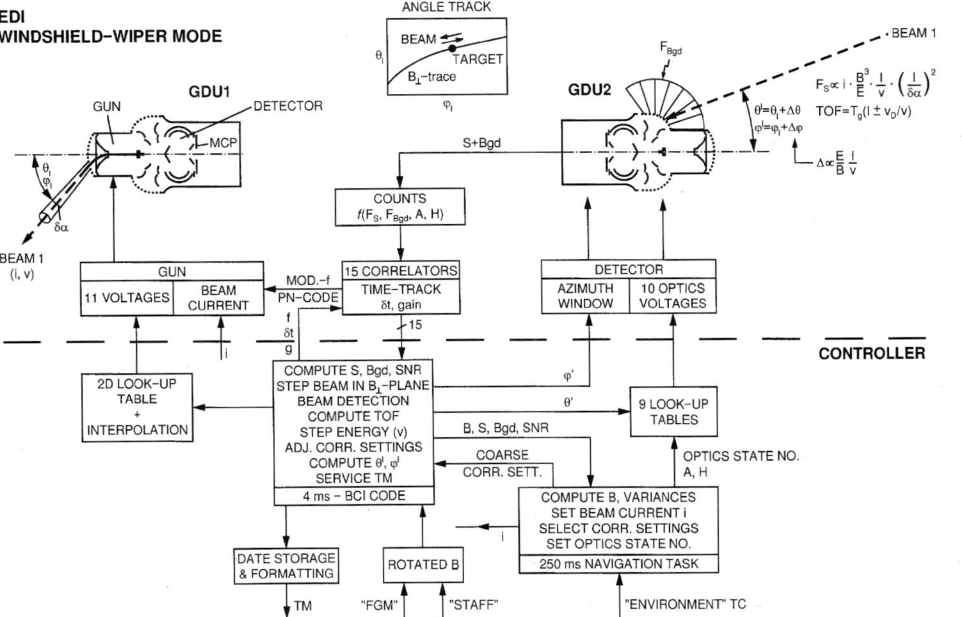

Figure 12. Illustration of the Windshield-Wiper mode of operation. The electron beam from GDU1, characterized by its current i, speed v, direction Θ and ϕ, width $\delta\alpha$, modulation frequency f and code-type, is swept in the plane perpendicular to **B**, as determined from the FGM and STAFF magnetometers. Signals recorded by the detector in GDU2 are analyzed to determine when the beam is on target. A similar link (not shown) exists between the gun in GDU2 and the detector in GDU1. The detector count rate consists of the background signal (Bgd) from the flux of ambient plasma electrons (F_{Bgd}) and the beam signal (S) from the incident flux of beam electrons (F_S) if the beam is on target. Correlators are used to distinguish beam and background electrons, as well as to measure the time-of-flight of the beam electrons. Once beam passage has occurred, the sweep direction of the beam is reversed and the process repeated, as indicated in the box at the top of the figure. For small E/B, the return beam is directed parallel to the emitted beam, but for large E/B there is a significant offset, Δ. Every 4 ms the controller exercises a software loop that directs beams and detectors, and sets the correlator parameters. A navigation task running every 250 ms selects beam currents, optics states, etc. Look-up tables are used to convert directions into deflection voltages.

(3) parallel ion flows. The pitch-angle sub-mode allows pitch-angle distributions to be measured for either electrons or ions at a sequence of directions spaced equally in pitch angle, starting with the $+\mathbf{B}$ and $-\mathbf{B}$ directions for each detector respectively. The number of pitch-angle directions and spacing between directions can be changed via mode initialization tables.

The perpendicular-ion sub-mode of PASIF permits surveys in the plane perpendicular to **B** at equally-spaced azimuthal directions. This sub-mode will be used with azimuthal spacings at or near 30° and 60° and can be used to infer ion flow velocities perpendicular to **B**.

The parallel-ion sub-mode of PASIF permits surveys in a family of cones with the **B**-field direction as their common axis. Such surveys will reveal the angular offset from the **B** direction of the distribution peak in the nearly field-aligned flows, thereby allowing for cases with perpendicular ion flows.

The maximum and minimum voltage limits available to the EDI optics were chosen originally to allow electron beams of energy 1 keV or less to be detected over all desired look directions in a region greater than a 2π steradian hemisphere. As we scale the optics voltages to observe ambient particles of energy greater than 1 keV, we may reach these limits on some electrodes, particularly the deflectors. This will cause the look directions at some extreme polar angles (near the pole and the equator) to become inaccessible at higher energies. Also, because the ion states achieve different polar look-angles by repulsively reflecting the ions entering the aperture grid, the sensitivity of the optics to ions decreases in the neighborhood of the pole and vanishes at the pole. In the relatively extreme case of observing ions at 10 keV, polar angles between 48° and 90° can still be achieved, providing access to two thirds of a hemisphere.

Acknowledgements

We acknowledge the contributions to the EDI proposal of the co-investigators A. Pedersen (ESA/ESTEC), V. Formisano and E. Amata (CNR), K. Tsuruda, M. Nakamura and H. Hayakawa (ISAS). We are indebted to the following individuals for the part they have played in instrument manufacture, testing and qualification: at MPE, P. Obermayer, K.-H. Kaiser, E. Künneth, W. Goebel, W. Stöberl, K. Sigritz, and the staff of the workshops headed by G. Pfaller, P. Reiss, and L. Pichl; at UCSD, G. Kendrick, J. Ritter, S. Beros, and K. Larson; at UNH, J. Googins, C. Kletzing, K. Mello, S. Turco, P. Kelly, W. Isaac, D. Hallmark, D. Bodet, and P. Vachon; and at LPARL, M. Gabriel, G. King, K. Strickler, and E. McFeaters.

References

Aggson, T. L. and Heppner, J. P.: 1977, 'Observations of Large Transient Magnetospheric Electric Fields', *J. Geophys. Res.* **82**, 5155.

Baumjohann, W. and Haerendel, G.: 1985, 'Magnetospheric Convection Observed Between 06:00 and 21:00 LT: Solar Wind and IMF Dependence', *J. Geophys. Res.* **90**, 6370.
Baumjohann, W., Haerendel, G., and Melzner, F.: 1985, 'Magnetospheric convection observed between 06:00 and 21:00 LT: Variations with Kp', *J. Geophys. Res.* **90**, 393.
Baumjohann, W., Sckopke, N., LaBelle, J. K., Klecker, B., Lühr, H., and Glassmeier, K. H.: 1987, 'Plasma and Field Observations of a Compressional Pc 5 Wave Event', *J. Geophys. Res.* **92**, 12203.
Cowley, S. W. H.: 1978, 'A Note on the Motion of Charged Particles in One Dimensional Magnetic Current Sheets', *Planetary Space Sci.* **26**, 539.
DeForest, S. E. and McIlwain, C. E.: 1971, 'Plasma Clouds in the Magnetosphere', *J. Geophys. Res.* **76**, 3587.
Escoubet, P., Schmidt, R., and Goldstein, M. L.: 1996, 'Cluster: Science and Mission Overview', *Space Sci. Rev.* , this issue.
Hones, E. W., Jr.: 1979, 'Transient Phenomena in the Magnetotail and Their Relation to Substorms', *Space Sci. Rev.* **23**, 393.
McIlwain, C. E.: 1972, in B. M. McCormac (ed.),*Earth's Magnetospheric Processes*, D. Reidel Publ. Co., Dordrecht, Holland, p. 268.
McIlwain, C. E.: 1981, in S.-I. Akasofu and J. R. Kan (eds.),*Physics of Auroral Arc Formation*, AGU, Washington, p. 173.
Melzner, F., Metzner, G., and Antrack, D.: 1978, 'The GEOS Electron Beam Experiment', *Space Sci. Rev.* **4**, 45.
Paschmann, G., Melzner, M., Haerendel, G., Bauer, O. H., Baumjohann, W., Sckopke, N., Treumann, R., McIlwain, C. E., Fillius, W., Whipple, E. C., Torbert, R. B., and Quinn, J. M.: 1993, in W. R. Burke (ed.),*Cluster: Mission, Payload and Supporting Activities*, ESA SP-1159, p. 115.
Pedersen, A., Grard, R., Knott, K., Jones, D., Gonfalone, A., and Fahleson, U.: 1978, 'Measurements of Quasi-Static Electric Fields Between 3 and 7 RE on GEOS-1', *J. Geophys. Res.* **22**, 333.
Pedersen, A., Cattell, C. A., Fälthammar, C.-G., Formisano, V., Lindqvist, P.-A., Manka, R. H., Mozer, F. S., and Torbert, R.: 1984, 'Quasistatic Electric Field Measurements with Spherical Double Probes on the GEOS and ISEE Satellites', *Space Sci. Rev.* **37**, 269.
Riedler, W., Torkar, K., Rüdenauer, F. *et al.*: 1996, 'Active Spacecraft Potential Control', *Space Sci. Rev.* , this issue.
Sheldon, R. B. and Gaffey Jr., J. D.: 1993, 'Particle Tracing in the Magnetosphere: New Algorithms and Results', *Geophys. Res. Letters* **20**, 767.
Sheldon, R. B. and Hamilton, D. C.: 1994, 'Ion Transport and Loss in the Earth's Quiet Ring Current: 2. Diffusion and Magnetosphere Ionosphere Coupling', *J. Geophys. Res.* **99**, 5705.
Tsuruda, K., Hayakawa, H., and Nakamura, M.: 1985, in A. Nishida (ed.),*Science Objectives of the Geotail Mission*, ISAS, Tokyo, p. 234.

ACTIVE SPACECRAFT POTENTIAL CONTROL

W. RIEDLER and K. TORKAR
Institut für Weltraumforschung der Österreichischen Akademie der Wissenschaften, Inffeldgasse 12, A-8010 Graz, Austria

F. RÜDENAUER and M. FEHRINGER
Österreichisches Forschungszentrum Seibersdorf, A-2444 Seibersdorf, Austria

A. PEDERSEN, R. SCHMIDT, R. J. L. GRARD and H. ARENDS
Space Science Department, ESA/ESTEC, NL-2200 AG Noordwijk, The Netherlands

B. T. NARHEIM and J. TROIM
Forsvarets Forskningsinstitutt, N-2007 Kjeller, Norway

R. TORBERT
University of New Hampshire, Durham, NH 03824, U.S.A.

R. C. OLSEN
Naval Postgraduate School, Monterey, CA 93943, U.S.A.

E. WHIPPLE
University of Washington, Seattle, WA 98195, U.S.A.

R. GOLDSTEIN
Jet Propulsion Laboratory, Pasadena, CA 91109, U.S.A.

N. VALAVANOGLOU
Technische Universität Graz, A-8010 Graz, Austria

HUA ZHAO
Center for Space Science and Applied Research, Chinese Academy of Sciences, Beijing 100080, P.R. China

Abstract. Charging of the outer surface or of the entire structure of a spacecraft in orbit can have a severe impact on the scientific output of the instruments. Typical floating potentials for magnetospheric satellites (from +1 to several tens of volts in sunlight) make it practically impossible to measure the cold (several eV) component of the ambient plasma. Effects of spacecraft charging are reduced by an entirely conductive surface of the spacecraft and by active charge neutralisation, which in the case of Cluster only deals with a positive potential. The Cluster spacecraft are instrumented with ion emitters of the liquid-metal ion-source type, which will produce indium ions at 5 to 8 keV energy. The operating principle is field evaporation of indium in the apex field of a needle. The advantages are low power consumption, compactness and high mass efficiency. The ion current will be adjusted in a feedback loop with instruments measuring the spacecraft potential (EFW and PEACE). A stand-alone mode is also foreseen as a back-up. The design and principles of the operation of the active spacecraft potential control instrument (ASPOC) are presented in detail. Flight experience with a similar instrument on the Geotail spacecraft is outlined.

1. Introduction

Spacecraft in the Earth's magnetosphere and in the solar wind environment charge electrostatically in response to various charging currents. The most important cur-

Space Science Reviews **79:** 271–302, 1997.

Table I
Typical floating potentials of spacecraft in sunlight

Plasma environment	Floating potential, V	Reference
Solar wind	+5 to +10	Pedersen *et al.* (1983)
Magnetosheath	+2 to +5	Pedersen *et al.* (1983)
Outer magnetosphere	+2 to +15	Pedersen *et al.* (1983)
Lobes	+15 to +100	Pedersen *et al.* (1983)
Plasmasheet (quiet)	+10 to +20	Lindqvist (1983) (ISEE-1)
Plasmasheet (geosynchronous, disturbed)	−70	Whipple *et al.* (1983) (ISEE-1)
	>−2000	Reasoner *et al.* (1976) (ATS-6)
Plasmasphere	0 to +1	Lindqvist (1983) (ISEE-1)
	−5.4 to 0	Whipple *et al.* (1974) (OGO 3)

rents are due to photoemission caused by sunlight, plasma currents due to the environmental electrons and ions, and secondary electron currents caused by the impact of primary electrons and ions. In steady state, a spacecraft will charge to an equilibrium potential where the sum of the currents vanishes so that there is no net transfer of charge between the spacecraft and the environment (for reviews, see Grard, 1973; Whipple, 1981).

The plasma densities in the Earth's ionosphere and plasmasphere are large enough that their plasma ion and electron currents by far exceed the currents from photoelectrons. Spacecraft in these regions charge to negative potentials of a few volts, because of the lower mass and consequently higher thermal velocities of the electrons as compared to the ions. Moderate negative potentials may also be found in the ring current, near-Earth plasma sheet and auroral zones, and transient charging events may occur there under storm or substorm conditions, when the plasma electron temperature is high. Negative potentials of several hundreds of volts in sunlight have been reported for spacecraft with exposed insulating areas such as ATS-5, ATS-6, and SCATHA (Olsen and Purvis, 1983; Mullen *et al.*, 1986), but spacecraft with conducting exposed surfaces such as ISEE and GEOS have experienced much smaller potentials (Wrenn, 1979).

Outside the plasmasphere, where the plasma density decreases from some 100 cm^{-3} to much lower values, and as long as the electron energy is not too high (less than a few keV), current balance to a spacecraft is usually between the plasma electron and photoelectron currents, and spacecraft potentials are a few volts positive. In extremely low density plasmas (<0.1 cm^{-3}), such as in the lobes of the Earth's magnetotail, potentials can reach values up to about 100 V. GEOS and ISEE observations for sunlight conditions have been presented in Pedersen *et al.* (1983) and Lindqvist (1983). A summary of spacecraft potentials in sunlight is provided in Table I.

The particularly disturbing and even dangerous local charging of surfaces by natural energetic electron beams or variable secondary emission characteristics of the surface material is avoided by an entirely conductive surface of the spacecraft. Taking this into account, the potential of a Cluster satellite should be similar to that of the GEOS and ISEE spacecraft: namely, the potential should be determined by a balance between photoemission and plasma electron currents, as long as the spacecraft is outside the plasmasphere and in sunlight, but not in the region of substorm or storm plasmas. Several authors have shown experimentally that, under these conditions, the potential of a spacecraft is determined by the quantity $n\sqrt{kT_e}$, where n is the plasma electron density, and kT_e the electron energy (for GEOS-2: Knott *et al.*, 1983 and Schmidt and Pedersen, 1987; for ISEE-1: Lindqvist, 1983), and the plasma electron current is nearly proportional to this quantity. In orbital-limited mode, when the Debye length is much larger than the body, which is a typical case in the outer magnetosphere, the random electron current I_{a0} from the plasma to a spherical satellite surface A is approximated for a Maxwellian velocity distribution to

$$I_{a0} = \frac{A}{4} n_e e \sqrt{(8kT_e)/(\pi m_e)} \,, \tag{1}$$

in the notation by Pedersen (1995) of the original equation developed by Mott-Smith and Langmuir (1926). For positive spacecraft potentials V_s the plasma electron current is further approximated by

$$I_a = I_{a0}(1 + V_s/V_e) \,, \qquad V_e = kT_e/e \,, \tag{2}$$

with e, n_e, m_e, T_e being the charge, density, mass, and temperature of the plasma electrons and k the Boltzmann constant. The area A for a spherical spacecraft of radius R is $4\pi R^2$. With $A = 8$ m^2 for a Cluster spacecraft area and $V_s = 0$, the plasma electron current of Equation (1) becomes numerically:

$$I_{a0} = 0.22 n_e \sqrt{kT_e} \quad (\mu\text{A}) \,, \tag{3}$$

where I_{a0} is in μA, n_e in electrons cm^{-3} and (kT_e) in eV.

The photoelectron current I_p is, to a large extent, a function of the material properties of the surface. The photoelectron saturation current density j_{ps} varies for different materials and between laboratory and *in-orbit* measurements. Feuerbacher and Fitton (1972) quote laboratory data of j_{ps} for several spacecraft materials in the range between 10 and 40 μA m^{-2}. For the ISEE-1 spacecraft a value of 20 μA m^{-2} was derived by Pedersen *et al.* (1984). Higher values up to 60–80 μA m^{-2} have been found in a study of ISEE-1, GEOS-1, GEOS-2, and Geotail potentials by Pedersen (1995). Part of this variability is explicable by different materials, surface finishes, and aging processes in the space environment.

The photoelectron current effectively emitted decreases with $V_s > 0$ when a fraction of the photoelectrons return to the surface. The total current from the sunlit cross-sectional area, A_s, is for positive spacecraft potential V_s:

$$I_p = A_s j_{ps} \exp(-eV_s/kT_p) \,. \tag{4}$$

The mean energy, kT_p, of the photoelectrons is about 1.5 eV (Grard, 1973). It can be considered as a typical value for a wide range of materials and has been derived using the appropriate solar spectrum together with laboratory data. Using experimental data from satellites, Pedersen (1995) found a best fit to the current-voltage characteristics with e-folding voltages of 2.0 to 2.5 V for low and moderate spacecraft potentials below about 20 V, and even higher e-folding voltages for fits at higher potentials.

If we use the value for j_{ps} of 80 μA m^{-2} quoted by Schmidt *et al.* (1995) and a projected area A_s of a Cluster spacecraft of 3 m^2 we obtain numerically

$$I_p = 240 \exp(-eV_s/kT_p) \quad (\mu\text{A}) \,. \tag{5}$$

The floating potential of the spacecraft in our approximation results from the equilibrium $I_a = I_p$.

The determination of an effective surface area for the plasma electron current is difficult. Pedersen (1995) uses the projected spacecraft area for practical reasons, while the total surface area of about 22 m^2 would overestimate the current because it would neglect focusing effects. For the numerical examples in this work a value between the projected and the total area, $A = 8$ m^2, has been used. With Equations (3) and (5) and $kT_p = 2$ eV one gets for Cluster

$$V_s = 14 - 2 \ln(n_e \sqrt{kT_e}) \quad (\text{V}) \tag{6}$$

with n_e in electrons cm^{-3} and (kT_e) in eV as an estimate of the floating spacecraft potential without operation of the ion emitter. Schmidt *et al.* (1995) report good agreement of a best fit with $kT_p = 2$ eV with Geotail data for $0 < V_s < 10$ V. Even this simple formula shows that the potential rapidly increases well above 10 V when the plasma density becomes smaller than 1 cm^{-3}, but it underestimates high spacecraft potentials. Pedersen (1995) derives a best fit for $V_s > 10$ V using $j_{ps} = 80$ μA m^{-2} and a correction term to I_p which transforms Equation (5) into the relationship

$$I_p = 240 \exp(-V_s/2) + c \exp(-V_s/7.5) \quad (\mu\text{A}) \tag{7}$$

with $c = 9$. Schmidt *et al.* (1995) show that some Geotail observations still exceed the I_p value given by Equation (7) and $c = 9$ and suggest $c = 24$, which fits best the Geotail data with $V_s > 10$ V in the absence of potential control. Figure 1 shows the floating potential of a Cluster spacecraft over the quantity $n\sqrt{kT_e}$ in three approximations based on the Equations (5) and (7) and therefore only valid for quiet, i.e., no substorm-type conditions. The solid curve according to $c = 24$ in Equation (7) is considered the most probable prediction. By the normalisation

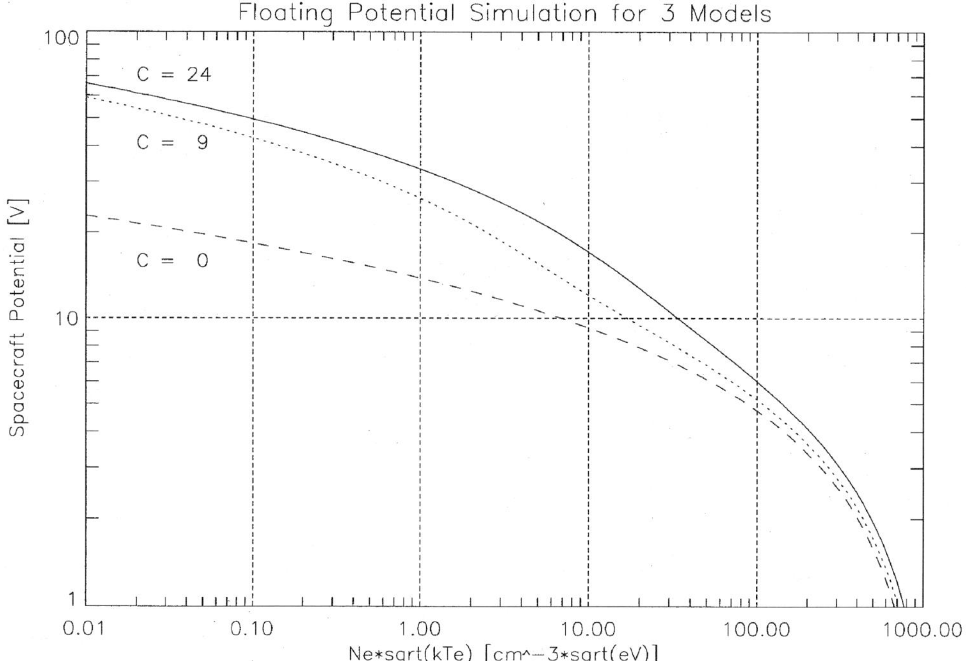

Figure 1. Floating potential of a Cluster spacecraft over the quantity $n\sqrt{kT_e}$ in three approximations: the solid line curve follows Equation (7) with $c = 24$ after Schmidt *et al.* (1995), the dotted and dashed lines are the overall best fit (Equation (7) with $c = 9$) and the best fit for $0 < V_s < 10$ V (Equation (6)), respectively, after Pedersen (1995).

of the abscissa the temperature variations of the electrons have been eliminated apart from corrections due to the term $(1 + V_s/V_e)$ of Equation (2). The figure is obtained for a plasma electron energy $kT_e = 100$ eV and remains valid for higher energies as long as the assumptions for the underlying equations hold. For smaller energies, this correction leads to a decrease of the spacecraft potential by factors of ≈ 1.1 and ≈ 1.5 for 10 and 1 eV, respectively, with respect to the normalised curves.

2. Scientific Objectives and Capabilities

2.1. Reasons for Satellite Charge Control

The primary motivation for active spacecraft potential control (ASPOC) on Cluster is to ensure effective, complete measurement of the ambient plasma distribution functions. On Cluster, the instruments that gain most are PEACE (Plasma Electron And Currents Experiment, Johnstone *et al.*, this issue) and CIS (Cluster Ion Spectroscopy, Rème *et al.*, this issue) for the electron and ion populations, respectively. Additional benefits may be found in improved data for the long-wire electric-field measurement (Electric Fields and Waves, EFW, Gustafsson *et al.*, this issue) and the electron-beam probe for electric fields (Electron Drift Instrument, EDI, Paschmann *et al.*, this issue).

Typical floating potentials for magnetospheric satellites of up to several tens of volts obscure, or render impossible, the measurement of the core of the ion-distribution function, which has a thermal energy comparable to the satellite potential. This problem was indicated by discrepancies in density calculations from satellites such as GEOS-1 (Décréau *et al.*, 1978). The densities inferred from ion spectrometers, for example, did not agree with the total electron density measurements obtained from wave techniques. Measurements in eclipse, made on ATS-6, SCATHA, and DE-1 have shown the appearance of previously 'hidden' ion populations, invisible in sunlight (Olsen, 1982).

Figure 2, from Olsen (1982), shows ion-distribution functions for the outer plasmasphere. The 90° measurements (upper panel) show the low-temperature plasma visible in an eclipse. The field-aligned component (lower panel) shows how deceptive the sunlight measurements can be – the high energy tail has a different temperature than the core of the plasma distribution. These unique measurements of the outer plasmasphere, at the equator, were possible only because the absence of the photoelectron current allowed the satellite potential to fall to near 0 V. Unfortunately, this situation is a rather rare occurrence, since it is only possible at local midnight, in the near-Earth regions. Negative charging to lesser degrees can occur throughout the near-Earth plasma sheet and auroral zones.

Other fundamental problems occur in the measurement of the anisotropic distributions outside the plasmasphere, such as the field-aligned flows which make up the polar wind (Olsen *et al.*, 1986). The bulk of the distribution is lost due to the satellite potential, and it is not possible to obtain fundamental parameters such as density, temperature and flow velocity for H^+.

Moreover, electron measurements gain by satellite potential control: usually, the low-energy portion of the electron spectra is contaminated by photoelectrons from the satellite surface, trapped by the positive satellite potential. Such effects can cause substantial errors in interpreting electron spectra. Automatic calculation of moments, in particular, is made more difficult.

Deleterious effects of spacecraft charging also extend to electric-field measurements. The double-probe technique for electric-field measurements (EFW, Gustafsson *et al.*, this issue) can respond to the local electric field induced by the satellite charge. This is one motivation for extending the booms to substantial distances from the satellite. By reducing the positive potential of the satellite and of the conductive long-wire booms, supporting the electric-field probes, the local field perturbation should be reduced, enhancing the accuracy of the electric-field measurement.

The electron-beam technique for electric-field measurements (EDI, Paschmann *et al.*, this issue) is also somewhat sensitive to satellite potential. If the beam energy is 1 keV and the spacecraft potential is 50 V, the perturbation to the measurement

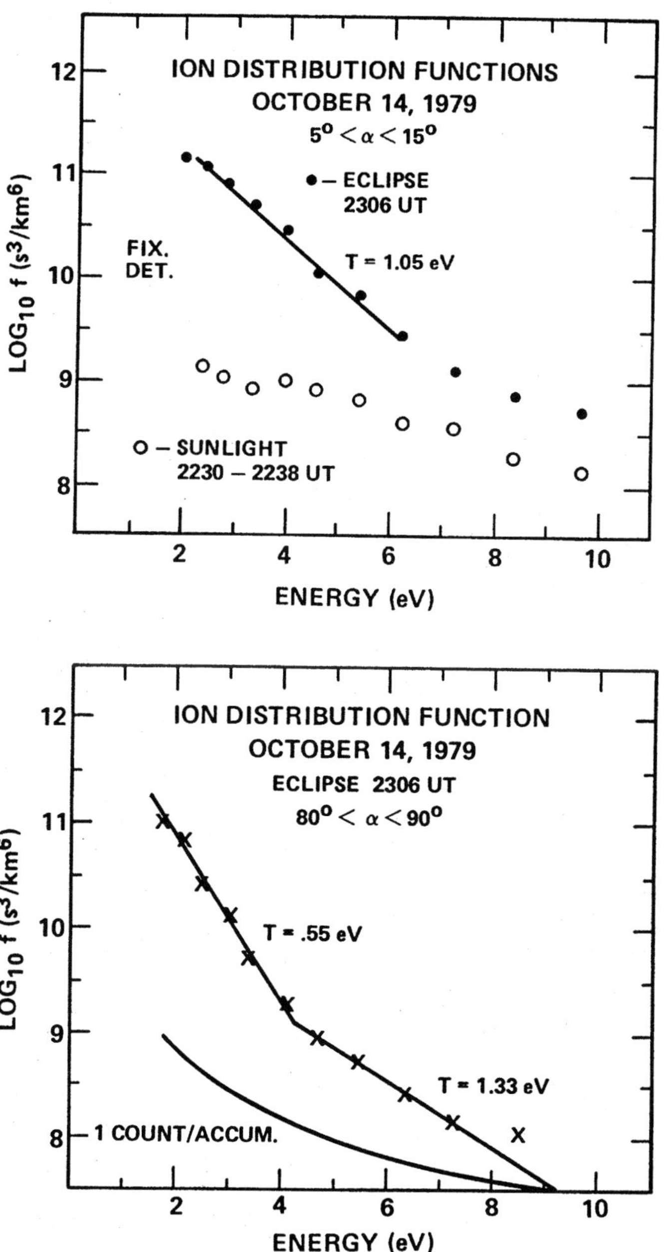

Figure 2. Data from the SCATHA satellite, in sunlight and eclipse, at local midnight, near geosynchronous orbit.

is about 5%. The disturbance should be kept to low values in order to be able to use the full capability of the beam technique to measure small electric fields.

Finally, by active spacecraft potential control the scientific database of spacecraft potential measurements will be enhanced and prepare the ground for a better understanding of the charging processes.

2.2. Technique of Spacecraft Potential Control

The basic approach to the active control of spacecraft potential involves the emission of charges from the spacecraft sufficient to balance the excess of charge accumulating on the vehicle from the environment. The instrument ASPOC is only capable of emitting positive charges, and hence cannot control negative spacecraft potentials. However, over most parts of the orbit with possible exceptions in the near-Earth plasma sheet and the auroral zones the photoemission of electrons is expected to drive the spacecraft potential positive relative to the plasma potential. Even in dense plasma the photoelectron emission prevents large negative excursions (>10 V) except in the most extreme magnetospheric environments; i.e., substorm or storm plasmas. An electron source, preferably emitting lower temperature electrons, would be necessary to prevent negative charging.

In the case of a positive spacecraft potential, which is of primary concern here and overlaps with the regions of primary scientific interest for the Cluster mission, it is necessary to emit positive ions in order to control the spacecraft potential. One approach to the technique has, for example, been described in Pedersen *et al.* (1983) and was most recently implemented in an experiment on the Geotail spacecraft (Schmidt *et al.*, 1993). The working principle of the technique is illustrated in Figure 3. The normal floating potential, indicated by 'V_{FO}', is established where the emitted photoelectron current balances the current collected from the ambient plasma (as shown above). If an additional positive ion emission current (I_e) is emitted, the equilibrium between all currents occurs at a lower potential (V_{FO}'); hence this arrangement reduces the floating potential of the spacecraft. By adjusting the emitted positive ion current, the spacecraft potential can be lowered to fall within the steep slope of the photoemission characteristic.

The effect of this technique, given the size of the Cluster spacecraft, is illustrated in Figure 4. It depicts the spacecraft potential as a function of the plasma parameter $n\sqrt{kT_e}$ for different ion emission currents between 0 and 40 μA. As in Figure 1 above, the solid and dotted lines refer to $c = 24$ and $c = 9$, respectively, in Equation (7). The curves labelled '$I = 0$' correspond to the curves '$c = 24$' and '$c = 9$' in Figure 1. One can recognise that the most prominent effect of active ion emission is a stabilisation of the spacecraft potential with respect to variations of ambient plasma parameters. Only at medium to high densities (e.g., $n_e > 3$ cm^{-3}) the emission of a constant ion emission current $I_e > 5$ μA would not stabilise the potential completely, but rather reduce the amplitude of natural variations further with respect to an already small range in the uncontrolled case. The energy (in this numerical example assumed as $kT_e \approx 10$ eV) is less important than the density. It

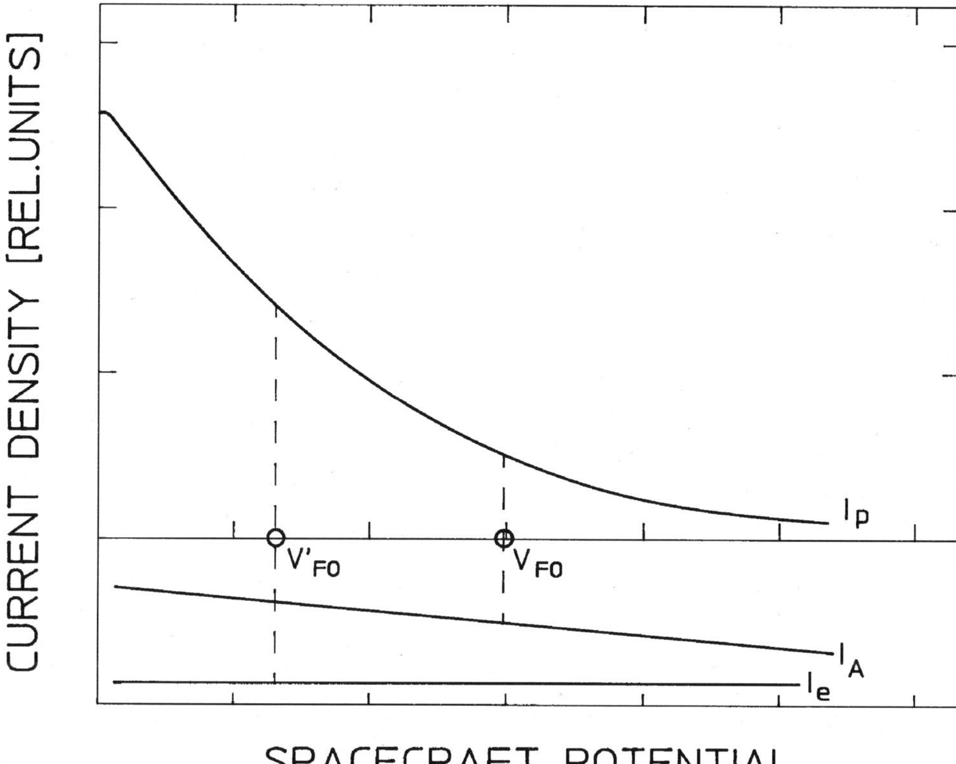

Figure 3. Current-voltage diagram of a sunlit surface in space. The floating potential V_{FO} is determined by the current balance $I_p = I_a$. If an additional ion current I_e is emitted, then $I_p = I_a + I_e$ and the surface potential would settle at V'_{FO}.

is mainly for this range of plasma parameters that the possibility of automated ion current control has been considered.

For ($n_e > 100$ cm^{-3}, $kT_e \approx 10$ eV) the electron current from the ambient plasma becomes sufficiently large to compensate for most of the photoemission. The floating potential lowers compared to the thin plasma case, and the ion emission becomes ineffective. Equation (6) yields the plasma density at which the potential turns negative with $n_e \approx 350$ cm^{-3} ($kT_e \approx 10$ eV). Figure 4 also illustrates that the range of plasma parameters, where the emission of an ion current could contribute to this change of polarity, is narrow.

Finally, in the discussion of Figure 4, one should note that the absolute values of the spacecraft potential, together with medium to high ion emission currents, are dominated by the photoelectron yield of the sunlit spacecraft surfaces. Although measurements from previous space experiments are slowly accumulating, they have not so far been fully conclusive. Within the range of photoelectron saturation current densities j_{ps} between 20 and 80 μA m^{-2} mentioned in the introduction,

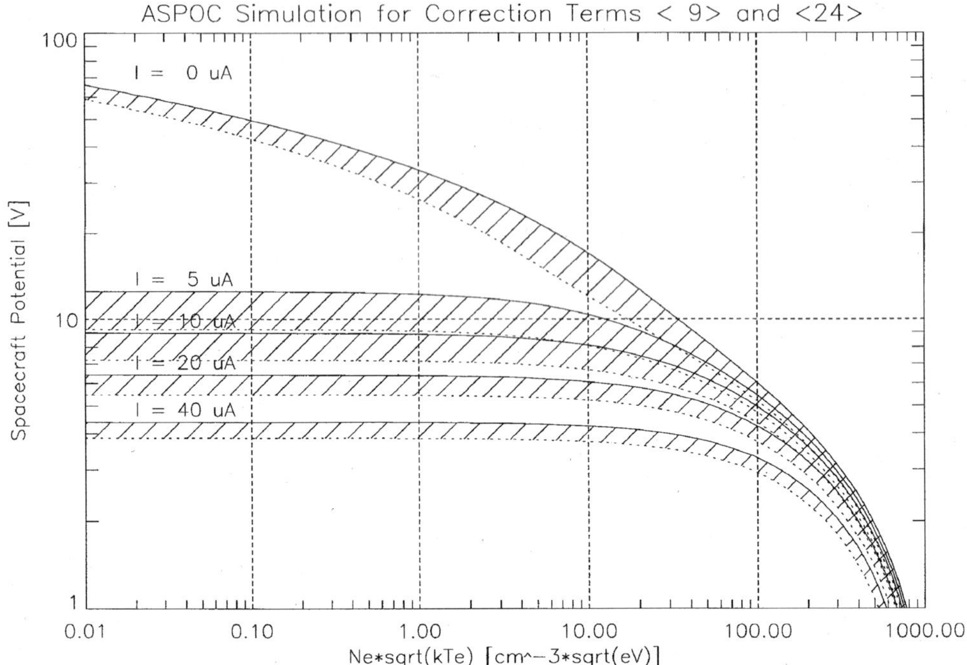

Figure 4. Spacecraft potential as a function of the plasma parameter $n\sqrt{kT_e}$ for different ion emission currents between 0 and 40 μA, calculated for a Cluster-size spacecraft. The solid and dotted lines refer to $c = 24$ and $c = 9$, respectively, in Equation (7).

the curves in Figure 4 correspond to the upper limit, thereby giving a worst-case estimate for the capabilities of the ion emission. Neglecting any contribution from plasma electrons and without the correction introduced in Equation (7) in order to explain very high observed potentials, the balance between photoemission and active ion emission results in a spacecraft potential V_s given by

$$V_s = \frac{kT_p}{e} \ln(A_s j_{ps}/I_e) . \tag{8}$$

Using the values for j_{ps} of 20 and 80 μA m^{-2}, an ion emission current $I_e = 10\,\mu$A would set the Cluster spacecraft potential within 3.6 and 6.4 V. The study of the photoemission characteristics and its development over the mission lifetime form another part of the ASPOC scientific objectives.

2.3. Precautions against Interference with Other Experiments

2.3.1. *Electromagnetic Compatibility and Beam/Spacecraft Interaction*

Any harmful effect on any other experiment by actively controlling the spacecraft potential is unlikely if the emission current is kept within reasonable limits with respect to beam-plasma instability thresholds. However, the necessity of operating the emitter in close consultation with all the other experimenters is recognised.

The emission of a charged particle beam into a plasma is a potential source of electrostatic and electromagnetic noise. The extent of this possibility in the case of ASPOC is not yet totally understood, especially since a thorough analysis of electrostatic and electromagnetic effects during the active potential control experiment on-board the *Geotail* spacecraft launched in 1992 is not yet available. However, effects being strong enough to raise concerns are clearly absent. Some weak electrostatic noise was observed in the frequency range below 20 Hz (Schmidt, personal communication), but very often the ion emitter was serving as neutraliser for an electron emitter which operated simultaneously and may have generated a large proportion of the observed emissions. More experience in orbit with Geotail and Cluster instruments is expected to shed more light on the understanding of the beam-plasma interaction.

There are numerous theoretical papers dealing with ion beam-plasma interactions (e.g., Kintner and Kelley, 1983; Miura *et al.*, 1983; Walker, 1986). However, work on the triggering of wave activity by ion beams with current characteristics comparable to those of ASPOC ($I_e \approx 10\ \mu$A, energy 5 to 8 keV) was practically non-existent in the literature prior to these technically similar investigations on Geotail and Cluster. Ion beams with several orders of magnitude higher current densities were injected into the ionosphere and generated waves around the lower hybrid resonance frequency (Jones, 1981). The injection of a non-neutralised proton beam (2 mA, 6.5 keV) from the Space Shuttle did never trigger any observable high frequency wave (experiment PICPAB, Mourenas *et al.*, 1989). Other experiments found increased emission in the magnetic field around the local electron gyro frequency.

A numerical simulation study (Schmidt *et al.*, 1992) dealt with the plasma response to the emission of very weak ion beams where both the background plasma and the ion beam parameters were adjusted to those of the instrument described here. It was found that the waves had a maximum growth when both the beam and the wave propagation direction were transverse to the ambient magnetic field. The growth rates were very sensitive to the temperature of the ions in the emitted beam and, depending on the background plasma, the instability could be reduced by increasing the beam current. The conditions prevailing in the central plasma sheet and the plasma sheet boundary layer produced the lowest current threshold for the beam-driven lower hybrid type instability, followed by the solar wind. In the lobe, the beam must be very cold for an instability to be excited. The results were based on exponential wave growth in a linear treatment. The level of self-stabilisation in nature owing to nonlinear effects will be subject of further studies.

Since the instabilities were found to be sensitive to ambient plasma conditions, the direction of the magnetic field and the temperature of the beam, only an adverse combination of parameters would lead to wave growth. In the lobes, where spacecraft potential control is most valuable, the injection only rarely occurs transverse to the magnetic field and the plasma conditions have the highest threshold for wave growth.

2.3.2. *Chemical Contamination by the Ion Beam*

Chemical interactions that could be considered as potential concerns to the spacecraft or instruments include:

- condensation of neutral indium in the vicinity of the ion emitter,
- return to the spacecraft of In^+ ions after one or more gyrations,
- interaction of the ion beam with spacecraft surfaces.

The condensation of this material on cold surfaces, however, is extremely unlikely. The vapour pressure of indium is only 1×10^{-15} Torr at 250 °C and the total surface from which indium evaporates is in the order of 1 mm^2 for the active, hot emitter and about 9 mm^2 for the surfaces at the environmental temperature. Also the measured mass efficiency (see Section 3.1.1.) sets an upper limit to neutral indium emission of 3%. Furthermore, a baffle is provided by the beam focusing electrodes together with the box orifice. Only neutrals within a cone of 17° may penetrate through the orifice and drift into the ambient plasma at thermal velocity. Their return to the spacecraft skin can only occur through two unlikely processes: the first possibility is that, due to collisions with the ambient gas, the path of the neutrals is reflected back towards the spacecraft. The typical mean free path length for neutral collisions is a couple of 100 m if the spacecraft-generated gas pressure in its close vicinity was as high as 10^{-6} Torr; it would be in excess of 10^4 m if the pressure dropped below 10^{-8} Torr. Hence, collisions are extremely unlikely to occur. Another possibility would be introduced by photo-ionisation of indium and attraction of these ions by negatively charged, non-conductive and shadowed surfaces. The photo-ionisation efficiency at 1 AU is in the order of 10^{-7}. The particle density, inferred from the indium partial pressure of 10^{-15} Torr at nominal operating temperature, is $n = 10^3$ m^{-3}, which yields a negligible amount of In^+ to be produced in the neighbourhood of the spacecraft. Other ionisation processes such as charge exchange or critical ionisation velocity effects are estimated to be negligible as well. Finally, a condensation of indium, even after lengthy laboratory tests, was never found inside the vacuum tank by visual inspection.

In the following discussion only In^+ is considered as other charge states of indium constitute only a small fraction (<10%) of the beam (see Section 3.1.1) and their results can easily be extrapolated. The return of the In^+ beam to the spacecraft can be assessed in the following way. The typical ion energy is about 6 keV. Ions come back to the spacecraft whenever the beam injection takes place perpendicular to the ambient magnetic field. In the Cluster orbit the magnetic field encountered in the near-perigee and apogee segments of the orbit will reach about 1000 nT and 10 nT, respectively. This yields, for the indium ion gyration period, τ and radius, R_G the values in Table II.

Table II
Gyro parameters for indium ions (In^+)

Location	B (nT)	τ (s)	R_G (km)
Near perigee	1000	7.5	120
Near apogee	10	750	12 000

The huge gyro radii near apogee, but even near perigee, in combination with the beam opening angle of 15° half width diffuse the beam to negligible densities after the first gyration period, even in the rare case of beam injection perpendicular to the magnetic field. For near-parallel injection the beam ions would spiral away from the spacecraft. For perpendicular injection at $B = 10$ nT the beam is spread over an area equivalent to a circle of about 40 000 km in diametre, corresponding to a flux of 5×10^{-6} cm^{-2} s^{-1} at $I_e = 10\ \mu$A.

2.3.3. *Negative Spacecraft Potential*

The working principle of ASPOC does not permit to control a negative potential of the spacecraft with respect to the ambient plasma. Therefore routine operation of ASPOC is not foreseen when the uncontrolled spacecraft potential is negative, to avoid driving the spacecraft potential more negative. In feedback mode, the on-board software of the instrument would switch off the ion beam. In stand-alone mode the beam current would be set to a value derived from theoretical considerations and in-flight tests during the commissioning phase. This value is set by time-tagged commands to vary according to the expected plasma environment along the trajectory, including a safety margin to avoid beam-induced negative charging under abnormal conditions.

As mentioned in the introduction, negative charging would be expected when the plasma temperature is very high or when the density of cold plasma is higher than several 100 cm^{-3} (see Figure 1). The latter condition prevails in the plasmasphere and may be fulfilled near perigee at 4 R_E geocentric distance. Only occasional Cluster payload operations are foreseen in this region, and no active ion emission will be scheduled there. It is expected that the outbound crossing of the plasmapause will be associated with a slow increase in the spacecraft potential from small negative to small positive values. Ion emission may be programmed to begin in that region of small positive potential and thus provide a smooth transition from uncontrolled negative to controlled, small positive potentials.

Another situation where negative charging is expected occurs during eclipses. If limited payload operations are performed during short eclipses, it is obvious that ASPOC would not normally be switched on at that time as the spacecraft potential would be driven even more negative.

A remaining possibility for transient negative charging is due to intense energetic electron fluxes, e.g., above auroral field lines during the passage of the near-Earth plasmasheet during a substorm. If active spacecraft potential control were operating during such an unpredictable event the spacecraft potential at which a balance of currents is established would be shifted to more negative values roughly in proportion to the ratio of the total current, which is increased by active ion emission, over the energetic electron current alone, neglecting secondary effects. During negative charging events the energetic electron current must be greater than (and will sometimes by far exceed) the total photoelectron current, for which Equation (5) gave an expected numerical value for Cluster of $I_p = 240\ \mu$A. One should relate this value to the typical ion current of ASPOC ($I_e \approx$ 10–20 μA) to find that the effect on the spacecraft potential will be small.

3. Technical Solution

3.1. Liquid-Metal Ion Source (LMIS)

3.1.1. *Operating Principle*

The ion emitter is a 'solid needle'-type Liquid-Metal Ion Source (LMIS), previously described in the literature, using indium as charge material (Mahoney *et al.*, 1969; Evans and Hendricks, 1972; Wagner and Hall, 1979; Dixon and v. Engel, 1980). A solid needle, usually made of tungsten (W), with a tip radius between 2 and 5 μm, is mounted in a reservoir with the charge material. The reservoir can be heated in order to melt the metal. A potential of 5 to 8 kV is applied between the needle and an extractor electrode. If the needle is well wetted by the metal, the electrostatic stress at the needle tip pulls the liquid metal towards the extractor electrode. This stress is counteracted by the surface-tension forces of the liquid. The equilibrium configuration assumed by the liquid surface is that of a Taylor cone (Taylor, 1964) with a total tip angle of 98.6°. The apex of the Taylor cone in practice reaches a diameter of 1 to 5 nm (Kingham and Swanson, 1984). The field evaporation of positively charged metal atoms in the strong apex field leads to emission of a high-brightness external ion beam from this cone apex with a beam brightness of the order of 10^6 A cm^{-2} sr^{-1} at maximum beam energy.

Since the emission zone is in the liquid state, ions leaving the surface can be continuously replenished by hydrodynamic flow of liquid metal from the reservoir to the needle apex, so that a stable emission can be maintained. Owing to its extremely high source brightness, the LMIS is particularly well-suited for formation of microfocused ion beams and has found wide application in microelectronic technology. The basic technology is well developed on an industrial basis. The present experimenter team has developed the LMIS for space applications for spacecraft potential control and scanning ion microscopy (Schmidt *et al.*, 1993; Rüdenauer *et al.*, 1992; Riedler *et al.*, 1992). The advantages of the LMIS principle are: low power consumption – mostly determined by the heater power to keep the indium reservoir at about 520 K; high mass efficiency; compactness and low mass.

Indium (stable isotopes with 115 amu and 113 amu to 95.7% and 4.3%, respectively) has been chosen as the ion-source charge material because of its low vapour pressure, which prevents contamination of the source insulators and ambient spacecraft surfaces. On the other hand, the melting point at 156.6 °C is high enough to prevent melting of an unheated source charge, even at the maximum expected elevated environmental temperature.

The current-voltage diagram of the LMIS is characterised by a well-defined emission onset voltage and a slope of typically 0.5 to 1.0 μA per 10 V increase of the extraction voltage.

The energy spread and the composition are important properties of the ion beam. Laboratory tests (Rüdenauer *et al.*, 1987) revealed that the beam consists of >90% single-charged In^+ with minor fractions of double- and triple-charged ions and single-charged indium clusters. The average charge number per emitted charged indium particle is 0.98, and the mass efficiency, i.e., the fraction of the total mass taken by single-charged ions, at 10 μA is about 80%. The energy spectrum of the main In^+ contribution is composed of a quasi-Gaussian component centred a few volts below the electric potential of the tip and of a lower energy tail extending more than 1 keV below the centre width. The emitted In^+ ion current shows some broadening of the energy spectrum with increasing currents at the extraction electrode. At an emission current of 10 μA, the energy width is 150 eV, but a low-intensity, low-energy tail down to more than 500 eV below nominal beam energy can be expected.

3.1.2. *Design of the Emitter Module*

Figure 5 shows the cross-section of an ion emitter module. One such module consists of an array of four individual ion emitters, which are operated one at a time. The individual emitters are of cylindrical geometry. The indium reservoir and the needle sitting on top are kept at high voltage. The ion sources are individually and indirectly heated from below by a Pt100 wire resistor embedded into a ceramic insulator tube. This scheme enables the source to be heated from a grounded power supply, with the tip itself still being kept at high voltage. The individual emitters are mounted in a slab of porous ceramic with extremely low heat conduction ($<5 \times 10^{-4}$ W K^{-1} cm^{-1}). The thermal isolation of the source has been designed to minimise the heater power consumption, which is below 0.6 W in the present design. The mass of an individual emitter is 1.2 g including the heater. Each reservoir contains 200 mg of indium which suffices for about 2000 hours of operation per emitter at 10 to 15 μA. Three emitters are more than sufficient to achieve the design goal of 5000 hours per module. The fourth emitter serves as a back-up. In fact, three or four emitters per module, or 7 to 8 emitters per instrument, are operational in the final configuration of the flight units.

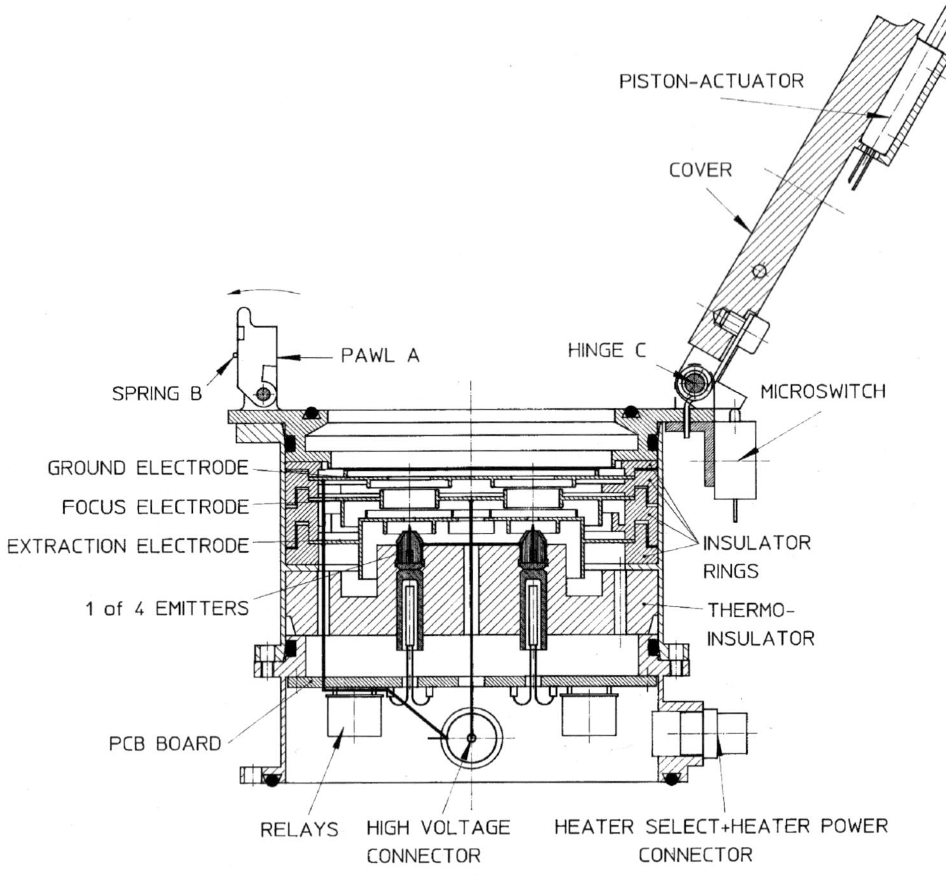

Figure 5. Schematic of LMIS module (see text).

All emitters have a common extraction and focusing lens arrangement with individual openings for each tip. It consists of a grounded extractor electrode, a focusing electrode at beam potential and a second ground electrode. These three electrodes constitute unipotential lenses for individual tips with the tip apex located in one focal point. The divergent ion beam (opening angle $\approx 90^{\circ}$) emitted from the tip is focused by this lens into a nominally parallel beam. The cold secondary side of the high-voltage supply is connected to the extraction and outer electrodes of the focusing lens and to metal tubes around the heater elements. These tubes protect the heaters and their power supply, which is connected to ground, from disastrous high-voltage strokes. With this grounding scheme, all possible paths for high-voltage discharges are confined within the high-voltage loop.

The relevant currents flowing in this system are:

– The current carried by the emitted ion beam. This current loop is closed via the spacecraft surface by the ambient plasma. This current is referred to as ion current or beam current.

– The total current delivered by the high-voltage supply to the emitter. This current includes the beam current and internal loss currents (e.g., the current to the extraction and beam-focusing electrodes), and is therefore always larger than the beam current. The percentage of loss currents within the total current is small (<10 to 20%) for small to medium currents and may increase to 30 to 50% near the maximum total current (about 65 μA).

Owing to the wide-angle nonparaxial rays entering the electrode system, the lens aberrations actually will produce a finite divergence of $\pm 15°$ (half maximum) in the outgoing beam. The tip and the focusing electrode are at the same potential, so no additional power supply or voltage divider is required for beam focusing. Since the beam shaping and focusing optics is purely electrostatic, the lens properties and the beam shape remain unchanged if the tip voltage (identical to focusing voltage) changes.

The insulating slab and focusing system is rigidly mounted onto an aluminium support structure. The electrical connections to the resistive heater elements are made via 125 μm-diameter nickel wires from a printed-circuit board, rigidly mounted to the back of the support structure.

3.1.3. *Protective Covers and Late Exchange of Modules*

In order to avoid oxidation, the indium in the liquid-metal ion source should never become exposed to air or water vapour. An almost hermetically closed volume has been designed (leak rate $\approx 8 \times 10^{-8}$), in which the emitters can be stored in a protective gaseous atmosphere. It will be opened after launch once a reasonable outgassing of the satellite has taken place. The cover system (see Figure 5) consists of a hinged plate which tightens the LMIS module on an O-ring. The cover plate is held down and locked by a special hook, which is held in position by a helically-wound bending spring. To open the cover, a pyrotechnic piston actuator, mounted perpendicular to one of the faces of the hook, is actuated. A pin pushes the hook away from its locking position and the spring-loaded cover plate will open.

Because of the small volume of gas and the inevitable leakage, the initial internal overpressure of 0.2 bar would disappear within about one year during storage on the ground and cause air to enter from outside, contaminating the indium. As the delivery time of the ASPOC flight units was about 1.5 to 2.5 years before the predicted launch date, a two-step solution has been implemented. For the initial deliveries the flight units were equipped with dummy emitter modules to be used during most of the test activities before launch. They are identical to real emitters except that no indium is present and the beam current paths are simulated by electrical resistors. These dummies were replaced by the real emitter modules in the late exchange period before the spacecraft were shipped to the launch site. After about one year the gas filling of the emitter modules has to be renewed.

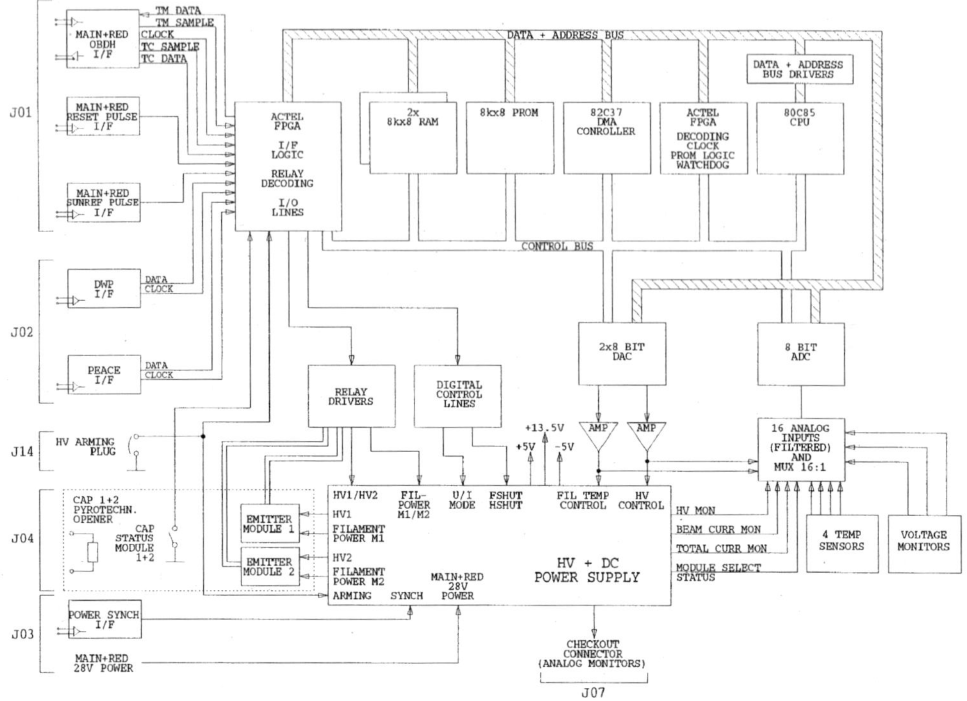

Figure 6. Electrical block diagram of ASPOC.

3.2. Control Electronics

The instrument utilises a microprocessor for controlling the experiment and for data handling. It basically operates and controls the ion-emitting system (high voltage and heater power), performs the start-up procedure of the emitters, and serves the interfaces to the on-board data-handling and telecommand units; the electric-field instrument and low-energy electron spectrometer (EFW and PEACE) provide spacecraft potential data in real-time. Special attention is paid to the monitoring and safety of the high-voltage unit.

Because of the low data rate (108 bit/s) both housekeeping and science data are transmitted through the housekeeping channel. Complete status information is given within two housekeeping data frames ($\approx$ 10.3 s).

Figure 6 shows an electrical block diagram of ASPOC with emphasis on the DPU. The flight software is downloaded from PROM into CMOS RAM when the instrument is turned on. The processor is an 80C85. A large part of the logic circuits that are not interfacing the spacecraft is contained in a programmable gate array (Actel). The DPU and the spacecraft interface logic cover two printed-circuit boards. The DPU has a watchdog timer. If a counter is not reset regularly by the program running in the DPU, it will perform a full reset of the DPU and a reload of the programme from the PROM into the RAM after 8 s. The impact on telemetry is a reset of all parameters to the power-on state, a loss of data for up to 18 s, followed by standby operation of the instrument.

The DC converter and the high-voltage unit take up one printed-circuit board each. The DC converter provides three fixed voltages (+5 V, +13.5 V, −5 V) and a variable output for the heater elements in the emitters. The high-voltage unit can power one of the two emitter modules at a time in voltage- or current-controlled modes. Analogue monitors of the high voltage, the total output current at high voltage and the effective ion-beam current are provided. The latter measurement necessitates a special grounding concept for the emitter supply unit. The power consumption consists of an almost constant component of (1.5 W for the DPU and the heating of one emitter filament and a variable part which is largely proportional to the emitted ion current. While 3 W peak primary power can be reached by design, this value should not be sustained for longer periods because of lifetime-limiting effects on the ion emitters. The maximum, technically optimum power during nominal operations is 2.7 W, and an average value of 2.4 W primary power complies with the requirements on the ion current for spacecraft potential control in the expected plasma environment.

3.3. Mechanical Concept

The ASPOC instrument is a single unit consisting of an electronics box and two cylindrical ion-emitter modules. Figure 7 shows a unit with dummy emitter modules and both covers in the open position. The thermal blanket of the spacecraft is attached to the plate on top of the unit. The viewing direction is parallel to the spin axis to avoid spin modulation of the equilibrium current. The wall thickness of the box is 0.8 mm. It carries four printed-circuit boards and a motherboard. The electronic components have been selected in general to withstand the predicted dose of ionising radiation of $\approx$ 15 krad (Si) at the wall thickness of 0.8 mm. Additional spot shielding has been applied to a few parts.

4. Instrument Data Summary

The main instrument data are summarised in Table III.

5. Flight Operations

5.1. Feedback Modes

In the basic operations mode a measurement of the spacecraft potential is supplied to ASPOC by either the electric-field experiment (EFW, Gustafsson *et al.*, this issue) or the electron analyser (PEACE, Johnstone *et al.*, this issue) and this

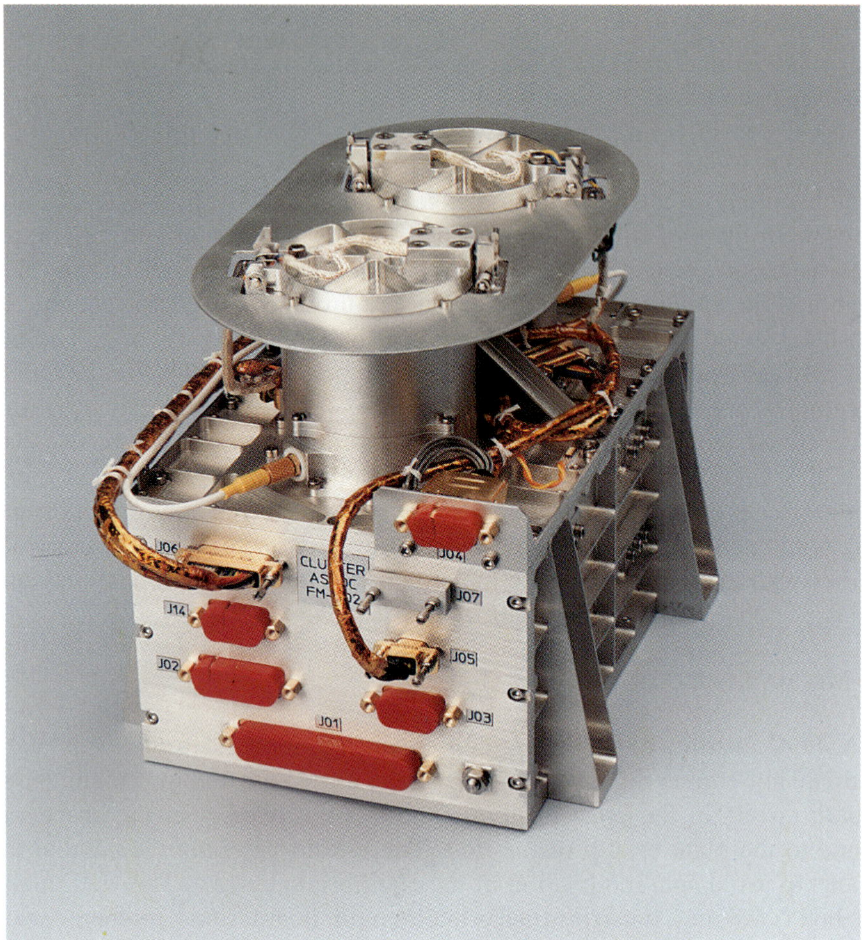

Figure 7. One of the ASPOC flight models.

information is then used to adjust the emission current sufficiently to reduce the spacecraft potential to some predetermined value, as described in the discussion of Figure 3, in a closed-loop scheme. This mode is called feedback mode from EFW or PEACE, with acronyms 'FEFW' and 'FPEA', respectively. The measurements of the spacecraft potential are updated once every spin and sent to ASPOC via dedicated serial, digital Inter-Experiment Link (IEL) interfaces. Data from EFW consist of the voltage measured between one pair of spheres and the spacecraft body when operating in voltage mode. The value is sampled every second and sent to the Digital Wave Processor (DWP, Woolliscroft *et al.*, this issue) instrument, which combines it with operating mode information from the WHISPER sounder (Décréau *et al.*, this issue) and transmits the product to ASPOC. The PEACE instrument calculates on-board the spacecraft potential from the distribution function of

Table III
Instrument data summary

Quantity	Value
Mass	
Electronics box	1410 g
2 emitter modules with covers	360 g
Harness	50 g
Flange for thermal blanket	30 g
Total	1850 g
Size	
Electronics box	187 × 157 × 95 mm
Emitter modules, closed	60 mm diam. × 75 mm
Overall	187 × 157 × 170 mm
Power	
Average power	2.4 W
Peak power	2.7 W
Telemetry rate	108 bit s^{-1}
Design lifetime	10000 hours at 10 μA
Beam characteristics	
Species	In^+
Atomic mass	113, 115 amu
Energy	5 to 8 keV
Current	max. 50 μA, design: 10 μA
Opening angle (half maximum)	±15°
Direction	along spin axis

electrons in the range below about 25 V and transmits a value if the calculation was successful.

Referring to Equation (6), we see that over most of the plasma regime of interest, photoemission dominates the current balance to the spacecraft. The required ion emission current to control the spacecraft potential would be the order of a few tens of μA, as illustrated in Figure 4. In the case when the plasma current dominates, the required emission current would be smaller. The control algorithm has been optimised to the expected current-voltage characteristics of the spacecraft. As the basic objective is the reduction of the average potential, the updates of the ion current in response to potential variations are performed slowly, with time constants of the order of several spin periods, and will be correlated with the spin period as the basic time period on-board the spacecraft, i.e., updates of the ion current will occur at most once per spin period.

The presence of two independent data sources necessitates the implementation of a priority scheme. It is envisaged to analyse the first orbits and to change the scheme used by the on-board processor if required. By default, priority is given to EFW data because of the higher resolution than PEACE (0.034 V vs $\approx$ 1.4 V) and the more straightforward way in which the satellite potential is derived. It is expected that the algorithm used in the PEACE instrument may be misled occasionally by 'exotic' energy distributions of the low-energy electrons.

The current and acceleration voltage of the ion beam will be available to all experimenters, which will enable them to interpret their data also with respect to any remaining effects of the spacecraft potential.

5.2. Stand-alone Modes

The feasibility of the active spacecraft potential control loop will finally be determined during the in-orbit commissioning phase. Cases where the emission of a constant ion current is preferred to maintaining a constant spacecraft potential by the science community, or where technical reasons prevent the implementation of the control loop in orbit, must be foreseen as alternative operational modes.

Where no signal from EFW or PEACE is used, a stand-alone mode of operation involves setting the ion-emission current to some predetermined value based on the spacecraft current-voltage characteristics and experience gained in flight. The name of this mode is Constant Beam Current Mode (code 'IION'). This value can be set by ordinary time-tagged commands to vary according to the expected plasma environment along the orbit. A maximum of four different current settings, each optimised for one of the main regions, namely the magnetospheric lobes, plasma sheet, magnetosheath, and solar wind, should be sufficient. At the same time the current level must be sufficiently low to ensure that the spacecraft potential is not driven negative. Equation (6), Figure 4 and the considerations in Section 2.3.3 show, however, that unpredicted negative charging events are unlikely. In the constant beam current mode, the control of the potential would not be as good as in the feedback mode, but could still be used to reduce the spacecraft potential to a few volts positive relative to the ambient plasma potential according to the steeper part of the photoemission characteristic.

There is another stand-alone mode where the total current fed into the ion emitter (rather than the ion beam current) is kept constant. The purpose of this mode (code 'ITOT') is to provide a fallback solution in case, however unlikely, of electromagnetic interference problems with the beam current monitor on-board. The measurement of the total current is probably less sensitive to electromagnetic interference than the measurement of the beam current, the latter being closed over a huge loop through the ambient plasma and the spacecraft structure.

5.3. Test and Commissioning Mode

For calibration and 'active experiment' purposes a Test and Commissioning Mode (code 'T&C') has been defined, covering instrument operations which are not related to maintaining a constant spacecraft potential. From a considerable amount of previous experience (Knott *et al.*, 1983; Schmidt and Pedersen, 1987; Lindqvist, 1983), Pedersen, 1995; Schmidt *et al.*, 1995), the current-voltage characteristics of typical spacecraft are already well-known, but the necessity for a measurement of this characteristic for each Cluster spacecraft at the beginning of the mission, and regular re-measurements to account for changes in the photoemission properties of the surface, remains. This measurement is simply carried out by sweeping the ion emission current in incremental steps over some convenient range, allowing simultaneous measurements of the spacecraft potential. The length of each step is 8 or 16 s, the equivalent of about 2 to 4 spin periods, and the total duration of this mode is 2 to 5 min depending on the step size and the emission current range, which will be adjusted to meet the scientific objectives of the particular measurement. A thorough operation in this mode, i.e., with small current steps over a wide current range for a duration of about 5 min, is planned at the beginning of each planning period (three orbits). A short verification sweep is intended at some suitable time, typically immediately after the turn-on of ASPOC at the beginning of a data-acquisition interval, in every orbit.

Apart from providing a highly improved environment for other experiments, scientific investigations of the photoelectric characteristics, scientific investigations of dependence of the spacecraft potential on plasma parameters, and of spacecraft charging in different plasma environments, can be carried out in a so-called active mode. In agreement with the scientific community, the ion current can be varied in a defined way for a short time to enable the cooperating plasma experiments to calibrate their response to spacecraft potential variations. Such experiments will be carried out at long, regular intervals, preferably in sections of the orbit which are not of prime interest to the mission as a whole. Stepping through different current levels must occur in synchronisation with scans, sampling intervals, etc. of other experiments.

5.4. Hot Standby Mode

The ion-beam emission may have to be turned off in a pre-planned manner during time intervals varying from a few seconds up to fractions of hours or more, depending on the expected ambient plasma conditions or because of operational requirements on the payload, for instance during interference testing between instruments. During such pauses, the ion emitter may be kept at elevated temperature to ensure immediate re-start capability. This mode, which is associated with pre-programmed emission pauses is called hot standby mode (code 'HOT'). It is worth mentioning here that a similar state of the instrument (emitters heated, but the ion beam turned off) is reached when the instrument is forced to turn off the ion beam temporarily due to a failure condition in one of the feedback modes. This state is called 'feedback wait status' (code 'WAIT').

5.5. Standby Mode

The standby mode (code 'STDB') leaves the instrument in a completely passive state: the high voltage, the ion-beam emission, and the heater filaments of the emitters are turned off. The instrument only serves the interfaces to the spacecraft and the Inter-Experiment Link and produces housekeeping data. Standby mode is the power-on mode and also the contingency mode in case of problems with the ion-beam emission.

The instrument does not have permanent memory. Standby mode is therefore useful to keep parameters in internal memory, which otherwise would have to be uplinked again after power-off. Standby mode will be used between data-taking periods within one orbit, in which ASPOC ion-beam operations are scheduled. ASPOC will be turned off during passes through the radiation belts.

5.6. Start-up

The description of modes would be incomplete without reference to the start-up of an ion emitter. It has been mentioned in previous sections that the ion emitters must be heated. Depending on the ambient temperature, it takes up to 15 min to reach a temperature inside the emitters that is sufficient to ignite the ion beam. The period from the beginning of the heating until a few seconds after the ignition of the beam is defined as the start-up time. Immediately after the emission has stabilised the instrument switches into the desired operating mode (feedback or stand-alone). Due to limited resources, only one emitter can be heated at a time. Whenever a change of active emitter is required, the previous emitter has to be turned off before the start-up of the new emitter can begin.

5.7. Technical Mode

Finally, the instrument features a technical mode (code 'TECH'). In this mode some additional low-level commands for the control of the emitter filaments, for the internal D/A converter, and for switching between emitter modules are enabled. This mode is suitable for low-level check-out of the instrument. A summary of operational modes is given in Table IV.

5.8. Contingencies

Possible failure conditions from external sources (e.g., an unexpected absence of spacecraft potential data in feedback mode) or technical problems with the instrument (e.g., failure of an emitter) may be grouped into three categories:

(1) Failure of an ion emitter.

(2) Failure of one of the inter-experiment links from the instruments EFW/DWP or PEACE.

(3) Disturbances by the WHISPER instrument in its active mode.

Table IV
ASPOC mode summary

Mode	Description	Ion current	Typical duration
Standby	Passive mode	None	No limit
Start-up	Warm up, then ignite ion emitter		15 min
Feedback from EFW	Control loop with spacecraft potential data from EFW	Variable	No limit, emitter change after several hours
Feedback from PEACE	Control loop with spacecraft potential data from PEACE	Variable	No limit, emitter change after several hours
Stand-alone	Emission not dependent on spacecraft potential	Constant, >0	No limit, emitter change after several hours
Calibration	Measure current-voltage characteristics	Step function	2 to 10 min
Active experiment	Ion beam experiments	Step function	No limit

Case 1 is covered by contingency procedures which will be activated from ground. Case 2 is covered by the instrument's capability to switch into a back-up operational mode which does not rely on the inter-experiment link data. If any unexpected interruption of the spacecraft potential data flow occurs while ASPOC is in feedback mode the ion emission is kept at the last value for a few spin periods before the current is either turned off or set to constant value (standby or stand-alone mode).

As for case 3, it is still an open question as to whether the WHISPER instrument in its active sounding modes will have a significant effect on the measurements of the spacecraft potential and the potential itself. In its active mode as a relaxation sounder, the instrument WHISPER emits pulses from the wire boom antennae with amplitudes up to 200 V_{pp} (Décréau *et al.*, this issue) and analyses the echoes returned from the plasma for resonances. Many instruments, including EFW and PEACE, have implemented blanking periods during the sounder pulses in order to ensure clean measurements. The presence of any effects will be established during the test and commissioning phase at the beginning of the mission. If it is found necessary to discard the spacecraft potential by EFW or PEACE, or to turn off the ion beam altogether, the ASPOC instrument can be commanded into a suitable mode whenever WHISPER is operating in a mode that has been found to be disturbing. This can be achieved by time-tagged ASPOC mode commands at the same time as WHISPER mode changes or, in the case of frequent changes, the

ASPOC on-board software can be commanded to react autonomously to WHISPER mode settings. This is accomplished by transmitting three-bit WHISPER operating mode information in every spin period through DWP to ASPOC. The ASPOC software reacts according to a configurable decision table.

6. Inter-Experiment Calibration and Test Plans

The ion emitters of ASPOC constitute one of several active experiments on the Cluster payloads. The sounder instrument WHISPER has already been mentioned. The Electron Drift Instrument (EDI) operates with a steered electron beam with about 1 keV energy and up to 1 μA (typically several 100 nA) beam current. Finally, the double-probe electric-field instrument EFW applies bias currents of some 100 nA to each of the four probes in order to clamp their potential close to the ambient plasma potential. This technique is necessary for accurate electric-field measurements. The main inter-experiment effect of the bias current consists of its influence on the current equilibrium between the spacecraft and the plasma and thus of the spacecraft potential. All these methods are potentially disturbing for other measurements. Therefore the inter-experiment calibration and interference measurements are highly important. A major part of the commissioning campaign before the beginning of nominal payload operations is filled with tests of the active modes of ASPOC, WHISPER, EDI, and EFW.

The instrument ASPOC supports the inter-experiment tests by three modes of operation:

(1) The ion beam is turned on and off in intervals of about 10 min. All other instruments operate in their most sensitive modes and perform comparative measurements between the ON and OFF conditions. At the same time, other active instruments (mainly WHISPER) also cycle between different operating parameters. If available, different techniques to suppress eventual disturbances are tested, for example the synchronisation of measurement intervals with the WHISPER pulses.

(2) The ion beam is turned on to a maximum value for some 10 min. Instruments measuring waves in the plasma are turned on to highest sensitivity. It is expected that interference – if present at all – would show in the electrostatic wave spectrum. The instruments EFW, with its capability to resolve up to 8 kHz using the internal memory, and WBD (Wide Band Data, Gurnett *et al.*, this issue) are the most important co-players in this test, which will be repeated in different plasma environments.

(3) The active instruments ASPOC, WHISPER, EDI and EFW influence the current balance to the spacecraft in different ways, and the effect will not only depend on the emitted currents, but also on the plasma region. The detailed definition of spacecraft potential campaigns is in progress. It involves the operation of ASPOC in many current steps while plasma instruments (mainly EFW, PEACE and

CIS) monitor primary and secondary effects on the spacecraft potential. Similar tests are carried out while EDI slowly increases the electron current.

The commissioning phase does not only deal with inter-experiment interference. Prior to this, a comprehensive check-out of all operating modes of ASPOC will be carried out.

(1) The stand-alone mode, emitting a constant ion current, is the most simple mode. Nevertheless, it provides an opportunity to test the satisfactory behaviour of the ion emitters in regular operations. There is a theoretical possibility of contamination of unused ion emitter needles by sputtered material from the active emitter. A very thin layer of contaminating substance on a needle would increase the voltage needed to initiate ion emission from an emitter. Only after the emitter has been ignited does the liquid flow of the indium remove the contaminants. Therefore it is planned to ensure a regular cycling of emitters in order to limit the maximum turn-off time. The commissioning phase will be used to establish the best strategy for this cycling.

(2) The feedback modes, using on-board data of the spacecraft potential for the ion emission, could not be tested on the ground from end to end. Part of the commissioning will be dedicated to establishing the response of the spacecraft and the instruments actually determining the spacecraft potential, EFW or PEACE, to variations of the ion beam. From this experience the optimum control loop parameters for the feedback modes will be calculated and up-linked to the instrument. Moreover, this test must be carried out in different plasma environments, at least once in tenuous plasma in the magnetotail and once elsewhere, for instance in the solar wind.

7. Previous In-Orbit Experience

A similar type of liquid-metal ion emitter has been operated in an experiment onboard the space station MIR (Rüdenauer *et al.*, 1992). It was shown that the emitters can be operated for extended periods under microgravity conditions. Operating parameters are similar to those observed on ground. Also post-flight analysis of the emitters in the laboratory did not exhibit any deviations, e.g., of the wetting of the needles with liquid indium, from tests on ground. A scanning ion microscope which utilises a similar ion source has also been built for the station MIR (Riedler *et al.*, 1992) and has been operated successfully several times since 1991, most recently within the Euromir project.

A dedicated active potential control experiment using an ion beam has been flown on-board the Japanese Geotail spacecraft. First results show that spacecraft potential had been reduced from values ranging up to 70 V – the maximum value observed – down to about +2 V with respect to one of the electric-field probes which are current-biased and thus kept at a potential slightly above the ambient plasma potential (Schmidt *et al.*, 1995). Very good stability of the controlled potential was

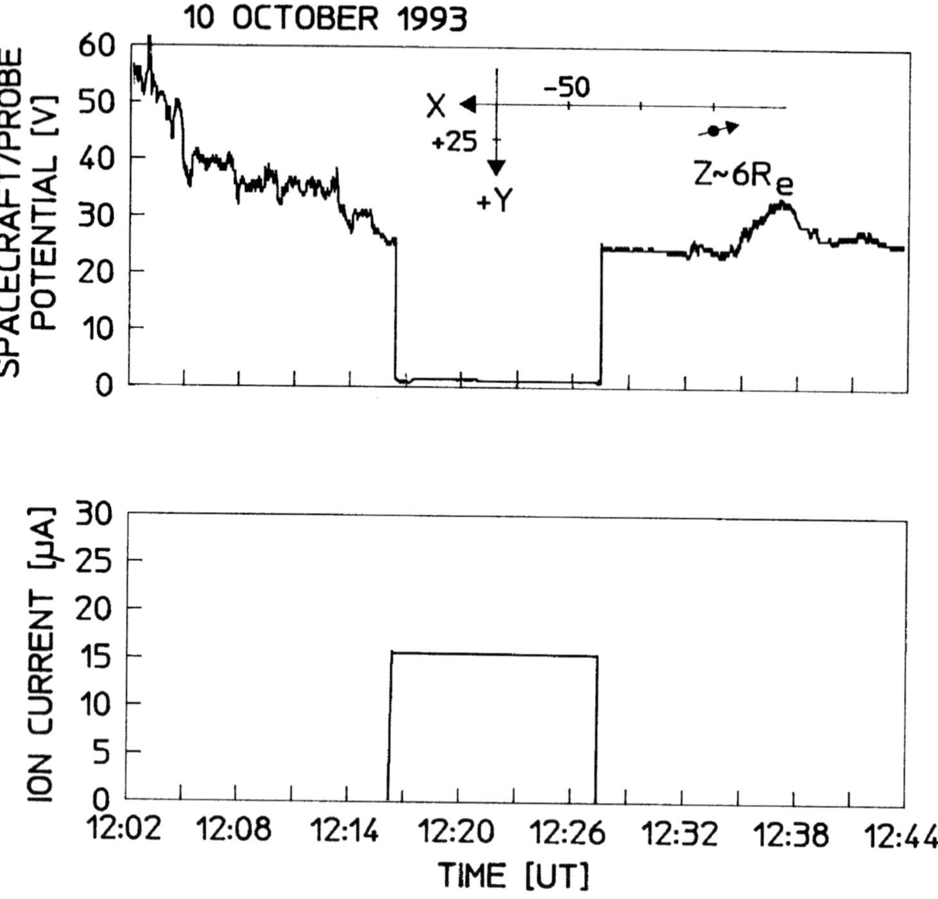

Figure 8. Ion emitter operations on Geotail on 10 October, 1993. The ion emitter was turned on at 12:16 UT with 15 μA emission current (*lower panel*). The spacecraft potential (*upper panel*) drifted between +56 V and +24 V in the uncontrolled state, but remained at about +1.7 V with the ion emitter on. The location of the spacecraft in GSM coordinates is indicated.

observed in stand-alone mode. Neither variations of the ambient plasma conditions nor operations of two electron emitters on Geotail produce significant variations of the controlled surface potential, as long as the resulting electron-emitter currents remain much smaller than the ion-emitter current. Figure 8 from Schmidt *et al.* (1995) demonstrates the effect of the ion emitter operations on Geotail.

Another attempt to apply ion beams for potential control will be undertaken on the Soviet Interball Auroral spacecraft planned to be launched in 1996 (Schmidt *et al.*, 1988). This instrument also includes a similar liquid-metal ion emitter and a saddle-field ion source operating with nitrogen.

8. Planned Science Studies Including Four-Spacecraft Aspects

The scientific objectives of the ASPOC investigations are twofold: at first, the ion emitter provides uniform reference conditions, i.e., a low spacecraft potential which is constant over time and on all four spacecraft. Thereby the main goal of Cluster, the measurement of spatial gradients of the plasma parameters, is supported. In this respect the scientific objectives of ASPOC are embedded into the objectives of the Cluster community. Emphasis is laid on supporting studies of the low-energy plasma.

For the main objective of keeping the spacecraft potential at a constant, low value, the importance of active control is greatest where the uncontrolled potential is likely to be extremely high, for instance in the lobe regions. A second interesting objective is linked to the various plasma boundary crossings, where ASPOC will maintain equal measurement conditions on either side of the boundary.

Secondly, active experiments with the ion beam and studies of spacecraft charging are intended. They will cover a very small fraction of the total time distributed over all regions of the orbit. Typically at the beginning of data-acquisition intervals, when some instruments perform their in-flight calibration cycles, ASPOC will also carry out short sweeps of the ion current in order to establish the current-voltage characteristic of the spacecraft and its variation over the mission lifetime and under different plasma conditions in the outer cusp region, in the magnetotail and in the solar wind.

Studies of interference from the ion beam (for instance by electrostatic noise) will be carried out in the commissioning phase and, if any interference were found, the operating parameters would subsequently be adjusted to minimise the effects. It may, however, be scientifically interesting to study the ion beam(plasma interactions in more detail than the commissioning phase permits. For this purpose ASPOC operations which trigger beam-plasma interaction phenomena could be scheduled occasionally in selected plasma regions within special campaigns. There is a preference for performing these active experiments in regions with stable ambient plasma conditions throughout the duration of the experiment.

The archiving of ASPOC data will go hand in hand with the scientific exploitation of the instrument. The modest data volumes reduce the work load and facilitate the build-up of a comprehensive data archive. Apart from the raw data a set of processed files in Common Data Format will be generated.

9. Conclusion

Charging of the outer surface or entire structure of a spacecraft in orbit can have a severe impact on the scientific output of the instruments which measure the low energy plasma or the electric field. The ASPOC instrument is designed to control the surface potential of the Cluster spacecraft to near-zero values for all of their operational lifetimes, if the uncontrolled spacecraft potential would be positive. As well as providing an improved environment, scientific investigations of the photoelectric characteristics and spacecraft charging can be carried out. The instrument's small size, mass and power consumption qualify it for the special requirements of the Cluster mission. First results from a similar instrument on the Geotail spacecraft show an effective reduction of uncontrolled spacecraft potentials ranging up to 70 V down to a few volts and a very good stability of the controlled potential even with constant ion currents.

Acknowledgements

We gratefully acknowledge the work of the following persons on hardware and software related to ASPOC: G. Fremuth, K. Fritzenwallner, F. Giner, H. Jeszenszky, Ch. Kürbisch, K. Kvernsveen, B. Kyrkjedelen, G. Laky, S. Neukirchner, U. Nischelwitzer, R. Wallner.

References

Décréau, P. M. E., Etcheto, J., Knott, K., Pedersen, A., Wrenn, G. L., and Young, D. T.: 1978, 'Multi-Experiment Determination of Plasma Density and Temperature', *Space Sci. Rev.* **22**, 633.

Décréau, P. M. E. *et al.*: 1997, this issue.

Dixon, A. J. and v. Engel, A.: 1980, 'Studies of Field Emission Gallium Ion Sources', *Inst. Phys. Conf. Ser.* **54**, 292.

Evans, C. A., Jr. and Hendricks, C. D.: 1972, 'An Electrohydrodynamic Ion Source for the Mass Spectrometry of Liquids', *Rev. Sci. Instr.* **43**, 1527.

Fahleson, U.: 1967, 'Theory of Electric Field Measurements Conducted in the Magnetosphere with Electric Probes', *Space Sci. Rev.* **7**, 238.

Feuerbacher, B. and Fitton, B.: 1972, 'Experimental Investigation of Photo-Emission From Satellite Surface Material', *J. Appl. Phys.* **43** (4), 1563.

Grard, R. J. L.: 1973, 'Properties of the Satellite Photoelectron Sheath Derived From Photoemission Laboratory Results', *J. Geophys. Res.* **78**, 2885-2906.

Gurnett, D.: 1997, *et al.,* this issue.

Gustafsson, G.: 1997, *et al.,* this issue.

Johnstone, A.: 1997, *et al.,* this issue.

Jones, D.: 1981, 'Xe^+-Induced Ion-Cyclotron Harmonic Waves', Active Experiments in Space Plasmas, *Adv. Space Res.* **1** (2), 103.

Kingham, D. R. and Swanson, L. W.: 1984, 'Mechanics of Ion Formation in Liquid Metal Ion Sources', *Applied Phys.*. **A34**, 123.

Kintner, P. M. and Kelley, M. C.: 1983, 'A Perpendicular Ion Beam Instability: Solutions to the Linear Dispersion Relation', *J. Geophys. Res.* **88**, 357.

Knott, K., Korth, A., Décréau, P., Pedersen, A., and Wrenn, G.: 1983, 'Observations of the GEOS Equilibrium Potential and its Relation to the Ambient Electron Energy Distribution', in *Spacecraft Plasma Interactions and Their Influence on Field and Particle Measurements, Proceedings of the 17th ESLAB Symposium*, ESA SP-198, p. 19.

Lindqvist, P.-A.: 1983, 'The Potential of ISEE in Different Plasma Environments', in *Spacecraft Plasma Interactions and Their Influence on Field and Particle Measurements, Proceedings of the 17th ESLAB Symposium*, ESA SP-198, p. 25.

Mahoney, J. F., Yahiku, A. Y., Daley, H. L., Moore, R. D., and Perel, J.: 1969, 'Electrohydrodynamic Ion Source', *J. Applied Phys.* **40**, 5101.

Miura, A., Okuda, H., and Ashour-Abdalla, M.: 1983, 'Ion-Beam Driven Electrostatic Ion Cyclotron Instabilities', *Geophys. Res. Letters* **10** (4), 353.

Mott-Smith, H. and Langmuir, I.: 1926, 'The Theory of Collectors in Gaseous Discharges', *Phys. Rev.* **28**, 727.

Mourenas, D., Béghin, C., and Lebreton, J. P.: 1989, 'Electron Cyclotron and Upper Hybrid Harmonics Produced by Electron Beam Injection on Spacelab 1', *Ann. Geophys.* **7** (5), 519.

Mullen, E. G., Gussenhoven, M. S., Hardy, D. A., Aggson, T. A., Ledley, B. G., and Whipple, E. C.: 1986, 'SCATHA Survey of High-Level Spacecraft Charging in Sunlight', *J. Geophys. Res.* **91**, 1474.

Olsen, R. C.: 1982, 'The Hidden Ion Population of the Magnetosphere', *J. Geophys. Res.* **87**, 3481.

Olsen, R. C., Chapell, C. R., and Burch, J. L.: 1986, 'Aperture Plane Potential Control for Thermal Ion Measurements', *J. Geophys. Res.* **91**, 3117.

Olsen, R. C. and Purvis, C. K.: 1983, 'Observations of Charging Dynamics', *J. Geophys. Res.* **88**, 5657.

Paschmann, G. *et al.*,: 1997, this issue.

Pedersen, A.: 1995, 'Solar Wind and Magnetosphere Plasma Diagnostics by Spacecraft Electrostatic Potential Measurements', *Ann. Geophys.* **13** (2), 118.

Pedersen, A., Chapell, C. R., Knott, K., and Olsen, R. C.: 1983, 'Methods for Keeping a Conductive Spacecraft Near the Plasma Potential', in *Spacecraft Plasma Interactions and Their Influence on Field and Particle Measurements, Proceedings of the 17th ESLAB Symposium*, ESA SP-198, p. 185.

Pedersen, A., Cattell, C. A., Fälthammar, C. G., Formisano, V., Lindqvist, P. A., Mozer, F., and Torbert, R.: 1984, 'Quasistatic Electric Field Measurements with Spherical Double Probes on the GEOS and ISEE Satellites', *Space Sci. Rev.* **37**, 269.

Reasoner, D. L., Lennartsson, W., and Chappell, C. R.: 1976, in A. Rosen (ed.), 'Relationship Between ATS-6 Spacecraft Charging Occurrences and Warm Plasma Encounters', in *Spacecraft Charging by Magnetospheric Plasmas, Prog. Astron. Aeron.* **47**, 89.

Rème, H., *et al.*: 1997, this issue.

Riedler, W., Rüdenauer, F. G., Beck, P., Berzhatyi, V., Fehringer, M., Finsterbusch, R., Neznamova, L., Pammer, R., Pürstl, F., and Steiger, W.: 1992, 'MIGMAS/A: Test of a Scanning Ion Microscope Onboard the Soviet Space Station MIR', *Proc. International Space Year Conference*, Münich, Germany, ESA ISY-4 (COSY-8), p. 127.

Rüdenauer, F. G., Steiger, W., Studnicka, H., and Pollinger, P.: 1987, *Int. J. Mass Spectr. Ion Proc.* **77**, 63.

Rüdenauer, F. G., Riedler, W., Berzhatyi, V., Fehringer, M., Göschl, E., Kropiunig, C., Neznamova, L., Steiger, W., and Torkar, K.: 1992, 'LOGION: Operation of a Liquid Metal Ion Emitter Module Under Microgravity', *Proc. International Space Year Conference*, Münich, Germany, ESA ISY-4 (COSY-8), p. 121.

Schmidt, R. and Pedersen, A.: 1987, 'Long-Term Behaviour of Photo-Electron Emission from the Electric Field Double Probe Sensors on GEOS-2', *Planetary Space Sci.* **35**, 61.

Schmidt, R., Arends, H., Nikolaizig, N., and Riedler, W.: 1988, 'Ion Emission to Actively Control the Floating Potential of a Spacecraft', *Adv. Space Res.* **8** (1), 187.

Schmidt, R., Schriver, D., and Ashour-Abdalla, M.: 1992, 'Plasma Response to the Emission of Very Weak Ion Beams for Spacecraft Potential Control', *J. Geophys. Res.* **97**, 14959.

Schmidt, R., Arends, H., Pedersen, A., Fehringer, M., Rüdenauer, F., Steiger, W., Narheim, B. T., Svenes, R., Kvernsveen, K., Tsuruda, K., Hayakawa, H., Nakamura, M., Riedler, W., and Torkar, K.: 1993, 'A Novel Medium-Energy Ion Emitter for Active Spacecraft Potential Control', *Rev. Sci. Instr.* **64** (8), 2293.

Schmidt, R., Arends, H., Pedersen, A., Rüdenauer, F., Fehringer, M., Narheim, B. T., Svenes, R., Kvernsveen, K., Tsuruda, K., Mukai, T., Hayakawa, H., and Nakamura, M.: 1995, 'Results from Active Spacecraft Potential Control on the Geotail Spacecraft', *J. Geophys. Res.* **100** (A9), 17253.

Taylor, G. I.: 1964, 'Disintegration of Water Drops in an Electric Field', *Proc. Royal Soc. London* **A280**, 383.

Wagner, A. and Hall, T. M.: 1979, 'Liquid Gold Ion Source', *J. Vac. Sci. Tech.* **16**, 1871.

Walker, D. N.: 1986, 'Perpendicular Ion Beam-Driven Instability in a Multicomponent Plasma: Effects of Varying Ion Composition on Linear Flute Mode Oscillations', *J. Geophys. Res.* **91**, 3305.
Whipple, E. C.: 1981, 'Potentials of Surfaces in Space', *Rept. Prog. Phys.* **44**, 1197.
Whipple, E. C., Krinsky, I. S., Torbert, R. B., and Olsen, R. C.: 1983, 'Anomalously High Potentials Observed on ISEE', in *Spacecraft Plasma Interactions and Their Influence on Field and Particle Measurements, Proceedings of the 17th ESLAB Symposium*, ESA SP-198, p. 35.
Whipple, E. C., Warnock, J. M., and Winkler, R. H.: 1974, 'Effect of Satellite Potential on Direct Ion Density Measurements Through the Plasmapause', *J. Geophys. Res.* **79**, 179.
Woolliscroft, L. *et al.*: 1997, this issue.
Wrenn, G. L.: 1979, 'Spacecraft Charging', *Nature* **277**, 11.

THE CLUSTER ION SPECTROMETRY (CIS) EXPERIMENT

H. RÈME, J. M. BOSQUED, J. A. SAUVAUD, A. CROS, J. DANDOURAS,
C. AOUSTIN, J. BOUYSSOU, Th. CAMUS, J. CUVILO, C. MARTZ, J. L. MÉDALE,
H. PERRIER, D. ROMEFORT, J. ROUZAUD and C. d'USTON
CESR, BP 4346, 31028 Toulouse Cédex 4, France

E. MÖBIUS, K. CROCKER, M. GRANOFF, L. M. KISTLER and M. POPECKI
UNH, Durham, U.S.A.

D. HOVESTADT, B. KLECKER, G. PASCHMANN and M. SCHOLER
MPE, Garching, Germany

C. W. CARLSON, D. W. CURTIS, R. P. LIN and J. P. McFADDEN
SSL, Berkeley, U.S.A.

V. FORMISANO, E. AMATA, M. B. BAVASSANO-CATTANEO, P. BALDETTI,
G. BELLUCI, R. BRUNO, G. CHIONCHIO and A. DI LELLIS
IFSI, Frascati, Italy

E. G. SHELLEY, A. G. GHIELMETTI and W. LENNARTSSON
Lockheed, Palo Alto, U.S.A.

A. KORTH and H. ROSENBAUER
MPAE, Lindau, Germany

R. LUNDIN and S. OLSEN
SISP, Kiruna, Sweden

G. K. PARKS and M. McCARTHY
U.W., Seattle, U.S.A.

H. BALSIGER
Bern University, Bern, Switzerland

Abstract. The Cluster Ion Spectrometry (CIS) experiment is a comprehensive ionic plasma spectrometry package on-board the four Cluster spacecraft capable of obtaining full three-dimensional ion distributions with good time resolution (one spacecraft spin) with mass per charge composition determination. The requirements to cover the scientific objectives cannot be met with a single instrument. The CIS package therefore consists of two different instruments, a Hot Ion Analyser (HIA) and a time-of-flight ion COmposition and DIstribution Function analyser (CODIF), plus a sophisticated dual-processor-based instrument-control and Data-Processing System (DPS), which permits extensive on-board data-processing. Both analysers use symmetric optics resulting in continuous, uniform, and well-characterised phase space coverage. CODIF measures the distributions of the major ions (H^+, He^+, He^{++}, and O^+) with energies from ~0 to 40 keV/e with medium (22.5°) angular resolution and two different sensitivities. HIA does not offer mass resolution but, also having two different sensitivities, increases the dynamic range, and has an angular resolution capability (5.6° × 5.6°) adequate for ion-beam and solar-wind measurements.

Space Science Reviews **79:** 303–350, 1997.

1. Scientific Objectives and Experiment Capabilities

The prime scientific objective of CIS is the study of the dynamics of magnetised plasma structures in the vicinity of the Earth's magnetosphere, with emphasis on the physics of the Earth's bow shock, the magnetopause boundary, the polar cusp, the geomagnetic tail and the plasma sheet. Past experience has demonstrated that the study of the macrophysics and microphysics requires that the local orientation and the state of motion of the plasma structures be determined as accurately as possible. The four Cluster spacecraft with relative separation distances that can be adjusted to spatial scales of the structures (a few hundred kilometers to several thousand kilometers) give for the first time the unambiguous possibility to distinguish spatial from temporal variations. These scientific objectives require the investigation of many different phenomena, including solar-wind/magnetopause interactions, substorms and auroras, reconnection, generation of field-aligned currents, polar cusps and upstream foreshock dynamics.

The Cluster spacecraft will encounter ionic plasma of vastly diverse characteristics in the course of one year (Figure 1). A highly versatile and reliable ionic plasma experiment is therefore needed.

The variety of conditions encountered in the various magnetospheric regions sets a number of requirements in order to provide scientifically valuable products everywhere.

(1) A great dynamic range is necessary in order to detect fluxes as low as those of the lobes, but also those as high as solar-wind fluxes, throughout the solar cycle.

(2) Hot populations are present in vast regions of the magnetosphere and of the magneto-sheath. In order to provide a satisfactory and uniform coverage of the phase space with sufficient resolution, a broad energy range and a full 4π angular coverage are necessary. The angular resolution should be sufficient to be able to separate multiple populations, such as gyrating or transmitted ions from the main population downstream of the bow shock, and to be able to detect fine structures in the distributions.

(3) Cold beams, such as the solar wind, require a high angular and energy resolution in a limited energy and angular range. Because of the limited energy range required, a beam tracking algorithm should be implemented in order to be able to follow the beam in velocity space. Moreover, for example in the foreshock regions, any study of backstreaming ions requires the simultaneous observation of the solar-wind cold beam and of the backstreaming particles. Therefore, together with the solar-wind coverage described above, a coverage of the entire phase space excepting the sunward sector, with broad energy range, is also required.

(4) In the case of sharp boundaries, such as discontinuities or boundary crossings, it is necessary not to miss any information at the discontinuity, thus a very efficient means of mode change, which allows adaptation to the local plasma conditions, should be provided.

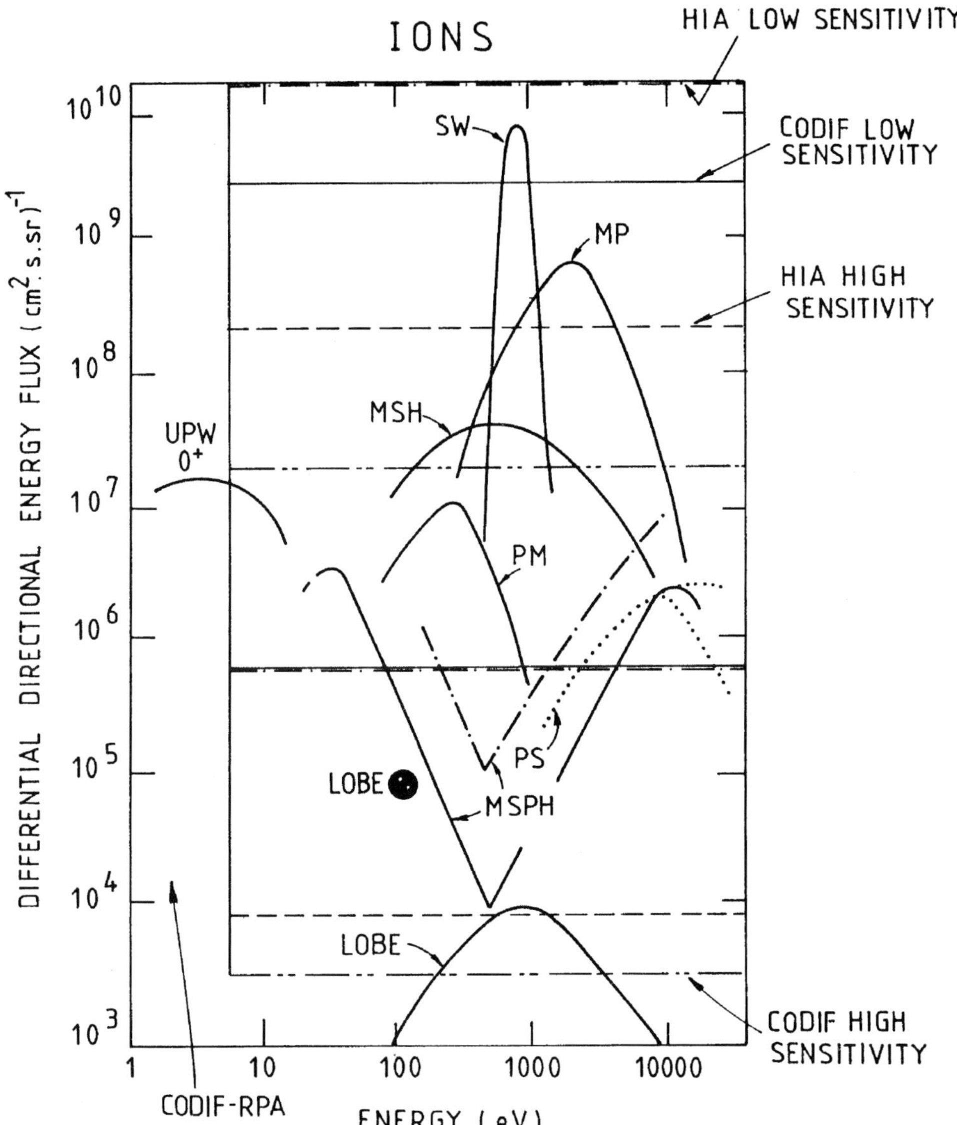

Figure 1. Representative ion differential directional energy fluxes to be encountered in the solar wind (SW), the magnetopause (MP), the magnetosheath (MSH), the plasma mantle (PM), the magnetosphere (MSPH), the plasma sheet (PS), the lobe and upwelling ions (UPW). The range studied by the low sensitivity of HIA is limited by - - - - - - - - - - - - - - - - -, the range of the high sensitivity of HIA by -, the range of low sensitivity of CODIF by full lines, and the range of high sensitivity of CODIF by - ·· - ·· - ·· - ·· - ·· - ·· - ·· - ··.

(5) Moments of the whole three-dimensional (3D) distribution (and of the sunward sector, in solar-wind mode) should be computed on-board , with high time

resolution, to continuously generate key parameters, necessary for event identification.

(6) To study detailed phenomena of magnetospheric plasma physics all the particle populations must be identified and characterized, therefore a 3D distribution is needed. In order to transmit the full 3D distribution, while overcoming the telemetry rate limitations, a compression algorithm must be introduced, which allows an increased amount of information to be transmitted.

So, to meet the scientific objectives, the CIS instrumentation has been designed to satisfy the following criteria, simultaneously on the 4 spacecraft:

– Provide uniform coverage for ions over the entire 4π steradian solid angle with good angular resolution.

– Separate the major mass ion species, i.e., those which contribute significantly to the total mass density of the plasma (generally H^+, He^{++}, He^+, and O^+).

– Have high sensitivity and large dynamic range ($\geq 10^7$) to support high-time-resolution measurements over the wide range of plasma conditions to be encountered in the Cluster mission (Figure 1).

– Have high ($5.6^\circ \times 5.6^\circ$) and flexible angular sampling resolution to support measurements of ion beams and solar wind.

– Have the ability to routinely generate on-board the fundamental plasma parameters for major ion species and with one spacecraft spin time resolution (4 s). These parameters include the density (n), velocity vector ($\mathbf{V}$), pressure tensor (P), and heat flux vector ($\mathbf{H}$).

– Cover a wide range of energies, from spacecraft potential to 40 keV e^{-1}.

– Have versatile and easily programmable operating modes and data-processing routines to optimize the data collection for specific scientific studies and widely varying plasma regimes.

– Rely as much as possible on well-proven sensor designs flown successfully on the AMPTE and *Giotto* missions.

To satisfy all these criteria, the CIS package consists of two different instruments: a Hot Ion Analyser (HIA) sensor and a time-of-flight ion COmposition and DIstribution Function (CODIF) sensor.

The CIS plasma package is versatile and is capable of measuring both the cold and hot ions of Maxwellian and non-Maxwellian populations (for example, beams) from the solar wind, the magnetosheath, and the magnetosphere (including the ionosphere) with sufficient angular, energy and mass resolutions to meet the scientific objectives. The time resolution of the instrument is sufficiently high to follow density or flux oscillations at the gyrofrequency of H^+ ions in a magnetic field of 10 nT or less. Such field strengths will be frequently encountered by the Cluster mission. Oscillations of O^+ at the gyrofrequency can be resolved outside 6–7 R_E. So this instrument package will provide all of the ionic plasma data required to meet the Cluster science objectives (Escoubet and Schmidt, 1997).

Cluster has been conceived as a global instrument which allows, using four spacecraft, vorticity, gradients, divergences, ... to be determined and thus enables macroscopic quantities, for example the electric current from the Curl-B to be measured. Electric currents perpendicular to the magnetic field can also been determined from the pressure gradient of particles. In the simplest case of an ideal magneto-hydromagnetic equilibrium, the perpendicular current density can be written $J_{\perp} = (B \times \nabla P)/B^2$. More details about the ∇P method accuracy can be found in Martz (1993) and in Martz and Sauvaud (1995). For a satellite distance which is small compared to the current filament transverse dimensions, the ∇P method gives good results. The Curl-B method can give less accurate results due to errors on the magnetic field and the satellite separation. On the contrary for a satellite distance which is comparable to the current filament dimensions, the mathematical error introduced by the non-linear variations of the pressure drives an error which becomes very large when a satellite is located outside the current filament. The Curl-B method is free of this latter error. Thus the Curl-B method and the pressure-gradient methods are complementary for estimating the electric currents.

With its capability to provide three-dimensional distribution functions simultaneously for several major species with high time resolution, the CIS instrument will make substantial contributions to the study of the solar-wind magnetosphere interaction, the dynamics of the magnetosphere, the physics of the magnetopause boundary, the polar cusp and the plasma sheet boundary layer, the upstream foreshock and solar-wind dynamics, the magnetic reconnection and the field-aligned current phenomena. For example an important contribution will be made to the understanding of the formation of the bow shock and its role in the heating and acceleration of incoming ion populations, to take just an example.

At the quasi-perpendicular bow shock the specular reflection of ions, their subsequent energy gain in the upstream $\mathbf{V}_{\mathrm{sw}} \times \mathbf{B}$ convection electric field, and their final escape into the downstream region, are known to be important in the dissipation of energy at the shock. The scale length on which the scattering of the original ring distribution and the final thermalisation occurs can at best be guessed (Sckopke *et al.*, 1990). With separation distances from a few 100 km to a few 1000 km the vital scales of several ion gyro radii can be covered with the spacecraft configuration. In addition, CIS will provide the ion distribution functions separately for all major species with a time resolution of one spacecraft spin. Therefore, it will be possible to study the behaviour of He^{++} at the shock. Because of their higher energy all He^{++} ions penetrate the shock. Their bulk velocity is larger than that of the protons, and the entire He^{++} population therefore gyrates in the downstream region. This difference is particularly important. Although the He^{++} ions make up only $\sim 6\%$ of the solar wind density, their contribution to the heating downstream of the shock must be comparable to that of the few percent of protons which are reflected and then start to gyrate. Finally, He^{+} pick-up ions of interstellar origin in the solar wind (Möbius *et al.*, 1985, 1988; Gloeckler *et al.*, 1993) present another important source for ion reflection and downstream heating and thermalization. Because pick-up ions fill a sphere in velocity space with a radius equal to the solar wind speed, centred around the solar wind, ions will interact differently with the shock depending on their origin in velocity space. CIS will provide the resolution to determine the original pick-up distribution and the fate of the reflected ions.

The quasi-parallel bow shock is known to be the source of a diffuse energetic ion population and low-frequency waves. The recent success with hybrid simulations has demonstrated that the energetic ion and wave activity are necessary ingredients of the shock formation which itself is a high dynamic (or even quasi-cyclic) process (e.g., Quest, 1988; Burgess, 1989; Scholer and Terasawa, 1990). This simulation work has paved the way for a combined in-depth study of the evolution of the ion distributions across the shock and their temporal variation with the CIS instrument. As has been shown in a modelling with simulated spacecraft (Giacalone *et al.*, 1994), a close collaboration between the data analysis and simulations will be needed for this task. It can also be expected that the association of ion density enhancements with magnetic pulses upstream of the shock can be identified with CIS and the multi-spacecraft capabilities (Scholer, 1995) as opposed to statistical studies which only yield an average spatial distribution (Trattner *et al.*, 1994). These new measurements will significantly further our understanding of the wave-particle interactions at the bow shock and their importance for ion acceleration and shock structure.

2. The Hot Ion Analyser (HIA)

The Hot Ion Analyser (HIA) instrument combines the selection of incoming ions according to the ion energy per charge ratio by electrostatic deflection in a symmetrical quadrispherical analyser which has a uniform angle-energy response with a fast imaging particle detection system. This particle imaging is based on microchannel plate (MCP) electron multipliers and position encoding discrete anodes.

2.1. Electrostatic Analyser Description

Basically the analyser design is a symmetrical quadrispherical electrostatic analyser which has a uniform 360° disc-shaped field of view (FOV) and extremely narrow angular resolution capability. This symmetric quadrisphere or 'top hat' geometry (Carlson *et al.*, 1982) has been successfully used on numerous sounding rocket flights as well as on the AMPTE/IRM, Giotto and WIND spacecraft (Paschmann *et al.*, 1985; Rème *et al.*, 1987; Lin *et al.*, 1995).

The operating principles of the analyser are illustrated by cross-section and top views in Figure 2. The symmetric quadrisphere consists of three concentric spherical elements. These three elements are an inner hemisphere, an outer hemisphere which contains a circular opening, and a small circular top cap which defines the entrance aperture. This analyser is classified as quadrispherical simply because the particles are deflected through 90°. In either analyser a potential is applied between

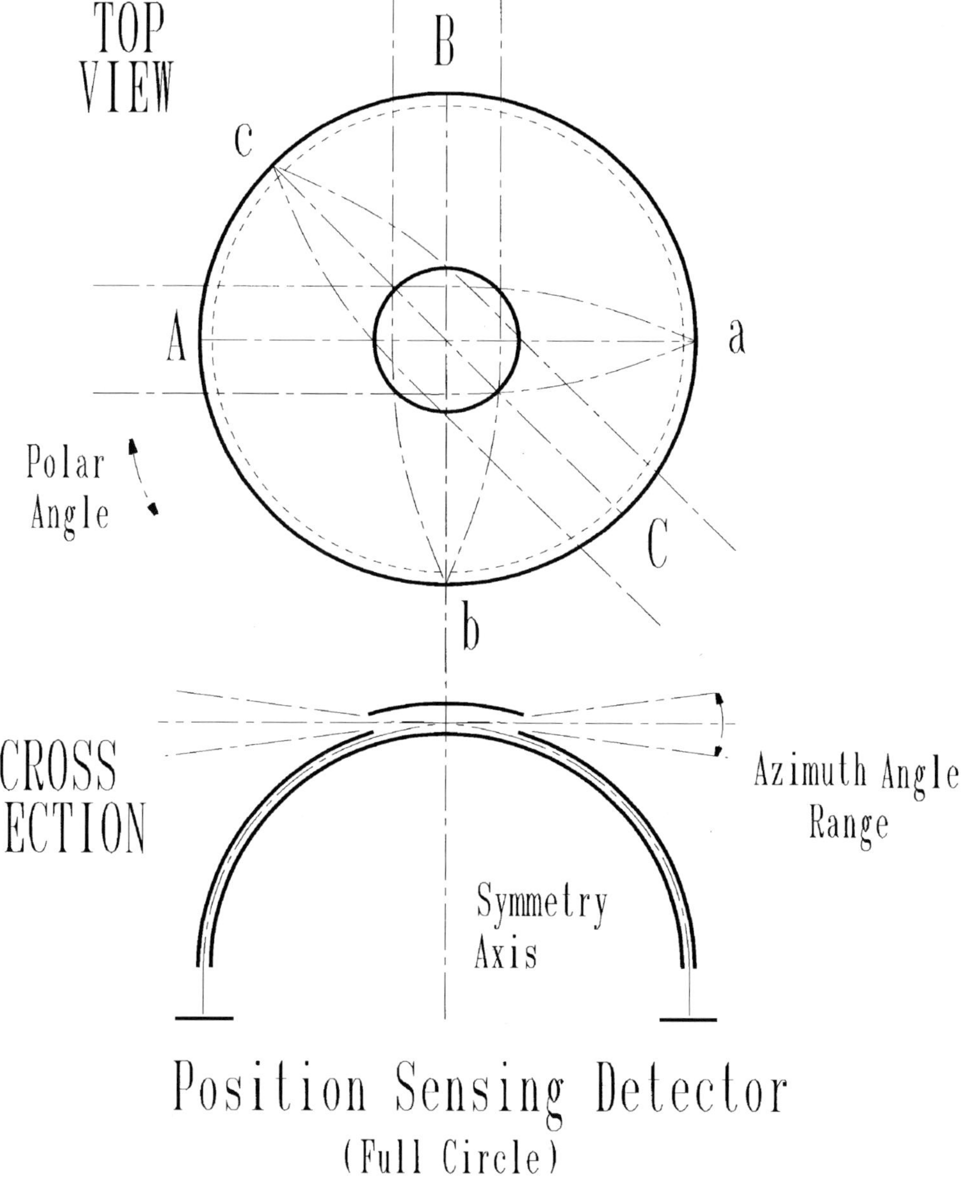

Figure 2. Particle orbits in a normal quadrisphere and in a symmetrical quadrisphere.

the inner and outer plates and only charged particles with a limited range of energy and initial azimuth angle are transmitted. The particle exit position is a measure of the incident polar angle which can be resolved by a suitable position-sensitive detector system. The symmetric quadrisphere makes the entire analyser, including the entrance aperture, rotationally symmetric. Trajectories are shown to illustrate

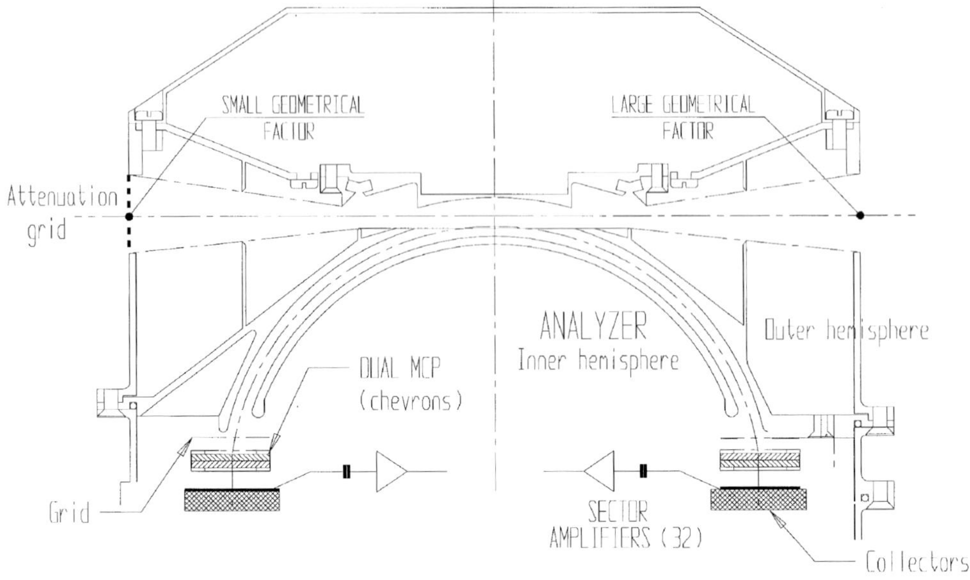

Figure 3. Principle of the HIA electrostatic analyser.

the focusing characteristics which are independent of polar angle. Throughout the paper we will use the following convention: the angle about the spin axis is the azimuth angle whereas the angle out of the spin plane is called polar angle.

In conclusion the symmetrical quadrispherical analyser has good focusing properties, sufficient energy resolution, and the large geometrical factor of a quadrisphere. Because of symmetry, it does not have the deficiencies of the conventional quadrisphere, namely limited polar angle range and severely distorted response characteristics at large polar angles, and it has an uniform polar response.

The HIA instrument has $2 \times 180°$ FOV sections parallel to the spin axis with two different sensitivities, with a ratio 20–30 (depending of the flight model and precisely known from calibrations), corresponding respectively to the 'high G' and 'low g' sections. The 'low g' section allows detection of the solar wind and the required high angular resolution is achieved through the use of $8 \times 5.625°$ central anodes, the remaining 8 sectors having in principle 11.25° resolution; the 180° 'high G' section is divided into 16 anodes, 11.25° each. In reality, sectoring angles are respectively $\sim 5.1°$ and $\sim 9.7°$, as demonstrated by calibrations (see Section 2.5). This configuration provides 'instantaneous' 2D distributions sampled once per 62.5 ms ($\frac{1}{64}$ of one spin, i.e., 5.625° in azimuth, which is the nominal sweep rate of the high voltage applied to the inner plate of the electrostatic analyser to select the energy of the transmitted particles. For each sensitivity section a full 4π steradian scan is completed every spin of the spacecraft, i.e., 4 s, giving a full 3D distribution of ions in the energy range ~ 5 eV e^{-1} to 32 keV e^{-1} (the analyser constant being ~ 6.70).

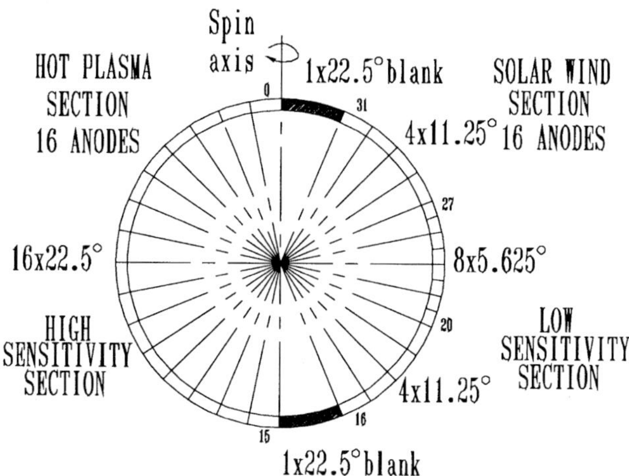

Figure 4. Principle of the HIA anode sectoring.

Figure 3 provides a cross-sectional view of the HIA electrostatic analyser. The inner and outer plate radii are 37.75 mm and 40.20 mm, respectively. The analyser has an entrance aperture which collimates the field of view, defines the two geometrical factors and blocks the solar UV radiation.

2.2. Detection System

A pair of half-ring microchannel plates (MCP) in a chevron pair configuration detects the particles at the exit of the electrostatic analyser. The plates form a $2 \times 180°$ ring shape, each 1 mm thick with an inter-gap of $\sim$ 0.02 mm, with an inner diameter of 75 mm and outer diameter of 85 mm. The MCPs have 12.5 μm straight microchannels with a bias angle of 8° to reduce variations in MCP efficiency with azimuthal direction. The chevron configuration with double thickness plates provides a saturated gain of 2×10^6, with a narrow pulse height distribution. The plates have a high strip current to provide fast counting capability. For a better detection efficiency ions are post-accelerated by a $\sim$ 2300 V potential applied between the front of the first MCP and a high-transparency grid located $\sim$ 1 mm above. The anode collector behind the MCPs is divided into 32 sectors, each connected to its own pulse amplifier (Figure 4).

2.3. Sensor Electronics

Signals from each of the 32 MCP sectors are sent through 32 specially designed very fast A121 charge-sensitive amplifier/discriminators (Figure 5) that are able to count at rates as high as 5 MHz. Output counts from the 32 sectors are accumulated in 48 counters (including 16 redundant counters for the solar wind), thus providing the basic angular resolution matrix according to the resolution of the anode sectoring.

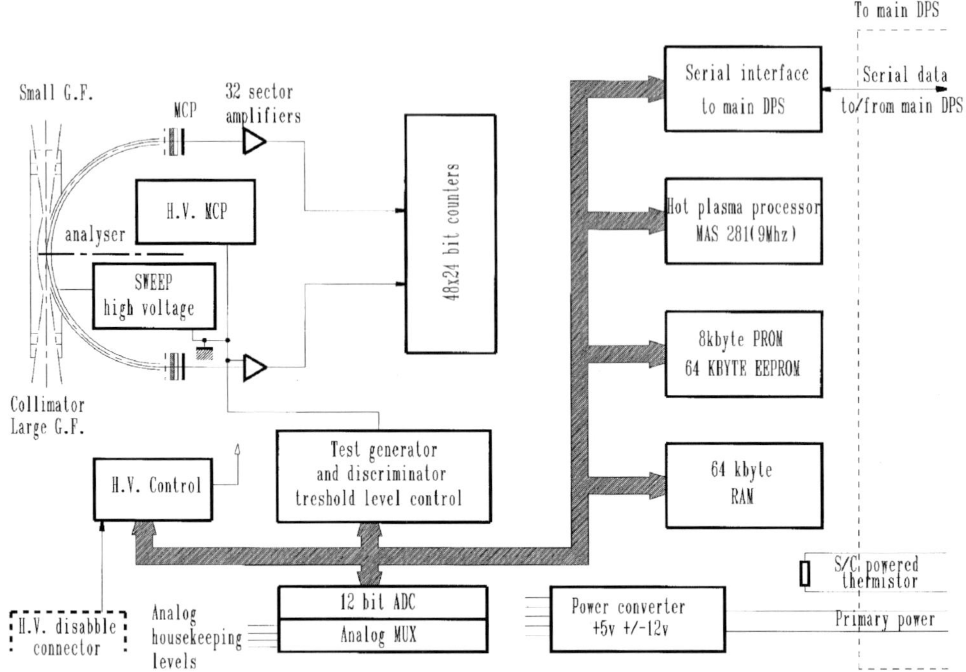

Figure 5. HIA functional block diagram.

According to the operational mode several angular resolutions can be achieved:

– in the normal resolution mode, the full 3D distributions are covered in $\sim 11.2°$ angular bins ('high G' geometrical factor); this is the basic mode inside the magnetosphere;

– in the high-resolution mode the best angular resolution, $\sim 5.6° \times 5.6°$, is achieved within a 45° sector centred on the Sun direction, using the 'low g' geometrical factor section; this mode is dedicated to the detection of the solar wind and near-ecliptic narrow beams.

Low Power Converter

The primary 28 V is delivered through the CODIF/DPS box to the HIA low-voltage power converter which provides +5 V and ±12 V for digital and analog electronics. In order to reduce the power consumption, HV power supplies are directly powered from the primary 28 V with galvanic insulation between the primary ground and the secondary ground.

High Voltage Power Supplies

HIA needs a high-voltage power supply to polarise MCPs at $\sim$ 2300–2500 V and a sweeping high voltage applied on the inner plate of the electrostatic analyser. The high voltages to polarise the MCPs are adjustable under control of the data processor system (DPS) microprocessor.

The energy/charge of the transmitted ions is selected by varying the deflection voltage applied to the inner plate of the electrostatic analyser, between 4800 and 0.7 V. The exponential sweep variation of the deflection voltage is synchronised with the spacecraft spin period. The sweep should consist of many small steps that give effectively a continuous sweep. The counter accumulation time defines the number of energy steps, i.e., 31 or 62 count intervals per sweep. The covered energy range and the sweeping time are controlled by the onboard processor through a 12-bit DAC and a division in two ranges for the sweeping high voltage. So the number of sweeps per spin, the amplitude of each sweep and the sweeping energy range can be adjusted according to the mode of operation (solar wind tracking, beam tracking, etc.). In the basic and nominal mode the sweep of the total energy range is repeated 64 times per spin, i.e., once every 62.5 ms, giving a $\sim 5.6^\circ$ resolution in azimuth resolution.

The HIA block diagram is shown in Figure 5.

2.4. In-flight calibration test

A pulse generator can stimulate the 32 amplifiers under processor control. This way all important functions of the HIA instrument and of the associated on-board processing can easily be tested. A special test mode is implemented for health checking of the microprocessor by making ROM checksums and RAM tests. The sweeping high voltage can be tested by measuring the voltage value of each individual step and the MCP gain can be checked by occasionally stepping MCP HV and by adjusting the discrimination level of charge amplifiers.

Performances of the HIA sensor are shown in Table I and in Figure 1. The full sensor is shown in Figure 6.

2.5. HIA performances

Pre-flight and extensive calibrations of all four HIA flight models were performed at the CESR vacuum test facilities in Toulouse, using large and stable ion beams of different ion species and variable energies, detailed studies of MCPs and gain level variations, MCP matching, angular-energy resolution for each sector from a few tens of eV up to 30 keV. Typical performances of the HIA instrument are reproduced in Figure 7. The analyser energy resolution $\Delta E/E$ (FWHM) is $\sim$18%, almost independent of anode sectors and energy; thus the intrinsic HIA velocity resolution is $\sim$9%, only about half of the average solar wind spread value. This is equivalent to an angular resolution of $\sim 5^\circ$ and is thus quite consistent with the angular resolution capabilities of the instrument, i.e., $\sim 5.9^\circ$ (FWHM) in azimuthal angle, as indicated in Figure 7, and $\sim 5.6^\circ$ in polar angle. The polar resolution stays, as expected, almost constant, $\sim 9.70^\circ$, over the 16 sectors (anodes 0 to 15) constituting the 'high G' section (Figure 8). Anodes 16 to 31 correspond to the 'low g' section and their response transmission is attenuated by a factor of 20–30

Table I

Main-measured parameters

- Full 3D ion distribution functions
- Flux as a function of time, mass and pitch angle
- Moments of the distribution functions: density, bulk velocity, pressure tensor, heat flux vector
- Beams

ANALYSERS	ENERGY RANGE	ENERGY DISTRIBUTION (FWHM)	TIME RESOLUTION 2D ms	TIME RESOLUTION 3D s	MASS RESOLUTION $M/\Delta M$	ANGULAR RESOLUTION	GEOMETRICAL FACTOR (TOTAL)	DYNAMICS $(cm^2 \text{ sec sr})^{-1}$
HOT ION ANALYSER HIA	~ 5eV/e - 32 keV/e	18%	62.5	4	-	~ 5.6° x 5.6°	cm^2.sr.kev 3.5×10^{-4} E(keV) for one half 7.10^{-3} E(keV) for the other half	10^4 - 2×10^{10}
ION COMPOSITION AND DISTRIBUTION FUNCTION ANALYSER CODIF	~ 0 - 40 keV/e Mass range 1 - 32 amu	16%	125	4	~ 4 - 7	~ 11.2° x 22.5°	2.16×10^{-3} cm^2 sr for one half 2.3×10^{-5} cm^2 sr for the other half 3.5×10^{-2} cm^2 sr for the RPA	3.10^3 - 3.10^9

ANALYSERS	FULL INSTANTANEOUS FIELD OF VIEW	MASS	POWER (Nominal Operations)
HOT ION ANALYSER HIA	8° x 360°	2.49 kg	3.36 watts
ION COMPOSITION AND DISTRIBUTION FUNCTION ANALYSER CODIF	8° x 360°	8.30 kg (including Data Processing System)	7.28 watts (including DPS)

CIS Total Weight: 10.79 kg without harness
Average CIS total raw power: 10.64 watts

CIS Telemetry: ~ 5.5 kbit/s
Expected total bit number (for the 4 spacecraft): 10^{12} bits

Figure 6. The HIA Sensor.

(depending of the flight model) due to the presence of a pin-hole grid placed in front of the 180° collimator; the polar resolution of sectors 20 to 27 is $\sim 5.1°$. Thus, when compared to the basic sectoring, $\sim 5.6°$ and $\sim 11.2°$, all effective polar resolutions are reduced, due to existence of an insulation space between the discrete anodes, as well as by the presence of support posts within the field of view. Finally, experimental energy, angle resolutions and transmission factors are introduced in the geometrical factor used to compute moments of the distribution function.

2.5.1. *UV Rejection*

A number of very interesting events are expected to occur when the HIA spectrometers face the Sun (2 times/spin): of course the intense solar wind, but also, for example, tailward ion beams flowing along the PlasmaSheet Boundary Layer (PSBL). Also a number of measures are applied in order to suppress or limit the solar UV contamination. Part of the UV is rejected by the entrance collimator; moreover, the inner surface of the outer sphere is scalloped and both spheres (and all internal parts) are treated and coated with a special black cupric sulfide. Extensive vacuum chamber tests of the HIA analysers were performed, using a calibrated continuous discharge source for extreme UV at He-584 Å and Lα 1215 Å lines. Reduction of the solar UV light reflectance at the Lα line is demonstrated in Figure 9. The resulting maximum count rate recorded by the sunward looking

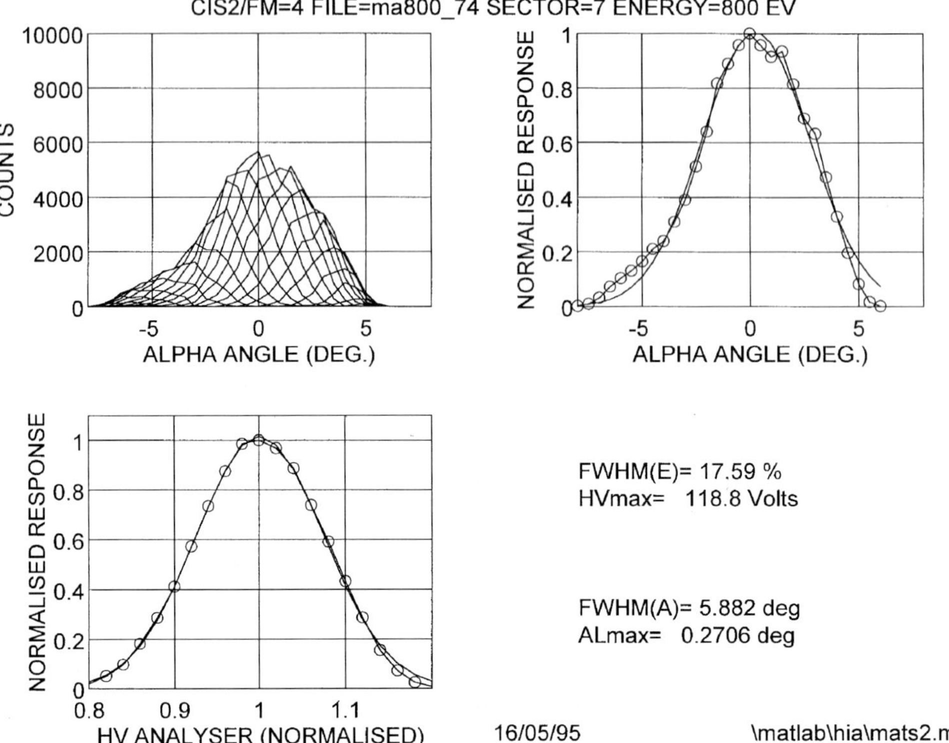

Figure 7. Typical energy and angular resolutions of the HIA analyser (flight model FM4), for an energy beam of 15 keV; the energy resolution is $\sim$ 18% and the intrinsic azimuthal resolution $\sim$ 5.9°.

sector (11.2° wide) is about 80 counts s^{-1} (for an intensity equivalent to 3 Sun intensity units) and the UV contamination is distributed over about $\sim$ 100° in polar angle; such a contamination is acceptable in the solar wind as well as in the magnetosphere.

3. The Ion Composition and Distribution Function Aanalyser (CODIF)

The CODIF instrument is a high-sensitivity mass-resolving spectrometer with an instantaneous 360° $\times$ 8° field of view to measure full 3D distribution functions of the major ion species (as much as they contribute significantly to the total mass density of the plasma), within one spin period of the spacecraft. Typically these include H^+, He^{++}, He^+, and O^+.

The sensor primarily covers the energy range between 0.02 and 40 keV/ charge. With an additional Retarding Potential Analyser (RPA) device in the aperture system of the sensor with pre-acceleration for the energies below 25 eV e^{-1}, the range is extended to energies as low as the spacecraft potential. So, CODIF will cover the core of all plasma distributions of importance of the Cluster mission.

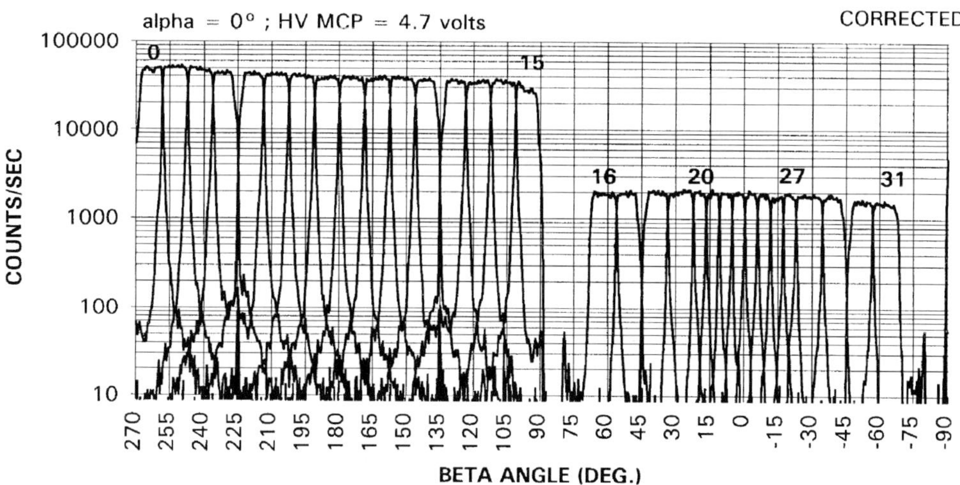

Figure 8. Calibrated relative transmission of the HIA polar sectors (beta angle in the vacuum chamber). Sectors 0–15 have $\sim 9.7°$ (FHWM) angular resolution; transmission of sectors 16–31 is attenuated by a factor of ~ 25 and equatorial sectors 20–27 have $\sim 5.1°$ (FWHM) angular resolution. 0–180° axis corresponds to the spacecraft spin axis.

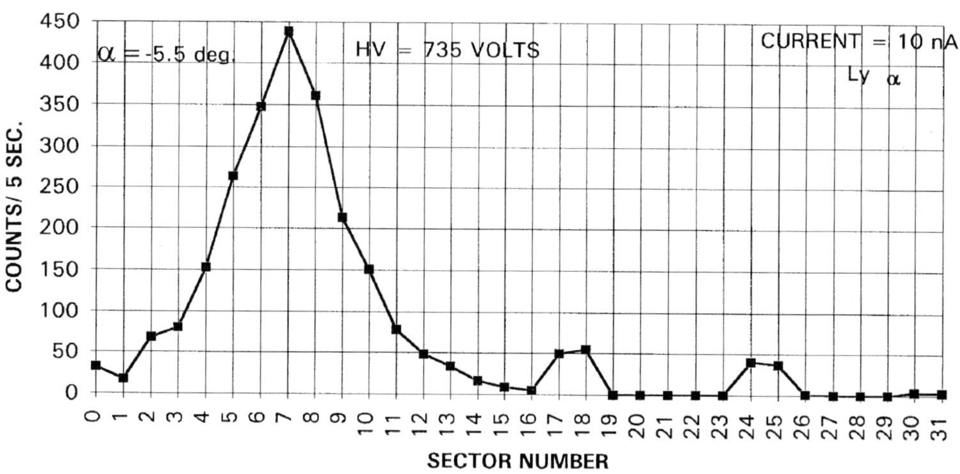

Figure 9. Azimuthal distribution of the solar UV count rate for each sector; the UV source is centered on sector 7 and the polar angle is $-5°$ (when the UV sunlight hits the inner hemisphere). Lα intensity measured by a windowless Au photodiode is equivalent to $\sim$3 Sun units.

To cover the large dynamic range required for accurate measurements in the low-density plasma of the magnetotail on the one hand and the dense plasma in the magnetosheath/cusp/boundary layer on the other, it is mandatory that CODIF employs two different sensitivities. The minimum number of counts in a distribution needed for computing the basic plasma parameters, such as the density, is about 100. These must be accumulated in 1 spin to provide the necessary time resolution. However, the maximum count rate which the time-of-flight system can handle is $\sim 10^5$ counts s^{-1} or 4×10^5 counts $spin^{-1}$. This means the dynamic range achievable with a single sensitivity is only 4×10^3.

Figure 1 shows the covered fluxes ranging from magnetosheath/magnetopause protons to tail lobe ions (which consist of protons and heavier ions); fluxes from $\sim 10^3$ to over 10^8 must be covered, requiring a dynamic range of larger than 10^5. This can only be achieved if CODIF incorporates two sensitivities, differing by a factor of 100. CODIF therefore will consist of two sections, each with 180° field of view, with different (by a factor of 100) geometrical factors. This way one section will always have count rates which are statistically meaningful and at the same time can be handled by the time-of-flight electronics. The exception is solar wind H^+ which will often saturate the instrument, but will be measured with HIA.

The CODIF instrument combines ion energy per charge selection by deflection in a rotationally symmetric toroidal electrostatic analyser with a subsequent time-of-flight analysis after post-acceleration to $\geq$20 keV e^{-1}. A cross section of the sensor showing the basic principles of operations is presented in Figure 10.

The energy-per-charge analyser is of a rotationally symmetric toroidal type, which is basically similar to the quadrispheric top-hat analysers. It has an uniform response over 360° of polar angle. The energy per charge selected by the electrostatic analyser E/Q, the energy gained by post-acceleration eU_{ACC}, and the measured time-of-flight through the length d of the time-of-flight (TOF) unit, τ, yield the mass per charge of the ion M/Q according to $M/Q = 2(E/Q + eU_{\mathrm{ACC}})/(d/\tau)^2\alpha$. The quantity α represents the effect of energy loss in the thin carbon foil ($\sim 3\ \mu\mathrm{g\ cm}^{-2}$) at the entry of the TOF section and depends on particle species and incident energy.

3.1. Electrostatic Analyser Description

The electrostatic analyser (ESA) has a toroidal geometry which provides optimal imaging just past the ESA exit. This property was first demonstrated by Young *et al.* (1988). The ESA consists of inner and outer analyser deflectors, a top-hat cover and a collimator. The inner deflector consists of toroidal and spherical sections which join at the outer deflector entrance opening (angle of 17.9°). The spherical section has a radius of 100 mm and extends from 0 to 17.9° about the z-axis. The toroidal section has a radius of 61 mm in the poloidal plane and extends from 17.9° to 90°. The outer deflector covers the toroidal section and has a radius of 65 mm. The top-hat cover consists of a spherical section with a radius of 113.2 mm, which extends from 0 to 16.2°. It thus fits inside the entrance aperture of the outer deflector. The top-hat cover contains an O-ring outside the spherical section, which together with a lip on the outer deflector provides a seal of the sensor interior during integration and launch activities when the protection retractable cover will be closed. The outer

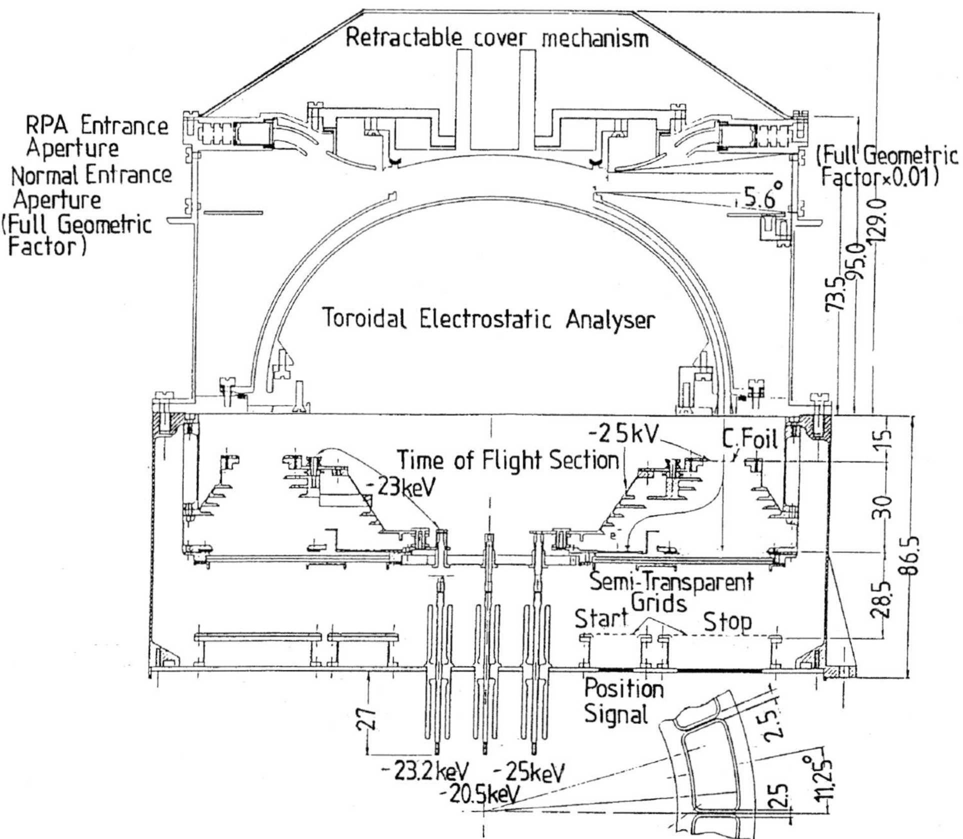

Figure 10. Cross-sectional view of the CODIF sensor. The voltages in the TOF section are shown for a 25 kV post-acceleration.

deflector and the top-hat cover will be at signal ground under normal operation, but will be biased at about −100 V during RPA operation. The inner deflector will be biased with voltages varying from −1.9 to −4950 V to cover the energy range in normal ESA operation and set to about −113 V for the RPA.

The fact that the analyser has complete cylindrical symmetry provides the uniform response in polar angle. A beam of parallel ion trajectories is focused to a certain location at the exit plane of the analyser. The exit position, and thus the incident polar angle of the ions, is identified using the information from the start detector (see Section 3.2) . The full angular range of the analyser is divided into 16 channels of 22.5° each. The broadening of the focus at the entrance of the TOF section is small compared to the width of the angular channels.

As illustrated in Figure 10, the analyser is surrounded by a cylindrical collimator which serves to define the acceptance angles and restricts UV light. The collimator consists of a cylindrical can with an inner radius of 96 mm. The entrance is covered by an attenuation grid with a radius of 98 mm which is kept at spacecraft ground. The grid has a 1% transmission factor over 50% of the analyser entrance and >95% transmission over the remaining 50%. The high transmission portion extends over the azimuthal angle range of 0 to 180° where 0° is defined along the spacecraft spin axis. The low transmission portion, whose active entrance only extends from 22.5° to 157.5° in order to avoid the counting of any crossover from the other half, has a geometric factor that is reduced by a factor of $\simeq$ 100 in order to extend the dynamic range to higher flux levels. On the low-sensitivity half, the collimator consists of a series of 12 small holes vertically spaced by approximately 1.9° around the cylinder. These apertures have acceptance angles of 5° FWHM, so there are no gaps in the polar angle coverage. The ion distributions near the polar axis are highly over-sampled during one spin relative to the equatorial portion of the aperture. Therefore, count rates must be weighted by the sine of the polar angle to normalise the solid-angle sampling for the moment calculations and 3D distributions.

The analyser has a characteristic energy response of about 7.6, and an intrinsic energy resolution of $\Delta E/E \simeq 0.16$. The entrance fan covers a viewing angle of 360° in polar angle and 8° in azimuth. With an analyser voltage of 1.9–4950 V, the energy range for ions is 15–40 000 eV e^{-1}. The deflection voltage is varied in an exponential sweep. The full energy sweep with 30 contiguous energy channels is performed 32 times per spin. Thus a partial two-dimensional cut through the distribution function in polar angle is obtained every $\frac{1}{32}$ of the spacecraft spin. The full 4π ion distributions are obtained in a spacecraft spin period. Including the effects of grid transparencies and support posts in the collimator each 22.5° sector has a respective geometric factor $A\Delta E/E\Delta\Theta\pi/8 = 2.16 \times 10^{-3}$ cm^{2}sr in the high sensitivity side and 2.3×10^{-5} cm^{2}sr where A denotes the aperture area and $\Delta\Theta$ the acceptance angle in azimuth. The acceptance in polar angle is $\pi/8$. The detection efficiencies of the time-of-flight system vary with particle energy, species and individual MCP assembly and therefore are treated separately in Section 3.5.

The outer plate of the analyser is serrated in order to minimize the transmission of scattered ions and UV. For the same reason the analyser plates are covered with a copper black coating. Behind the analyser the ions are accelerated by a post-acceleration voltage of 16–25 kV, such that also thermal ions have sufficient energy before entering the TOF section.

Retarding Potential Analyser

In order to extend the energy range of the CODIF sensor to energies below 15 eV e^{-1}, an RPA assembly is incorporated in the two CODIF apertures (see Figures 11 and 12). The RPA provides a way selecting low-energy ions as input to the CODIF analyser without requiring the ESA inner deflector to be set accurately near 0 V. The RPA collimates the ions, provides a sharp low-energy cutoff at a normal incident grid, pre-accelerates the ions to 100 eV after the grid, and deflects the ions into the ESA entrance aperture. The RPA will pass a 1 mm wide beam with

$\pm 5°$ of polar angle range and to about 10° of azimuthal angle about the center axis of each of the entrance apertures over the 360° field of view, giving a geometric factor of 0.035 cm^2 sr. Collimation in azimuthal angle to $\pm 10°$ of normal incidence to the RPA grid limits the entrance area to about half. The energy pass of the ESA is about 5–6 eV at 100 eV of pre-acceleration, assuming all deflection voltages are optimised. This energy pass is very sensitive to the actual RPA deflection optics, so that deflection voltages will have to be determined at about the 1% level.

The RPA consists of a collimator, RPA grid and pre-acceleration region, and deflection plates. The collimator section is kept at spacecraft ground.

The deflection system provides a method of steering the RPA low-energy ions into the CODIF ESA.

The RPA grid and pre-acceleration region consist of a pair of cylindrical rings, sandwiched between resistive ceramic material. Both inner and outer cylindrical rings contain apertures separated by posts every 22.5°, similar to the ESA collimator entrance, to allow the ions to pass through the assembly. The RPA grid is attached to the inner surface of the outer cylindrical ring. This outer ring has a small ledge which captures the RPA grid and which also provides the initial optical lens that is crucial to the RPA operation. Both inner and outer cylindrical rings are in good electrical contact with the resistive kapton (silver epoxy). During RPA operation the outer cylindrical ring is biased from spacecraft ground to about +25 V and provides the sharp low-energy RPA cutoff. This voltage is designated V_{rpa} in Figure 11. The inner cylindrical ring tracks the outer ring voltage and is biased at -100 V $+V_{rpa}$. The inner cylindrical ring, the ESA outer deflector, and the ESA top-hat cover are electrically tied to the RPA deflector.

The RPA deflection plates consist of three toroidal deflectors located above the ESA collimator entrance and one deflector disk located below the collimator entrance. The three toroidal deflectors are used to deflect the ions into the ESA. The deflector disk is used to prevent low-energy ions from entering the main aperture and to collect any photoelectrons produced inside the analyser, while in RPA mode.

3.2. Time-of-Flight and Detection System

The CODIF sensor uses a time-of-flight technology (Möbius *et al.*, 1985). The specific parameters of the time-of-flight spectrometer have been chosen such that a high detection efficiency of the ions is guaranteed. High efficiency is not only important for maximizing the overall sensor sensitivity, but it is especially important for minimising false mass identification resulting from false coincidence at high counting rate. Too thin a carbon foil would result in a significant reduction in the efficiency of secondary electron production for the 'start' signal, while an increase in thickness does not change the secondary electron emission significantly (Ritter, 1985). Under these conditions a post-acceleration of $\geq$20 kV is necessary for the mass resolution of the sensor.

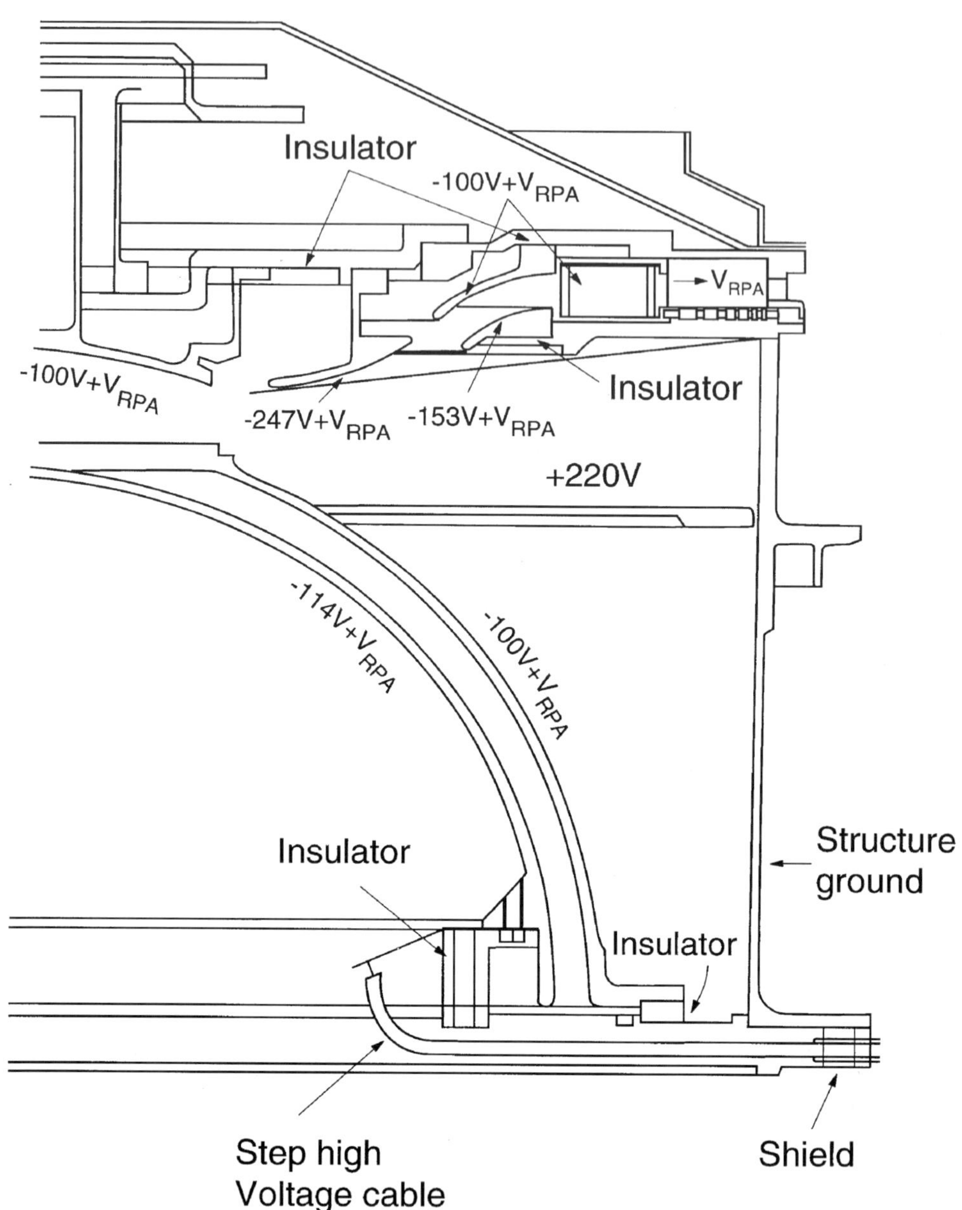

Figure 11. Geometry of the CODIF RPA.

After passing the ESA the ions are focussed onto a plane close to the entrance foil of the time-of-flight section (Figure 12). The TOF section is held at the post-acceleration potential (between -16 kV and -25 kV) in order to accelerate the ions into the TOF section with a minimum energy greater than 20 keV charge^{-1}. With this potential configuration the ion image on the foil extends from $r = 70$ mm

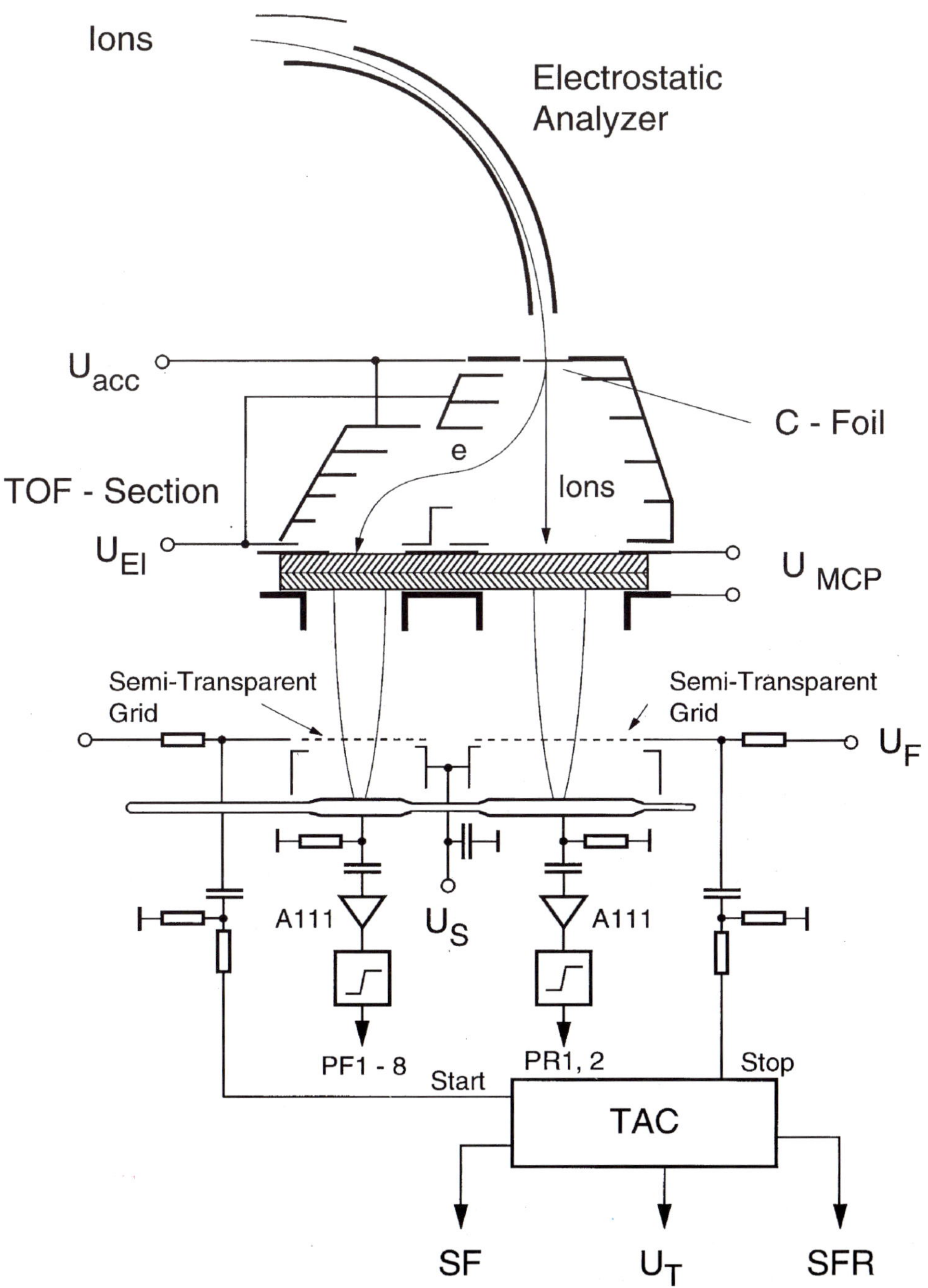

Figure 12. Schematics of CODIF sensor.

to $r = 80$ mm at high energies, and the image diameter is reduced to 3 mm or less at energies lower than 5 keV.

In the TOF section, the velocity of the incoming ions is measured. The flight path of the ions is defined by the 3 cm distance between the carbon foil at the entrance and the surface of the 'stop' microchannel plate (MCP). The start signal is provided by secondary electrons, which are emitted from the carbon foil during the passage of the ions. The entrance window of the TOF section is a 3 μg cm^{-2} carbon foil, which is an optimum thickness between the needs of low-energy loss and straggling in the foil and high efficiency for secondary electron production. The electrons are accelerated to 2 keV and deflected onto the start MCP assembly by a suitable potential configuration.

The secondary electrons also provide the position information for the angular sectoring. The carbon foil is made up of separate 22.5° sectors, separated by narrow metal strips. The electron optics are designed to strongly focus secondary electrons originating at a foil onto the corresponding MCP start sector.

Between the two sections with different geometrical factors a non-conducting plate with conducting strips in the appropriate equipotential configuration will ensure that no electrons and ions can penetrate into the other section.

The MCP assemblies are ring-shaped with inner and outer radii of 6×9 cm and 3×5 cm for the stop and start detectors, respectively. In order to achieve a high efficiency for the detection of ions at the stop MCP, a positive bias voltage is applied to collect secondary electrons from the 40% dead area of the MCP and the carbon foil.

The signal output of the MCPs is collected on a set of segmented plates behind the start MCPs (22.5° each), behind the stop MCPs (90° each), and on thin wire grids with ≈50% transmission at a distance of 10 mm in front of the signal plates, all being at ground potential (see Figure 10). Thus almost all of the post-acceleration voltage is applied between the rear side of the MCPs and the signal anodes. All signal outputs are coupled to 50 Ω impedance preamplifier inputs via a capacitor-resistor network. The timing signals are derived from the 50% transmission grids, separately for the high- and the low-sensitivity TOF section. The position signals, providing the angular information in terms of 22.5° sectors, are derived from the signal plates behind the start MCP.

Table I summarises the main performances of the HIA and CODIF sensors.

3.3. Sensor Electronics

The sensor electronics (Figure 13) of the instrument comprise two time-to-amplitude converters (TACs) to measure the time-of-flight of the ions between the start carbon foil and the stop MCPs, two sets of eight position discriminators at the start MCPs, two sets of two position discriminators at the stop MCPs, and the event selection logic. Each individual ion is pulse-height-analysed according to its time-of-flight,

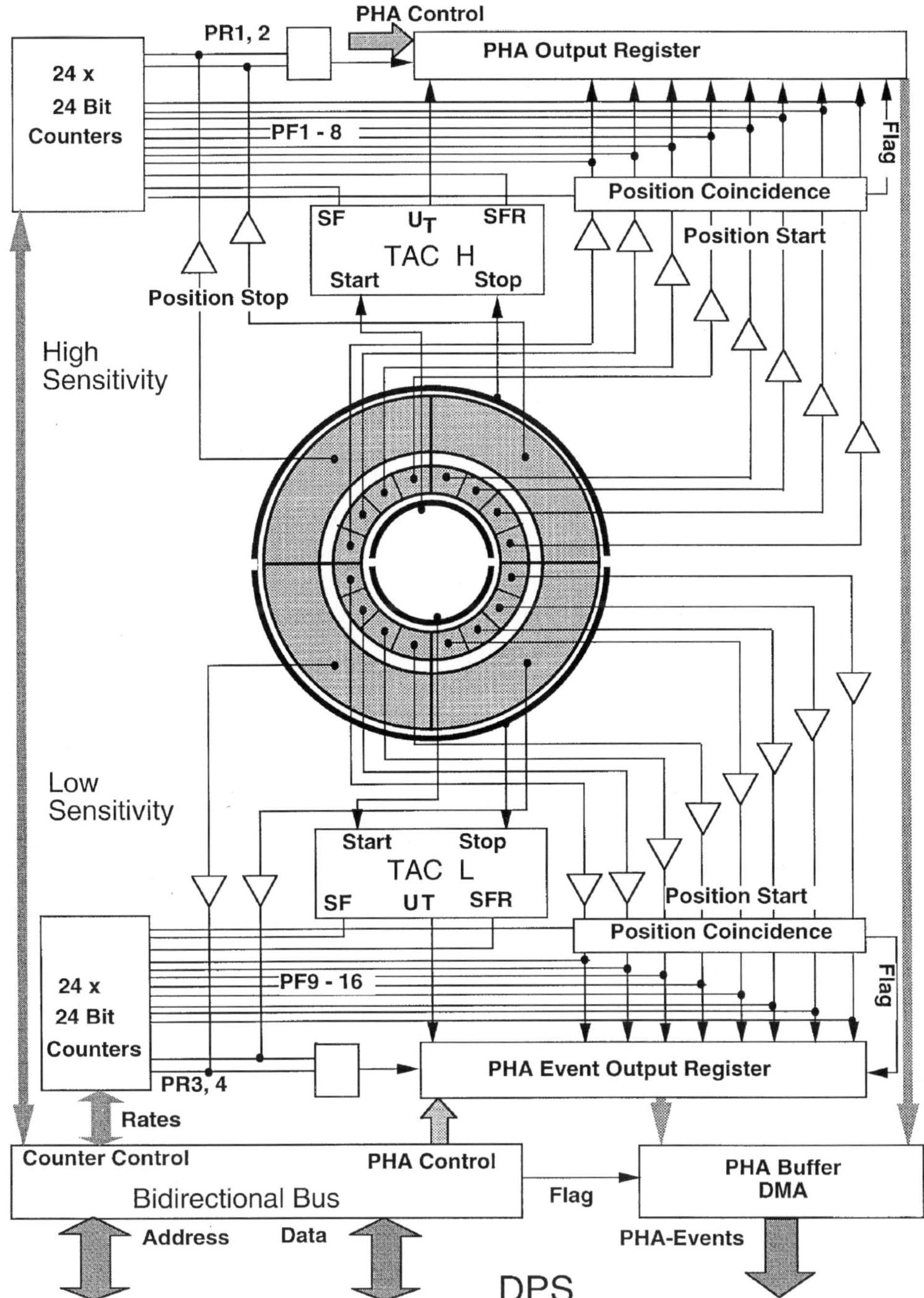

Figure 13. Functional block diagram of the CODIF sensor electronics.

incidence in azimuthal (given by the spacecraft spin) and polar angle (given by the start position), and the actual deflection voltage.

The eight position signals for each TOF section (in order to achieve the 22.5° resolution in polar angle) are independently derived from the signal anodes, while the timing signals are taken from the grids in front of the anodes. Likewise, the stop MCPs, consisting of four individual MCPs, are treated separately to carry along partial redundancy. By this technique the TOF and the position signals are electrically separate in the sensor. The position pulses are fed into charge-sensitive amplifiers and identified by pulse discriminators, the signal of which is directly fed into the event selection logic.

The TOF unit is divided into two TOF channels. The outputs of the 50% transmission grids in front of the signal anode of the MCPs are capacitively coupled to two input stages of the TOF electronics. These input stages consist of a preamplifier (rise time <0.9 ns) and a fast timing discriminator using a tunnel diode, and are contained in custom-made hybrids. The outputs of these hybrids are used to drive the TAC. The TAC provides an output signal whose amplitude is proportional to the time delay between the start and stop pulses. In addition, the TAC generates logical output signals for each start and stop pulse. The measured overall timing accuracy of the electronics for an amplitude of 100 mV of the start and stop pulse is 0.2 ns. The TAC output pulse is pulse-height analysed by a fast analog-to-digital converter (ADC) with a conversion time of <6 μs.

The conditions for valid events are established in the event-selection logic. The respective coincidence conditions can be changed via ground command. Several count rates are accumulated in the sensor electronics. There are monitor rates of the individual start and stop detectors to allow continuous monitoring of the carbon foil and MCP performance. The total count rates of TOF coincidence show the valid events accumulated for each TOF section. These rates can be compared with the total stop count rates in order to monitor in-flight the efficiency of the start and stop assemblies. There is a digital monitoring of all essential housekeeping values, like the deflection HVs, the post-acceleration voltage, the MCP HVs, and all the supply voltages of the electronics. In addition, the temperatures of the detector and electronics compartment are monitored. A simplified block diagram of the CODIF sensor electronics is presented in Figure 14.

In order to protect the MCPs, the solar-wind protons and the solar-wind alpha particles will be blocked from detection by a simple scheme during the sweeping cycle, as shown in Figure 15. The sweep, starting at high energies, is shown for the high-sensitivity section in the upper panel and for the low-sensitivity section in the lower panel in log E and azimuthal angle. The location of solar-wind protons and alphas in phase space is known from the HIA. By the use of this information the voltage sweep, which starts at high energies, is stopped above the alphas when the high-sensitivity section is facing the solar wind, and above the protons when the low-sensitivity section is facing the solar wind. The result is a small data gap for both sections of the sensor simultaneously. The primary purpose introducing

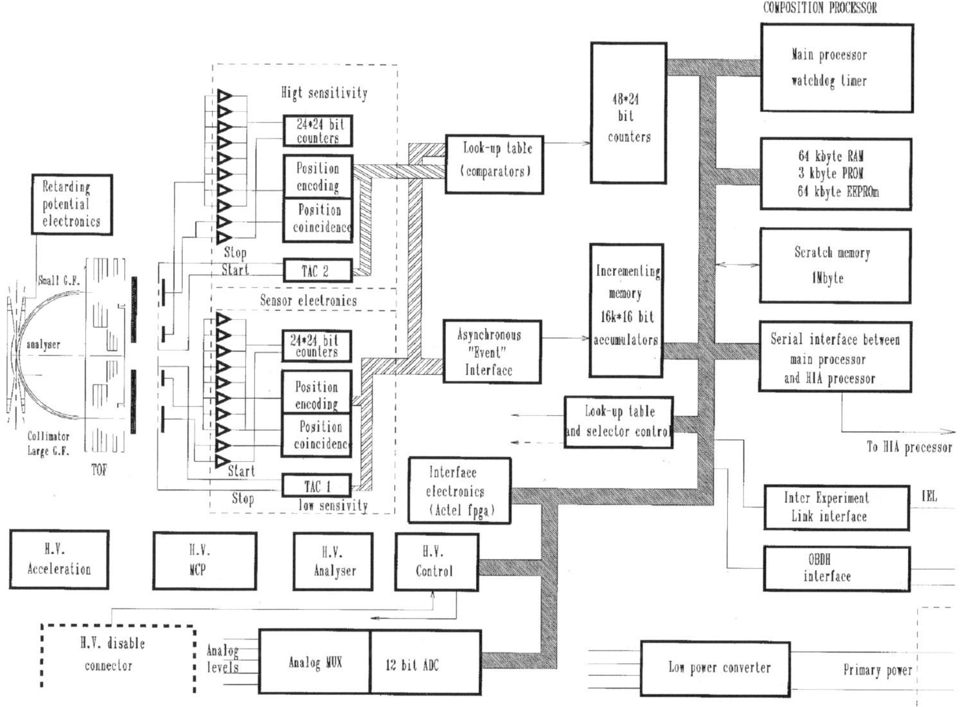

Figure 14. CODIF overall functional block diagram.

this scheme is to avoid short-time gain depression of the MCP area which would otherwise persist for the order of 1 s after the impulsive high count rate that would result from the solar wind.

3.3.1. *Counters and Incrementing Memory*

Each half of the CODIF sensor (high and low sensitivity) has eight angular bins and at least 64 TOF bins. A look-up table is used to combine TOF and energy into four mass species. The 32 combined mass/angle signals from each of the two parts of the analyser are sent into a selector circuit. The CODIF processor selects which part of the analyser is used. The 32 selected mass/angle signals feed the 48 counters, which are read out by the CODIF processor once per energy step. These counter values are used for moments, distribution functions, etc.

In addition to the counters, the CODIF has an incrementing memory accumulator used to make high-mass-resolution spectrograms. This memory can hold a full distribution (with limited angular resolution), so that long time averages can be made without using processor resources (memory, read-out time). According to the actual count rates, the CODIF processor selects which sensivity range (hence which half of the analyser) to accumulate. The TOF data from the selected half of the analyser are combined with the energy step and a look-up table and accumulated

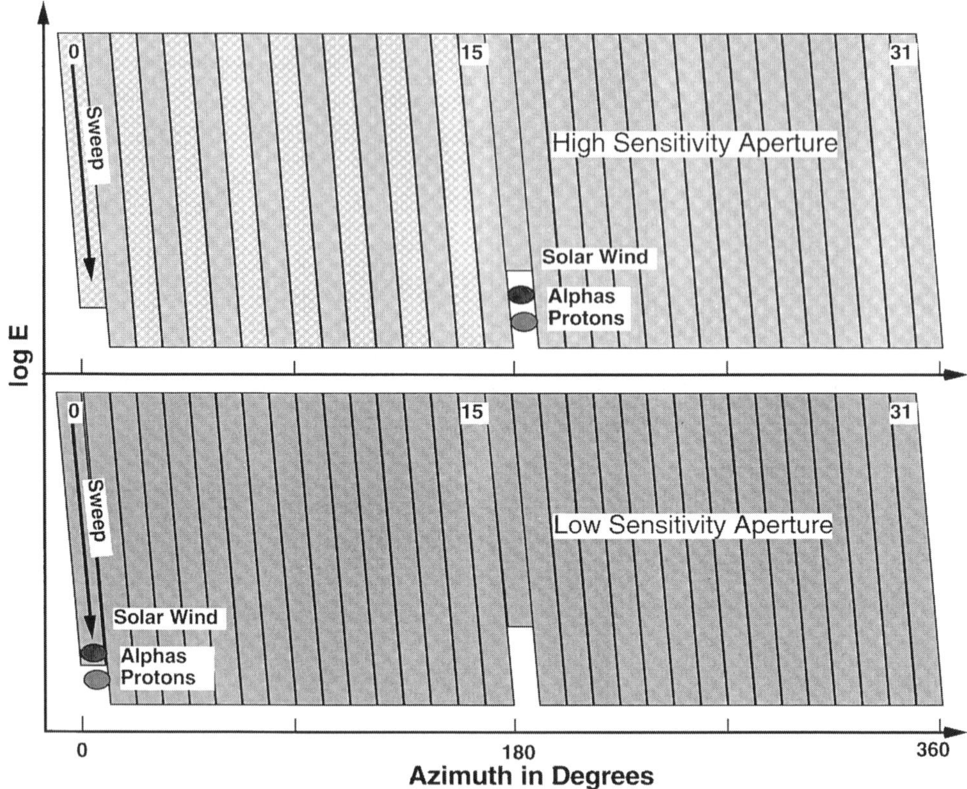

Figure 15. Energy sweeping scheme of CODIF in the solar wind. The sweep is shown in log E versus azimuthal angle for the high-sensitivity section (*upper panel*) and low-sensitivity section (*lower panel*), starting at the high energy end. When looking into the solar wind, the sweep stops above the alpha particles for the high-sensitivity section and above the protons for the low-sensitivity section.

into 64 mass channels. The eight angle bins from the selected half are combined with the spacecraft rotation sweep angle with another look-up table into 16 angular sectors with about 45° resolution. The 16 angles and 64 masses are combined with 16 energy bins to address the incrementing memory. The incrementing memory requires $16 \times 64 \times 16 = 16\,384$ accumulators.

3.3.2. *High Voltage System*

A sweep-voltage high-voltage power supply generates an exponential voltage waveform from 1.9 to 4950 V for the electrostatic analyser. A $\geq$20 kV static supply feeds the post-acceleration voltage, which can be adjusted via ground command. Another adjustable supply is used for the MCPs and the collection of secondary electrons. It supplies up to 5 kV and is floated on top of post-acceleration voltage. All high-voltage power supplies are run from the spacecraft raw power instead of from the low-voltage power converter. Thus the efficiency factor of the low-voltage power converter does not apply to the corresponding power consumption. Our design still garantees a galvanic separation of secondary and primary power.

3.4. Other elements

3.4.1. *Retractable Cover*

A retractable cover is used for CODIF, mainly to avoid carbon foil damage during launch operations.

3.4.2. *In-flight Calibration*

Routinely an in-flight-calibration (IFC) pulse generator stimulates the two independent TOF branches of the electronics according to a predefined program. Within this program all important functions of the sensor electronics and the subsequent on-board processing of the data can be automatically tested. Temporal variations of calibration parameters can be measured. The in-flight calibration can also be triggered by ground command in a very flexible way, e.g., for trouble shooting purposes. In addition, the known prominent location of the proton signal can, if necessary, serve as a tracer of changes in the sensor itself.

3.5. CODIF performances

3.5.1. *Resolution in Mass per Charge*

The instrumental resolution in mass per charge is determined by a combination of the following effects:

– energy resolution of the electrostatic deflection analyser ($\Delta E/E = 0.16$);

– TOF dispersion caused by the angular spread of the ion trajectories because of the characteristics of the analyser and the straggling in the carbon foil (the angular spread of $= 13°$ leads to $\Delta\tau/\tau = 0.03$);

– TOF dispersion caused by energy straggling in the carbon foil ($\Delta\tau/\tau$ up to $= 0.08$ for 25 keV O^+);

– electronic noise in the TOF electronics and secondary-electron flight time dispersion (typically 0.3 ns).

The resulting TOF dispersion amount $\Delta\tau/\tau \leq 0.1$, which finally leads to a M/Q resolution between 0.15 for H^+ and 0.25 for low-energy O^+. A sample TOF spectrum for various 25 keV ions is shown in Figure 16 from calibration measurements with the CODIF sensor. It is demonstrated that all major ions are well separated by the sensor.

3.5.2. *Time-of-Flight Efficiency Curves*

The TOF efficiency is a function of the ion species and the total energy, the sum of the original ion energy plus the energy gained in the post-acceleration potential. The efficiency was also found to be a function of the position at which the ion enters the instrument. An efficiency versus energy curve is determined for each species, with an overall normalisation factor necessary for each azimuthal position. The

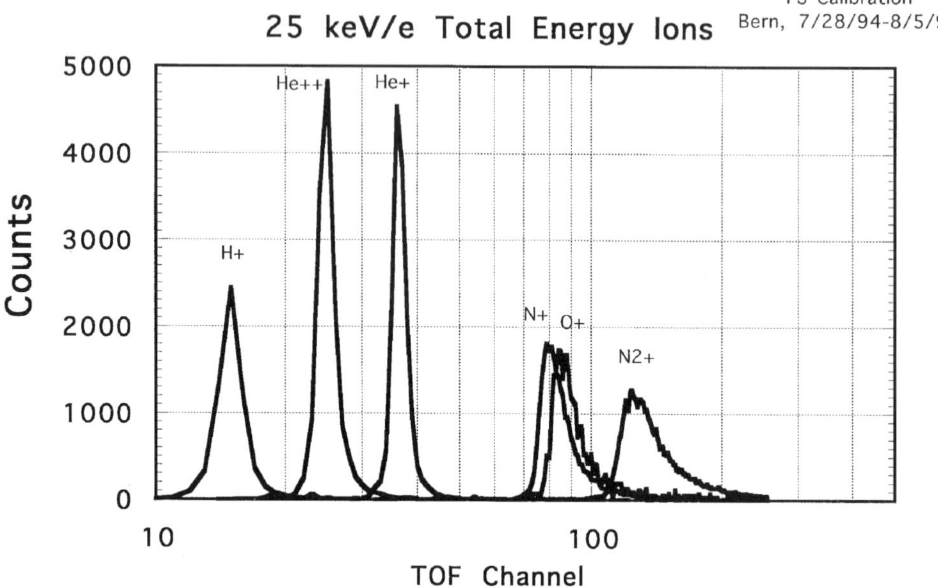

Figure 16. Time-of-flight spectrum of 25 keV H^+, He^{++}, He^+, N^+, O^+, and $N_2{}^+$ ions, as measured with the Flight Spare model of CODIF during a calibration at the University of Bern.

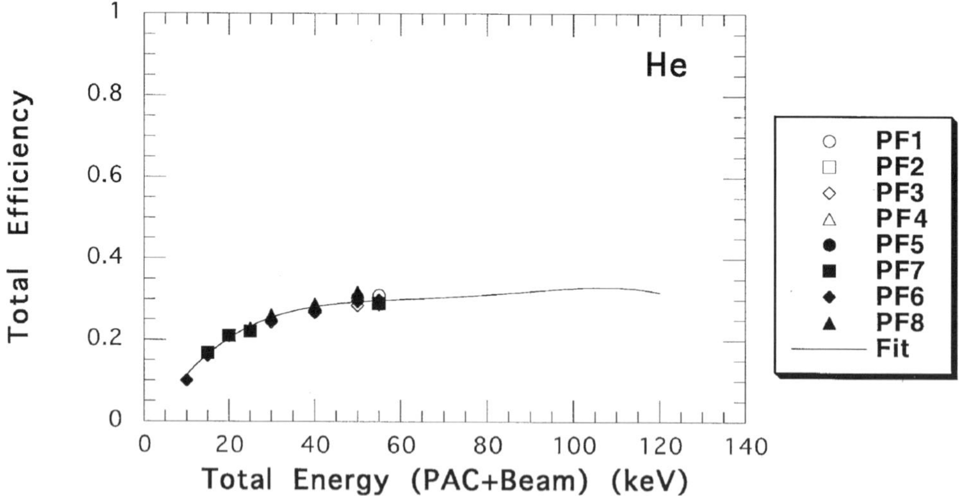

Figure 17. CODIF time-of-flight efficiency for Helium. PF1 through 8 indicate the individual angular sectors of 22.5° each of the high resolution side of the sensor.

efficiencies take into account the probability of getting a 'start' and 'stop' signal, plus the probability that an event will satisfy the valid event conditions. Figures 17 and 18 show the efficiency curves for He and for N^+. These curves were determined using data from Flight Model 1. Data from all positions on the high-sensitivity side

Table II
Total position adjustment for CIS flight model 1

Pixel	H^+	He	N^+
1.000	0.424	0.496	0.560
2.000	0.826	0.848	0.767
3.000	0.882	0.951	0.892
4.000	0.397	0.512	0.565
5.000	0.905	0.944	0.918
6.000	1.101	1.056	1.086
7.000	1.199	1.042	0.950
8.000	0.911	0.775	0.860
10.000	0.705	0.776	0.760
11.000	0.997	0.980	0.951
12.000	1.255	1.227	1.012
13.000	2.838	2.406	1.688
14.000	0.470	0.524	0.739
15.000	0.949	0.919	0.969

of the instrument are shown. The He curve is applicable to both He^+ and He^{++}. The N^+ and O^+ efficiencies are quite close, so the N^+ curve is used to represent the N^+ and O^+ group measured by the instrument. The final H^+curve has not yet been determined. Table II shows the position factors needed for each pixel and species for Flight Model 1. Note that this information is model-dependent, and this Table is shown just as an example of the amount of variablity that is observed.

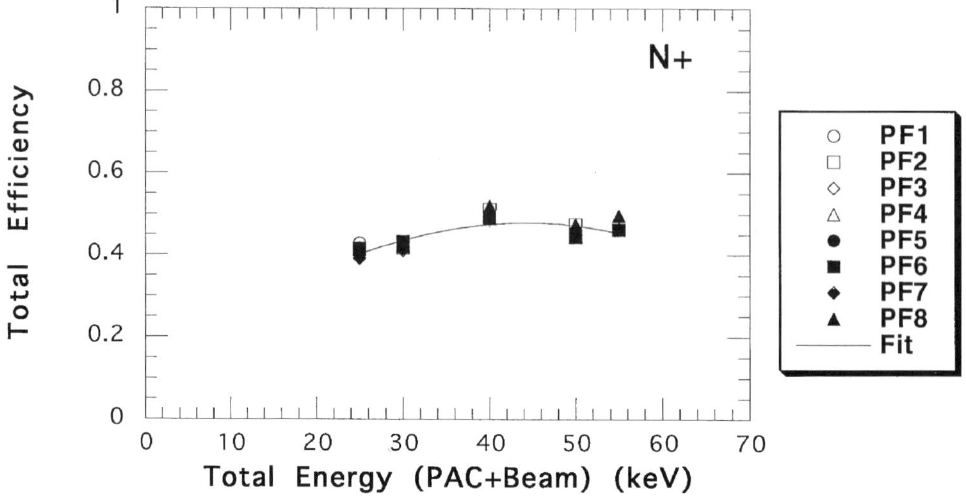

Figure 18. CODIF time-of-flight efficiency for N^+ ions.

3.5.3. *Immunity to Background*

Compared to a single detector instrument, the TOF sensors have an inherently better immunity to background, because of the start-stop coincidence requirement. The main source of background in these instruments is chance coincidence counts due to penetrating radiation and to events not detected by either start or stop MCP because detection efficiencies are less than 100% according to the relation:

$$R_{\mathrm{CHANCE}} = R^2_{\mathrm{SIGNAL}} \mu_1 (1 - \mu_2) \mu_2 (1 - \mu_1) \Delta\tau \; , \tag{1}$$

where R_{SIGNAL} is the input rate of the ion species with maximum flux, $\Delta\tau$ is the total TOF window, and μ_1 and μ_2 are the efficiencies of the start and stop assemblies, respectively. It can be easily seen that low efficiencies reduce the signal-to-noise ratio significantly. Therefore, great care has been applied to ensure as high efficiencies as possible for both start and stop detectors. The stop detector has an efficiency exceeding 90%. The efficiency of the start detector assembly is limited by the secondary electron emission of the foil. For protons of 25 keV the emission efficiency is reduced to $\sim$ 50% (Ritter, 1985). Similar values have been reached with the laboratory model of the sensor. Figure 19 shows the background-to-signal ratio versus incident proton flux level for 25 keV O^+ (the TOF window for the integration of counts is $\Delta\tau/\tau = 0.25$). It is demonstrated that the mass density of the plasma can still be determined with an accuracy of 10% (O^+/H^+ ratio better than 0.006) for proton fluxes as high as $= 10^8$ (10^{10}) ions s^{-1} cm^{-2} sr^{-1} with the full (or reduced) aperture, respectively.

A possible background due to UV photons is negligible, since the deflection plates are covered with copper black and photons undergo at least three scatterings before reaching the TOF system.

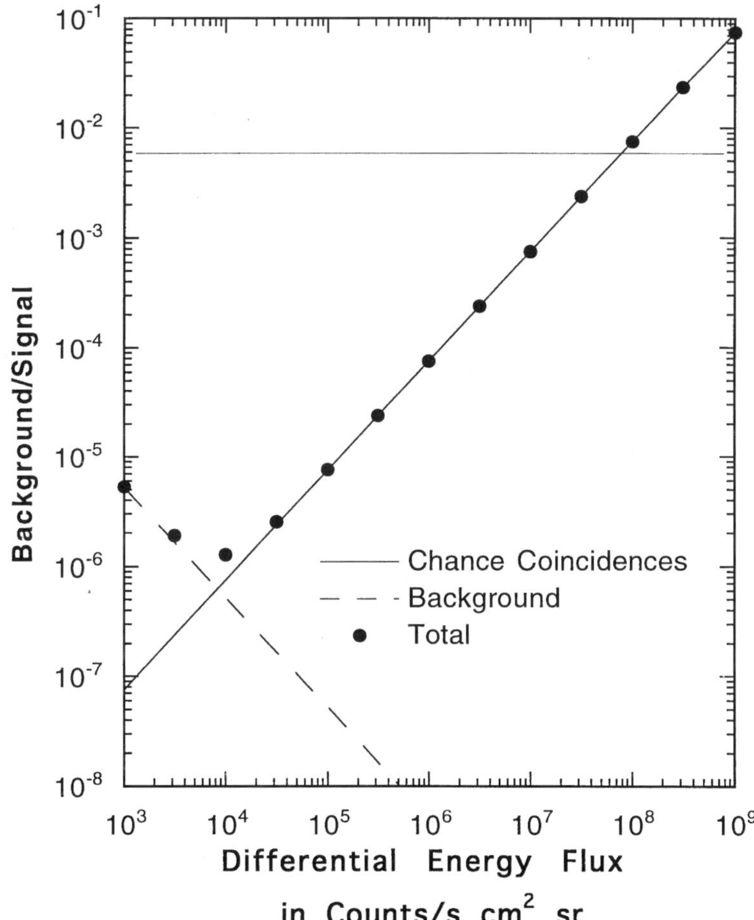

Figure 19. Background-to-signal ratio versus absolute proton flux for background due to chance coincidence of events creating only a start or a stop pulse (full line, computed for the actual start and stop efficiencies of CODIF). At low flux levels the background due to penetrating radiation and intrinsic effects of the MCPs prevails (broken line, taken from AMPTE SULEICA flight data (Möbius *et al.*, 1985)). The combination of both effects is shown by the solid circles. The horizontal line represents the limit for 10% accuracy of the mass density, if O^+ is the minor species.

3.5.4. *Dynamic Range*

The design of the electrostatic analyser guarantees a large geometrical factor in the high-sensitivity section $A\Delta E/E\Delta\tau\pi = 0.025$ cm^2 sr. The energy bandwidth is $\Delta E/E = 0.16$. The efficiency of the TOF unit is about 0.5. Differential energy fluxes as low as $\sim 3 \times 10^3$ ions s^{-1} cm^{-2} sr^{-1} can be detected by the instrument with the full time resolution of 1 spin period and about 5 counts energy^{-1} channel. The sensitivity is increased for longer integration time accordingly. Therefore the dynamic range reaches seven decades. The upper flux limit of the instrument amounts to 3×10^9 ions s^{-1} cm^{-2} sr^{-1}, which leads to a count rate of 10^5 counts s^{-1}

Figure 20. The CODIF sensor.

in one TOF unit (near saturation of the analysing electronics) and still guarantees a mass density determination to better than 10% accuracy for the reduced aperture geometry.

Performances of the CODIF sensor are summarised in Table I and in Figure 1. The full CODIF sensor is shown in Figure 20.

4. Data Processing System

4.1. On-board Data-processing System

Because of the high sensitivity and high intrinsic velocity-space resolution of the CIS instruments, continuous transmission of the complete 3D ion distributions sampled at the full time and angular resolution would require impossibly large bit rates. So, extensive on-board data-processing is a fundamental aspect of the CIS experiment. The CIS flight software has been designed to meet the scientific requirements of the mission even in limited transmission bit-rate allocation conditions.

First, the instrument data system (DPS) controls the operation and data collection of the two CODIF and HIA instruments, formats the data for the telemetry channel, and receives and executes commands. In addition, the DPS analyses and compresses on-board the tremendous amount of data to maximise the scientific return despite the limited CIS telemetry allocation. The DPS and the CODIF instrument are integrated in one box called CIS-1 and HIA is integrated in another box called CIS-2.

4.1.1. *Moments*

Moments of the distribution functions measured by the analysers will be computed by the DPS and continuously transmitted with maximum time resolution (1 spin period or 4 s). Main on-board calculated moments for CODIF and HIA instruments, to within a multiplicative factor dependent on the analyser geometrical factors, are given by the sums of Table III. These moments include particle density N_i (including partial densities over several energy ranges for CODIF, and sunward and anti-sunward densities for HIA), the three components of the flow vector V_i, the six unique components of the momentum flux tensor, and the heat flux vector. From these, the full pressure tensor can be deduced as well as the temperature anisotropies $T_{\|}/T_{\perp}$. We anticipate that full 4π space coverage of the analysers and their clean response function will guarantee a high accuracy for the on-board computed moments. To calculate moments, integrals over the distribution function are approximated by summing products of measured count rates with appropriate energy/angle weighting over the sampled distribution. As already said, moments for HIA and for CODIF (for four masses) are computed once per spin.

Besides instrument sensitivity and calibration, the accuracy of computed moments is mainly affected by the finite energy and angle resolution, and by the finite energy range. The requirement on instrumental accuracy is best demonstrated in the measurements of mass flow through the magnetospheric boundary and in the computation of the current density in current layers like the magnetopause and the Flux Transfer Events (FTEs). Directional errors in the bulk velocity of less than 2° and relative errors less than 5% in the product of bulk velocity times number density of the different species are highly desirable. As for the mass flow, quantitative tests of other conservation laws (stress and energy balance) require measurements of plasma moments with uncertainties less than 5%. Paschmann *et al.* (1986) tested the capability of the AMPTE/IRM plasma instrument in a simulation study. For parameters typically observed in high-speed flow events, the simulation shows that density, velocity, temperature and pressure are accurately measured to within 5%. With the better azimuthal coverage and resolution of the CIS instruments, improved accuracy (in comparison to AMPTE/ IRM) of the plasma moments is expected (Martz, 1993). The accuracy requirements concerning the analysis of two- and three-dimensional current structures as well as shear and vortex flows, i.e. measurements strongly related to the four spacecraft aspect, are fulfilled by the capability of the instrument.

Table III

Moment definitions

To within a multiplicative factor dependent on the analyser geometric factor, the moments are given by the following sums:

Density:

$$N = \sum_E 1/V(E) \times \sum_\phi \sum_\theta C(\theta, \phi, E)$$

Bulk Velocity:

$$NV_x = \sum_E \sum_\phi \cos(\phi) \times \sum_\theta \cos(\theta) \times C(\theta, \phi, E)$$

$$NV_y = \sum_E \sum_\phi \sin(\phi) \times \sum_\theta \cos(\theta) \times C(\theta, \phi, E)$$

$$NV_z = \sum_E \sum_\phi \sum_\theta \sin(\theta) \times C(\theta, \phi, E)$$

Heat Flux Vector:

$$NH_x = \sum_E V^2(E) \times \sum_\phi \cos(\phi) \times \sum_\theta \cos(\theta) \times C(\theta, \phi, E)$$

$$NH_y = \sum_E V^2(E) \times \sum(\phi) \sin(\phi) \times \sum_\theta \cos(\theta) \times C(\theta, \phi, E)$$

$$NH_z = \sum_E V^2(E) \times \sum_\phi \sum_\theta \sin(\theta) \times C(\theta, \phi, E)$$

Pressure Tensor:

$$NP_{xx} = \sum_E V(E) \times \sum_\phi \cos^2(\phi) \times \sum_\theta \cos^2(\theta) \times C(\theta, \phi, E)$$

$$NP_{yy} = \sum_E V(E) \times \sum_\phi \sin^2(\phi) \times \sum_\theta \cos^2(\theta) \times C(\theta, \phi, E)$$

$$NP_{zz} = \sum_E V(E) \times \sum_\phi \sum_\theta \sin^2(\theta) \times C(\theta, \phi, E)$$

$$NP_{xy} = \sum_E V(E) \times \sum_\phi \cos(\phi) \times \sin(\phi) \times \sum_\theta \cos^2(\theta) \times C(\theta, \phi, E)$$

$$NP_{xz} = \sum_E V(E) \times \sum_\phi \cos(\phi) \times \sum_\theta \cos(\theta) \times \sin(\theta) \times C(\theta, \phi, E)$$

$$NP_{yz} = \sum_E V(E) \times \sum(\phi) \sin(\phi) \times \sum_\theta \cos(\theta) \times \sin(\theta) \times C(\theta, \phi, E)$$

where:

E is energy

$\vec{V}$ is velocity

Θ, ϕ are the analyzer viewing angles

$C(\theta, \phi, E)$ are the measured counts

4.1.2. *Reduced Distributions*

Other reduced distributions, including pitch-angle distributions, averages (over 2 to 5 spin periods) or snapshots of the 3D distributions, will be computed with resolutions dependent upon the specific scientific objectives and telemetry rate. The two-dimensional pitch-angle distribution requires far less telemetry than the full distribution, thus allowing higher time resolution. Pitch-angle distributions can be transmitted when the magnetic field direction (provided on-board by the magnetometer) is in the field of view of the detector.

4.1.3. *On-Board Processing Unit*

Accomplishing these computations in real time is a heavy processing burden, and requires a sophisticated data system, both in terms of hardware and software. The data system is based on a set of two microprocessors (Figure 21). The main processor, located in the CIS-1 box, interfaces with the spacecraft On-Board Data Handling System (OBDH), the magnetometer, the plasma wave experiments (DWP), and the CIS-2 processor. It is in charge of formatting telemetry data, receiving and executing commands or passing them to the other processor, and controlling the burst memory. It also controls, collects and analyses data from the CODIF. The second processor is included in CIS-2 box and controls, collects and analyses data from the HIA. The main processor is interfaced with the second one by a serial data line; the HIA processor will compress the data so that the serial link can transmit at the highest data rates.

Each processor system consists of a Marconi MAS281 microprocessor, together with RAM, EEPROM and ROM memories, and a 'watchdog' timer. The 'watchdog' timer is used to monitor the health of the operation, software and reset the processor if the operation is abnormal. This allows the processors to recover automatically from transient problems such as radiation-induced single-event upsets. The software is designed to come up in a safe operating mode on reset. All hardware is radiation-hardened to the greatest extent possible, with spot shielding being used on any parts that are not sufficiently hard to withstand the Cluster environment.

4.1.4. *Scratch Memory*

The CIS experiment acquires data at nearly the fastest useful rate. In order to store a series of many two- and three-dimensional distributions at full time resolution, a 1 Mbyte memory is included in the instrument, so that discontinuities can be studied in detail.

4.2. TELEMETRY

4.2.1. *Data Products*

Tables IV and V give HIA and CODIF scientific telemetry products, respectively. Products consist of on-board computed moments, one-, two-, and three-dimenstional distributions and pitch-angle distributions. The high flexibility in

Table IV
HIA scientific telemetry products

Quantity	Product no.	Accum. Size	Basic Time (spin)	Total (bits)	bit/s
I. HOT POPULATIONS (large geometrical factor section)					
Moments **3Δϕ(n,3v,6P,3H)**	**P2**	**468 words 16 spins 1 packet**	**1**	**468 x 16 + 32**	**117.5**
Hot 3D Max Resolution (Large G Section) 62E x 8θ x 16ϕ	**P5**	**3968 words 4 packets**	**1**	**63488 + 192**	**15920**
3D 31E x 88Ω	**P6**	**1364 words 2 packets**	**1**	**21824 + 96**	**5480**
3D 31E x 42Ω	**P7**	**651 words 1 packet**	**1**	**10416 + 48**	**2616**
1D 62E	**P9**	**8 spins 1 packet**	**1**	**496 x 8 + 32**	**125**
1D 31E	**P18**	**8 spins 1 packet**	**1**	**248 x 8 + 32**	**63**
2D Azim. Distribution (integrated over polar angles) 31E x 16ϕ	**P10**	**496 words 2 spins**	**1**	**3968 x 2 + 48**	**998**
2D Polar Distribution (integrated over azim. angles) 31E x 16θ	**P11**	**496 words 2 spins**	**1**	**3968 x 2 + 32**	**996**
2D Polar Distribution 31E x 16θ for 3 sectors (solar wind, antisolar and flanks)	**P20**	**1488 words 2 spins**	**1**	**11904 x 2 + 32**	**2976**
2D Pitch Angle Distrib. Cut (2 slices/spin when *B* is in the field of view) **31E x 16θ x 2 slices**	**P12** **P19**	**496 words 1 spin** **2 spins**	**0.5 x 2** **1**	**3968 x 2 + 48** **3968 x 2 + 64**	**1996** **1008**
3D 16E x 88Ω	**P15**	**704 words 1 packet**	**1**	**11264 + 48**	**2828**
3D 30E x 88Ω	**P16**	**1320 words 2 packets**	**1**	**21120 + 96**	**5304**
3D 62E x 88Ω	**P17**	**2728 words 3 packets**	**1**	**43648 + 144**	**10948**
3D 31E x 8θ x 16ϕ (*)	**P21**	**1984 words 2 packets**	**1**	**31744 + 96**	**7960**
3D 31E x 8θ x 16ϕ compressed (**)	**P23**	**992 words 2 packets**	**1**	**15872 + 5 x 2 x 16**	**4008**

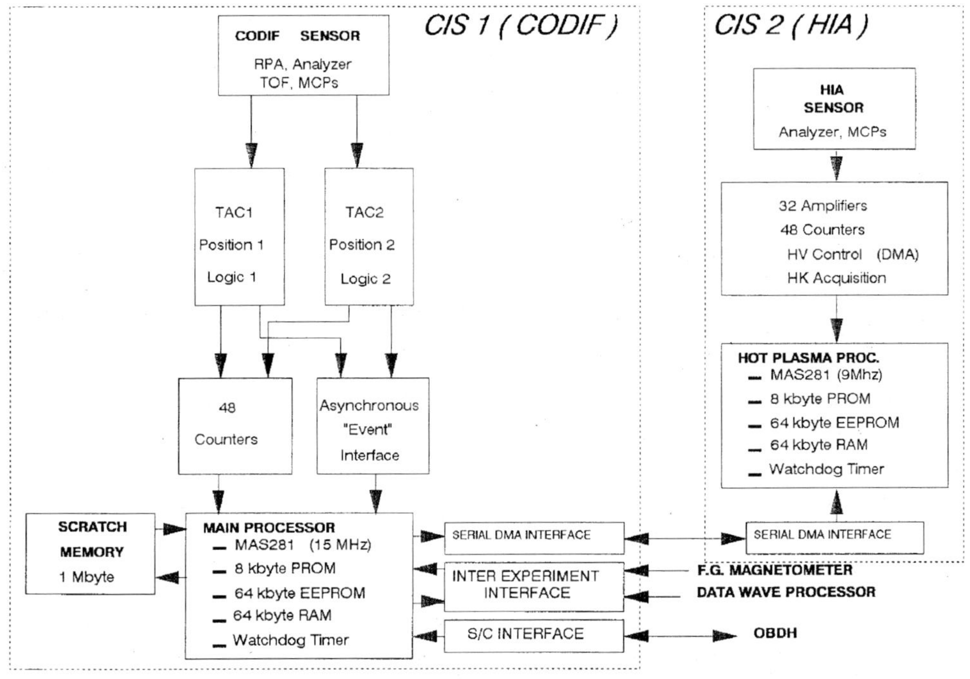

Figure 21. CIS data-processing system.

selecting data products to be transmitted at a given period depends upon the telemetry mode, bit rate sharing between CIS-1 and -2, and of course of the plasma environment; energy, angle, and time resolutions can be optimised to extract maximum information relevant to the scientific objectives. Data format changes are programmed within the instrument and do not require any reformatting of the spacecraft or ground data systems.

For example HIA produces typically a data volume of 32 polar sectors times 62 energies times 32 azimuth sectors, 16 bit-words, sampled in one spin period (4 s). Such a very high data rate has to be handled by a real-time operating system in order to elaborate and compress data to a few kbit s^{-1} telemetry stream output. All information is transmitted as log-compressed 8-bit-words, except moments which are transmitted with 12 bits. Pitch-angle distributions are instantaneous measurements when **B** is in the field of view of the instruments, and typical full 3D distributions are reduced to 88 Ω (solid angles) by taking into account oversampling in the polar regions.

A linear compression scheme is implemented as a part of the on-board CIS software, which allows the possibility to transmit compressed 3D distributions more often. The compression factor can be adjusted by setting new values to the compression parameters. A number of simulations have proven that a factor of

Table IV
Continued

Quantity	Product no.	Accum. Size	Basic Time (spin)	Total (bits)	bit/s
II. COLD POPULATIONS - SOLAR WIND (small geometrical factor section)					
Cold moments for solar wind 2ΔE (n,3v,6P,3H)	**P4**	**156 words 16 spins**	**1**	**312 x 16 + 16 x 18**	**82.5**
3D 31E x 8θ x 8ϕ (cold 3D)	**P8**	**992 words 1 packet**	**1**	**15872 + 32**	**3976**
2D Cold Azim. Distrib. (θ integration) 31E x 8ϕ (5.6° each)	**P13**	**496 words 4 spins**	**1**	**1984 x 4 + 32**	**498**
2D Cold Polar Distrib. (ϕ) integration) 31E x 8θ (5.6° each)	**P14**	**496 words 4 spins**	**1**	**1984 x 4 + 32**	**498**
3D 31E x 8θ x 16ϕ (*)	**P22**	**1984 words 2 packets**	**1**	**31744 + 32**	**7944**
3D 31E x 8θ x 16ϕ compressed (**)	**P24**	**992 words 1 packet**	**1**	**15872 + 4 x 16**	**3984**

ϕ : azimuthal angle
θ: polar angle
Ω: solid angle

Packet header: 2 x 16 bits = 32 bits
Frame Header: 9 x 16 bits = 144 bits / 5.1522 sec
(duration independent of the TM mode)

1 word = 16 bits

*: calibration products
**: compression ≥ 2 (2.5 should be expected; 2 assured)

two in the compression factor can easily be reached without any loss of data. The chosen algorithm for this compression is based on the evaluation of the dispersion of the maximum of a data token around the average of the data token itself. If the maximum (Max) satisfies the following:

$$\text{Max} - k^* \sqrt{(\text{Max})} < \text{Token}_{\text{Average}} ,$$

where k is an ajustable parameter factor to set the dispersion, the data are assumed to be equal to the $\text{Token}_{\text{Average}}$ which is transmitted as representative of the whole token. Otherwise the token length is scaled by a factor two and the above inequality is applied until the relation is satisfied or the token length has been reduced to one. If k is assumed to be 0 the compression becomes error-free.

Basically, for HIA, the high-sensitivity section has full 180° coverage and hot population data are computed using data from this section. When there is a cold

Table V

CODIF scientific telemetry products

Quantity	Product no.	Packet numbers	Basic Time (spins)	Total (bits)	bit/s
I. HOT POPULATIONS					
Moments **3ΔE(n,3v,6P,3H) x 4M**	P7	1	1	1872 + 32	476
3ΔE (Heat Flux Tensor, 7 more components) x 1M	P8	1	1	252 + 32	71
3D 64M x 8E x 6Ω *(6Ω : 2 polar, 4 perpendicular)*	P11	2	2	24576 + 64	3080
3D 4M x 31E x 88Ω (*) **	P13 + P16 + P18	7	1	87296 + 224	21880
3D 4M x 16E x 88Ω (*) **	P12 + P15 + P17	4	1	45056 + 128	11296
3D 1M x 31E x 24Ω	P14	1	1	5952 + 32	1496
2D PAD Cut 4M x 31E x 8θ *(2slices/ spin when B is in the field of view)* *	P24	1/slice	0.5	7936 x 2 + 2 x 32	3984
2D PAD Cut 4M x 16E x 8θ *(2slices/ spin when B is in the field of view)***	P23	1/slice	0.5	4096 x 2 + 2 x 32	2064 (1032/slice)
2D 4M x 31E x 16ϕ * **2D 2M x 16E x 16ϕ** (pro ons + α) ** or **4M x 16E x 8ϕ** **	P19 P21 P20	1 1 1	1 1 1	15872 + 32 4096 + 32 4096 + 32	3976 1032 1032
2D 1M x 31E x 32ϕ	P22	1	1	7936 + 32	1992
90° PAD 2M x 4β x 4E	P25	1	1	256 + 32	72
Live Pulse Height Data **Time of flight: 8 bits** **Azim. Position: 5 bits** **Energy Step: 7 bits** **Sector: 3 bits** **Proton mode: 1 bit**	P28	1		24 x k + 32 every 2 spins k > 1	depending of k value
Monitor Counting Rates **16 signals x 16E x 16ϕ**	P27	4	32 spins	32768 + 128	257
II. RPA MODES					
RPA diagnostic product **4M x 31E x 88 Ω x 4 sweeps**	P29	1	75	349184 + 32	1164
RPA Science product 4M x 8E x 9Ω **+ 1 start energy for each mass 5 bits** **+ 1 start angle for each mass 7 bits**	P30	1	1 spin every N spins	2316 + 32	587/N
RPA Moments	P31	1	1	624 + 32	164
III. COLD POPULATIONS					
Cold Populations Moments **3ΔE(n,3v,6P,3H) x 4M** **3ΔE** (Heat Flux Tensor, 7 more components) **x 1M**	P9 P10	1 1	1 1	1872 + 32 252 + 32	476 71
Solar Wind Helium **+ Proton Tail 2M x 8Ω x 8E**	P26	1	1	1024 + 32	264

ϕ: azimuthal angle (spin phase angle)
θ: polar angle
Ω: solid angle
β: near 90° pitch angles (the highest possible angle) for gyrotropic distributions (for 4 high energies)
4M: protons, a + 2 other masses

* best possibility
** basic use
(*) possibility to have different time of resolution for the different masses

Packet Header: 32 bits

population like the solar wind, the rest of the spin (360°–45°) is not be ignored, but, using the large geometrical factor section, data are taken and transmitted.

For CODIF, 4M stands for the four major species: H^+, He^{++}, O^+, and He^+. 64M 3D distributions can be read out at a slow rate. They give more detailed information about the presence of minor species. 4M, 88 Ω (solid angles), 3D distributions should be read out as often as possible, after all the other data types have been accommodated. A priority scheme for the time resolution is given according to the abundance of the species:

H^+ highest resolution,
He^{++} or O^+ highest resolution or slower by a factor of 2,
He^+ or other species factor of 2 or factor 4 slower.

4.2.2. *Remote-sensing Distribution*

Close to boundaries a distribution of four angles at 90° pitch-angle (phase 0°, 90°, 180°, 270°) is accumulated for two species (H^+ and O^+) in the four highest energies. This allows the boundary motions to be traced.

4.2.3. *Live Pulse Height Data*

For each particle CODIF measures the following parameters:

Time-of-flight:	8 bits (giving 256 values)
Azimuthal position:	5 bits (32 sectors)
Proton mode:	1 bit
Energy step:	7 bits (one between 128 elementary steps)
Pixel number:	3 bits
Total:	24 bits each

4.2.4. *Monitor Rates*

To check the performance and the counting efficiency of CODIF certain monitor rates have to be accumulated and transmitted with the science data:

2 Start	(each time-to-amplitude converter)
2 Coincidence	(each time-to-amplitude converter)
16 Start position	
4 Stop position	

To cut down in bit rate a specific scheme is proposed by which only every fourth energy step and every eighth sector are transmitted at a time. A cycle is completed after 32 spins.

4.2.5. *Telemetry Formats*

Instrument science and housekeeping data are read out over a single serial interface; the two types are differentiated by separate word gates. Telemetry is collected as a series of blocks, a fixed number per telemetry frame. Telemetry frames are always 5.152222 s in duration independent of telemetry mode, and are synchronised by a 'Reset' pulse that occurs at the beginning of each frame. Housekeeping data consists of 54 bytes per telemetry frame. Science can be collected in a variety of modes with different bit-rates; these modes are subdivided into 'Normal' and 'Burst' Modes, differentiated by the number of blocks per frame (10 for normal and 62 for burst). The different bit rates for Normal Mode are generated by changing the number of words per block.

Mode	Name	Bit s^{-1}	Block size	Blocks/frame	Bytes/frame
NM1	Normal mode	5 527	356	10	3 560
NM2	Ion mode	6 521	420	10	4 200
NM3	Electron mode	4 503	290	10	2 900
BM1	Normal burst mode	26 762	278	62	17 236
BM2	WEC/WBB TR mode	6 546	68	62	4 216
BM3	Event memory readout	29 456	306	62	18 972

BM3 is a special mode to dump the instrument scratch memory only; it is not an ordinary operating mode.

Two contingency modes exist in which all available data will go either to CIS-1 (CODIF) or to CIS-2 (HIA).

The four Cluster spacecraft will fly through a number of different plasma environments, and there must be a mechanism to change the mode of the instrument with a minimum number of commands when moving from one region to another. The CIS instruments have a large amount of flexibility either in the selection of the operating mode and in the reduction of the data necessary to fit the available telemetry bandwith. The instrument must be capable of making many changes to the operational details in response to a few commands.

Table VI shows the 16 CIS basic operation modes with the bit-rate sharing between CODIF and HIA, defined for each spacecraft bit-rate mode. The CIS instruments will operate in the different regions of the Earth's environment in these 16 operative modes that, for the five telemetry regimes foreseen (forgetting HK and BM3 modes), give a total amount of 80 science data transmission schemes. Each basic scheme corresponds to a given sequence of products, spanning from the momenta of the ion distributions to the 3D.

Roughly speaking, all these 16 operative regimes can be grouped into solar-wind tracking oriented modes, solar-wind study modes with the priority on the backstreaming ions, magnetospheric modes, an RPA mode and a calibration mode. Moreover, part of these solar-wind and magnetospheric modes are duplicated in a similar mode in which 3D compression is introduced.

For the HIA instrument two basic modes of operations, mixing basic products defined in Table IV have been implemented according to the plasma populations

Table VI
CIS operation modes

MODE		Mode Name	TELEMETRY MODE CIS-2 BITRATE (bit/s)				TELEMETRY MODE CIS-1 BITRATE (bit/s)			
			NM1	NM2	NM3	BM1	NM1	NM2	NM3	BM1
0	SW-1	SOLAR WIND / SW tracking	1 272	1 272	1 272	7 000	4 255	5 252	3 231	19 762
1	SW-2	SOLAR WIND / 3D backstreaming ions	1 272	1 272	1 272	7 000	4 255	5 252	3 231	19 762
2	SW-3	SOLAR WIND / SW tracking	2 135	2 135	2 135	13 162	3 392	4 386	2 368	13 600
3	SW-4	SOLAR WIND / 3D backstreaming ions	2 135	2 135	2 135	13 162	3 392	4 386	2 368	13 600
4	SW-C1	COMPRESSION SW-3 (+3Ds) solar wind tracking	2 135	2 135	2 135	13 162	3 392	4 386	2 368	13 600
5	SW-C2	COMPRESSION SW-4 (+3Ds) backstreaming ions	2 135	2 135	2 135	13 162	3 392	4 386	2 368	13 600
6		RPA								
7	PROM	PROM OPERATION								
8	MAG-1	MAGNETOSPHERE 1	1 272	1 272	1 272	7 000	4 255	5 252	3 231	19 762
9	MAG-2	MAGNETOSPHERE 2	2 135	2 135	2 135	13 162	3 392	4 386	2 368	13 600
10	MAG-3	MAGNETOSPHERE 3	3 124	4 148	2 135	13 162	2 403	2 373	2 368	13 600
11	MAG-4	MAG-1 SHEATH/TAIL	1 272	1 272	1 272	7 000	4 255	5 252	3 231	19 762
12	MAG-5	MAG-2 SHEATH/TAIL	2 135	2 135	2 135	13 162	3 392	4 386	2 368	13 600
13	MAG-C1	COMPRESSION MAG-1 + 3Ds	1 272	1 272	1 272	7 000	4 255	5 252	3 231	19 762
14	MAG-C2	COMPRESSION MAG-4 + 3Ds sheath/tail	1 272	1 272	1 272	7 000	4 255	5 252	3 231	19 762
15	CAL.	CALIBRATION								

CIS (HIA + CODIF):
NM1: 5527 bit/s
NM2: 6521 bit/s
NM3: 4503 bit/s
BM1: 26762 bit/s

Normal modes: NM1 or BM1

Baseline is to have identical modes on the 4 spacecraft

encountered along the Cluster orbit: so-called (a) 'magnetospheric' modes, and (b) 'solar-wind' modes. In both modes moments are systematically transmitted, computed every spin from the data acquired on the high-sensitive half-hemisphere ('high G' section) when the spacecraft are inside the magnetosphere, from the attenuated half-hemisphere section ('low g') when the spacecraft are in the interplanetary medium. This way one of the goals of the mission, i.e., to be able to produce high-resolution (4 s) moments by on-board computation, has been fullfilled for all the listed regimes apart from the calibration mode. The computed moments are used on-board to drive automatic operative mode changes (when this option has been remotely enabled) to better follow fluctuations which require fast-sensitivity-adapting capabilities or to select the best energy sweep regime to cover the local solar wind distribution.

'Magnetosphere' basic modes stay relatively simple, i.e., the full energy-angle ranges are systematically covered and the different data products (including moments) are deduced from the 62E $\times$ 88Ω energy solid angle count rate matrices accumulated on the 'high G' section.

'Solar wind' modes allow a precise and fast measurement (4s) of the ion flow parameters (H^+, He^{++}). For that, in the solar wind, the sweep energy range is automatically reduced and adapted every spin, centred on the main solar wind velocity by using a criterion based on the H^+ thermal and bulk velocities computed during the previous spin. Moreover, detailed 3D distributions (e.g., for upstreaming ions and/or for interplanetary disturbances) are included in the basic products transmitted to the telemetry.

In both regions, and within the HIA telemetry allocation, a maximum bit rate has been allowed for transmission as often as possible of full size (or reduced) 3D distributions. From Table VII it can be anticipated that Burst Modes should be considered as those having the highest expected scientific return.

Finally, to best fit the sampling activity to the plasma environment, if an auto-switching variable has been asserted, it is possible to run the instrument in the auto-change mode from magnetospheric to solar-wind configuration and *vice versa*. The switching criteria are presently based on locally measured plasma Mach number checks.

Science data packets include a number of data products from both HIA and CODIF in a flexible format. Data are time-tagged in such a way as to allow absolute timing of the data on the ground. The format allows bit rate allocations to the various data products to be changed relatively easily with minimal impact on ground processing. All auxiliary data necessary to analyse the data, such as instrument operational mode and timing information, are included in science data products, as it could be difficult to recombine housekeeping packets with the science packets.

Finally, housekeeping data (81 bit s^{-1}), extensively used during spacecraft development tests, give all the information needed to follow the health and safety of the instrument.

Table VII

Examples of basic operational modes of the CIS-2 (HIA) experiment grouped in 4 tables according to the space region of interest. For each telemetry mode (NM1, NM2, ...) different combinations of scientific products (see their definition in Table IV) are defined, computed from count rates provided by the high-sensitive half analyser ('high G' section) and/or the low geometrical factor half ('low g' section). Each basic mode (MAG-1, MAG-2, ..., SW-2, SW-3, ...) in a given telemetry mode refers to Table VI.

MAGNETOSPHERIC MODES

TELEMETRY MODE / OPERATION MODE				HIA Bit rate (bit/s)		HIGH G SECTION							
						M	1D	2D			3D		
						P2	P9	P10	P11	P12	P6	P15	P17
NM1	NM2/BM2	NM3	BM1	Alloc.	HIA	Mom.	62E	2Dϕ AZ	2Dθ POL	2DαPAD	31Ex88Ω	16Ex88Ω	62Ex88Ω
5527	6521/6546	4503	26762			117.5	125	998	996	1996/1008	5480	2828	10948
MODES 8-11				1272	1186							3 sp	
MODES 6-9-12				2135	2070						3 sp		
MODE 7				2135	2112								
10				3124	3071							1 sp	
	MODE 10			4148	4079					1 sl.			
			6-7-8-11	7000	6731					1 sl.			
			9-10-12	13162	13062					2 sl.			

SOLAR WIND MODES

TELEMETRY MODE / OPERATION MODE				HIA Bit rate (bit/s)		HIGH G SECTION					Low g SECTION			
						1D	2D		3D		M	2D		3D
						P18	P10	P20	P6	P15	P4	P13	P14	P8
NM1	NM2/BM2	NM3	BM1	Alloc.	HIA	31E	2DϕAZ	2DθPOL	31Ex88Ω	16Ex88Ω	M	2Dθ POL	2Dϕ AZ	31Ex8θx8ϕ
5527	6521/6546	4503	26762			63	998	2976	5480	2828	82.5	498	498	3976
MODE 0				1272	1275		5 sp							/4 sp
MODE 2				2135	2141									/2 sp
			0-6	7000	6869				2 sp					
			MODE 2	13162	12531									
MODE 1				1272	1088					3 sp				/18 sp
MODE 3				2135	2074				3 sp			/2 sp		/18 sp
			MODE 1	7000	6307									/5 sp
			MODE 3	13162	6464									/15 sp

COMPRESSION MAGNETOSPHERE

TELEMETRY MODE / OPERATION MODE				HIA Bit rate (bit/s)		HIGH G SECTION				
						M	3D			
						P2	P6	P15	P17	P23
NM1	NM2/BM2	NM3	BM1	Allocated	HIA	Moments	31Ex88Ω	16Ex88Ω	62Ex88Ω	31Ex8θx16ϕ
5527	6521/6546	4503	26762			117.5	5480	2828	10948	3206 [COMP=2.5]
MODES 13-14				1272	~1270					~ 3 sp

COMPRESSION SOLAR WIND

TELEMETRY MODES / OPERATION MODES				HIA Bit rate (bit/s)		HIGH G SECTION	Low g SECTION		
						3D	M	2D	3D
						P23	P4	P13	P24
NM1	NM2/BM2	NM3	BM1	Allocated	HIA	31Ex8θx16ϕ	Moments	2Dθ POL	31Ex8θx8ϕ
5527	6521/6546	4503	26762			3206 [COMP=2.5]	82.5	498	1992 [COMP=2]
MODE 4				2135	~2135				
MODE 5				2135	~2135	2 sp.			/16 sp.

3 sp.: integrated over 3 spins /3 sp.: once every 3 spins 1,2 sl.: 1 or 2 slices

Table VII shows the scientific products of HIA transmitted nominally in the various telemetry modes.

4.3. Modes of Operation

4.3.1. *Telemetry*

One of the decisive variables which affects the instrument operation is the telemetry mode; when the telemetry mode changes, the CIS instrument receives a single command and changes accordingly its bit rate allocation and data product collection mechanism to match the available telemetry. Some instrument parameters stay mode-independent and are programmable, such as MCP voltage.

The DPS is made of a small PROM, some EEPROM, and some RAM memories. The non-volatile EEPROM memory contains most of the on-board code and parameter tables, the RAM memory is used primarily to hold data blocks and some operational parameters and the PROM memory contains the bootstrap code needed to load or change the EEPROM. The EEPROM memory cannot be read while it is being programmed, and programming takes several millisec per block; it contains most of the operational parameters so that they do not have to be reloaded on power-up.

As a basic philosophy the default operational parameters are kept in EEPROM memory, while the current operational parameters are in RAM memory. The telemetry mode independent parameters are copied from the defaults on processor reset (this is called the 'Fixed Table'). The 'Operational Mode Table' is copied from the default table to set up a new mode after commanding. Sometimes it may also be desirable to follow automatic operational mode changes based only on science data (e.g., moments) collected by the instrument. The 'Telemetry Allocation Table' is a subset of the Operational Mode default Table; when the telemetry rate changes, the appropriate Telemetry Allocation Table is copied from the default table for the new rate and the current operational mode.

The CIS-1 and CIS-2 instruments have separate tables, but of course are controlled by the same telemetry rate and operation mode commands.

4.4. Ground Science Data Processing

The CIS raw telemetry will be pipeline-processed at the French Cluster Data Centre at CNES, Toulouse, where CESR-developed software will be running. Level-1 and Level-2 data products will thus be systematically generated. Level-1 files correspond to decommutated and decompressed data, organised in flat files, in full time resolution, one file per spacecraft-day-data product. Level-2 files are CDF files in physical units, and they include density for the major ion species, bulk velocity, parallel and perpendicular temperature. These files will be organised following the Cluster Science Data System (CSDS) recommendations, and they will populate two data bases: the Prime Parameter Data Base (PPDB: 4 spacecraft, 4 s resolution) and the Summary Parameter Data Base (SPDB: 1 spacecraft, 1 min resolution). The contents of these data bases will be distributed to other National Data Centres on a daily basis. The PPDB will be accessible to the whole Cluster community, and the SPDB will be public domain. Due to their broad accessibility, and to the quality of their data products, these data bases will permit joint analysis of plasma parameters from several instruments, further enhancing the science return of the Cluster mission.

The health and the performance of the CIS instrument will be monitored at various levels, by using files retrieved via the network from the Cluster Data Disposition System (DDS) , both at JSOC and at CESR.

5. Conclusion

The general characteristics of the two CIS instruments, including scientific performances, weight and raw power, are summarised in Table I. Note that the entrance of each sensor is put about 10 cm outside the spacecraft platform in order to have an unobstructed field of view and to minimise the effect of the spacecraft potential on the trajectories of the detected low-energy particles. The two planes of view of CODIF and HIA are parallel and tangential to the spacecraft body. The free field of view of the two sensors is $15° \times 360°$ (Figure 22).

In summary, by their unique features, the CIS instruments will provide fast measurements of the major plasma ion species with greatly improved accuracy and resolution. The inherent flexibility of the instrument control will allow a permanent optimisation of the scientific operation according to the various situations encountered all along the Cluster mission. The extensive on-board data processing will not only improve the time resolution of the measurements and significantly reduce data ground-processing costs, but will also make the plasma fundamental parameters quickly and directly available in an usable form to other investigators.

Acknowledgements

The CIS instrument was supported by many institutions. At CESR financial support came from CNES grant No. 208. The contribution of MPE Garching was supported by the Bunderministerium für Forschung und Technologie, Germany, under grant number 50 OC 89069. The high voltage supply of CIS-2 was provided by MPAE Lindau with the support of the Max-Planck-Gesellschaft zur Förderung der Wissenschaften and DARA grant 50 OC 89030. For IFSI CIS experiment was supported by Italian Space Agency (ASI). The work at the University of Washington, UNH, LPARL and UC Berkeley was supported by NASA contract. The Swedish participation was funded by the Swedish National Space Board and the Swiss participation by the Swiss National Science Foundation and the State of Bern.

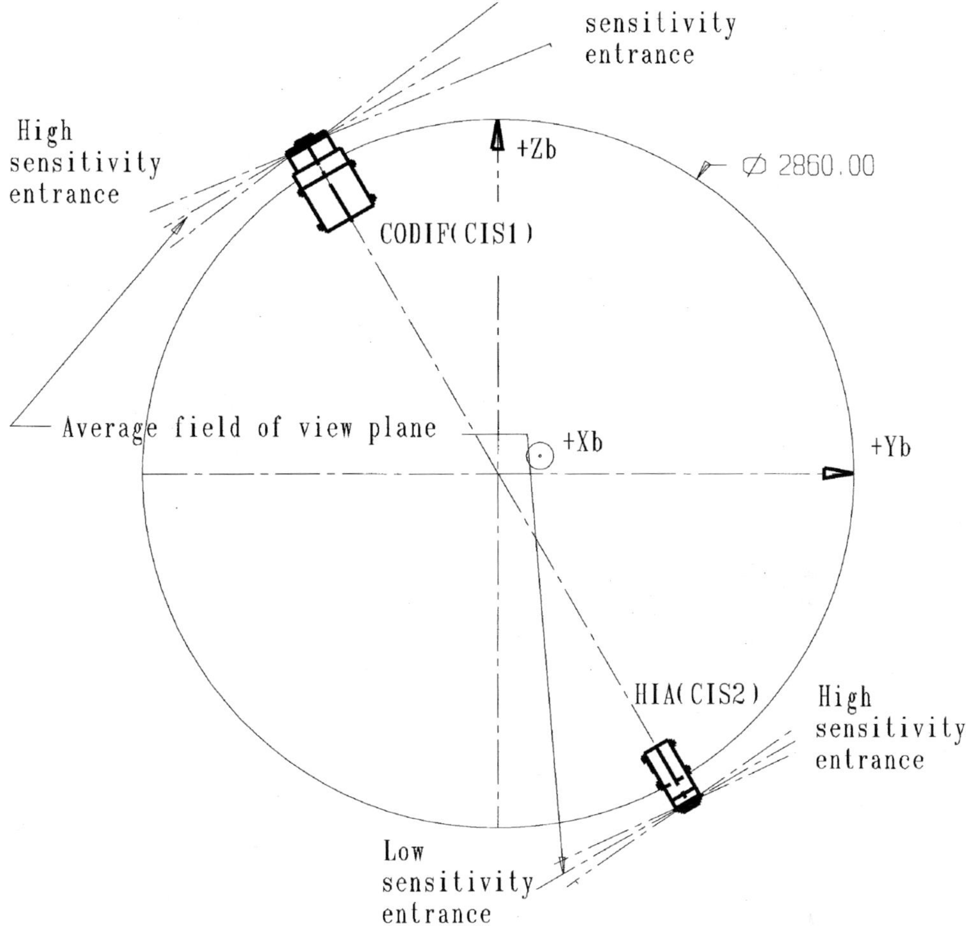

Figure 22. Position of the two sensors on the spacecraft and their fields of view.

We would like to thank J. P. Thouvenin for his permanent support and J. Y. Prado, H. Poussin, and M. Nonon-Latapie at CNES Toulouse, all the Cluster ESA Project staff at ESTEC and at ESOC for their excellent cooperation and the JSOC staff and particularly P. Chaizy in charge of the CIS instrument at JSOC.

The authors thank the 2 referees for their very efficient work.

References

Burgess, D.: 1989, 'Cyclic Behaviour at Quasi-Parallel Collisionless Shocks', *Geophys. Res. Letters* **16**, 345.

Carlson, C. W. *et al.*: 1982, 'An Instrument for Rapidly Measuring Plasma Distribution Functions with High Resolution', *Adv. Space Res.* **2** (7), 67.

Escoubet, P. C. and Schmidt, R.: 1997, 'Cluster Science and Mission Overview', *Space Sci. Rev.*, this issue.

Giacalone, J. *et al.*: 1994, 'Artificial Spacecraft in Hybrid Simulations of the Quasi-parallel Earth's Bow Shock: Analysis of Time Series Versus Spatial Profiles and a Separation Strategy for Cluster', *Ann. Geophys.* **12**, 591.

Gloeckler, G. *et al.*: 1993, 'Detection of Interstellar Pick-up Hydrogen in the Solar System', *Science* **261**, 70.

Lin, R. P. *et al.*: 1995, 'A Three-dimensional Plasma and Energetic Particle Investigation for the Wind Spacecraft', *Space Sci. Rev.* **71**, 125.

Martz, C.: 1993, 'Spectrométrie Ionique dans la Magnétosphère et le Vent Solaire. Simulation et Précision des Mesures Coordonnées au Moyen des 4 Satellites de la Mission Cluster', Thesis, Paul Sabatier Toulouse University.

Martz, C. and Sauvaud, J. A.: 1995, Accuracy of Ion Distribution Measurements and Related Parameters Using Top-hat Ion Spectrometers; Application to the CIS Cluster Experiment, submitted to *Rev. Sci. Instruments*.

Möbius, E. *et al.*: 1985, 'The Time-of-Flight Spectrometer SULEICA for Ions of the Energy Range 5-270 keV/charge on AMPTE IRM', *IEEE Trans. Geosc. Remote Sens.* **GE-23**, 274.

Möbius, E. *et al.*: 1985, 'Direct Observation of He^+ Pick-up Ions of Interstellar Origin in the Solar Wind', *Nature* **318**, 426.

Möbius, E. *et al.*: 1988, 'Interaction of Interstellar Pick-up Ions with the Solar Wind', *Astrophys. Space Sci.* **144**, 487.

Paschmann, G. *et al.*: 1985, 'The Plasma Instrument for AMPTE IRM', *IEEE Trans. Geosci. Remote Sens.* **GE-23**, 262.

Paschmann, G. *et al.*: 1986, 'The Magnetopause for Large Magnetic Shear: AMPTE/IRM Observations', *J. Geophys. Res.* **91**, 11099.

Quest, K. B.: 1988, 'Theory and Simulation of Collisionless Parallel Shocks', *J. Geophys. Res.* **93**, 9649.

Rème, H. *et al.*: 1987, 'The Giotto Electron Plasma Experiment', *J. Phys. E: Sci. Instrum.* **20**, 721.

Ritter, H.: 1985, 'Sekundärelektronenemission von Kohlenstoffolien beim Durchgang von Ionen im Energiebereich von 40 bis 500 keV', MPE Report 190.

Scholer, M.: 'Interaction of Upstream Diffuse Ions with the Solar Wind', *Adv. Space Sci.*, in press.

Scholer, M. and Terasawa, T.: 1990, 'Ion Reflection and Dissipation at Quasi-parallel Collisionless Shocks', *Geophys. Res. Letters* **17**, 119.

Sckopke, N. *et al.*: 1990, 'Ion Thermalization in Quasi-perpendicular Shocks Involving Reflected Ions', *J. Geophys. Res.* **95**, 6337.

Trattner, K. *et al.*: 1994, 'Statistical Analysis of Diffuse Ion Events Upstream at the Earth's Bow Shock', *J. Geophys. Res.* **99**, 13389.

Young, D. *et al.*: 1988, '2π-radian Field-of-View Toroidal Electrostatic Analyzer', *Rev. Sci. Instr.* **59**, 743.

PEACE: A PLASMA ELECTRON AND CURRENT EXPERIMENT

A. D. JOHNSTONE and C. ALSOP
Mullard Space Science Laboratory

S. BURGE
Rutherford Appleton Laboratory

P. J. CARTER, A. J. COATES, A. J. COKER and A. N. FAZAKERLEY
Mullard Space Science Laboratory

M. GRANDE
Rutherford Appleton Laboratory

R. A. GOWEN
Mullard Space Science Laboratory

C. GURGIOLO
Southwest Research Institute

B. K. HANCOCK
Mullard Space Science Laboratory

B. NARHEIM
Norwegian Defence Research Establishment

A. PREECE
Rutherford Appleton Laboratory

P. H. SHEATHER
Mullard Space Science Laboratory

J. D. WINNINGHAM
Southwest Research Institute

R. D. WOODLIFFE
Mullard Space Science Laboratory

Abstract. An electron analyser to measure the three-dimensional velocity distribution of electrons in the energy range from 0.59 eV to 26.4 keV on the four spacecraft of the Cluster mission is described. The instrument consists of two sensors with hemispherical electrostatic energy analysers with a position-sensitive microchannel plate detectors placed to view radially on opposite sides of the spacecraft. The intrinsic energy resolutions of the two sensors are 12.7% and 16.5% full width at half maximum. Their angular resolutions are 2.8° and 5.3° respectively in an azimuthal direction and 15° in a polar direction. The two sensors will normally measure in different overlapping energy ranges and will scan the distribution in half a spacecraft rotation or 2 s in the overlapped range. While this is the fastest time resolution for complete distributions, partial distributions can be recorded in as little as 62.5 ms and angular distributions at a fixed energy in 7.8 ms. The dynamic range of the instrument is sufficient to provide accurate measurements of the main known populations from the tail lobe to the plasmasheet and the solar wind. While the basic structure of the instrument is conventional, special attention has been paid in the design to improving the precision of the instrument so that a relative accuracy of the order of 1% could be attained in flight in order to measure the gradients between the four spacecraft accurately; to decreasing the minimum energy covered by this technique from 10 eV down to 1 eV; and to providing good three dimensional distributions.

Space Science Reviews **79:** 351–398, 1997.

1. Scientific Objectives

1.1. Mission Objectives

In-situ measurements from spacecraft in the magnetosphere suffer from a fundamental ambiguity. A time change in the measurement made by the spacecraft along its orbit may have been due to either a temporal change in the magnetosphere or a spatial change along the spacecraft track, or indeed a combination of the two. A single spacecraft has no way to distinguish these possibilities. Furthermore, many of the important physical quantities in the magnetosphere are vector gradient quantities which cannot be measured from a single spacecraft without making unverifiable assumptions. The Cluster mission aims to resolve magnetospheric fine structure which is inherently 3-dimensional and time-dependent, and to determine vector gradients by deploying four identical spacecraft in nearby, non-coplanar orbits. Four spacecraft are the minimum necessary to achieve these objectives. To measure the gradients the instruments must be capable of measuring small differences between the four spacecraft which therefore demands more accuracy and precision from the instruments than previous missions.

1.2. Electron Instrument Objectives

The electron instrument is one of a set of complementary instruments monitoring the particle distributions and fields within the plasma. Its role is to measure the distribution function of the electrons as accurately and frequently as possible and to provide the data at the best resolution that the spacecraft telemetry will allow. The measurements will be used to determine the nature and causes of processes such as acceleration, scattering, and diffusion which modify the electron distribution.

Several of Cluster's instruments are designed to measure parts of the velocity distribution of the electrons. At energies up to $\sim$20 eV, the temperature and density can be measured by the Langmuir probes which are part of the electric field experiment (Gustafsson *et al.*, 1996). The density can also be measured by the resonance sounding experiment (Decreau *et al.*, 1996). The PEACE instrument covers a higher energy range, from 0.59 eV up to 26.4 keV, using the technique of differential electrostatic analysers with electron multipliers. At even higher energies ($E > 20$ keV) semiconductor detectors measure the energy of the electrons directly (Wilken *et al.*, 1996).

The energy range covered by the PEACE instrument is a key one in the dynamics of the magnetosphere because it carries a substantial proportion of the energy density. The electrons are generally more dynamic than the ions because of their higher mobility, and they respond more quickly to changes. It can also be anticipated that they are the most likely electric current carriers. Since their thermal velocities are much higher than the bulk velocities in the medium they can often be used for tracing the behaviour of field lines during dynamic processes. The expected distributions are predominantly gyrotropic and can, to first order, be represented by a pitch angle/energy distribution. Although bulk flow velocities perpendicular to the magnetic field are much smaller than the thermal velocities of the distribution they can be measured with acceptable accuracy demonstrating that the distributions are not completely symmetrical about the magnetic field. Higher order moments of the distribution are important for the physics and can often be meaningfully obtained. At sharp boundaries and in rapidly varying situations non-gyrotropic features may be found. Three-dimensional distributions are required to understand the nature of such kinetic processes.

The advantages of the electrostatic analyser/electron multiplier technique are that:

(a) it provides differential energy measurements;

(b) it provides angular resolution of a few degrees;

(c) it provides complete coverage of all directions when its field of view is rotated by the spacecraft.

In principle the volume of phase space which it samples can be made as small as desired. In practice the resolution is limited by external constraints.

There have been a number of previous missions which have used a similar technique to measure electron distributions in three-dimensions and at high speed in the outer magnetosphere and solar wind.

On ISEE, the Fast Plasma Experiment (Bame *et al.*, 1978) had a limited three dimensional capability with four polar angle sectors covering the range from -55° to $+55^\circ$ with respect to the spacecraft equator and was able to acquire a complete distribution in 24 s. Its energy resolution was approximately $(\Delta E/E) = 60\%$.

On Ampte-IRM the Plasma Instrument had full three-dimensional capability (Paschmann *et al.*, 1985) an angular resolution of $22.5^\circ \times 22.5^\circ$ and an energy resolution of 20%. The data was returned in the form of velocity moments of the distribution at one-spin (4.35 s) intervals and reduced-resolution three-dimensional distributions at 3 spin intervals.

The Ampte-UKS Electron Instrument (Shah *et al.*, 1985) had three-dimensional capability but limited time resolution and azimuthal angular resolution since it only made one energy sweep per second, or five sweeps per spin. The data was returned as 3D distributions. With each new instrument the capabilities have been improved and new features of the electron distribution have been revealed. Therefore we naturally seek to improve the capabilities of this instrument with respect to earlier investigations. In particular, we have concentrated on three aspects which are important for the Cluster mission:

(a) improving the precision of the instrument, in order to measure the gradients accurately;

(b) decreasing the minimum energy covered by differential analysers from 10 eV to 1 eV;

(c) providing good three-dimensional distributions.

2. Requirement and Constraints

2.1. Plasma Populations

The plasma characteristics to be sampled during the Cluster mission will be highly variable. The electron populations with temperatures in the range covered by this instrument include plasmaspheric electrons, polar wind, solar wind, plasmasheet and auroral electrons. Their densities range from $\sim 10^3$ m^{-3} in the tail lobes to more than 10^8 m^{-3} in the disturbed solar wind; flow speeds range from nearly stagnant (less than a few tenths of km s^{-1}) in the inner magnetosphere and central plasma sheet to more than 1000 km s^{-1} in the disturbed solar wind and plasma sheet boundary layer. In many regions more than one particle population may be present. For example, magnetospheric electron populations within the dayside magnetosphere commonly contain both a cold (<10 eV) ionospheric component and a hot (>1000 eV) plasma sheet component. Other complexities are likely to be present including field-aligned beams, loss cones and gyrophase-bunched distributions. The instrument must have sufficient sensitivity and dynamic range to cover the range of densities and be able to resolve multiple populations.

2.2. Low-Energy Electron Measurements

Although low-energy electrons can be measured by probes, a differential technique made by an instrument like that described here is more sensitive and provides more detailed information, particularly about the angular distribution. Few previous differential instruments have measured below 10 eV. An exception was the S302 instrument on GEOS-1 and GEOS-2 (Wrenn *et al.*, 1981) which had a lower limit of 0.5 eV. The experience gained with that instrument has been applied in this design. Observations in the energy range from 1 eV to 10 eV are important for several reasons:

(a) the range contains most of the electron distribution in the solar wind and the bow shock;

(b) the current-carrying particles in current sheets are likely to be low-energy electrons;

(c) low-energy electrons play an important role in wave-particle interactions;

(d) low-energy electrons control the spacecraft/plasma interaction.

The Cluster mission provides an important opportunity to improve our knowledge about this energy range because it includes an ion emitter (Riedler *et al.*, 1996) to control the spacecraft potential. When it is operating the spacecraft-potential is clamped to a low positive potential and allows spacecraft photo- and secondary electrons to escape from the sheath. They no longer make a background to observations of the ambient plasma. This will improve the quality of measurements in the low-energy range significantly. The instrument should be able to make reliable observations of electrons with energies as low as 1 eV, which under most circumstances will be less than the energy corresponding to the spacecraft potential. Such coverage is necessary to separate electrons in the spacecraft sheath from ambient plasma electrons.

2.3. Angular and Energy Resolution

Electron distributions in space contain structures in energy and angle down to the finest resolution which has been measured in space. Such structures often provide important information about the processes which are affecting the distributions and are keys to the physics of the medium. For example, the loss cone throughout much of the region of interest to Cluster is of the order of a few degrees and contains not only reduced fluxes but sometimes intense field aligned beams. It is impossible to set an appropriate value for the resolution required; it is important to achieve the best possible.

2.4. Complete and Uniform Coverage of Viewing Direction

The sensors must be able to obtain the details of the plasma population, however it is moving with respect to the spacecraft and its spin axis. It is important that the instrument, by its design or deployment, does not introduce a bias into the coverage.

2.5. Relative Accuracy

The essence of using the data from four identical spacecraft means taking the differences between the individual measurements to find the gradients. If the difference is small, to obtain the gradient with modest accuracy requires the individual measurements to be made with very small errors relative to each other. We set as a target for the design that the measurements from two sensors in the same plasma should differ by less than 1%, which means that a gradient with a difference of 10% would be measured with an accuracy of 10%. This does not mean an absolute accuracy of 1% but a relative accuracy of 1%. This distinction is important in this type of instrument where the absolute calibration for intensity cannot yet be achieved at that level.

2.6. Time Resolution

The separation of the spacecraft will vary from 200 km to 18 000 km during the mission. Plasma bulk velocities in magnetospheric structures are typically of order 100 km s^{-1} but may be much higher, sometimes more than 1000 km s^{-1}. To follow the fastest motion between two spacecraft at their closest requires time resolution of order 0.1 s or less. Some of the important boundaries such as the bow shock and the magnetopause will pass over the spacecraft within a second, well within the spin period of the spacecraft. If these structures are to be fully investigated the instruments should be fast enough to resolve a number of points on the gradient as the boundary is crossed.

2.7. Reliability and Redundancy

It is essential for the scientific objectives to have operational sensors on all four spacecraft. If we assume that the probability of failure is constant, and the average sensor lifetime is 6 years then there is only a 30% probability that all four instruments will be operating after the two years of the Cluster main mission. To increase the probability to 70% would require a detector lifetime of 21 years. However if there is redundancy in the instrumental capability on each spacecraft the probability of retaining the four-spacecraft capability for two years increases. For example with pairs of independent sensors the probability of at least one sensor on each spacecraft surviving two years returns to 70% even with the same detector lifetime of 6 years. It is not likely that the probability of failure is constant and the detector lifetime cannot be predicted on the basis of a normal testing programme, so figures like those given here are completely hypothetical. The argument shows that it is important to build the detectors with attention to reliability and that the most effective measure is to include redundancy in the operations on each spacecraft.

2.8. Mission Constraints

The main constraints which impacted on the design of the instrument were:

(a) Keeping the instrument from exceeding the mass allocation by too much which meant that there had to be some descoping in the planned instrument characteristics including, reducing the field-of-view of HEEA from 360° to 180°; reducing the number of angular sectors in each sensor from 16 to 12; not including an instrument burst memory.

(b) Compressing the data to fit within the telemetry allocation which meant that 3D distributions can only be returned with reasonable time resolution when burst mode telemetry is being used.

(c) Staying within the power allocation limited the processing speed which could be employed in the onboard data processor which in turn limited the resolution with which the velocity moments could be calculated.

(d) Using the rotation of the spacecraft to scan azimuthal angles which imposed a minimum time for a complete distribution of half a spin period. It is only possible to obtain partial distributions in shorter periods.

3. Technical Description

3.1. Overview

There were two types of electrostatic analyser which could have been used for this experiment. They were the 'Top Hat' type (Carlson *et al.*, 1983) used on Ampte IRM

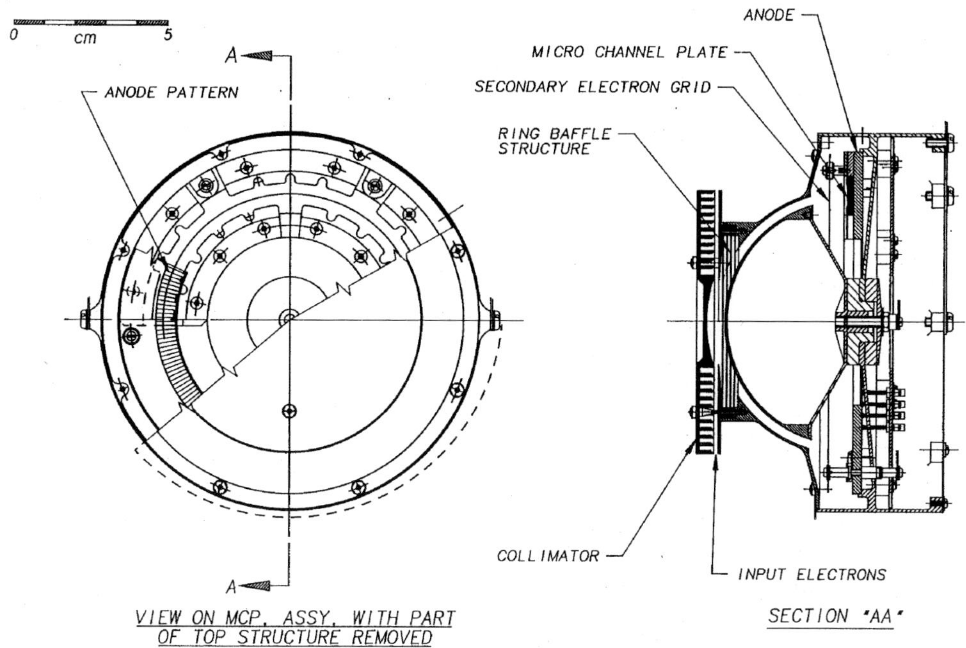

Figure 1. A cross-section of the LEEA analyser showing the structure of the input collimator, the electrostatic analyser and the anode pattern.

(Paschmann *et al.*, 1985) and Giotto (Reme *et al.*, 1987), and the Triquadrispherical type (Johnstone *et al.*, 1985) used on Ampte-UKS, Giotto (Johnstone *et al.*, 1987), CRRES (Hardy *et al.*, 1993), Tether (Oberhardt *et al.*, 1994), Suisei (Mukai and Miyake, 1986), and Geotail (Mukai *et al.*, 1994). The Top Hat has a 360° field of view and a turning angle inside the analyser of less than 90°. The Triquadrispherical has a 120° field of view but a turning angle of nearly 270°. A comparison of the properties of the two types of analyser is not simple and a determination of the 'best' design depends on the requirements. We estimated that the Top Hat would be better for this application because the 'flythrough' geometry of the input aperture promised better low-energy performance and because one sensor could cover the whole field of view. The advantage of flythrough geometry is that particles whose energy is too high to pass through the analyser plates should go straight through the analyser without impacting any structure. This should have minimised the production of secondary electrons inside the analyser and thus enhanced its performance for the measurement of low-energy electrons. Likewise we anticipated that the input aperture could be designed so that photons would also pass through without generating photoelectrons. As is shown below this turned out not to be possible to the degree we required.

The instrument consists of two sensors of the same basic Top Hat type (Carlson *et al.*, 1983; Sablik *et al.*, 1990) with hemispherical electrostatic energy analysers and an annular microchannel plate with a position-sensitive readout as detector (Figure 1). The acceptance energy of the detector is set by the voltage applied to the inner hemispherical plate and the direction of arrival of the electron in the meridian plane of the spacecraft is determined from the position at which the electron strikes the detector. Both sensors can cover the full energy range from 0.59 eV to 26.4 keV. The main difference between them is in their geometric factor. The Low Energy Electron Analyser (LEEA) which is designed to specialise in coverage of the very lowest electron energies (0.59 eV–9.45 eV) has the smaller (by a factor 4) geometric factor appropriate for the high fluxes to be found at low energy. The smaller geometric factor is achieved by reducing the field of view at the entrance collimator, and reducing the size of the input aperture. These features also help to enhance the performance for the measurement of low energy electrons. The High Energy Electron Analyser (HEEA) specialises in the measurement of the upper end of the electron energy spectrum and because it has a larger geometric factor extends the dynamic range of the combination of sensors. While it is usual for such analysers to have a field of view of 360°, these only have a field of view of 180°, which enables them to be mounted so that they view radially from the spacecraft. The pair of sensors are mounted opposite to each other on the spacecraft so that together they cover the complete angular range in a half rotation of the spacecraft.

3.2. Electron Optics

The first stage in the detailed design of the instrument was to set up a numerical analysis (Woodliffe and Johnstone, 1996) of the performance of the analyser which could then be confirmed by calibration. The objectives for the numerical simulation were:

– to optimize the analyser electron optics;
– to study the effect of manufacturing accuracy on gradient measurements;
– to prove, using calibrations, that the analysers perform as predicted.

A cross-section of an analyser showing the important parameters to be defined is shown in Figure 2. The relation between various parameters has been determined by Carlson *et al.* (1983) and further analysed by Sablik *et al.* (1985, 1990). However, to meet the scientific performance requirements in this design it was necessary to go outside the range of values which they had specified. In particular by changing the relative spacing between the inner and outer deflection plates the relation specifying the Top Hat angle was no longer valid. The Top Hat angle controls the input deflection of the beam and must be adjusted so that the angular response in the plane of the diagram is symmetric about the plane of the entrance aperture. The focussing in azimuthal angle is affected by the truncation angle, the position of the grid and the detector.

The optimum analyser performance specified in terms of geometric factor, field-of-view, and energy response, was obtained by adjusting the following analyser

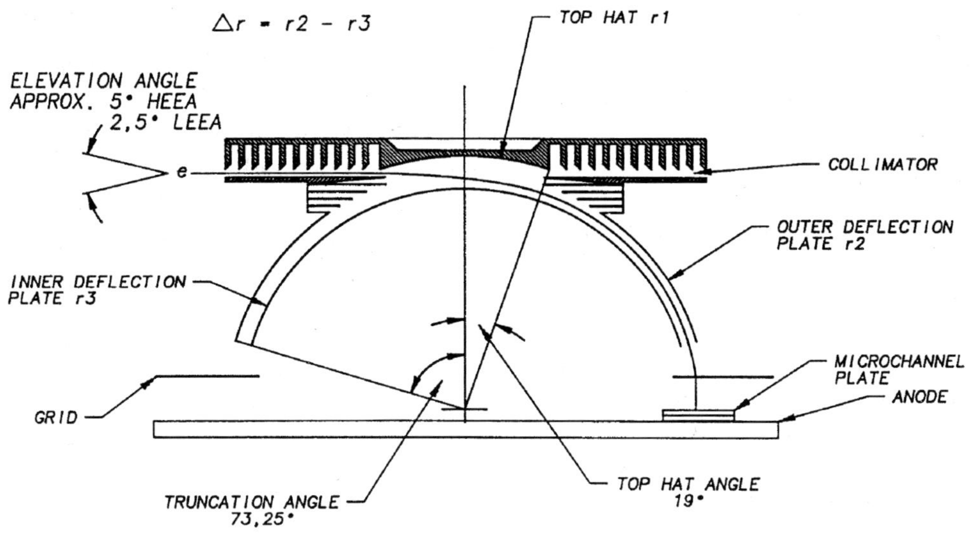

Figure 2. A diagram of the electron optical elements of the sensors.

dimensions in the simulations: the Top Hat angle, the inner, outer and Top Hat spherical section deflection plate radii, the truncation angle of the inner and outer deflection plates, and the length and width of the collimator at the analyser entrance aperture. In order to test the flythrough performance of the analyser we set a criterion that when the analyser was set to measure 20 eV electrons, at the upper end of the secondary electron spectrum, that electrons with energies of 50 eV and above, the ones with the highest secondary electron production rate, would not be able to enter the plates. We found from the simulations that it is not possible to design an input collimator which is restrictive enough to completely prevent the access of high-energy electrons and photons unless the response of the analyser itself is reduced unacceptably. The 270° turning angle of the Triquadrispherical analyser may in fact provide better discrimination against secondary electrons and photons than does the Top Hat's flythrough geometry with a much smaller turning angle. Therefore to ameliorate the effect of secondary electrons in this region a ring baffle structure was constructed around the Top Hat (Figures 1 and 2) so that most of the electrons striking the outer plate in the entrance region would be scattered into the baffle structure and would not pass around the track followed by the particles being measured. The simulation program was used to determine what effect this new structure would have on the response before it was implemented. The results showed that the only effect was that it reduced the peak energy of the response by a negligible 2.6% in energy. Later calibrations (Figure 15) made on the modified analysers, confirmed this result.

HEEA and LEEA have identical electron optical designs except for the input aperture and collimator. The Top Hat plate radii are 40, 43, 46 mm, respectively. The Top Hat angle is 19° (Figure 2). The measured performance figures are shown

Table I
Sensor characteristics

Sensor	LEEA	HEEA	
Energy range	0.59 eV–26.4 keV	0.59 eV–26.4 keV	
Energy resolution (FWHM)	0.127 ± 0.006	0.165 ± 0.007	
Energy sweeps per spin	16, 32, or 64	16, 32, or 64	
Field of view, polar	179.4°	179.4°	
azimuthal	$2.79° \pm 0.14°$	$5.27° \pm 0.20°$	
Angular resolution polar	3.75°, 15°	3.75°, 15°	
Geometric factor, per 15° zone	1.6×10^{-8}	6.0×10^{-8}	m^2 sr eV/eV
Maximum total count rate over all anodes	$>10^7$	$>10^7$	s^{-1}

Geometric factors include grid transparency and microchannel plate efficiency factors. The term eV/eV indicates that the relative energy bandpass has been included and that to obtain the total geometric factor the value in the table should be multiplied by the centre energy of the bandpass.

in Table I. Where a spread is given the spread is the standard deviation of the measured values obtained from the calibration of 60 discrete anodes in 5 sensors.

The effect of sunlight in the sensor has been minimised by incorporating baffles in the input collimator, and by applying a highly absorbent, diffusely reflecting surface coating to all internal surfaces. The baffles reduce the amount of ultraviolet light which can reach the detector after reflecting off internal structure. The diffuse coating cuts down the amount of solar Lyman alpha which reached the detector after multiple reflections. In this way the stimulation of the detector by photoelectrons generated by the action of the ultraviolet light on the internal surfaces or by direct illumination of the detector by the photons themselves is minimised.

The collimator aperture, which defines the angular response of the analyser (Figure 1) has a 'saw tooth' baffle design incorporated into it to reduce scattering of particles and sunlight, with the central part having a spherical profile in order to maintain the desired electric field in the electrostatic analyser. The second element in the protection is the series of concentric 'ring' baffles already mentioned which form the surface of the outer hemisphere near the entry point for the electrons. They also scatter photons away from the direct path to the detector. The baffle structure ensures that they are scattered into the baffle structure and lost. Thus there is is no direct line of sight from the aperture entrance to the hemisphere solid surface.

The diffusely-reflecting black coating, which has been applied to the aperture structures and deflecting plates, is made from copper oxide crystals grown using the EBONOL-C process. The coating is electrically conducting, has good adhesion, is capable of coating the fine geometry of the baffle structures and is sufficiently thin (<8 microns) and uniform to maintain the mechanical accuracy requirements of the electrostatic analyser.

By configuring the sensors such that high energies are being analysed when the sensors are viewing the Sun, the deflecting field within the electrostatic analyser can be used to minimise the amount of the photoelectron component, which has energy less than 10 eV, reaching the microchannel plate.

3.3. Energy Coverage

The energy range of the instrument is from 0.59 eV to 26 400 eV in 88 levels. The first 16 energy levels are equally spaced linearly in the range from 0.59 eV to 9.45 eV. The remaining levels are equally spaced logarithmically by a factor of 1.165 over the rest of the range. The energy is swept downwards and may be started from any level over the full range of the sweep. The first few accumulation bins in each sweep form the flyback period when the voltage is being charged up to its starting level. The sweep rate is synchronized to the spin period to ensure an integral number of sweeps during each spin. The starting point for the spin can be controlled by varying the delay between the Sun pulse from the spacecraft and the initiation of the spin in the instrument. This allows the data collection from the electron detectors to be synchronized with the data collection of the ion instrument CIS (Reme *et al.*, 1996) and the sweep to be adjusted so that high energies are being detected when the Sun is shining in the instrument aperture (Section 3.2).

There are four sweep modes whose purpose is to vary the azimuthal angular resolution of the instrument (Figure 3):

– Low Angular Resolution LAR; energy range 60 levels; 16 sweeps/spin; 22.5° resolution.

– Medium Angular Resolution MAR; energy range 60 levels; 32 sweeps/spin; 11.25° resolution.

– High Angular Resolution HAR: energy range 30 levels; 64 sweeps/spin; 5.625° resolution.

– Fixed Energy FE; a constant energy for as long as required for energies up to 1800 eV.

Both sensors may use all four sweep modes and they may operate in different sweep modes, or start from different levels simultaneously.

3.4. Swept High Voltage Supplies

To cover the energy range 0.59 eV to 26 400 eV requires voltages between 100 mV and 4200 V to be applied to the inner hemisphere. This is a dynamic range of a factor of 42 000. Although in normal operation the LEEA sensor will cover a low-energy range and HEEA will cover a high-energy range, the sweep high voltage generator which has been developed is able to provide stable voltages over the whole voltage range. The same design of supply can therefore be used in both sensors and both sensors can be operated over the full energy range. In this way we obtain redundancy of operation between the sensors.

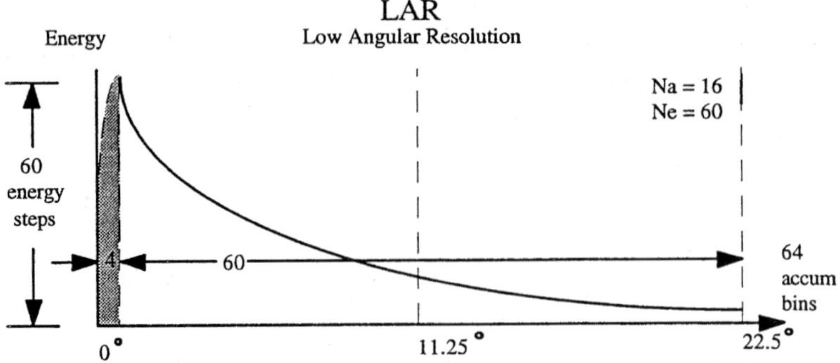

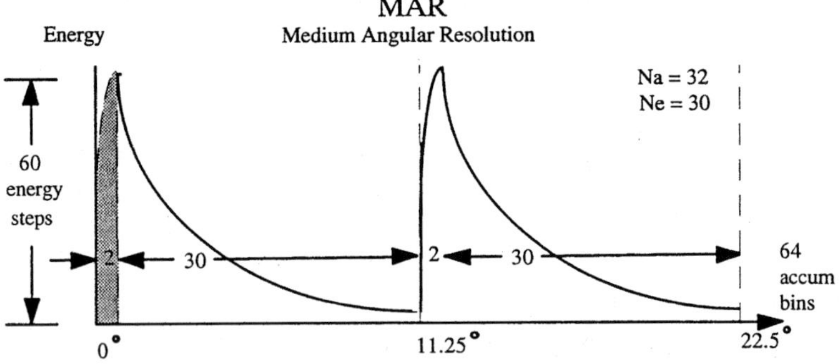

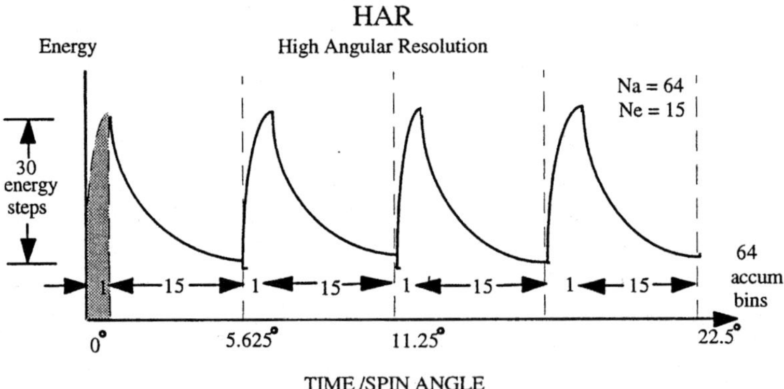

Figure 3. The pattern of the energy sweeps in the three main modes with flyback period in grey. Only one such period has been highlighted in each sweep mode.

The wide dynamic range introduces many problems. The dynamic range is 4.2×10^4 in voltage, more than 2×10^9 in power into a resistive load for the normal sweep and more than 10^{10} in power during the flyback period. The sweep supply was built to a specified accuracy of $\pm 1\%$ or ± 100mV, whichever is greater over the complete range. A novel system using an expanded Digital-to-Analogue Converter

(DAC) method was used, where the output from the DAC (which provides the input to a multiplier chain) was used to modify the reference voltage which in turn increases the step size as the output increases. The variability at the low-energy end means that the 1% accuracy cannot be maintained at the lowest energies although the voltages have been measured and will be recorded in flight.

3.5. Detectors

The electron detectors are double-thickness chevron-pair microchannel plates. These plates give a saturated pulse-height distribution with a full-width-half maximum of 70% at a modal gain of the order of 2×10^6. The resistivity of the glass is low enough to allow the plate to respond to count rates up to 1×10^4 per sq mm per second or approximately 10^6 electrons per anode per second, without significant gain degradation due to saturation effects. To ensure optimum performance of the microchannel plates during the full lifetime of the mission, the power supply operating the microchannel plate of each sensor has 16 possible levels (off, 60V for ground testing, and 14 in-flight operating levels). This allows the microchannel plate bias to be increased to recover gain lost as a result of degradation with use. The bias voltage at the input to the microchannel plate has an average value of approximately 100V to ensure all electrons have enough energy to be efficiently detected.

The background generated by penetrating radiation can be measured by biassing the grid in front of the microchannel plate negatively by 8 V to repel electrons and setting the plate voltage to the minimum of 0.59 eV. With these settings no electrons can reach the detector through the proper channel so that any counts which are registered must be the result of penetrating particles or multiply-scattered secondaries.

Electrons leaving the microchannel plate traverse a gap of 200 microns before striking a conducting anode. A voltage of +150 V is applied between the anode surface and the back surface of the microchannel plate to optimize the spreading of the charge cloud leaving the microchannel plate. This ensures that both fine zone and the coarse zone electrodes collect enough charge to be stimulated as required by the coincidence circuitry used for selection of fine and coarse zones (see Section 3.6). Each anode consists of a number of interleaved radial electrodes, each ~80 microns wide and separated from the adjacent electrode by a 20 micron insulator gap. The electrodes are connected alternately (Figure 4) to fine zone and coarse zone pre-amplifiers and discriminators. This arrangement is required to ensure that the fine zone anode collects the same amount of charge as the corresponding coarse zone. The structure used to hold the microchannel plate and the anode has been designed to minimise the area of insulator exposed to electron clouds between the microchannel plate and the anode. Such exposure has been shown to cause noise and, sometimes, to damage the microchannel plate.

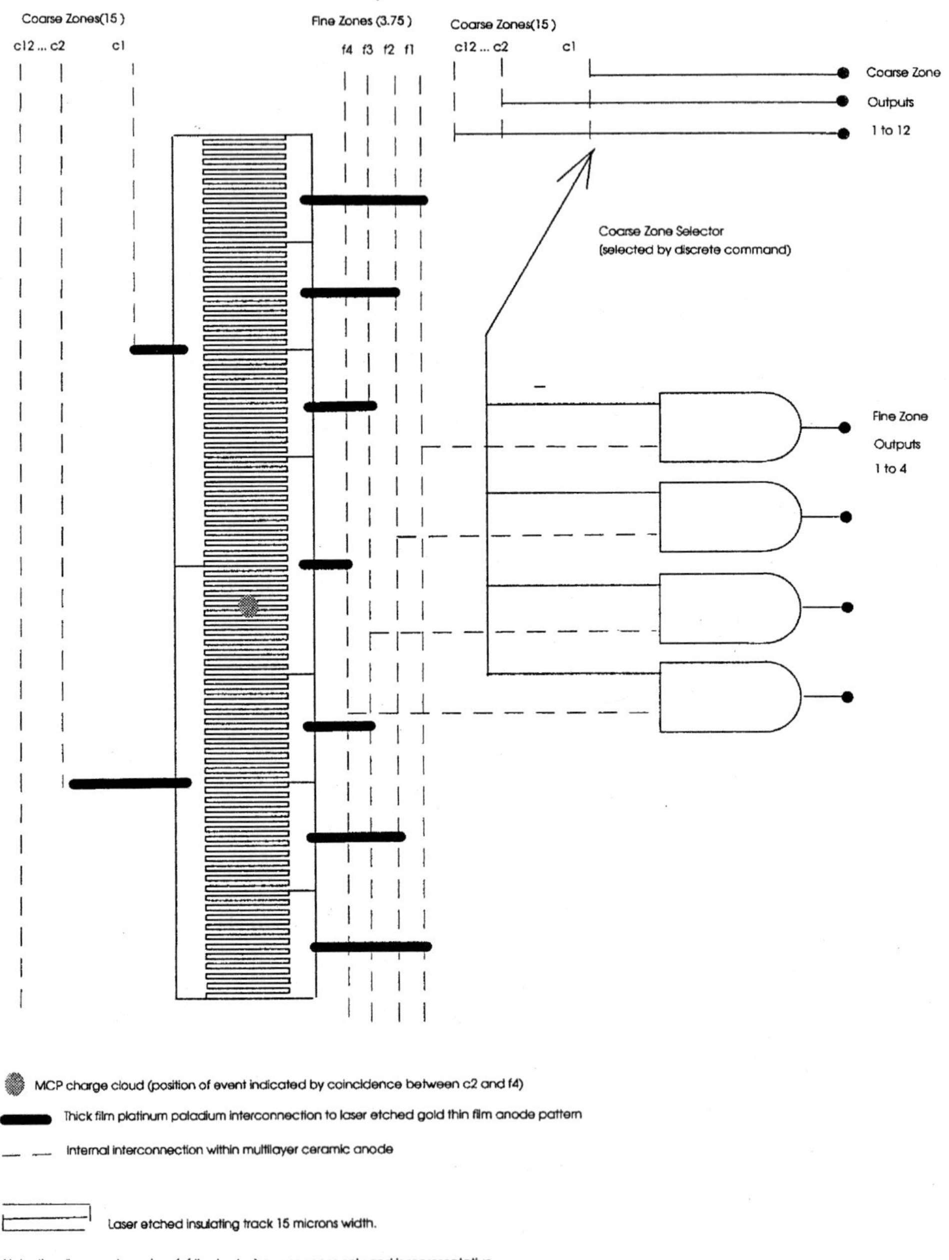

Figure 4. The structure of the anode showing coarse and fine zone interconnections.

The anode was constructed by sputter-coating a thin film of copper on a ceramic surface and then cutting a pattern using laser etching. The complex interconnection scheme for the fine and coarse zones was implemented by using a multilayer ceramic printed-circuit board of which the anode was the top layer. To achieve the necessary level of electromagnetic screening of the anode from the analyser structure, a signal ground plane was incorporated into the bottom surface of the multilayer ceramic.

The discriminator threshold levels were chosen carefully to be above the noise generated by spacecraft electronic subsystems, whilst being low enough to maintain good counting efficiency without excessive microchannel plate gain – which would reduce their lifetime. The level of the noise below the threshold was measured indirectly by measuring the count rate with an externally generated input of varying amplitude. The maximum amplitude was found to be typically equivalent to 300 000 microchannel plate electrons. The threshold has been set at a level corresponding to 400 000 microchannel plate electrons to allow some margin for gain depreciation or increase in electronic noise. The noise levels below threshold have been checked on each sensor when installed on the spacecraft and found to be equivalent in all cases to the figures given above.

Each sensor has a built-in pulse generator which can be used to check the performance of the amplifiers and discriminators during ground-testing and in-flight. This generator provides a sequence of pulses at different amplitudes and at different frequencies. Figure 5 is a block diagram of the sensor electronics.

3.6. Angular Coverage

The only position-sensitive readout which can achieve the high count rates necessary for a wide dynamic range is the discrete-anode. In this scheme each anode, or resolution element, is connected to its own amplifier and counter. The fields-of-view of both sensors are divided into 12 coarse zones 15° wide in order to meet the basic requirements. Each coarse zone is then subdivided into 4 fine zones 3.75° wide to provide the more detailed resolution sometimes needed (Figure 6). However, only one coarse zone of each sensor can be selected for fine zone readout at any given time because the fine zone counts are accumulated in coincidence with counts in the selected coarse zone. There are therefore 12 coarse zone amplifier/counter chains and four fine zone amplifier/counter chains.

3.7. Data Accumulation

The data are accumulated in 16 bit counters in the sensors and then passed to the Data Processing Unit for processing. There are 1024 accumulation periods per spin in all modes. At the nominal spin period of 4 s, the accumulation period will be 3.9 ms.

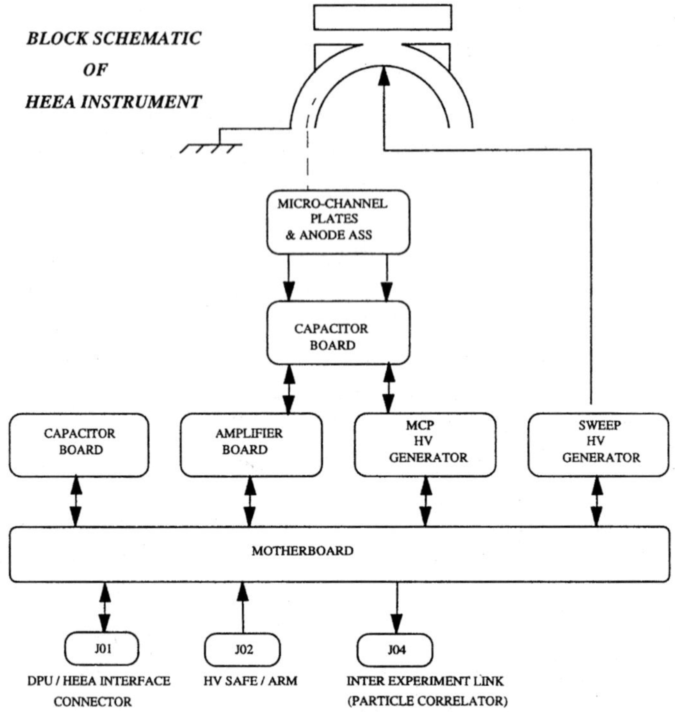

Figure 5. A block diagram of the sensor electronics.

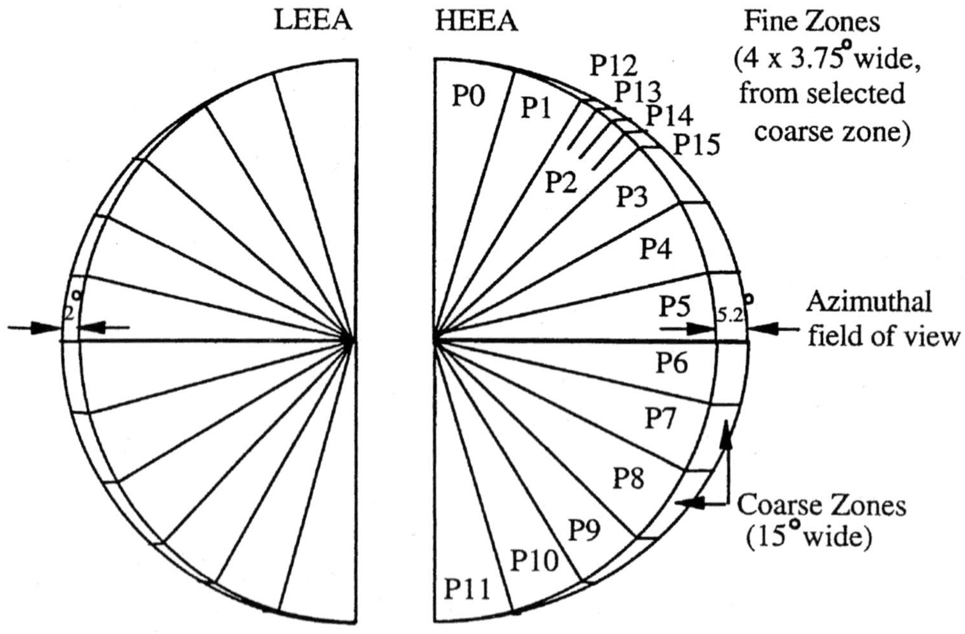

Figure 6. The angular response of the two analysers.

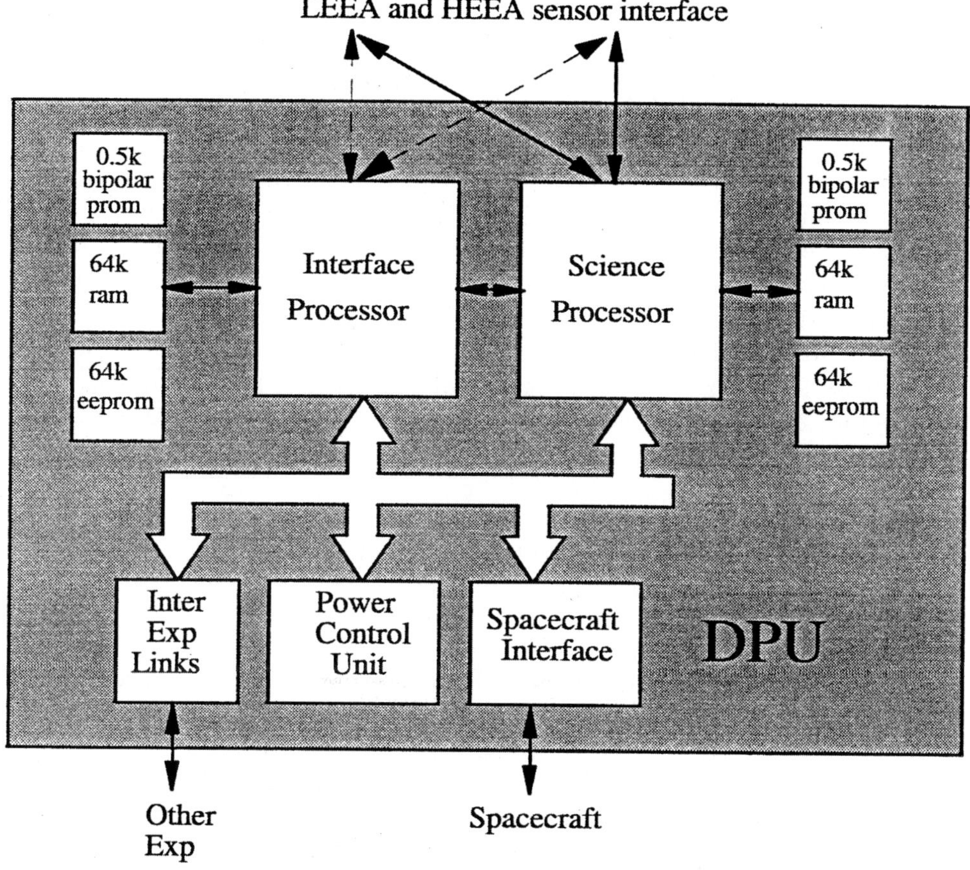

Figure 7. Block diagram of the Data Processing Unit.

3.8. Data Processing Unit

The tasks of the Data Processing Unit (DPU) include the following:

(a) Interface to spacecraft: power conditioning and distribution, command processing and distribution, telemetry interface to spacecraft, signal (e.g., Sun pulse) interface to spacecraft, data compression.

(b) Interface to sensors: sensor control, data processing.

(c) Autonomous control of the instrument: on-board house-keeping monitoring, experiment control. The above tasks are handled in a central unit (Figure 7) which is split into three main parts: the interface processor, the science processor and the power conditioning unit. The science processor carries the main data processing load. Both processors are based on the Transputer T222 microprocessor which combines high throughput with an architecture designed for parallel and distributed processing.

3.9. Inter Experiment Links

The PEACE instrument receives data onboard from three instruments, the magnetometer FGM (Balogh *et al.*, 1996), the electron beam instrument EDI (Paschmann *et al.*, 1996) and the radio sounder WHISPER (Decreau *et al.*, 1996). The magnetic field direction is used to determine when the magnetic field is within the field of view of the analysers for pitch angle distribution collection and to determine which coarse zone should be subdivided for fine resolution. EDI and WHISPER are both capable of causing interference to PEACE and the information received warns of potential interference.

PEACE provides data to DWP (Wooliscroft *et al.*, 1996), and ASPOC (Riedler *et al.*, 1996). DWP takes a pulse stream from one anode of the HEEA detector as one input for the wave-particle correlator and ASPOC takes the PEACE estimate of the spacecraft potential (Section 5.3).

4. Meeting the Requirements

4.1. Plasma Populations

The relation between the counting rate of the sensor and the phase space density of the electron distribution is given by (Johnstone *et al.*, 1987):

$$\frac{\mathrm{d}N}{\mathrm{d}t} = [\mathrm{GF}]v^4 f(\mathbf{v}) \,, \tag{1}$$

where [GF] is the geometric factor of the sensor, in units of m^2 sr eV/eV as given in Table I, and $\mathbf{v}$ is the velocity at the centre of the response of the electrostatic analyser. If the distribution can be described as a Maxwellian with temperature T and density n then $f(\mathbf{v})$ is given by

$$f(\mathbf{v}) = n\left[\frac{m}{2\pi kT}\right]^{3/2} \exp\left[-\frac{mv^2}{2kT}\right] . \tag{2}$$

The maximum count rate in the sensor for such a distribution occurs when

$$v^2 = \frac{4kT}{m} . \tag{3}$$

Therefore if we take as the ability to characterize the distribution the criterion that this peak counting rate must be within the energy range of the analyser (0.59 eV–26.4 keV) then the sensors described here can measure electron distributions with temperatures in the range from $\sim$4000 K to $\sim 1.5 \times 10^8$ K. The flow velocities anticipated, say up to 5000km s^{-1}, correspond to energies up to 70 eV and will be observed as anisotropies in the distribution. The maximum counting rate in such a distribution, at the energy given in Equation (3) is

Table II
Dynamic range of sensors

Temperature	4000 K	1.5×10^8 K
Maximum	2×10^9	9×10^6
1% accuracy (Section 4.5.1)	1.5×10^5	800
Minimum	9×10^3	50

Tabulated values are densities in m^{-3}.

$$\left[\frac{dN}{dt}\right]_{max} = 0.097n\left[\frac{2kT}{m}\right]^{1/2}[GF] . \tag{4}$$

If we specify that the working range of the instrument is given by the ability to measure a distribution both at the maximum counting rate given in (4) and at rates a factor of 10 below the maximum then we can determine the dynamic range of the instrument.

The maximum counting rate is determined by the speed of the electronics and by the saturation characteristics of the microchannel plate. The electronic dead-time for the anodes has been measured to be of the order of 1 μs (Section 7.2). Including the dead time correction (Equation (8) Section 5.2) it is possible to obtain corrected count rates up to 10^6 per anode s^{-1}. This count rate is within the maximum count rate density of which the microchannel plates are capable without saturation (Section 3.5).

The minimum useable count rate is governed by the background noise count rate. This depends on a number of factors such as; the natural background of penetrating radiation from cosmic rays or the radiation belts; and the noise level from spacecraft interference. The former is beyond our control except for the amount of shielding around the detector. We estimate that there is approximately 2 mm Al equivalent shielding so the threshold for penetration is approximately 1 MeV. The penetrating radiation background is variable and can be measured by switching on the retarding potential grid (-8 V) in front of the microchannel plate and setting the voltage on the analyser plates to pass only electrons with much lower energy than 8 eV. The measured count rate then gives a measure of the background from penetrating particles. The measured level of background noise in the laboratory is 2 counts s^{-1} per 15° anode. We take this to be the minimum useable count rate, although to make a measurement with reasonable accuracy at this level would require integration over a number of spins.

Putting these values in equation (4), taking HEEA to measure the minimum intensity and LEEA to measure the maximum gives the following dynamic range for the product $n\sqrt{T}$:

$$6 \times 10^5 > n\sqrt{T} > 1.2 \times 10^{11} .$$

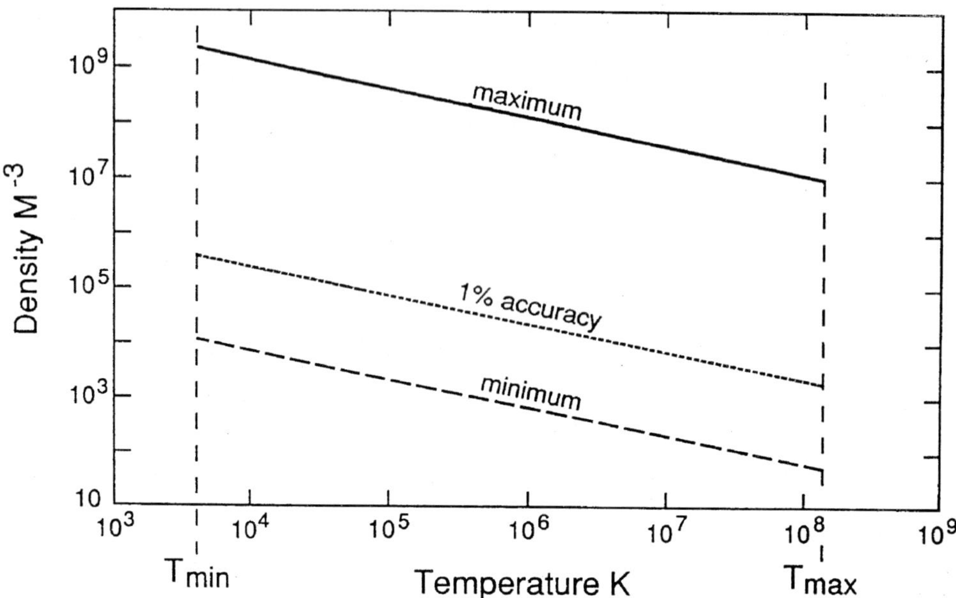

Figure 8. Dynamic range of the complete instrument. It can achieve 1% relative accuracy for observations above the middle line.

The measurement range is summarized in Table II (Figure 8).

4.2. Low-Energy Electron Measurements

Low-energy electron measurements are difficult to make because they are strongly affected by the interaction between the spacecraft and the ambient plasma environment and by the generation of secondary electrons within the analyser itself.

4.2.1. *Internal Secondary Electrons*

There are two sources of secondary electrons inside the analyser which can produce erroneous measurements. The first source consists of electrons with higher energy than the electrostatic analyser is set to measure. Such electrons strike the structure near the entrance aperture because they are not deflected sufficiently by the electric field. In doing so they generate secondary electrons with energy in the range from approximately 1 eV to 10 eV. If the analyser has been set to measure electrons in this energy range then they are able to pass through the analyser and register on the detector where they create an undesirable background to the ambient electrons being measured. The second source produces electrons in the same energy range; it consists of photoelectrons caused by the influx of solar ultraviolet radiation into the analyser. Since such electrons are only generated when the Sun is within the field of view of the analyser, the effect can be eliminated from the observations.

The effects of both sources of unwanted electrons have been minimised by the inclusion of the input baffle which consists of a series of thin parallel plates (Figure 2). The baffles are designed to cover the angular range near the analyser entrance which is directly accessible to photons or high-energy electrons entering the LEEA sensor. The only place where the production of internal secondaries can lead to detected electrons is when they are produced by particles or light striking the knife-edges of the baffle elements. In addition the smaller field of view and aperture of the LEEA analyser reduces the amount of solar ultraviolet, and high-energy particles which are able to penetrate between the analyser plates and generate secondaries in the first place.

4.2.2. *Spacecraft-Plasma Interactions*

Both types of secondary electrons are also produced on the surface of the spacecraft and form a sheath around the spacecraft which may be tens of metres thick. The effect they have on the observations depends on the spacecraft potential relative to the plasma and the potential depends in turn on the nature of the distributions. The spacecraft potential is determined by achieving a current balance so that the net current to the spacecraft is zero, i.e.,

$$I_i + I_e + I_{pe} + I_{se} = 0 \, , \tag{5}$$

where I_i is the ambient ion current; I_e is the ambient electron current; I_{pe} is the photoelectron current; I_{se} is the secondary electron current.

Outside the plasmapause the dominant term is the photoelectron current so the spacecraft potential Φ_{sc} is a few volts positive. Thus only those photoelectrons with more energy than $e\Phi_{sc}$ are able to overcome the potential barrier and leave the spacecraft. This is just sufficient to balance the current due to the flux of ambient electrons and ions to the spacecraft. The remaining photoelectrons form a cloud or sheath of electrons around the spacecraft which is electrostatically trapped. Some of these electrons return to the spacecraft and some of them enter the apertures of the detectors. When ASPOC ion emitters are turned on the emitted ion current will then have to be balanced by extra photoelectron current. The overall effect will be to allow almost all the photoelectrons to leave the spacecraft and to allow the potential to fall to a small positive value. When the spacecraft goes into eclipse the potential is determined by the temperature of the ambient electrons and the secondary electron current. In the outer magnetosphere where the effective temperature (kT/e) may reach the equivalent of 20 keV, the spacecraft potential may correspondingly reach -20 kV. In these circumstances the ion emitters of ASPOC will not reduce the potential. However, on Cluster, there will normally be no operations in eclipse.

The effect of these various influences on the observations are summarized in Figure 9. It shows how the various electron populations might be observed in a spectrum obtained by a sensor as an illustration of the processes involved. The spectrum is divided into two parts at the energy corresponding to the spacecraft

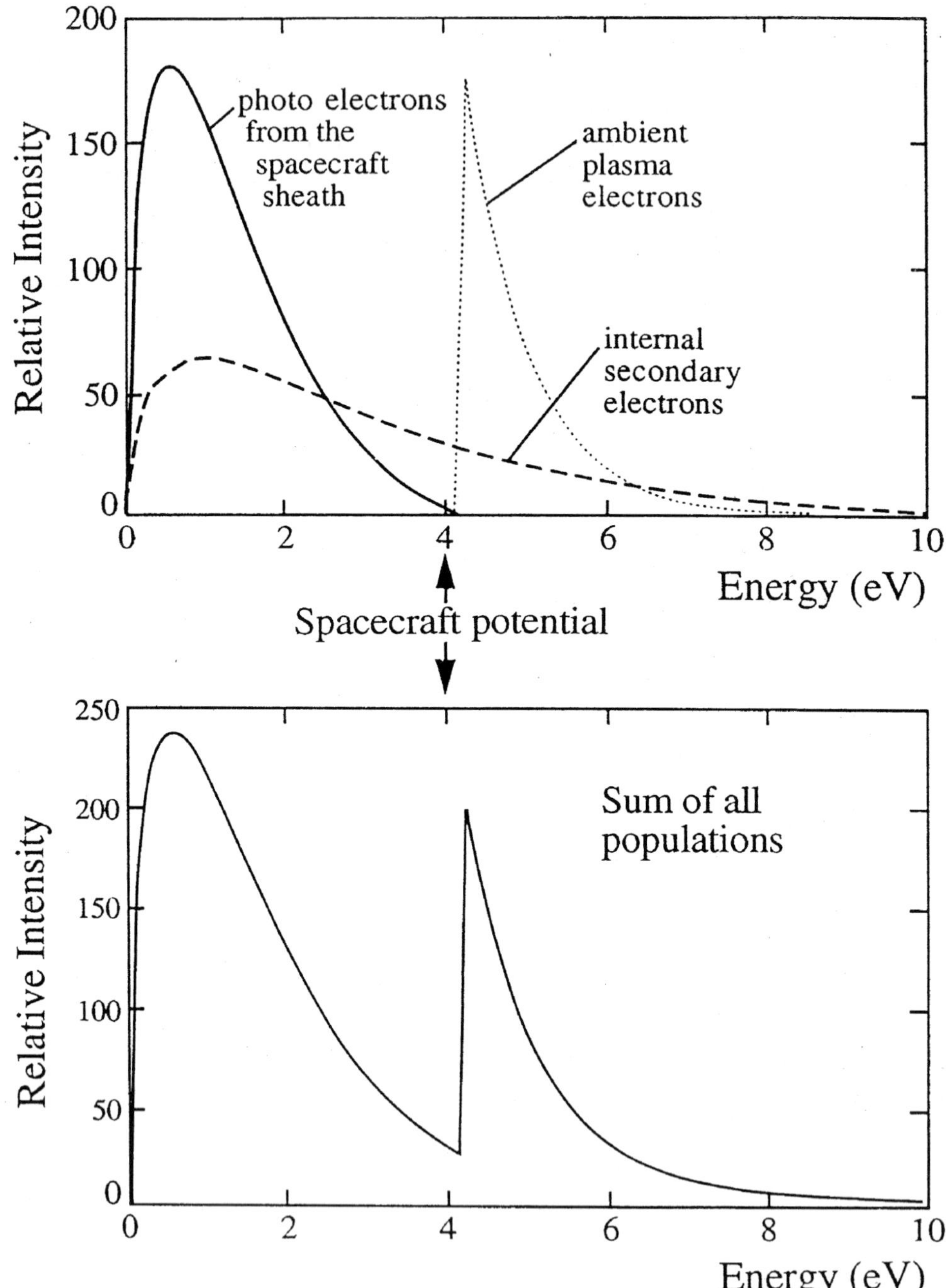

Figure 9. A diagram of the secondary electron populations which might be encountered at various times by the instruments. The relative magnitude and temperatures of the different populations are likely to be extremely variable and as a result the appearance of the total distribution could be considerably different from this sketch.

potential. Below this energy the spectrum comprises internally-produced secondary electrons and electrons from the photoelectron and secondary electron sheath around the spacecraft. The intensity of the sheath electrons drops to zero at the spacecraft potential, since at that point they are able to leave the neighborhood of the spacecraft and will not return to it. All the electrons from the ambient plasma are observed at energies above the spacecraft potential, because they have been accelerated into the sensor by the spacecraft potential. Thus if the internally-produced secondary electrons have been kept at a low level by the analyser design then the spacecraft potential will mark a minimum in the energy spectrum. The operation of the ASPOC instrument should keep the spacecraft potential below 2 V thereby minimizing the electrons returning from the sheath and minimizing the correction to the intensities of the ambient electrons.

The sensors have been mounted so that their fields-of-view are directed radially from the spacecraft's cylindrical surface and not tangentially as is sometimes done (Figure 10). This is to allow measurement of electrons which approach the sensor along paths which are as nearly normal to the surface as possible and not electrons which travel across the surface. This should minimise the effect of any surface charge on small insulating areas.

External surfaces of the spacecraft close the sensor apertures are all conducting and grounded and there are no exposed voltages nearby. There is a potentially serious problem with the star mapper which has a large mirror with an insulated surface which is always shielded from sunlight. It is capable of charging to negative potentials in high temperature plasma and affecting the potential of the whole spacecraft.

4.3. Angular and Energy Resolution

The angular resolution is determined by the number of discrete anodes forming the detector readout system and the sweep mode. In a compromise between resolution and instrument mass and power consumption we used 15° anodes sizes over the full field of view of each sensor but included the possibility to examine one of the anodes at a time with 3.75° resolution. This enables, for example, a study of fine structure in the pitch angle distribution near the loss cone. In the normal sweep mode MAR, the azimuthal angular resolution is 11.25°; the other two modes, HAR and LAR allow this to be halved or doubled, respectively.

The high voltage supplies can set the electrostatic energy analysers at 88 different levels over the complete range of the detector from 0.59 eV to 26.4 keV. The levels are logarithmically spaced, except for the linear part (Section 3.3) at the low-energy end. The spacing is therefore 11.65% of the mean energy. This is just within the energy resolution of the analysers. The energy bandpasses of the LEEA and HEEA are 12.7% and 16.5% respectively. These two latter figures give the intrinsic, and hence the best, resolution of the analysers but the actual resolution of the data depends also on the nature of the distribution which is transmitted (Section 5).

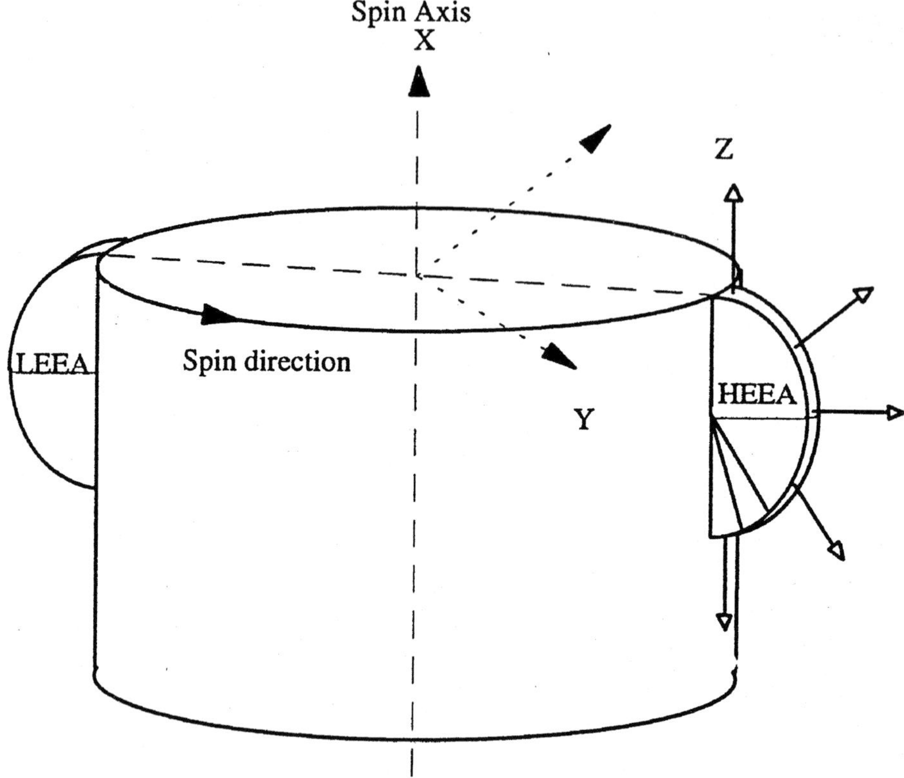

Figure 10. The deployment of the analysers on the spacecraft.

The high voltage supplies are swept and not stepped which broadens the energy resolution according to the sweep rate and the accumulation time.

4.4. Complete and Uniform Coverage of Viewing Direction

Both analysers have a uniform field of view over 180°. It is arranged to be in a meridian plane relative to the spacecraft spin axis so that as the spacecraft rotates each analyser will cover all directions. The Full-Width-Half-Maximum azimuthal angle of acceptance for the analysers is 5.27° for HEEA and 2.79° for LEEA (Table I) but the full width of the response is approximately double that value (Figure 15). Therefore, in the LAR sweep mode (16 sweeps per spin) there are angular gaps between the measurements at the same energy on successive sweeps for both sensors for polar angles in the range 14° to 166° (LEEA) and 30° to 150° (HEEA). For LEEA there is also a gap in the MAR mode (32 sweeps per spin) from 30° to 150°. Some extremely narrow field-aligned distributions, which are also sharply peaked in energy, might be missed intermittently on some occasions. Such a possibility must be checked when distributions of this type are present.

4.5. Relative Accuracy

4.5.1. *Statistical Fluctuations*

The statistical fluctuation in the measurement due to the finite number of counts collected by the detector is an unavoidable source of error. Its magnitude can be estimated from simulations of the operation of a sensor and the on-board moment calculations. Kessel *et al.* (1989) found that the amplitude of the fluctuations in the density value is given by the Poisson value $1/\sqrt{N}$ where N is the total number of counts in the distribution. They also found that the velocity and temperature fluctuations are of the same relative order of magnitude. To achieve 1% relative accuracy therefore requires a minimum of 10 000 counts in the distribution. For a distribution acquired in 2 s the average count rate would therefore have to be of the order of 330 counts anode^{-1} s^{-1} on HEEA and the corresponding value, taking into account the different geometric factor of 70 counts anode^{-1} s^{-1} on LEEA. Since the maximum intensity entails a count rate per anode of approximately 10^6 on LEEA then it is only possible to achieve 1% accuracy over the upper part of the dynamic range, i.e., a factor 15 000 from the peak down. The corresponding density is indicated in Table II in the row labelled '1% accuracy'.

4.5.2. *Reproducibility of Sensors*

The geometric factor of the sensors is a combination of two main factors; the geometrical response of the analyser with its collimator; and the efficiency of the microchannel plate. The former is controlled by the mechanical structure of the analyser and should therefore be stable with time; the latter is dependent on the nature of the surface of the microchannel plate and is liable to change with use. Since the microchannel plate's behaviour cannot be predicted it is necessary to have methods of calibrating the performance in flight. One advantage of the four satellite mission is that it provides a straightforward way of doing this. When the four spacecraft are in nearly uniform plasma conditions a simple statistical comparison of the count rate of different sensors and different anodes at the same energies will quickly reveal whether one detector, or part of a detector is developing a different response. The design of the instrument allows for changes in high voltage bias to correct the differences or to adjust the sensitivity parameters used in the moment calculations. However this process can only be carried out if the different sensors are making their measurements at the same energy. This is determined by the mechanical structure and requires that the sensors are constructed with a high degree of reproducibility.

The effect of mechanical variations within the machining tolerances on the performance of the analysers was assessed by the numerical model of the analysers. It showed that the most serious problems were caused by eccentric offsets between

Table III
Hemisphere offsets

Spacecraft	LEEA (microns)	HEEA (microns)
1	25	18
2	53	32
3	13	39
4	11	10
Flight spare	10	56
Target value	38	50

the hemispheres of the electrostatic analysers. If there is an offset then the mean energy of the analyser becomes a quasi-sinusoidal function of the azimuthal angle because the gap between the analyser plates is changing, and the energy bandwidth also varies with position. If an inflight intercalibration is attempted with too large an energy difference between the detector elements then the relative values will become dependent on the electron spectrum which is naturally variable. We set a target that the energy response curves had to overlap to within 10% of their width in order to achieve the target for relative accuracy.

The fractional error in the mean energy is given by

$$\frac{E_R - E_N}{E_N} = \frac{\Delta d}{d}, \tag{6}$$

where E_N is the nominal mean energy and E_R the real mean energy of the analyser. The analyser plate spacing is d and the offset is Δd. Since the energy bandwidth is 0.165 (HEEA) or 0.127 (LEEA) the offset must be less than 50 or 38 microns, respectively. A number of precautions had to be taken to achieve this accuracy. The radius of the insulator for the inner hemisphere was minimised in order to reduce the thermal expansion gaps; the inner hemisphere assembly was mounted in its structure before machining to ensure alignment; the hemispheres were machined instead of using the lighter but less rigid spinning method of construction. The responses of all anodes on all flight sensors have been measured by calibration in a beam system (Section 7) and the results are summarized in Figure 11. This shows a histogram of the distribution of centre energies and of energy bandpasses for both sensors. The standard deviation over the centre energy values obtained from measurements of 60 anodes is 7% of the energy bandpass for LEEA and 6% for HEEA. From the variation of the mean energy with azimuth around an individual sensor it is possible to estimate the offset of the inner hemisphere. The values are given in Table III.

These measurements show that the target has been achieved for all but two sensors.

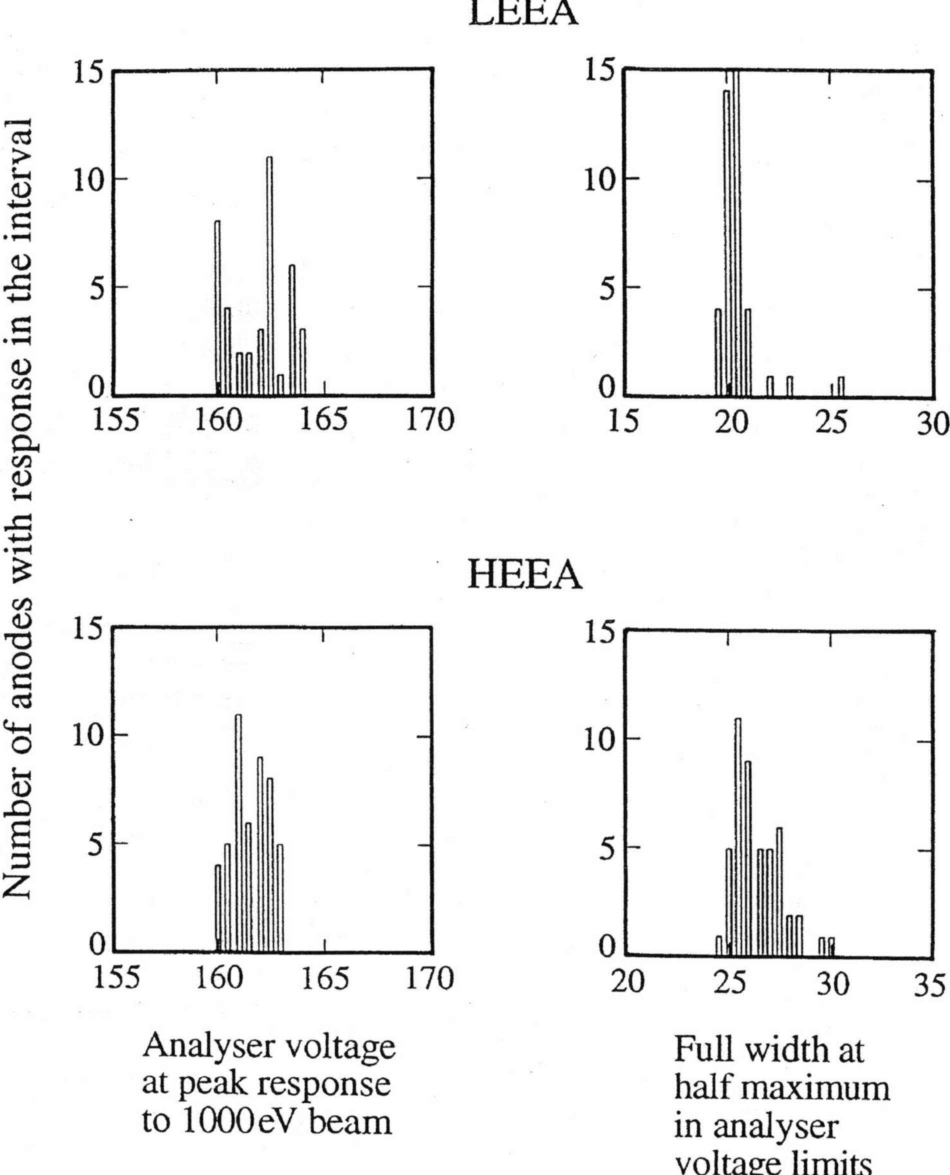

Figure 11. A summary of calibration results showing a histogram of the values of the centre and the width of the energy bandpass for each anode as measured. The units on the abscissa are volts applied to the analyser's inner hemisphere.

4.5.3. *Other Sources of Error*

The two factors mentioned above are the ones which have most influence on the relative accuracy of the measurements. Another factor which affects the comparison between the spacecraft is the resolution of the moment calculations and the precision with which the values are transmitted. As long as the angular and energy structures in the distribution are not narrower than the resolution elements ($30^\circ \times 22.5^\circ \times 0.4E$; see Section 5.2) then the moment summations are accurate both absolutely and relatively (Kessel *et al.*, 1989). In the presence of features whose angular width is less than the size of the elements in the moment summations then the differences between moments calculated on the different spacecraft will not be representative of gradients in the physical conditions. However, this is a fundamental problem with the use of highly compressed data in the form of moments in the presence of multiple populations. It is essential to have more detailed information, in the form of pitch-angle distributions and three-dimensional distributions, even if at lower time resolution, to check the validity of the moments.

The summation values are transmitted to the ground (Section 5.2) with sufficient precision to achieve 1% accuracy on the individual values. However some of the derived values, such as the temperature, involve taking what might be a small difference between two large values. In these circumstances the 1% accuracy might not be maintained. A further possibility of increased error occurs with the dead time correction when large. However the difference in the sensitivities of LEEA and HEEA means that the size of the correction is different and therefore LEEA can be used to check the accuracy of the correction to HEEA when its count rates are high. Nevertheless, in the top of the dynamic range of the sensors, within a factor of ten of the maximum in Table II and Figure 8, the statistical variations will increase above the 1% level.

4.6. Time Resolution

The timing of three-dimensional plasma measurements is coupled to the spin period because the rotation of the spacecraft is used to provide an angular scan. This limits measurements of the complete distribution with our pair of sensors with a combined field-of-view of 360° to twice per spin for those parts of the energy range which are covered by both sensors. This provides a time resolution of 2 s. For those parts of the energy range covered by one sensor only, the time resolution is one spin period or 4 s. Such resolution is not adequate to resolve the most rapidly-varying structures but has to be accepted as a limitation on current techniques of measurements.

The time resolution of the sensors can be improved under some circumstances. Most electron distributions are nearly gyrotropic so that one energy sweep of the sensors can provide a good characterization of the distribution in terms of an energy/pitch angle distribution. If the magnetic field is not within the field of view during the sweep then the distribution will not cover all pitch angles but it may be enough to determine the relative timing between the four spacecraft as they pass through a magnetospheric structure. The best available time resolution in this case 62.5 ms which is constrained by the data output. Even faster time resolution is available when one sensor can be operated in the fixed energy mode. Then the

Table IV
Peace telemetry allocation

Telemetry mode	Bits s^{-1}	Bytes/spin 4 s spin	Bytes/spin 3.6 s spin	
Normal mode 1	2 515.42	1257.7	1131.9	
Normal mode 2	1 521.67	760.8	684.7	CIS priority
Normal mode 3	3 540.22	1770.1	1593.1	Peace priority
Burst mode 1	15 980.68	7990.3	7191.3	
Burst mode 2	3 658.23	1829.1	1646.2	
Burst mode 3	1 926.00	963.0	866.7	

resolution can be reduced to 7.8 ms for a polar angular distribution at the chosen energy. This time resolution is only available when the spacecraft telemetry is in Burst mode 1. This is sufficient for the variations mentioned in Section 2.6 even though complete moments are not obtained at this speed.

4.7. Reliability and Redundancy

The main redundant feature of the instrument is the use of two completely independent sensors, LEEA and HEEA. If there should be a failure in one sensor it will not affect the other which can continue to collect most of the information. There will be some loss of time resolution, energy coverage and dynamic range. The telemetry used by the failed sensor can be re-allocated to the remaining sensor. The redundancy between the two sensors was made possible in part by the design of the high voltage sweep supply which can operate over an extremely wide dynamic range of output voltages.

The DPU has built-in redundancy since it contains two separate processors, the science processor and the interface processor. Both are needed to be able to carry out all the processing functions of the instrument but if one should fail then software is included onboard which will enable either processor to take control of the whole experiment, albeit with some loss of functionality.

The extensive testing carried out on the engineering model and five flight models has also had the result that more potential problems have been discovered than would have been possible in the usual three-model programme. This should increase the overall reliability since the lessons learned in testing the final model were applied in the refurbishment of earlier models.

5. Data Products

The instrument must be able to adapt to the 6 different science telemetry output rates allocated to PEACE shown in Table IV.

Table V
Science data format

		Size bytes per spin	Abbreviation	Resolution	
CORE	moments	168	MOM-D	16 bit	±0.4%
CORE	s/c potential, etc.	7	SCP-S	not compressed	
CORE	pitch angle distribution	390	PAD-D	8 bit	±1.5%
	low-energy reduced	192	LER-S	8 bit	±3.1%
	high-polar resolution	480	HIP-D	8 bit	±1.5%
	3D reduced resolution	2 880	3DR-D	8 bit	±3.1%
	3D full resolution	23 040	3DF-D	8 bit	±1.5%
	3D single sensor	variable	3DX	8 bit	either

Notes: S indicates data from a single sensor; D indicates data from both sensors; products are listed in order of priority.

Since the collection of scientific data is directly related to the rotation of the spacecraft, the information output has to be related to the spin period. However the telemetry allocation is time synchronized. The spin period is specified to be 4 ± 0.4 s so that there is no fixed relationship between the telemetry allocation in bits per second and the science output in bits per spin. The science data format, which is based on one spin of data, has been designed to float within the spacecraft format and to have its own synchronising pattern.

Since the rate at which data is collected by the instrument is more than 3 Mbits s^{-1}, a considerable amount of data compression is required to fit within the telemetry allocation. The most efficient method is to calculate the velocity moments of the distribution onboard, essentially treating the electron population as a fluid and obtaining values for the density, velocity, pressure tensor and heat flux. The electron distribution can be characterized by a relatively small number (13) of values. If the distributions to be encountered were all in, or near, thermal equilibrium, such an approach would be all that was required but this is not the case. In most magnetospheric regions more than one population may be present and the individual populations may contain non-thermal features which are essential for an understanding of the physics of the situation. Therefore we have developed a hierarchy of data products (Table V), which provide increasing amounts of information, and which can be transmitted if the telemetry capacity is available. While the CORE is always transmitted in the order given in Table V, the order of the optional distributions can be selected on command according to the scientific priorities of the investigation.

The count values in all matrices are telemetered as single byte, quasilog-compressed numbers. The eight bits are divided either into a 4-bit mantissa and a 4-bit exponent which gives a maximum error of ±3.125% or 5-bit mantissa and a

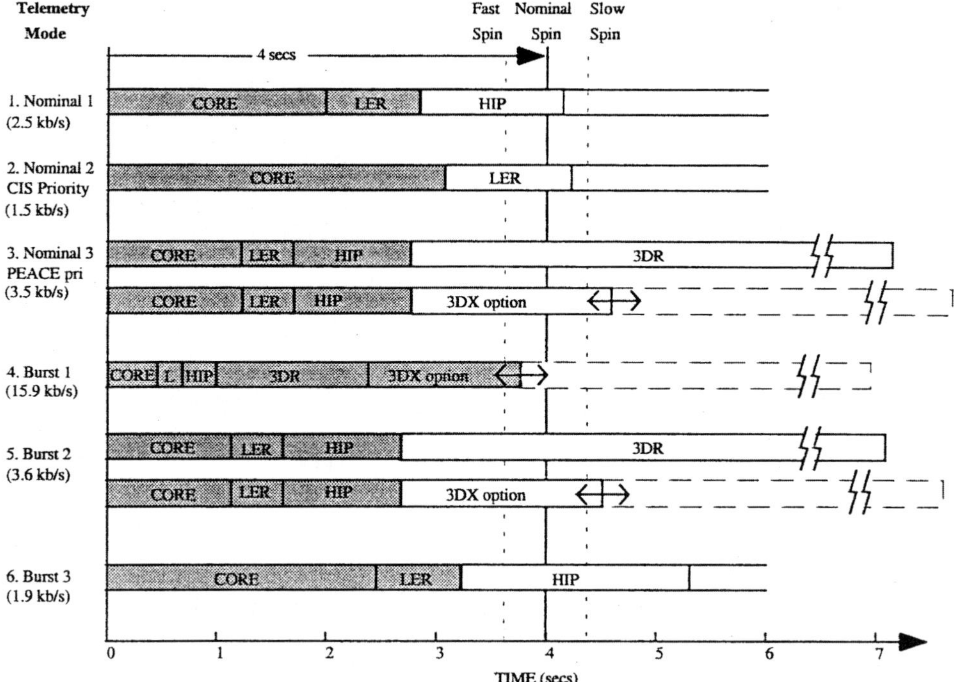

Note that 3DF-D is very large. Even in BM1, 5 spins (about 20 seconds) are required to transmit a single 3DF-D data product. Also note that the size of a 3DX distribution can be varied over a large range. CORE contains MOM-D, SCP-S and PAD-D

Figure 12. Transmission of distributions for the six telemetry rates.

3-bit exponent which gives a maximum error of $\pm 1.5\%$. The 4-bit exponent gives a wider range.

The CORE data product (containing moment sums and pitch angle distribution data) has the highest transmission priority. CORE has been sized so that it can be transmitted every spin even in the least favorable scenario, that of the CIS priority, and the fastest spin rate (3.6 s). In the standard operational scenario the smallest data allocation will be Normal mode 1, which allows transmission of CORE together with the low-energy range distribution LER at spin rate. In fact, for all telemetry allocations other than CIS priority, there is a useful portion of the telemetry allocation remaining after transmission of CORE and LER, allowing transmission of additional data products, although it is only in the Burst Mode 1 case that transmission of detailed three dimensional distributions becomes possible every spin (Figure 12).

The data product transmission scheme ensures that the telemetry allocation is always fully used, whatever the data rate. Crucial features of the scheme are the capability to start the transmission of a data product during one spin and complete it during the next spin, and the capability to automatically decide whether or not to

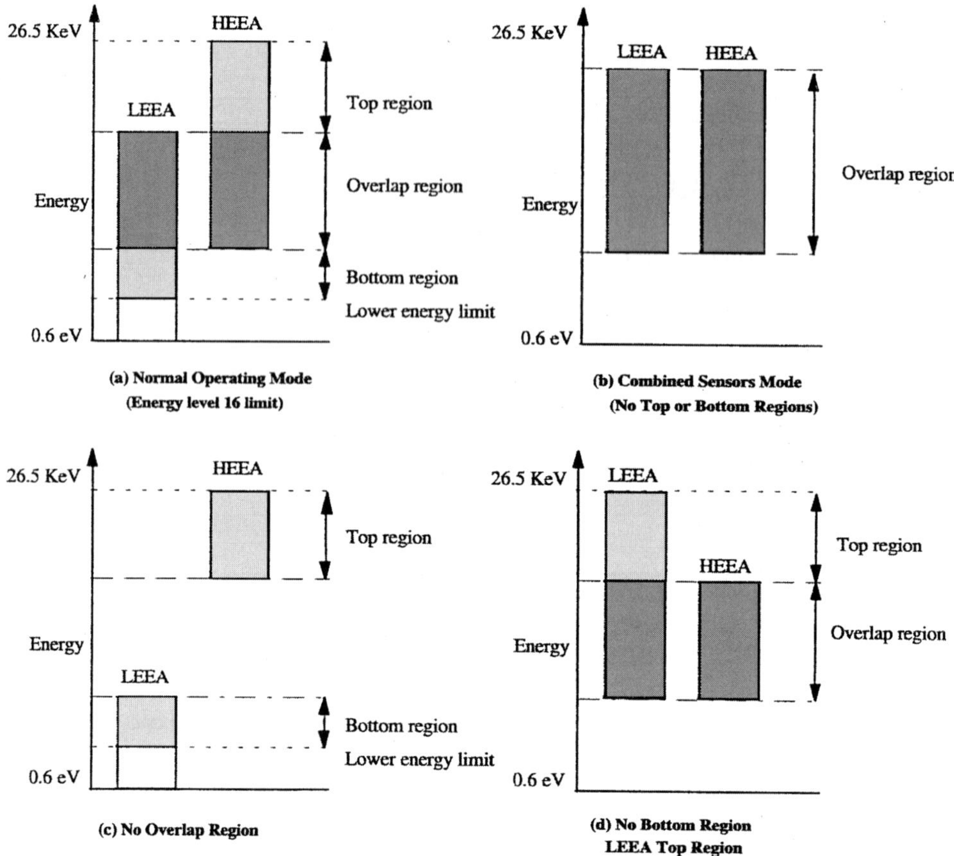

Figure 13. Possible energy sweep schemes for the combination of two analysers.

exercise this option. Consider a system in which all data products from a given spin must be transmitted within a spin period. A simple transmission scheme would start transmitting data at the beginning of each new spin, implying the truncation of data products from the previous spin which were in the process of being transmitted when the previous spin came to an end. The PEACE scheme decides which data products can be transmitted at spin resolution, and ensures that they are transmitted at that rate. It also transmits an additional data product (the one which would be truncated in the previous scenario) as often as possible, while ensuring that the transmission of the other data products at spin rate is not affected.

This is best illustrated by example. For a 4-s spin, in Normal mode 1, it is possible to transmit 1257 bytes per spin. Transmission of CORE and LER every spin requires only 806 bytes, leaving 451 bytes unused. The next data product on the priority list is HIP, which, at 494 bytes is slightly too large to be completely transmitted during the spin. The PEACE system transmits HIP in full, 'stealing' 43 bytes from the following spin. Thus, after transmission of CORE and LER on the second spin, only 408 bytes remain, and transmission of HIP in full takes a bigger bite out of the following spin. After several spins, the number of bytes left in the current spin after transmission of HIP from the previous spin (which has spilled over into the current spin) is barely enough to transmit CORE and LER for the current spin. If the HIP for the current spin were to be transmitted, it would not be possible to transmit both CORE and LER for the next spin during the next spin. In order to guarantee the transmission of CORE and LER every spin, the HIP for the current spin is not transmitted and the CORE and LER for the next spin are transmitted directly after the CORE and LER for the current spin. Thus CORE and LER are transmitted every spin, and HIP is transmitted as often as possible using the remainder of the telemetry allocation.

Larger data products, such as 3DR, are broken down into several segments for transmission and can be reassembled in ground processing. Thus a single 3DR distribution of 2928 bytes can be transmitted as segments of 732 bytes over several spins in Normal mode 1 (which allows only 1257 bytes to be sent per spin) together with spin rate transmission of CORE and LER. Data products may be removed from the data product priority list by command, but their order in the list may not be changed.

5.1. Moment Computation

The values calculated on board are:

$$
\begin{aligned}
&\int f(v)\,\mathrm{d}v && 1 \text{ value}\,, \\
&\int v\,f(v)\,\mathrm{d}v && 3 \text{ values}\,, \\
&\int vv\,f(v)\,\mathrm{d}v && 6 \text{ values}\,, \\
&\int v|v^2|\,f(v)\,\mathrm{d}v && 3 \text{ values}
\end{aligned}
\tag{7}
$$

and the total counts making 14 values per distribution. These functions are converted to the usual plasma parameters on the ground.

The moment calculations are carried out onboard for the two sensors and for several energy ranges separately as follows, assuming that HEEA is set at a higher energy than LEEA (Figure 13):

- HEEA from its preset level to the start of the LEEA sweep once/spin,
- HEEA levels overlapping the LEEA sweep, twice/spin,
- LEEA levels overlapping the HEEA sweep, twice/spin,
- LEEA from the bottom of the overlap to level 17, once/spin.

The integrations are carried out by making a summation over the elements of a matrix of counts. The size of the matrix, and hence the resolution of the integration is independent of the sweep mode and is as follows:

15 energies × 6 polar angles × 16 azimuthal bins .

All the count values in the matrix are corrected for the dead-time appropriate to the amplifier/counter chain which collected the data before being included in the integration. The correction term is

$$C_{\text{true}} = C_{\text{obs}} \frac{C_{\text{max}}}{C_{\text{max}} - C_{\text{obs}}} , \tag{8}$$

where C indicates a count rate and C_{max} which is the maximum possible count rate is given by

$$C_{\text{max}} = \frac{T_{\text{acc}}}{\tau_d} , \tag{9}$$

where τ_d is the measured dead time and T_{acc} is the accumulation time. The values can be changed in flight by uplinking new values if determined to be necessary from an analysis of the data. The DPU adjusts the dead-time algorithm to account for variations in the spin period which affect the accumulation time T_{acc}. The maximum count output from the dead-time correction is limited to approximately 8000 counts per accumulation. The integration also takes into account differences in the efficiency or sensitivity of each anode. The efficiency of each anode has been determined by calibration in the laboratory, but if it is found in flight that the efficiency of one anode has changed relative to the others then new values can be uplinked to ensure that unbiassed moment values are obtained.

The moment data are transmitted as 16-bit values with a sign, a mantissa of 7 bits and an exponent of 8 bits. This gives a range of more than $10^{\pm 38}$ and a precision of 1 in 256.

The integrations given in Equation (7) when carried out on board will include any background noise, from the detectors or from penetrating radiation. If the distribution of noise, and its intensity is known from measurements or can be deduced from the nature of the three-dimensional distributions, then an integration over the noise alone can be made on the ground and can be subtracted from the transmitted values, thereby correcting them.

5.2. Low-Energy Data (LER, 3DX-LEF)

The low-energy range is taken to be the linear range of the sweep from levels 1 to 16 (0.59 eV $< E <$ 9.45 eV). Data collected in this range must be treated differently because they are likely to be affected by the interaction between the spacecraft and its local environment. Moments calculated in this energy range by the technique outlined above would risk being in error because the distribution will have been distorted by acceleration through the spacecraft potential and because there will be additional local electron populations generated by the spacecraft. The data are treated in two ways; first of all, the simplest parameter to describe the interaction, the spacecraft potential, is estimated from the energy spectrum and then second, in order to provide as much detailed information as possible, reduced 3D distributions are telemetered.

The spacecraft potential will be obtained by the following algorithm:

(i) A reduced resolution matrix will be produced from a spin of data by taking:

energy range – the lowest 16 bins in the sweep whatever the preset;

azimuth – 8 successive sweeps whose the phasing relative to the solar direction pulse will be selected by command;

polar – 4 successive polar zones selected by command.

(ii) From the 32 spectra of counts so obtained the algorithm searches from the highest energy to find the energy at which the slope becomes negative. It then searches in the energy range from there to the lowest energy to find the energy with the absolute minimum count rate. This gives 64 values for the energy of the minimum. From these 64 energy values the average, the variance, and the maximum and minimum values are telemetered. There are the following control options for the calculation:

- the starting sweep position relative to the solar direction can be selected.
- either LEEA or HEEA can be selected;
- the computation can be commanded off.

The objective of the algorithm is to search for the minimum in the distribution corresponding to the break between electrons from the sheath and ambient electrons. However, as described in Section 4.2 the observation can be obscured by internally-generated secondary electrons. Therefore, the algorithm must be regarded as experimental until flight data have confirmed its validity. As much flexibility as possible has been included so that it can be adjusted in the light of flight experience. For the most detailed low-energy data, whenever possible a complete 3D distribution for the low-energy range is transmitted (3DX-LEF) for either LEEA or HEEA. The size of the matrix is given in Table VI. If there is not enough telemetry capacity for the full distribution, distributions with reduced resolution (LER) will be telemetered. The energy range remains the same but the distribution is reduced (i) by summing four adjacent polar zones to give 3 broad polar zones and (ii) by taking, in LAR Mode, every fourth sweep on each spin and then shifting in phase for the next spin so that, after four spins, all sweeps will have been covered. This scheme effectively reduces the time resolution, but gives a means of checking whether there have been time variations during the acquisition period. In MAR mode the same scheme is used with every fourth sweep taken, while in HAR it is every eighth.

5.3. Pitch Angle Distributions (PAD)

The pitch-angle distribution is obtained by collecting the data from the polar field of view of the two sensors on the two occasions each spin when the magnetic field direction is within the azimuthal field of view of a sensor. Within the spin, data from the full range of pitch angles from 0 to 180° will be collected from both sensors with the basic 15° resolution. In order to select the data in this way the magnetic field direction is required onboard. Two alternative methods will be used:

– the FGM magnetometer;
– the symmetry direction of the 3D electron distribution calculated internally.

The value measured during the previous spin is used so that it is possible during intervals with a rapidly-varying magnetic field that there will be an error.

Only alternate energy bins can be transmitted within the telemetry allocation for one spin in the normal sweep mode MAR. Thus data collected from even bins are transmitted during one spin; data from the alternate odd bins are collected and transmitted during the next. In the HAR mode all data are transmitted every spin; in LAR only every fourth bin is transmitted.

5.4. High-resolution polar distributions (HIP)

The HIP distributions contain data from the fine zones. Since they are normally selected to be from the coarse zone and sector through which the magnetic field direction passes they provide information about narrow field-aligned distributions. The HIP distribution is collected from a single spin of data. There are three control modes for selecting the coarse zone as follows:

(i) select data from the azimuth sector and the polar coarse zone through which the magnetic field passed in the previous spin;
(ii) select data from an azimuth sector and polar coarse zone by command;
(iii) as for (ii) except that the polar coarse zone increments by one each spin.

5.5. Three-dimensional distributions (3DF, 3DR and 3DX)

The 3DF distribution contains both LEEA and HEEA coarse zone data at the full resolution of which the sensor is capable. For the 3DR distribution the resolution is halved in each dimension to produce a distribution with one eighth of the size of the full distribution.

The 3DX distribution was added at a late stage in the development as the nature of the software allowed for the possibility. It provides additional flexibility in the construction of reduced distributions from the initial full three-dimensional distribution with full measurement resolution. It allows for distributions to be constructed which will fit within the telemetry allowance in one spin. The distribution is usually, though not necessarily, constructed from the data from a single sensor. There are three ways of reducing the size:

(a) by selecting a fraction of the sweeps per spin;
(b) by summing over either polar or energy bins;
(c) by applying a window to the energy range.

Table VI
Output distributions

Distribution	Sweep energy mode	Polar bins	Azimuth zones	Sensors sectors	
LER	LAR	16	3	4	1
	MAR	8	3	8	1
	HAR	8	3	8	1
PAD	All	15	13	1	2
HIP	LAR	60	4	1	2
	MAR	30	4	2	2
	HAR	15	4	4	2
3DF	LAR	60	12	16	2
	MAR	30	12	32	2
	HAR	15	12	64	2
3DR	All	15	6	16	2
3DX	All	Selectable	Selectable	Selectable	1
3DX-LEF	LAR	16	12	16	1
	MAR	8	12	32	1
	HAR	4	12	64	1

Table VII
Instrument mass

Low-energy electron analyser	1758 ± 5 gm
High-energy electron analyser	1738 ± 5 gm
Data processing unit	1999 ± 14 gm
Total	5490 ± 10 gm

6. Resource Summary

6.1. Mass

The range of values given in Table VII covers all the five flight units.

6.2. Power consumption

The values are given in Table VIII.

Table VIII
Power consumption

Mean secondary power consumption	
Low-energy electron analyser	694 mw
High-energy electron analyser	774 mw
Data processing unit	1831 mw
Total	3299 mw
Total mean primary power consumption	4713 mw
Maximum power (>8 ms)	8457 mw
Inrush power	9250 mw

7. Calibration

7.1. Calibration System

Figure 14 shows the electron calibration system built to test the instrument. Ultraviolet light from a mercury lamp generates photoelectrons from a gold film deposited on a quartz disk. The quartz disk and its guard ring structure are mounted on insulating ceramic rods, rated up to 40 kV, inside a grounded mu-metal case. A negative bias voltage on the gold produces a nearly-uniform electric field between the gold film and a stainless steel grid mounted on the mu-metal case which encloses the source structure. Electrons are accelerated from the gold surface through the grid whose opening defines the exit aperture. The circular cross-section of the resulting electron beam is 150mm in diameter and is nearly-uniform over the PEACE aperture. There is negligible angular divergence of the beam electron trajectories at energies above 30 eV where the transverse velocity components from the photon interaction are small. A channel electron multiplier is mounted on the rotary table to provide an independent assessment of the beam intensity and the beam current is measured directly by a picoammeter.

A reliable intercalibration between the 10 PEACE flight sensors (which includes the 2 flight spare sensors) was provided by a measurement of the flux from tritium beta-particle source. This measurement was incorporated in the standard calibration sequence to provide a quick and simple test of the relative response. Tritium emits electrons with a range of energies up to a maximum energy of 18 keV. The rate of emission from the source is well known and even though it decays with a half life of 12.33 years it provides a convenient and reliable reference for each sensor.

The instrument being calibrated is mounted on a two-axis rotary table which allows movement of the instrument over the complete azimuthal and elevation angle response range. The whole system (the electron gun, the rotary table and base plate) is contained within an outer mu-metal enclosure for magnetic shielding which is 700 mm long and 640 mm diameter and is mounted in the vacuum chamber

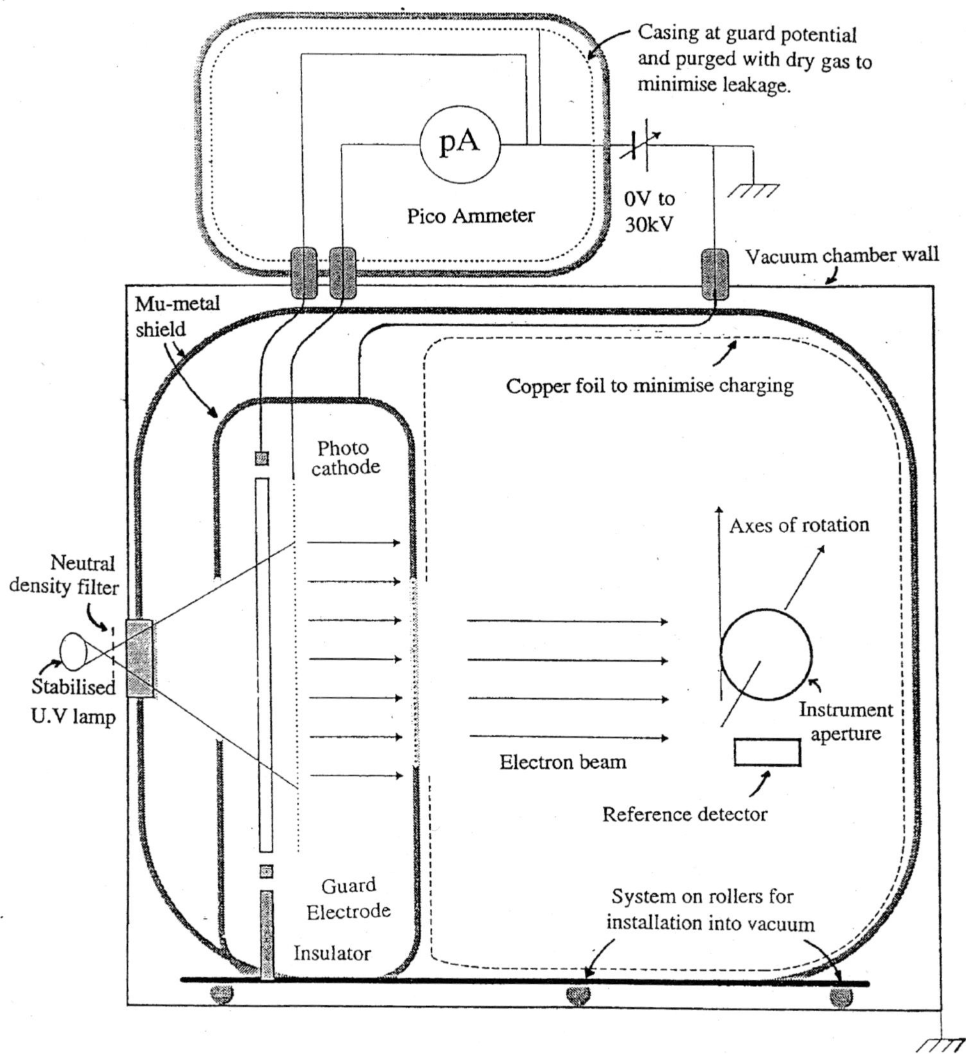

Figure 14. A diagram of the calibration system.

on teflon rails. The residual magnetic field in the chamber is less than one-tenth of the Earth's field and the beam divergence is less than 1° at 1000 eV.

7.2. Calibration Sequence

Using the tritium source, the operational voltage of the microchannel plate is selected using the following criteria:

– the microchannel plate should be in the 'plateau' region of its gain versus voltage curve where the change of count rate with microchannel plate voltage is a minimum;

– the sum of the fine zone counts should equal the counts from the corresponding coarse zone;

– the noise level should be less than 2 counts per anode s^{-1};

– the count rate should be within a factor of two of the average for sensors of that type (LEEA or HEEA).

When these conditions have been established the sensor is mounted on the two axis rotary table so that energy/angle scans can be carried out. The voltage on the electron gun is set to one of the standard sequence of 9 energies, ranging from 30 eV to 25 000 eV, at which the detailed calibration is carried out. Then the following scans are made:

(a) the energy is varied over the range $E \pm 0.2E$ in 10 equal steps where E is the energy at the peak of the response;

(b) the elevation angle is varied from $-5°$ to $+5°$ in $1°$ steps (HEEA);

(c) the polar angle is varied from $-90°$ to $+90°$ taking a set of measurements at the centre of each anode.

Then setting the elevation angle and the energy where the peak count rate was recorded the response is measured in steps of $1.5°$ from $-90°$ to $+90°$ over the complete polar angle range and the performance of the fine zones is assessed.

The light output from the mercury ultraviolet lamp is reduced by a known amount by using neutral density filters, and the dead time is estimated by comparing the changes in the observed count rate.

7.3. Calibration Results

Figure 15 shows a comparison between the response of one of the anodes as measured in a calibration and the response obtained from the numerical analysis. The differences between them are negligibly small. The numerical model has proved to be an accurate representation of the performance of the analyser.

Further calibration results are displayed in Figure 11 which presents a histogram of performance parameters for the complete set of sensors and in Table I which lists the parameters and their standard deviations.

7.4. Solar Ultraviolet Response

The response of the sensors to sunlight entering the aperture is measured with a krypton resonance lamp. The main solar ultraviolet component is the Lyman alpha line at 121 nm which has an intensity of approximately 2×10^{11} photons $cm^{-2}\ s^{-1}$ at 1 astronomical unit (AU). The lamp emits photons at a wavelength of 123 nm in $30°$ wide cone at a rate of $2.5 \times 10^{15}\ s^{-1}$. The analyser is mounted on the two axis rotary table in the vacuum chamber, allowing the azimuthal and elevation angle response to be measured, when the aperture is illuminated by the lamp.

The main results of the measurements are:

(a) The energy response peaks at 3 eV and goes to zero at energies >10 eV. Switching the internal grid to -8 V reduces the count rate significantly and applying

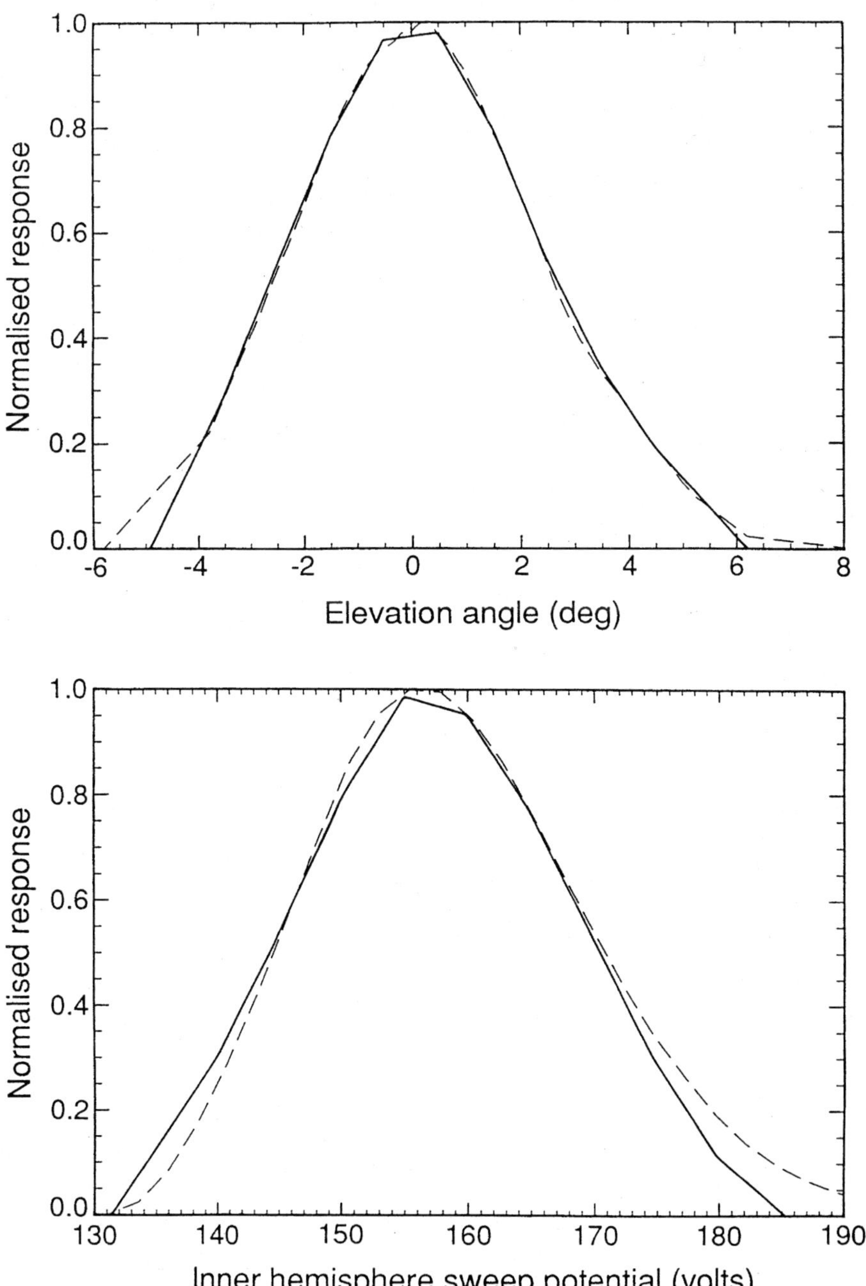

Figure 15. A comparison of the measured response (dashed line) of an analyser with the theoretical response (solid line).

a voltage to the inner hemisphere reduces the count rate to negligible levels. This proves that internally-generated photoelectrons are the main source of the solar background counts and direct illumination of the microchannel plate surface by the ultraviolet light has been prevented. If a voltage is applied to the inner hemisphere the photoelectrons are eliminated. Therefore, with appropriate settings on the grid and inner hemisphere, it will be possible to measure the background rate due to other sources such as penetrating radiation in flight.

(b) The width of the response in elevation angle is $\sim 5°$ the same as the opening angle of the collimator. This indicates that negligible levels of light are being reflected from the aperture baffles. The response is sufficiently narrow to allow the timing of the sweep to be configured such that energies >10 eV are being sampled when the analyser aperture is pointing within 5° of the solar direction in LAR and MAR sweep modes. Hence in these sweep modes the sensors will not be affected by solar photons even when they sweep down to the lowest energies.

(c) In HAR sweep mode (set to its lowest possible energy range), since the entire sweep only lasts for 5.625° of the spacecraft rotation one sweep will be affected. The measurements indicate that in these conditions we can expect approximately 500 counts from the solar Lyman alpha flux at 1 AU.

(d) The ratio of the number of photon counts in the microchannel plate (without applying the grid voltage or the sweep voltage) to the number of such photons entering the aperture is $\sim 10^{-8}$.

7.5. Intercalibration with Other Instruments

The flight spare sensors will be maintained in operating configuration for post flight checks and for intercalibration with other sensors in different beam systems should unexplained differences become apparent in operations.

8. Flight operations

8.1. Instrument Operation

8.1.1. *Startup, Shutdown*

The instrument will turn on in a programmed sequence which checks the instrument health (i.e., temperatures, voltages, current levels) for out-of-limit values, and take the appropriate action if any are detected. Then it will check experiment software.

8.1.2. *Uplink New Parameters*

The on-board software will need to refer to various numerical parameters, e.g., out-of-limit values; high voltage settings, amplifier thresholds; calibration values for on-board numerical software. As these values change with time new values will occasionally require uplinking.

8.1.3. *Uplink New Software*

The new software could include new autonomous procedures developed to take account of malfunctions or new scientific operating modes developed in response to discoveries.

8.2. Scientific Modes

8.2.1. *Normal Mode Telemetry*

The instrument will usually be operated in a standard mode which provides the optimum unbiassed coverage. In this mode the LEEA sensor covers the energy range from 0.59 eV to 1324 eV and HEEA covers 34 eV to 26.4 keV. Thus there is overlap and the potential for half spin resolution over the energy range from 34 eV to 1324 eV. Both sensors will be in the MAR sweep mode which gives angular resolution of 11.25° (azimuthal) × 15° (polar) and energy resolution of two basic levels 24.5%. The CORE data products are designed to provide standard baseline data on the electron distribution and so will be transmitted from all four spacecraft on all standard orbits. These data will allow the interspacecraft differences to be obtained. Some detailed distributions will only be transmitted from a single spacecraft with different spacecraft transmitting different parts of the distribution in order to provide the most detailed picture of the entire distribution. This enables the moment values to be validated or conversely may indicate that they are unreliable parameters for characterizing the distribution.

The following data products are available from all four spacecraft:

$$\text{MOM} - \text{D},\ \text{SCP} - \text{S},\ \text{PAD} - \text{D},\ \text{LER} - \text{S}\,.$$

In addition the following products are only available from the spacecraft indicated once every few spins where the value of 'few' depends on the spin period and the telemetry rate:

Spacecraft 1 – 3DX with full energy, and polar resolution but only two azimuthal segments which is effectively an alternative pitch angle distribution from a single sensor.

Spacecraft 2, 3, 4 – 3DX with full energy and angular resolution but a restricted range of energies. A different range of energies will be selected for each spacecraft with 2 low (LEEA), 3 mid (LEEA), and 4 (HEEA) high-energy range.

8.2.2. *Burst Mode Telemetry*

In Burst Mode 1, the main high rate mode, the following data products will always be transmitted each spin from all four spacecraft:

$$\text{MOM} - \text{D},\ \text{SCP} - \text{S},\ \text{PAD} - \text{D},\ \text{LER} - \text{S},\ \text{HIP} - \text{D},\ \text{3DR} - \text{D}\,.$$

In addition each of the four spacecraft will transmit the same extra data product as in Normal Mode operations above but at one spin resolution.

8.3. In-Flight Calibration

The performance of the sensors will be checked in operations by in-flight calibration procedures carried out on a routine basis. Two types of procedure are foreseen at this stage. First, checks on the performance of each sensor to establish proper operating voltages for microchannel plates, gains or thresholds of amplifiers. Secondly, detailed statistical intercomparison of the distributions obtained by different sensors in a relatively uniform plasma environment. One of the unusual features of the operational programme is that by comparing the four instruments when in a uniform plasma it will be easy to detect if the performance of one of them is deteriorating relative to the others. Corrective action can then be taken quickly so that the data output is not degraded.

9. Data Analysis and Archiving Plan

Data from the instrument will be processed and archived using SDDAS (Southwest Data Display and Analysis System). When the data arrives on the CD-ROM, it is in the form of telemetry values. The first step will be to separate PEACE Science and Housekeeping data from the data from other instruments on the spacecraft and reorganize it into the data products listed in Table VI. The data will be stored in the IDFS format. This consists of three types of file; a header file which contains metadata about the source of the data such as spacecraft, timing, etc.; the data file in which the data is stored as a time series of the telemetry values; and a VIDF (Virtual Instrument Description File) which contains all the information required to convert the data file into geophysical values. The data will be archived in this form so that any time it is used, the raw telemetry data will be accessed and converted to geophysical values using the most up-to-date version of the VIDF. The VIDF will be maintained by the operations team and by data analysts who will continually check the sensor calibrations and examine the data for any errors. If an error is discovered, perhaps through data analysis activities, later in the mission then the VIDF can be corrected so that any later use of the data will employ the 'best' calibration data. The system can be maintained as long as scientists continue to use the data.

When the raw data arrives on CD-ROM it will be ingested on to the Operations Team Server in the IDFS format (Figure 16). At this stage a thorough validation of the data will take place and if there are any problems with the data it will not be released for exploitation before it has been corrected. If there are any reasons to be careful in the use of the data caveats will be listed in the header data. Once validated it will be transferred to the Exploitation Server to be accessed by scientists. Any necessary corrections will be put into the current VIDF.

To produce scientific data for display and analysis IDFS read routines are applied which combine data and VIDF files. They can be used to extract the data in any

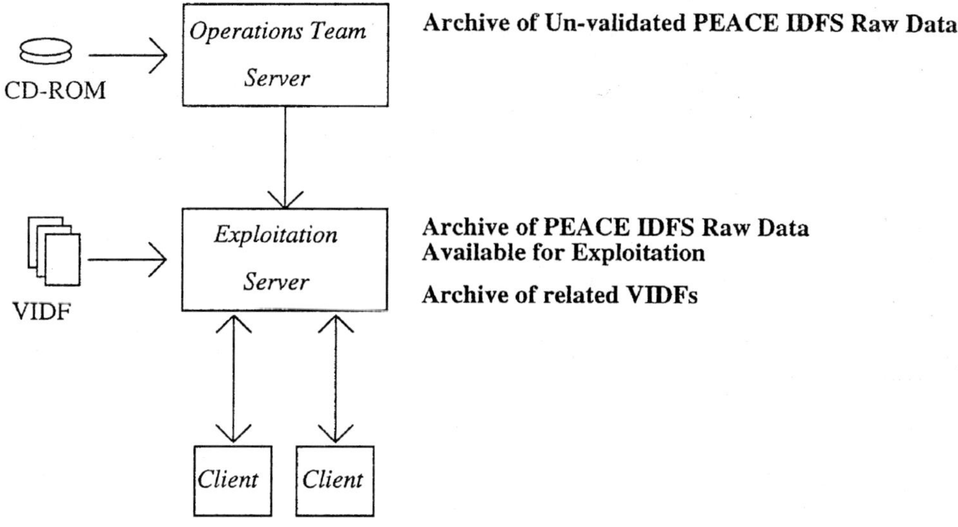

Figure 16. PEACE data preparation.

convenient form as shown in Figure 17. User's output routines can be written in the C language and incorporate IDL routines. The necessary software to access the data via networks can be made available.

A Data User's Manual will be produced and maintained current to the operations. It will contain:

(1) description of the instrument, its settings, modes and data outputs;

(2) description of the onboard data analysis techniques;

(3) report of the laboratory calibrations and the test history;

(4) record of the operations in orbit;

(5) record of instrument performance including inflight calibrations, changes in the onboard calibration settings, operational anomalies, performance checks, microchannel plate performance, external interference sources;

(6) description of the spacecraft/plasma interactions and their effect on low-energy data;

(7) description of the data processing on the ground, and common data analysis software;

(8) explanation of the limitations and potential sources of error in the moment data;

(9) catalogue of the data in the archive.

This will provide the main guide to data user's during the mission and for any future use of the data. Access to the full resolution data will be maintained as long as possible.

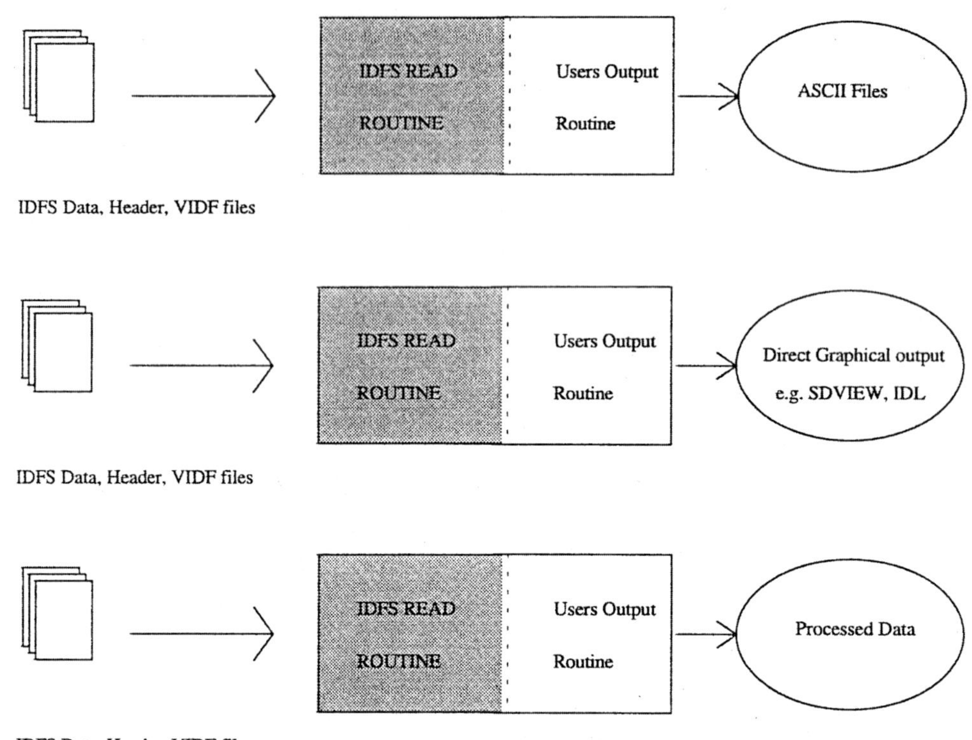

Figure 17. Production of science data from IDFS raw data.

Acknowledgements

We acknowledge the contributions of many engineers and scientists to the PEACE investigation in the development of scientific objectives, instrument design, construction, testing and calibration. They include: S. J. Schwartz (Queen Mary Westfield College), D. S. Hall (Rutherford Appleton Laboratory), J. Sharber (Southwest Research Institute), P. Reiff (Rice University), M. Goldstein (Goddard Space Flight Center), M. Thomsen (Los Alamos National Laboratory), T. E. Kennedy, A. Murrell, A. Spencer, J. Raymont, J. E. Hall (Mullard Space Science Laboratory), J. J. Berthelier, Y. Meyerfeld (CRPE).

References

Balogh *et al.*: 1997, this volume.

Bame, S. J., Asbridge, J. R., Felthauser, H. E., Glore, J. P., Paschmann, G., Hemmerich, P., Lehmann, K., and Rosenbauer, H.: 1978, *IEEE Trans. Geosci. Electr.* **GE-16**, 216.

Carlson, C. W., Curtis, D. W., Paschmann, G., and Michael, W.: 1983, 'An Instrument for Rapidly Measuring Plasma Distributions Functions with High Resolution', *Adv. Space Res.* **2**, 67.

Décréau, P. M. R. *et al.*: 1997, this volume.

Gustaffson *et al.*: 1997, this volume.
Hardy, D. A., Walton, D. M., Johnstone, A. D., Smith, M. F., Gough, M. P., Huber, A., Pantazis, J., and Burkhardt, R.: 1993, 'Low Energy Plasma Analyzer', *IEEE Trans. Nuc. Sci.* **NS-40**, 246.
Johnstone, A. D., Kellock, S., Coates, A. J., Smith, M. F., Booker, T., and Winningham, J. D.: 1985, 'A Spaceborne Plasma Analyser for the Three-Dimensional Measurements of the Velocity Distribution', *IEEE Trans. Nuc. Sci.* **NS-32**, 139.
Johnstone, A. D., Coates, A. J., Wilken, B., Studemann, W., Weiss, W., Cerulli-Irelli, R., Formisano, V., Borg, H., Olsen, S., Winningham, J. D., Bryant, D. A., and Kellock, S. J.: 1987, 'The Giotto Three-Dimensional Positive Ion Analyser', *J. Phys. (E)* **20**, 795.
Kessel, R. L., Johnstone, A. D., Coates, A. J., and Gowen, R. A.: 1989, 'Space Plasma Measurements with Ion Instruments', *Rev. Sci. Inst.* **70**, 3750.
Mukai, T., Machida, S., Saito, Y., Hirakawa, M., Terasawa, T., Kaya, N., Obara, T., Ejiri, M., and Nishida, A.: 1994, 'The Low-Energy Particle (LEP) Experiment Onboard the Geotail Satellite', *J. Geomag. Geoelect.* **46**, 669.
Mukai, T. and Miyake, W.: 1986, 'Transmission Characteristics and Fringing Field Effect of a 270° Spherical Electrostatic Analyser', *Rev. Sci. Inst.* **57**, 49.
Oberhardt, M. R., Hardy, D. A., Slutter, W. E., McGarity, M. O., Sperry, D. J., Everest III, A. W., Huber, A. C., Pantazis, J. A., and Gough, M. P.: 1994, 'The Shuttle Potential and Return Experiment (SPREE)', *Nuovo Cimento* **17**, 67.
Paschmann, G., Loidl, H., Obermayer, P., Ertl, M., Laborenz, R., Sckopke, N., Baumjohann, W., Carlson, C. W., and Curtis, D. W.: 1985, 'The Plasma Instrument for AMPTE IRM', *IEEE Trans. Geosci. Remote Sensing*, **GE-23**, 262.
Paschmann, G. *et al.*: 1997, this volume.
Reme, H., Cotin, F., Cros, A., Medale, J. L., Sauvaud, J. A., d'Uston, C., Anderson, K. A., Carlson, C. W., Curtis, D. W., Lin, R. P., Korth, A., Richter, A. K., Loidl, A., and Mendis, D. A.: 1987, 'The Giotto Electron Plasma Experiment', *J. Phys. (E)* **20**, 721.
Reme, H. *et al.*: 1997, this volume.
Riedler *et al.*: 1997, this volume.
Sablik, M. J., Winningham, J. D., Gurgiolo, C., and Johnstone, A. D.: 1985, 'Computer Simulation of an Electrostatic Spherical Analyser Used as an Energy Spectrograph', *Rev. Sci. Inst.* **56**, 1320.
Sablik, M. J., Scherrer, J. R., Winningham, J. D., Frahm, R. A., and Schrader, T.: 1990, 'TFAS (A Tophat for All Species): Design and Computer Optimisation of a New Electrostatic Analyser', *IEEE Trans. Geosci. Remote Sensing* **GE-28**, 1034.
Shah, H. H., Hall, D. S., and Chaloner, C. P.: 1985, 'The Electron Instrument on the AMPTE UKS', *IEEE Trans. Geosci. Remote Sensing*, **GE-23**, 293.
Wilken *et al.*: 1997, this volume.
Woodliffe, R. D. and Johnstone, A. D.: 1996, in J. Borovsky, R. Pfaff, and D. Young (eds.), *The Use of Numerical Simulation in the Design of the Cluster/Peace 'Top Hat' Analyser Electron Optics, 'Measurement Techniques for Space Plasmas'*, AGU Geophysical Monograph XX, American Geophysical Union, Washington DC, submitted.
Wooliscroft *et al.*: 1997, this volume.
Wrenn, G. L., Johnson, J. E. F., and Sojka, J. J.: 1981, 'The Suprathermal Plasma Analysers on the ESA GEOS Satellites', *Space Sci. Inst.* **5**, 271.

Acronyms

ASPOC	Active SPacecraft POtential Control
CIS	Cluster Ion Spectrometer
DWP	Digital Wave Package
EDI	Electron Drift Instrument
EFW	Electric Field and Wave experiment
FGM	Flux Gate Magnetometer
HAR	High Angular Resolution
HEEA	High Energy Electron Analyser
IDFS	Instrument Data File Set
LAR	Low Angular Resolution
LEEA	Low Energy Electron Analyser
MAR	Medium Angular Resolution
PAD	Pitch Angle Distribution
PEACE	Plasma Electron And Current Experiment
SDDAS	Southwest Data Display and Archiving System
VIDF	Virtual Instrument Description File
WHISPER	Waves of HIgh frequency and Sounder for Probing of Electron density by Relaxation

RAPID

The Imaging Energetic Particle Spectrometer on Cluster

B. WILKEN, W. I. AXFORD, I. DAGLIS, P. DALY, W. GÜTTLER, W. H. IP, A. KORTH, G. KREMSER, S. LIVI, V. M. VASYLIUNAS and J. WOCH
Max-Planck-Institut für Aeronomie, D-37191 Katlenburg-Lindau, Germany

D. BAKER
Laboratory for Astrophysics and Space Physics, University of Colorado, Boulder, CO 80309, U.S.A.

R. D. BELIAN
Los Alamos National Laboratory, Los Alamos, NM 87545, U.S.A.

J. B. BLAKE, J. F. FENNELL and L. R. LYONS
The Aerospace Corp., Los Angeles, CA 90009, U.S.A.

H. BORG
Institut för Rymdfysik, Swedish Institute of Space Physics, 90187 Umeå, Sweden

T. A. FRITZ
Center of Space Physics, Boston University, MA 02215, U.S.A.

F. GLIEM and R. RATHJE
Institut für Datenverarbeitungsanlagen, Techn. University, D-38023 Braunschweig, Germany

M. GRANDE and D. HALL
Science and Engineering Research Council, Rutherford Appleton Laboratory, Chilton, 0X11 OQX, U.K.

K. KECSUEMÉTY
Central Research Institute for Physics, Academy of Sciences, H-1525 Budapest, Hungary

S. MCKENNA-LAWLOR
Experimental Physics Department, St. Patrick's College, Maynooth, Ireland

K. MURSULA and P. TANSKANEN
Department of Physics, University of Oulu, 90570 Oulu, Finland

Z. PU
Department of Geophysics, Peking University, Beijing 100871, China

I. SANDAHL
Institutet för Rymdfysik, Swedish Institut of Space Physics, 98128 Kiruna, Sweden

E. T. SARRIS
Department of Electronic Engineering, University of Thrace, Xanthi, Greece

M. SCHOLER
Max-Planck-Institut für Extraterr. Physik, 85748 Garching, Germany

M. SCHULZ
Lockheed Research Laboratory, Palo Alto, CA 94304, U.S.A.

F. SØRASS and S. ULLALAND
Physics Department, University of Bergen, 5007 Bergen, Norway

Space Science Reviews **79:** 399–473, 1997.

Abstract. The RAPID spectrometer (Research with Adaptive Particle Imaging Detectors) for the Cluster mission is an advanced particle detector for the analysis of suprathermal plasma distributions in the energy range from 20–400 keV for electrons, 40 keV–1500 keV (4000 keV) for hydrogen, and 10 keV $nucl^{-1}$–1500 keV (4000 keV) for heavier ions. Novel detector concepts in combination with pin-hole acceptance allow the measurement of angular distributions over a range of 180° in polar angle for either species. Identification of the ionic component (particle mass A) is based on a two-dimensional analysis of the particle's velocity and energy. Electrons are identified by the well-known energy-range relationship. Details of the detection techniques and in-orbit operations are described. Scientific objectives of this investigation are highlighted by the discussion of selected critical issues in geospace.

Table of Contents

1. Introduction

Over many years of intense research the Earth's magnetosphere has emerged as a highly structured and dynamic, magnetically contained body of plasma. At times or permanently, parts of the magnetosphere seem to be connected with interplanetary magnetic field lines. The field topology in the outer regions of the magnetosphere and its time dependence is by and large a result of currents carried by the thermal plasma. The suprathermal component,on the other hand, may be less important for most of the macroscopic plasma quantities but it plays an important role on its own right due to peculiarities in the physics of energetic particles. Acceleration processes in the magnetosphere of still unknown nature energise particles in the magnetosphere to hundreds of keV. The relatively fast motion of these particles can carry information about the energisation process over significant distances to an observing platform. Studies of the intensity profile, the energy distribution and the ionic mass and charge composition can provide important clues on the nature of the process. Furthermore, the kinetic properties of these particles can be used as a tool to trace out plasma structures over distances as large as an Earth radius by utilising the particle's gyroradius. Information can even be transmitted over global distances by the rapid drift of energetic particles in field gradients or, even more important, by field-aligned swift electrons travelling with speeds comparable with the speed of light. In tail-like field configurations these particles can transmit over very large distances almost instant information on changes in the field topology.

The Cluster polar orbit ($4 \times 19.6\ R_E$) provides excellent opportunities for energetic particle studies. The physics at the magnetopause, the bow shock, and the near-Earth magnetotail are key regions of interest for the RAPID investigation. The state-of-the-art detection techniques, the large energy range for nuclei and electrons, and the complete coverage of the unit sphere in velocity space lead to the following capabilities:

– Remote-sensing of local density gradients over distances comparable with particle gyroradii. Species-dependent structures in gradients can be studied, gradient motions can be resolved to one spin period ($T = 4$ s).

– Identification of major ion species (H, He, CNO) in the energetic plasma component. A special operational mode allows the identification and analysis of Energetic Neutral Atoms (ENA). The importance of ENAs in the magnetosphere and the detectability with RAPID is briefly described in Section 3.5.

– Characterisation of magnetic-field-line topologies using the fast motion of energetic electrons.

These observational features allow detailed studies in most regions of geospace visited by Cluster. Among a vast number of topics a selection discussed in Section 3 illustrates how RAPID can scrutinise as yet unexplained phenomena.

The RAPID instrument uses two different and independent detector systems for the detection of nuclei and electrons. The Imaging Ion Mass Spectrometer (IIMS) identifies the nuclear mass of incident ions or neutral atoms from the kinetic energy equation: a time-of-flight and energy measurement determines the particle mass. One-dimensional images of spatial intensity distributions result from the projection principle. The Imaging Electron Spectrometer (IES) is dedicated to electron spectroscopy. The well-known energy-range relationship is used to identify electrons over a limited energy range.

Close collaboration with the other Cluster teams is essential to bring to bear the wealth of information expected from this multi-spacecraft mission which indeed offers an unprecedented scientific tool for studies of long-standing problems in the magnetosphere.

An essential complement to the Cluster space segment is the Cluster science data system (CSDS) which is a pre-requisite for bringing together multiple data sets from the 44 Cluster instruments. A successful analysis of the enormous volume of data returned by Cluster requires support by a data network such as CSDS.

Section 2 gives an account of the detection techniques employed in IIMS and IES and expands on specific aspects of the signal processing and data generation in the two segments of RAPID. This Section also provides useful information on the flight operations and the data analysis. Section 3 addresses in some detail selected scientific objectives of the RAPID investigation.

2. Instrumentation

2.1. The RAPID Spectrometer

Outer envelopes of the RAPID spectrometer with some principal dimensions are shown in Figure 1. The instrument is physically a single structure which contains all major elements of the instrument as sketched in Figure 2: the sensor systems IIMS and IES, the front-end electronics or Signal Conditioning Unit (SCU), and the Digital Processing Unit (DPU) with the the Low-Voltage Power-Supply (LVPS) and the spacecraft interface in the back of the box. Figure 3 is a photograph of the EM unit.

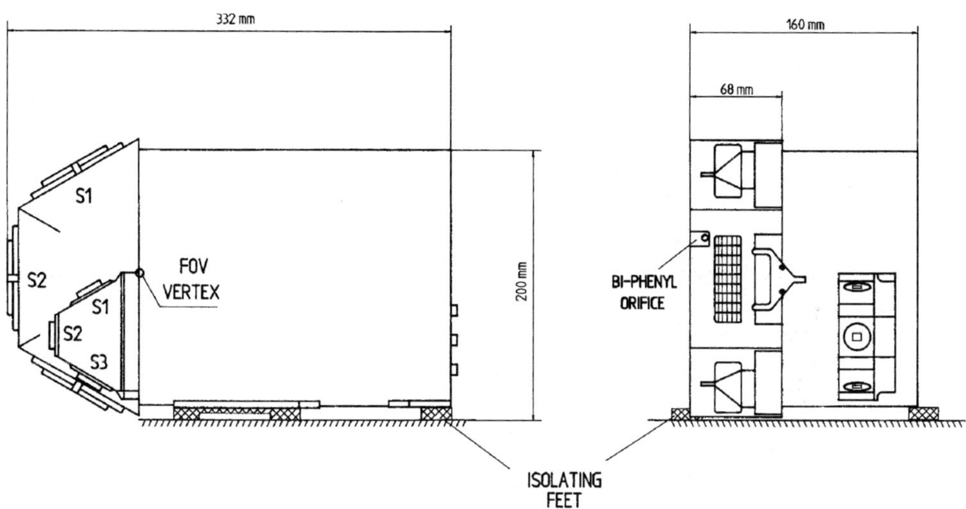

Figure 1. Mechanical configuration of the RAPID instrument showing the sensor system IIMS (on the left side of the unit in the right panel) and IES (small system in the right panel). Individual detector heads are identified by Sn (n = 1, 2, 3).

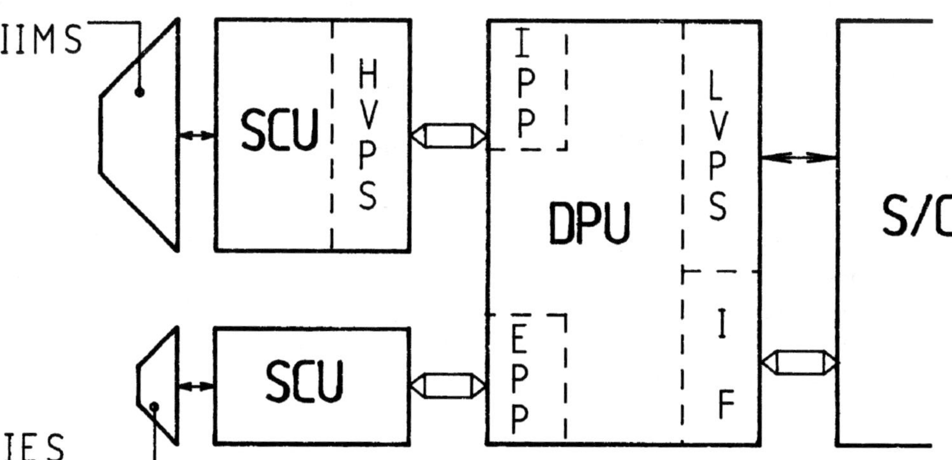

Figure 2. Key components of the dual sensor spectrometer integrated into the single RAPID box structure.

The sensor system for nuclei, the Imaging Ion Mass Spectrometer (IIMS), is composed of three identical SCENIC heads (Spectroscopic Camera for Electrons, Neutral and Ion Composition). The positions of the three systems S1, S2, and S3 in the instrument reference system are marked in Figure 1. Each spectrometer head Sy (y = 1, 2, 3) is protected by a mechanical door. After insertion into the vacuum of space the individual door latches are released by bi-phenyl ($C_{12}H_{10}$) operated mechanisms and the doors are rotated into the open position by the action of a

Figure 3. A photograph of the RAPID unit. The sensor system IIMS is on the left side.

spring. An opened door and the orifice for the bi-phenyl evaporation is sketched in Figure 1 for head S2 (right panel in Figure 1). This rather straightforward scheme for an one-shot actuator is based on the vast difference in evaporation speed of large bi-phenyl molecules in air and in vacuum. However, the obvious simplicity of such a device is somewhat offset by the difficulty to predict with reliability the accurate release time due to the uncertainty in the knowledge of the bi-phenyl temperature during the launch phase. Measurements in the laboratory suggest a delay time of about 30 to 40 hours for the doors to be released after launch.

The electron detector, IES, is composed of three identical sensor heads as well, however, the detection technique differs entirely from the principle used in SCENIC. Again the apertures are protected by bi-phenyl operated closures but in this case the 'mechanism' is rather simplified: the tiny entry holes of the IES heads are closed by bi-phenyl 'plugs' which leave the holes open after evaporation in space. The positions of the heads Sn are also shown in Figure 1.

2.1.1. *The Nuclei Detector SCENIC*

The centre piece of the IIMS sensor system is the SCENIC detector head. In essence SCENIC is a miniature telescope composed of a time-of-flight (TOF) and energy (E) detection system. The novel aspect is the imaging of flux distributions and the capability to identify Energetic Neutral Atoms (ENA) in a certain energy band.

The particle-identifying function of the SCENIC spectrometer is obtained from a two-parameter measurement: the particle velocity (V) and the energy (E) are measured as independent quantities, the particle mass A is then uniquely determined either by computation ($A \sim EV^{-2}$) or by statistical analysis in two-dimensional (V, E) space with the mass A as the sorting parameter. Actually the velocity detector measures the flighttime (T) the particle needs to travel a known distance in the detector geometry. The time-of-flight/energy measuring technique is described in detail, e.g., by Wilken (1984).

Figure 4 shows cross-sections of the SCENIC detector telescope drawn to scale. A particular feature is the triangular structure with a 60° opening angle. The energy measuring solid-state detectors (SSD) are mounted in the apex at the rear of the system. A group of two SSDs (an energy detector ED and a back detector BD) is combined in an anticoincidence condition for high-energy-electron detection. The flighttime (T) measuring system is the entry element of the telescope. It is essentially composed of a thin foil (10 μg cm^{-2} Lexan foil with aluminium coating) and the front surface of detector ED. The distance between the foil and detector ED along the line of symmetry is the nominal flight path for the T-measurement.

Particles passing through the telescope release secondary electrons (SE) from the entry foil. The SE are accelerated and directed to a microchannel plate (MCP) for detection. The MCP output signal constitutes the START signal for the T-measurement. Details of the isochronous SE transfer to the START-MCP are shown in the upper cross-Section of Figure 4. Upon impact of the particle on detector ED secondary electrons are ejected from its surface as well. These SE are transferred to the STOP-MCP by a technique similar to the start electrons. The STOP signal completes the T-measurement.

The energy E of the incident particle, reduced by the loss in the START foil, is measured in detector ED. For sufficiently high energies the particle is able to penetrate detector ED and to strike the back detector BD. This leads to the elimination of the event from analysis as described in Section 2.2.1.

Figure 4 shows the START foil as an elongated rectangle. The design of the START-system is such that the SE transfer to the MCP is not only isochronous but also position-preserving: Four read-out anodes behind the START-MCP (not shown in Figure 4) correspond to four contiguous segments on the entry foil and each of these forms a 12° by 15° viewing cone with the ED detector in the back of the system. In a sense this geometry can be considered a degenerated case of a 'projection camera' with only one pixel in the back plane. In this special case the 'virtual' image plane coincides with the entrance foil.

Incident particles are strongly collimated before they reach the T/E-telescope. Two microchannel collimators (COLL1 and COLL2 in Figure 4) define a highly anisotropic field-of-view with 6° lateral and 60° polar opening. A set of plates between the collimating elements with alternating potentials (0 and $+U_{\mathrm{def}}$) forms a linear electrostatic deflector (DEFL). The primary purpose of DEFL is to protect the instrument from overloads due to large fluxes of low-energy particles (e.g.,

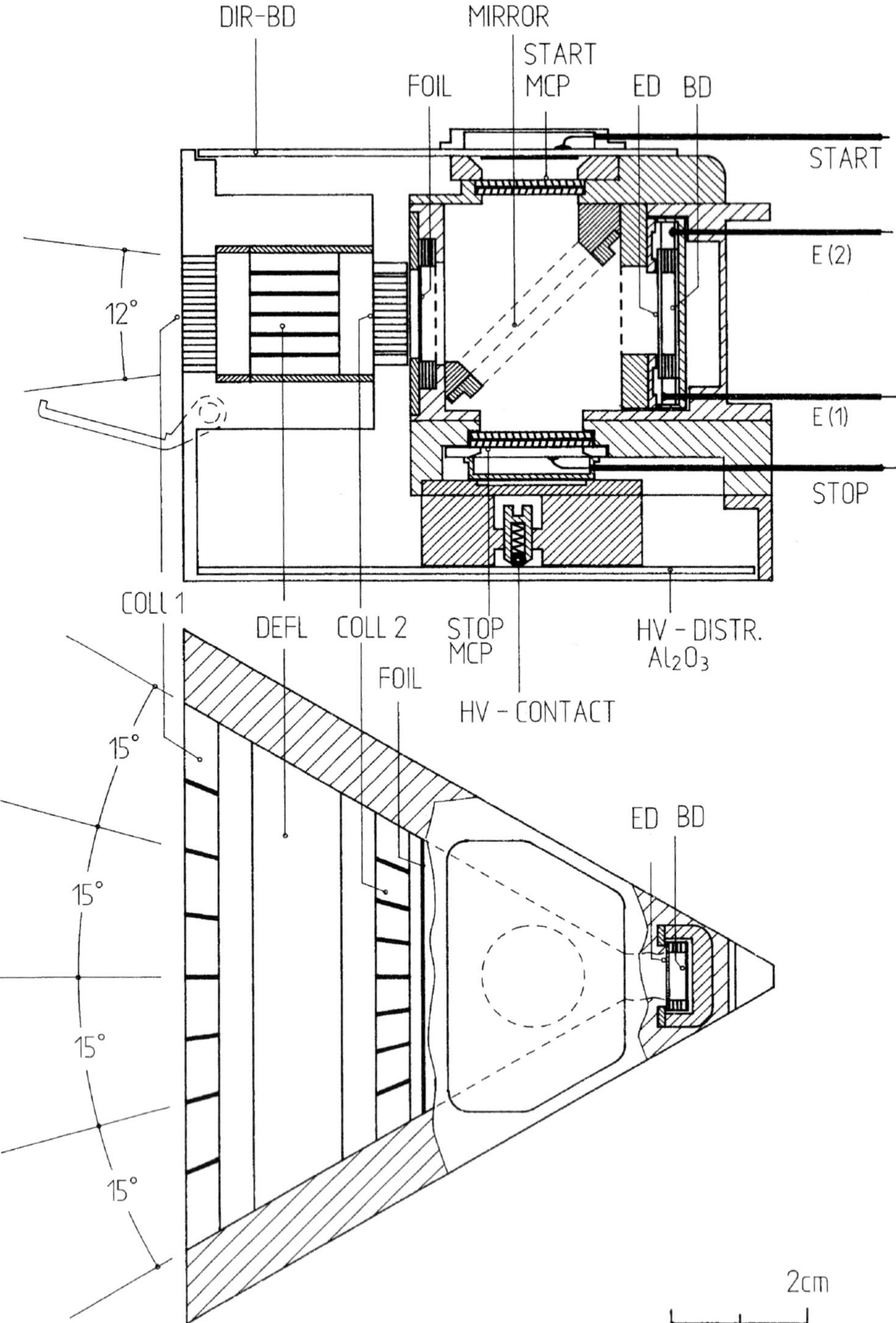

Figure 4. Cross-sections of the SCENIC head. Two narrow collimators (COLL1 and COLL2) and a set of deflection plates (DEFL) form the entry element. The foil (FOIL) and the solid-state detector (ED) define the time-of-flight geometry. The detector BD is in anti-coincidence to ED · Microchannel plates (MCP) detect 'START' and 'STOP' secondary electrons.

Table I
Technical parameters of the SCENIC head

Flightpath s (mean)	34 mm
Field-of-view (total)	6° × 60°
Polar angles	4 × 15°
E-detector (ED)	
Area/thickness	5 × 15 mm^2/300 μ
B-Detector (BD)	
Area/thickness	5 × 15 mm^2/300 μ
START foil[a]	
Al/Lexan/Al	63 nm/83 nm/63 nm
Deflection voltage	0–10 kV

[a] The high-quality foils were manufactured by the Luxel Corporation, Friday Harbor, WA (U.S.A.).

solar wind plasma). Some selected technical parameter of the SCENIC head are listed in Table I.

The relatively high gain $g = 10$ achieved with the active collimator with the deflection voltage DEFL set to 10 kV allows efficient separation of energetic neutral atoms (ENA) from ions for energies up to 150 keV (this operational configuration of RAPID is called ENA mode). The energy band between 20 and 150 keV is generally considered important for magnetospheric neutrals produced in the ring current region. RAPID, in the ENA mode, covers the high energy end of the ENA distribution (compare also Section 3.5).

The deflection voltage DEFL can be set to 16 different levels between 0 and 10 kV. Experience with an instrument on GEOTAIL similar to RAPID shows that a rather low level for DEFL (about 1 kV) is sufficient to avoid excessive counting rates in the inner magnetosphere and thus provides adequate protection. There is no plan to operate DEFL at any other level than described above.

The fraction in (E, T) space covered by the SCENIC head is shown in Figure 5 together with calculated loci of major magnetospheric nuclei or groups of nuclei. The width of the particle traces reflects the effect of flight path variations over the 60° opening of the SCENIC head. The energy (E) scale from 0 keV up to 4000 keV is partitioned by electronic thresholds A and C which define the lower (30 keV) and the upper (4000 keV) limit of the *measured* energy range. The upper end of the linear energy range B is defined by the last channel of the 8-bit amplitude-to-digital-converter (ADC). The time (T) range extends essentially from 0 ns to 80 ns.

Particles (electrons or nuclei) with sufficient energy to penetrate the ED-detector create a veto signal in the BD-detector which results in the elimination of this event from subsequent analysis.

With reference to Figure 5 the following definitions are used to identify hydrogen, helium, particles of the CNO and Si-Fe groups (carbon, nitrogen, and oxygen

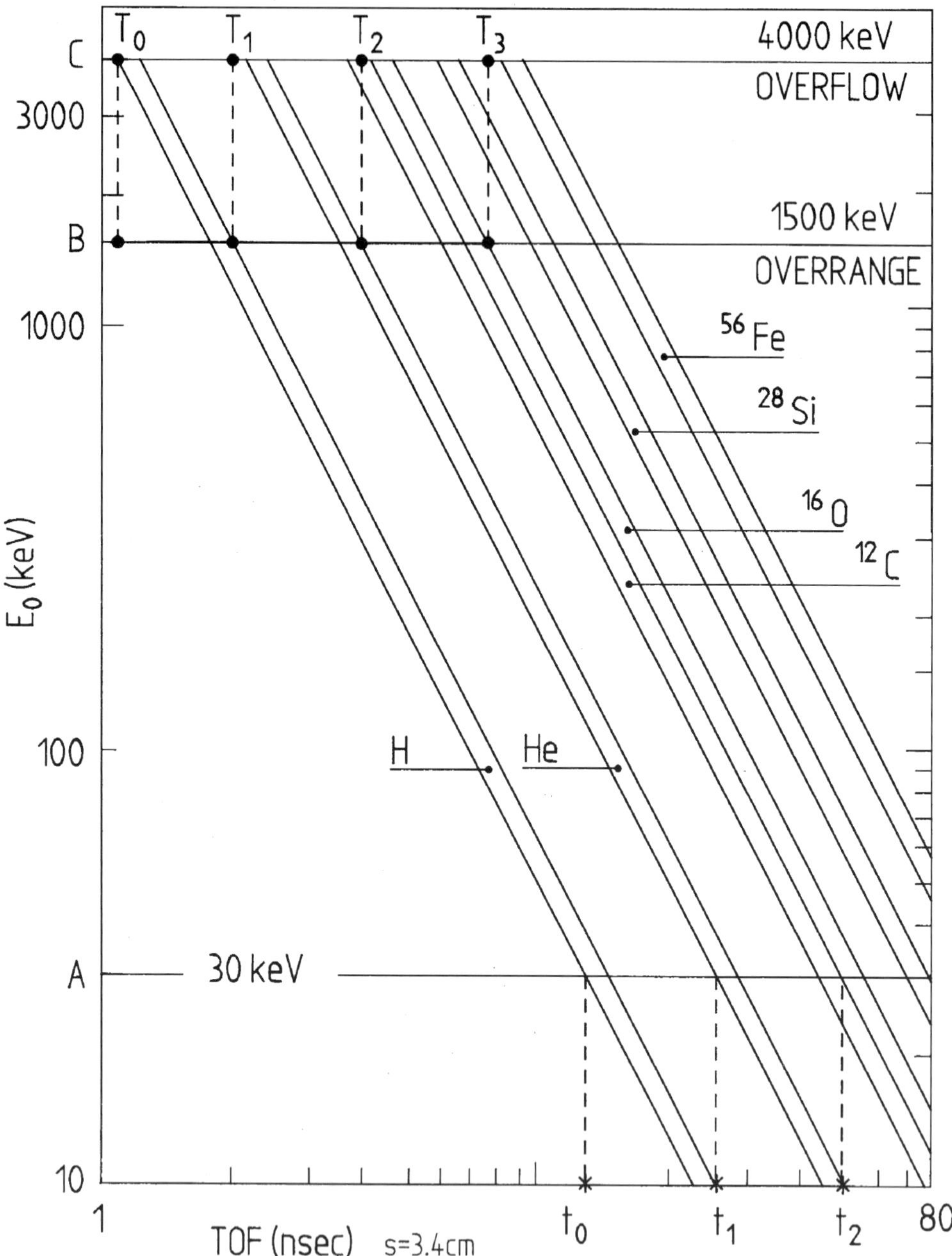

Figure 5. The energy-time plane covered by the IIMS detector. The width of the particle traces reflects the variation of the flight-path in the SCENIC geometry. Energy thresholds A and C define the lowest energy value accepted and over-range, respectively. The level B denotes channel 256 in the energy ADC which is equivalent to the upper limit for the linear energy response.

in the CNO group and particles with masses between silicon and iron in the Si–Fe group cannot be resolved as individual species):

Logic state in the E-channel	T−range	Species	Remarks
$\overline{ABC}$	$t0 - t1$	H	Reduced resolution
$\overline{ABC}$	$t1 - t2$	He	Reduced resolution
$A\overline{BC}$	2–80 ns	H, He, CNO, Si–Fe	Nominal resolution
$AB\overline{C}$	$T0 - T1$	H	Unique identification
$AB\overline{C}$	$T1 - T2$	He	Unique identification
$AB\overline{C}$	$T > T2$	CNO and heavier	No mass resolution

Comment:
A and C denote physical threshold, B is defined by the last channel of the energy ADC (upper limit of the linear range).

As mentioned earlier the IIMS sensor system is composed of three identical SCENIC heads in a configuration such that contiguous coverage over 180° in the polar angle is achieved (the polar angle is defined with respect to the Cluster spin axis). The sectored rotation of the spacecraft provides the completing azimuthal coordinate. The design features of the SCENIC head combined with the specific layout of the electronic system lead to performance parameters summarised in Table II. It should be noted that hydrogen and helium particles are identified by the time-of-flight T only if the *measured* particle energy is outside the nominal range, i.e., below 30 keV (measured energy = 0 keV) or above 1500 keV. In the latter case the measured energy value is constant and equal to E_M (see Table II) which corresponds to the highest ADC channel.

The response functions of the SCENIC head (and likewise of the IIMS sensor system) are generally describe by a complex family of energy-dependent functions parameterised by the particle mass A and the selected type of science data (science data (SD) are defined in Section 2.4.). The conversion from observed counting rates n (cts s^{-1}) to particle flux j in physical units is therefore represented by functions of the form

$$j = [GF\ \varepsilon(E, A; SD)]^{-1} n$$

with GF denoting the geometric factor and ε describing the detection efficiency as a function of particle energy (E), particle mass (A) and selected Science Data (SD).

2.1.2. *The Electron Detector IES*

Electrons with energies from 20 keV and 400 keV are measured with the novel Imaging Electron Spectrometer (IES). Advanced microstrip solid-state detectors

Table II
Characteristic parameters of the IIMS and IES sensor systems

	IIMS	IES
Energy range (keV)		
Hydrogen $(0, T)$	40–75	–
Hydrogen (E, T)	75–1500	–
Hydrogen (E_M[a], T)	1500–4000	–
Helium $(0, T)$	40–100	–
Helium (E, T)	100–1500	–
Helium (E_M[a], T)	1500–4000	–
CNO	210–1500 (oxygen)	–
Electrons	–	20–400
ENA	40–200	–
Mass classes (amu)	1, 4, 12–16, 28–56	–
Mass resolution (A/dA)	4 (oxygen)	–
Field-of-view	±3° × 180°	±17.5° × 180°
Angular coverage		
Polar (range/intervals)	180°/12	180°/9
Azimuthal (range/sectors)	360°/16	360°/16
Deflection voltage (kV)		
Range/steps	0–10/16	–
Geometric factor (cm^2 sr)		
Total (180°)/per pixel	$2.7 \times 10^{-2}/2.2 \times 10^{-3}$	$2.0 \times 10^{-2}/2.2 \times 10^{-3}$

[a] The value E_M corresponds to channel 256 in the energy ADC.

having a 0.5 cm × 1.5 cm planar format with three individual elements form the image plane for three acceptance 'pin-hole' systems. Each system divides a 60° segment into three angular intervals. A schematic cross-Section of an IES pin-hole camera is presented in Figure 6. Three of these detectors arranged in the configuration shown in Figure 1 provide electron measurements over a 180° fan in polar angle. Figure 7 shows an exploded view of the complete IES sensor head with the integrated monolithic Switched Charge/Voltage Converter (SCVC) chip and boards in the stack containing bias resistors and coupling resistors required by the charge sensitive pre-amplifier/detector operation.

The 800-micron-thick ion-implant solid-state devices are covered with a 450 μg cm^{-2} (Si eq) absorbing window which eliminates ions up to 350 keV through the mass dependent range-energy relationship. The principle energy range for IES is shown in Figure 8: The curves labeled e and p show the amount of energy deposited in an 800 μ silicon detector by electrons and protons, respectively. Electrons with energies above E_D penetrate the detector and, as a result, the deposited energy is no longer proportional to the incident energy. Protons, on the other hand, can traverse an absorbing foil in front of the detector only for energies above E_A.

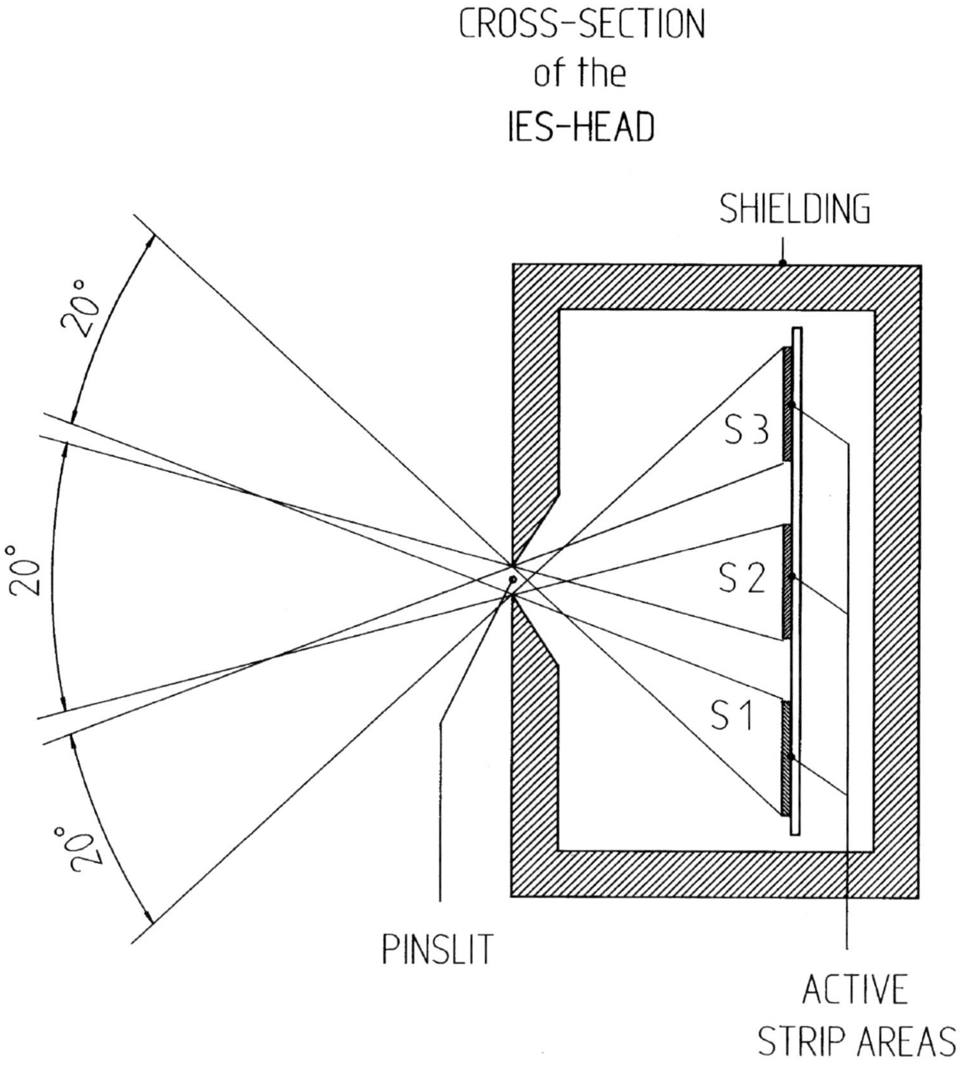

Figure 6. The IES Sensor Concept. Multiple look directions are achieved using a single detector with multiple elements placed behind a pin slit.

The separation of electrons in the nominal range 20–400 keV from high energy electrons $E \leq E_D$ and from protons between E_A and the proton 'veto' discriminator relys on the observational fact that the particle spectra are steeply decreasing with energy.

Figure 9 demonstrates the very good angular resolution obtained with an IES camera for three electron energies (65, 90, and 114 keV). The nine individual strips on the three focal plane detectors are interrogated by the multichannel SCVC circuit.

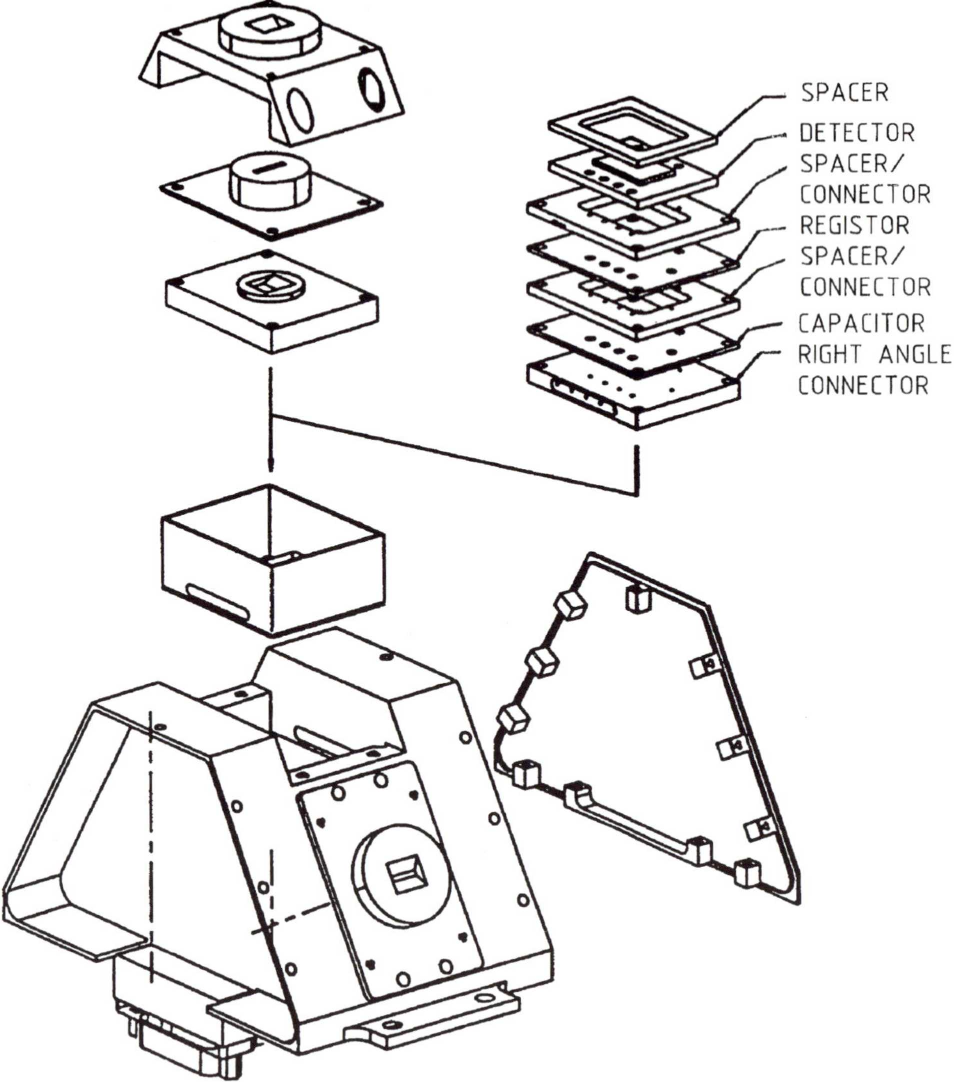

Figure 7. The IES sensor head in an exploded configuration showing the orientation of the three pin slits and the proximity of the microelectronics pre-amplifier board to the sensors. Boards contained in the stack actually contain the bias resistor and coupling capacitors required by the charge sensitive pre-amplifier/detector operation.

The SCVC provides for each particle coded information on the strip number and particle energy. This primary information is transferred to the DPU for further evaluation.

The SCVC is laid out for a total of 16 charge-sensitive pre-amplifier channels of which nine are used in this IES application. The circuit diagram of a single channel in the SCVC chip is shown in Figure 10 (the chip was developed and qualified by

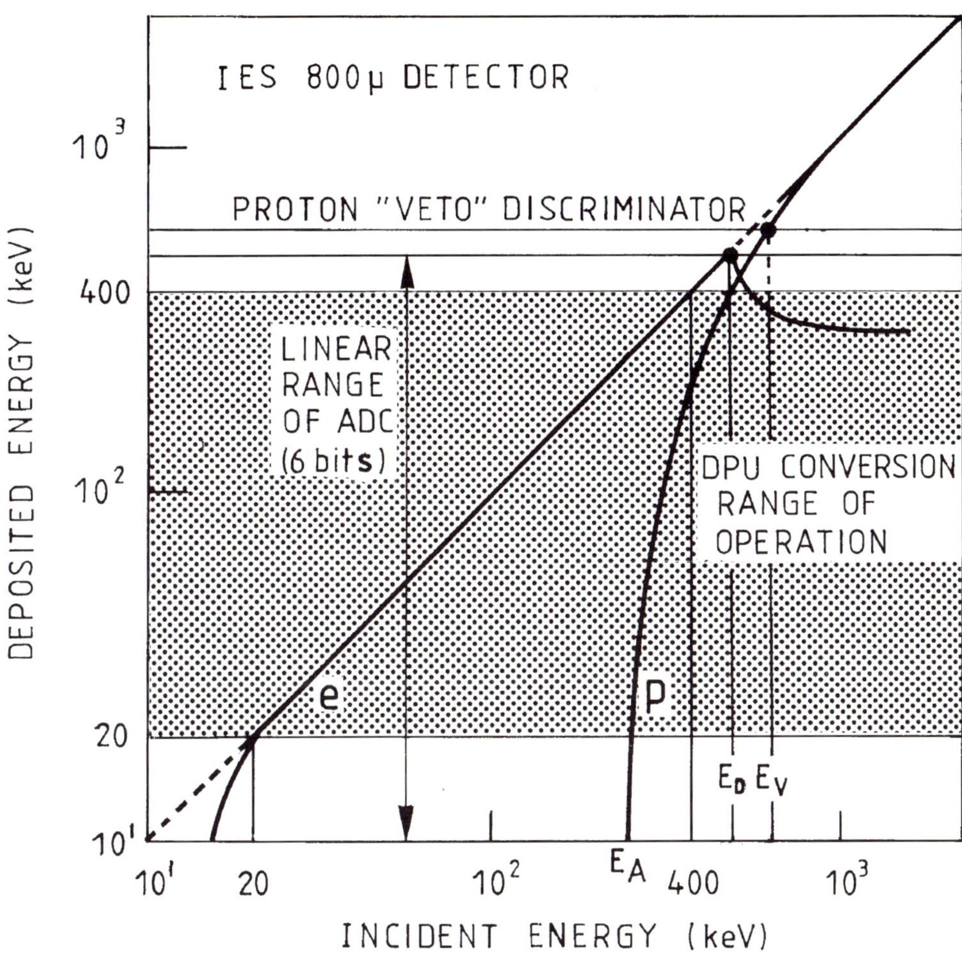

Figure 8. Energy range of the IES detector system. The shaded area indicates quasi-unique electron identification. An absorber in front of the detector transmits protons only for energies exceeding E_A. Electrons with energies above E_D and protons in the narrow energy and E_A onto E_V can confuse the electron measurement but the rather low intensity makes this background insignificant under all preactical conditions.

the Rutherford Appleton Laboratory, Oxford (UK)). Each pre-amplifier channel has typically a noise at zero input capacitance of 700 electrons r.m.s. equivalent to 6 keV Full-Width-at-Half-Maximum (FWHM). The total power consumed by the chip is about 10 mW.

According to Figure 10, the charge placed in the solid-state detector by an incident particle is stored on a capacitor (e.g., C2) following the pre-amplifier. This stored charge is compared to a background value stored in a companion capacitor (e.g., C1). The difference between these two values of charge is then strobed out and fed into a comparator for further evaluation (see Section 2.2.2).

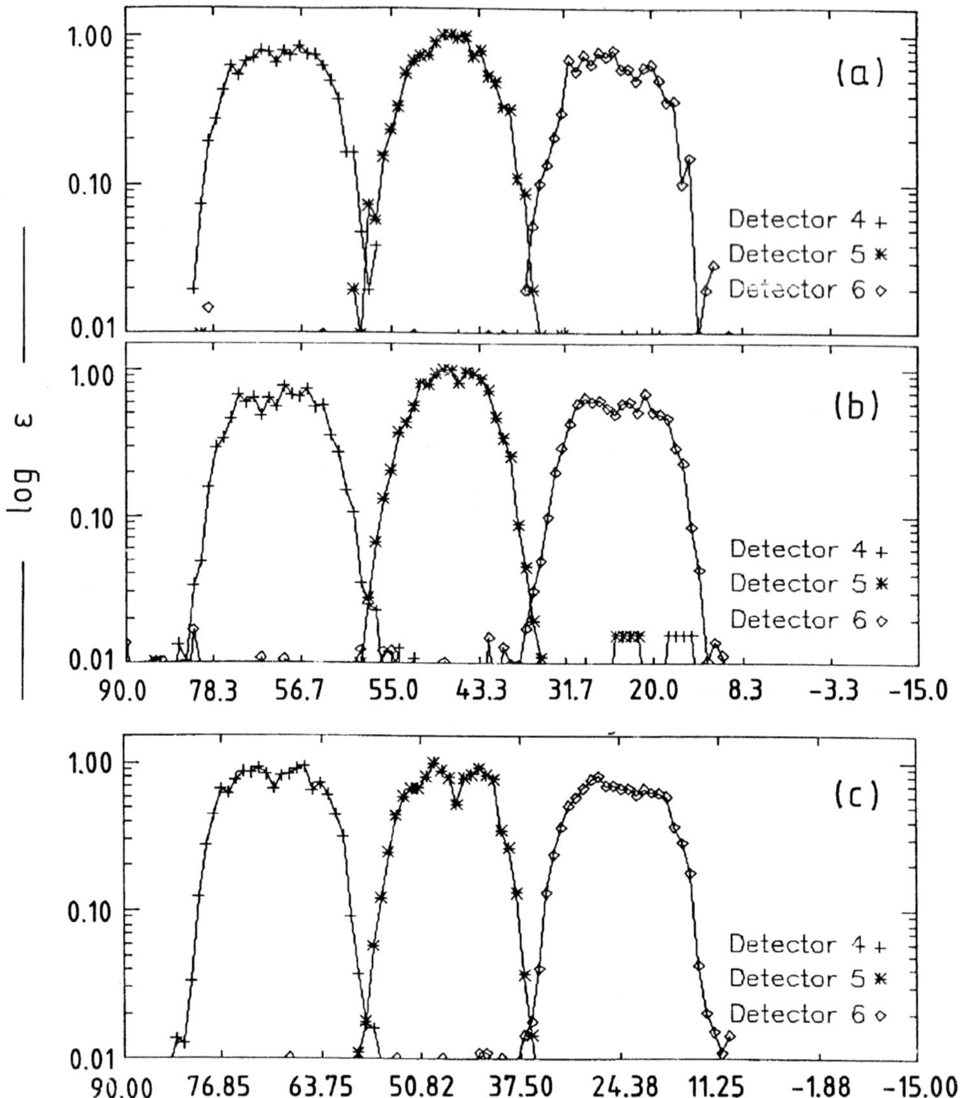

Figure 9. Angular response of an IES head with three detector strips. The panels (a), (b), and (c) show angular scans for 65, 90, and 114 keV electrons.

The SCVC chip operates the nine detector strips in a parallel/sequential fashion:

(1) At the time $t = 0$ all nine detector/charge sensitive pre-amplifier systems are activated for an integration time t_{int} (t_{int} can be set to 2 μs, 10 μs, 50 μs, and 100 μs).

(2) After closing the integration window, the nine channels are strobed-out in 47 μs (a 1–2 μs deadtime between the integration interval and the beginning of the read-out process is insignificant).

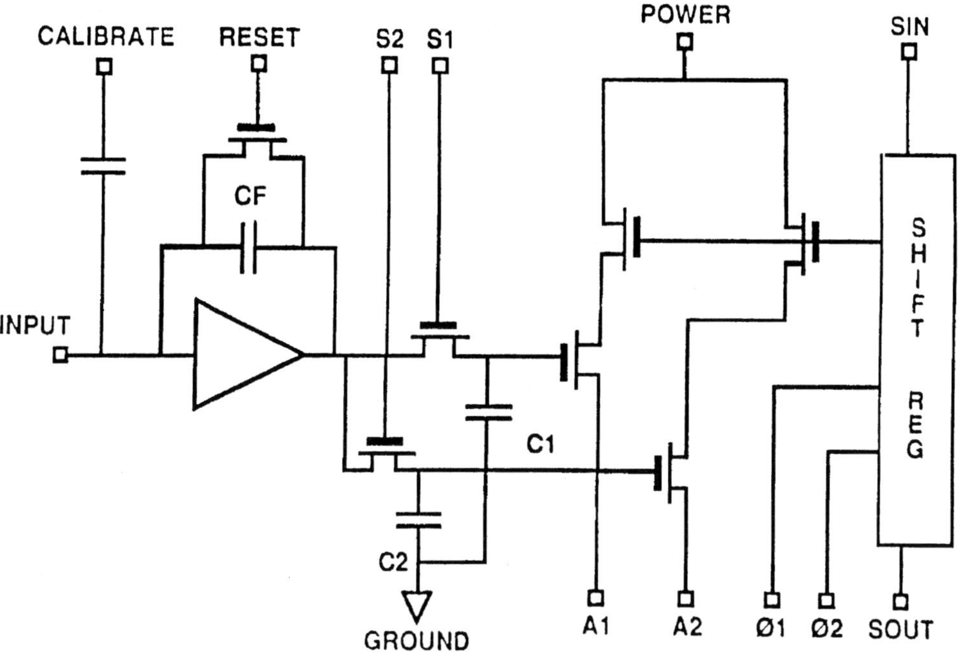

Figure 10. Schematic of the Switched Charge/Voltage Converter (SCVC) showing one of 16 identical channels. The switches S1 and S2 control the charging of the capacitors C1 and C2.

As a result the duty cycle of the system varies from 4% ($t_{\text{int}} = 2\ \mu$s) to 68% (for $t_{\text{int}} = 100\ \mu$s); the long integration times are particularly useful for low electron fluxes. Particle pile-up effects in the SCVC present a serious problem and, therefore, the actual counting rates have to be monitored carefully and the integration time t_{int} has to be adjusted according to the flux level. IES has the built-in feature to switch automatically to a shorter integration time t_{int} if the counting rate exceeds a certain limit. The limit value is set such that pile-up has a rather low probability.

The SCVC design principle leads to the existence of residual charge values at the entrance of the comparator. These offsets or 'pedestal' values are different for each channel but they are stable over time and temperature; the analysis of digitised charge values must correctly handle these pedestals. An example for the variation of pedestal values is given in Figure 11 for the flight unit RAPID-5: the pedestal positions for the nine detector strips, measured with the 8-bit ADC in the instrument, varies in the extreme by as much as a factor of 10 to 15. The impact of the pedestal value on instrument performance and data interpretation is discussed in Section 2.2.2.

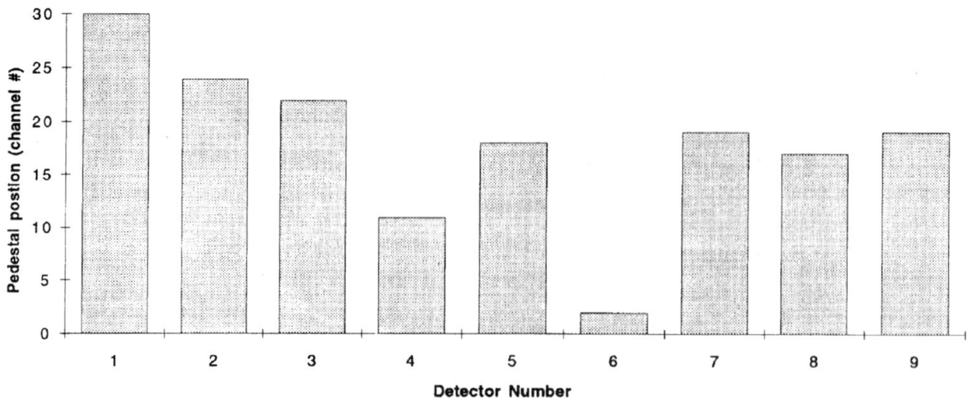

Figure 11. The mean pedestal positions of the 9 IES detector strips in the RAPID unit F5. The positions are given as channel numbers out of total of 256 CH.

2.2. Signal-conditioning units (SCU)

The general lay-out of the RAPID instrument in Figure 2 shows that either of the two sensor systems is followed by dedicated circuitries called the Signal Conditioning Unit (SCU). The primary task of the SCU is to provide proper analogue amplification and signal shaping, event-definition logic, control functions for configuring the detector system and to interface with the Digital Processing Unit or DPU.

2.2.1. *The IIMS Signal-Conditioning Unit*

Figure 12(a) is a simplified representation of the IIMS Signal Conditioning Unit (SCU) which is intended to show essential components and their role in the complex signal generation and signal processing on different levels. An important task of the SCU is to ensure that signals generated in the IIMS sensor system are indeed caused by a single incident particle. Within limits the SCU detects and excludes events which involve two or more particles arriving at the sensor system within a defined time window. The following is a description of major electronic SCU components and the sequential signal-processing on different levels:

Level 0

A particle passing through a SCENIC head Sy is analysed with respect to its flight time T (or equivalently to its velocity), to its energy E and to its direction-of-incidence (DIR). Accordingly, the SCENIC head generates a set of signals (analogue and digital) in the energy, time and direction channels with the following definitions:

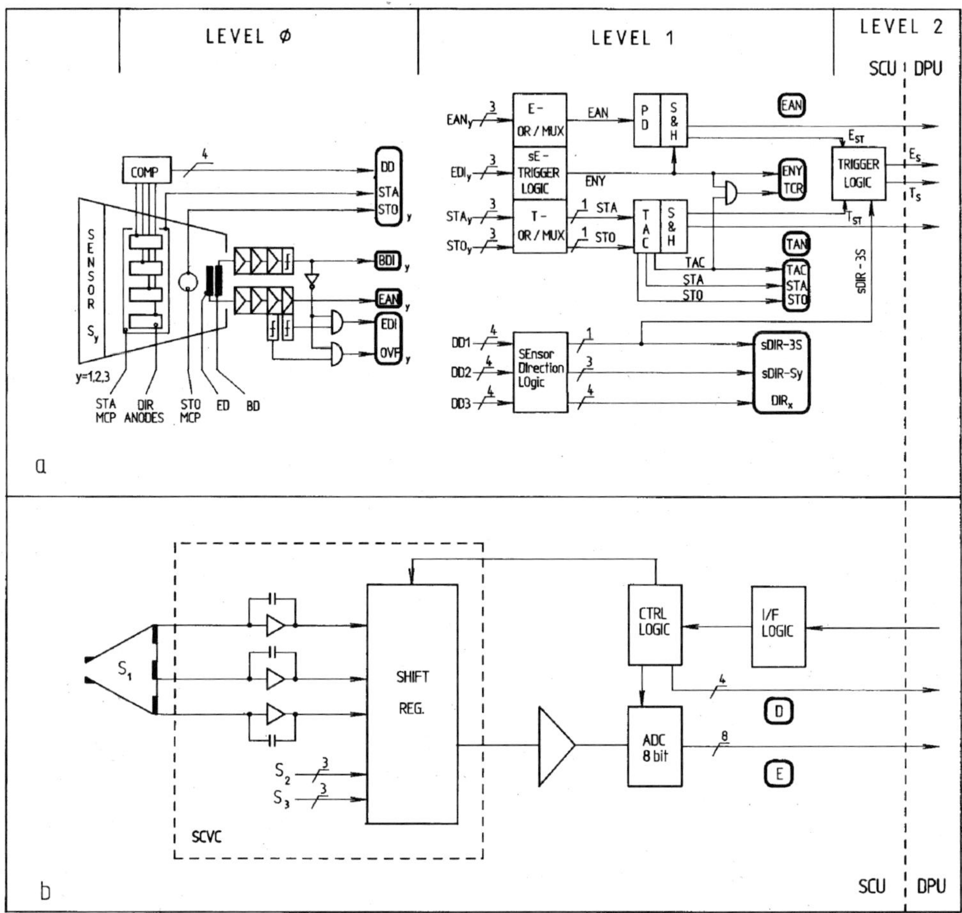

Figure 12. Simplified diagrams of (a) the IIMS and (b) IES signal conditioning units with emphasis on the signal and data formation (data are shown in heavy rounded rectangles).

CH	Signal	Definition
E	EAN	Analogue signal from the energy detector ED.
	EDI, OVF	Digital pulses from the lowest (A) and highest (C) thresholds in the ED amplifier chain. Overflow indicator.
	BDI	Digital pulse from the back detector BD, BDI inhibits EDI and OVF.
T	STA, STO	START and STOP signals.
DIR	DD	One-out-of-four direction signals.

The signal multiple (EAN, EDI, OVF, BDI, STA, STO, DD)y obtained from the head Sy represents an event at level 0. The subset (EAN, EDI, STA, STO,

DD)y is used for further processing in the SCU whereas the duple (BDI, OVF)y is transferred directly to scalers in the DPU.

Level 1

The signals (EAN, EDI, STA, STO, DD)y are offered to the next stage in the SCU for a first evaluation. As sketched in Figure 12 the analogue signals EANy and the time signals (STA, STO)y are connected to so-called OR/MUX devices for selection. The DPU can specify the sensor operation by setting the E and T multiplexer to the OR or MUX mode. In the OR mode signals are accepted on a first-come-first-served basis; in operational terms this mode is called Parallel Mode (PM). In the MUX mode sensors Sy are selected sequentially and only signals from a given sensor are accepted at any one time; in operational terms this mode is called Serial Mode (SM).

An analogue EANy signal in the E-channel is processed only if the digital EDIy meet some constraints. The EDIy signals are evaluated in the sE-TRIGGER LOGIC (sE stands for 'single EAN') to ensure that a single Sy sensor was active. The peak detector PD in the analogue path can operate on the EAN signal only if exactly one out of the three EDIy lines carries a pulse. If this condition is met sE-TRIGGER LOGIC creates an ENY pulse to indicate that the (EAN, EDI)y combination concurs with the above-mentioned requirements. In general, an EAN signal will not be accepted by the PD circuit if:

– its amplitude was below the lower threshold A (no EDI created);

– the particle energy was sufficiently high to stimulate the back detector BD and the resulting BDI signal disabled the EDI pulse;

– more than one EDI pulse was detected by the sE-TRIGGER LOGIC.

The T-OR/MUX circuit in the T-channel selects a (STA, STO) pair in a similar manner as the EAN signal is selected in the E-channel. The time-to-amplitude converter (TAC) transforms the pair into an analogue signal called TAN (T analogue) with an amplitude proportional to the observed flighttime T, and issues a digital TAC pulse to indicate the detection of a valid TAN signal.

In the DIR-channel a device marked SEDILO (Figure 12) converts the pulse pattern on the DDy lines (three times four lines) into signals and codes characterising the status and contents of the direction measurement:

(1) A sDIR-3S pulse is issued if only a single active x-direction was found in all three detector heads.

(2) The sDIR-Sy pulse defines the stimulated detector head Sy ($y = 1, 2, 3$).

(3) The DIR-x ($x = 1, 2, 3, 4$) code defines the four look-directions in Sy.

In summary the Level 1 processing leads to the following products: a new multiple of digital pulses (STA, STO, TAC, ENY) is established, reflecting that START and STOP pulses are selected and a valid TAN signal is produced; an EAN signal from a single detector is received and the sampling process in the Sample-and-Hold circuitry (S & H) is initiated. As indicated in Figure 12(a) ENY and TAC signals are combined to form a TCR pulse (functionally TCR is formed at this stage but the logic is actually located in the DPU). This pulse proves that valid analogue energy (EAN) and time (TAN) signals are present. However, it is important to note that the compliance of the (EAN, TAN) pair with the trigger condition specified in the TRIGGER LOGIC has yet to be demonstrated in the Level 2 processing phase.

The DIR-channel has found a single direction, a corresponding (sDIR-3S) signal is created and the direction is characterised by the duple (sDIR-Sy, DIR-x) with the ranges $y = 1, 2, 3$ and $x = 1, 2, 3, 4$.

As discussed above, a particle's direction of incidence within 180° polar angle is defined by a coarse direction Sy (one of the three 60° SCENIC heads Sy) and a fine direction DIR-x (division of Sy in four 15° intervals). The coarse direction Sy can be obtained from three different channels: From the E-channel (EDI-Sy), from the DIR-channel (sDIR-Sy) and from the multiplexer system (E-, T-, DIR-MUX) by selecting a sensor head Sy. The fine direction DIR-x, on the other hand, is extracted only from the DIR-channels. Of particular interest is the susceptibility of this system to particle pile-up in the DIR-channel and its dependence on the sensor mode (parallel or serial):

Serial Mode

The DPU activates a single sensor head Sy at any given time and cycles through the sensor heads Sy ($y = 1, 2, 3$) in a preprogrammed sequence. A unique coarse direction y is therefore imposed for all events. A fine direction DIR-x will be issued by SEDILO only if the DDy code from the DIR-channel shows unambiguous direction information. Invalid DDy codes (two or more lines show high levels due to multi-particle interaction or charge splitting between read-out modes) disable SEDILO and the triple (sDIR-3S, sDIR-Sy, DIR-x) will not be generated. However, the DPU accepts the remaining digital signals for accumulation in COUNTER ARRAY but the classification is restricted to MTRX data as will be discussed in Section 2.3.1.

Parallel Mode

All three sensor heads are active and particles are accepted on a first-come-first serve basis. The direction of incidence (Sy, DIR-x as defined in Figure 20) is obtained from measured quantities only. The unbiased sensitivity over 180° makes this mode attractive in low flux environments but this advantage is increasingly qualified by the susceptibility to multi-particle events if the flux exceeds a certain critical level.

The response to ambiguous directional measurements can be presented in a convenient form by using the definitions:

– DDu(2): DIR-detector in sensor head Su shows two active anodes due to multiparticle interaction or charge splitting; the DIR-detectors from the other heads are assumed to be inactive;

– 2DD: DIR-detectors from two different heads, Su and Sv, show valid single directions; the third detector is assumed to be inactive;

– EDI-y: single (s) or multiple (m) y-directions active, and by referring to the DPU description in Section 2.3.1:

Sensor		SEDILO			DPU	
EDI-y	DDy	sDIR-3S	sDIR-Sy	DIR-x	Class.	CTR. array
s, m	DDu(2)	–	–	–	I-MTRX	Yes
s	valid	Yes	$\neq$EDI-Sy	Yes	no	Yes
s, m	2DD	–	sDIR-Su;v	–	no	Yes

Level 2

The digitisation of the EAN and TAN amplitudes in the analogue-to-digital converters (actually located in the DPU) depends on precise coincidence conditions. The circuitry TRIGGER LOGIC imposes commandable trigger modes and time windows on the event pattern (EAN, TAN, sDIR) with sDIR-3S abbreviated to sDIR:

Trigger mode	Accepted event type
1. $E + (T\star$ sDIR$)$	$(E, T$, sDIR$)$, $(0, T$, sDIR$)$
2. $E + T$	(E, T), $(E, 0)$, $(0, T)$
3. $E \star T\star$ sDIR	$(E, T$, sDIR$)$
4. $E \star T$	(E, T)
5. E	(E, T), $(E, 0)$
6. T	(E, T), $(0, T)$

The above trigger conditions lead obviously to a digital filter function with effects on the overall detection efficiency of the spectrometer. The following is a qualitative assessment of the filter effect:

– The triple coincidence required in Modes 1 and 3 has generally a rather low rate of occurrence. This is even amplified by the relatively low probability of creating a sDIR-3S pulse. As a result little practical value will be put on these modes.

– The double coincidence in Mode 2 has a reasonable efficiency but the accepted event structures have clearly a mixed distribution.

– The double coincidence in Mode 4 is lower in occurrence rate than Mode 2, but it creates clean distributions with acceptable efficiency. This mode is therefore the preferred operational mode.

– Modes 5 and 6 are implemented merely for the case of drastic failures in either the energy or the time channel.

The E- and T-ADC in the DPU start the conversion in binary codes if the analogue pair (EAN, TAN) meets the coincidence requirement set in TRIGGER

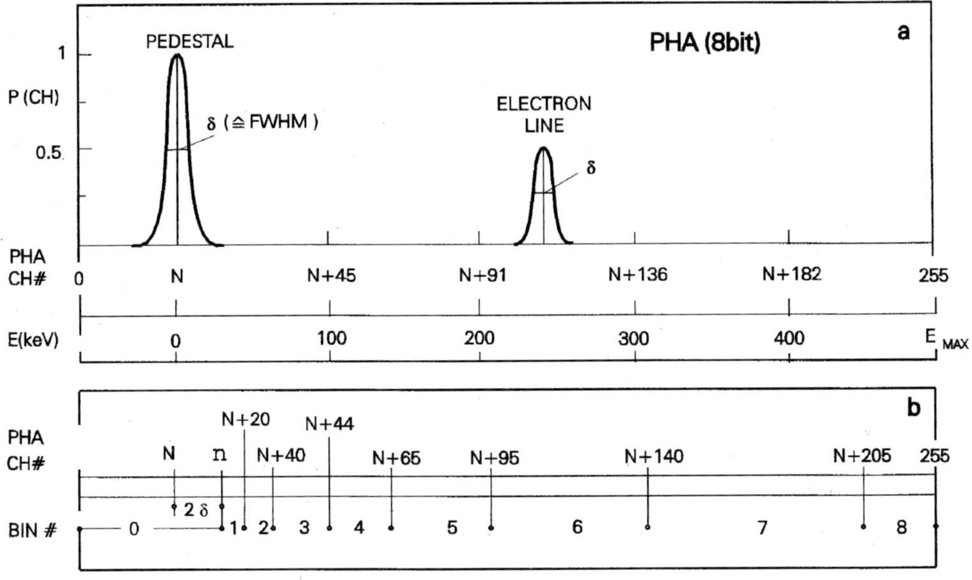

Figure 13. Principle of the pulse-height analysis for an IES detector strip/pre-amplifier channel. The pedestal distribution with the width δ reflects the variation of the minimum charge about its mean value at channel $= N$. If an incident electron deposits additional charge on the strip the pulse amplitude is found in channel $N + \Delta$ ('electron line'). The position of the pedestal ($CH = N$) corresponds therefore to the particle energy $E = 0$ keV. The lower part of the figure shows the prescription for mapping the 256 channels into 9 bins: a single pedestal bin ($b0$) and eight energy bins $b1$ to $b8$).

LOGIC and the event, now represented in an all-digital form, is ready for the classification process in the DPU.

2.2.2. *The IES Signal-Conditioning Unit*

Figure 12(b) deliniates the principal features of the IES signal conditioning unit (SCU). The initial amplifier stages for the nine energy channels and a multiplexer (comparator/shift register) are implemented in monolithic technology. This SCVC chip (see also Section 2.1.2) is physically integrated into the sensor housing. The second part of the SCU, designed with standard electronic components, accepts the serialised output signals from the chip for amplification in a single amplifier chain and subsequent digitisation in an 8-bit ADC. The resultant digitised signal represents the 'pulse height' or energy E of the incident particle.

More precisely, the measured signal amplitude corresponds either to the pedestal value (explained in Section 2.1.2) or to the pedestal-plus-energy value if a particle was detected. Figure 13 is an illustration of the peculiar effect of the pedestal on the electron energy range. The pulse amplitude offered to the 8-bit ADC is proportional to the charge found by the SCVC on the strip capacitor. However, there is always a minimum charge in the detector-pre-amplifier channel (pedestal) with a mean value, e.g., in ADC channel N, the fluctuations of this charge are characterised by the FWHM (full width at half maximum) value δ. If a particle-related additional charge was present, the signal amplitude is shifted to a higher value called 'electron line' in Figure 13; this line has the same width δ as the pedestal distribution.

Two consequences are related to the existence of the pedestal:

(1) The total number of counts per time interval R found in the pulse-height distribution (obtained with a pulse-height analyser (PHA) is a constant. The value R is equal to the SCVC sampling frequency, i.e., a count is either found in the pedestal (no particle present) or in the appropriate ADC channel (particle energy E).

(2) In energy space the lower limit $E = 0$ keV is defined by the PHA channel N; the upper limit of the energy range is $E_{\text{max}} = (255 - N)g$, with g denoting the system gain (nominally $g = 2.0$ keV CH^{-1}).

The conclusion from the above is that each single detector strip is associated with a specific energy range defined by its characteristic pedestal value N, a fact that has to be accounted for in the data analysis.

The output of the IES/SCU is a set of nine ADC values corresponding to the signal-plus-pedestal recorded since the previous readout. As indicated in Figure 12(b), each of the energy measurements is associated with a four-digit direction number D to form an (E, D) address pair for the information processing in the follow-on Electron Pre-Processor (EPP) in the DPU and accumulation by the microprocessor (see Section 2.3.2). This permits the DPU to handle the IES data at a sample time equivalent to strobing-out the SCVC channels.

As stressed in Section 2.1.2, the IES is a very compact sensor system. Because of this compactness, it has not been possible to monitor the system in the standard manner of electronic pulse stimulation through a charge terminator. Instead, the full system can be stimulated with a set of radioactive sources which produce a series of gamma- or X-ray lines in the 20 to 100 keV region. When a set of spectra is recorded from these sources, it is possible to verify the IES system performance very accurately, demonstrate the linearity of the amplifier gains, and observe the effect of the pedestal variation from channel-to-channel.

Figure 14 displays, as an example, the X-ray spectrum obtained from detector strip No. 2 in the RAPID flight unit F4 when a standard $_{63}\text{Eu}^{155}$ source was mounted on the IES head. The radioactive source emits a series of X-ray lines with energies at 18.9, 26.5, 45.3, 60.0, 86.5, and 105.3 keV. The distribution in Figure 14 shows clearly a very strong peak at the pedestal position (PO) and peaks (P1 to P4) for the four most intensive X-ray lines. A fitting routine with gaussians allows the extraction of the characteristic values of the peaks (position and width). The obtained channel numbers, when referenced to the pedestal position, correspond to the X-ray lines at 45.3, 60.0, 86.5, and 105.3 keV. The FWHM (full width at half maximum) values, which should be equal for all lines, reflect the noise in the selected detector strip No. 2 (the obvious variation in the FWHM values is probably due to the fitting procedure). The line positions allow a very accurate check of the system linearity and slope (equivalent to the gain factor g in keV CH^{-1})

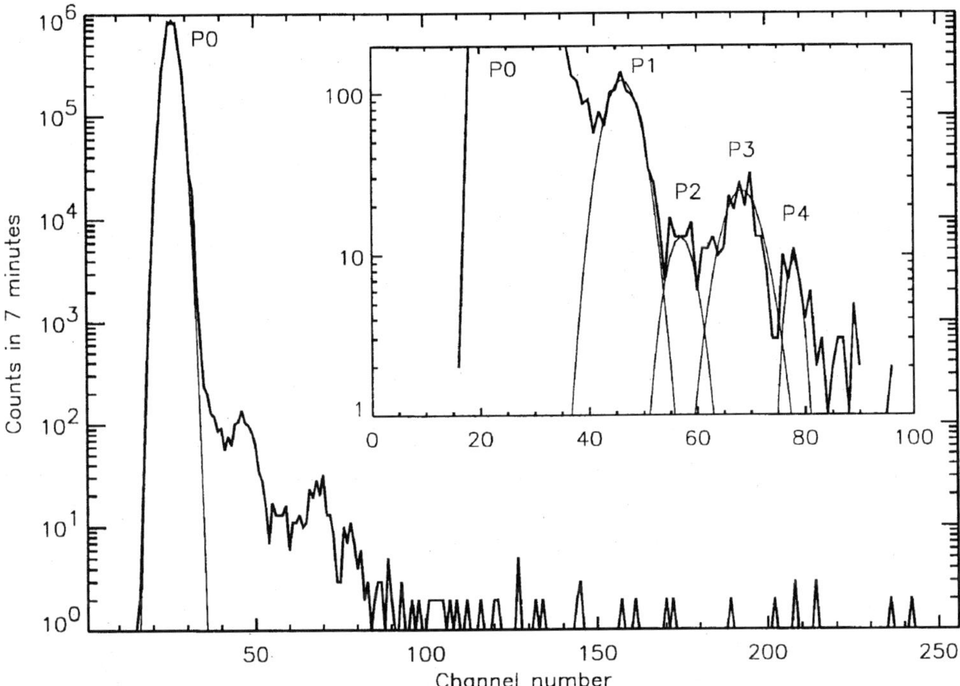

Figure 14. X-ray spectrum from an EU[155] radioactive source obtained with detector strip number 2 in the RAPID unit F4 (integration time 2 μs). The peaks P0 through P4 correspond to the pedestal and four X-ray lines at 45 keV, 60 keV, 80 keV, and 105 keV.

as demonstrated in Figure 15. This test and verification concept with a radioactive source as an external stimuli has the obvious advantage of providing truly front-to-end information with no electronic interference. The shortfall, on the other hand, is that the source can only be used for ground tests, in-orbit IES has rather limited test capabilities.

2.3. The Digital Processing Unit (DPU)

The internal digital processing unit (RAPID-DPU) serves the SCUs and sensor systems (IIMS and IES), evaluates and compresses the primary event data rate to a level which is compatible with the telemetry capacity and arranges the output data in the format of an Experiment Data Block (EDB). The present description of the DPU and its functions is a rather brief extract with strong emphasis on the data manipulation and the final construction of 'Science Data'.

The simplified DPU block diagram in Figure 16 amplifies the following key elements:

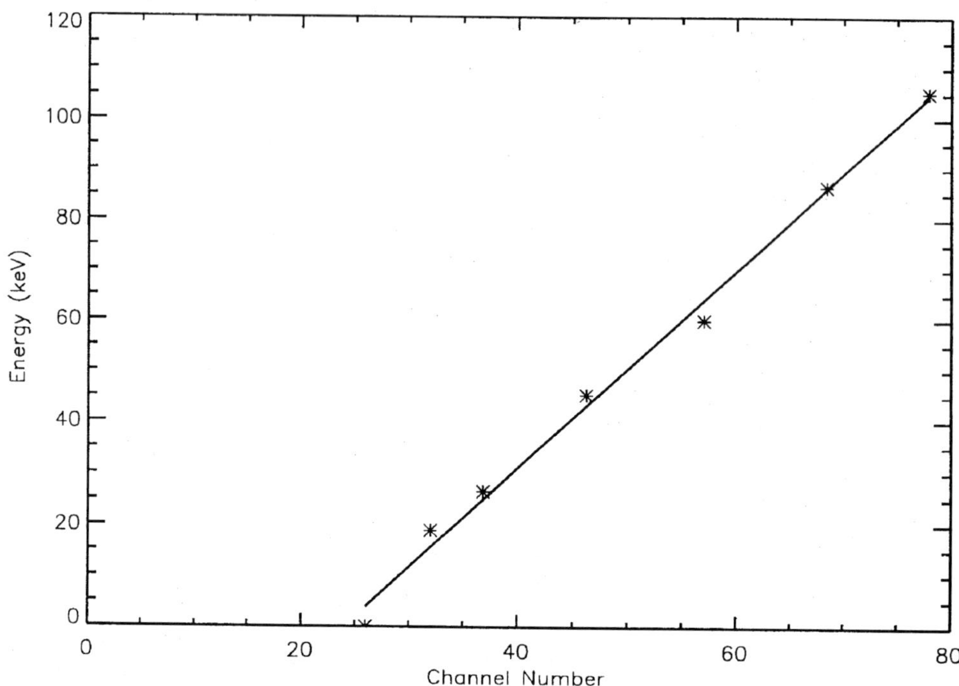

Figure 15. Energy versus channel plot produced from the pedestal and x-line positions in Figure 14. This technique monitors the channel gain (slope of the straight line) and linearity.

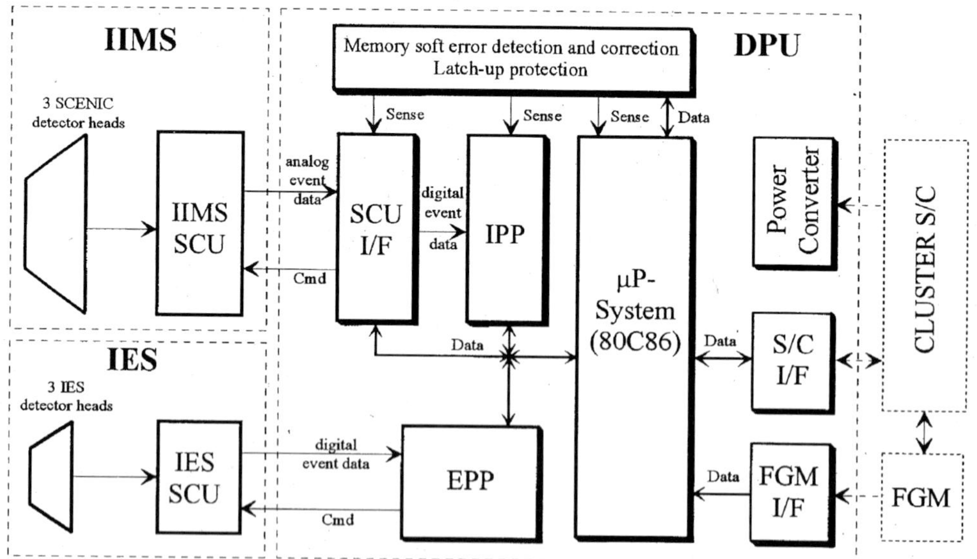

Figure 16. Block diagram of the RAPID digital processing unit (DPU) and the interfaces to the sensors, the spacecraft, and the magnetometer (FGM).

- Interface to the IIMS SCU.
- Ion Pre-Processor (IPP).
- Electron Pre-Processor (EPP).
- Microprocessor system (80C86-based).
- Memory protection and latch-up detection safeguard electronics.
- Interface to the Cluster spacecraft.
- Inter-experiment-link (IEL) to the magnetometer instrument (FGM).
- Low-voltage power converter.

2.3.1. *IIMS Event Processing*

A main task of the DPU is the compression of the enormous quantity of data received from the IIMS/SCU system. It was mentioned in the previous section that each fully defined particle event is described by two analogue signals (EAN, TAN) and a set of 18 digital pulse channels. These data are processed in the DPU and eventually transformed into ‘Science Data’ for nuclei. The various data types created in the SCU on the different levels of processing are transferred to the DPU as schematically shown in Figure 17. The multitude of IIMS signal channels, shown on the left side of Figure 17, is divided into three groups. The first group includes the 18 digital pulse channels. A subset (TAC, EDI-y, sDIR-Sy, DIR-x) is passed through a logic to expand the TAC and EDI-y pulses into channels with higher directional resolution. This process leads to pulse types of the form TAC-y, TAC-yx and EDI-yx. These pulses and the set (STA, STO, BDI-y, OVF-y, ENY, sDIR-3S) are offered to COUNTER ARRAY (Figure 17) to increment appropriate scalers.

The second group is essentially a duplication of direction relevant information from the DIR- and E-channels (sDIR-Sy, sDIR-3S, DIR-x, EDI-y) and the overflow indicator OVF-y. This subset is used for consistency checks in a precursing process before the analogue signals EAN and TAN in the third group are accepted for digitisation and classification. Events showing overflow in the energy channel ($E \geq$ 4 MeV) and/or inconsistent direction information in EDI-y and sDIR-Sy are discarded. For all other events the two ADCs can be enabled by the trigger signals Es and Ts. At the same time the DPU processes the event-related direction information (sDIR-Sy, DIR-x) and synchronises it with the digitised (E, T) pair to remove dynamic phase shifts. The DPU also converts the SCU direction code (Sy, DIR-x) into a serial number $D = 0, \ldots, 11$ which defines twelve unique directions within the 180° polar range of the instrument with the counting convention specified in Section 2.5.1. (Figure 20). In case the sDIR-3S pulse fails to indicate the presence of a high resolution directional measurement (positive identification of a single direction out of the 12 polar angular intervals) the DPU determines a coarse y direction from the EDI-y pattern and assignes to these the direction numbers D = 12, 13, 14 (inspect Figure 20 for the counting convention).

The event (E, T, D) is now prepared for the classification process in the follow-on ion pre-processor or IPP in Figure 18. The objective is to extract from the (E, T)

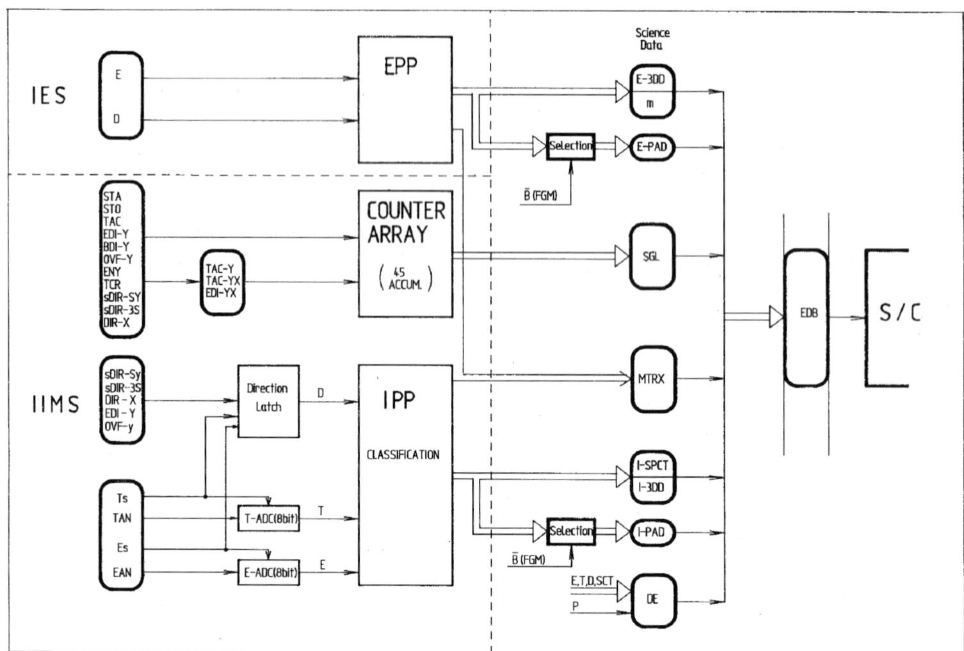

Figure 17. Schematic representation of the data processing in the DPU. Input data from the SCUs are shown on the left (heavy rounded rectangles). The sorting and classification processes eventually result in the Science Data which, in turn, are organised in Experiment Data Blocks (EDB) for transmission. The vertical dashed line separates the Micro-Processor supported data evaluation (right side) from the hardware logic and the pre-processing systems EPP (Electron Pre-Processor) and IPP (Ion Pre-Processor).

pair the particle mass A and the energy per mass ratio E/A with high precision. In addition the $A-(E/A)$ space is subdivided into a coarse bin field which is then combined with high-resolution direction information. This leads to a substantial reduction in the required data rate without the necessity to reduce resolution in time and direction.

The DPU initiates the classification process upon the appearance of at least one of the trigger signals Es and Ts (compare Figure 12(a)). The E/A ratio is established in a straightforward single table look-up technique, since this quantity is obtained from the measured flighttime T directly. The particle mass A, on the other hand, depends on energy E and flighttime T. A five-step successive approximation in an $E=f(T,A)$-table is applied to obtain the mass A. This process can handle a maximum event rate of roughly 50 000 s^{-1}.

The final product of the classification is the construction of high-resolution $(A, E/A)$ vectors which are used to address:

(1) a (32×64) matrix counter field and

(2) a bin-definition field.

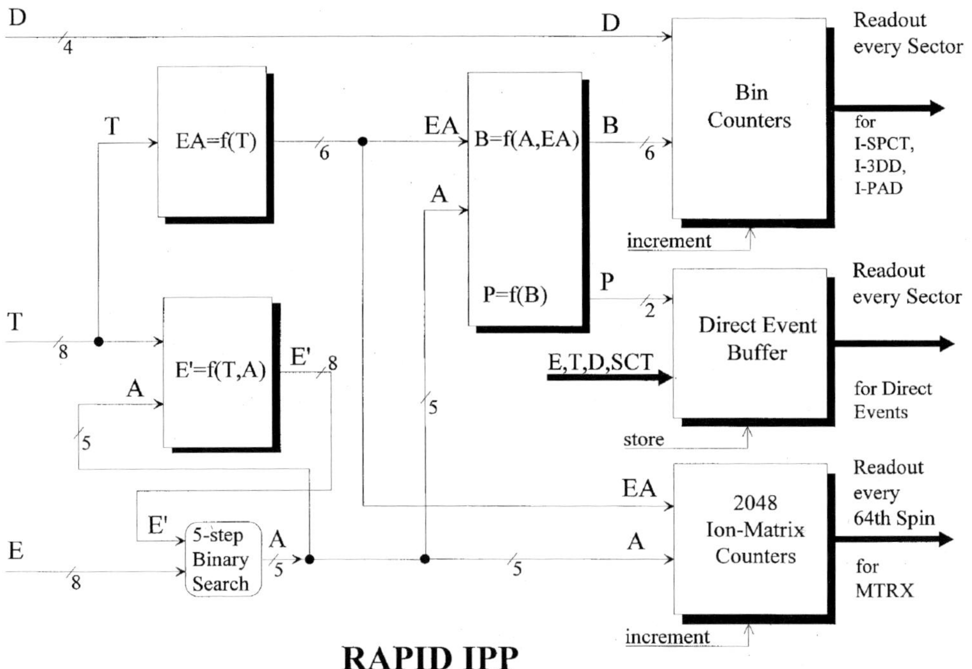

Figure 18. Schematic diagram of the ion pre-processor IPP showing the data flow in the classification process.

The selected matrix counters are incremented, the contents of this counter field represents the IIMS part of the 'Science Data' MTRX. The output of the bin-definition field is a bin number B which defines 35 bins in the $(A - (E/A))$ plane. The bin number, combined with the direction number D, addresses the bin counter array and the respective bin counter is incremented. The contents of the bin counter array is the basis for the 'Science Data' I-SPCT, I-PAD, and I-3DD described in Section 2.4. This process reduces the $(A - (E/A))$-matrix from a total of 2048 to 35 elements which cover the same area in $(A - (E/A))$ space with a coarser resolution.

The above description of the classification process refers essentially to (E, T, D) events with valid values for all three parameters. Events with missing parameters are processed according to the following scheme (missing parameters are shown as 0):

E	T	D	Classification
0	T	D	yes
E	0	D	no
E	T	0	MTRX only
0	T	0	MTRX only

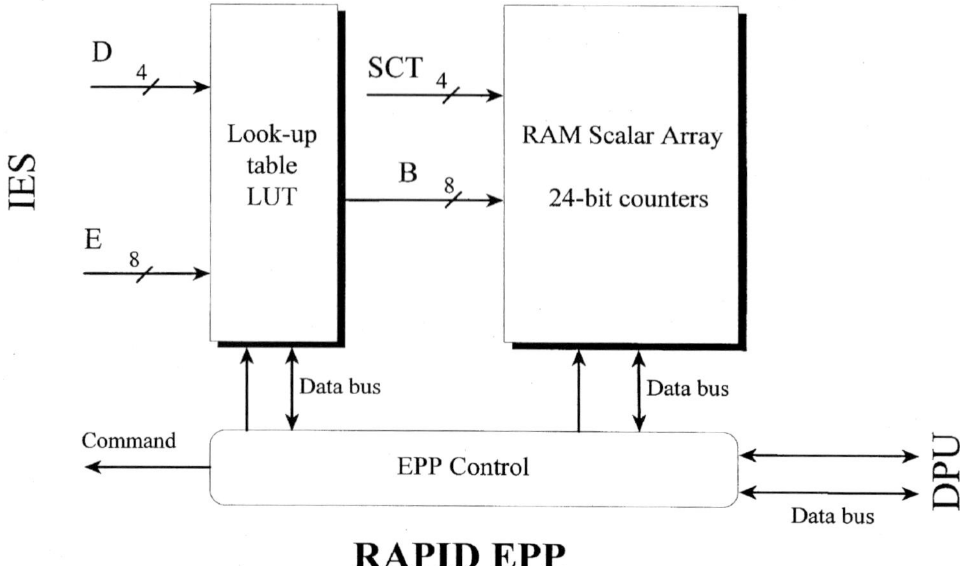

Figure 19. Schematic diagram of the electron pre-processor (EPP). Electron input data (energy E and direction number D) are passed through a sorting process to reduce the data volume.

2.3.2. *Electron Pre-Processor (EPP)*

According to Figure 17 signals from the IES /SCU are transferred to the Electron Pre-Processor (EPP) in the DPU. The EPP tasks are the provision of a serial command interface to IES and the pre-processing of IES event data. A simplified block diagram of the electron pre-processor is presented in Figure 19. A valid electron event at the EPP input is described by a digital signal duple (E, D), with E (8 bit) and D (4 bit) denoting the electron's energy and direction of incidence, respectively. The (E, D) pair serves as an input vector for a bin definition look-up table (LUT) which defines a bin number B (8 bit). The bin number B and the current sector number SCT (SCT defines 16 azimuthal sectors) are concatenated to form another vector pointing to a RAM SCALER field (a RAM-based counter array) and the selected counter is then incremented. The contents of the RAM SCALER serves as the basis for the generation of the electron Science Data E-3DD and E-PAD. These data types, with reduced binning in energy space (8 bins and 2 bins, respectively) form by far the most important data for routine operation in-orbit.

As mentioned earlier, the IES SCU uses a 256-channel pulse-height-analyser or PHA (the PHA consists of an 8-bit ADC and the associated RAM scaler field which stores the pulse-height distribution). The limited telemetry capacity requires information reduction by mapping the original 256 PHA channels to 8 energy bins (or 2 bins for E-PAD). The principle of the bin-mapping concept is illustrated in Figure 13(b) for E-3DD (8 bins). The bin boundaries b are referenced to the pedestal position ($CH = N$ corresponding to $E = 0$ keV) by using the algorithm $b = (N + ch)$, with ch = 20–205, etc., denoting predefined fixed channels as shown in Figure 13. The seven boundaries $b = (N + 20)$ to $b = (N + 205)$ define six energy bins ($b2$ to $b7$) with a quasi-logarithmical increase in width. These bins move obviously in 'channel space' but they are fixed in energy space (!) since they are referenced to the pedestal position N ($E = 0$ keV).

However, inspection of Figure 13(b) reveals:

(1) The width of the first bin ($b1$) is defined by the system noise δ: $\Delta(b1) = (20 - 2\delta)$. This follows from the prescription for the lower boundary n of $b1$: $n = (N + 2\delta)$.

(2) The width of the last bin ($b8$) is defined by the pedestal position N: $\Delta(b8) = (50 - N)$.

As illustrated in Figure 13(a), the channel boundaries can be recast in energy space by applying the mean value g = 2.0 keV channel^{-1} for the system gain factor (the variation of g for the nine detector strips is no more than $\pm 5\%$).

Actually the above mapping process creates a total of 9 bins in energy space:

– Bin 0 (between ch 0 and $ch\ n = (N + 2\delta)$ contains the vast majority of the pedestal counts plus counts of the energy spectrum between $E = 0$ keV and $E = (2\delta g)$ keV. The counts in this bin are transmitted in the housekeeping data (HK-data). Bin 0 monitors in essence the pedestal count rate and allows a judgement on the performance of the read-out system for the channel in question.

– Bin 1 through 8 form the 8 energy bins used in the Science Data for E-3DD (the two energy bins for E-PAD are generated in the same spirit), they are essentially free of noise counts from the pedestal distribution. An accurate knowledge of the pedestal position N and noise width δ is required for the channel-to-energy transformation.

The DPU constructs the bin boundaries in the 256-channel-to-8-bin mapping process from look-up tables (LUTs). This implies the following steps:

(1) The LUT for a given IES unit (with nine detector strips) contains pairs $(Nn, \delta n)$, with n = 1–9, describing the pedestals for the nine detector strips and the values of 7 fixed bin boundaries. From this, as delineated above, the DPU computes the 9 bin boundaries required for the description of the 'pedestal bin' bin 0 and the 8 energy bins (bin 1 to bin 8). Since the noise values δn are dependent on temperature T and integration time t_{int} ($t_{\text{int}} = 2, 10, 50, 100\ \mu$s) some fine tuning is necessary. Therefore a quadruple of LUTs (representing the four integration times t_{int}) is used for each of the two expected operational temperatures $T = -10$ °C and $T = +20$ °C. This amounts to a total of 8 LUTs for a given IES unit.

(2) To increase the operational flexibility each RAPID DPU stores the LUTs of all five flight units (including the spare unit), i.e., each DPU can choose from a total of 40 LUTs.

(3) The compressed format of the 40 LUTs (described in (1)) is stored in a ROM. During commissioning or at certain phases along the operational orbit a LUT quadruple which is appropriate for the IES unit of interest and the temperature T

is selected by telecommand (TCMD). This quadruple is then copied into the non-volantile RAM (NVRAM). At this stage some editing of the $(Nn, \delta n)$ values is possible. In a next step the LUT quadruple is decompressed to generate a quadruple with explicit bin boundaries stored in the NVRAM. With a final TCMD the LUT with the optimum t_{int} is selected from the quadruple, fully expanded, and written into the EPP memory (the selected rad-hard memory components have very low SEU susceptability; periodic reloading of the EPP memory limits the duration of corruption potential of IES measurements to fractions of an orbit). It is possible to combine the selected optimum choice LUT with a different value for the integration time t_{int} if this leads to better results.

The bin defining LUTs can be exchanged by telecommand thus allowing arbitrary schemes for binning the energy (E) and direction (D) ranges. A set of pre-defined LUTs is permanently available in the DPU. From either these resident LUTs or from up-linked new tables the DPU generates the electron Science Data (consult Table III for definitions):

(1) E-3DD (8 energy bins), E-PAD (2 energy bins) and the detector index m;
(2) Histogram mode (256 energy bins); this is not a routine mode.

Look-up tables are dedicated to either group (1) or group (2). This implies that, in contrast to IIMS, the two groups are mutually exclusive since only a single LUT is active at any given time. A set of about 16 different LUTs is required to cover the entire range in energy and direction for the high-resolution data in group (2).

2.3.3. *On-Board Pitch-Angle Computation*

The RAPID spectrometer is connected to the magnetic-field instrument FGM via the Inter-Experiment Link (IEL). FGM sends 64 uncorrected magnetic-field vectors $\mathbf{B} = (B_x, B_y, B_z)$ per spacecraft rotation ($T = 4$ s). Vector components are offered in digital form with a width of 12 bits each. The objective is to determine for each of the 16 azimuthal sectors which look direction in the IIMS and IES fan, respectively, is perpendicular to the $\mathbf{B}$-vector (the DPU uses the second $\mathbf{B}$-vector received in a given sector as the reference vector).

The DPU implements this 90° pitch-angle determination by applying the following algorithm in each sector: Unit vectors $\mathbf{D}_\nu$, ($\nu = 0, \ldots, 11$ for IIMS and $\nu = 0, \ldots, 8$ for IES) are introduced to describe the boresights of the detector look directions. A software routine calculates the 12 (9) vector products ($\mathbf{D}_\nu \cdot \mathbf{B}$) and determines the vector $\mathbf{D}_n$ for which the product assumes a minimum value, i.e., the direction $\mathbf{D}_n$ corresponds to 90° pitch angle. For each $\mathbf{D}_n$ the DPU assigns by means of a look up table two D_ν corresponding to look directions closest to the magnetic field direction (parallel and antiparallel). The number of events accumulated with these direction numbers form the I-PAD (E-PAD) Science Data.

2.4. The IIMS and IES Science Data

The final data products resulting from the DPU are called 'Science Data'. According to Figure 17 the main body of the Science Data contains ion data (I-SPCT, I-PAD, I-3DD), electron data (E-PAD, E-3DD) and MTRX data. Each one of these data types is obtained from the classification process in IPP and the bin sorting in EPP. Four additional types of Science Data are provided:

– SGL-Data (IIMS): the DPU samples the 45 accumulators in COUNTER ARRAY with specified frequencies and forms single parameter rates called SGL-Data.

– DE-Data (IIMS): a fraction of unprocessed socalled direct events (DE) is selected to bypass the classification for transmission to the ground. A DE event is characterised by the 3 parameters (E, T, D) denoting energy (8 bit), time-of-flight (8 bit), and direction (4 bit). The selection of DE events is based on a four-step priority P which is assigned to the bin number B (see Figure 18). The P assignment can be changed by telecommand; default is $P =$ const. for all bin numbers B of a given particle species. Priority $P \doteq 3$ refers to high-priority particles. With this definition priorities are assigned as follows:

Priority P	0	1	2	3
Species	e, p	He	CNO	Si-group

A maximum of 16 DEs per priority is accepted in each of the 16 azimuthal sectors on a first-come-first-served basis and written into a $(4 \times 16 \times 16)$ event buffer. Events are selected for addition to an EDB by applying the following prescription:

(a) At any given time the four least significant bits of the spin counter (INDEX) designate an azimuthal sector with the sector number SCT_n ($\mathrm{SCT}_n =$ INDEX MOD 16).

(b) The DPU starts the transmission in sector SCT_n by reading out the contents with decreasing priority $P\mu$ ($\mu = 3, 2, 1, 0$); the DPU sequences through the sectors until the number of events equals the maximum number S allowed per EDB. More precisely this can be written as:

$$\sum_{\nu=\mathrm{SCT}_n}^{\mathrm{SCT}_n+15} \sum_{\mu=3}^{0} P_\mu(\nu \mathrm{mod} 16) \leq S$$

with $P\mu$ designating the number of events per priority P and sector ($P\mu \leq 16$); the total number of DE events S per EDB is given in Table IV. At this stage a DE event is represented by ($E = 8$ bit, $D = 4$ bit, $\mathrm{SCT} = 4$ bit).

– Detector index m (IIMS and IES): defines for each sector the direction perpendicular to the magnetic field vector (m = 0–15). The directions in I-PAD and E-PAD are determined from tables with 'm' as the input parameter.

– **B**-vector polarity m-signs (IIMS and IES): indicates the polarity of the **B**-field in each sector.

Definitions of the Science Data together with a brief scientific description are given in Table III. It should be noted that RAPID/IIMS generates the same Science Data in energetic neutral atom (ENA) mode as in the ion mode.

The energy channels in Table III ($8d$ and $2w$) represent a contiguous coverage of the complete energy range for a given species. A precise position of the channel boundaries result from beam calibrations and will be compiled in the RAPID Data Analysis Reference Document (DARD) together with other calibration data.

The DPU samples the above data types and constructs an Experiment Data Block (EDB) as the basic unit for the data transmission to the ground. The EDB period $T = 4$ s is defined by the spacecraft telemetry system, however, the data structure varies with the telemetry mode. Tables IV(a) and IV(b) show the distribution of the Science Data in an EDB for the nominal (NM1 to NM3) and burst mode (BM1 and BM2) telemetry.

The principle difference between NM EDBs and BM EDBs is the higher sampling rate in the BM mode except for the electron data E-PAD and E-3DD. The former is omitted from the BM EDB because the pitch-angle distribution can be computed on the ground from the E-3DD distribution which covers the sphere with high angular resolution. The E-3DD data in NM mode are integrated over a spin whereas these data are subdevided in 16 assimuthal sectors in BM mode.

The socalled single rates SGL-0, SGL-1, SGL-2, and SGL-3 represent groups of counting channels from IIMS:

(a) One-dimensional parameter counters such as STOP rates and rates from the energy detectors.

(b) Two-dimensional parameters such as time-of-flight event rates (TAC), direction resolved rates from the energy detectors.

(c) Three-dimensional parameters such as time-of-flight/energy event rates (TCR) and direction-resolved TAC rates.

The frequency of transmission of a given SGL rate reflects the importance for the data analysis.

2.4.1. *Scratch Memory and Telemetry Mode BM3*

RAPID features a 64 kByte scratch memory which is organised as a ring buffer. It is used to collect BM1 mode EDBs as described in Table IV(b) in a continuous fashion. Upon the appearance of a trigger signal provided by the magnetic field instrument (FGM) via the IEL (Section 2.3.3) the instrument continous to collect data for one half of the ring buffer size. In this way the buffer captures high resolution data symmetric about the time at which the trigger pulse was issued. The total contens of the buffer corresponds to an observation time of approximately 2 min.

Table III
The RAPID science data

Data type	Particle species	No. of E-ch.	Polar intervals	Azimuthal sectors	Scientific description
H-SPCT	H	8 d	i	i	Proton energy spectrum
I-SPCT	He, CNO	8 d	i	i	He, CNO energy spectra
I-PAD	H	2 w	$3 \times 15°$	16	Selected pitch-angles (0°, 90°, 180°)
I-3DD	H, He, CNO	8 d	$12 \times 15°$	16	3D distribution
MTRX	all	all	i	i	$(A - (E/A))$ distribution
E-PAD	Electrons	2 w	$3 \times 20°$	16	Selected pitch-angles (0°, 90°, 180°)
E-3DD(NM)	Electrons	8 d	$9 \times 20°$	i	3D distribution, 9 pixels
E-3DD(BM)	Electrons	8 d	$9 \times 20°$	16	3D distribution, 144 pixels
m	all	–	–	16	Detector index
m-signs	all	–	–	16	**B**-vector polarity
DE	all	256	max $12 \times 15°$	16	High resolution direct events
SGL	na	na	na	variable	Digital pulse rates

d = narrow differential energy channels, w = wide-energy channels, i = integral

The BM1 EDBs read into the ring buffer are protected by check bytes appended to the EDB. Transmission to the ground is performed exclusively in telemetry mode BM3.

2.4.2. *Housekeeping Data and In-Flight Calibration*

Finally, it should be noted without details that SCU and DPU also provide valuable information in two other data fields which support the interpretation of the Science Data in a significant way. The first set comprises analogue and digital housekeeping data which reflect the actual operational configuration of the instrument and the health of all subsystems in an engineering sense. These data are transmitted in a dedicated telemetry channel. The second data set comes from the IIMS built-in precision pulse generator used to monitor and characterise the performance of the IIMS/SCU and, to some extent, of the DPU as well. The calibrator system operates in two different modes:

– The In-Flight Particle Simulator (IFPS): a calibrating (E, T) pulse pair is injected into the front-end electronics of the SCU once per spin and with a fixed phase. The DPU varies the pulse amplitudes by cycling through a pre-programmed sequence which simulates particle events such that an even coverage in $(A-(E/A))$ space is achieved. The IFPS is permanently active and cannot be switched off. A comparison of the simulated event pattern with the image returned to the DPU reveals irregularities in the analogue and digital signal processing. IFPS results are transmitted with full resolution in a dedicated area of the EDB (not included in Table IV) and contribute to science data I-SPCT, MTRX and SGL rates.

– The In-Flight Functional Test (IFFT): this routine must be initiated by telecommand; the instrument resumes the pre-test operational mode automatically after completing the test . The IFFT monitors threshold positions and amplifier linearity in the SCU. It should be noted that the transmission of science data is incomplete during execution of the IFFT test routine.

2.5. Flight Operations

2.5.1. *Coverage in Phase Space*

RAPID is mounted on the spacecraft platform such that the field-of-view covers a range of 180° in the polar direction of a spherical system defined by the satellite's spin vector and spin plane. The sensor systems, IIMS and IES, scan the full range of the solid angle as the satellite rotates (16 azimuthal sectors). The data returned are used to construct intensity distributions of distant particle sources on a spherical or plane image area with 192 pixels for IIMS and 144 pixels for IES. Figure 20 shows the orientation of the instrument in the spacecraft reference system, defines the IIMS and IES look directions, and indicates the convention for the direction numbers D.

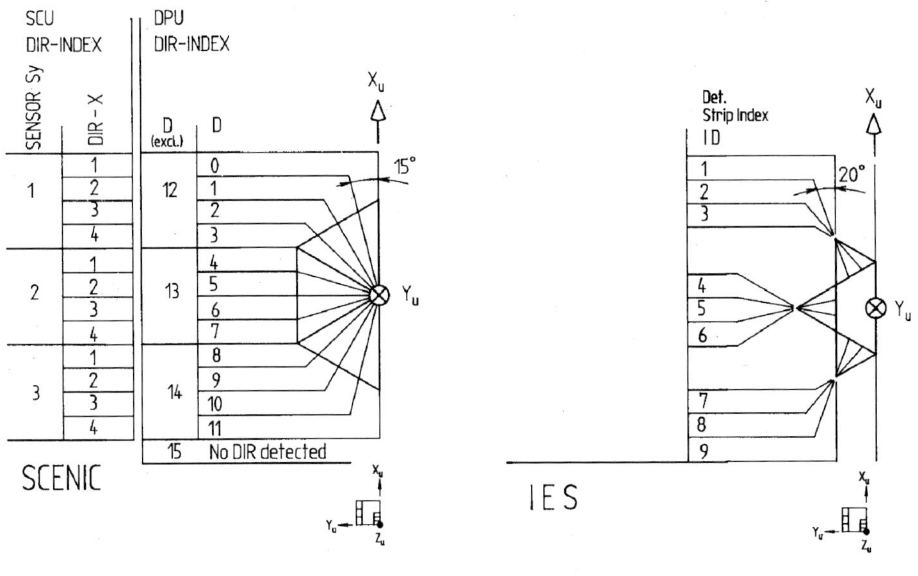

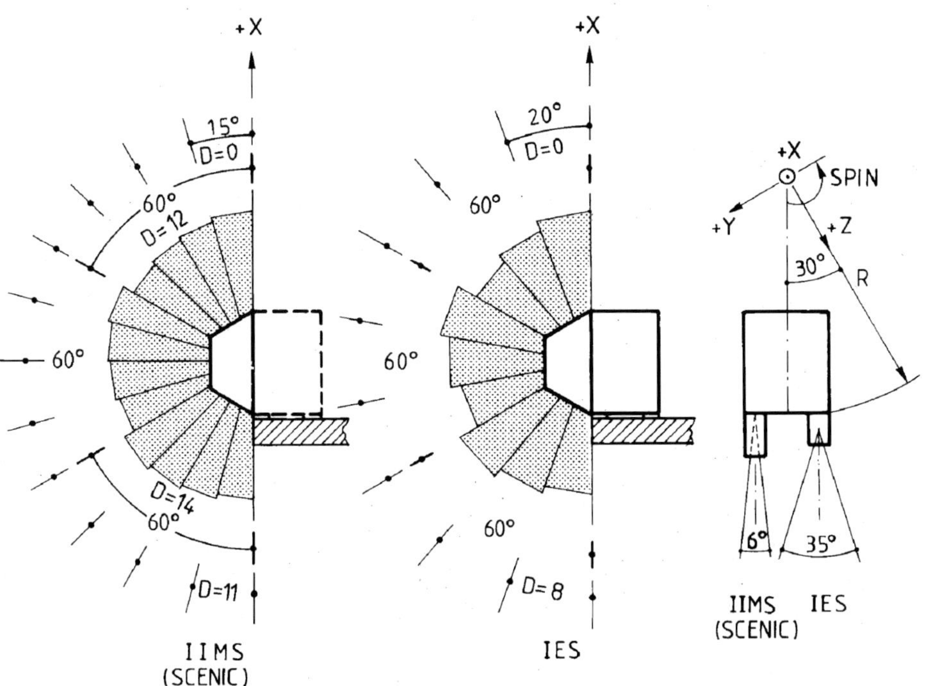

Figure 20. Orientation of RAPID on the Cluster spacecraft. The convention for the direction numbers D is shown for the two sensor systems.

Table IV(a)

Distribution of the science data in an EDB (1024.8 bit s^{-1}) (Telemetry modes: NM1, NM2, NM3, and BM2); Telemetry rate: 1024.8 bit s^{-1}

Science data	M–CH	E–CH	D–CH	Bytes per set	Bytes/ spin	Acc. period (spins)	Transfer time (spins)	Duty cycle
DE 20 ions, 3 bytes each	–	–	–	60	60	1	1	100%
SGL-0 (2 counters)	–	–	–	4	1	4	4	100%
SGL-1 (3 counters)	–	–	–	3	1	4	4	100%
SGL-2 (18 counters)	–	–	–	18	3	4	8	50%
SGL-3 (22 counters)	–	–	–	22	1	4	32	12.5%
H-SPCT	1	8	–	8	8	1	1	100%
I-SPCT	2	8	–	16	4	4	4	100%
I-PAD	1	2	3	6	96	$\frac{1}{16}$ [b]	1	100%
I-3DD	3	8	12	288	144	$\frac{1}{16}$ [b]	3 2	3%
I-MTRX	32	64	–	2048	8	64	256	25%
E-PAD	–	2	3	6	96	$\frac{1}{16}$ [b]	1	100%
E-3DD(NM)	–	8	9	72	72	1	1	100%
m	–	–	–	$\frac{1}{2}$	8	$\frac{1}{16}$	1	100%
m-signs	–	–	–	$\frac{1}{8}$	2	$\frac{1}{16}$	1	100%
Sync marker				3	3	–	1	
Subcommutation INDEX (EDB number)				1	1	–	1	
Content descriptors				2	2	–	1	
E/T-CAL				2	2	–	1	
				$\sum$	512			

[a] In units of the nominal spin period $T = 4$ s.
[b] 16 samples per spin ($16 \times \frac{1}{16}$).

2.5.2. *Spin Sectorisation and Spin Phase Offset*

The RAPID data are binned into 16 sectors, numbered 0–15, during the time of one rotation about the spin axis. The start of a spin is determined by the detection of a Sun pulse, plus an offset that allows the spin boundary (start of sector 0) to be located at any desired azimuth.

For comparison of spin-averaged data, it is desirable that all experiments start their spins simultaneously. This is considered more important than having all the spin boundaries pointing in the same azimuth. Thus spin-averaged data from RAPID, PEACE, and CIS will all be taken over the same time interval.

Bearing in mind various instrumental restrictions, it has been agreed by the Cluster teams to define a spin rotation as the time from one Sun-reference-pulse plus *offset*, to the next Sun-reference-pulse plus *offset*, where *offset* is defined as a rotation angle, with the same value for all instruments. The agreed value is

Table IV(b)
Distribution of the Science Data in an EDB (4620.92 bit s^{-1}) (Telemetry mode: BM1)

Science data	M–CH	E–CH	D–CH	Bytes per set	Bytes/ spin	Acc. period (spins)	Transfer time (spins)	Duty cycle
DE 106 ions, 3 bytes each	–	–	–	318	318	1	1	100%
SGL-1 (3 counters)	–	–	–	3	80	$\frac{1}{16}$	1	100%
SGL-2 (18 counters)	–	–	–	18	9	2	2	100%
SGL-3 (22 counters)	–	–	–	22	3	2	8	25%
H-SPCT	1	8	–	8	8	1	1	100%
I-SPCT	2	8	–	16	4	4	4	100%
I-PAD	1	2	3	6	96	$\frac{1}{16}$[b]	1	100%
I-3DD	3	8	12	288	576	$\frac{1}{16}$[b]	8	12.5%
I-MTRX	32	64	–	2048	32	64	64	100%
E-PAD	–	–	–	–	–	–	–	–
E-3DD(BM)	–	8	9	72	1152	$\frac{1}{16}$[a]	1	100%
m				$\frac{1}{2}$	8	$\frac{1}{16}$	1	100%
m-signs				$\frac{1}{8}$	2	$\frac{1}{16}$	1	100%
Sync marker				9	9	–	1	
Subcommutation INDEX (EDB number)				1	1	–	1	
Content descriptors				2	2	–	1	
E/T-CAL				2	2	–	1	
Spare				2	2	–	1	
				$\sum$	2304			

[a] In units of the nominal spin period $T = 4$ s.
[b] 16 samples per spin ($16 \times \frac{1}{16}$).

$$\text{offset} = 75/1024 \times 360^\circ \approx 26.367^\circ \,.$$

For RAPID and the Sun Sensor (which replies the Sun-reference-pulse), the relevant data for this calculation are:

	Location on spacecraft relative to $+Y$ axis	Desired angle to Sun at start of spin
Sun sensor	26.2°	26.367°
RAPID	60.0°	60.167°

From these data one calculates that the Sun is to appear at $360^\circ - 60.167^\circ = 299.833^\circ$, or in sector 13.326 (see Figure 21).

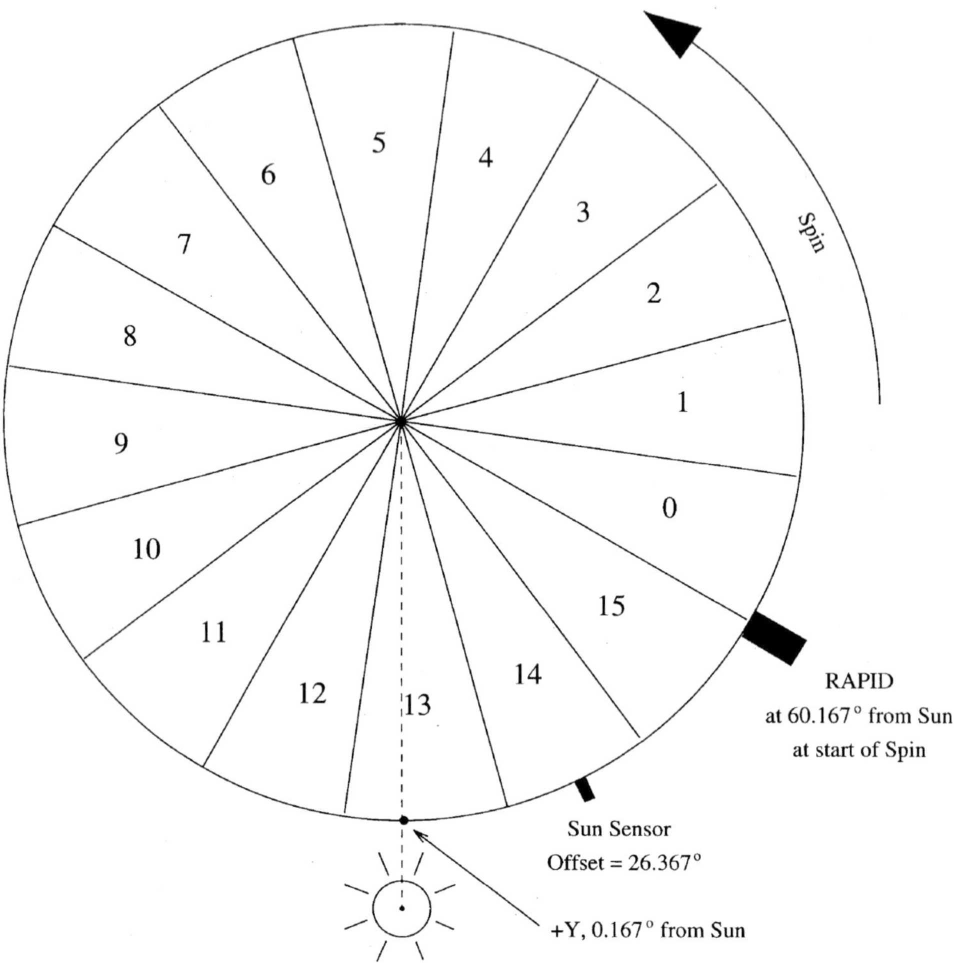

Figure 21. Definition of the RAPID sectors in the spin plan and the angular position of RAPID and the Sun Sensor relative to the Sun-satellite line at the start of a spin. Incidentally the spacecraft $+y$ axis only 0.167° away from the S/C-δm line at this instant in time.

For programming the DPU to accomplish this, we require the sector in which RAPID is looking when the Sun pulse occurs. This is the sector corresponding to the angle $360° - 26.367°$ which is sector 14 plus $\frac{212}{256}$. This result corresponds exactly to the precise definition of *offset*.

The azimuth of the centre of sector n is then given by

$$71.417° + 22.5° n \, .$$

The azimuth is the angle of rotation from the Sun counterclockwise when viewed looking down the spin axis (from the north). Figure 21 illustrates the position of the Sun sensor and RAPID at the time when a spin starts. The sector pattern then follows.

It should be noted that the above formula applies only if the plane in which the Sun is sensed is located 26.2° from the $+Y$ axis. In the spin and attitude information given in the auxiliary files, the exact location of this plane is specified, including alignment and orientation errors. Any such deviation from the nominal value will not affect the time of the spin boundary, but will alter the centre positions of the sectors, in that their values will be *reduced* by the amount that the effective Sun sensor location is *increased*.

2.5.3. *Operational Modes*

The two RAPID spectrometers IIMS and IES, are to a large degree independent and each system can be put into specific 'configuration modes' by telecommand. Combinations of IIMS and IES configuration modes are called 'operational modes'. With just a few exceptions all 'operational modes' can be used in both telemetry modes (Science Nominal Modes NM-1 = NM-2 = NM-3 = 1024.80 bit s^{-1} and Science Burst Mode BM-1 = 4620.92 bit s^{-1}; BM-2 and BM-3 are insignificant for operational purposes).

2.5.4. *Configuration Modes*

IIMS and IES are largely independent subsystems of RAPID with different internal set-up structures called Configuration Modes (CM). Configuration modes (emergency modes are not discussed) are conveniently represented by Mode Matrices(AB) and (LMN) for IIMS and IES, respectively.

2.5.4.1. *IIMS Configuration Modes (CM)*

The IIMS CM Matrix (AB) describes two independent functional levels in the internal control system of the instrument. The parameter A defines sensor/DPU settings and the parameter B defines HV conditions in the instrument:

A	B			
0 = OFF	0 = HV OFF	(relay OFF)		
$1 = S$	1 = HV ON	STA = 0V	STO = 0 V	DEF = 0 V
$2 = P$	2 = HV ON	STA = r	STO = s	DEF = t
3 = SWG	3 = HV ON	STA = R	STO = S	DEF = T
4 = IFFT ON	4 = HV ON	STA = R	STO = S	DEF = 0 V
5 = DPU test	5 = HV ON	STA = R	STO = S	DEF = 10 kV

Explanations for the functions A: Serial (S), parallel (P) and arbitrary switching (SWG) operation of the three IIMS heads; IFFT: In-flight functional tests generator; DPU: Memory dump.

Explanations for the functions B: HV, High-voltage relay; STA, STO, DEF, high-voltage supplies for Start MCP, Stop MCP and deflection voltage DEFL; r, s, t, arbitrary step numbers (1 to 16) for STA, STO, and DEF; R, S, T, fixed step numbers for STA, STO, and DEF defined by pre-launch tests.

From these definitions follows the IIMS mode matrix $\mathbf{AB}$:

A \ B	0	1	2	3	4	5
0	00	–	–	–	–	–
1	10	11	12	13	14	15
2	20	21	22	23	24	25
3	30	31	32	33	34	35
4	40	41	42	43	44	45
5	50	51	52	53	54	55

The CM matrix $\mathbf{AB}$ describes the multitude of IIMS modes available in flight operations.

2.5.4.2. *IES Configuration Modes (LMN)*

The operation of the IES subsystem requires three parameters L, M, and N which refer to three independent functional levels:

L: In-flight calibration.

A testpulse with constant amplitude is fed into the SCVC multiplexer.

This function can be activated in parallel to the current operational mode of the instrument.

M: Energy binning parameter.

This parameter transforms the 8-bit primary accuracy of energy signals into an output signal with lower or equal accuracy. The transformations are performed by look-up tables (LUTs); different LUTs are used for different instrument temperatures:

(1) LUT-1 Temperature $T = +20$ °C, 256 to 9 binning (8 energy bins plus 1 pedestal bin).

(2) LUT-2 Temperature $T = -10$ °C, 256 to 9 binning (8 energy bins plus 1 pedestal bin).

(3) LUT-3 Temperature $T =$ any, 1 to 1 binning (8-bit accuracy, socalled histogram mode).

This LUT is called if full energy resolution is required.

N: Integration time parameter.

This parameter defines the integration time t_{int} (2, 10, 50, 100 μs) in the detector read-out system. Selection of an integration time t_{int} depends on the ambient particle flux.

With these definitions the parameters M and N describe the IES status (the parameter L can be set to 0 or 1):

M	N	IES status
0	0	OFF
1 LUT1 (+20 °C)	1 $t = 2\ \mu$s	ON
2 LUT2 (– 10°C)	2 $t = 10\ \mu$s	ON
3 LUT3	3 $t = 50\ \mu$s	ON
4	4 $t = 100\ \mu$s	ON
5	5	MDM

MDM = Memory Dump Mode.

The resulting IES Configuration Matrix **LMN** is ($L = 0$ or 1):

M N	0	1	2	3	4	5
0	00	–	–	–	–	–
1	–	11	12	13	14	–
2	–	21	22	23	24	–
3	–	31	32	33	34	–
4	–	–	–	–	–	–
5	–	51	52	53	54	55

2.5.5. *RAPID Operational Modes (OP Modes)*

RAPID operational modes are constructed from IIMS and IES configuration modes. Operational Modes are coded OP (AB.LMN):

Mode	AB (IIMS)	LMN (IES)
POWER ON (default mode)	10	014
Stand-by	10	0MN
Hot stand-by	11	0MN
Nominal operation		
Low flux mode	24	014 or 024
High flux mode	14	011 or 021
In-flight Calibration	4B (64 s)	1MN
ENA	15 or 25	LMN

The Energetic Neutral Atoms (ENA) mode is a special IIMS mode in which the deflection voltage DEF in the collimator is set to 10 kV. With this setting all ions with $E \leq 200$ keV are excluded from the detection system, i.e., detected particles with energies between 75 keV and 200 keV are considered neutral atoms.

The detection of ENAs requires the RAPID unit on only one spacecraft to be configured in the ENA mode (OP 15.LMN or 25.LMN). Orbital segments in the lobes/polar cap are ideal for this purpose (low background from ions).

2.6. Routine In-Orbit Operations

The routine operations of RAPID are largely driven by the ambient particle flux (i.e., the region in geospace). Two selected categories of orbit may serve to illustrate anticipated sequences of operational modes (OP) along a trajectory:

(1) Apogee in the magnetotail inside the magnetopause

Region in geospace	OP	Mode
Plasma sheet	14.011 or 14.021	nominal high flux
Magnetopause (skimming)	14.011 or 14.021	nominal high flux
Lobe/polar cap	24.014 or 24.024	nominal low flux
Cusp	24.014 or 24.024	nominal low flux
Inner magnetosphere (inside $L = 5$)	11.011 or 11.021	hot stand-by (IIMS)

(2) Apogee in the solar wind (outside the bow shock)

Region in geospace	OP	Mode
Solar wind	24.014 or 24.024	nominal low flux
Bow shock	24.014 or 24.024	nominal low flux
Magnetosheath	24.014 or 24.024	nominal low flux
Magnetopause (crossing)	14.011 or 14.021	nominal high flux
Cusp	24.014 or 24.024	nominal low flux
Polar cap/lobe	24.014 or 24.024	nominal low flux
Inner Magnetosphere (inside $L = 5$)	11.011 or 11.021	hot stand-by (IIMS)

Transitions between Operational Modes require predefined groups of telecommands (socalled procedures Pn). A major concern in the RAPID commanding is to take proper account of the high voltage switching involved in a chosen sequence of telecommands (e.g., HV relay position, HV limit value, HV target value, etc.). The 'road map' in Figure 22 provides an overview for transitions between 'adjacent' operational modes (major modes are highlighted by heavy lines) and safe routes to more 'distant' OPs.

2.7. Data handling

The Onboard Data Handling (OBDH) system of CLUSTER integrates RAPID data into the CLUSTER telemetry format and prepares the transmission to a ground station. After placing the CLUSTER data on a suitable storage medium (CD-ROMs) the complete data set is made available to each PI institute and the national institutions participating in the CLUSTER Science Data System (CSDS). It is the task of the German CLUSTER Data Center (GCDC) to extract the RAPID data from the storage medium (CD-ROM) and convert selected subsets to physical units and make them, depending on the time resolution, accessible by the CLUSTER teams and the general scientific community, respectively. Spin-averaged time profiles of fluxes for electrons, protons, helium, and CNO group ions for two energy ranges each and field aligned anisotropies for electrons and protons from four satellites are provided for coordinated data analysis by CLUSTER Co-Investigators, while one-minute averages of the same parameters from a single spacecraft are made available to the general scientific community. For both data sets, publication may be undertaken only with the approval of the RAPID PI. These GSCS data products are complemented by the data analysis effort at the institutions participating in the RAPID investigation.

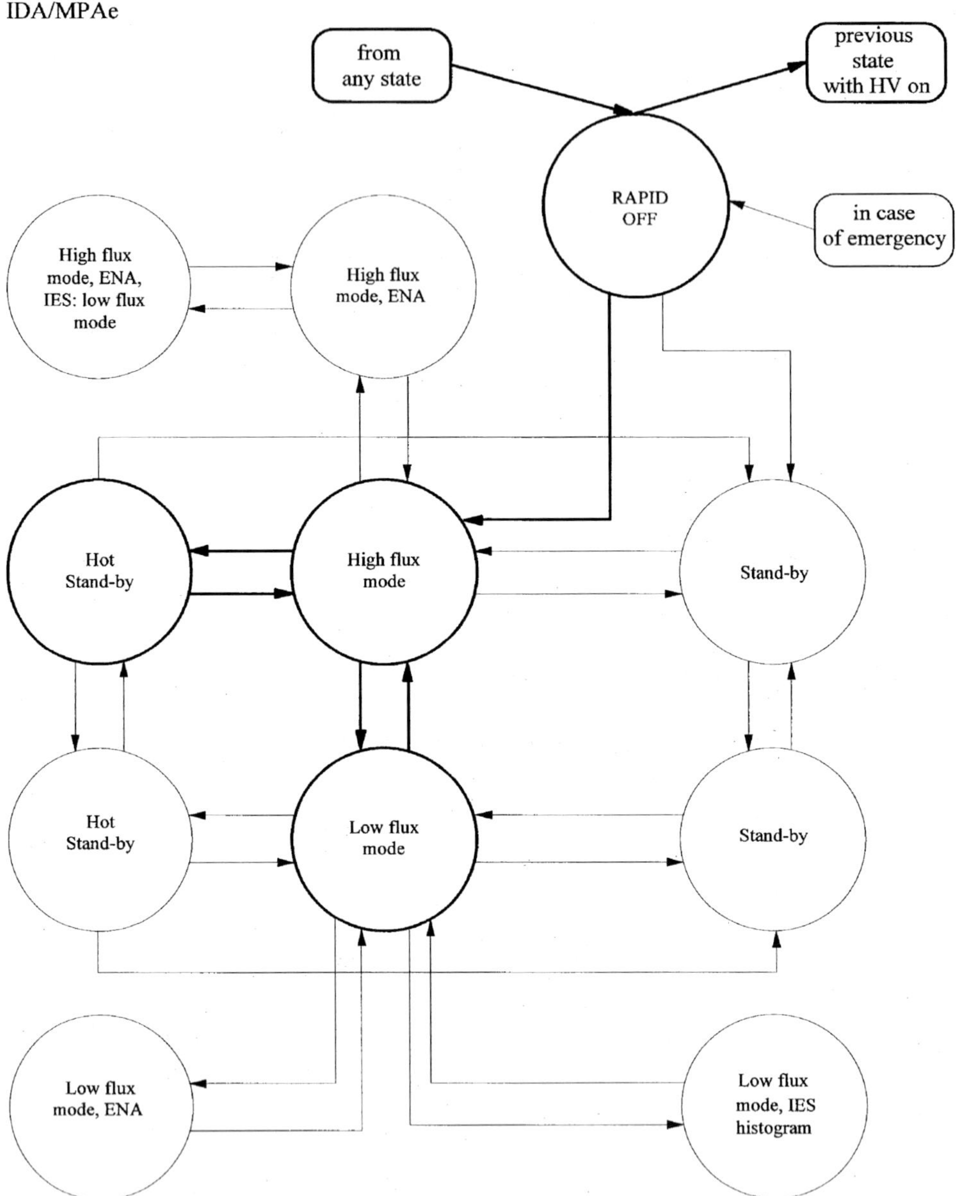

Figure 22. Road map for transitions between different configuration modes CM illustrating the principle. Major modes are connected by heavy lines.

CLUSTER passes through regions in geospace with vastly different field topologies and plasma properties. Reliable region identification is therefore essentual for a successful interpretation of the RAPID measurements. Identification of characteristic regions in the magnetosphere such as plasmasheet, lobes, cusp, and

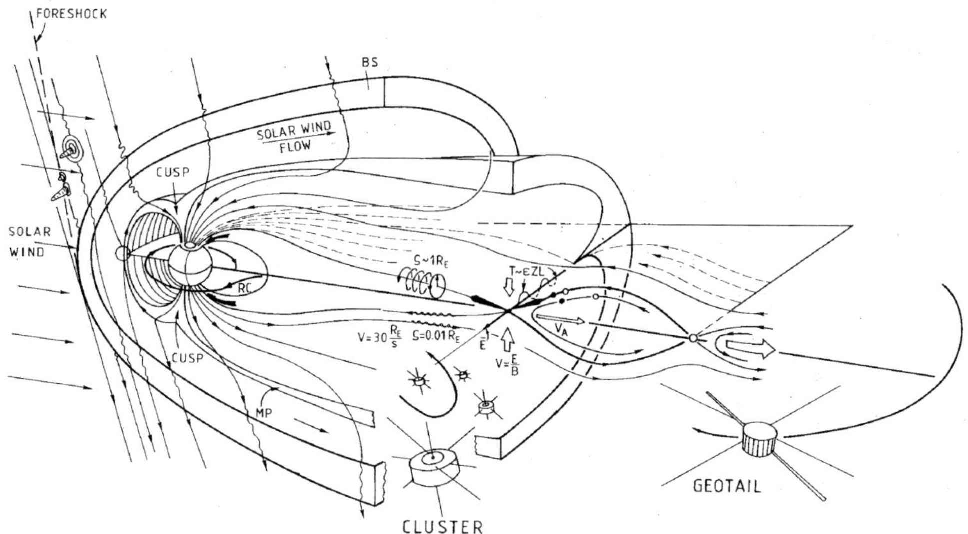

Figure 23. Three-dimensional 'tail-view' of the Earth's magnetosphere with the Cluster fleet of spacecraft and *Geotail*.

magnetopause make it mandatory to combine the RAPID data with magnetic field and plasma observations. An important contribution from RAPID is the coverage of the high energy tail in the particle distribution functions and the detection of boundaries by remote sensing techniques (e.g., deviations from gyrotropy).

3. Scientific Objectives

3.1. Introduction

The energetic-particle measurements with RAPID will be able to expand significantly our understanding of solar-terrestrial physical processes in most principal regions of the magnetosphere. These are in particular the: (1) middle magnetosphere, (2) low latitude boundary layer, (3) auroral zone extension to high altitude, (4) plasma sheet and its associated boundary layer, (5) plasma mantle/tail lobes, (6) magnetopause, (7) cusp/magnetosheath, and (8) bow shock region.

Figure 23 illustrates a tail-view of the magnetospheric system, showing most of the plasma regions and boundaries presently believed to exist in geospace. The microphysical processes occurring within these key regions – especially during the times of major dynamical reconfigurations – are areas in which RAPID will provide essential information. These studies are particularly dependent upon the four-spacecraft Cluster configuration.

The following topics are meant to highlight the scientific contributions that can be made with RAPID but they can by no means give a complete picture of the return expected from this instrument.

3.1.1. *Energetic Particles as Scientific Tools*

The energetic tail of the plasma distribution is often considered insignificant in plasma dynamics, though in many of the magnetospheric regions it can contribute significantly to the total plasma energy density. Furthermore, the high speed and substantial gyroradii of energetic particles can probe with high time resolution plasma properties and magnetic-field topologies over large distances. For example, 50 keV electrons travel along field lines with a speed of nearly 20 R_E s^{-1}. These swift electrons trace-out field lines in the magnetosphere and in the magnetotail in a rather short time. They can provide nearly instantaneous information about changes in the field configuration in distant regions of geospace.

Energetic ions, on the other hand, have large gyroradii which sense the existence or the approach of boundaries over thousands of kilometres. Figure 24 delineates the remote-sensing capabilities of 100 keV protons (90° pitch-angle assumed) in the context of a Cluster tetrahedron (the separation distance between the four spacecraft is assumed to be 1 R_E). The approximate dimensions of the tetrahedron/gyroradius systems are shown schematically for five selected key regions in geospace. The five insets illustrate the length of the local gyroradius in relation to the spacecraft separation distance.

It is obvious from Figure 24 that large separation distances between the Cluster spacecrafts are particularly important in the upstream region and in the magnetotail (in particular in the plasma sheet). For short separation distances (1000 km or less) the remote-sensing ranges of the four spacecrafts overlap strongly and the performance of the tetrahedron is essentially reduced to that of a single platform. It should be noted, however, that electron gyroradii (100 keV) are always small compared to the size of the tetrahedron.

Finite gyroradius effects in the ion angular distribution function have been successfully used on ISEE to probe the dayside magnetopause (Fritz *et al.*, 1982), FTE motion (Daly and Keppler, 1982), and plasma sheet motion in the geotail (Andrews *et al.*, 1981). The remote-sensing method can be applied in many ways, both for step function gradients and for finite density gradients.

Although this method has the advantage that boundary orientations and motions can be determined by a single spacecraft, it does contain some problems and ambiguities. The application of the step function gradient is often complicated, being tailored to each event individually. The finite gradient permits an analytical application by means of particle anisotropy measurements, but the additional Compton-Getting (motional) anisotropies must also be corrected for. In both cases, a rotation of the magnetic field during the analysis period introduces intractable complications.

A necessary assumption in all applications of remote-sensing is that the particle boundaries have a radius of curvature much greater than the ion gyroradius, i.e.,

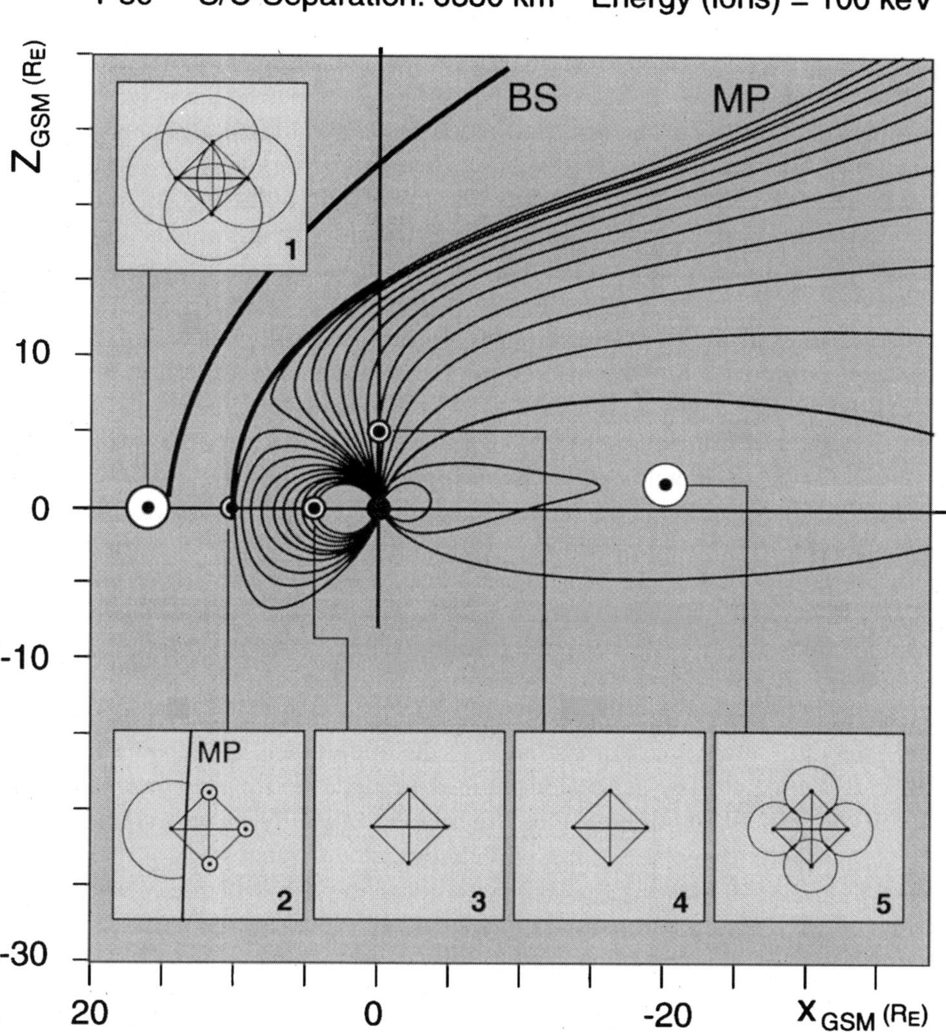

Figure 24. The Cluster tetrahedron (separation 1 R_E) and the remote-sensing distance of energetic particles in geospace illustrated for 100 keV protons and 90° pitch angle. Circles in five selected regions in the noon-midnight meridian indicate the approximate size of the local tetrahedron-plus-gyroradius geometry (the circles are drawn in the paper plane but they are actually perpendicular to the local magnetic field vector). The five insets show the relative length of the gyroradius for the different locations. In the inner magnetosphere 3 and 4 the gyroradii are too small to be resolved.

that they be planar. Departures from this assumption have been found at the magnetopause (Fahnenstiel, 1981) and in the geomagnetic tail (Andrews *et al.*, 1981). In such situations only multi-spacecraft analyses will be able to determine the complicated surface-wave motion of the boundary in question.

3.1.2. *The Role of Energetic Particles in Space Physics*

As we consider the orginal exploration of the terrestrial magnetosphere, we see that energetic particles (along with magnetic-field measurements) first revealed the existence of the Van Allen radiation belts, the magnetopause, the magnetosheath, the bow shock, the plasma sheet and much about the length and persistence of the geomagnetic tail. In a very real sense, the exploration of the rest of the Solar System has also been led by the detection of energetic particles. The remarkable discoveries and interpretations of the Pioneer-10 and -11 missions to Jupiter and to Saturn basically relied upon energetic ($\geq$ 50 keV) particle measurements: the discoveries made and the models formed based upon these measurements remain very much in place today, even after other missions to these planets have occurred. Similarly, in the case of Mercury – explored thus far only by Mariner-10 – we basically only have knowledge of the $\geq$ 35 keV electron population and a portion of the thermal electron population. Nonetheless, we have a remarkably persuasive picture of the basic elements of the Hermean system (Baker *et al.*, 1986).

The IMP-8 spacecraft had on-board sensors of large geometric factor that provided very sensitive measurements of $\geq$ 200 keV electrons at $\sim$ 35 R_E geocentric distance. With these nearly continuous measurements, Baker and Stone (1976, 1977a, b, 1978) were able to map and probe the complete structure of the middle magnetotail as never before. Energetic electron layers were discovered carrying significant energy flows outside the magnetopause and this energy flux was seen to vary with the nature of the solar wind-magnetosphere interactivity. In an even more dramatic demonstration of the importance of energetic electrons, such early work clearly delineated regions of open and closed magnetic field lines in the midtail region and thus contributed crucially to understanding and resolving an internal debate going on between groups measuring only the lower energy ($\leq$ 30 keV) plasma. Subsequent work using the IMP-8 electron sensors (Bieber *et al.*, 1982) demonstrated in a clear way that energetic electrons, and only energetic electrons, can be used to fully assess magnetic-field-line topology and thus distinguish between competing magnetotail dynamics models. The work referred to here remains definitive and similar methods were used extensively on ISEE-1 and -2 (Fritz *et al.*, 1984) to assess where, when and how acceleration and transport processes occur in and around the magnetosphere.

In another context, a large data set dominated by 3D electron measurements for energies $\geq$ 30 keV has been obtained for nearly two decades at geostationary orbit by the Los Alamos spacecraft (Baker *et al.*, 1982). Using these data as a basis, it has been possible to understand many of the issues related to the acceleration, transport and loss of energetic particles in the near-tail region. Indeed, it is the improved understanding of the microphysics of near-tail (substorm) acceleration that will be in the focus of Cluster. The beauty and strength of energetic-particle measurements lies in their high rectilinear speed and their relative immunity to large-scale magnetospheric convection electric fields. With such particles we can see clearly separated plasma regions, we can uniquely assess field-line topologies, we can map connectivity from the magnetospheric equatorial plane to the ionosphere, and we can sense the global changes in magnetospheric configuration that lead inevitably to substorm onset (Baker *et al.*, 1981, 1984).

Quite recently our team has had the experience of ISEE-3 and *Geotail* in the deep-magnetotail, as a case in point of the importance of energetic particles. A major result of the ISEE-3 tail mission was the confirmation of the release of plasmoids during substorms and the $\geq$ 70 keV electron have proven to be crucial in establishing the field structure, topology and scale size of plasmoids (Scholer *et al.*, 1984, 1985; Wilken *et al.*, 1995).

There can be a tendency in the study of plasma systems to look only in the region of maximal number density and ignore the tail of the plasma distribution function. However, in an astrophysical sense, most of what we know about the Universe comes from energetic-electron-associated processes. It is the synchrotron emission in active galactic nuclei, in radio jets and in remote stellar systems that tell us of the magnetic, electrical and plasma properties of these systems. It is the synchrotron and bremsstrahlung X-ray emissions that come from electrons that tell us of the environments of distant pulsars. It is energetic electrons that give us Type II and Type III solar radio bursts and it is electrons that let us diagnose distant supernova shock fronts as they propagate through the interstellar medium. Thus, we are reliant upon energetic electrons in the cosmic sense, and this emphasises even more the importance of understanding their acceleration and transport in our best and greatest local opportunity, namely, the Earth's magnetosphere.

What, then, will emerge as the principal science foci of Cluster? Undoubtedly there will be great attention to magnetotail acceleration regions and to plasma boundary regions. Key to making progress will be assessing field-line mapping and topology, the nature of the acceleration process, the mechanisms of particle transport in the tail and detailed study of the boundaries separating distinctly different plasma regions. We can sensibly carry out these studies only by using complete energetic-particle measurements on Cluster. Only energetic particles have the high speeds ($\geq 0.2c$) to allow assessment of magnetic reconnection regions. Only electrons counterpointed with ion measurements can be used to diagnose the nature of the tail acceleration region. Only electrons, specifically, have the small gyroradii to permit the rapid probing while preserving the clear separation between topologically distinct plasma regimes.

3.1.3. *RAPID and the Plasma Instruments CIS and PEACE*

The Cluster scientific payload is a carefully balanced set of instruments covering the full range of plasma waves, fields and particle parameters. All of these instruments are needed to generate a comprehensive description of static and dynamic plasma conditions. Among these instruments the energetic-particle spectrometer RAPID and the plasma instruments CIS (Cluster Ion composition Spectrometer) and PEACE (Plasma Electron and Current Experiment) form a triad of particle instruments in which the inherent low pass filter functions of the two plasma analysers is complemented by the RAPID high pass filters for ions and electrons. The combined CIS/PEACE/RAPID measurements effectively eliminate 'window effects' which often lead to misinterpretations.

The ion composition instrument CIS is capable of obtaining full 3D ion distributions with mass per charge composition determination. It covers the energy range between about 0 and 40 keV e^{-1}. The electron analyser PEACE measures electrons between 0.7 eV and 26 keV. The energy ranges of these instruments connect directly to the energy ranges of the RAPID instrument (20 keV to 450 keV for electrons and 40 keV to 1500 keV for protons).

Comprehensive energy coverage is fundamental in studying for example boundaries and boundary layers in space plasmas. This may help to answer the following selected questions:

– How does the plasma cross the magnetopause and enter onto closed field-lines?

– How can the reconnection sites on both the equatorward and poleward edges be identified? Where does reconnection most preferentially take place?

– What are the widths of magnetopause boundary layers? How do they vary along the flanks and how do they depend on the IMF polarity (northward/southward)? How important are these boundary layers in driving large-scale magnetospheric convection?

– How do the magnetopause (dayside) currents couple to the cusp field-aligned currents?

– What are the general characteristics of current sheets (density, flow, temperature, heat fluxes, beta)?

– What is the structure of the cross-tail current sheet?

– Which feature in the magnetosphere corresponds to the Harang discontinuity?

3.2. Bow Shock and Upstream Phenomena

One prominent feature in the solar wind interaction with the magnetosphere concerns the various particle-acceleration processes occurring upstream of and at the bow shock. The ISEE-1 and -2 observations have shown that these regions are very rich in small- and large-scale phenomena directly related to the generation of plasma turbulence and particle energisation, and hence are most important for testing theories of different shock acceleration mechanisms and wave-particle interactions. Some of the ISEE results and the advances expected from the Cluster mission are outlined in the following.

Particle acceleration at MHD shocks has become one of the key topics in space plasma physics during the last decade, stimulated by both extensive theoretical studies and in-situ measurements within the Solar System. In this field, the Earth's bow shock has attracted particular attention because of its proximity. The energetic-particle population near the bow shock has been related to diffusive Fermi acceleration (Gosling *et al.*, 1978; Scholer *et al.*,1979; Wibberenz *et al.*, 1985) and compared with model calculations (Lee, 1982; Ellison, 1985). Recently, the bow shock origin of the diffusive upstream ions has been questioned (Anagnostopoulous *et al.*, 1986; Sarris *et al.*, 1986) and it has been proposed that this energetic-particle population is due to escape of magnetospheric ions into the upstream region. The question is of considerable importance.

Diagnostic tools to distinguish these two origins are the following:

– simultaneous measurements of ions and electrons at different positions inside and outside the Earth's magnetosphere,

– check for the simultaneous occurrence of an electron component (the bow shock acceleration in the energy range above 20 keV is restricted to ions),

– correlation with the momentary directions of the interplanetary magnetic field and flow direction of the solar wind,

– measurement of the energy spectrum over a wide range, throughout various phases of individual events,

– measurement of the chemical composition,

– correlation with the simultaneous occurrence of an increased level of turbulence in the magnetic field,

– study of anisotropies in three dimensions.

Apart from the question of the origin of upstream particles, a number of fundamental problems in the interaction between the bow shock and energetic particles are still unsolved. These include: the injection of thermal solar wind ions into the diffusive shock acceleration mechanism, in relation to the shock structure; the efficiency of shock acceleration as a function of the shock Mach number, the shock normal angle, the position on the bow shock etc.; the escape mechanism (upstream, downstream, perpendicular to the average field) which finally determines the shape of the energy spectrum; conditions for the growth of wave turbulence; the scattering of particles by hydromagnetic turbulence, in particular, the variation of the mean free path with particle parameters like velocity and/or energy/charge; the conditions for the appearance of limiting fluxes in a steady state situation; the influence of the accelerated particles on the ambient medium.

Previous studies established, depending on the location around the shock front, two separate populations of suprathermal ions. The so-called beam ions, observed predominantly in the quasi-perpendicular bow shock regime (Asbridge *et al.*, 1968; Gosling *et al.*, 1978; Paschmann *et al.*, 1979) were identified as beams of particles streaming along the interplanetary magnetic field, with energies that seldom exceed 20 keV. However, at higher energies, a reflected diffuse ion population exists in the quasi-parallel shock regime (Lin *et al.*, 1974; Gosling *et al.*, 1978). The diffuse ions (protons, alphas, and CNO) have energies in the range 30–100 keV, and exhibit rather isotropic distribution, indicating a process of random scattering. Diffuse shock acceleration might be at work here. Another population, intermediate between the beam ions and the diffuse ions, has also been identified (Paschmann *et al.*, 1981).

It is possible that, while the beam ions have their origin in the specular reflection of solar wind ions at the quasi-perpendicular shock front (Sonnerup, 1969; Paschmann *et al.*, 1980) some of them could act as seed particles for the diffuse ions via further acceleration. The energy density of the reflected diffuse ions is about one third of the solar-wind-energy density (Ipavich *et al.*, 1981b), the shock acceleration is thus a very efficient mechanism. Mitchell and Roelof (1983) have made the interesting suggestion that the acceleration of the diffuse ions is self-limiting in the sense that the parallel energy ($E_{||}$) of the 30–200 keV energy ions cannot exceed the magnetic-field-energy density ($B^2/2\mu_0$) by more than a factor of 2. This effect can be investigated in detail with the RAPID instrument as it has comprehensive angular (4π) as well as energy (20 keV to 1 MeV) coverage for different ion species (p, α, and CNO). Observations at the four different positions relative to the shock will specifiy the boundary conditions for models of the processes occurring.

The appearance of the diffuse energetic ions is usually accompanied by large-amplitude low-frequency waves in the magnetic field as well as the solar wind plasma. The presence of ULF waves is essential in the theoretical model of first-order Fermi acceleration (Axford, 1981; Lee, 1983). The Fermi acceleration mechanism requires that the ions be resonantly scattered between the converging upstream magnetic irregularities and the shock front or the downstream waves. Several ingredients are important in the diffuse shock acceleration process. First of all, the diffusion coefficient ($K_{||}$) should have a certain energy dependence such that the scale length λ for the e-folding distances for the upstream distribution of the diffuse ions should be different for ions of different energies. While Ipavich *et al.* (1981a) have determined $\lambda = 7.2\ R_E$ for 30 keV-protons, Wibberenz *et al.* (1985) have found λ = const. ($6.5 \pm 1.5\ R_E$) for 25–36 keV protons without significant energy dependence. With the energy window of the RAPID experiment covering the full energy width of the upstream diffuse ions and with multiple measurements at the different positions of the spacecraft, this outstanding issue will be thoroughly addressed by the Cluster mission.

Cluster will offer the advantage of simultaneous multipoint composition measurements of complete spectra from thermal energies up to hundreds of keV, with coordinated RAPID and CIS measurements. In addition, both RAPID and CIS are capable of obtaining full 3D ion distributions with high time resolution.

Other issues that shall be thoroughly addressed by RAPID measurements are:

– exact determination of the bow shock and its motion through detailed examinations of small-scale structures,

– energy dependence of the diffusion coefficient K through the multipoint measurements of RAPID that cover the full energy width of the upstream diffuse ions,

– comparative importance of parallel and perpendicular diffusion,

– specification of the boundary conditions for the models of the processes leading to beam ions and diffuse ions.

The temporal and directional variations of the interplanetary magnetic field have obviously strong effects on the upstream particle population. Simultaneous observations of the undisturbed solar wind by the SOHO and WIND spacecraft will be of great importance in correlating the various magnetospheric phenomena with interplanetary conditions.

3.3. The Magnetopause, Magnetosheath, and Cusp Regions

After deceleration at the bow shock, the solar wind plasma continues to travel through the magnetosheath around the magnetopause, the actual separation between the solar wind and terrestrial magnetic field. The main questions regarding this region are:

– Which is the dominant source for energetic ions in the magnetosheath?

– What processes occur during the propagation of the plasma from the bow shock to the magnetopause?

– To what extent, where and under what conditions is the magnetopause penetrated by the solar wind?

Several studies have shown that both the bow shock and the magnetosphere are sources of energetic ions in the magnetosheath (Anagnostopoulos and Kaliabetsos, 1994). Multispecies measurements from the AMPTE/CCE spacecraft covering the wide energy range 1.5 keV–2 MeV (Paschalidis, 1994) indicated that the leakage of magnetospheric particles is the most important contributor to the energetic ($E \geq 50$ keV) population of the magnetosheath, as compared to Fermi acceleration and to shock drift acceleration. Fermi acceleration predicts the opposite local time asymmetry than that observed by AMPTE/CCE (spectra harder at the duskside than at the dawnside), and shock drift acceleration should be a minor contributor to the energetic population because of upper-energy cutoff limitations.

Cluster will provide multi-point composition measurements over a wide energy range which are essential for the assessment of the issues regarding the magnetosheath and magnetopause region.

3.3.1. *Magnetic-Field-Line Reconnection and Flux Transfer Events (FTE)*

Over a quarter century ago it was suggested that the magnetic field lines in the solar wind could combine with those of the Earth to provide direct entry and acceleration of solar wind plasma into the magnetosphere. This process, known as 'reconnection' or 'merging', is thought to be the basic mechanism responsible for the input of solar wind energy into the magnetosphere, and formation of the plasma convection in the magnetosphere and ionosphere. Dungey (1961) first suggested that magnetic reconnection between the Earth's magnetic field and the IMF may occur at the dayside magnetopause when the IMF has a southward orientation. Fast plasma observations from ISEE-1 and -2 provided the first convincing *in-situ* evidence for quasi-steady-state reconnection at the dayside magnetopause (Paschmann *et al.*, 1979; Korth *et al.*, 1982; Sonnerup *et al.*, 1981). By analysing data of ISEE-1 and -2, Russell and Elphic (1978, 1979) found that reconnection in the dayside magnetospheric boundary region appears to be more frequently associated with the FTEs. Since then various theoretical transient reconnection models have been developed to explain the formation and observed features of FTEs (Lee and Fu, 1985; Scholer, 1988; Southwood *et al.*, 1988; Liu and Hu, 1988).

The reconnection process has turned out to be far more complicated than originally imagined. The multi-point measurements of the Cluster experiments will be able to probe the complex structure of reconnection events in terms of spectra, composition and flow directions, thus yielding information on energisation, sources and motion of the particles. In addition, the ability to use finite gyroradius effects is a powerful tool to detect the motion of the structure as a whole.

Principal observational features of so-called FTEs at the magnetopause are:

– a sinusoidal fluctuation in B_N, the component of the magnetic-field normal (and outwards) to the magnetopause,

– isotropic electrons and streaming ions in the magnetosheath.

Field-aligned anisotropies of energetic ions uniquely identify the magnetospheric hemisphere to which the magnetosheath field line is connected.

In the magnetosphere, the energetic-particle intensities are reduced during FTEs, to values that are only slightly higher than in the magnetosheath FTEs. This indicates that the FTEs have a continuity through the magnetopause permitting magnetospheric energetic particles to escape into the magnetosheath, and possibly into interplanetary space. Conversely, it has also been demonstrated by Paschmann *et al.* (1982) that magnetosheath plasma is found within the magnetospheric FTEs. These phenomena have also been observed by GEOS and STARE (Sofko *et al.*, 1979).

Surveys of the occurrence and properties of FTEs (Rijnbeek *et al.*, 1984; Daly *et al.*, 1984) demonstrate that they are a persistent feature of the dayside magnetopause during times when the interplanetary field is directed southward.

Using ISEE-1 and -2 to make two-point measurements of FTEs, Saunders *et al.* (1984) have shown that the events are highly structured, possessing twisted fieldlines and vortex plasma motion. In fact, based on these measurements alternate interpretations of FTE signatures have been proposed by Saunders (1983) and Southwood *et al.* (1988). According to the suggestion by Saunders (1983) FTE signals will be observed if a flux bulge due to transient changes in the rate of quasi-steady reconnection passes across the satellite. It is quite clear that four-point measurements will greatly add to our understanding of these complex features by probing in detail their cross-sectional structure and temporal variations.

That FTEs are spatially and temporally localised reconnection events is now widely (although not universally) accepted. It is even believed that there is a continuous spectrum of reconnection event sizes and time scales covering the range from quasi-stationary reconnection to the fastest FTEs. Another possibility is that FTEs are very large scale features of the magnetopause that only appear to be localised due to wave motion of the magnetopause surface and the (at best) two-point measurements previously available. The four-point measurements of Cluster will be able to answer this question in ways that the two-point measurements of ISEE could not. Energetic-particle measurements are vital to any investigation of reconnection events because:

– they are the only reliable indicator in the magnetosheath of the hemisphere to which the flux tube is connected (necessary for source determination);

– the fact that the particles exist for longer than their bounce time (about 1 min for 25 keV protons) means that there is a resupply process, which is still one of the mysteries of reconnection observations for which multi-point spectral measurements are required;

– composition measurements are essential for the resupply and source determination;

– the use of energetic particles as remote-sensing probes can determine orientation and motion of the events, needed to discover their ultimate destination and four-point measurements will be necessary if the radius of curvature of the event surface is comparable to the ion gyroradius, as appears most likely.

So far reconnection at the dayside magnetopause for southward IMF has been studied extensively. It has also been suggested that when the IMF has a northward component, reconnection proceeds in high latitudes behind the cusps. Dungey (1963) first inferred that magnetic reconnection could occur at the nightside magnetopause while the IMF is northward. Following Dungey's expectation some authors discussed the possible reconnection site, processes and associated effects during northward IMF periods (Russell, 1972; Maezawa, 1976; McDiarmid, 1978; Crooker, 1979). ISEE-3 data showed that plasmas in the magnetosheath can enter the magnetosphere through a large area on the magnetopause, implying that in general the magnetotail has an open configuration (Gosling *et al.*, 1984, 1985). Song and Russell (1992) suggested that high-latitude reconnection behind the polar cusps may lead to the formation of the low-latitude boundary layer for northward IMF. Stasiewicz (1991) discussed a global model of magnetic merging in more general situations.

A recent two-dimensional MHD simulation studied the transient reconnection in the high-latitude boundary layer (HLBL) and proposed a global reconnection model for the period of northward IMF (Liu *et al.*, 1994; Liu *et al.*, 1995; Zhang and Liu, 1995; Zhang *et al.*, 1995). It was found that the Kelvin-Hemholtz instability generated by the flow shear in the boundary layer causes generator effects which may increase the growth rate of the tearing mode instability, thus leading to reconnection in the nightside HLBL.

It has been shown that when the IMF is northward, in the auroral magnetosphere, the electric-field patterns, field-aligned currents, particle precipitation and auroral phenomena are considerably different to those in southward IMF periods (Schunk, 1988). We suggested that magnetic reconnection in HLBL plays a crucial role in leading to these complex phenomena.

Signatures of reconnection in the HLBL which RAPID measurements will help to identify include:

– Generally energetic particle flows in the nightside HLBL are tailward. If there are Earthward energetic particle flows somewhere behind the cusps, they may be regarded as clear evidence of nightside high-latitude boundary-layer reconnection.

– When magnetic reconnection proceeds, the magnetic line in the reconnection region must be open. Thus it is expected that the particles in the HLBL are a mixture of magnetospheric and magnetosheath populations, albeit with occasional heating.

– Magnetic reconnection at HLBL should produce plasmoids as well.

– High-speed plasma flows with a magnitude greater than 400 km s^{-1} persisting in various time scales can also be regarded as a signature of transient and quasi-steady-state reconnection.

– Heavy ions from the ionosphere may be observed in both the HLBL and the adjacent nightside magnetosheath.

3.4. Magnetic Storms and Substorms

3.4.1. *The Storm Time Magnetosphere*

RAPID will be an ideal instrument to continue the line of research in the investigation of Geomagnetic Storms suggested by work with the CRRES and Viking satellites. Viking found a polar cap dominated by He^{+2} ions, and a transition to a more equatorial, trapped belt of He^{+} ions. This feature can also be seen by CRRES, and is compatible with the charge exchange process. There is a need to compare observations with detailed models for charge exchange processes, in order to establish how much of this behaviour is due to charge exchange, and how much to the injection of new populations.

Solar wind alpha-particle entry to the magnetosphere during quiet times was initially investigated using *Viking* data. It was found that the variation of the mean charge state with invariant latitude is significantly different from that of the anisotropy. This indicates the existence of a transition region where particles diffusing to lower latitude are trapped. They subsequently diffuse to still lower latitudes, and undergo charge exchange, thus providing a major source of inner magnetosphere singly-charged helium. There appears to be a very low abundance of oxygen in this region, again suggesting that the singly-charged helium population is not extracted from the ionosphere.

Studies with calculations of ion trajectories have shown that high cross-tail electric fields at very disturbed times allow direct access of ions from the solar wind into the midnight region. The four-spacecraft configuration of Cluster and the full angular coverage available from RAPID will be a powerful tool for testing the entry model calculations, as well as investigating flows along the lobe boundaries, in order to investigate the connectivity at this time.

The composition ratios of energetic ions measured by RAPID, combined with data from the upstream monitors (Wind and SOHO), can help to identify changes in the solar wind source populations during storms. The speed of entry into the magnetosphere will be compared with estimates from existing models (Fennell *et al.*, 1993; Yeoman *et al.*, submitted) to assess the required magnitude and variation of the cross-tail electric field. During quiet times we can use the particle pitch-angle distributions and composition ratios to examine the inward diffusion of solar wind material from polar cap entry to trapping region and subsequently to lower latitudes in order to assess this mechanism as a major source of magnetospheric ions.

An interesting aspect will be to study large scale changes and the calorimetry of the ring current during active storm and substorm periods. This will be performed by using a relatively novel technique which was previously used on *Geotail*: a special operating mode of the RAPID instrument (ENA mode, Section 2.5.5) allows the detection of Energetic Neutral Atoms (ENAs). ENAs are produced as a result of charge exchange reactions between energetic ring current ions and low-energy neutral hydrogen. As the resulting neutrals are unaffected by magnetic or electric fields they provide a useful mechanism for remote-sensing dominant source region enabling calorimetry of the energy transfers to the ring current during storms and substorms to be performed. The detection of ENAs with RAPID is delineated in Section 3.5.

3.4.2. *Substorms and Their Near-Earth Manifestations*

Magnetospheric substorms have been investigated for several decades (Akasofu, 1991). Substorm phenomena are related to the interaction between the solar wind, the magnetosphere and the ionosphere. Current phenomenological models describe substorms in terms of different phases: growth phase, expansion phase, and recovery phase (Rostoker *et al.*, 1980). The sequence of these phases is reflected in the development of the cross-tail current (Kaufmann, 1987). The growth phase begins typically about one hour before the onset of the expansion phase and results in a stretching of the geomagnetic field-lines towards a more tail-like configuration (McPherron, 1972) attributed to an increase in the cross-tail current in the near-Earth plasma sheet. The stretched magnetic field in the tail represents a large reservoir of magnetic energy which can drive magnetospheric substorms. An outstanding feature of substorms is the sudden onset of the expansion phase (the so-called substorm onset) characterised by auroral breakup phenomena and strong spatial and temporal variations of the relevant magnetospheric parameters. During this short time interval (a few minutes) the magnetosphere recovers from the stretched configuration towards a more dipolar-like configuration with strong variations of the particle fluxes (Sauvaud and Winkler, 1980). The reconfiguration (the so-called dipolarisation) is associated with a partial diversion (i.e., decrease) of the cross-tail current into the ionosphere (Kaufmann, 1987). Increase in the electron fluxes (electron injection) and the accompanying dipolarisation may be delayed by up to a few minutes with respect to substorm onsets (Korth *et al.*, 1991; Pu *et al.*, 1992).

Over the last decade, new and better observations are gradually leading to a more consistent model of the interrelations between these various magnetospheric phenomena. Despite these advancements it has not yet been possible to establish a model that fits all sets of observations. Fundamental arguments related to basic substorm physics persist and different competing substorm models describing the observations are appearing in the literature. Summaries of substorm models were given by Lui (1991) and Siscoe (1993).

Most of the controversies in connection with substorm models are associated with the processes that are thought to trigger the expansion phase, in which region of geospace the expansion is initiated, and the relationship between near-Earth observations and high-speed ion flows further down the tail (Lopez, 1993). The suddenness of the expansion onset led to the assumption of an instability which diverts (or disrupts) the cross-tail current. Auroral morphology studies performed with data from the Viking satellite (Elphinstone *et al.*, 1991) demonstrate that the first arcs to develop at breakup are conjugate in the Tsyganenko magnetic field model (Tsyganenko, 1987) with the inner edge of the plasma sheet being typically at 6–9 R_E from the Earth. Jacquey *et al.* (1991) concluded on the basis of the magnetic signatures observed in the tail lobes by ISEE-1 that the dipolarisation starts close to the Earth ($R < 15\ R_E$) and propagates tailward. In a multi-point study based on observations in the near-Earth tail, Jacquey *et al.* (1993) found indications that the current disruption associated with the substorm onset starts at 6–9 R_E and propagates tailward after the onset of the substorm expansion. These results, as well as the investigations of substorm phenomena using observations in the auroral zone and GEOS-2 observations (Korth *et al.*, 1991; Roux *et al.*, 1991a, b; Kremser *et al.*, 1992; Ullaland *et al.*, 1993), are consistent with the inner edge of the plasma sheet being the location of expansion phase onsets. Therefore, expansion phase is very likely triggered on closed field-lines by the near-Earth ($R < 10\ R_E$) magnetotail processes in agreement with Lui (1991) and Fairfield, (1992), who further summarised arguments for a near-Earth triggering region.

A most prominent characteristic of the near-Earth Neutral Line (NL) model of substorms (Hones, 1976) is the formation of a plasmoid by magnetic reconnection at substorm onset. In this model an X-type neutral line is formed close to the Earth implying tailward ion flows. McPherron and Hones (1991) found observational evidence for the existence of the near-Earth X-type neutral line to be located between 10 and 20 R_E downtail. On the other hand Baumjohann *et al.* (1989) and Baumjohann *et al.* (1990) analysed a large ion flow data set and magnetic field measurements in the mid-tail ($-9\ R_E < X_{\rm GSM} < -19\ R_E$; AMTE-IRM). Their observations of the plasma flows in the plasma sheet show that jet-like high-speed flows are often present and that they are dominantly Earthward, which would imply that a possible NL forms somewhere tailward of $-20\ R_E$. In a study based on *Geotail* and IMP-8 data Mitchell *et al.* (1994) found indications of NL formation Earthward of $X = -35\ R_E$. A conclusion based on these observations will be that the distance between the near-Earth NL and the location of expansion phase onset often is too large for a NL formation to be the proximal cause of the expansive phase (Siscoe, 1993).

The four Cluster spacecrafts will cross the current sheet most frequently at large distances (about 19 R_E) which is at the outer edge of the region of interest for local substorm breakup phenomena. Complementary measurements from ground-based instrumentation and from space platforms such as geosynchronous, polar-orbiting, and solar-wind satellites should be used in addition to the RAPID, CIS, and FGM data in order to resolve fast-moving substorm phenomena.

3.4.3. *Substorms and Magnetotail Dynamics*

A major topic to be addressed using energetic-particle measurements is that of acceleration processes in the tail. Energetic particles have been observed continuously over time periods of hours with impulsive time variations on time scales < 1 min. However, because of the motion of the plasma-sheet over a single spacecraft and the limited intensity thresholds in the instrument, it has not been possible to distinguish the continuous plasma-sheet energetic-particle population and its sources from the population produced by impulsive acceleration associated with substorms. Such distinctions are vital to theories of substorms and related induced electric fields such as those associated with tearing mode instability theory. A group of four spacecrafts will make it possible to address this problem by simultaneously sampling nearby regions within the relatively narrow plasma sheet and its boundary layer.

The best direct evidence for particle acceleration and probably the most important energetic phenomena in the magnetosphere are the high-intensity field-aligned 'impulsive' bursts, displaying Inverse Velocity Dispersion (IVD). These events are usually detected at the high-latitude plasma-sheet boundary during thinnings of the plasma sheet. However, a number of cases have been found where 'impulsive' bursts with inverse velocity dispersion were observed well inside the plasma-sheet or during plasma-sheet expansions. These observations suggest that for some events the IVD may be a temporal rather than a spatial effect. However, the most straightforward explanation is that the events are caused by accelerations at a neutral line (Sarris and Axford, 1979). Cluster is a perfect mission for resolving spatial from temporal effects. The RAPID high-resolution electron measurements can provide an extra 'fingerprint' of the acceleration mechanism, which is missing from all previous observations of impulsive bursts because of instrumental limitations.

The passage of a 'source' of energetic particles over a spacecraft inside the plasma-sheet has been inferred on the basis of reversals in both the streaming of the energetic particles and the plasma flow. However, severe objections have been raised to such interpretations of the observations. Cluster must search specifically for such events with good time resolution, since the expected 'reversal' must be observed simultaneously at all energies (within $\sim$ few seconds.) at each spaceceraft. If such an encounter with the 'source' is observed, then the RAPID measurements on the multiple Cluster spacecrafts will be able to measure its size and speed.

Single-spacecraft measurements of the time evolution of the angular distributions of energetic particle intensities during cases of counterstreaming between energetic ions and electrons have been used to infer the presence of transient ($\sim$ 10–20 s) $\mathbf{E}$-fields parallel to the magnetic field. From simultaneous measurements of the variations in the energy spectra of the ambient energetic proton and electron populations under the influence of the $\mathbf{E}_{||}$-field, the 'effective' potential traversed by the energetic particles has been estimated to be a few tens of keV. The region over which the inferred transient $\mathbf{E}_{||}$-field extends may be determined using multi-point observations of such events. Furthermore, such multi-point measurements of various species with different charges will provide an excellent self-consistency test for the assumed existence of the field-aligned $\mathbf{E}$-field.

The $\mathbf{B}$-field microstructure (i.e., magnetic loops or islands with lengths of a few R_E) which develops in the magnetotail following acceleration events, has been inferred mostly on the basis of single-spacecraft measurements. Suspected neutral line formation in the near-Earth tail region and the subsequent plasma acceleration has led Baker *et al.* (1979) to propose the substorm sequence reproduced in Figure 25. Energetic electron anisotropies were used to indicate whether field-lines were open or closed, but such mapping cannot be certain given the spatial scale sizes and temporal variation. Multiple-spacecraft measurements of energetic particles, plasma and field with good time resolution ($\leq$ few seconds) can contribute to resolving these uncertainties.

Pulsations of the plasma-sheet boundary with periodicities of approximately 2 to 9 min have been detected from measurements of energetic-particle intensities. Such oscillations are not apparent in the measurements of the $\mathbf{B}$-field alone. There is strong evidence that these pulsations are due to a standing (and not a travelling) wave at the plasma-sheet boundary.

Energetic particle measurements will also be useful in the tail lobes during periods of observable solar particles. Regions of open lobe field-lines can be identified, and using timing comparisons with SOHO or Wind, the length of open field-lines can be estimated. In the past, incorporating two spacecrafts inside and outside the magnetosphere, the time delay in the entry of solar particles in the lobes of the magnetotail was used in order to infer the 'effective' length of the geotail (Van Allen *et al.*, 1971; Fennell *et al.*, 1973). However, the entry of solar particles into and transport inside the plasma-sheet has not been studied in detail. Recent results have indicated that the access of solar particles into the plasma-sheet is prompt near the neutral sheet and delayed at high plasma-sheet latitudes, being always considerably faster than the particle access in the lobes of the magnetotail. With simultaneous multi-spacecraft observations by Cluster (with spacecrafts located at various distances from the neutral sheet as well as inside the lobes) it will be possible to examine directly the access of solar particles to the different regions under distinct configurations of the upstream field. Furthermore, the study of the expected dawn-dusk asymmetries in the time delay for the entry of solar particles into the plasma-sheet will become possible by multi-spacecraft observations across the magnetopause.

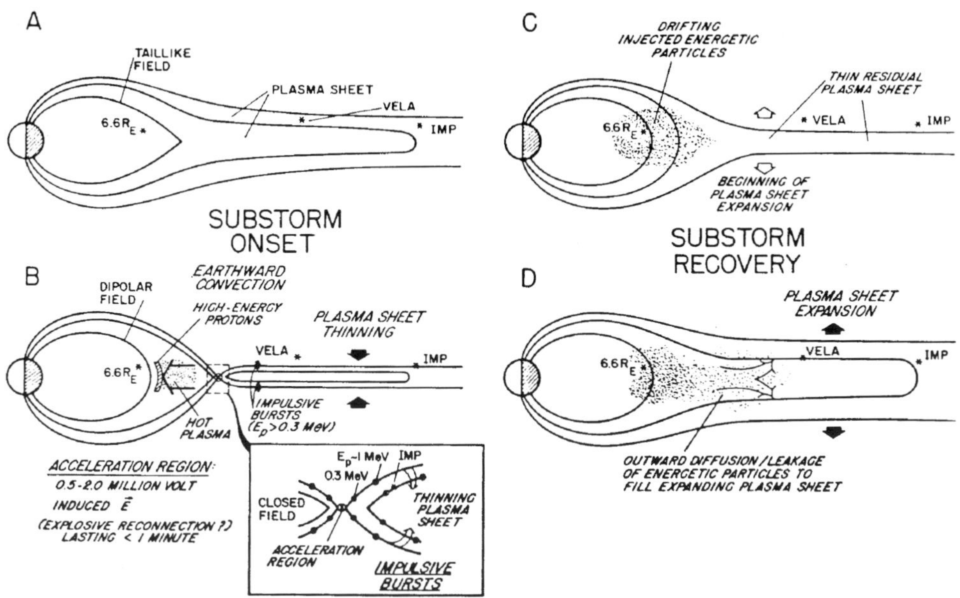

Figure 25. Schematic sequence of energetic particle events predicted by the substorm model of Baker *et al.* (1979). (a) The inner magnetosphere just prior to substorm onset showing the buildup of stress evidenced by the tail-like field. (b) The magnetosphere just after onset showing a dipolar field configuration and the accelerated ion bunches streaming sunward toward the trapped radiation zones and antisunward along the thinning plasma sheet. (c) Conditions just prior to substorm recovery at the beginning of the plasma sheet expansion. (d) Expansion of the plasma sheet and the subsequent filling of the expanding sheet with energetic protons diffusing out of the trapped region.

3.5. Detection of Energetic Neutral Atoms (ENA)

Energetic neutral atoms are produced in the Earth's magnetosphere through charge exchange between hot ($E \geq 1$ keV) magnetospheric or outflowing ionospheric plasma and the geocorona. The geocorona is an exospheric extension of relatively cold ($\sim$ 1000 K) neutral atoms (mainly hydrogen), which resonantly scatter solar Lα radiation. The produced neutrals leave the interaction region with essentially the same energy and velocity direction as the incident ion. In a way the geocorona acts like an imaging screen for magnetospheric/ionospheric ions. Thus, imaging the neutral atom distribution is a means to make the magnetospheric plasma visible. Remote-sensing of energetic neutral atoms (ENA) can provide line-of-sight integrated observations of source populations. The extraction of physical parameters will, however, require unfolding of the product ($j_{\text{ion}} \times n_{\text{H}}$) from the line-of-sight integrals (Roelof *et al.*, 1992).

Following the pioneering work of Roelof on the observation of the terrestrial ring current through energetic neutral atom imaging (Roelof *et al.*, 1985; Roelof, 1987), growing attention has been paid by the scientific community to the potential importance of neutral atom imaging for magnetospheric research (Williams *et al.*,

1992; Daglis and Livi, 1995), and several modelling studies were conducted during the last few years (Hesse *et al.*, 1993; Orsini *et al.*, 1994).

The polar orbit of Cluster is particularly favourable for neutral atom imaging of the ring current and near-Earth magnetotail because of the low ion background in the lobes and the polar region. Its relatively long orbital period is ideal for the continuous monitoring of the evolution of dynamical processes in the magnetosphere (storms, substorms).

In order to detect energetic neutral atoms, the RAPID unit on one spacecraft is operated in the ENA Mode as described in Section 2.5.5. In this mode the 10 kV deflection voltage in the collimator is activated. The electric field reduces the transmission of ions with energies below 200 keV by 3 to 4 orders of magnitude. Particles observed in the critical energy band 40–200 keV are then considered neutral atoms. Figure 26 illustrates, as an example, the angular resolution obtainable in the equatorial magnetosphere when Cluster passes over the pole. Although the pixel size is coarse the 'image' can probably detect the location and resolve the spatial extent of plasma injections into the active magnetosphere with sufficient resolution.

The geometry shown in Figure 26 refers clearly to a rather favorable situation in terms of angular resolution and radial distance to the dominant ENA sources. The storm time ENA differential flux at this location can be of the order of 50 part./(cm^2 s sr keV) for energies above 30 keV. At apogee (19.6 R_E) an integral flux of $\prec$1 part./(cm^2 s keV) is expected.

3.6. HEP/*Geotail* – A PRECURSOR OF RAPID

The energetic-particle package HEP on the Japanese spacecraft *Geotail* (launched 24 July, 1992) contained a first version of an imaging ion spectrometer which is identical to RAPID in all relevant components, performance, and Science Data discussed in this report (except for IES). In a first phase *Geotail* studied the deep tail in double lunar swing-by ecliptic trajectories which, unlike Cluster, allowed extended observations in the plasma sheet. Figure 27 illustrates the IIMS technique for particle identification. Each particle is characterised by an energy (E) and time-of-flight (T) measurement. The sorting parameter in the (E, T) plane is the particle mass A (from the left side the mass lines refer to hydrogen, helium, CNO, and Si–Fe group ions, respectively).

The four frames in Figure 27 document an impulsive oxygen event observed at about $X = -60\ R_E$ in the magnetotail. The pre-event chemical composition consisted mainly of hydrogen and helium with only little oxygen present. At 19:00 UT the oxygen flux suddenly increased for a short time to a substantial level. These frames are constructed from the Science Data 'Direct Events' (DE) which represent the full resolution of the spectrometer. As mentioned earlier in this report, the majority of these events are reduced on-board to a lower level of precision to meet the telemetry capacity.

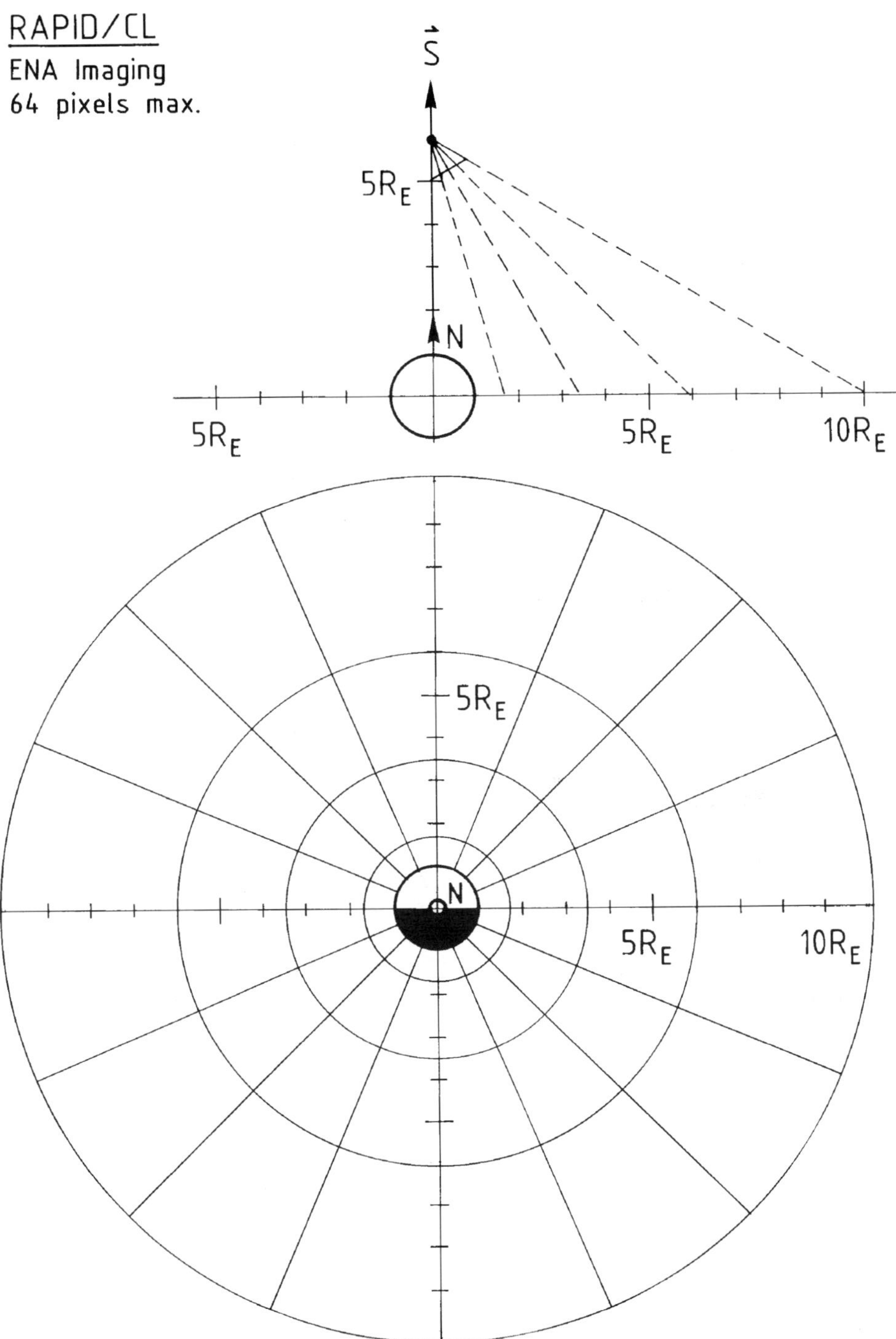

Figure 26. Imaging of energetic neutral atoms in the equatorial magnetosphere. The Cluster spacecraft with RAPID in the ENA mode is assumed to travel over the Northpole. The lower part shows the angular resolution.

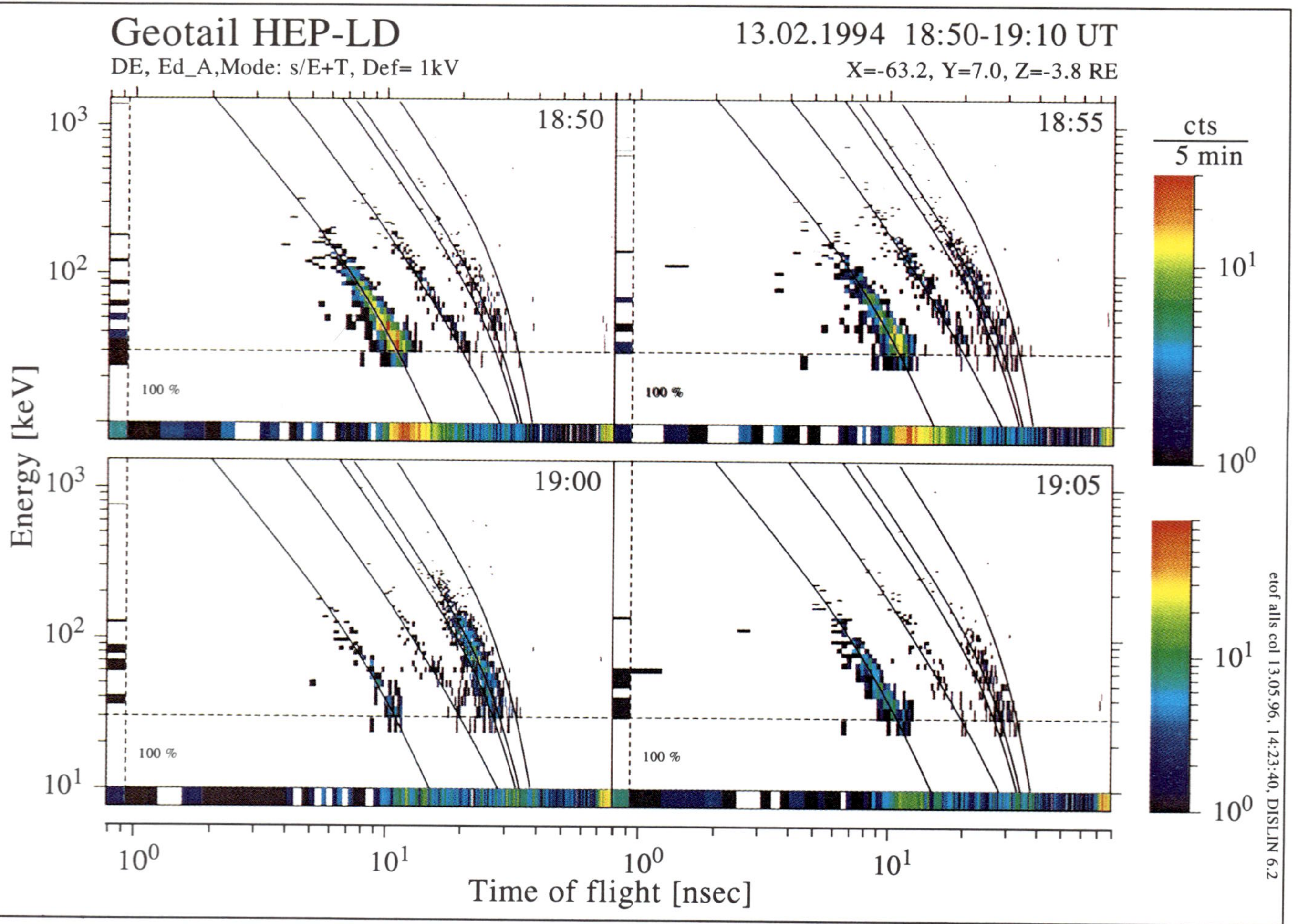

Figure 27. A sequence of two-parameter scatter plots from the spectrometer HEP-LD on *Geotail* demonstrate the mass sorting in the (E, T)-plane. The mass lines identify hydrogen, helium, CNO, and Si–Fe (from the left). A strong oxygen burst occured in the interval 19:00–19:05 UT in the deep magnetotail (GSE coordinates on the upper right).

Figure 28 shows the evolution of the three-dimensional intensity distributions for the same time intervals. The oxygen (19:00–19:05 UT) appears mainly at or near 90° pitch angle. This is evidence for strong plasma convection potentially driven by a reconnection process Earthward of the spacecraft.

Although *Geotail* observed these oxygen-rich events well tailward of the Cluster nightside apogee distance, the above representations may serve as guidelines for the type of measurements RAPID is expected to return. Most regions of geospace visited by Cluster are expected to show strong angular anisotropies either due to gradients in the plasma or to streaming distributions.

3.7. Derivation of Plasma Parameters from RAPID Measurements

The orientation of RAPID on the spacecraft platform and the coverage in phase space (unit sphere) is described in Section 2.5.1. The conversion from observed counting rates n_c to the differential flux intensity j can then be completed through functions of the form

$$j = [GF\epsilon(E, M; SD)]^{-1} n_c \,, \tag{1}$$

where GF indicates the geometric factor, ϵ denotes the detection efficiency as a function of particle energy E, particle mass M and selected science data channel SD. By definition, the differential directional flux intensity $j(E, \theta, \phi)$ represents the number of particles per energy unit centred at the energy E, per unit area in the direction of (θ, ϕ), per unit time interval, and per unit solid angle; θ and ϕ denote the polar angle and azimuthal angle in the local satellite coordinate system, respectively. It can be shown that $j(E, \theta, \phi)$ is related to the particle-distribution function f in the phase space (E, θ, ϕ) as (Pu *et al.*, 1992)

$$f(E, \theta, \phi) = (m/2)^{1/2} \frac{j(E, \theta, \phi)}{E^{1/2}} \,, \tag{2}$$

or,

$$f(E_k, \theta_i, \phi_l) = (m/2)^{1/2} \frac{j(E_k, \theta_i, \phi_l)}{E_k^{1/2}} \,, \tag{3}$$

where $k = 1, 2, \ldots, N_k$ indicates energy channels of the spectrometer, $i = 1, 2, \ldots, N_i$ and $l = 1, 2, \ldots, N_l$ denote the polar and azimuthal looking directions of detectors, respectively. Furthermore, a number of fundamental plasma quantities such as the number density N, the mean velocity $\mathbf{U}$ and the thermal pressure tensor P, etc., can be obtained by calculating moment integrals of $f(E, \theta, \phi)$ (Lundin *et al.*, 1982). Therefore, the observed function $j(E, \theta, \phi)$ may be considered a basic element to derive a variety of plasma parameters (the technique to extract plasma parameters is described in detail by Pu *et al.* (1996)).

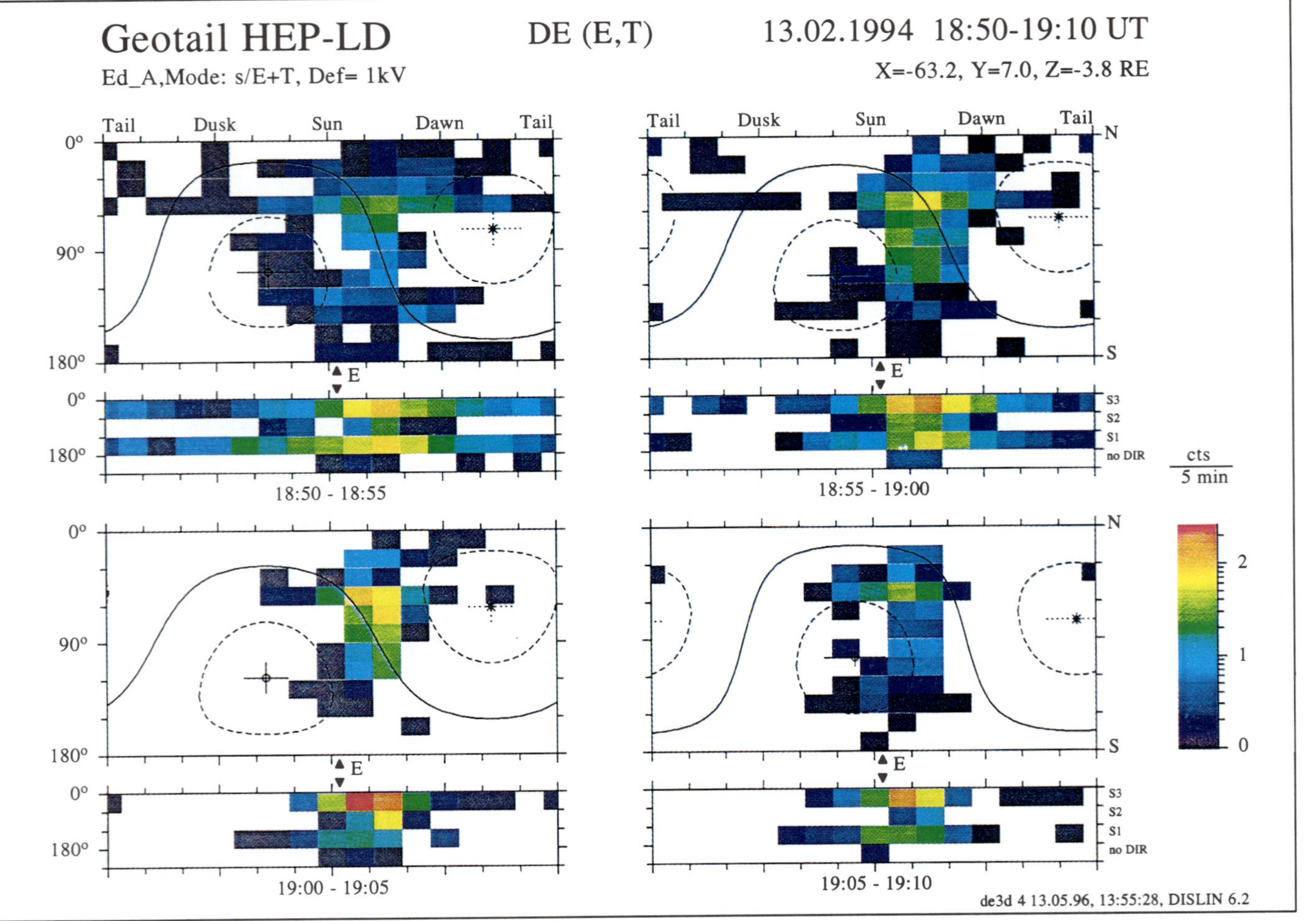

Figure 28. Polar-versus-azimuthal angle displays for the same time intervals as in Figure 27. The color-coded counts per angular bin represent the integral rate (summed over energy and particle mass). Each frame shows the direction of the B-vector and the loci of selected pitch angles (45° and 135° dashed, 90° heavy line). The small frames are similar displays with reduced resolution in the polar angle (60° only). The intensity increase in the 19:00–19:05 UT frame is mainly caused by CNO group particles (presumably oxygen) as indicated in Figure 27.

It is possible to transform from the Local Satellite Coordinate System to Boundary-Normal Coordinate Systems, such as the Magnetopause-Normal Coordinate System, the Shock-Normal Coordinate System and the Principle-Axes Coordinate System, by knowing the rotation matrices. We may also define a Local Magnetic Field Line Coordinate System $(x'y'z')$ in which the z' axes is along the local magnetic field. The latter can be obtained either by *in-situ* measurements, or from current magnetic-field models.

Energetic particles measured by RAPID at the same instance in time may originate from different sources. These different groups of particles are heated or accelerated through different mechanisms, and hence may possess different number density, mean velocity and thermal pressure tensor. In general they lie in different energy ranges, and are characterised by different types of distribution functions. Thus, it makes more sense for the derivation of plasma parameters to begin with distinguishing different populations of particles, and then to calculate moment integrals for them separately.

4. Summary

The dual sensor system in the RAPID spectrometer identifies energetic electrons and ions in the energy range 20–400 keV and 10 keV nucl^{-1}–4000 keV, respectively. The moderate ion mass resolution of about 4 (for oxygen ions) allows the identification of all ions and ion groups with significance in geospace plasmas.

Novel concepts are employed in the sensor systems to achieve angular resolution over a 180° range in polar angle. The imaging ion mass spectrometer IIMS uses two-dimensional time-of-flight/energy analysis to determine the particle mass. Addition of position sensing on the entrance foil (START foil) is a major evolutionary step in the detection technique which, in combination with a geometrically small STOP detector (at the end of the flight path), provides direction sensitivity. This approach divides the 180° polar range in 12 contiguous angular intervals. Sectoring the satellite spin plane with a maximum of 16 azimuthal intervals covers the unit sphere in velocity space completely with a total of 192 angular bins.

The imaging electron spectrometer IES achieves directional sensitivity by using novel microstrip solid-state devices in combination with a pin-hole acceptance. This technique divides the 180° polar range into 9 angular intervals which, with 16 sectors in the spin plane, corresponds to a total of 144 bins on the unit sphere. A major thrust of the paper is to show the signal processing and the formation of so-called Science Data which contain the ultimate return from RAPID.

The advanced RAPID concept represents an important scientific tool in the complement of plasma instruments on-board the four Cluster spacecraft. New capabilities of the novel spectrometer type permit a variety of studies in all regions of geospace visited by Cluster. Three selected areas, the bow shock and upstream region, reconnection at the magnetopause and FTEs, and substorms and the dynamics in the magnetotail are discussed in detail to highlight outstanding physical problems to be addressed by the RAPID investigation, and to emphasise specific contributions expected from studies of the suprathermal plasma component. On a broader scale, the three instruments RAPID, CIS, and PEACE form a set of spectrometers which covers an energy range from 0.7 eV to 400 keV for electrons and approxemately 0 keV to 4000 keV for ions. This wide energy range is a pre-requisite for advanced studies in geospace.

Acknowledgements

The RAPID investigation is the effort of an international consortium with experienced individuals from many European countries and the United States. RAPID is technically a combination of two self-contained spectrometer systems, the nuclei detector IIMS and the electron detector IES, which share the service of a common data processor. The IIMS spectrometer and the common DPU are constructed by MPAe and IDA (Technische Universität, Braunschweig) with the assistance of European participating institutions. The IES system is a contribution from LANL/Los Alamos (USA) and The Aerospace Corporation/Los Angeles (USA). It is worth noting that the RAPID subsystems IIMS/DPU benefitted greatly from a precursing similar spectrometer (LD) built for the Japanese spacecraft *Geotail*. Exclusive of the authors of this paper we list the following persons because of their significant contribution to the LD/RAPID effort: H. Böker, N. Meyer, E. Schäfer, O. Matuschek, K. H. Otto and H. Wirbs (MPAe); C. Herbert and P. O'Leary (St. Patrick's College, Maynooth, Ireland); B. Gerlach (IDA, Braunschweig), L. Ersland and Å. Sandvand (University of Bergen, Norway), R. Rasinkangas and T. Braysy (University of Oulu, Oulu/Finland), W. Ford and B. Laubscher (LANL, Los Alamos/USA), J. Osborn (The Aerospace Corporation, Los Angeles/USA), W. Gerwien (MBB, Munich/Germany), and K. Wurth and K. Leißle (Dornier System, Friedrichshafen/Germany), S. Brown and C. Smith (GSFC, Greenbelt/U.S.A.). We also thank I. Güttler and S. Siebert-Rust (MPAe) for their assistance in project management and administration. The enormous task to build, test and calibrate a total of six flight units for the two rather demanding missions Cluster and *Geotail* is not conceivable without the skill and support of these professionals.

The work at participating institutions is supported by national grants. Funding to MPAe and IDA is provided by the Deutsche Agentur für Raumfahrtangelegenheiten (DARA) under Grant Number 50 OC 89097; the Rutherford Appleton Laboratory contribution is supported by the UK Science and Engineering Research Council; the University of Bergen received grants from the Norwegian Research Council for Science and Humanities; and the effort in the United States has been supported by NASA at three laboratories: with Order Number S-47197E at the Los Alamos National Laboratory (Los Alamos internal technology development funding was used also); with Grand Numbers S-86525E and NAG5-2266 at the Boston

University Center for Space Physics; and with Contract Number NAS5-31452 at The Aerospace Corporation, Los Angeles.

References

Akasofu, S. I.: 1991, in J. R. Kan, T. A. Potemra, and T. Iijima (eds.), 'Development of Magnetospheric Physics', *Magnetospheric Substorms*, Geophysical Monograph 64, American Geophysical Union, p. 3.

Anagnostopoulos, G. C. and Kaliabetsos, G. D.: 1994, 'Shock Drift Acceleration of Energetic ($E \geq$ 50 keV) Protons and ($E \geq$ 37 keV nucl^{-1}) Alpha Particles at the Earth's Bow Shock as a Source of the Magnetosheath Energetic Ion Events', *J. Geophys. Res.* **99**, 2335.

Andrews, M. K. *et al.*: 1981, 'Plasma Sheet Motions Inferred from Medium-Energy Ion Measurements', *J. Geophys. Res.* **86**, 7543.

Asbridge, J. R. *et al.*: 1968, 'Outward Flow of Protons from the Earth's Bow Shock', *J. Geophys. Res.* **73**, 5777.

Axford, W. I.: 1981, 'Acceleration of Cosmic Rays by Shock Waves', workshop on Plasma Astrophysics, organised by the International School of Plasma Physics, held at Varenna, Italy, *ESA SP-161*, p. 425.

Baker, D. N. and Stone, E. C.: 1976, 'Energetic Electron Anisotropies in the Magnetotail: Identification of Open and Closed Field-Lines', *Geophys. Res. Letters* **3**, 557.

Baker, D. N. and Stone, E. C.: 1977a, 'The Magnetopause Electron Layer Along the Distant Magnetotail', *Geophys. Res. Letters* **4**, 133.

Baker, D. N. and Stone, E. C.: 1977b, 'The Relationship of Energy Flow at the Magnetopause to Geomagnetic Activity', *Geophys. Res. Letters* **4**, 395.

Baker, D. N. and Stone, E. C.: 1978, 'The Magnetopause Energetic Electron Layer. 1. Observations Along the Distant Magnetotail', *J. Geophys. Res.* **83**, 4327.

Baker, D. N. *et al.*: 1979, 'High-Energy Magnetospheric Protons and Their Dependence on Geomagnetic and Interplanetary Conditions', *J. Geophys. Res.* **84**, 7138.

Baker, D. N., Hones, E. W., Jr., Higbie, P. R., Belian, R. D., and Stauning, P.: 1981, 'Global Properties of the Magnetosphere During a Substorm Growth: A Case Study', *J. Geophys. Res.* **86**, 8941.

Baker, D. N., Higbie, P. R., Belian, R. D., Hones, E. W., and Klebesadel, R. W.: 1982, in C. T. Russel and D. J. Southwood (eds.), 'The Los Alamos Synchronous Orbit Date Set', *The IMS Source Book*, AGU, Washington DC, p. 82.

Baker, D. N., Akasofu, S. I., Baumjohann, Bieber, J. W., Fairfield, D. H., Hones, E. W., Jr., Mauk, B. H., McPherron, R. L., and Moore, T. E.: 1984, 'Substorms in the Magnetosphere, Chapter 8 of Solar Terrestrial Physics - Present and Future', *NASA Publ.*, Washington DC, p. 1120.

Baker, D. N., Simpson, J. A., and Eraker, J. H.: 1986, 'A Model of Impulsive Acceleration and Transport of Energetic Particles in Mercury's Magnetosphere', *J. Geophys. Res.* **91**, 8742.

Baumjohann, W., Paschmann, G., and Cattell, C. A.: 1989, 'Average Plasma Properties in the Central Plasma Sheet', *J. Geophys. Res.* **94**, 6597.

Baumjohann, W., Paschmann, G., and Lühr, H.: 1990, 'Characteristics of High-Speed Ion Flows in the Plasma Sheet', *J. Geophys. Res.* **95**, 3801.

Bieber, J. W., Stone, E. C., Hones, E. W., Jr., Baker, D. N., and Bame, S. J.: 1982, 'Plasma Behavior During Energetic Electron Streaming Events: Further Evidence for Substorm-Associated Magnetic Reconnection', *Geophys. Res. Letters* **9**, 664.

Crooker, N. U.: 1979, 'Dayside Merging and Cusp Geometry', *J. Geophys. Res.* **84**, 951.

Daglis, I. A. and Livi, S.: 1995, 'Merits for Substorm Research from Imaging of Charge-Exchange Neutral Atoms', *Ann. Geophys.* **13**, 505.

Daly, P. W. and Keppler, E.: 1989, 'Observation of a Flux Transfer Event on the Earthward Side of the Magnetopause', *Planetary Space Sci.* **30**, 331.

Daly, P. W. *et al.*: 1984, 'The Distribution of Reconnection Geometry in Flux Transfer Events Using Energetic Ion, Plasma, and Magnetic Data', *J. Geophys. Res.* **89**, 3843.

Dungey, J. W.: 1961, 'Interplanetary Magnetic Field and Auroral Zones', *Phys. Res. Letters* **6**, 47.

Dungey, J. W.: 1963, in C. DeWitt, J. Hiebolt, and A. Lebeau (eds.), 'The Structure of the Ionosphere, or Adventures in Velocity Space', *Geophysics: The Earth's Environment*, Gordon and Breach, New York, p. 526.

Ellison, D. C.: 1985, 'Shock Acceleration of Diffusive Ions at the Earth's Bow Shock: Acceleration Efficiency and A/Z Enhancement', *J. Geophys. Res.* **90**, 29.

Elphinstone, R. D., Murphree, J. S., Cogger, L. L., Hearn, D., Henderson, M. G., and Lundin, R.: 1991, in J. R. Kan, T. A. Potemra, and T. Iijima (eds.), 'Observations of Changes to the Auroral Distribution Prior to Substorm Onset', *Magnetospheric Substorms*, Geophys. Monogr. Ser. 64, AGU, Washington, D.C., p. 257.

Fahnenstiel, S. C.: 1981, 'Standing Waves Observed at the Dayside Magnetopause', *Geophys. Res. Letters* **11**, 1155.

Fairfield, D. H.: 1992, 'Advances in Magnetospheric Storm and Substorm Research: 1989–1991', *J. Geophys. Res.* **97**, 10 865.

Fennell, J. F. *et al.*: 1973, 'Access of Solar Protons to the Earth's Polar Caps', *J. Geophys. Res.* **78**, 1036.

Fennell, J. F., Roeder, J., Spence, H., Livi, S., and Grande, M.: 1993, *Composition Changes in Association with Energetic Particle Flux Drop Outs*, IAGA, Buenos Aires.

Fritz, T. A. *et al.*: 1982, 'The Magnetopause as Sensed by Energetic Particles, Magnetic Field, and Plasma Measurements on November 22, 1977', *J. Geophys. Res.* **87**, 2133.

Fritz, T. A., Baker, D. N., McPherron, R. L., and Lennartsson, W.: 1984, in E. W. Hones Jr. (ed.), 'Implications of the 11:00 UT March 22, 1979 CDAW-6 Substorm Event for the Role of Magnetic Reconnection in the Geomagnetic Tail', *Magnetic Reconnection in Space and Laboratory Plasmas*, Vol. 203, AGU, Washington DC.

Gosling, J. T. *et al.*: 1978, 'Observations of Two Distinct Populations of Bow-Shock Ions in the Upstream Solar Wind', *Geophys. Res. Letters* **5**, 957.

Gosling, J. T., Baker, D. N., Bame, S. J., Hones, E. W., Jr., McComas, D. J., Zwickl, R. D., Slavin, J. A., Smith, E. J., and Tsurutani, B. T.: 1984, 'Plasma Entry Into the Distant Tail Lobes: ISEE-3', *Geophys. Res. Letters* **11**, 1078.

Gosling, J. T., Baker, D. N., Bame, S. J., Feldman, W. C., and Zwickl, R. D.: 1985, 'North-South and Dawn-Dusk Plasma Asymmetries in the Distant Tail Lobes: ISEE 3', *J. Geophys. Res.* **90**, 6354.

Hesse, M., Smith, M. F., Herrero, F., Ghielmetti, A. G., Shelley, E. G., Wurz, P., Bochsler, P., Gallagher, D. L., Moore, T. E., and Stephen, T. S.: 1993, 'Imaging Ion Outflow in the High-Latitude Magnetosphere Using Low-Energy Neutral Atoms', *Society of Photo-Optical Instrumentation Engineers, Title: Instrumentation for Magnetospheric Imagery II*, **2008**, 83.

Hones, E. W., Jr.: 1976, in D. J. Williams (ed.), 'The Magnetotail: Its Generation and Dissipation', *Physics of Solar Planetary Env.*, AGU, Washington, D.C., p; 559.

Ipavich, F. M. *et al.*: 1981a, 'Temporal Development of Composition, Spectra and Anisotropies During Upstream Particle Events', *J. Geophys. Res.* **86**, 11153.

Ipavich, F. M. *et al.*: 1981b, 'A Statistical Survey of Ions Observed Upstream of the Earth's Bow Shock: Energy Spectra, Composition and Spatial Variation', *J. Geophys. Res.* **86**, 4337.

Jacquey, C., Sauvaud, J. A., and Dandouras, J.: 1991, 'Location and Propagation of the Magnetotail Current Disruption During Substorm Expansion: Analysis and Simulation of an ISEE Multi-Onset Event', *Geophys. Res. Letters* **18**, 389.

Jacquey, C., Sauvaud, J. A., Dandouras, J., and Korth, A.: 1993, 'Tailward Propagating Cross-Tail Current Disruption and Dynamics of Near-Earth Tail: A Multi-Point Measurement Analysis', *Geophys. Res. Letters* **20**, 983.

Kaufmann, R. L.: 1987, 'Substorm Currents: Growth Phase and Onset', *J. Geophys. Res.* **92**, 7471.

Korth, A., Pu, Z. Y., Kremser, G., and Roux, A.: 1991, in J. R. Kan, T. A. Potemra, and T. Iijima (eds.), 'A Statistical Study of Substorm Onset Conditions at Geostationary Orbit', *Magnetospheric Substorms*, Geophysical Monograph 64, American Geophysical Union, p. 343.

Korth, A. *et al.*: 1982, 'Two-Satellite Study of Proton Drift on Quiet Days', *Adv. Space Res.* **1**, 173.

Kremser, G., Tanskanen, P., Ullaland, S., Perraut, S., Roux, A., and Korth, A.: 1992, 'Influence of Storm Sudden Commencements on Magnetospheric Substorms', *Proc. International Conference on Substorms ICS-1*, March 23–27 1992, Kiruna, Sweden, ESA SP-335, p. 291.

Lee, M. A.: 1982, 'Coupled Hydromagnetic Wave Excitation and Ion Acceleration Upstream of the Earth's Bow Shock', *J. Geophys. Res.* **87**, 5093.
Lee, M. A.: 1983, 'Coupled Hydromagnetic Wave Excitation and Ion Acceleration at Interplanetary Traveling Shocks', *J. Geophys. Res.* **88**, 6109.
Lee, L. C. and Fu, Z. F.: 1985, 'A Theory of Magnetic Flux Transfer Events at the Earth's Magnetopause', *Geophys. Res. Letters* **12**, 105.
Lin, R. P. *et al.*: 1974, '30-to-100 keV protons Uplasma-Sheettream of the Earth's Bow Shock', *J. Geophys. Res.* **79**, 489.
Liu, Z. X. and Hu, Y. D.: 1988, 'Local Magnetic Field Reconnection Caused by Vortices in the Flow Field', *Geophys. Res. Letters* **15**, 752.
Liu, Z. X., Mu, S., and Pu, Z. Y.: 1994, 'Magnetic Field Configuration at Magnetopause for Different IMF Conditions', *Chinese Sci. Bull.* **39**, 1759.
Liu, Z. X., Zhu, Z., and Pu, Z. Y.: 1995, 'Transient Reconnection at High Latitude Boundary Layer', *Acta Geophys. Sinica* **39**, 1.
Lopez, R. E.: 1993, 'Auroral and Related Phenomena', *Adv. Space Res.* **4** (13), 189.
Lui, A. T. Y.: 1991, 'A Synthesis of Magnetospheric Substorm Models', *J. Geophys. Res.* **96**, 1849.
Lundin, R., Hultqvist, B., Pissarenko, N., and Zackarov, A.: 1982, 'The Plasma Mantle', *Space Sci. Rev.* **31**, 247.
McDiarmid, I. B., Burrows, J. R., and Wilson, M. D.: 1978, 'Magnetic Field Perturbation in the Dayside Cleft and Their Relationship to the IMF', *J. Geophys. Res.* **83**, 5753.
McPherron, R. L.: 1972, 'Substorm Related Changes in the Geomagnetic Tail: the Growth Phase', *Planetary Space Sci.* **20**, 1521.
McPherron, R. L. and Hones, E. W., Jr.: 1991, 'Relation of the Substorm Current Wedge to Dropout of the Near-Earth Plasma Sheet', *EOS* **72** (17), 251.
Maezawa, K.: 1976, 'Magnetic Convection Induced by the Positive and Negative Z-Components of the Interplanetary Magnetic Field: Quantitative Analysis Using Polar Cap Magnetic Records', *J. Geophys. Res.* **81**, 2289.
Mitchell, D. G. and Roelof, E. C.: 1983, 'Dependence of 50-keV Uplasma-Sheettream Ion Events at IMP 7&8 Upon Magnetic Field Bow Shock Geometry', *J. Geophys. Res.* **88**, 5623.
Mitchell, D. G., Angelopoulus, V., McEntire, R. W., Williams, D. J., Lui, A. T. Y., Decker, R. B., Krimigis, S. M., Roelof, E. C., Christon, S. P., Kokubun, S., Yamamoto, T., Paterson, W., Frank, L. A., Lepping, R. P., and Reeves, G. D.: 1994, 'Two-Point Correlations of Energetic Particle Beams in the Magnetotail', *EOS Trans. AGU* **75** (43), 575.
Orsini, S., Daglis, I. A., Candidi, M., Hshieh, K. C., Livi, S., and Wilken, B.: 1994, 'Model Calculation of Energetic Neutral Atoms Precipitation at Low Altitudes', *J. Geophys. Res.* **99**, 13 489.
Paschalidis, N. P., Sarris, E. T., Krimigis, S. M., McEntire, R. W., Levine, M. D., Daglis, I. A., and Anagnostopoulos, G. C.: 1994, 'Energetic Ion Distributions on Both Sides of the Earth's Magnetopause', *J. Geophys. Res.* **99**, 8687.
Paschmann, G. *et al.*: 1979, 'Plasma Acceleration at the Earth's Magnetopause: Evidence for Reconnection', *Nature* **282**, 243.
Paschmann, G *et al.*: 1980, 'Energization of Solar Wind Ions by Reflection from the Earth's Bow Shock', *J. Geophys. Res.* **85**, 4689.
Paschmann, G. *et al.*: 1981, 'Characteristics of Reflected and Diffuse Ions Upstream from the Earth's Bow Shock', *J. Geophys. Res.* **86**, 4355.
Paschmann, G. *et al.*: 1982, 'Plasma and Magnetic Field Characteristics of Magnetic Flux Transfer Events', *J. Geophys. Res.* **87**, 2159.
Pu, Z. Y. and Zong, Q-G.: 1996, *Plasma Parameters Derived from Data obtained with an Energetic Particle Spectrometer Onboard a Satellite*, MPAE-W-056-96-03.
Pu, Z. Y., Korth, A., and Kremser, G.: 1992, 'Plasma and Magnetic Field Parameters at Substorm Onsets Derived from GEOS 2 Observations', *J. Geophys. Res.* **97**, 19 341.
Rijnbeek, R. P. *et al.*: 1984, 'A Survey of Dayside Flux Transfer Events Observed by ISEE, 1 and 2 Magnetometers', *J. Geophys. Res.* **89**, 786.
Roelof, E. C., Mitchell, D. G., and Williams, D. J.: 1985, 'Energetic Neutral Atoms ($E \sim 50$ keV) from the Ring Current: IMP 7/8 and ISEE, 1', *J. Geophys. Res.* **90**, 10 991.

Roelof, E. C.: 1987, 'Energetic Neutral Atom Image of Storm-Time Ring Current', *Geophys. Res. Letters* **14**, 652.

Roelof, E. C., Mauk, B. H., and Meier, R. R.: 1992, 'Instrument Requirements for Imaging the Magnetosphere in Extreme-Ultraviolet and Energetic Neutral Atoms Derived from Computer-Simulated Images', *Society of Photo-Optical Instrumentation Engineers, Title: Instrumentation for Magnetospheric Imagery* **1744**, 19.

Rostoker, G., Akasofu, S. I., Forster, J., Greenwald, R. G., Kamide, Y., Kawasaki, K., Lui, A. T. Y., McPherron, R. L., and Russel, C. T.: 1980, 'Magnetospheric Substorms – Definition and Signatures', *J. Geophys. Res.* **85**, 1663.

Roux, A., Perraut, S., Robert, P., Morane, A., Pedersen, A., Korth, A., Kremser, G., Aparicio, B., Rodgers, D., and Pellinen, R.: 1991a, 'Plasma Sheet Instability Related to the Westward Traveling Surge', *J. Geophys. Res.* **96**, 17 697.

Roux, A., Perraut, S., Morane, A., Robert, P., Korth, A., Kremser, G., Pedersen, A., Pellinen, R., and Pu, Z. Y.: 1991b, in J. R. Kan, T. A. Potemra, and T. Iijima (eds.), 'Role of the Near-Earth Plasmasheet at Substorms', *Magnetospheric Substorms*, Geophysical Monograph 64, American Geophysical Union, p. 201.

Russell, C. T.: 1972, 'The Configuration of the Magnetosphere', in E. R. Dyer, Jr. (ed.), *Critical Problems of Magnetospheric Physics*, National Academy of Sciences, Washington D.C., p. 1.

Russell, C. T. and Elphic, R. C.: 1978, 'Initial ISEE Magnetometer Results: Magnetopause Observations', *Space Sci. Rev.* **22**, 681.

Russell, C. T. and Elphic, R. C.: 1979, 'ISEE Observations of Flux Transfer Events at the Dayside Magnetopause', *Geophys. Res. Letters* **6**, 33.

Sarris, E. T. and Axford, W. I.: 1979, 'Energetic Protons Near the Plasma Sheet Boundary', *Nature* **277**, 460.

Sarris, E. T. *et al.*: 1986, 'Simultaneous Measurements of Energetic Ion ($\geq$ 50 keV) and Electron ($\geq$ 220 keV) Activity Upstream of Earth's Bow Shock and Inside the Plasma Sheet: Magnetospheric Source for the November 3 and December 3, 1977 Upstream Events', preprint.

Saunders, M. A.: 1983, 'Recent ISEE Observations of the Magnetopause and Low Latitude Boundary Layer: A Review', *J. Geophys.* **52**, 190.

Saunders, M. A. *et al.*: 1984, 'Flux Transfer Events: Scale Size and Interior Structure', *Geophys. Res. Letters*, **11** 131.

Sauvaud, J. A. and Winkler, J. R.: 1980, 'Dynamics of Plasma, Energetic Particles, and Fields Near Synchronous Orbit in the Nighttime Sector During Magnetospheric Substorms', *J. Geophys. Res.*, **85** 2043.

Scholer, M.: 1988, 'Magnetic Flux Transfer Events at the Magnetopause Based on Single X-Line Burst Reconnection', *Geophys. Res. Letters* **15**, 291.

Scholer, M. *et al.*: 1979, 'Pitch Angle Distributions of Energetic Protons Near the Earth's Bow Shock', *Geophys. Res. Letters* **6**, 707.

Scholer, M., Gloeckler, G., Hovestand, D., Ipavich, F. M., Klecker, B., Baker, D. N., Baumjohann, W., and Tsurutani, B. T.: 1984, 'Simultaneous Observation of the Plasma Sheet in the Near Earth and Distant Magnetotail, ISEE-1 and ISEE-3', *Geophys. Res. Letters* **11**, 1034.

Scholer, M., Baker, D. N., Bame, S. J., Baumjohann, W., Gloeckler, G., Smith, E. J., and Tsurutani, B. T.: 1985, 'Correlated Observations of Substorm Effects in the Near Earth Region and the Distant Magnetotail', *J. Geophys. Res.* **90**, 4021.

Schunk, R. W.: 1988, 'Magnetosphere-Ionosphere-Thermosphere Coupling Processes', in *Solar-Terrestrial Energy Program: Major Scientific Problems*, Proc. of a SCOSTEP Symposium, held during the XXVII COSPAR Planetary Meeting, p. 52.

Siscoe, G.: 1993, 'Recent Activity in Substorm Research', *Adv. Space Res.* **4** (13), 165.

Sofko, G. J. *et al.*: 1979, in B. Battrick (ed.), 'STARE Ionospheric Electron Flows During the August 28, 1978 GEOS-2 Magnetopause Crossing', *Proc. Magnetospheric Boundary Layers Conf.*, Alpbach, ESA SP-148, p. 183.

Song, P. and Russell, C. T.: 1992, 'Model of the Low-Latitude Boundary Layer for Strongly Northward Interplanetary Magnetic Field', *J. Geophys. Res.* **97**, 1411.

Sonnerup, B. U. Ö.: 1969, 'Acceleration of Particles Reflected at a Shock Front', *J. Geophys. Res.* **74**, 1301.

Sonnerup, B. U. Ö., Paschmann, G., Papamastorakis, I., Sckopke, N., Haerendel, G., Bame, S. J., Asbridge, J. R., Gosling, J. T., and Russell, C. T.: 1981, 'Evidence for Magnetic Field Reconnection at the Earth's Magnetopause', *J. Geophys. Res.* **86**, 10049.

Southwood, D. J., Farrugia, C. J., and Saunders, M. A.: 1988, 'What Are Flux Transfer Events', *Planatary Space Sci.* **36**, 503.

Southwood, D. J. *et al.*: 1988, 'What Are Flux Transfer Events?', *Planetary Space Sci.* **36** (II), 503.

Stasiewicz, K.: 1991, 'A Global Model of Gyroviscous Field-Line Merging at the Magnetopause', *J. Geophys. Res.* **96**, 77.

Tsyganenko, N. A.: 1987, 'Global Quantitative Models of the Geomagnetic Field in the Cislunar Magnetosphere for Different Disturbance Levels', *Planetary Space Sci.* **35**, 1347.

Ullaland, S., Kremser, G., Tanskanen, P., Korth, A., Roux, A., Torkar, K., Block, L. P., and Iversen, I. B.: 1993, 'On the Development of a Magnetospheric Substorm Influenced by a Storm Sudden Commencement: Ground, Balloon, and Satellite Observations', *J. Geophys. Res.* **98**, 15 381.

Van Allen, J. A. *et al.*: 1971, 'Asymmetric Access of Energetic Solar Protons to the Earth's North and South Polar Caps', *J. Geophys. Res.* **76**, 4262.

Wibberenz, G. *et al.*: 1985, 'Dynamics of Intense Upstream Ion Events', *J. Geophys. Res.* **90**, 283.

Wilken, B.: 1984, 'Identification Techniques for Nuclear Particles in Space Plasma Research and Selected Experimental Results', *Rep. Prog. Phys.* **47**, 767.

Wilken, B., Zong, Q. G., Daglis, I. A., Doke, T., Livi, S., Maezawa, K., Pu, Z. Y., Ullaland, S., Yamamoto, T.: 1995, 'Tailward Flowing Energetic Oxygen Ion Bursts Associated with Multiple Flux Ropes in the Distant Magnetotail: *Geotail* Observations', *Geophys. Res. Letters*, accepted.

Williams, D. J., Roelof, E. C., Mitchell, D. G.: 1992, 'Global Magnetospheric Imaging', *Rev. Geophys.* **30**, 183.

Yeoman, T., Lühr, H., Friedel, R. W. H., Coles, S., Grande, M., Perry, C. H., Lester, M., Smith, P. N., Orr, D., and Singer, H.: 1997, CRRES: Ground-Based Multi-Instrument Observations of an Interval of Substorm Activity', *Ann. Geophys.*, submitted.

Zhang, H. and Liu, Z. X.: 1995, 'Manifestation of Disturbance at Magnetopause in Plasmasheet', *Chinese Sci. Bull.*, submitted.

Zhang, H., Liu, Z. X., and Pu, Z. Y.: 1995, 'Transient Reconnection at Nightside Magnetopause', *J. Geophys. Res.*, submitted.

CLUSTER MISSION OPERATIONS

P. FERRI and M. WARHAUT
Mission Operations Department, European Space Operations Centre, Darmstadt, Germany

Abstract. The Cluster ground segment design and mission operations concept have been defined according to the basic mission requirements, namely, to allow the transfer of the four spacecraft from the initial geostationary transfer orbit achieved at separation from the launcher into the final highly elliptical polar orbits, such that in the areas of scientific interest along their orbits, the four spacecraft will form a tetrahedral configuration with pre-defined separation distances, to be changed every six months during the mission.

The Cluster mission operations will be carried out by ESA from its European Space Operations Centre; the task of merging the Principal Investigators' requests into coordinated, regular scientific mission planning inputs to ESOC will be undertaken by the Joint Science Operations Centre. The mission products will be distributed to the scientific community regularly in form of CD-ROMs. Principal Investigators will also have access to quick-look science, housekeeping telemetry and auxiliary data via an electronic network.

1. Introduction

The development and implementation of the Cluster ground segment and its post-launch operation in conjunction with the space segment has been delegated to the European Space Operations Centre (ESOC) (Warhaut, 1995). This task includes: the transfer of the four Cluster satellites into their operational orbits in the correct relative positions, following their launch on the first flight of the new Ariane-5 launcher; the combined operation of the four spacecraft and their payload throughout their in-orbit lifetime (a minimum of two years for science-product generation) according to the mission requirements and the requests generated by the Principal Investigators (PIs) during the mission itself; the collection and transfer to the scientific community of all the raw science data generated by the payload instruments, in conjunction with selected spacecraft and mission auxiliary data.

The challenge of Cluster from the mission operations' point-of-view is the simultaneous launch and control of four identical spacecraft which have to be configured and manoeuvred into a complex set of trajectories with a series of time-critical orbit changes. During the routine phases of the mission the Cluster ground segment has to support and coordinate the simultaneous operations of the four spacecraft with careful utilisation of the available ground contact with the spacecraft.

Space Science Reviews **79:** 475–485, 1997.

2. Ground Segment

The Cluster ground segment (Figure 1) will provide capabilities for monitoring and control of the four satellites in all mission phases, as well as for reception, archiving and distribution of the payload instrument data. It is composed of a Ground Stations and Communications Network (GSCN), an Operations Control Centre (OCC), a dedicated communications network for data distribution to the scientific community and the Cluster Science Data System (CSDS).

The GSCN (Figure 2) performs telemetry, telecommand and tracking operations within the S-band frequencies, via the two ESA ground stations located in Odenwald (Germany) and in Redu (Belgium), and Wideband Data (WBD) acquisition via the NASA Deep-Space Network (DSN). In the critical Launch and Early Orbit Phase (LEOP) and Transfer Orbit Phase (TOP), the ground station network will be augmented with stations located in Kourou, Malindi and Perth. NASA/DSN will provide back-up support during LEOP and, for about 10 days of the TOP phase, during which the spacecraft will be too distant to be properly acquired by the ESA stations, NASA/DSN will become the prime stations.

The Operations Control Centre (OCC) located at ESOC in Darmstadt, performs mission control of the spacecraft and payload, compatible with satellite safety, scientific requirements and overall system resources. It also carries out all activities related to mission planning and scheduling of the overall system (ground segment, spacecraft and payload), taking into account the requests received from the PIs, as well as all system constraints and capabilities.

The core of the OCC facilities is the Cluster Mission Control System (CMCS, see Sorensen *et al.*, this issue) which supports all mission control tasks, including real-time/near real-time data- processing tasks essential for controlling the mission; acquisition and interim storage of raw scientific data, to be accessible together with raw housekeeping and auxiliary data by the scientists at remote locations, and to provide for distribution of all data on a hard data medium; payload command request handling and the planning and scheduling of satellite operations.

The mission control system is supplemented by the Flight Dynamics System (FDS), which controls all activities related to orbit and attitude determination and control, and a number of auxiliary hardware and software tools.

The dedicated communications network provides the support services for retrieval of quick-look raw science, housekeeping and auxiliary data stored at the OCC, submission of command requests to the OCC and electronic data exchange of scientifically processed data and related software between the individual CSDS Data Centres (see also Schmidt and Escoubet, this issue).

The CSDS, which includes the Joint Science Operations Centre (JSOC, see Hapgood *et al.*, this issue) at Didcot (UK), supports scientific mission planning and instrument command request preparation for submittal to the OCC, and a set of national Data Centres (DCs) to support scientific data-processing and distribution activities.

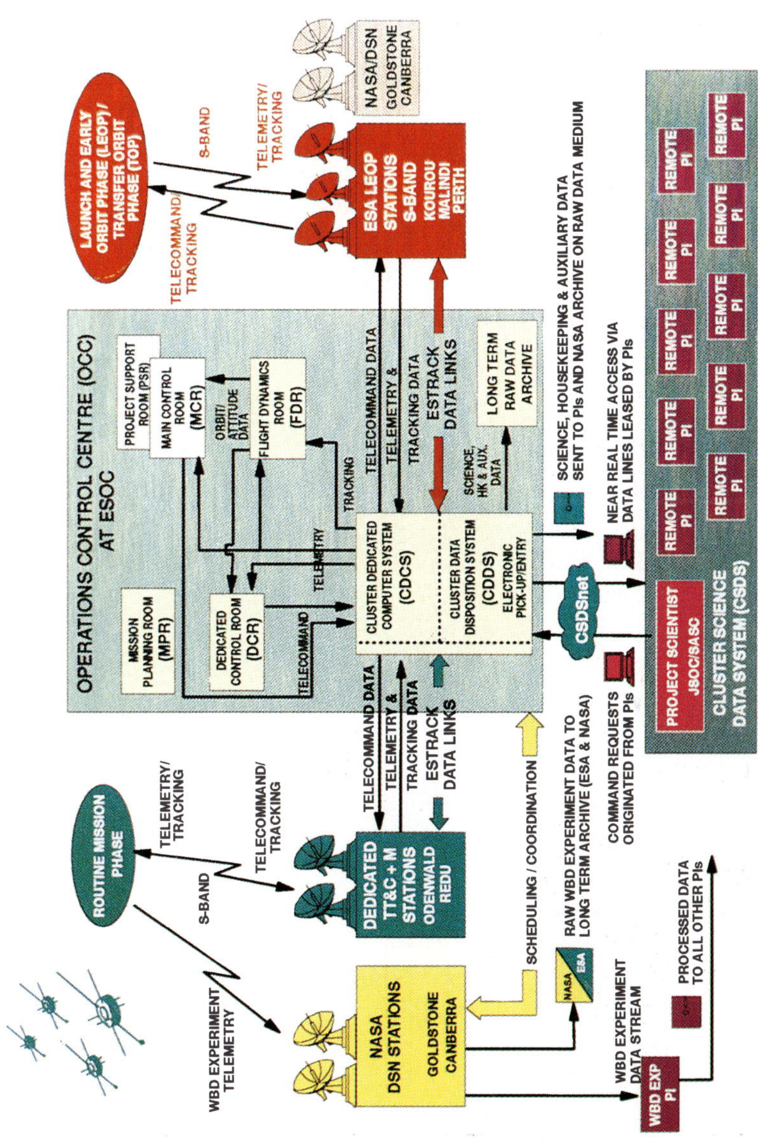

Figure 1. Cluster Ground Segment.

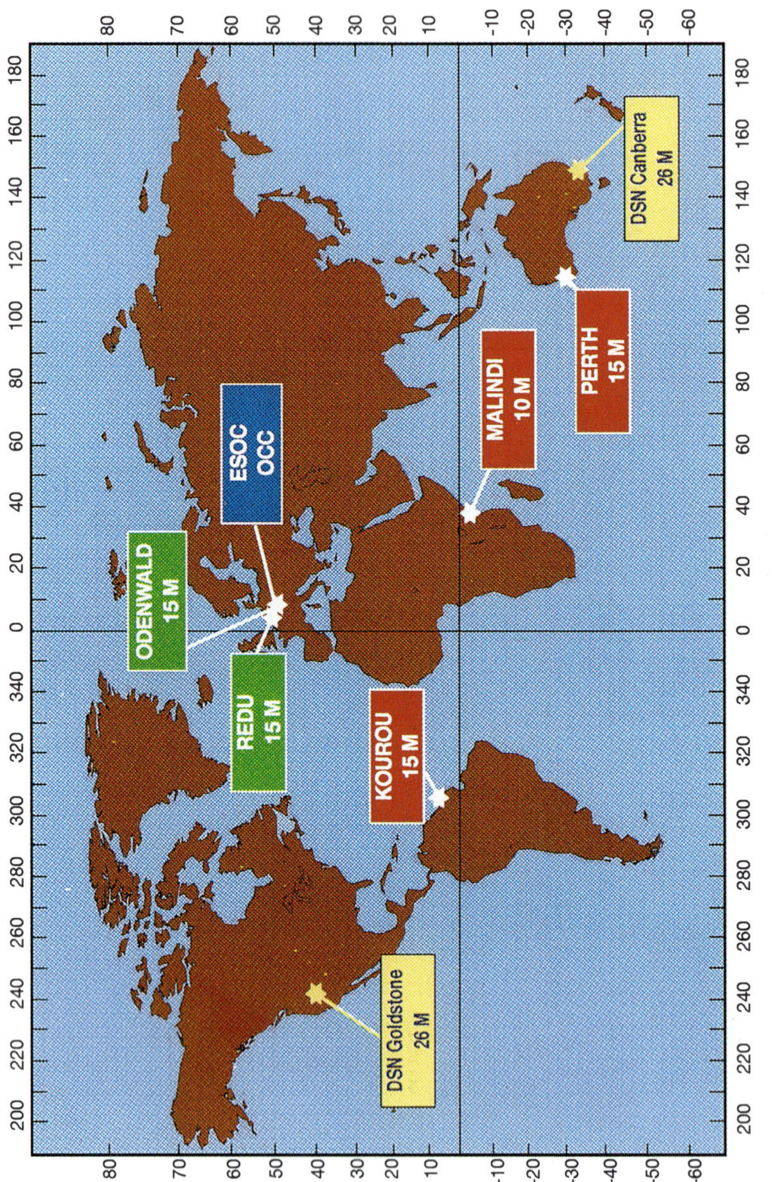

Figure 2. Cluster Ground Station Network.

3. Mission Operations

The Cluster mission can be broadly divided into three phases:

– the Launch and Early Orbit Phase/Transfer Orbit Phase (LEOP/TOP), covering the activities from launch and first acquisition of the four spacecraft throughout a four-week period of orbit manoeuvres until delivery of the four spacecraft into the desired mission orbits;

– the Commissioning and Verification Phase (CVP), a period of about ten weeks in which the initial payload activation will be performed, and the spacecraft subsystems and payload instruments will be commissioned and calibrated to verify their in-orbit performance and to prepare for the scientific measurements proper;

– the Mission Operations Phase (MOP), consisting of all routine operations in support of payload activities and scientific data acquisition. This phase has a nominal duration of 24 months, interrupted once every six months by orbit manoeuvres aimed to adjust the separation distances between the four spacecraft.

3.1. Launch and transfer to mission orbit

All four Cluster spacecraft will be launched on a single Ariane-5 launcher, which will inject each of the spacecraft into a geostationary transfer orbit approximately 30 min after lift-off, in a timed sequence lasting a few minutes. The ground tracking network in this phase will consist of five ESA stations supported by the two NASA DSN stations at Goldstone (U.S.A.) and Canberra (Australia), all of which will be connected to a single Operations Control Centre located at ESOC. During the first revolution in the injection orbit the most time-critical activities will be carried out in rapid succession on all four spacecraft, under coverage of two of the ESA stations. These activities include initialisation of the on- board thermal- and reaction-control subsystems, and the initial verification of the health of the spacecraft subsystems. A spin-up manoeuvre from the initial five to the nominal fifteen revolutions per minute (rpm) will conclude the initial activities on all spacecraft within eight hours from launch.

After collection of tracking measurements from the ground stations and acquisition of the desired attitude, a sequence of manoeuvres will be initiated to reach the pre-defined mission orbits, a set of highly elliptical (28 000 km perigee, 125 000 km apogee) orbits with inclination close to 90° and the line of absides close to the equatorial plane. The spacecraft will be manoeuvred in pairs, to minimise operational complication in particular in case of contingencies. The nominal transfer scenario foresees six orbit manoeuvres for each spacecraft. A number of attitude- and spin-correction manoeuvres will also be carried out during this phase.

Most of these activities will be performed in real-time contact with one of the ground stations, although for some orbit manoeuvres ground coverage cannot be guaranteed. A subset of the telemetry data generated during the long non-coverage periods will be recorded in a dedicated area of the on-board memory. Use of the solid-state recorders designed for the science data collection is also foreseen in critical periods of this phase. Owing to the large distances from Earth reached by the spacecraft during this phase, up to 380 000 km, the RF-link margins of the 15 m antennae of the ESA ground stations will not be sufficient to guarantee full coverage, which will therefore for several days be provided by the 26 and 34 m antennae of the DSN stations.

During the launch and early orbit phases the Mission Control Team at ESOC will be on duty around the clock, in two 12-hour shifts. A complement of about 150 people will be involved full or part-time in Cluster operations during this critical phase of the mission.

3.2. Spacecraft and Payload Commissioning

Once achievement of the nominal mission orbit is confirmed, the spacecraft will be configured for scientific measurements. This requires first of all the deployment of the two rigid booms and the four 50 m long wire booms which carry some of the instrument sensors. In addition, the second RF antenna, located on the bottom of the spacecraft, will be deployed at the beginning of this phase. These activities involve critical pyrotechnics operations and attitude and spin manoeuvres. They will start once all four spacecraft have reached the desired operational orbit and the end of the Transfer Orbit Phase (TOP) has been declared. Deployment of the wire booms will be carried out in successive steps with long pauses in between to allow collection and analysis of scientific data from the sensors on the top of the wires, with a total duration of about two weeks for each spacecraft. By overlapping the activities on different spacecraft, wire boom deployment operations will be completed in about one month.

Payload instrument commissioning will start immediately after the completion of the rigid-boom-deployment activities, and continue throughout the following eight to ten weeks. Once all the instruments have been commissioned a measurement campaign will be carried out, aiming to identify all sources of interference between different instruments. Finally, in the intercalibration phase, all the instruments will be operated together in specific regions of the magnetosphere. This final phase also serves as a commissioning exercise for the ground-segment mission-planning cycle, involving the JSOC at Didcot and the OCC at ESOC.

Ground coverage during this phase will be ensured by two ESA ground stations at Redu and Odenwald; the DSN ground stations used during the previous mission phases will also be available but only for a few hours per orbit. The OCC will provide full and continuous operations support during the entire phase although, due to the long duration and the low time-criticality of the commissioning activities, the Mission Control Team will be smaller than during the LEOP/TOP phases. However, a team of experts for each instrument being commissioned will be required to be on hand at the OCC during this period. Instrument teams and their ground support equipment will be located in a dedicated area of the OCC and linked to the operations control rooms via voice and data links. All operations will be carried out following pre-defined procedures, which may involve 'go-nogo' inputs to the mission control team by the instrument science teams, based on their evaluation of the scientific telemetry data acquired and processed in real-time. In the case of unforeseen on-board software maintenance activities on the payload, a direct link will be provided for transmission of the relevant software updates from the instrument ground-support equipment to the OCC mission-control system, which uses the data to construct the relevant telecommands.

3.3. Routine Mission Operations

The mission-control concept during the routine phase is based on the use of a single control centre in conjunction with two dedicated ESA ground stations (Redu and Odenwald). All payload operations will be pre-planned and executed according to an agreed plan produced in a cyclic mission-planning exercise. No real-time payload operations are foreseen during this phase, except for special operations like complex instrument software maintenance activities or in the case of contingencies. Real-time operations with the spacecraft will be normally limited to the acquisition of stored telemetry data of all types (housekeeping, memory dumps, science), recorded during the long non- coverage periods and dumped to the ground station during the real-time contact periods, and the uplink to the on-board memory of time-tagged commands for the execution of all pre-planned spacecraft and payload operations.

With the four spacecraft in the nominal operations orbit the ground coverage is about 50% of the total orbital period of about 57 hours, divided into a long visibility period (pass) of up to 20 hours and one or two short passes of a few hours every orbit. The relatively short distances between the four spacecraft and between the two ground stations result in a practically identical (within a few minutes) ground coverage for all spacecraft and stations. Typically one station will track one spacecraft for half of the available geometrical visibility time, and a second spacecraft for the rest of the pass. The second ground station will be similarly assigned to the second pair of spacecraft, resulting in an average coverage for each spacecraft of about 25% of the mission time.

The Cluster spacecraft generates fixed-frame telemetry in a number of different possible combinations of frame types/telemetry modes (Table I). Simultaneous downlink of real-time housekeeping and low-rate science data is possible during on-board storage dump activities, allowing continuous monitoring of spacecraft health and independence between routine spacecraft control activities and science data acquisition.

The baseline approach for science operations is to activate the payload instruments and acquire science data – in one of six possible combinations of sampling rates – only in specific areas of the orbit, when crossing the regions of the magnetosphere that are most interesting from a scientific point-of-view. If the spacecraft is

Table I
Cluster telemetry modes

Telemetry modes	Bit rate (kbit s^{-1})	Contents
1, 2	2	housekeeping
3, 4	22	housekeeping, normal mode science
5, 6	131	housekeeping, burst mode science
7	131/262	housekeeping, normal mode science, recorder dumps
8	262	WBD science data
9, 10	131/262	housekeeping, normal mode science, memory dumps

in real-time contact with the ground upon entering one of the regions of scientific interest, then the science data are directly downlinked to the station with a bit rate of 22 kbit s^{-1} (normal mode) or 131 kbit s^{-1} (burst mode); otherwise all data are stored in one of two on-board recorders of the capacity of 2.25 Gbit each.

Considering the available resources, including on-board storage, downlink, ground segment storage and processing capabilities, continuous scientific data collection in normal mode will be possible over the entire orbit, whilst the use of burst mode data-acquisition periods will reduce the scientific measurements to a minimum of six hours per orbit, in the burst mode only case.

In addition to the above modes of science data acquisition the Wideband Data (WBD) instrument will be capable of transmitting data at a rate of 262 kbit s^{-1} for short periods each orbit to a DSN ground station, with the only constraints being availability of the DSN station and non-simultaneity of coverage from an ESA station.

Dumping of on-board recorded data of all kinds will be possible at two speeds, depending on the distance of the spacecraft from the ground station (131 or 262 kbit s^{-1}). As indicated above, all dumping activities are incompatible with real-time collection of science data in burst mode and with the special dump of WBD data to the DSN stations.

3.3.1. *Mission Operations Planning and Implementation*

All routine mission operations will be performed in accordance with an integrated space and ground segment schedule produced by a periodic planning exercise (Charrat *et al.*, 1996), which can be broadly divided into a long-term baseline planning cycle and a short-term detailed planning cycle.

The long-term planning is based on a planning period of six months, generally the period over which the spacecraft separation distances remain unchanged, and is completed three months before the start of the planning period at the latest. This activity involves the Project Scientist and all PIs who together define the areas of scientific interest and the desired data acquisition modes for each orbit in the planning period. The plan is then presented at the Science Operations Working Group (SOWG), a small working group comprising the Project Scientist, representatives of each of the payload instruments, ESOC and JSOC, for comments and approval. ESOC is also requested to analyse the plan for feasibility in terms of spacecraft and ground-segment resources availability and operability, and to provide detailed comments. The finalised version of the baseline plan is then completed by all PIs with indications, for each of the data acquisition periods contained in the plan, of the required operating mode for their instrument. The plan is then passed to JSOC to form the basis of the detailed short-term planning cycles.

The short-term planning is based on planning periods of three orbits (about one week) and begins eight weeks before the start of each planning period, when JSOC submits electronically to ESOC a complete set of operations requests for the payloads of all four spacecraft, based on the agreed baseline plan and on eventual further inputs directly received from the PIs. At ESOC this request is analysed via a software mission planning tool which checks the correctness of the request and its feasibility in terms of resources (on-board power and data storage). If the request is acceptable, a further check is carried out on the availability of ground segment resources (ground station data storage and link capacity between the station and the OCC), If the response is positive the request is accepted and confirmed back to JSOC. If not, the plan is rejected with a detailed explanation of the resource violation and JSOC is requested to modify it accordingly. At this stage ESOC also prepares, based on the accepted request from JSOC, the necessary request for Deep-Space Network support for the downlink of data from the WBD instrument.

Further iterations of the request for science operations related to a specific planning period are allowed, including possible changes to already accepted requests, until the week before the start of the planning period. At this point all requests are frozen and the final implementation cycle begins at ESOC. Only minor change requests involving operations which have no impact on the space- or ground-resources utilisation are allowed in this phase.

Final implementation of the science operations works on a cycle of nominally three days, in which the finalised plan is translated into time-tagged operations schedules for the four spacecraft and the two ground stations. The schedules are then released to the relevant control computers.

Shortly before the beginning of a visibility period the ground station operations schedule is activated and the station operations are performed, in the absence of anomalies, completely automatically. The time-tagged operations schedule is then uplinked to each spacecraft at the next opportunity during the visibility period; in general a period of 48 hours ahead will be covered by each uplink of time-tagged commands to guarantee continuity of science operations, bearing in mind that each spacecraft will undergo long periods (up to 30 hours) without ground contact.

The other major activity during the periods of contact with the spacecraft is the dump of all data (housekeeping and science) recorded during the previous non-coverage period. The data are dumped in reverse order, stored at the ground station and transmitted off-line to the OCC. Depending on the amount of data dumped and on the real-time activities underway at the time of processing, the transfer of the dumped data to the OCC may take up to 24 hours from the end of the dump. UTC time-stamping of the real-time frames is carried out at the station, allowing a correlation of on-board events and the UTC within 2 ms. The on-board time to UTC correlation, derived from the real-time telemetry, is then applied at the OCC to the recorded telemetry, using the on-board-time field of the recorded frames.

All real-time routine operations are carried out by two spacecraft controllers, in charge of the control and monitoring of two spacecraft and one ground station each. It should be noted that the Cluster-dedicated ground stations of Redu and Odenwald will be unmanned and remotely controlled from ESOC, where the main Station Computers are located. Engineering support is provided on-call within the Flight Control Team in case of unexpected behaviour of one of the spacecraft, whilst network controllers and engineers are available in case of non-nominal behaviour of the automatic ground station control schedule. Mission planning activities are carried out during normal working hours, five days a week by the Flight Control Team engineers.

3.3.2. *Mission Product Distribution*

At the OCC science data and housekeeping data contained in the telemetry stream will be extracted and stored as raw data, chronologically ordered and sorted by spacecraft and experiment, and will include timing data so as to permit correlation with respect to UTC. Further processing of science data and verification of the correct functioning of the experiments will be performed by the Principal Investigators.

Auxiliary data such as spacecraft orbital positions, separation distances, attitude, spin phases and rates, will be provided to the Principal Investigators to allow full analysis of the scientific measurements.

The data acquired from the Cluster spacecraft, including the auxiliary data, will amount to approximately 1 Gbyte day^{-1}. The Data Disposition System allows quick access to the most recent data (the last 10 days), which will be available over communications lines on a request basis. This access is restricted to PIs and JSOC only and to a fixed maximum amount per day (a few Mbytes, user-dependent). The complete data set is distributed in the form of CD-ROMs to all PIs and Cluster Science Data Centres within three weeks of generation of the data, on a daily basis. A long-term archive of all mission data will be kept at ESOC for ten years from the end of the mission.

4. Conclusions

The Cluster mission operations will be supported from ESOC in the last years of the century, forming a bridge between the scientific missions of the late 1980s and early 1990s like Hipparcos and Eureca and the missions of the next millennium such as XMM, Integral and Rosetta. The design of the ground segment and the operations concept anticipate many of the characteristics of these future missions in terms of high usage of distributed architecture and local- and wide-area networks for data-processing and distribution, automatic remote control of the ground stations, reduction in operations costs by concentration of mission-control responsibilities into two 24-hour positions. The simultaneous launch of four satellites and the related activities in low Earth orbit will also demonstrate the capability of ESOC to support this type of parallel critical operations with a single control centre and mission-control team.

References

Charrat, B., Ferri, P., and Sweeney, M. A.: 1996, 'The Cluster Mission Planning Concept', *ESA Bull.* **87**.

Hapgood, M.A. *et al.*: 1997, 'The Joint Science Operations Centre', *Space Sci. Rev.*, this issue.

Schmidt, R. and Escoubet, C. P.: 1997, 'The Cluster Data System (CSDS) – a New Approach to the Distribution of Scientific Data', *Space Sci. Rev.*, this issue.

Sorensen, E. M., Merri, M., and Di Girolamo, G.: 1997, 'The Cluster Data Processing System: a Distributed System in Support of a Challenging Scientific Mission', *Space Sci. Rev.*, this issue.

Warhaut, M.: 1995, 'Cluster Mission Implementation Plan', ESOC Internal Paper CL-ESC-PL-010, Issue 4, February 1995.

THE JOINT SCIENCE OPERATIONS CENTRE

M. A. HAPGOOD, T. G. DIMBYLOW, D. C. SUTCLIFFE, P. A. CHAIZY,*
P. S. FERRON,** P. M. HILL*** and X. Y. TIRATAY‡
Cluster Joint Science Operations Centre, Space Science Department
Rutherford Appleton Laboratory, Chilton, Didcot
Oxfordshire, OX11 0QX, U.K.

Abstract. The Joint Science Operations Centre (JSOC) has been established to provide the operational interface between the Instrument Principal Investigators (PIs) and the European Space Operations Centre (ESOC). Its key task will be to merge inputs from the Cluster instrument teams and to generate the coordinated command schedule for operation of the scientific payload. In addition, it will collect and process data needed to plan those operations and will monitor the performance of the mission and individual instruments. This paper outlines the JSOC subsystems that have been built to carry out these tasks and highlights points of scientific or technical interest within these systems.

1. Introduction

The Joint Science Operations Centre (JSOC) has been established to support the Cluster Project Scientist in coordinating the complex science operations of the Cluster mission (Dunford, 1993). It will form part of the operational ground segment for the Cluster mission (Ferri and Warhaut, this issue) and will work under the direction of the Cluster Project Scientist to provide the operational interface between the Instrument Principal Investigators (PIs) and the European Space Operations Centre (ESOC), which is responsible for the day-to-day operation of the four spacecraft.

The services to be provided by JSOC may be divided into four main areas:

– Commanding. This is the preparation and review of the coordinated command schedule for the Cluster science payload during nominal mission operations. JSOC will implement this schedule in accordance with the coordinated mission planning agreed by the Cluster Science Working Team (SWT) and the known scientific and technical constraints on the mission and individual instruments.

– Planning. This is the supply of information to support mission planning by the Science Working Team and to support the JSOC commanding tasks as outlined above.

– Monitoring. This is the monitoring of mission and instrument performance on behalf of SWT and of the PIs.

* On secondment from Centre National de la Recherche Scientifique.
** On secondment from Institutet för Rymdfysik, Uppsala.
*** On secondment from Max Planck Institut für extraterrestrische Physik, Garching.
‡ Visiting from Ecole Nationale Supérieure de Physique, Grenoble.

Space Science Reviews **79:** 487–525, 1997.

– Information dissemination. This is the dissemination of information to the scientific community in support of ground-based solar-terrestrial physics programmes and of other related missions.

JSOC has been built by, and will be operated by, a small international team based at the Rutherford Appleton Laboratory (RAL) under contract to ESA. JSOC comprises four main subsystems, each of which deals with one of the four tasks described above. These subsystems are described in Sections 2–5 of this paper. Following this, Section 6 describes the hardware and software environments within which JSOC has been implemented. Finally, Section 7 outlines an enhancement to JSOC which has recently been funded by ESA – the provision of support to help coordinate ground-based solar-terrestrial physics observations with Cluster operations.

This paper does not aim to give a complete detailed description of JSOC. That objective is already met by the formal design documents produced during the JSOC implementation phase (see Dimbylow and Hill, 1995, and other references given therein). Instead we aim here to give an overview of the various JSOC subsystems and then to highlight points of major scientific and technical interest.

2. Commanding Subsystem

2.1. Overview

The key task of JSOC is to produce the coordinated command schedule for the Cluster science payload. This schedule is then passed to the Operations Control Centre at ESOC, where it is processed by the Cluster Mission Planning System (Sørensen *et al.*, this issue) to generate the individual command schedules for the four Cluster spacecraft and the ground segment.

Within this scheme, both JSOC and ESOC will check for potential conflicts in the command schedule. JSOC is responsible for checking 'science conflicts', i.e., detecting instrument modes which are likely to reduce the overall scientific return from the payload; ESOC is responsible for checking 'engineering conflicts', i.e., commands that could endanger the spacecraft or the instruments. The JSOC checking for 'science conflicts' is implemented via PI-defined constraints as discussed below.

The details of the ESOC checking are outside the scope of this paper. However, it is appropriate to note here that the command schedules prepared by JSOC are specified in terms of 'command sequences'. These sequences are blocks of time-tagged commands which have a well-understood functionality. Many of the sequences to be used in JSOC command schedules have already been verified in spacecraft tests on the ground; the other 'JSOC' sequences will be tested during the commissioning phase in the first three months after launch. By restricting JSOC command schedules to use only these well-understood sequences, the risk of dangerous commands being sent to the payload is reduced. Thus when new sequences are defined during mission operations, their safe operation must be verified as soon as possible.

In generating the payload command schedule, JSOC will process various inputs as follows:

– The Master Science Plan generated by the Cluster Science Working Team (Paschmann and Sckopke, 1996). This Plan specifies the agreed timeline of the spacecraft telemetry mode (e.g., normal mode data-taking, burst mode data-taking) and of the instrument modes to be set up on each of the four spacecraft. JSOC will prepare a computerised version of the Plan; this will be held in a set of files termed Top-Level Instrument Schedule (TLIS) files. There will be one file for each JSOC planning period, which will typically be three orbits long (perigee-to-perigee) – giving a duration of about one week.

– PI-supplied constraints on instrument modes, e.g., between modes of different instruments, between instrument modes and the plasma regions in which they are used. JSOC will verify that the Master Science Plan (coded in TLIS files) meets these constraints and issue warnings to SWT when exceptions are found. This is the mechanism by which JSOC will identify science conflicts and liaise with the SWT in order to resolve them.

– PI-supplied procedures describing how to command transitions between the modes listed in the Master Science Plan. JSOC will use these procedures to construct an initial command schedule for each instrument. The schedule for an instrument will be made available for review by the PI team. It will be held in a file termed a PI Observations Request (PIOR) file. These procedures (and hence the PIOR files) may contain names which represent instrument parameters that will be resolved to numerical values at a later stage of the command schedule preparation, e.g., instrument gain settings. These parameter names are termed 'henceforth parameters'. They must be fixed within a planning period but may be revised between periods.

– PI requests for revision of his or her instrument command schedule. These should refine details of instrument operation. The PI is responsible for ensuring that such requests do not alter the instrument mode agreed by SWT. The revised command schedule will be made available to the PI and can be subject to further review and revision, as required by the PI.

– PI-supplied values for henceforth parameters.

The coordinated command schedule is generated by merging the command schedules of the individual instruments and resolving values of henceforth parameters. The coordinated schedule is then transferred to ESOC for processing by the Cluster Mission Planning System (Sørensen *et al.*, this issue). This processing can generate either one or three responses from ESOC to JSOC thus:

– A validation report is always returned. This indicates whether or not the ESOC software has successfully checked the syntax and content of the command schedule. If the validation is unsuccessful, JSOC must identify the incorrect data, take action to correct them, generate a revised schedule and submit it to ESOC.

– A spacecraft operations file (SCOP). This is returned only if the schedule has been validated successfully. It indicates whether or not ESOC has been able to accept all payload operations requested in the coordinated schedule. If so, the SCOP reports the detailed payload operations planned on the four spacecraft. If not, JSOC must identify ways of resolving the problem and ask the SWT to decide which solution should be implemented. JSOC will implement this solution, generate a revised schedule and submit it to ESOC.

– A ground segment operations file. This is returned together with the SCOP. It indicates details of the operations planned in the Cluster ground segment.

Figure 1 outlines the relationships between JSOC and other parts of the mission. We use this figure to emphasise three points:

(1) JSOC will work under the direction of the Cluster Project Scientist. In particular, the coordinated command schedule, generated by JSOC, is formally subject to his approval before transmission to ESOC. Thus JSOC will seek his instructions in the case of critical problems in the preparation of the schedule.

(2) JSOC will interact with the PIs in two distinct ways: (a) collectively in the form of the Science Working Team for mission-wide issues; (b) individually for issues affecting the detailed operation of a single instrument.

(3) The coordination of Cluster operations with ground-based STP experiments and with other space missions will be determined via the Science Working Team. The output of this coordination will be included in the Master Science Plan delivered to JSOC. To support this coordination task the formal membership of the SWT includes the Chairman of the Cluster Ground-Based Working Group. JSOC will support this SWT activity by collecting and making available information on ground-based experiments and on other space missions.

2.2. Command Timing

An important feature of the JSOC commanding subsystem is that command timings are expressed as offsets from the times of events chosen by the Science Working Team. The Master Science Plan (and hence the TLIS files) are divided into data acquisition periods. The start of each period is expressed in the Plan as a time offset from one of the following orbit and scientific events:

- perigee,
- apogee,
- magnetopause crossing,
- bow shock crossing,
- neutral sheet crossing.

The times of changes in the spacecraft telemetry mode within a data acquisition period are then expressed as times from the start of the period. The end of each

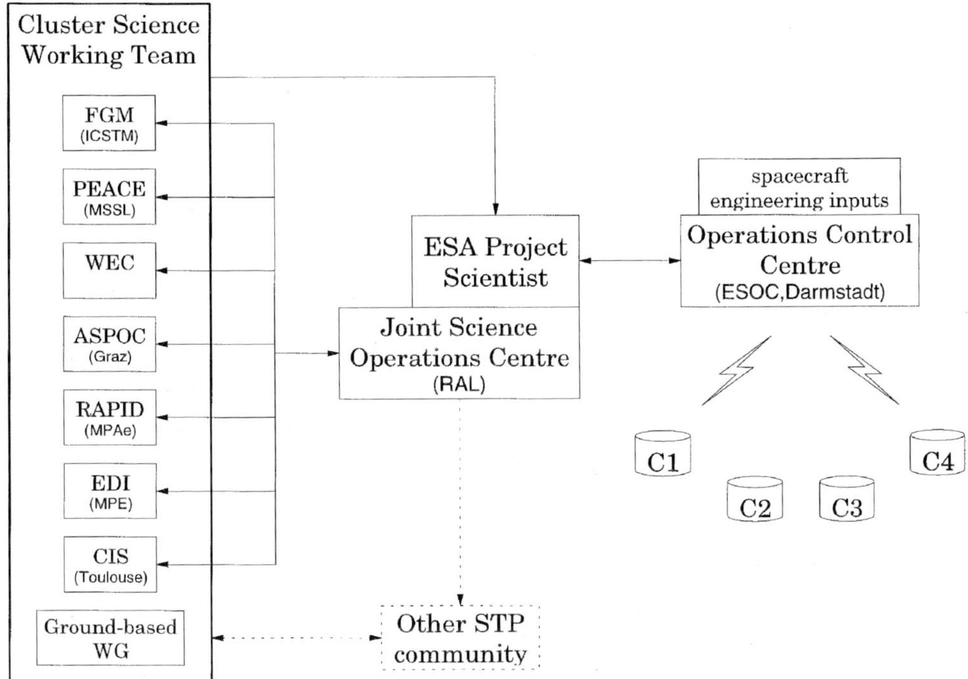

Figure 1. Relationship between JSOC and other parts of the Cluster mission. WEC is the Wave Experiment Consortium comprising the five instruments DWP, EFW, STAFF, WBD, and WHISPER. GBWG is the Cluster Ground-Based Working Group.

datâ acquisition period is expressed as a change to a telemetry mode OFF (i.e., no data-taking).

The times of instrument mode transitions are expressed relative to the times of changes in the spacecraft telemetry mode or relative to the orbit and scientific events above. This gives important flexibility in setting instrument modes. Most instruments require a mode transition when the spacecraft telemetry mode is altered; these mode transitions are naturally timed with respect to telemetry mode changes. However, all instruments require some of their mode transitions to be associated with crossings of magnetospheric boundaries; these mode transitions are naturally timed at offsets relative to those crossings.

The command timing is resolved to absolute time (Universal Time) only when JSOC generates the coordinated command schedule for delivery to ESOC. The Universal Times associated with the orbit and scientific events above are therefore an input to the software which generates the coordinated command schedule. These times are generated in the JSOC planning subsystem as described in the next section and are made available to members of SWT via the information dissemination subsystem.

Note that the orbit and scientific events relate to spacecraft motion around an orbit. But there are four spacecraft. So it is necessary to consider which spacecraft's motion is used to define these events. The choice of spacecraft is made by the Science Working Team from within the following possibilities supported by the JSOC software:

– For a crossing of a plasma boundary (i.e., magnetopause, bow shock, neutral sheet) the Master Science Plan may use the time associated with either: (i) a crossing by one of the operational Cluster spacecraft; or (ii) a crossing by a fictitious average spacecraft, whose position is the centroid of the operational Cluster spacecraft. The choice of spacecraft can be altered at any point in the Plan. The centroid approach is particularly appropriate when the spacecraft are relatively close together (separations $< 1R_E$) as it will yield an average crossing time for the set of operational spacecraft. The individual spacecraft approach may be more appropriate when the spacecraft are at large separations ($> 1R_E$).

– For crossings of the magnetopause and bow shock, the JSOC software provides several different predicted times based on different levels of solar-wind ram pressure (see detailed discussion in Section 3.1.5). For commanding we always use times based on median values of ram pressure because these provide a good estimate of the average crossing time (Hapgood, 1994).

– For perigee and apogee, the average spacecraft concept is not very meaningful. So here the Master Science Plan may use times associated with a perigee or apogee of any of the spacecraft. The choice of spacecraft can be altered at any point in the Plan.

2.3. Commanding Database

A vital component of the commanding subsystem is the JSOC commanding database. This contains various data which are used in the production of the payload command schedule:

– The procedures used to command mode transitions. These procedures are entered by JSOC staff using information gathered from the Instrument User Manuals and from discussions with the PI teams.

– The values of henceforth parameters supplied by PIs.

– A cross-reference table mapping the PI alias names of command sequences to ESOC names.

– A cross-reference table identifying the command sequences that call 'on-board macros'. These are sets of telecommands, without time-tags, that may be held in the on-board computer. Their execution is started via a separate time-tagged telecommand uplinked to the spacecraft. JSOC can execute a macro only by calling a command sequence that contains the latter telecommand. Thus this cross-reference table allows JSOC to identify these command sequences. This information is vital. When a PI requests revision of a macro, JSOC must check its impact on any command schedules and mode transition procedures which implicitly call the macro. JSOC will use the table to identify those schedules and procedures.

The accuracy of the payload command schedule is critically dependent on the integrity of these data. To help maintain this we have exploited the power of modern database technology. The JSOC commanding data are stored in a relational database conforming to the SQL database standard (see Section 6.2). This standard allows database designers to specify an extensive range of integrity constraints on the data held in the database. When these constraints are enabled, the database software will reject any data that violate the constraints.

The constraints may take the form of simple conditions, e.g., check that a number lies in a range or that a label is a member of a set. They may also take the form of references from one database table to another. In this case values of a field in one table must lie in the set of values given in another table, where the latter table is the master definition of those values. To give a simple concrete example, the list of Cluster instrument names is given in a master table called INS_LIST. This contains just seven entries – one for each instrument, where for commanding purposes the five WEC instruments are treated as one. All other tables which contain instrument names are constrained so that they can only contain instrument names defined in INS_LIST.

These cross-table constraints are used extensively to ensure the integrity of the JSOC commanding data. The constraints can be represented graphically in an 'entity-relationship' or ER diagram (Chen, 1976; Oracle, 1992), which is a very useful tool in database design. Figure 2 shows the ER diagram for the JSOC database. The rectangular boxes represent tables containing data (entities in database jargon) and the lines between these boxes represent relationships between tables. Each relationship is described in words given in a diamond-shaped box and has a sense of direction indicated by the arrow. The ER diagram may be regarded as a model of the dataset being stored in the database.

This approach also helps to ensure that the tables within the database have been 'normalised'. This is an important concept in database design. A discussion of its formal definition is beyond the scope of this paper – please consult any textbook on databases, e.g., Date (1986). For the purposes of this paper, the concept may be taken to mean that the dataset must be broken down so that each table represents a distinct and meaningful component of the dataset. Figure 2 shows how this has been done for the JSOC commanding data.

2.4. Command Preparation

The preparation of the detailed command schedule for each planning period is illustrated in Figure 3. The process will begin eight weeks before execution when JSOC issues the initial set of instrument command schedule files (PIOR files). These files will be reviewed by PIs and, where necessary, revised by JSOC according to PI instructions. This process will be repeated, separately for each instrument, until

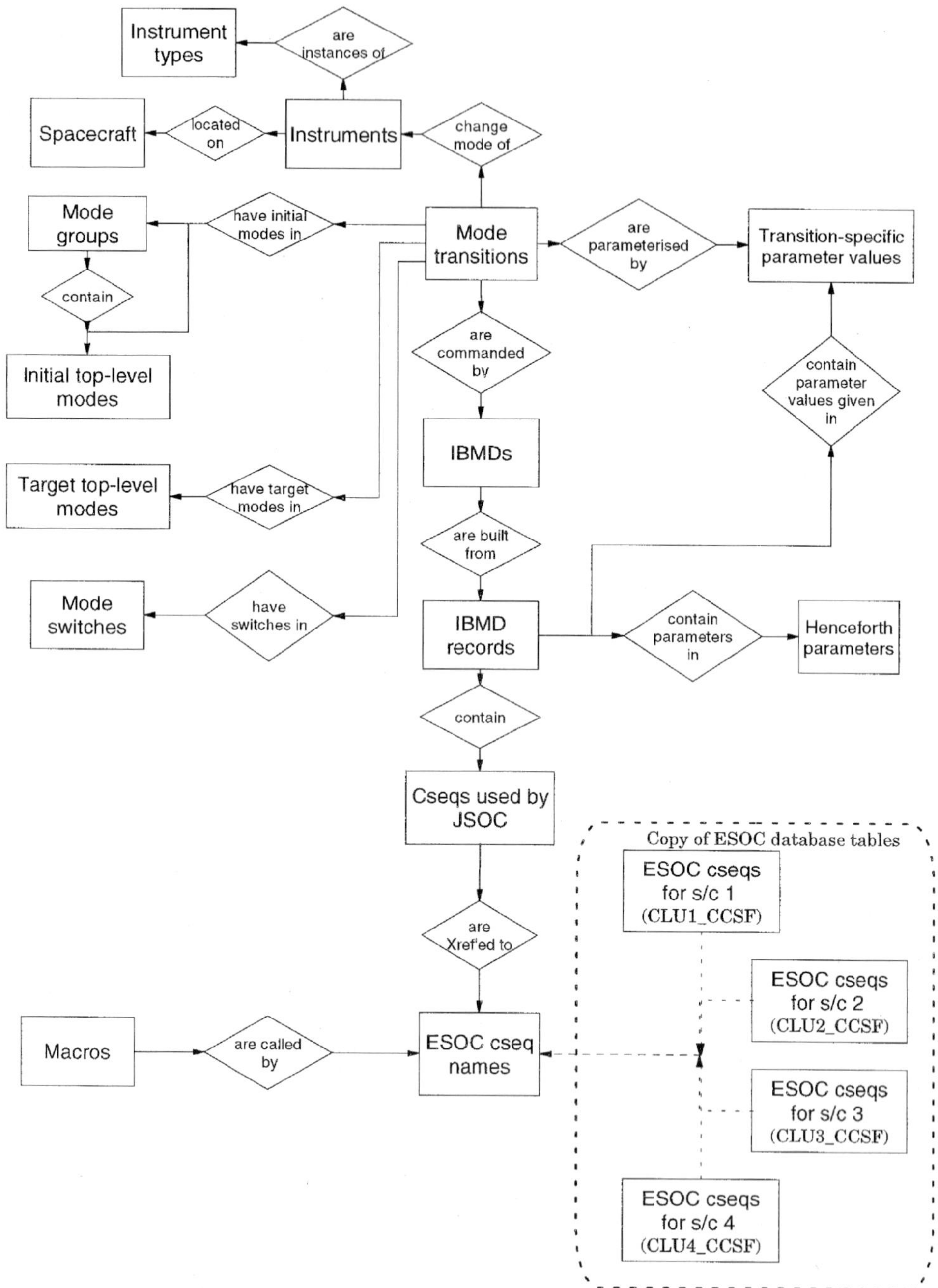

Figure 2. Entity-relationship diagram for the JSOC commanding database.

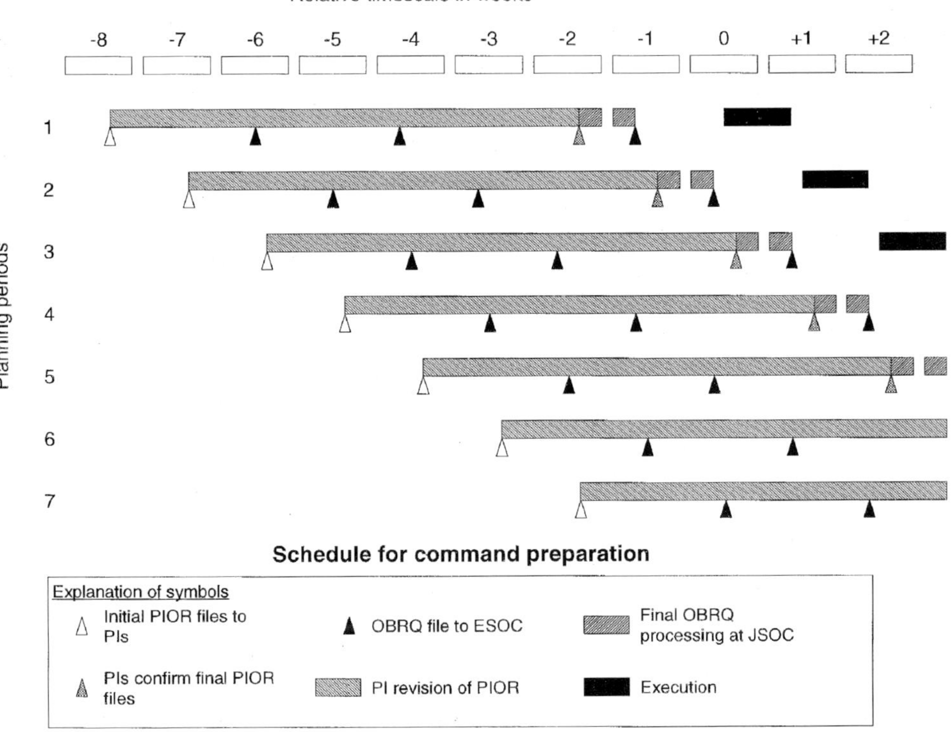

Figure 3. The schedule for command preparation.

JSOC has a set of acceptable instrument schedules, each of which should have been formally approved by the appropriate PI.

After the instrument schedule is approved, the PI remains free to request further changes. On receipt of a change request on an 'approved' schedule, JSOC will automatically cancel its approved status.

Six weeks before execution, JSOC will merge the latest schedules for each instrument to produce the initial version of the payload command schedule (an observational requests or OBRQ file). This initial payload schedule will be sent to ESOC for processing. Any problem found by ESOC will be reported to JSOC. JSOC will resolve these problems, in consultation with the Cluster Project Scientist and the PIs, and revise the instrument schedules as necessary. The new set of instrument schedules (including any other changes requested by PIs) will be merged to produce a new payload schedule which will be sent to ESOC.

Four weeks before execution, JSOC will again merge the existing instrument schedules to generate another payload schedule, which will be submitted to ESOC. This schedule will be used by ESOC to generate the requests to the NASA Deep Space Network (DSN) for downlink from the WBD instrument on Cluster. Hence, the file containing this schedule is termed the 'DSN OBRQ'. If ESOC report any problems with this version of the payload schedule, these must be resolved by JSOC, as for the initial payload schedule, prior to generation of the formal request to DSN.

Note that DSN downlink will be scheduled within JSOC from the beginning of command preparation. The DSN OBRQ indicates the point at which that information is passed to ESOC to be used in the formal request for DSN downlink; it is not the point at which JSOC starts to plan DSN activities. Note also that only data from WBD are downlinked to DSN ground-stations. Data from the other Cluster instruments are always downlinked to ESA ground-stations. The formal interface for requesting DSN downlink is always via ESOC to DSN (NASA, 1996). The role of JSOC is to prepare the payload schedule from which ESOC will generate the formal request to DSN. However, JSOC will receive information, direct from DSN, on the predicted load levels at the DSN ground stations. The processing and display of this information by JSOC is discussed in Section 3.3. We note here that it will allow the WBD team, in consultation with JSOC, to optimise the scheduling of WBD downlink with respect to scientific requirements, ground-station visibility and loading.

The final payload schedule is generated on the Tuesday of the week preceding the week (Monday to Sunday) which contains the start of execution. At this time JSOC will merge the latest instrument schedules to generate the required payload schedule, which will then be sent to ESOC for execution.

The preparation of the initial and DSN OBRQs does not prevent PIs from requesting changes to existing instrument schedules. PIs may request such changes at any time up to the Thursday preceding production of the final payload schedule.

At all deadlines (initial, DSN, and final), JSOC will use the set of latest instrument schedules for each instrument to generate the payload schedule, irrespective of the approval status of those instrument schedules - except in cases where there is a known serious problem with the latest instrument schedule. In the latter case, JSOC will use the latest instrument schedule that is known to command acceptable instrument operations.

2.5. USER INTERFACE

The commanding subsystem supports a wide range of functions to process the various files associated with the subsystem. To allow easy operation of the subsystem, these functions have been integrated into a single piece of software which is called the JSOC commanding controller or jcc (note that this acronym is always given in lower case to match the name of the executable code). This software is formally described in the JSOC Detailed Design Document (Dimbylow, 1995). We outline here its principal characteristics.

The user interface of jcc is based on the Motif windowing system. This allows the operator to select functions by pointing and clicking with a mouse. Text input is entered via pop-up windows. The main window of jcc is divided into columns, each of which deals with a different planning period. Each column is then divided into blocks, each of which deals with one type of file in the commanding subsystem. Buttons within each block allow the operator to examine the files and to execute whatever operations are necessary on those files. In cases where there is one file per instrument (rather than one file for the whole), the operator can select the instrument by clicking on a check box. Critical operations are protected by requesting the operator to confirm the action by clicking on a button in a pop-up box.

A very important attribute of the user interface is the use of colour to indicate the status of the different files. The possible status levels are:

– NONE. No processing has been undertaken so no status information for this period.

– READY. File ready for next action. Light green.

– WAITING. Incoming file is awaiting processing. Green.

– URGENT. Action on this file due soon. Orange.

– TODAY. Action on this file is due today. Yellow.

– LATE. Action on this file is overdue. Red.

– WARNING. Processing on this file has generated a warning. The operator must view the warning and decide if action is required. Violet.

– ERROR. Processing on this file has generated an error. The operator must fix this. Cyan.

– OK. Processing on this file was successful. Blue.

This modern user interface makes for considerable ease of operation. The operator can directly select the planning period to be processed and the actions to be performed. There is no requirement for the operator to learn a special command language - as was required in older styles of user interface. The use of colour to indicate file status allows the operator to assess the current state of command preparation by visual examination.

3. Planning Subsystem

This subsystem generates various data which are required to plan and command the mission:

(1)–Event prediction - the prediction of the orbit and scientific events used as reference times in the commanding subsystem, as described in Section 2.2.

(2)–Data recording constraints - the information and plots that will be used at JSOC to support the allocation and adjustment of data acquisition intervals. This is a critical factor in Cluster science planning.

(3)–Geometric position. This is the calculation of the spacecraft positions and separations. It also includes the calculation of quality factors that characterise the tetrahedron formed by the four spacecraft.

(4)–Magnetic position – the calculation of spacecraft position with respect to magnetic local time and invariant latitude/L-value. The time series of L-values will be further analyzed by JSOC to predict the crossings of critical L-shells. These predictions will be used to indicate when the Cluster spacecraft are expected to be in the radiation belt. They will also predict crossings of auroral field lines.

The software used to generate these planning data is discussed in detail in the rest of this section.

However, one important item is excluded from the JSOC planning subsystem as it has already been provided elsewhere. JSOC will not provide a three-dimensional graphics tool for viewing the Cluster orbit. This facility has already been made available to the Science Working Team via the Orbit Visualisation Tool (OVT), which has been developed under contract to ESA (Stasiewicz, 1990). OVT was developed specifically to assist Cluster planning by providing a three-dimensional view of the Cluster orbit, the configuration of the four spacecraft, and the scientific regions through which Cluster passes.

3.1. Event Prediction

3.1.1. *Overview*

The set of events to be predicted may be divided into two classes: (a) orbit events, namely perigee and apogee plus eclipse entry and exit, and (b) scientific events, namely crossings of the magnetopause, bow shock and tail neutral sheet.

These events will be predicted using data from event and orbit files supplied by ESOC (Smith and Robertson, 1995). These files are supplied in two forms (long-term and short-term), which have identical formats but different content. The long-term files will be generated after each major manoeuvre to adjust the separations of the Cluster spacecraft (every six months) and will predict their orbits for the rest of the the mission (including the effects of future manoeuvres). The short-term files will be generated at least once per week and will predict the orbits of the four spacecraft from 10 days before generation to 3.5 months after generation.

The orbit events will be taken from the long-term and short-term event files. JSOC will extract the time and spacecraft id of each event and then calculate its position and orbit number using the corresponding (long-term or short-term) orbit file together with orbit decompression software provided by ESOC (Smith and Robertson, 1995). This software gives spacecraft positions in geocentric inertial coordinates (mean epoch of J2000.0). These positions will then be converted to geocentric solar ecliptic (GSE) coordinates as described in Hapgood (1992) and Hapgood (1995).

In contrast, JSOC will have to predict the science events. The procedure to be used is as follows:

Table I
Plasma regions identified in the JSOC Planning Subsystem

Region code	Description
1	Solar wind
2	Magnetosheath
3	Inner magnetosphere
4	North tail lobe
5	South tail lobe

Table II
Plasma boundaries identified in the JSOC Planning Subsystem

Initial region code	Final region code	Boundary	Direction of crossing
2	1	Bow shock	Outbound
1	2	Bow shock	Inbound
3,4,5	2	Magnetopause	Outbound
2	3,4,5	Magnetopause	Inbound
4	5	Neutral sheet	Southbound
5	4	Neutral sheet	Northbound

– JSOC will generate spacecraft positions in GSE coordinates at five minutes time resolution – using long-term and short-term orbit files together with the decompression software as described for orbit events.

– This software also gives the orbit number at each spacecraft position.

–Each GSE position then will be analyzed to place it in one of the five regions in Table I. The criteria for assigning positions to regions are shown as a decision tree in Figure 4. The functions used to describe magnetospheric boundaries are presented in Figure 4 as generic forms, which are independent of the particular model used. The detailed implementation of these functions is described in subsequent subsections.

– Once every position has been assigned to one of the five regions, the time of each boundary crossing is determined by finding the two times between which the region code changes as shown in Table II.

– The crossing time is taken as the first of the two times between which the region code changes. This gives the predicted boundary crossing time to a resolution of five minutes, which is quite adequate for the purpose of Cluster operations; previous work at JSOC has shown that the one sigma uncertainty in such predictions, due to short-period ram-pressure fluctuations, may be of order of an hour (Hapgood, 1994). The first time is chosen, in preference to the mean of the two times, so that the event times *exactly* match times used in tables of predicted positions prepared by JSOC. This exact matching is important because the event and position tables are all stored in a relational database (see Section 6.2). It is envisaged that, at some point in the mission, it may be desirable to build multi-table queries on these tables, e.g., to list configuration quality parameters at magnetopause crossings. The exact time matching will make it easy to build such queries using standard database join functions. Thus the choice of first time gives us flexibility in the future use of the database.

– Similarly, the spacecraft position and orbit number at the crossing are taken as the position and orbit number at the first of the two times bounding the crossing.

For all events, orbit and scientific, the magnetic latitude and magnetic local time are calculated from the GSE position using the transformations given by Hapgood (1992). We use two different coordinate systems described in that paper: (i) geocentric-solar-magnetospheric (GSM) and (ii) solar-magnetic (SM) coordinates. The key features of these systems are the orientation of the X and Z axes: (a) in the SM system the Z axis is exactly the geomagnetic dipole axis and the X axis lies in the SM meridian containing the Sun and is normal to Z axis; but (b) in GSM the Z axis is the projection of the dipole axis on a plane normal to the Earth–Sun line while the X axis is the Earth–Sun line.

For magnetic latitude we take the latitude in the geocentric-solar-magnetospheric (GSM) coordinate system. This system was selected for JSOC applications, in preference to SM latitude, as it was considered to give a better ordering of phenomena on the dayside magnetopause. This is because the stagnation point for magnetosheath flow past the magnetopause lies at the sub-solar point, which has zero GSM latitude. This selection was made on dayside magnetopause phenomena because they are intimately associated with the exterior cusp, which is the prime objective of the Cluster mission.

The magnetic local time (MLT) is taken as the angle between two SM meridians – namely those containing the point of interest and the anti-sunward direction. We make no correction for non-dipole terms in the geomagnetic field as these are negligible for all points on the baseline orbit for Cluster (perigee = four Earth radii). Thus MLT is derived from the longitude ϕ in the solar-magnetic (SM) coordinate system thus:

$$\mathrm{MLT} = 12.0 + \phi/15.0\,, \tag{1}$$

where MLT is measured in hours between 0 and 24 and longitude in degrees between -180 and $+180$ (positive east).

The predictions of orbit and scientific events will be stored in a table in the JSOC planning database. The table will be initially populated using the first set of long-term orbit and event files and will cover a period of one year starting at the current date. This table will then be updated once per week (normally on a Friday) during mission operations using the latest short-term orbit and event files supplied by ESOC. Thus each weekly update will affect a four month window in the table. The table will be completely repopulated every time a new set of long-term orbit

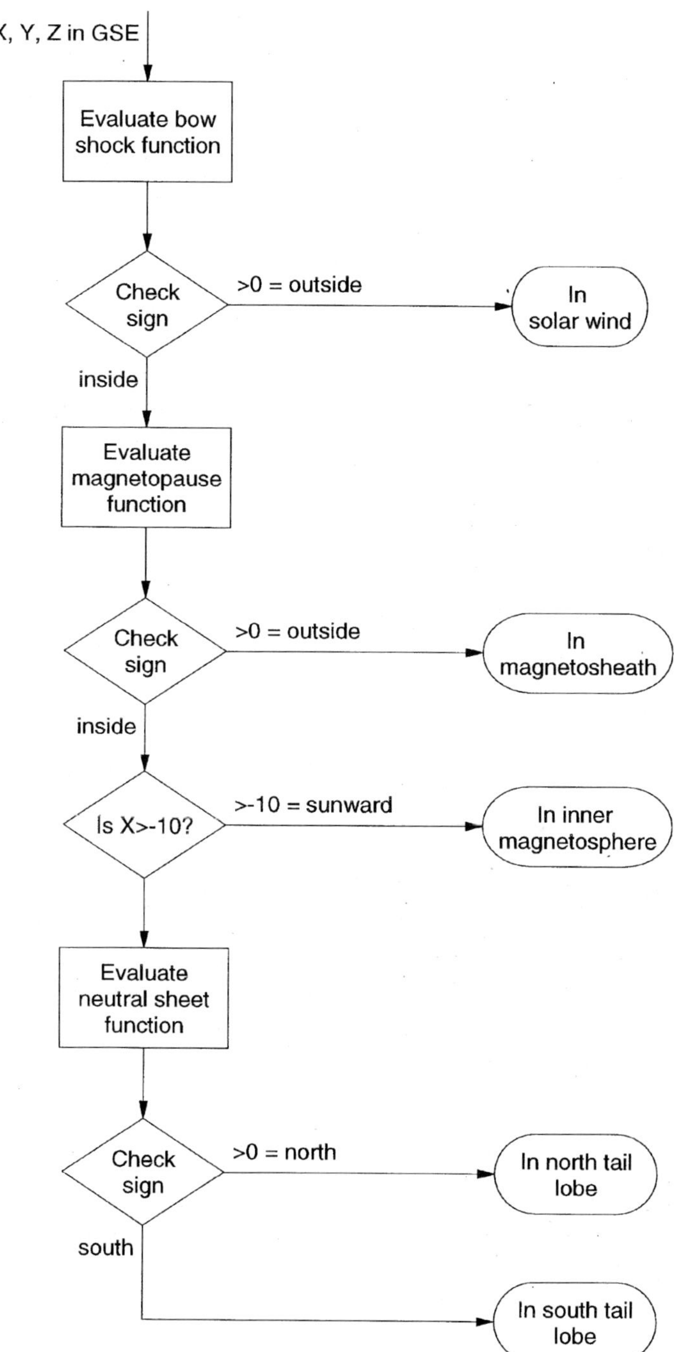

Figure 4. Decision tree for region identification.

and event files is received. It will again cover a period of one year ahead starting at the new current date.

Each weekly update is expected to modify the times of predicted events by of order a few seconds (S. Pallascke, private communication). Thus the events used as reference times for commanding may alter by about a minute in the course of the 10-week command preparation cycle described in Section 2.4. This shift is not expected to have any significant impact on the command timeline.

3.1.2. *Magnetopause*

This will be represented by the empirical model of Sibeck *et al.* (1991). This model is strongly recommended because it incorporates the basic understanding of magnetopause location, namely that the location is determined by a balance between solar wind ram pressure and magnetic pressure in the outer magnetosphere. In the Sibeck *et al.* model the mathematical functions fitted to the data are explicitly chosen to satisfy the pressure balance condition at the sub-solar magnetopause. Thus this model allows the prediction of Cluster magnetopause crossings in terms of ram pressure statistics: (i) the crossing time can be predicted using the 'average' ram pressure, and (ii) the uncertainty in the predicted crossing time can be estimated from a suitable measure of the short-term (periods of hours) variability of ram pressure.

Note that the recent empirical magnetopause model by Roelof and Sibeck (1993) describes magnetopause location as a function of both ram pressure and the z component of the interplanetary magnetic field, B_z. However, it does not incorporate pressure balance; it simply fits the data to a bivariate expansion about the mean values of ram pressure and B_z. It provides a good description of magnetopause location for its region of validity, namely ram pressure and B_z close to the mean (deviation < 1 sigma) but, as noted by Roelof and Sibeck, it can fail dramatically outside this region. It cannot be used to predict the effect of the full range of ram pressure variations on magnetopause crossing times.

Work at JSOC (Hapgood, 1994) has shown that the median is the best average of ram pressure for predicting magnetopause crossings (the distribution of ram pressure values is skew-symmetric so the mean is biased towards higher pressures). In addition, this work has shown that variability of ram pressure is dominated by short-term variations ($\leq$ 12-hour period), so a measure of the total variability can be used as a measure of the short-term variability. The upper and lower sextiles are used, i.e., one sixth of all ram pressure values are greater than the upper sextile and one sixth are less than the lower sextile. Two-thirds of values lie in the inter-sextile range so this range is equivalent to $\pm$ one standard deviation in a Gaussian distribution.

The use of median and sextiles is convenient because crossing time (measured with respect to perigee) is a monotonic but non-linear function of ram pressure. Thus the median and sextiles of ram pressure will map to the median and sextiles of crossing time, but the mean and standard deviation of ram pressure will not map to meaningful values.

Thus JSOC will predict Cluster magnetopause crossings with the Sibeck et al. model. The average crossing time will be estimated using a predicted median value of ram pressure; the uncertainty in the estimated crossing time will be derived from predictions of the sextiles of ram pressure. These predictions are discussed below in Section 3.1.5. The model magnetopause will be adjusted to allow for the aberration of the magnetopause due to the Earth's motion about the Sun. The adjustment will be a fixed four degree rotation about the GSE Z axis such that the tail is rotated towards the evening sector.

3.1.3. *Bow Shock*

This will be represented by the empirical model of Howe and Binsack (1972), which has previously been used in the mission analysis for Cluster (Pellón-Bailón and Rodríguez-Canabal, 1993). The empirical model of Fairfield (1971) was also considered, but we encountered problems with mathematical discontinuities in that model. Owing to pressure of the implementation schedule, it was decided not to pursue these problems and to implement the simpler Howe and Binsack model.

However, we adjusted the stand-off distance in this model – to implement the recommendation of the Inter-Agency Consultative Group (IACG) that the model bow shock should be adjusted according to ram pressure, so that the sub-solar bow shock maintains a stand-off distance three Earth radii outside the sub-solar magnetopause (Fairfield, 1990).

Thus, JSOC will predict the magnetopause stand-off distance using the Sibeck *et al.* model together with a predicted median value of ram pressure as discussed below in Section 3.1.5. This will be used to set the stand-off distance in the Howe and Binsack bow shock model. The latter model will then be used to predict Cluster crossings of the bow shock.

The model bow shock will be adjusted to allow for the aberration of the bow shock due to the Earth's motion about the Sun. The adjustment will be a fixed four degree rotation about the GSE Z axis such that the tail is rotated towards the evening sector.

3.1.4. *Neutral Sheet*

This will be represented by the neutral current sheet model which Tsyganenko used within his 1989 geomagnetic field model (Equation (11) of Tsyganenko, 1989). This model was used in the mission analysis for Cluster (Pellón-Bailón and Rodríguez-Canabal, 1993) and gives very similar results to the model of Fairfield (1980). However, the Tsyganenko model is preferred here because it explicitly includes the curvature of the neutral current sheet, around $x = -10R_E$, where the sheet changes its alignment from the dipole equator to the ecliptic. JSOC will use a fixed set of parameters for this model, i.e., independent of the level of solar-terrestrial activity, because Tsyganenko was unable to find a consistent or a scientifically realistic dependence on that activity.

3.1.5. *Solar-Terrestrial Activity*

The magnetopause and bow shock models used by JSOC will require predictions of solar wind ram pressure as inputs. JSOC will use predictions based on the state of the solar cycle.

To check the solar cycle variation we have calculated 12-month running values of the median and sextiles of ram pressure for each month in the period January 1973 to July 1993. The raw values of ram pressure were hourly values derived from hourly means of density and velocity in the 'OMNI' dataset prepared by the US National Space Science Data Center (NSSDC). The median and sextiles for each month, m, were calculated over the period from day 15 of month $m - 6$ to day 14 of month $m + 6$. Thus the 12-month running medians and sextiles are similar to the 12-month running means used in predictions of sunspot number and solar 10.7 cm radio flux. The median and sextiles were used only if $\geq$ 500 hourly values contributed to the calculation.

The resulting medians and sextiles are plotted in Figure 5(a) in a superimposed epoch format where the epoch is the month of maximum sunspot number in each solar cycle. The planned launch date of Cluster relative to the maximum of Cycle 22 in July 1989 is shown by the vertical line. Maximum is used as the epoch (rather than minimum) because it places the Cluster launch in mid-cycle.

Figure 5(a) shows that there is substantial agreement between the medians and sextiles from the two previous solar cycles from 4 to 9 years after maximum, i.e., the cycle phase around the expected launch of Cluster. This agreement is, perhaps, not surprising since it corresponds to solar minimum when the solar wind is relatively undisturbed. Similarly, the lack of consistency close to maximum is not surprising as it is the period when the solar wind is dominated by impulsive events.

Thus the ram pressure variations from the previous two solar cycles can be used to predict values for the period of Cluster operations. We take the weighted means of the pairs of curves shown in Figure 5(a) to predict the median and sextiles as a function of solar-cycle phase, where the weights are the number of hourly values contributing to each data point. We then convert the phase to absolute time, taking zero phase to be the solar maximum of July 1989. The resulting predictions (of median and sextiles) are shown in Figure 5(b).

These ram pressure predictions will be used as input to the magnetopause and bow shock models described in Sections 3.1.2 and 3.1.3 above. Thus JSOC will predict Cluster crossings of these two boundaries for three different levels of ram pressure. The predictions based on the median ram pressure will provide an estimate of when the spacecraft will cross the 'average' boundary position. The predictions based on the sextiles will provide an estimate of the uncertainity in the time (and position) of this boundary crossing.

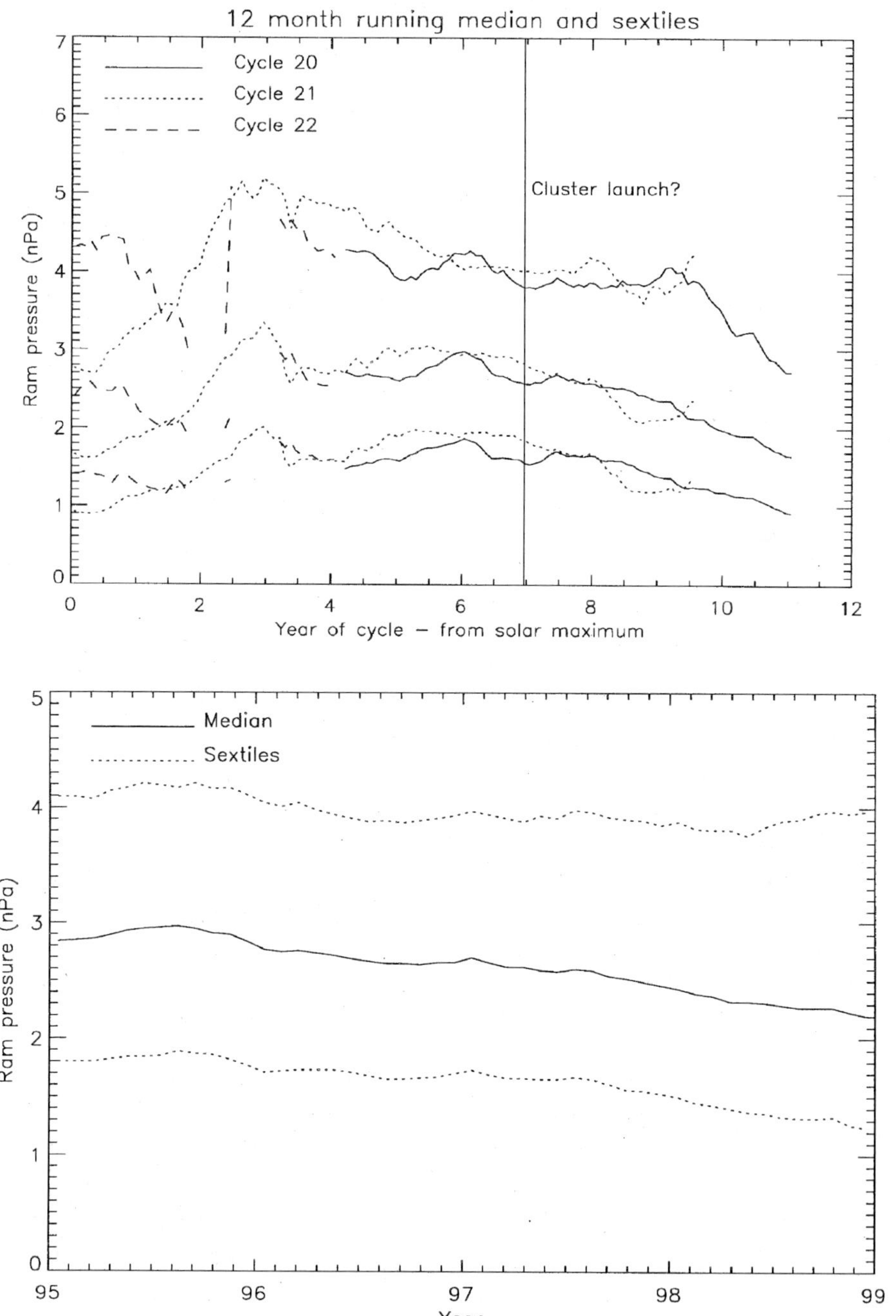

Figure 5. Observed (a) and predicted (b) ram pressure. The figure assumes a Cluster launch date in June 1996.

It is planned to review these predictions after the first Cusp constellation of Cluster and before the second Cusp constellation. This review will take account of the actual magnetopause locations measured by Cluster during the first Cusp constellation.

3.2. Bryant Plots

The event predictions will also be made available in graphical form using the Bryant plot format that was developed for the *AMPTE-UKS* mission (Bryant *et al.*, 1984). Figure 6 shows a example Bryant plot for Cluster.

– The main (top) panel shows the passage of the mission through the various plasma regions. The horizontal axis represents absolute time and the vertical axis time relative to perigee (also known as orbit phase). Thus the track of the mission is a set of near-vertical lines, sloping slightly to the right. The track is omitted to aid clarity in this example which shows over one year of Cluster operations.

- The solid curves show the times of magnetopause crossings. The outer curve corresponds to the upper sextile of ram pressure and the inner to the lower sextile. Thus there is a two-thirds probability of Cluster encountering the magnetopause between the curves.
- The dashed curve shows the times of bow shock crossings with curves corresponding to ram pressure sextiles as for the magnetopause.
- The magnetopause and bow shock predictions include the effect of aberration due to the Earth's motion around the Sun (as discussed in Sections 3.1.2 and 3.1.3).
- The dotted curve shows the times of the neutral sheet crossing. The neutral sheet model used has no dependence on solar-terrestrial activity so a single curve is plotted. The curve shows orbit-phase variations of around 5 hours due to the diurnal motion of the magnetic dipole.

– The middle panel shows the GSM latitude of the two magnetopause crossings.

– The lower panel shows the magnetic local time of apogee – to allow these plots to be related to the Master Science Plan, which specifies objectives by apogee local time.

3.3. Data Recording Constraints

One key planning task for JSOC is the provision of advice on the allocation and adjustment of data recording periods. To aid consideration of this problem, JSOC will generate plots which can display the orbit of any Cluster spacecraft with respect to:

(1) The model magnetopause and bow shock.

(2) Any Boolean combination of:

– data acquisition intervals,

– ground-station visibility from Redu,

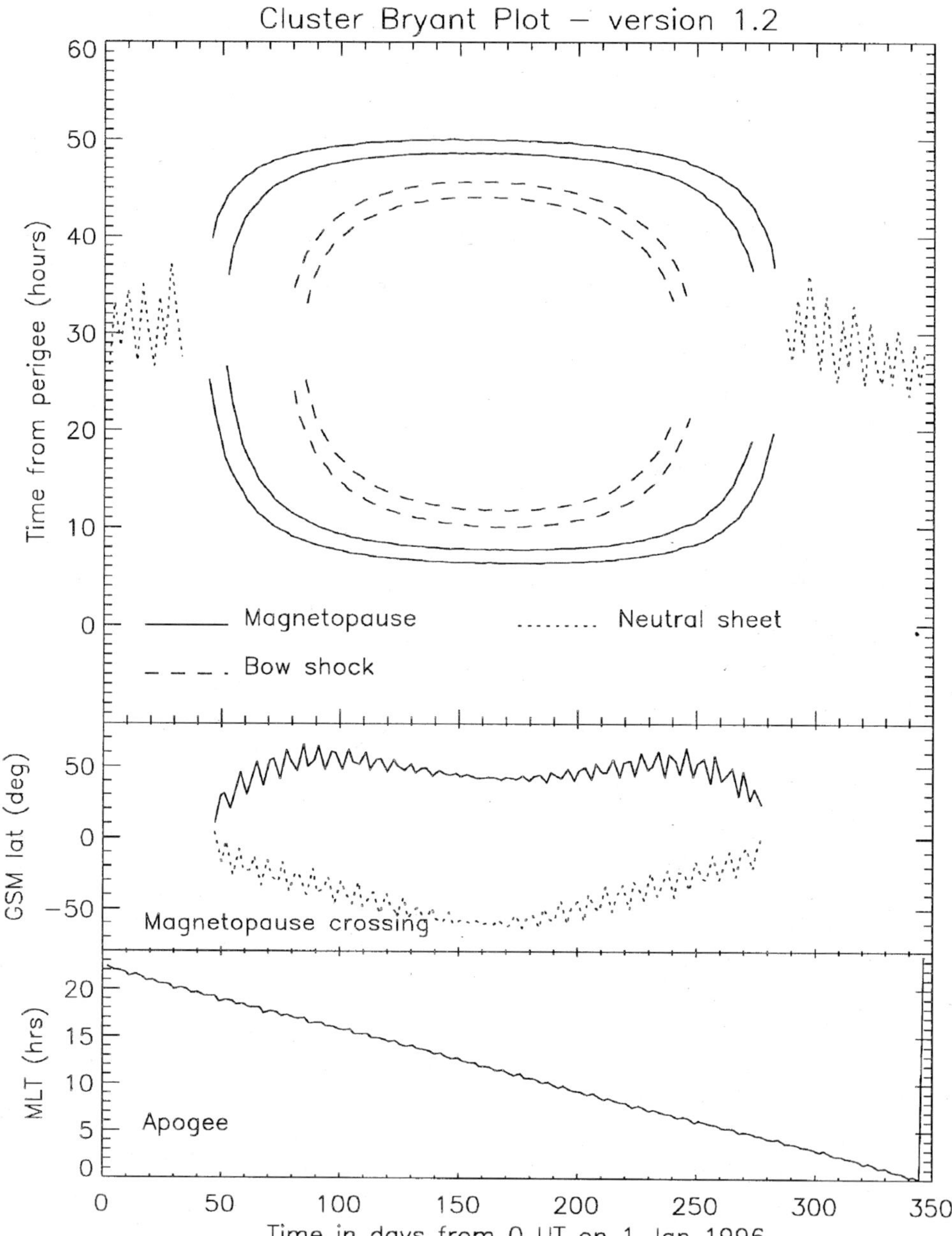

Figure 6. Example of a Bryant plot. This plot was produced using predicted orbit data for the original launch date of 1 December, 1995.

– ground-station visibility from Odenwald,
– ground-station visibility from Canberra,
– ground-station visibility from Goldstone,

– periods of predicted low, moderate or heavy load levels at Canberra,
– periods of predicted low, moderate or heavy load levels at Goldstone.

(3) Predicted crossings of the magnetopause, bow shock, neutral sheet and the auroral zone (as described in Sections 3.1.1 and 3.5.3).

(4) Entry to, and exit from, eclipses.

The ground-station visibility is calculated by transforming the spacecraft position $\mathbf{r}_S$ to geocentric coordinates (i.e., rotating with the Earth), calculating its position $\Delta\mathbf{r}_S = \mathbf{r}_S - \mathbf{r}_G$ relative to the fixed geocentric position of the ground station $\mathbf{r}_G$, and then obtaining the zenith distance of the spacecraft as seen from the ground station as $\theta = \arccos((\Delta\mathbf{r}_S \cdot \mathbf{r}_G)/(|\Delta\mathbf{r}_S||\mathbf{r}_G|))$. The spacecraft is taken to be visible if $\theta \leq 85°$. This method ignores the small difference between geocentric and geodetic coordinates (due to the oblateness of the Earth); this will generate small errors $\approx 0.1°$ in the zenith distance.

The predicted load levels at Goldstone and Canberra are taken from monthly reports to be supplied to JSOC by DSN.

The predictions of entry to, and exit from, eclipses are taken from the event files supplied by ESOC as described in Section 3.1.1. The crossings of the magnetopause, bow shock, neutral sheet and the auroral zone are predicted by the JSOC planning sub-system as described in Sections 3.1.1 and 3.5.3.

These 'visibility plots' shall be made available to PIs as Postscript files via the information dissemination subsystem. Figure 7 shows an example. It covers one orbit with apogee near dusk and shows the following features:

– The projection of the Cluster orbit on to a plane defined by the GSE Z axis and the apogee of that orbit (dotted curve). The r coordinate is equal to $\sqrt{(x_{GSE}^2 + y_{GSE}^2)}$ and is set positive in the sunward direction.

– Tick marks at one hour intervals around the orbit, starting at perigee.

– The intersection of the Sibeck magnetopause (for 2.8 nPa ram pressure - the solar cycle median value) with the projection plane is shown by the inner dashed line.

– The intersection of the Howe and Binsack bow shock (adjusted to stand-off $3R_E$ outside the magnetopause) with the projection plane is shown by the outer dashed line.

– The periods of data acquisition with visibility from the NASA/DSN ground stations at Canberra and Goldstone and no visibility from the ESA ground stations at Redu and Odenwald are shown by the thick continuous curve. These are prime periods for downlink of data from the WBD instrument.

– The other periods of planned data acquisition are marked as a thin continuous curve.

– The date and time of the first tick mark is shown in CCSDS ASCII time code A (CCSDS, 1990), which is the Cluster Science Data System standard for display of date and time.

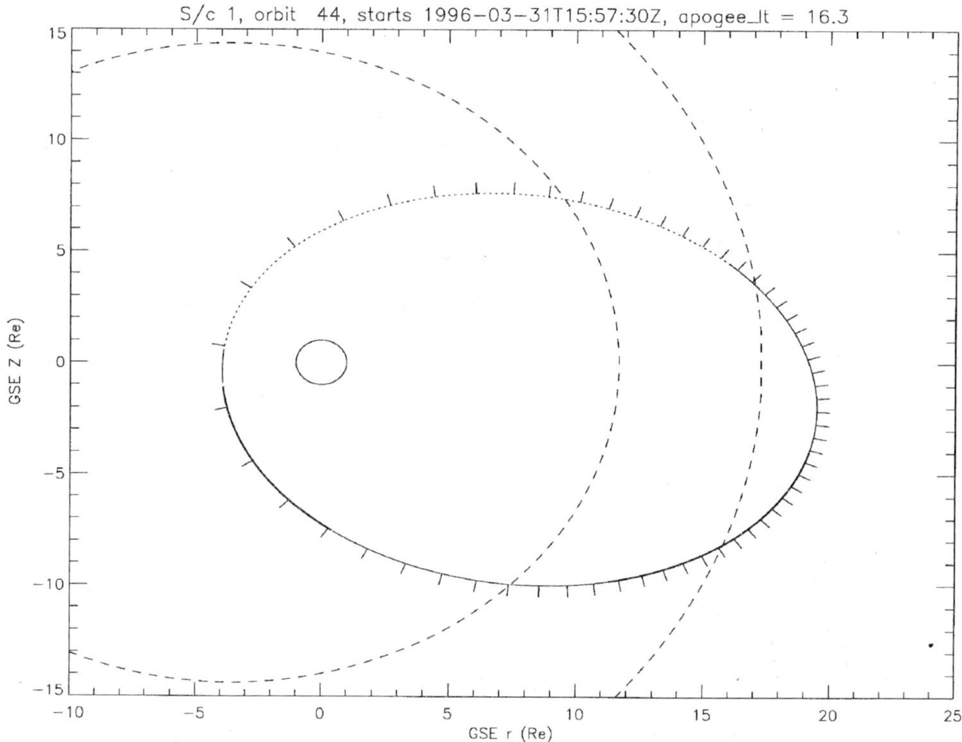

Figure 7. Example of an orbit plot.

3.4. GEOMETRIC POSITION

As already discussed in Section 3.1.1, JSOC will calculate the predicted positions and orbit numbers for all four Cluster spacecraft in inertial coordinates, using the orbit data files and decompression software provided by ESOC (Smith and Robertson, 1995). These positions will be converted to the GSE coordinate system (Hapgood, 1992, 1995). These data shall be placed in a table in the JSOC database.

However, there is also a requirement for JSOC to provide information that allows users to assess the quality of multi-spacecraft measurements for any predicted configuration of the four Cluster spacecraft. This may be assessed by a quality factor, or factors, based on the shape of the tetrahedron formed by the four spacecraft. Examples of such factors are those proposed by vom Stein *et al.* (1993) and by Robert *et al.* (1993), both of which have been adopted for use in the representation of the actual spacecraft configuration by the Cluster Science Data System (Schmidt and Escoubet, this issue). However, following the decision by the Cluster Science Operations Working Group, at its meeting in Paris on 5–6 October, 1994, JSOC will not use these parameters for the representation of the predicted spacecraft configuration but will instead use those proposed by Schoenmaekers (1994). The

Schoenmaekers parameters are based on the ellipsoid that exactly fits the positions of the four Cluster spacecraft; the three principal axes of this ellipsoid describe the shape of the tetrahedron and the direction of the major axis describes the orientation of the tetrahedron.

These parameters will be calculated from the decompressed orbit data using software supplied by ESOC and will be placed in a table in the JSOC database.

A future development of this work could include an assessment of the predicted orientation of the tetrahedron with respect to key plasma boundaries, e.g., the orientation at which the tetrahedron crosses the magnetopause could be calculated by determining the angle between: (i) the major axis of the Schoenmaekers ellipsoid, and (ii) the normal to the Sibeck model magnetopause. However, this task lies outside the current scope of JSOC.

3.5. Magnetic Position

3.5.1. *Magnetic Latitude*

JSOC will calculate the invariant latitude for each geometric position of each spacecraft. The results shall be stored in the 'scientific position' table in the JSOC planning database.

The invariant latitude will be calculated using the FLGALP subroutine from the MAGLIB software library developed by the Centre National d'Etudes Spatiales (CNES) in Toulouse (Kosik, 1994). This subroutine calculates the invariant latitude (and L-value) using a modified algorithm, due to Y. Galperin, that takes account of external as well as internal field terms. This algorithm is as follows: (i) starting from the point of interest, trace the field line to its footprint in the Northern hemisphere – using the complete field (internal + external), (ii) calculate the invariant latitude of the footprint using the McIlwain algorithm with the internal field only, and (iii) take this result to be the invariant latitude of the point of interest. For the internal field this subroutine uses IGRF90, the 1990 version of the International Geomagnetic Reference Field (Langel, 1992). For the external field it provides an option to select either: (a) the 'Long' version of Tsyganenko (1987) model, or (b) the Tsyganenko (1989) model. JSOC will use the 1989 model because that version includes an improved representation of the warped neutral current sheet in the tail. The footprint for the field-line tracing is taken at one Earth radius from the centre of the Earth. Note that the precise altitude of the footprint is not signficant so long as it is within the region dominated by the internal field.

It is anticipated that this software will be updated to use the newly-released IGRF95 internal field model (IAGA, 1996). JSOC will update its software (under careful configuration control) as soon as practicable.

Note that the invariant latitude, Λ, is directly related to the McIlwain L-value thus: $L = \sec^2 \Lambda$.

3.5.2. *Magnetic Local Time*

JSOC will also calculate the geomagnetic local time for each geometric position of each spacecraft. The TGML subroutine from the CNES/MAGLIB software library will be used. This calculates geomagnetic local time using the algorithm given in Kosik (1993), which is identical to that given in Equation (1) of the present paper. The results shall be stored together with the invariant latitude in the 'scientific position' table in the JSOC planning database.

3.5.3. *Critical L-Shells*

The time-line of L-values, derived according to Section 3.5.1, will be further processed to detect Cluster crossings of 'critical L-shells'. These are surfaces within the magnetosphere which are described by a McIlwain L-value and magnetic local time. Such surfaces may have a special significance for Cluster science or for the operation of the Cluster spacecraft. For example, these shells:

– May indicate the outer edge of the radiation belt. Thus we may use predictions of critical L-shell crossings to check that those Cluster instruments sensitive to the radiation belt environment are put in a safe state (e.g., PEACE, RAPID, ASPOC).

– May predict the location of auroral field lines. Thus we may use predictions of critical L-shell crossings to predict when the Cluster spacecraft cross auroral field lines.

We represent each L-shell as an array of 24 values - one L-value for each hour of magnetic local time. The spacecraft will be considered to be inside the shell when $L_{\mathrm{sc}} < L_{\mathrm{crit},x}(\mathrm{nint}(M_{\mathrm{sc}}))$ where L_{sc} and M_{sc} are the L-value and the magnetic local time of the spacecraft, nint is the nearest integer function and $L_{\mathrm{crit},x}$ is the array describing L-shell x. This approach has two practical advantages:

– That the JSOC planning system can handle any number of critical L-shells, each with its own array $L_{\mathrm{crit},x}$. Thus it is possible to specify different radiation belt safety limits for different instruments (and for the same instrument on different spacecraft).

– That it is straightforward to modify each $L_{\mathrm{crit},x}$ array during operations, e.g., if a PI decides to change the radiation belt safety limit in the light of experience with his or her instrument.

The L_{crit} array for each instrument will be specified by the PI and may be revised by him or her at any time, e.g., to take account of experience gained during the mission. The array describing the auroral field lines will be specified by JSOC using the auroral oval model of Holzworth and Meng (1975). These crossings of critical L-shells will be recorded as scientific events in the predicted events table of the JSOC planning database.

4. Monitoring Subsystem

The JSOC monitoring tasks are intended to detect problems that may affect the overall scientific performance of the mission. To carry out this monitoring promptly,

Table III
Science parameters to be derived from instrument housekeeping

Instrument	Parameter
ASPOC	ion current
EFW	spacecraft potential
FGM	B_{spin}, magnetic field component along the spacecraft spin axis $\lvert B_{equator}\rvert$, magnetic field magnitude perpendicular to the spacecraft spin axis
PEACE	particle counts per last spin
RAPID	particle count rates
STAFF	magnetic wave amplitudes, electric and magnetic wave gain levels at various frequencies
WBD	gain level
WHISPER	wave 'power'

JSOC will process and examine small amounts of instrument data (both housekeeping and science packets) retrieved from the Cluster Data Disposition System (CDDS) (Sørensen *et al.*, this issue) at ESOC. These data will be available within a few days of being recorded on the spacecraft. They will be retrieved from CDDS using software (CDDSREQ program) provided by the UK Cluster Data Handling Facility (Allen *et al.*, 1995).

The monitoring subsystem can be divided into three main tasks which are described in the following subsections. Note that these tasks are complementary to (a) the instrument health and safety monitoring undertaken at ESOC and (b) the detailed instrument performance monitoring undertaken by PIs.

4.1. Mission progress monitoring (MPM)

This is the task of verifying that the Cluster spacecraft are passing through the expected scientific regions at approximately the expected times. In particular, JSOC will try to identify and record the Cluster crossings of key plasma boundaries such as the magnetopause. This information is required: (a) to check that the mission is proceeding as planned, and (b) to support other monitoring activities described below. To do this JSOC will retrieve various science parameters which are stored in instrument housekeeping as shown in Table III.

These parameters should be sufficient to monitor the different plasma regions crossed by Cluster. Changes in the particle counts should provide a good indication of the gross changes in particle distributions which are seen across the magnetopause and the bow shock. The ion current and spacecraft potential will provide an indication of related changes in particle density at those boundaries. The magnetic field components are not converted to a scientific frame of reference, so the absolute values of components are not scientifically meaningful. However, we expect to detect the marked shear in the magnetic field which occurs at the neutral-sheet crossing and often at magnetopause. The wave parameters should provide an indication of the enhanced wave activity that is associated with most plasma boundaries.

The use of housekeeping data greatly reduces the amount of data that is required to monitor any given period (by a factor of 20 for normal mode data-taking and over 100 for burst mode). This approach allows JSOC to retrieve all housekeeping data taken on one spacecraft for the instruments listed in Table III. Thus, in principle, JSOC can monitor all regions crossed during Cluster data-taking operations.

The housekeeping data needed for mission progress monitoring will be retrieved from the CDDS as soon as they become available. This should normally be within two or three days of data-taking. The housekeeping data will then be processed to generate the parameters listed in Table III. Most of the processing software has been developed within JSOC using PI-supplied descriptions of the housekeeping data. However, the magnetic field data from the FGM instrument will be processed by software provided to JSOC by the PI team.

The parameters generated by this processing will be examined using a graphics tool developed within JSOC (Dimbylow, 1996). This allows the operator to select any data acquisition period and then to display any subset of the mission progress parameters available for that period. The parameters are displayed as time series in a multipanel plot. Thus the operator can look for related variations in several parameters, e.g., a high-shear magnetopause may be identified by a shear in the magnetic field with a change in particle counts slightly Earthward of the magnetic field shear.

The operator can mark the time of the boundary crossing with a cursor controlled by a mouse. On clicking the mouse, the graphics tool will calculate the time of the crossing (from the displayed data) and will prompt the user to identify the event type (from a list displayed in a pop-up window). The event time, event type and spacecraft id are then recorded in a table of mission progress events, which is stored in the JSOC planning database. The orbit number and GSE position of the event are also calculated (using routines from the planning subsystem) and stored in that table. Reports from the table will be made available to PIs.

The mission progress graphics tool will also be used to identify periods when Cluster is in a stable plasma environment, i.e., periods of at least 30 min when the mission progress parameters are subject to fluctuations $\leq 10\%$. These are periods suitable for the inter-experiment calibration (IEC) described in the next subsection. The mission progress tool will then be used to select these periods (typically one per plasma regime per orbit). The start and end of each IEC period will be recorded as events in the mission progress events table.

Table IV
Parameters for inter-experiment calibration

Parameter	Instruments
Electron density	CIS, PEACE, WHISPER
Plasma bulk velocity	CIS, PEACE
Ion current / spacecraft potential	ASPOC, EFW

4.2. INTER-EXPERIMENT CALIBRATION (IEC)

This is the task of routinely comparing similar quantities: (a) from different instruments on the same spacecraft, and (b) from the same instruments on different spacecraft. The aim of these comparisons is to give PIs early warning of any differences that develop between parameters which should be similar. Thus, PIs will be able to take remedial action as soon as possible and so maintain the consistency of the Cluster dataset. JSOC will look for two types of differences: (a) gross differences which suddenly and unexpectedly occur, (b) systematic long-term drifts in the relative values of similar parameters.

Note that this JSOC activity is not intended to check the absolute values of comparable parameters. That is a subject for dedicated inter-calibration campaigns agreed by the Science Working Team. The JSOC IEC task will supplement those campaigns by checking for unexpected differences arising between the campaigns.

This comparison of similar quantities is subject to two important constraints:

– That the four Cluster spacecraft are in a stable plasma environment, so that any differences in parameters arise from instrumental effects and not from spatial or temporal variations in the plasma. This constraint is met by the selection of stable periods (for IEC activities) as part of mission progress monitoring.

– That we make such comparisions in all the different plasma regimes crossed by Cluster, e.g., inner magnetosphere (ring current), tail lobes, magnetosheath and solar wind. Experience from the AMPTE mission suggests that the ratios between similar parameters are likely to vary between these regimes because of differences in instrumental response. Thus, when checking for long-term drifts, it is vital to check separately on several series of data - with each data series taken in a different plasma regime. This constraint is met by our objective of selecting one IEC period per orbit in each plasma regime.

The list of parameters to be compared is shown in Table IV. To generate these parameters JSOC will run software provided by the PI teams. These programs all require instrument science data as input. In some cases, housekeeping data are also required. JSOC will retrieve these data from the CDDS at ESOC for the IEC periods selected via the MPM tool.

These parameters will be examined using a graphics tool, similar to that for mission progress monitoring (Dimbylow, 1996).

4.3. Science Performance Monitoring (SPM)

This is the task of regularly examining small amounts of science data, on behalf of PIs, to check for indications of potential problems on individual instruments. This activity is intended to assist PIs by giving an early warning of any problems that may affect the scientific quality of the Cluster dataset. It supplements the checks on housekeeping data undertaken at ESOC. It is not intended to replace the detailed data checking undertaken by PI teams. Indeed, it may provide a trigger for such PI activities, by indicating cases that require detailed study.

The amount of data that can be retrieved from the CDDS at ESOC is necessarily limited by available resources (e.g., network bandwidth). Thus, to make best use of the finite amount of data which JSOC can retrieve from CDDS, SPM checks will, in general, be made on the same data samples as used for IEC monitoring. The only exceptions to this rule are a few cases where the SPM must be applied to data taken during execution of special test procedures on the instrument.

The data, including any necessary housekeeping, will be retrieved from the CDDS at ESOC and processed at JSOC using software provided by the PI teams. This software is, in some cases, the same as that used for IEC and, in other cases, is a separate package. The parameters generated by this software will be examined using a graphics tool which can display several parameters on a multi-panel plot (Dimbylow, 1996). The JSOC staff will use PI-supplied criteria to assess if the displayed data are acceptable or if they indicate potential problems. The results of this assessment will be reported to the appropriate PI.

4.4. Scientific Events Catalogue

In addition to the three main monitoring tasks described above, JSOC is also responsible for the production of a scientific events catalogue. This catalogue is intended to provide a detailed and definitive list of scientific events observed by the four Cluster spacecraft, e.g. magnetopause, bow shock and neutral sheet crossings, flux transfer events and auroral field line crossings. Experience from previous missions, e.g., AMPTE-UKS, indicates that a good event catalogue is a significant aid in the analysis of data from space plasma missions. For example, the catalogue may help to identify a good set of cases for event analysis.

The production of this catalogue is clearly similar to the mission progress monitoring but will take place on a longer time-scale using better quality and more comprehensive data products than the quick-look products used in mission progress monitoring. JSOC will identify events for this catalogue by the analysis of Cluster 'prime parameter' data from all four spacecraft. These parameters have been carefully defined within the Cluster Science Data System (Schmidt and Escoubet, this issue) to provide a set of high-quality data products at spin time-resolution. They are generated by the pipeline processing of science data on all four spacecraft using PI-supplied software operated at the CSDS National Data Centres. The data are disseminated within the Cluster Science Data System to ensure that they are available to all members of the Cluster science community. They will be made available to JSOC via the UK National Data Centre of CSDS, which is co-located with JSOC. They are expected to be available approximately six weeks after the raw data are taken on the spacecraft.

JSOC will examine prime parameters using a graphics tool similar to that used in mission progress monitoring and will identify scientific events (e.g., boundary crossings) using criteria to be agreed with the SWT. The full details of each event (event type, time, orbit number, spacecraft position, GSM latitude and magnetic local time) will be determined using the same mechanisms as for mission progress. These details will be recorded in a separate table in the JSOC planning database.

5. Information Subsystem

5.1. Content

This subsystem makes a variety of information available to members of the Cluster Science Working Team. The information made available includes:

– Predicted orbit. Postscript plots which show the location of data acquisition periods with respect to the scientific regions crossed on each orbit.

– Predicted orbit and scientific events as listed in Sections 3.1.1 and 3.5.3. SWT access to this dataset is very important as these data are used as references for command timing.

– Predicted geometric position. The GSE position of the reference spacecraft, the separation vectors of the other spacecraft and the quality factors for the spacecraft configuration.

– Predicted scientific position. The invariant latitude and magnetic local time of each spacecraft.

– Predicted solar cycle trends. The predicted monthly values of solar-wind ram pressure (medians and sextiles) used in the JSOC planning subsystem. The predicted monthly sunspot number is also provided.

– DSN visibility. Postscript plots which show spacecraft visibility from DSN ground stations with respect to the data acquisition periods and the scientific regions crossed on each orbit.

– Scientific events catalogue. Table of *observed* scientific events identified from examination of prime parameter data as described in Section 4.4.

– Help files describing the above.

– A copy of the Master Science Plan agreed by the SWT.

– JSOC documentation. Access to various formal documents prepared as part of JSOC implementation and testing.

– General information on the Cluster mission.

– Pointers to other sources of relevant information on other space missions and on ground-based solar terrestrial physics experiments.

5.2. Mechanisms

5.2.1. *World-Wide Web*

All the information described above will be made available via the World-Wide Web (WWW) server on the JSOC computer system. The WWW is an ideal tool for this JSOC task:

– A high-quality user interface is readily available to users in the form of various WWW 'browsers' such as Mosaic and Netscape. Thus there is no need for JSOC to develop or directly support the user interface.

– The user interface accesses JSOC via a client-server interaction. Thus the user interface is controlled by the computer on which the browser is located. This should be local to the user, so control of the user interface will not be affected by network problem. The user interface will respond promptly to user commands.

– Help files and other textual information can be formatted in the 'Hypertext Markup Language' (HTML). This will give a professional appearance to the result.

– HTML will also allow JSOC staff to provide hypertext links to other information - within the JSOC server, on other systems at RAL (e.g., Cluster ground-based data centre) and at other sites of interest to Cluster operations (e.g., International Solar-Terrestrial Physics mission planning and operations at NASA Goddard Space Flight Center (Mish *et al.*, 1995)).

– Tables of data in ASCII 'flat files' can be viewed via the browser's plain text viewer.

– Postscript plots (such as Figure 7) can be viewed if the browser supports an external Postscript viewer such as the public-domain product 'Ghostview'.

– The contents of directories of data and/or plot files can be listed by the viewer. The user can then examine the contents of any file by clicking on the filename with a mouse.

The JSOC WWW server also supports a second smaller set of pages. These provide information of general public interest about JSOC and are linked from the public information pages of JSOC's host institite (RAL).

5.2.2. *CSDS User Interface*

The WWW access to JSOC is complemented by access to some of the planning data via the Cluster Science Data System User Interface (CSDS-UI) (see Schmidt and Escoubet, this issue). This system uses the Oracle database management system to store and query various data tables generated by the CSDS. These include several tables in the JSOC planning database - namely the predicted events, geometric position, scientific position and solar cycle trends and the catalogue of observed scientific events.

The CSDS-UI does not directly access the tables held on the JSOC computer system. Instead, it accesses a copy, held at the user's CSDS National Data Centre, which is created as an Oracle 'snapshot'. This is a mechanism, provided in the Oracle software, which allows the database designer to specify: (a) that a table **A** is to be a subset of some other table **B**, and (b) the copy **A** is to be updated from **B** at regular intervals. Each update may be a complete replacement of **A** by a new copy from **B** – or it may be an incremental update changing only records in **A** that reflect changes in **B** since the previous update. Table **B** may be on a different computer to table **A**, provided these computers are linked by SQL*Net, the Oracle networking product.

The snapshots used by the CSDS-UI are complete copies of the JSOC tables and are held at the CSDS National Data Centres in Austria, France, Germany, Hungary, Scandinavia and the UK. These snapshots are updated as follows:

– The scientific events catalogue will be updated every working day during mission operations. Therefore, the snapshot copies of this table are updated every night just after midnight Universal Time (when the load on CSDS network links is expected to be low).

– The predicted orbit and scientific events, geometric position and scientific position tables are updated every week as revised orbit and events data are received from ESOC. Therefore the snapshot copies of these tables are updated every week (at present, early on a Saturday morning).

6. Implementation

6.1. Hardware

The JSOC system comprises three computers from the Sun SPARCstation series:

– A SPARCstation 5, which is the public front-end computer. This machine hosts PI accounts and the information dissemination subsystem.

– A SPARCstation 20, which is the operational computer. This machine hosts command preparation activities, the commanding and planning databases, and the communications with ESOC and with other parts of the Cluster Science Data System.

– A second SPARCstation 5, which acts a backup to the other two machines. All critical data and software are automatically copied and updated on this machine.

A large amount of disk space (12 gigabytes) is available on internal disks within the three computers and in an external disk pack. The disk space is configured so that the operational system can access all necessary files on the public front-end, but the main operational files are not accessible from the public front-end.

The computers are connected to the RAL local area network so that they are accessible from the X-terminals and PCs used by JSOC staff. This connection also gives access to the public Internet, which is used by PIs and other external users who wish to access the front-end computer. The operational computer is also connected to CSDSnet. This is a private TCP/IP network, funded by ESA, which links the various centres within the Cluster Science Data System (CSDS) – see Schmidt and Escoubet (this issue). JSOC's main use of CSDSnet is for transfer of command and monitoring data between JSOC and ESOC. In addition there is some exchange of planning data over CSDSnet to the various CSDS National Data Centres.

A fax modem is installed to allow transmission of computer-generated faxes from JSOC to PIs and to ESOC – as part of the commanding subsystem.

6.2. SOFTWARE

The JSOC computer runs under the Sun Solaris operating system (a Unix system), currently version 2.5. The JSOC implementation makes use of a large number of software packages, running under Solaris, as follows:

(1) Programming tools. A wide variety of tools are used:

– The SunProC compiler is used for compilation of C language programs. C is used in the commanding subsystem and most monitoring subsystem modules.

– 'Purify' from Pure Software Inc. is used to check C language programs for 'memory leaks', i.e., invalid access to computer memory.

– The Numerical Algorithms Group (NAG) 'f90' compiler is used for compilation of Fortran 90 programs. Fortran 90 is used in the planning subsystem and some monitoring subsystem modules.

– The standard Unix tool 'make' is used to construct procedures for building libraries and executables from C and Fortran source codes.

(2) Database. The JSOC commanding and planning databases are stored under the Oracle database management system. This system supports SQL, the standard database language. The main Oracle applications used are:

– SQL*Plus is used for creation of database objects, their routine updating and the execution of various standard queries required by JSOC subsystems. These functions are all executed via procedures (SQL scripts) stored in ASCII files. SQL*Plus is also used for interactive execution of ad-hoc queries.

– SQL*Loader is used to load commanding and planning data from ASCII files.

– SQL*Forms is used to examine and review the contents of many database tables.

– The Pro*C precompiler is used to access the commanding database from C language programs.

– SQL*Net is used to support remote network access to the planning subsystem from other CSDS National Data Centres.

– SQL*DBA was used for database system administration but is now superseded by Server Manager.

(3) Data visualisation. A number of applications in the planning and monitoring subsystem require graphical display of data.

– IDL from Research Systems Inc is used for all data visualisation.

(4) User interfaces. JSOC has made use of high-level tools for the production of graphical user interfaces in the commanding and monitoring subsystems. These run under the Motif windowing system.

– 'X-Designer' from Imperial Software Technology has been used to construct the commanding subsystem user interface (jcc).

– 'UIM/X' from Visual Edge Software Ltd has been used to construct user interfaces for various applications in the monitoring subsystem, e.g., the mission-progress-monitoring tool. UIM/X supports calls to other packages: (i) IDL is called to generate plots of monitoring data within windows created by UIM/X; (ii) SQL is called to store monitoring results in Oracle database tables.

(5) Configuration control. The ASCII source files for the JSOC system (e.g., C and Fortran codes, makefiles, SQL scripts, control data, etc) are kept under configuration control via the public domain product 'Concurrent Versions System' (CVS). This is a front end to the well-known Revision Control System (RCS).

(6) File transfer. A large number of files containing command and planning data are transferred between JSOC and ESOC - as described in Sections 2 and 3 above. These files are transferred using the Cluster File Transfer System (CFTS), which has been developed within ESOC (Hamer, 1995). This software is an application on top of the standard Internet protocols. It provides additional security, redundancy and monitoring for the transfer of these critical files.

(7) Fax generation. 'Trufax' from Versafax is used to send computer generated messages as faxes.

7. Ground-Based Support

The scope of JSOC has recently been extended to include this task. For this task the JSOC team works in close conjunction with staff in the World Data Centre C1 for Solar-Terrestrial Physics (WDC), who are supporting the Cluster ground-based data centre. The WDC is co-located with JSOC at the Rutherford Appleton Laboratory.

One key objective to facilitate coordinated observations by Cluster and ground-based solar-terrestrial physics (STP) experiments. Favourable configurations between Cluster and the ground-based experiments will be predicted using a tool already developed by the WDC team (Lockwood and Opgenoorth, 1995; Lockwood *et al.*, 1995). This requires, as input, predicted event data from the JSOC planning subsystem as described in Section 3.1.1 and 3.5.3. The predictions of favourable configurations will be made available to the potential user community via the Cluster ground-based data centre. This centre will also collect information on the plans for operations of ground-based STP experiments in conjunction with Cluster. JSOC will summarise the predictions and the plans of the ground-based community and will report this summary to the Cluster SWT.

The other main objective is the exchange of summary data between Cluster and the ground-based STP community. To aid the Cluster community, the WDC team will produce a quickly available index of auroral electrojet activity (Opgenoorth *et al.*, 1996). This will provide a clear indication of magnetospheric disturbances such as substorms. This would be a scientifically valuable complement to Cluster studies of the magnetosphere because substorm activity is difficult to identify from spacecraft observations. To generate this index, the WDC team will collect data from ground-based magnetic observatories in three latitude zones in the Northern Hemisphere:

– One zone similar to the traditional auroral electrojet observatories (Kamei and Maeda, 1981).

– Another zone at lower latitudes – to provide a better measure of electrojet activity at high levels of geomagnetic activity.

– A third zone at higher latitudes – to provide a better measure of electrojet activity at low levels of geomagnetic activity.

These data will be collected as quickly as possible - mainly via computer network links. They will then be processed to generate a set of electrojet indices (one for each zone), which will be made available to the Cluster SWT and the ground-based community via a World Wide Web server operated by the Cluster ground-based data centre.

Acknowledgements

This work was performed under ESA contract 10260/93/NL/CC. ESA also provided the licence for the Oracle database software. We thank all those who freely gave us copies of their software for use in JSOC: (a) J.C. Kosik at CNES for the MAGLIB library; (b) the ESOC Orbits and Attitude Department for the Cluster orbit decompression software; (c) J. Schoenmaekers at ESOC for the subroutine to calculate quality parameters; (d) the Cluster PI teams for software used in the monitoring subsystem to process data from their instruments; (e) G. Hall of the UK Cluster Data Handling Facility for the CDDS data retrieval software; (f) the ESOC Data Processing Division for the Cluster File Transfer System. The development of JSOC has benefited greatly from a close working relationship with the team building the Cluster Operations Control Centre at ESOC; we thank them for the friendly and professional way in which they have worked with us. We also thank all members of the Cluster Science Working Team and of the Cluster Project Team for the help and advice which they have provided. The support now to be provided by JSOC for ground-based STP experiments will build on work by the Cluster ground-based working group; we thank them for their efforts. Finally we thank the many other staff at RAL who have contributed to the JSOC development: E. Dunford, M. Lockwood, A. McDermott, D. Mac Randal, R. May, C. Perry, L. Sastry, G. Spalding, R. Stamper, D. Terrett, P. Vaughan and M. Wild.

References

Allen, A. J., Burgess, D., Dimbylow, T. G., Golton, E., Hall, G. W., Hare, A. R., Perry, C. H., Schwartz, S. J., and Spalding, G. H.: 1995, *UK Co-ordinated Data Handling Facility for Cluster: Detailed Design Document, CU-RAL-DD-0001, Issue 1.1*, Rutherford Appleton Laboratory: Chilton.

Bryant, D. A. *et al.*: 1984, *The AMPTE-UKS Mission Science Plan, UKS-510*, Rutherford Appleton Laboratory: Chilton, p. 29.

CCSDS: 1990, *CCSDS Time Code Formats. CCSDS 301.0-B-2, Blue Book, issue 2*, Consultative Committee for Space Data Systems: Washington DC, p. 2.

Chen, P. P.-S.: 1976, 'The Entity-Relationship Model – Towards a Unified View of Data', *ACM Trans. on Database Systems* **1**.

Date, C. J.: 1986, *An Introduction to Database Systems, 4th edition*, Addison-Wesley: Reading MA.

Dimbylow, T. G. and Hill, P. M.: 1995, *JSOC System Specification, DS-JSO-SS-0001, Issue 1.3*, Cluster Joint Science Operations Centre: Chilton.

Dimbylow, T. G.: 1996, *JSOC Detailed Design Document, DS-JSO-DD-0001, Issue 1.2*, Cluster Joint Science Operations Centre: Chilton.

Dunford, E., Vaughan, P., Golton, E., and Dimbylow, T.: 1993, in R. Schmidt (ed.), 'The Cluster Joint Science Operations Centre', *Cluster: Mission, Payload and Supporting Activitie*, ESA SP-1159, p. 291.

Fairfield, D. H.: 1971, 'Average and Unusual Locations of the Earth's Magnetopause and Bow Shock', *J. Geophys. Res.* **76**, 6700.

Fairfield, D. H.: 1980, 'A Statistical Determination of the Shape and Position of the Geomagnetic Neutral Sheet', *J. Geophys. Res.* **85**, 775.

Fairfield, D. H.: 1990, *Magnetospheric Boundaries for Use at the NSSDC Satellite Situation Center*, ,, Report to the Inter-Agency Co-ordinating Group, NSSDC, Greenbelt.

Ferri, P. and Warhaut, M.: 1997, 'Cluster Mission Operations', *Space Sci. Rev.* , this issue.

Hamer, P.: 1995, *Cluster Software User Manual*, ,, Vol. A0M38, CFTS - JSOC Supplement, ESOC: Darmstadt.

Hapgood, M. A.: 1992, 'Space Physics Coordinate Transformations: A User Guide', *Planetary Space Sci.* **40**, 711.

Hapgood, M. A.: 1994, *The Uncertainty in Predictions of Cluster Magnetopause Crossing Times*, Presentation at the first Cluster Science Operations Working Group, Paris, 15–16 February, 1994.

Hapgood, M. A.: 1995, 'Space Physics Coordinate Transformations: The Role of Precession', *Ann. Geophys.* **13**, 713.

Holzworth, R. H. and Meng, C.-I.: 1975, 'Mathematical Representation of the Auroral Oval', *Geophys. Res. Letters* **2**, 377.

Howe, H. C. and Binsack, J. H.: 1972, 'Explorer 33 and 35 Plasma Observations of Magnetosheath Flow', *J. Geophys. Res.* **77**, 3334.

IAGA: 1996, 'International Geomagnetic Reference Field, 1995 Revision', *Geophys. J. Int.* **125**, 318.

Kamei, T. and Maeda, H.: 1981, *Auroral Electroject Indices (AE) for July–December 1978, Data Book No. 4*, World Data Center C2 for Geomagnetism: Kyoto.

Kosik, J. C.: 1993, *Maglib – Lettre d'Information, No. 7, Coordinate Systems*, CNES, Toulouse.

Kosik, J. C.: 1994, *Maglib User's Guide, Version 2*, CNES, Toulouse.

Langel, R. A.: 1992, 'IGRF: 1991 Revision', *EOS, Trans. AGU* **73**, 182.

Lockwood, M., Stamper, R., Wild, M. N., and Opgenoorth, H. J.: 1995, 'Ground-Based Measurements in Support of Cluster: An On-Line Planning Procedure', Report **RAL-95-018**, Rutherford Appleton Laboratory, Didcot.

Lockwood, M. and Opgenoorth, H. J.: 1995, 'Opportunities for Magnetospheric Research Using EISCAT/ESR and CLUSTER', *J. Geoelect. Geomag.* **47**, 699.

Mish, W. H., Green, J. L., Reph, M. G., and Peredo, M.: 1995, 'ISTP Science Data Systems and Products', *Space Sci. Rev.* **71**, 815.

NASA: 1996, *Cluster Mission NASA/ESA Implementation Agreement, CL-EST-RS-0009.*,

Opgenoorth, H. J., Persson, M. A. L., Lockwood, M., Stamper, R., Wild, M., Pellinen, R., Pulkkinen, T., Kauristie, K., and Kamide, Y.: 1996, in M. Lockwood, H. J. Opgenoorth, and B. Battrick

(eds.), 'A New Family of Geomagnetic Disturbance Indices', *Ground-Based Observations in Support of Cluster*, ESA-SP Series, to be published.
Oracle: 1992, *Oracle7 Server Application Developer's Guide. Part No. 6695-70-1292*, Oracle Corporation: Redwood City, California, 1–2.
Paschmann, G. and Sckopke, N.: 1996, *Cluster Master Science Plan, CL-MPE-TN-0009*, Max-Planck-Institut für extraterrestrische Physik, Garching.
Pellón-Bailón, J.L and Rodríguez-Canabal, J.: 1993, *Cluster: Consolidated Report on Mission Analysis. CL-ESC-RP-0001, Issue 2.0*, ESA, Darmstadt.
Robert, P. and Roux, A.: 1993, in C. P. Escoubet (ed.), 'Influence of the Shape of the Tetrahedron on the Accuracy of the Estimate of the Current Density', *Spatio-Temporal Analysis for Resolving Plasma Turbulence (START)*, ESA WPP-047, p. 289.
Roelof, E. C. and Sibeck, D. G.: 1993, 'The Magnetopause Shape as a Bivariate Function of IMF B_z and Solar Wind Dynamic Pressure', *J. Geophys. Res.* **98**, 21421.
Schmidt, R. and Escoubet, C. P.: 1996, 'The Cluster Science Data System (CSDS) – a New Approach to the Distribution of Scientific Data', *Space Sci. Rev.*, this issue.
Schoenmaekers, J.: 1994, Presentation at the third Cluster Science Operations Working Group, Paris, 5–6 October, 1994.
Sibeck, D. G., Lopez, R. E., and Roelof, E. C.: 1991, 'Solar Wind Control of the Magnetopause Shape, Location and Motion', *J. Geophys. Res.* **96**, 5489.
Smith, S. and Robertson, C.: 1995, *Cluster Data Delivery Interface Document, CL-ESC-ID-0001*, ESOC, Darmstadt.
Sørensen, E. M., Merri, M., and Di Girolamo, G.: 1996, 'The Cluster Data Processing System: A Distributed System in Support of a Challenging Scientific Mission', *Space Sci. Rev.*, this issue.
Stasiewicz, K.: 1990, *Magnetic Field Modeling for Application to the Cluster Project: A Report to ESA Prepared Under the Contract ESOC 8635/90/D/MD(SC)*, IRF, Uppsala.
vom Stein, R., Glaßmeier, K. H., and Motschmann, U.: 1993, in C. P. Escoubet (ed.), 'Cluster as a Wave Telescope and a Mode Filter', *Spatio-Temporal Analysis for Resolving Plasma Turbulence (START)*, ESA WPP-047, p. 211.
Tsyganenko, N. A.: 1987, 'Global Quantitative Models of the Geomagnetic Field in the Cislunar Magnetosphere for Different Disturbance Levels', *Planetary Space Sci.* **35**, 1347.
Tsyganenko, N. A.: 1989, 'A Magnetospheric Magnetic Field Model with a Warped Tail Current Sheet', *Planetary Space Sci.* **37**, 5.

List of Acronyms

CLUSTER INSTRUMENTS

The eleven Cluster instruments are referred throughout this paper by their acronyms: ASPOC, CIS, DWP, EDI, EFW, FGM, PEACE, RAPID, STAFF, WBD and WHISPER. See the appropriate papers in this special issue for a description of these instruments.

The five wave instruments (DWP, EFW, STAFF, WBD and WHISPER) form the Wave Experiment Consortium, otherwise known as WEC.

OTHER ACRONYMS

Acronym	Description
AMPTE	Active Magnetospheric Particle Tracer Explorers
ASCII	American Standard Code for Information Interchange
CCSDS	Consultative Committee on Space Data Systems
CDDS	Cluster Data Disposition System
CMPS	Cluster Mission Planning System
CNES	Centre National d'Etudes Spatiales
CSDS	Cluster Science Data System
CSDS-UI	Cluster Science Data System - User Interface
CVS	Concurrent Versions System
DSN	Deep Space Network
ER	Entity-Relationship (diagram)
ESA	European Space Agency
ESOC	European Space Operations Centre
GBWG	Working group for the coordination of ground-based observations with Cluster
GSE	Geocentric-solar-ecliptic
GSM	Geocentric-solar-magnetospheric
HTML	Hypertext Markup Language
IACG	Inter-Agency Consultative Group
IAGA	International Association for Geomagnetism and Aeronomy
IDL	Interactive Data Language
IEC	Inter-experiment calibration
IGRF	International Geomagnetic Reference Field

Acronym	Description
jcc	JSOC commanding controller (acronym always in lower case)
JSOC	Joint Science Operations Centre
MAGLIB	CNES software library for magnetospheric physics
MLT	magnetic local time
MPM	Mission progress monitoring
MSP	Master Science Plan
NAG	Numerical Algorithms Group
NASA	National Aeronautics and Space Administration
NSSDC	US National Space Science Data Center
OBRQ	Observational Requests (file containing the command schedule for science payload on all four spacecraft)
OVT	Orbit Visualisation Tool
PI	Principal Investigator
PIOR	PI Observations Requests (file containing the command schedule for one instrument on all four spacecraft)
RAL	Rutherford Appleton Laboratory
RCS	Revision Control System
SCOP	Spacecraft Operations Plan (file)
SM	Solar-magnetic
SQL	Structured Query Language (database standard)
SPM	Science performance monitoring
STP	Solar-Terrestrial Physics
SWT	Science Working Team
TLIS	Top Level Instrument Schedule (a computerised representation of the Master Science Plan)
UKS	United Kingdom Sub-satellite (one of the AMPTE spacecraft)
WDC	World Data Centre C1 for Solar-Terrestrial Physics (RAL)
WWW	World Wide Web

THE CLUSTER DATA PROCESSING SYSTEM: A DISTRIBUTED SYSTEM IN SUPPORT OF A CHALLENGING SCIENTIFIC MISSION

E. M. SØRENSEN, M. MERRI and G. DI GIROLAMO
Flight and Control Systems Department, European Space Operations Centre, Darmstadt, Germany

Abstract. The Cluster mission is aimed at the study of small-scale structures that are believed to be fundamental in determining the behaviour of key interactive processes of cosmic plasma. The mission will be controlled from the European Space Operations Centre (ESOC). ESOC is also in charge of the commanding of the scientific payloads on-board the four Cluster spacecraft after negotiation with the Cluster Principal Investigators (PIs) and of collecting and distributing the mission's scientific results to the Cluster community.

This paper describes the process of translating the scientific requirements of the Cluster mission into a data-processing system supporting the mission via the definition of an appropriate operational scenario. In particular, the process of negotiation between the PIs and ESOC to command the spacecraft is mediated by the Joint Science Operations Centre (JSOC) and finalised by the Cluster Mission Planning System (CMPS) while the return of the data to the Cluster community is actuated by the Cluster Data Disposition System (CDDS). The Cluster Mission Control System (CMCS) provides the interface between these two systems and the spacecraft. These elements constitute the Cluster Data-Processing System (CDPS).

1. Introduction

Cluster is a scientific mission funded by the Solar Terrestrial Science Programme (STSP) whose purpose is to investigate key interactive processes of cosmic plasma. The other STSP mission is SOHO. Details on the SOHO ground segment can be found in (Cyr *et al.*, 1995). The Cluster mission comprises four identical spacecraft launched into large, elliptical polar orbits around the Earth. After transfer to their operational orbits the spacecraft will fly in formation to allow scientists to measure subtle changes in the interaction between the Earth and the Sun. The four spacecraft will look at how particles from the Sun interact with the Earth's magnetic field. Cluster will observe the magnetic and electrical interactions between the Earth and the Sun, by making direct measurements of the fields and particles trapped in the Earth's magnetic field. For the first time, small-scale fluctuations in interplanetary space will be measured between each of the four spacecraft as they orbit the Earth.

Responsibility for the definition, design and implementation of the Cluster ground segment has been assigned to the ESA's European Space Operations Centre (ESOC) in Darmstadt, Germany. ESOC is also responsible for the operation of the Cluster spacecraft (both platform and payload) and for the Cluster ground segment.

The Cluster Data-Processing System (CDPS) is an important part of the Cluster ground segment and it has been designed to meet the complex scientific objectives of the Cluster mission. The process of defining the CDPS started at ESOC in late

Space Science Reviews **79:** 527–555, 1997.

1990. The initial step was to carefully analyse the scientific objectives, to deduce an operational scenario, and then to define a system that would best fulfil them. This paper briefly introduces the Cluster operational scenario and describes in some detail the main components of the CDPS. To conclude, some lessons learned are discussed.

2. Cluster Operational Scenario

This section briefly summarises the mission operational scenario stemming from the Cluster scientific objectives and highlights the implications on the ground system design of these objectives as derived from the Cluster Mission Implementation Requirements Document (see MIRD 1991) which is the driving document for implementation of the ground segment. Details on Cluster operations can be found in (Ferri *et al.*, this issue).

2.1. The Cluster Mission Phases

The Launch and Early Orbit Phase (LEOP) and the Transfer Orbit Phase (TOP) are the earliest phases of the mission. Due to the criticality of these phases, five ESA ground stations, namely, Odenwald (Germany), Redu (Belgium), Kourou (French Guyana), Perth (Australia) and Malindi (Kenya) and two NASA Deep-Space Network stations, Goldstone (USA) and Canberra (Australia), are used. All stations are connected to a single Operations Control Centre (OCC) located at ESOC that provides spacecraft monitoring and control (housekeeping data only).

During the Commissioning and Verification Phase (CVP) and Mission Operations Phase (MOP), Cluster is operated from the ESA ground stations in Redu and Odenwald. Each station can support any one of the four Cluster spacecraft at the nominal data rates, that is, from 2 to 262 kbit s^{-1} for Telemetry (TM) and 2 kbit s^{-1} for Telecommands (TC). Ranging (to estimate the spacecraft positions) and Doppler (to estimate the spacecraft velocities) are also supported as required in parallel to commanding and telemetry reception.

The CDPS interfaces with the ground stations to receive and transmit telemetry and telecommands in a controlled and error-free manner. A no-break data-link connection is always established when the ground station is in real-time contact with a Cluster satellite. Transfer of non real-time telemetry is performed within 24 hours at most after reception at the ground station.

The CDPS interprets the Cluster telemetry and prepares telecommands in the accepted formats. Furthermore, it supports real-time telemetry (both housekeeping and science) and telecommands with the satellite currently on-line. At any time, there could be two spacecraft in real-time contact (one for each ground station).

2.2. Cluster operations

The CDPS controls and monitors the health of both spacecraft platform and payload, in accordance with agreed and pre-defined Flight Operations Procedures (FOP) under the specific Cluster operations constraints and in accordance with operational requirements. In particular, the CDPS schedules and implements satellite operations as required, including spacecraft recorder and RAM dumps, antenna switches, eclipse handling, ranging campaigns, network activities and support of the operations of the Wideband Data (WBD) payload. Real-time housekeeping data is accessible at the Operations Control Centre (OCC) within 30 s in support of payload commissioning and contingency situations. WBD science data is handled differently from that of the other payloads as, due to its high downlink bit rate, it requires the support of the NASA Deep-Space network stations of Goldstone and Canberra.

Mission operations are conducted so as to maximise the scientific return in a manner compatible with satellite safety and the overall ground segment system resources. This is achieved via a mission planning software facility based on operational constraints and scientific inputs from the PIs that are consolidated by the Joint Science Operations Centre (JSOC). Mission planning aims at obtaining the maximum amount of data acquired from simultaneous observations by the four spacecraft in orbital regions of scientific interest within the constraints imposed by station coverage, on-board resources and acquisition above 35 000 km. The regions of scientific interest and the corresponding requests for observations are electronically provided by the PIs via JSOC using a dedicated network (CSDSnet). Active control and management of on-board resources are also provided. The CDPS includes also a system for on-board software maintenance.

The CDPS makes available telemetry (housekeeping and science) and auxiliary data in near real-time (<5 min) to the Cluster community, that is the PIs, JSOC and the Cluster Science Data System (CSDS) community, via an ESOC-developed dedicated network (CSDSnet). CSDSnet allows inter-connection of ESOC, JSOC, and six Cluster data centres that are located in Austria, France, Germany, Hungary, Sweden, and United Kingdom. The United States and the Chinese data centres, although part of the CSDS, are connected to the CSDSnet via the United Kingdom and the Austrian data centres, respectively (Schmidt *et al.*, this issue). Following receipt or generation at the OCC, the data can be accessed remotely on a call-up basis. The CDPS supports simultaneous on-line operations of two Cluster satellites together with access by PIs to the mission products of all four.

The CDPS time-stamps the telemetry data by converting the on-board time to Universal Time Coordinated (UTC) with an accuracy below 2 ms for any one spacecraft. It also provides a commanding system capable of controlling any of the four Cluster spacecraft. During nominal operations, commands are decalibrated (i.e., put in a format that is understandable by the spacecraft), pre-transmission validated, and transmitted to the ground station(s) for uplinking in real-time. A complete history of all commands requested, transmitted and verified is maintained by the OCC. Hazardous operations on the spacecraft are, whenever possible, implemented during a ground station pass. Ranging and Doppler are also supported. The CDPS provides the capability to process housekeeping data in real-time for spacecraft control purposes. This includes command verification and out-of-limit detection. Housekeeping telemetry data, including the traditional binary and analog parameters and event messages, are processed by the OCC using calibration curves to convert them into the engineering and/or functional parameter values needed to monitor the status of the spacecraft platform and payload and to recover from anomalies, if any.

On average, a minimum of 95% of data originating simultaneously from the four spacecraft can be recovered and made available to PIs. Mission products include all payload science data (except the WBD telemetry data), all platform and payload housekeeping data and auxiliary data (e.g., telecommand history, station visibility, spacecraft orbital position, attitude and spin rate/phase, etc.). Housekeeping and science telemetry data (both real-time and playback) are extracted and stored as raw data, chronologically ordered and sorted by spacecraft and payload, and include quality data and additional timing data so as to enable the PI to correlate the data with respect to UTC. In addition to the near real-time access to mission data that is available via electronic network for the most recent data (ten days), all science (except WBD) data, housekeeping data and consolidated auxiliary data are copied and mailed to each PI and Co-Investigator (Co-I) within three weeks of its receipt at the OCC on a permanent storage medium. A common data interchange standard based on the Standard Formatted Data Unit (SFDU) concept is used in order to allow for efficient data distribution. Finally, data is archived for all payloads for a period of ten years from the end of the mission. Archived data includes all raw telemetry and auxiliary data.

3. Overview of the Cluster Data-Processing System

The Cluster Data-Processing System (CDPS) is part of the ESOC OCC, the central facility responsible for the operations of the four Cluster spacecraft. The CDPS has been developed under the responsibility of the Data-Processing Division in the Flight and Control Systems Department at ESOC. It is a distributed computer system whose main components are the Cluster Mission Control System, the Cluster Mission Planning System, the Cluster Data Disposition System and the Cluster Spacecraft Evaluation System. Figure 1 shows an overview of the CDPS and its location in the context of the Cluster ground and space segments.

The *Cluster Mission Planning System* (CMPS) is an off-line system that provides tools to plan in advance the mission operations based on inputs from the PIs (via JSOC) and from the operational staff at ESOC. The final output of this system

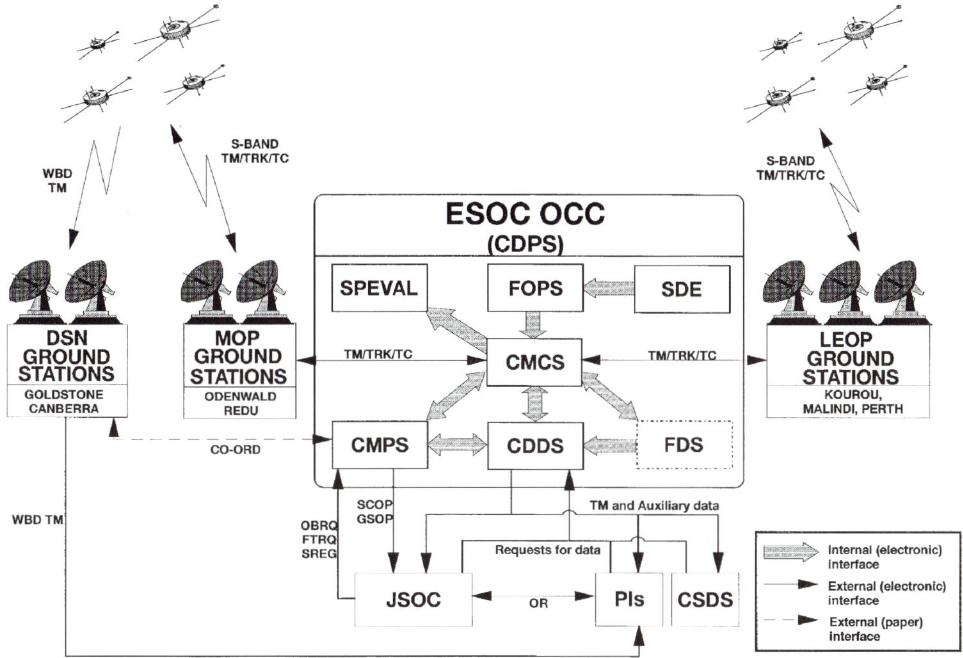

Figure 1. Overall Architecture of the Cluster Data Processing System. The main block represents the CDPS (except for the FDS block). Interactions with the Cluster community and the ground stations are shown for both MOP and LEOP.

is the generation of machine- and man-readable schedules to command the four spacecraft and the two ground stations.

The *Cluster Mission Control System* (CMCS) supports the exchange of telemetry, telecommand and ranging data between the OCC and the spacecraft via the dedicated communications lines and appropriate interfacing to the ground stations. It allows real- and near-real-time data-processing essential for monitoring and controlling the mission. The CMCS also has facilities for management and access of the operational database and for maintenance of the on-board software.

The CMCS relays in real-time the received housekeeping and science telemetry to the *Cluster Data Disposition System* (CDDS) which stores the last ten days of this data together with other auxiliary data to allow remote access by selected external computer nodes of the Cluster community for quasi-real-time inspection of mission data. Furthermore, it provides facilities for regular daily production of compact disks for distribution to the Cluster community and Co-Is and for permanent archive of both raw telemetry and auxiliary data.

The *Cluster Spacecraft Evaluation System* (SPEVAL) allows the archiving of all housekeeping telemetry and auxiliary data and their retrieval based on complex queries for spacecraft performance evaluation mostly for the benefit of the operational staff at ESOC. It supports the calculation of basic and complex statistics and sophisticated plotting of this information.

In addition, the *Flight Operations Procedure System* (FOPS) is an off-line system that is used by the mission operations team to prepare Cluster operations procedures and command sequences. The command sequences defined can be imported into the operational database.

These last two systems are not covered in this paper as they are part of the generic ESOC infrastructure, that is, they can be reused with minor modifications in support of other missions. During the system design, much emphasis has been placed on the concepts of back-up and redundancy to cope with emergency situations. The CDPS is a fully redundant system with no single-point failure.

4. The Cluster Mission Planning System

The planning of the Cluster mission is a complex task that involves four spacecraft each carrying eleven identical and independent payloads and all the traditional platform sub-systems. The main goal of the Cluster Mission Planning System (CMPS) is to schedule on-board and ground operations in order to maximise the scientific mission return within the constraints imposed by the on-board (e.g., storage and power capacities) and ground (e.g., ground station visibility and availability) systems. The CMPS is operationally used during the Mission Operation Phase.

The routine mission control concept is based on the use of a single control centre in conjunction with two dedicated ESA ground stations (Redu and Odenwald). All payload operations are pre-planned and executed according to an agreed plan produced in an iterative mission planning exercise. A slightly different handling is necessary for the WBD payload as it requires coverage from the NASA Deep-Space Network (DSN). The CMPS partially supports the coordination with the DSN authority for allocation of Cluster coverage periods and, upon their confirmation, it allows the scheduling of WBD operations. Real-time payload operations are not foreseen during the routine phase, except for special operations like payload software maintenance activities. In addition, most of the platform operations including dumps of the on-board recorders, are also included in the planning exercise.

To coordinate and consolidate the requests for scientific observations of the PIs, the Joint Science Operations Centre (JSOC) has been established in the United Kingdom (see Hapgood *et al.*, this issue). The process of planning Cluster operations involves several iterations between ESOC and JSOC. These have been formalised into four different planning levels:

- Long-term plan.
- Medium-term plan.
- Short-term plan.
- Operational plan.

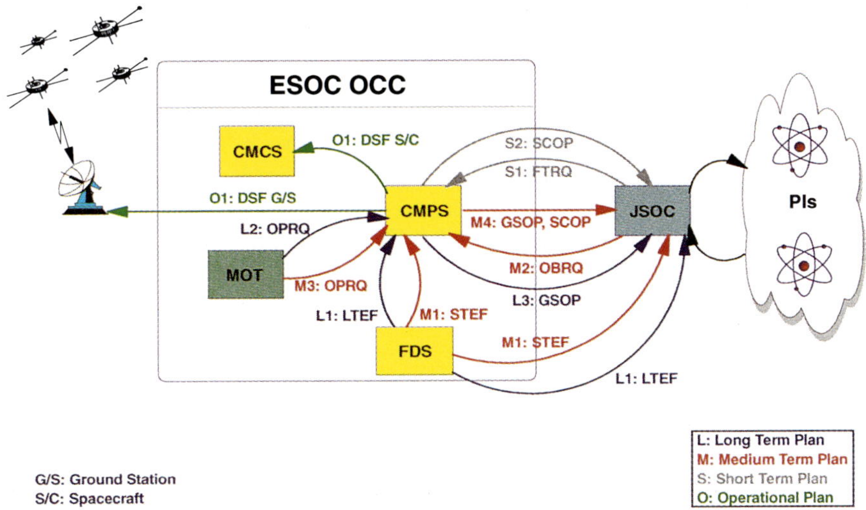

Figure 2. The Cluster Mission Planning System Environment. The various systems that interacts for mission planning purposes are shown together with a summary of the steps that are performed at the different planning levels.

Figure 2 describes and prioritises the different steps that are undertaken in the planning process. It is worth emphasising that these steps involve the exchange of files among several systems remotely located with respect to each other and using different hardware platforms. In order to simplify the interfaces and to allow human inspection of the exchanged information, all exchanged files are encoded in ASCII format. Comment lines can be inserted in these files and are preceded by an exclamation mark (!). The CCSDS ASCII Time Code A (Time Code Formats 1990) format is always used to specify dates and times and always refers to UTC. In addition, the content of any file that is sent from JSOC to the CMPS at ESOC is syntactically validated and an acknowledgement, the so-called File Validation Status Report (FVSR) file, is returned to JSOC. Where errors are detected in the file, the FVSR contains a detailed list of the problems found, together with their location in the file.

The CMPS has been developed using C language and runs on a DEC 4000-90 VAX/VMS computer. A sophisticated and intuitive man-machine interface based on Motif simplifies the interactions of the user with the system.

4.1. The long-term plan

The long-term plan initialises the planning process. It covers a period of about six months and has to be finalised three months before the start of the period itself. Inputs to this planning phase are:

```
! Standard File header
OBRQ 0321 1996-04-29T04:12:00Z

! File specific header
1996-06-29T01:00:00Z 1996-06-30T00:12:00Z     1

! OR for normal science with acquisition table "F" for spacecraft 1, 3 and 4.
! The OR contains 1 command sequence. The 1st time interval indicates the OR start-end
! time. Earliest/latest start/end are also given below
OBSREQLABEL1 1996-06-29T03:12:00Z 1996-06-29T04:59:00Z NORMAL F  134  1
 1           1996-06-29T03:10:00Z 1996-06-29T04:55:00Z
             1996-06-29T03:15:00Z 1996-06-29T05:03:00Z

! Execute Command Sequence SEQ12345 10 seconds before the OR start on spacecraft 1
! and 4. SEQ12345 has 12 parameters which are provided below. If a parameter is not
! provided, the default value is used
SEQ12345   14    -10  12
123.45,1272,ENABLE,STOP,X'23AD',,,127,1,2,3,4;
```

Figure 3. Example of Observational Request File (OBRQ). This file is produced by JSOC and contains Observation Requests that allow the scheduling of payload operations for science data taking. It is an input to the CMPS.

– The predicted flight dynamics events delivered by the ESOC Flight Dynamics System (FDS) to the CMPS (and JSOC) via the Long-Term Event Files (LTEF).

– The generic operation requests provided by the Mission Operation Team (MOT) at ESOC (e.g., eclipse handling, antenna switches, etc.) via the Operations Request (OPRQ) files.

Based on the above inputs, the CMPS generates a six-month plan with a preliminary list of ground segment operations. A report on this plan is electronically transmitted to JSOC in the form of a Ground Segment Operations (GSOP) file, discussed in more detail below.

4.2. The Medium-Term Plan

The long-term plan is used by the CMPS to initialise the medium-term plan which has a duration of six orbits (the average period of the nominal Cluster orbit is about 57 hours). In addition, payload and specialised platform operation requests are provided as input to the CMPS at this time together with more detailed and accurate flight dynamics events via the Short-Term Event File (STEF). The interactions between ESOC and JSOC start six weeks before the start of the medium-term plan.

JSOC is responsible for coordinating and merging PI Observational Requests (OR) into a single Observational Request (OBRQ) file that covers a period of about three orbits. OBRQs are electronically transmitted to the CMPS which automatically validates them. An example of the content of a valid OBRQ file is provided in Figure 3. The syntax of this file is described in (CRID 1995).

The medium term is the most complex level of the entire planning cycle and is schematically shown in Figure 4. The planning cycle foresees that each medium-term plan overlaps for 50% of its duration, i.e., three orbits, with the previously generated medium-term plan, thus avoiding discontinuity at the medium-term plan edges. As each OBRQ covers a period of three orbits, two of them are necessary

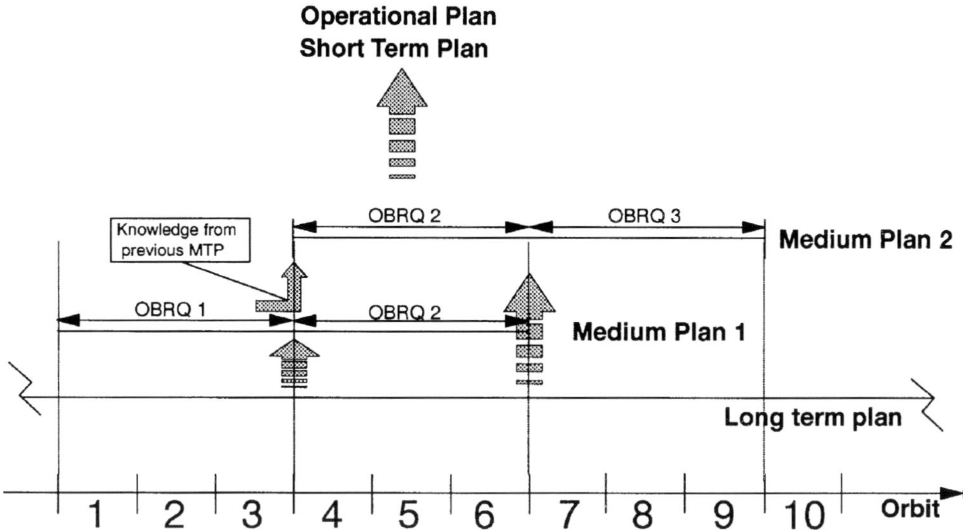

Figure 4. Schematic Overview of the Cluster Planning Cycle. This figure describes the process that leads to the generation of an operational plan and, consequently, to the production of the Detailed Schedule Files. The operational plan is directly derived from the short-term plan. Medium-term plan 2 is used to initialise the short-term plan based on the two OBRQs (2 and 3) included and the knowledge of the status of the previous medium-term plan (1) at the beginning of medium-term plan 2. All medium-term plans are derived from a unique long-term plan. This cycle is iterated every three orbits.

when planning at medium-term level. In addition to these two OBRQs, the CMPS has to retain knowledge about the status of the on-board resources (e.g., recorder fill levels, power usage, payload modes, etc.) at the beginning of the medium-term plan under consideration, as calculated from the previously generated and overlapping medium-term plan. To achieve this, the CMPS allows the user to initialise a medium-term plan with a previous, overlapping one. Manual initialisation is also supported.

Generic and plan-specific platform operation requests are also submitted to the CMPS by the Cluster Mission Operations Team in ESOC. This is done via the Operations Request file (OPRQ). The OPRQ allows the operational team to specify complex operations for both spacecraft and ground stations that can be linked to various events (e.g., acquisition of signal, eclipse, etc.) or absolute times by using a simple and flexible dedicated language. The CMPS supports the creation and modification of OPRQs that are syntactically validated before being accepted. An example of the content of a valid OPRQ file is provided in Figure 5. The syntax of this file is described in (MPS-ADD 1994).

Based on these inputs (STEF, OBRQs, OPRQs and the previous medium-term plan for initialisation) and on a number of configurable operational constraints, the CMPS is able to run its planning algorithm which, if all conflicts or resource clashes could be resolved, schedules spacecraft and ground stations operations including

```
operation pre_longecl 1234                       ! on each spacecraft
   TDA 20 1                                      ! housekeeping TM mode
   start on enter_eclipse -00:05:00Z always      ! to be executed 5 min before
   stop on  enter_eclipse  00:00:00Z             ! each long eclipse
commands
   00:00:00Z seq00000 enable,1;
   00:02:00Z seq00001 disable,1;
end

! Start on AOS and repeat each hour within passes. Assume 1 spacecraft only
operation ranging REDU
   start on REDU_AOS period 01:00:00Z till REDU_LOS always
   stop after 00:05:00Z
commands
   00:00:00Z start_track
   00:05:00Z stop_track
end
```

Figure 5. Example of Operations Request File (OPRQ). This file is produced by operational staff at ESOC and contains requests that allow the scheduling of platform and ground station operations. It is an input to the CMPS.

changes to the on-board data-acquisition modes, recorder dumps, period of real-time contact and ground station configuration. In cases of conflict, the CMPS warns the user of the cause of the conflict. For such cases, the CMPS supports a 'what-if' facility, the 'examine function', which allows deeper investigation and permits the user to try alternative solutions. The result of this scheduling exercise is graphically displayed on the screen with the possibility of generating printouts. Figure 6 gives an example of a scheduled medium-term plan as it appears on the CMPS monitor.

Regardless of the success or failure of the scheduling run, the CMPS allows the generation of two report files, the Spacecraft Operations file (SCOP) and the Ground Segment Operations file (GSOP), that are automatically transferred to JSOC. The SCOP reports on the status of each individual OR in the OBRQs used in the medium-term plan and identifies which OR is acceptable and which generates a conflict (the cause of the conflict is also explained). This type of reporting allows iterations between ESOC and JSOC that eventually lead to a conflict-free plan. An example of the content of a valid SCOP file is provided in Figure 7. The syntax of this file is described in (CRID 1995).

The GSOP contains a schedule for ground-segment-related events involving both spacecraft and ground stations. This includes Telemetry Data Acquisition (TDA) mode changes, spacecraft or ground station non-availability times due to scheduled maintenance, manoeuvres, or orbital events such as long eclipses, or NASA/DSN station availability. Moreover, information concerning resource consumption is provided. GSOP files are produced at different planning levels with slightly different contents. An example of the content of a valid GSOP file is provided in Figure 8. The syntax of this file is described in (CRID 1995).

4.3. Short-term plan

The CMPS uses the medium-term plan to initialise a short-term plan. This planning level covers a period of three orbits. The basic difference between the medium-

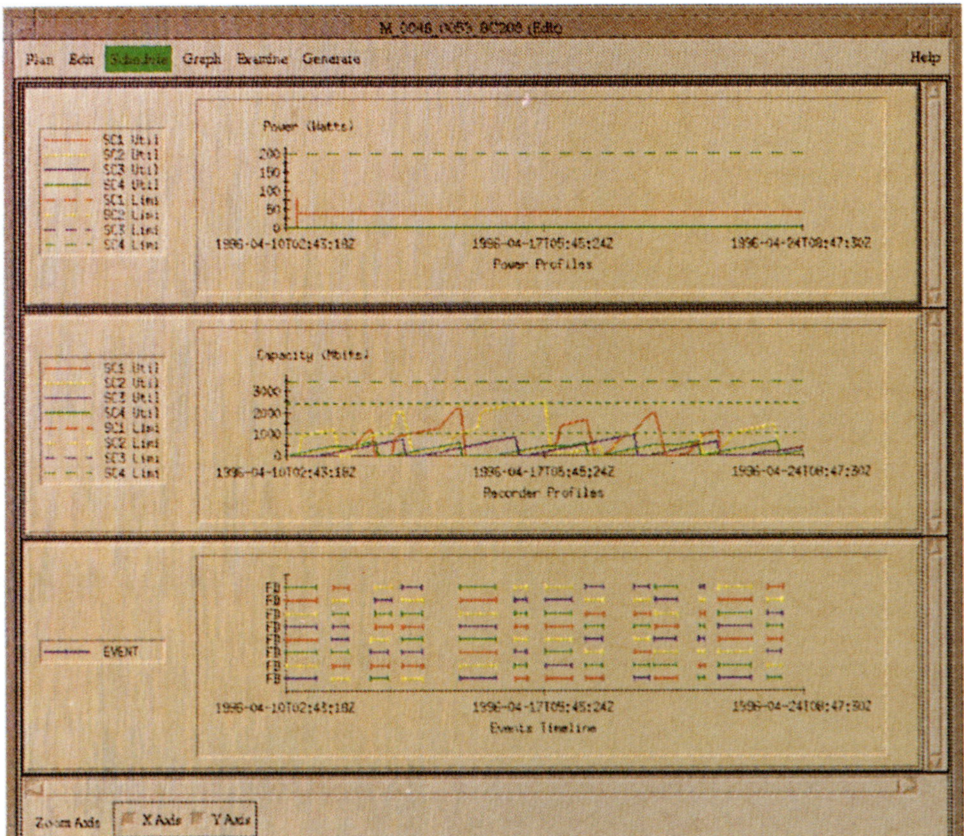

Figure 6. Example of a Scheduled Medium-Term Plan Screen. This figure is a snapshot of the CMPS screen for a scheduled medium-term plan. The screen is divided in three parts. The top panel gives the power profiles for each spacecraft. The dotted line (overlapping) represents the spacecraft-specific maximum power available. The middle panel gives the recorder profiles. There are three spacecraft specific dotted lines (overlapping) representing, from bottom to top, the total capacities of each of the two on-board recorders and the grand total capacity. The bottom panel allows the display of timelines of several event types including flight dynamics, region of scientific interest, scheduled OBRQs and OPRQs and spacecraft modes. In this panel, labels can be toggled on or off to provide more information. Colour coding is used to identify spacecraft specific information.

and the short-term planning levels is that, at the short-term level, the plan is frozen (no new scheduling can be performed) and only minor modifications are still possible. Minor in this context means that modifications must not affect the on-board power and the recorder usage profiles. Changes to the medium-term plan are actuated via the Fine-Tuning Request (FTRQ) files that may be submitted by JSOC electronically. The FTRQ may be used for requesting changes to the command sequences that were previously included in the medium- term plan via OBRQs. In particular, it is possible to insert new command sequences, to delete already existing command sequences and to update their parameters. The FTRQ, if any, must arrive at ESOC at least four days before the start of the short-term plan. An

```
SCOP 0034 1996-06-01T12:12:34Z

! File specific header including reference to the originating plan
1996-06-29T12:00:30Z 1996-06-29T15:09:00Z    2    1    1  M_0011_0014_MASTER

! First OR with two command sequences. OR accepted.
0121 LABEL_ABCDEF ACCEPTABLE      NORMAL 1  124    2
 1 1996-06-29T12:00:30Z 1996-06-29T03:15:00Z
   1996-06-29T03:12:00Z 1996-06-29T04:55:00Z
   1996-06-29T03:15:00Z 1996-06-29T04:59:00Z

SEQ12345    14     -10   12
123.45,1272,ENABLE,STOP,X'23AD',,,127,1,2,3,4;

ZCMD1234   124   -1800    0

! Second OR with one command sequence. OR not accepted because recorder capacity
! is exceeded.
0121 LABEL_123456 EXCEEDED REC.  BURST   3 1234    1
99 1996-06-29T04:12:00Z 1996-06-29T05:12:00Z
   1996-06-29T04:12:00Z 1996-06-29T05:12:00Z
   1996-06-29T04:12:00Z 1996-06-29T05:14:00Z
!
BURST001 1234     -60    3
FAST,ON,42;
```

Figure 7. Example of Spacecraft Operations File (SCOP). At the end of a planning session, the CMPS produces this file to provide JSOC with information on the results of the scheduling process.

```
! Standard file header
GSOP 0456 1996-06-10T12:00:00Z

! File specific header: schedule status
SCHEDULED    10  M_0011_0014_MASTER

! GSOP Data records
RECORDER CAPACITY (Mbit)    1040        2 1996-06-29T10:32:00Z
MAX POWER (W)                220        2 1996-06-29T10:32:00Z
RECORDER OCCUPANCY (%)     30.00        2 1996-06-29T10:32:00Z
POWER USAGE (%)            80.00        2 1996-06-29T10:32:00Z
TDA 401-r2                              2 1996-06-29T10:32:00Z
RECORDER OCCUPANCY (%)     50.00        2 1996-06-29T12:00:00Z
POWER USAGE (%)            81.00        2 1996-06-29T12:00:00Z
TDA 71-r1                               2 1996-06-29T12:00:00Z
UNAVAILABLE MAINTENANCE           RED     1996-06-29T12:45:30Z 1996-06-29T12:47:32Z
DSN                               CAN     1996-06-29T12:48:10Z 1996-06-29T13:15:00Z
```

Figure 8. Example of Ground Segment Operations File (GSOP). At the end of a planning session, the CMPS produces this file to provide JSOC with information on the status of the ground segment and of the spacecraft.

example of the content of a valid FTRQ file is provided in Figure 9. The syntax of this file is described in (CRID 1995).

The short-term plan allows the generation of a SCOP file that is automatically transferred to JSOC to report on the scheduling status.

4.4. Operational Plan

The CMPS uses a short-term plan to initialise an operational plan. Also this planning level covers a period of three orbits. The operational plan is completely frozen and can only be used to generate six Detailed Schedule Files (DSF), one for each of the four spacecraft and the two ground stations. The spacecraft DSFs are automatically transferred to the CMCS that expands them into the appropriate Cluster commands

```
! Standard file header
FTRQ 0456 1996-06-10T12:00:03Z

! File specific header: number of contained fine tuning requests
   3

! Delete command sequence SEQ45678 in OR OBSREQLABEL1
0123 OBSREQLABEL1 D
SEQ45678 1234      60    0

! Update parameters of command sequence SEQABCDE in OR OBSREQLABEL2
0123 OBSREQLABEL2 U
SEQABCDE    23       0    3
ON,FULL,-0.1;

! Insert command sequence SEQXYZ in OR OBSREQLABEL3
0123 OBSREQLABEL3 I
SEQXYZ       1    180    1
CLOSED;
```

Figure 9. Example of Fine Tuning Request File (FTRQ). This file is produced by JSOC and contains small modification to a previously submitted Observation. It is an input to the CMPS.

```
! Standard File Header
DSFS   44 1997-01-01T16:00:00Z

! Specific File Header
1997-01-02T12:30:00Z-1997-01-10T04:00:00Z     3

! Command call statement
200 0001 1TCOBDSON,,1997-01-02T12:30:00Z,,,,,,
-25.44,12AF,1997-01-02T12:35:00Z,START;

! Command sequence call statement
200 0002 SEQ 1STARTWEC,,1997-01-02T12:50:00Z,,IDLE,1000;
```

Figure 10. Example of Spacecraft Detailed Schedule File (DSF). At the end of a planning session at operational level, the CMPS produces six DSFs that allow the commanding of the four spacecraft and two ground stations.

that are uplinked to the on-board time-tagged command queue (Ferri *et al.*, this issue). The ground station DSFs are automatically transferred to the CMCS that relays them to the ground station where they are executed by the station computer. Due to a limitation in the number of commands that can be stored on-board, spacecraft DSFs are generated on a daily basis, while ground station DSFs are produced in one go for the entire three-orbit period. The CMPS allows the DSF generation period to be specified within the available planning interval and keeps track of those periods for which DSFs were already generated. An example of the content of a valid spacecraft DSF file is provided in Figure 10. The syntax of this file is described in (ICD 1995). Ground station DSFs largely resemble those of the spacecraft.

5. The Cluster Mission Control System

The CMCS forms part of the CDPS at ESOC (see Figure 1). It contains functions to support the real-time and near real-time data-processing tasks essential for controlling the mission. One of the main drivers for the design of the CMCS is the uniqueness of the Cluster Mission, being the first multiple spacecraft to be controlled by ESOC. In particular, the CMCS supports:

– Four spacecraft simultaneously during LEOP/TOP and two spacecraft simultaneously during MOP using the two unmanned ground stations;

– The collection and distribution of a large quantity of data.

The main functionalities to be performed by the CMCS (derived from MCS-SRD 1995) are:

– *Ground Station Interface and Control* to operate all ground station equipment interfaces. This includes facilities for telemetry reception, telecommand transmission, range and range-rate measurements and ground station scheduling for the two unmanned MOP stations.

– *Housekeeping Telemetry Processing* to monitor the spacecraft status in real-time and near-real time. The CMCS provides the capability to process housekeeping data including traditional binary and analog parameters and special Cluster event messages. Parameters can be processed using calibration curves to convert them into engineering and/or functional parameter values. The spacecraft housekeeping and science telemetry are stamped with UTC within 2 ms from any one spacecraft. This requires an allocation of the overall error budget to the ground segment of no more than 50 microsec, which includes errors for propagation and ground system delays. Spacecraft housekeeping and memory dump data are filed into local files (short-term history files), the contents of which are periodically transferred onto the SPEVAL system where all the spacecraft performance evaluation activities are performed.

– *Telecommand uplink and verification* to support real-time commanding and maintenance of the on-board time-tagged queue. All payload operations are pre-planned and executed from the on-board time-tagged queue. Real-time payload operations are used only for special operations like complex payload software maintenance activities and in case of contingencies. Real-time operations with the spacecraft are normally limited to the acquisition of stored telemetry data of all types (housekeeping, memory dumps, science), recorded during the non-coverage periods and dumped to the ground station during the real-time contact periods, and the uplink to the on-board memory of time-tagged commands for the execution of all pre-planned platform and payload operations.

– *On-board Software Maintenance facility* (OBSM) to provide management and configuration control of all changes performed to the on-board software. The CMCS also provides facilities to receive software patches for the payloads and uplink these patches.

– *Database subsystem* to support telemetry processing and telecommanding and to import the spacecraft manufacturer's database.

– *Man-Machine interfaces* to support all of the above functions.

The architectural design of the CMCS (see MCS-ADD 1993) is a distributed hardware architecture comprising six prime computers, each with a dedicated

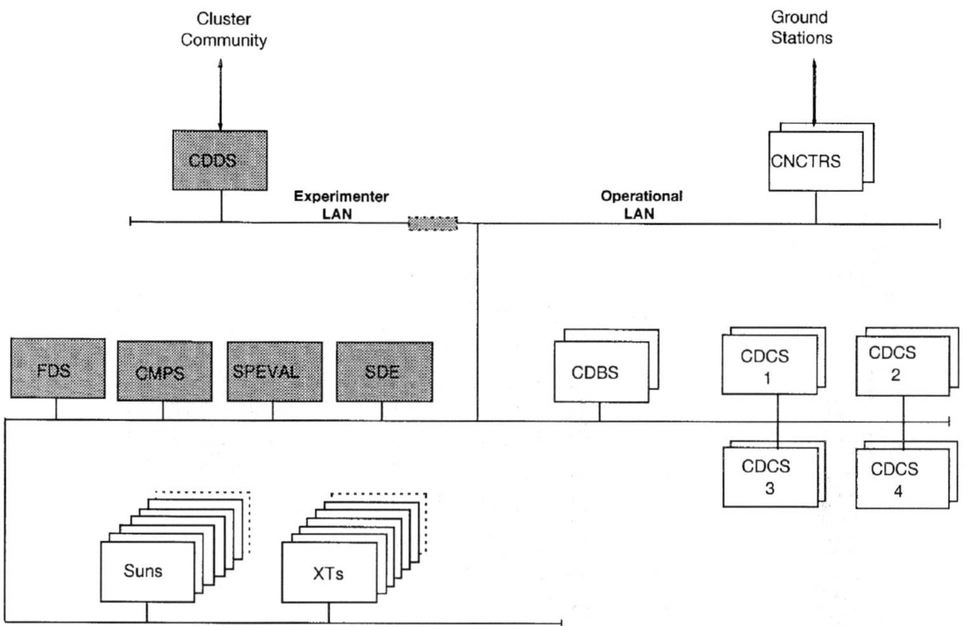

Figure 11. The Cluster Mission Control System Architecture.

standby machine (white boxes in Figure 11). The top-level architecture can be summarised as follows:

– *Cluster Controller and Telemetry Receiver System* (CNCTRS) to perform the ground station interface and control. It runs on a Digital VAXstation/VMS 4000–90A.

– *Cluster Spacecraft Dedicated Control System* (CDCSn, $n = 1, \ldots, 4$) to perform spacecraft monitoring/control and on-board software maintenance for a specific spacecraft. Each CDCS runs on a Digital VAXstation/VMS 4000–90.

– *Cluster Database System* (CDBS) to support Database functions. The CDBS is an off-line system using a single database for all four spacecraft and runs on a Digital VAXstation/VMS 4000–90.

The CMCS is on the operational LAN and this computer system can only be accessed by operations staff at ESOC (i.e., no remote access). The external access is via the CDDS. The CDDS allows secure connection between the external world (e.g., Cluster community) and the ESOC protected operational LAN in support, for instance, of payload command requests. The man-machine interface runs on SUN workstations for spacecraft monitoring and control and on X-window terminals for database maintenance, OBSM facility and CNCTRS applications. Each CDCS can support up to ten SUN workstations, each having three screens and a total of nine windows per Sun workstation. For monitoring and control, telemetry can be displayed in alphanumeric format, graphical format and as synoptic mimics. One example of mimic display is given in Figure 12.

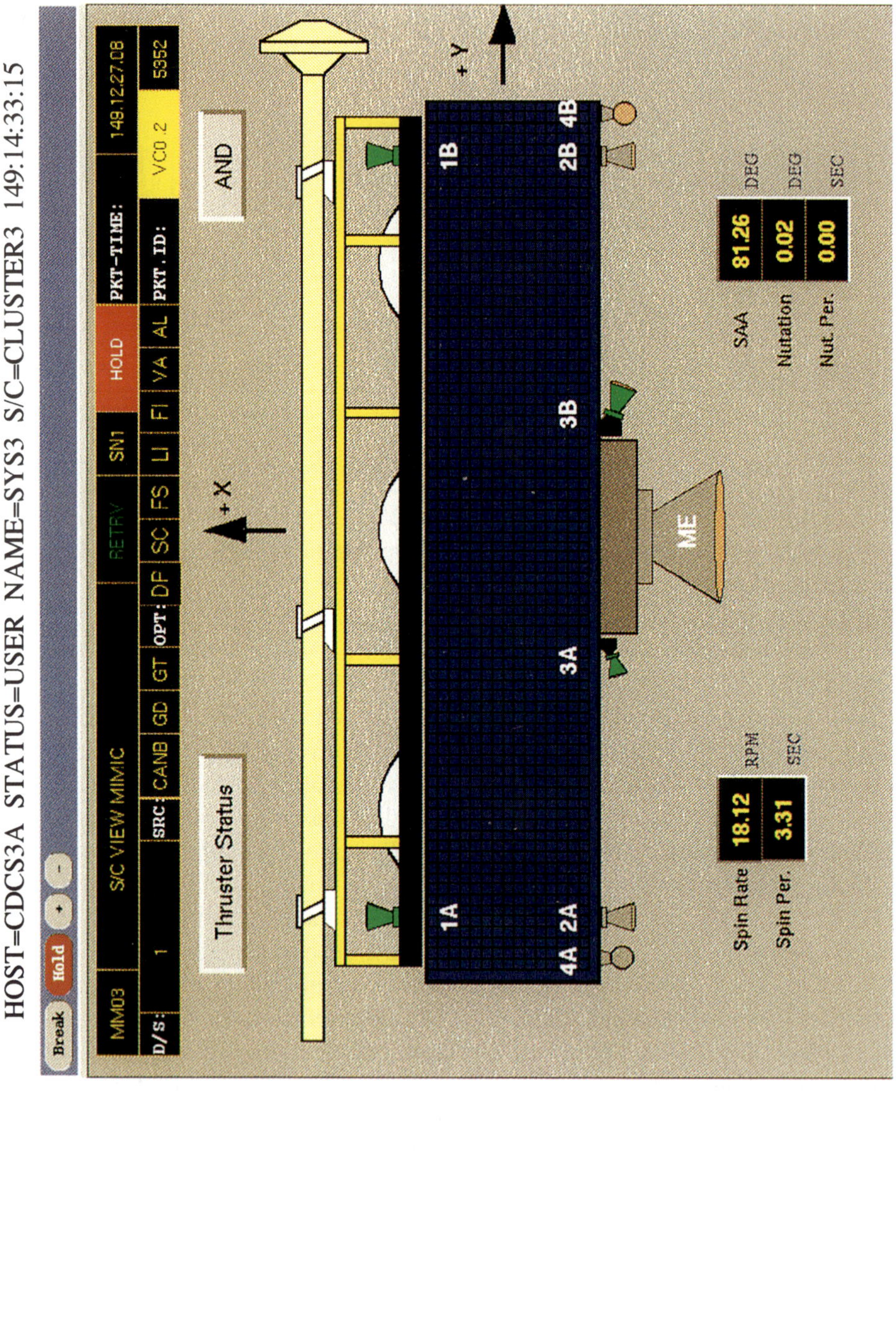

Figure 12. Example of CMCS Mimic Display. This mimic display represents the Attitude and Orbit Control Subsystem (AOCS) and it is used to monitor the Cluster thrusters.

6. The Cluster Data Disposition System

The main task of the Cluster Data Disposition System (CDDS) is to deliver the mission data to the Cluster community. This is performed by two distinct services that are provided by the system:

– *The On-Line Data Delivery Service* which allows on-line delivery on request of mission data for quick-look purposes via a dedicated electronic network (CSDSnet) as well as via the public Internet to PIs, JSOC and the CSDS data centres,

– *The Off-Line Data Delivery Service* which allows off-line production of recordable compact disks (CD-R) to be used as masters for mass production of CD-ROMs containing all Cluster data .

Both on-line and off-line delivered data are packaged in Standard Formatted Data Units (SFDU), more information on which may be found in SFDU – Tutorial 1992 and SFDU – Structure 1992. In order to allow for a smooth exchange of data among potentially different hardware platforms, all exchanged data is encoded in ASCII except for the telemetry that is left in binary format.

The types of mission data that are made available by these two services are identical (with the exception of two types of auxiliary data) and include house-keeping and science telemetry from all four Cluster spacecraft and all auxiliary data required to correctly process and interpret the telemetry. More specifically, the auxiliary data includes the following spacecraft-specific files (DDID 1995):

– *Long-Term Orbit File (LTOF):* this contains orbital state vectors for each major scientific phase of the Cluster mission, that is, Cusp 1, Tail 1, Cusp 2, and Tail 2 assuming a nominal configuration (see Escoubet *et al.*, this issue). It is generated by the Flight Dynamics System (FDS) at ESOC every six months for mission planning purposes. This data is only available via the on-line service.

– *Short-Term Orbit File (STOF):* this contains orbital state vectors at a higher resolution and hence accuracy than the LTOF based on the actually determined orbit (maximum error is ± 0.2 s per orbit). This file is generated by the FDS at least once per week and provides reconstituted data for the previous ten days prior to generation, and 3.5 months of predicted orbit data from the day of generation. This implies that the absolute cumulative error at the end of the 3.5 month period is less than 10 s.

– *Long-Term Event File (LTEF):* this contains events (e.g., start/end of eclipse, visibility from ground stations, start of revolution, start/end of antenna switch, perigee passes etc.) that, from the time of generation, cover the remaining period of the mission at a coarse resolution and accuracy. It is generated by the FDS every six months. This data is only available via the on-line service.

– *Short-Term Event File (STEF):* this contains events as in the LTEF, the only differences being in the quantity of events and the frequency of updates. This file is generated by the FDS at least once per week, with reconstituted events for the previous ten days and predicted events for the following 3.5 months.

– *Spacecraft Attitude and Spin Rates (SATT):* this contains actual and predicted data for the spacecraft attitude (errors are ±0.24°, actual, and ±0.64°, predicted), including spin axis right ascension and declination and the spin rate (errors are $\pm 1.3 \times 10^{-5}$ rpm at 15 rpm, actual, and $\pm 1.3 \times 10^{-3}$ rpm, predicted) and phase (errors are ±0.074°, actual, and ±0.85°, predicted) information for the specified validity times. It is generated by the FDS at least once per day.

– *Time Calibration File (TCAL):* this contains the relationship between the Spacecraft Event Time (SCET) and the On-Board Time (OBT), as calculated by the OCC in real-time. It is used to transform the OBT downlinked in the telemetry to SCET with an instrument-to-instrument error less than ±4 ms (the error portion introduced by the ground segment is less than $|\pm 50|$ ms). This file is generated by the CMCS every time a new calibration curve is calculated.

– *Command History File (CMDH):* this contains a log of all commands that have been uplinked to the spacecraft including their level of verification. It is generated by the CMCS every two hours.

– *Housekeeping Parameter Definition (HPD):* this contains the information needed to extract, decode and interpret the parameters in the housekeeping telemetry. It is generated by the CMCS every time the definition of any parameter is modified.

Figure 13 gives a schematic representation of the interactions of the CDDS with other systems and of its main products. The CDDS receives telemetry data in real-time from the front-end machine of the CMCS. The CMCS and the FDS provide auxiliary data that are automatically ingested in the CDDS. The CDDS is configured to keep on-line all the TM and auxiliary data that have been received in the last ten days. After that, they are archived on DAT tapes and are no longer available on-line.

In addition to this, the CDDS provides a safe gateway for secure communication between the highly protected operational environment in ESOC and the external world. This gateway is used by a number of ESOC systems including the CMCS, CMPS, SPEVAL and FDS.

The CDDS has been developed using C language and runs on a DEC 4000–105A VAX/VMS computer. A Personal Computer is connected via the network to the VAX and is responsible for the compact disk production.

6.1. On-line Data Delivery Service

As shown in Figure 13, the CDDS receives housekeeping and science telemetry data in real-time from the ground stations via the front-end computer of the CMCS, whilst the auxiliary data comes from both the CMCS and FDS. External users have no access to the latter two systems as they are protected by the ESOC network security system (firewall). The CDDS stores this data upon reception and makes it available a few seconds later to its users. Only the last ten days of data are kept on-line.

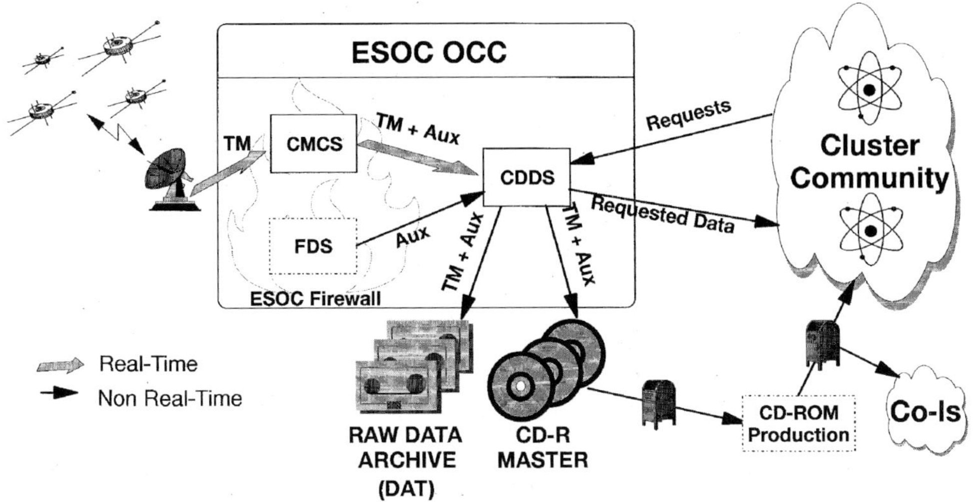

Figure 13. Overview of the Cluster Data Disposition System and its Environment.

Only authorised users (PIs and CSDS data centres) can access the on-line data delivery service. Each authorised user is given an account on the CDDS that is protected by a userid and a password and that maps into a user-dedicated directory. Authorised users must provide an account on their remote computers, also protected by userid and password, and allow access from the CDDS. Each user is also assigned a quota that indicates the maximum amount of data that he/she is authorised to request from the CDDS in one day.

The on-line data-delivery service works as a client-server model, with no interactive interface for the user, just the exchange of request files and resultant data transfers. A request for data is always initiated by the user. The user copies a request file from his own network node to his CDDS account. The CDDS processes these requests in turn, and if valid, extracts the data sets requested and copies them to the destination specified in the request. The transfers of the request file and the corresponding requested data are implemented as two independent file transfer sessions initiated by the user and the CDDS, respectively. Both the File Transfer Protocol (FTP) and DECnet are supported by the CDDS, although for Cluster only FTP is used. In case of transfer failure, the CDDS is programmed to retry the transmission at a later time. Both the number of retry attempts and the retry delay are configurable in the CDDS.

The request files are ASCII files that have always the name 'REQUEST.DDS' and that contain command statements expressed with a simple syntax. Submission of multiple request files at once is allowed and the file-name conflict is resolved by using the VMS file version number. There is only one command that can be used in the request file, the command 'REQUEST', the syntax of which is provided in Figure 14.

```
REQUEST [Qualifiers] Data_Identifier Target_Specification
```

Figure 14. Syntax of the Command Statement for the CDDS Request File.

```
! This is an example of a data request file, that would be sent to ESOC
! from the PI of Cluster payload EDI

! Request EDI HK Data since 24 March 1997 at 22:45 from all 4 spacecraft
REQUEST /SINCE=1997-03-24T22:45:00Z EDI.HKD "111.222.333.444:[EDI.DDS]EDI_HK.DAT"

! Request EDI burst science data of spacecraft 1 in the specified time
! range and transmit with compression
REQUEST /SINCE=1997-08-12Z/BEFORE=1997-08-13Z/SC=1/ZIP EDI.BSD
"111.222.333.444:[EDI.DDS]BURST.DAT"

! Request the auxiliary data LTEF for spacecraft 1 before 12 Jan 1997
REQUEST /SC=1/BEFORE=1997-01-12T LTEF.AUX
"111.222.333.444:[EDI.DDS]LTEF_Bef_120197.DAT"
```

Figure 15. Example of the Content of a CDDS Request File.

In addition to the command identifier 'REQUEST', a number of optional qualifiers can be provided to specify a particular spacecraft or subset of spacecraft ('/SC =‘), a time range ('/SINCE =' and '/BEFORE ='), the maximum number of bytes to be delivered in any single file ('/AMOUNT =') and whether the ZIP data compression algorithm should be used to compress the file to be returned to the user ('/ZIP'). The 'Data_Identifier' unambiguously specifies the data type of the request based on a pre-defined naming convention. The 'Target_Specification' gives the IP address of the target machine together with the name and the path of the file to which the requested data set shall be transferred. Any number of empty lines and comment lines (prefixed by an exclamation mark (!) in the first column) can be present between each REQUEST command. Figure 15 provides an example of the content of a valid request file. The syntax of request files is described in (DDID 1995).

The CCSDS ASCII Time Code A (Time Code Formats 1990) format is always used to specify dates and times and always refers to UTC. It has the format YYYY-MM-DDThh:mm:ssZ, where 'T' is the date-time separator and 'Z' is the mandatory terminator. The string '111.222.333.444' in Figure 15 represents the user- provided IP address of the target computer. After the IP address, the user must specify the name (including an optional path) of the file into which the requested data shall be copied.

Each CDDS response to a successful request for data is organised in three parts within a single file using the SFDU concept: the Acknowledgement, the Catalogue entry and the Requested data.

The *Acknowledgement* is expressed using the Parameter Value Language (PVL) (see PVLS 1992). This is a simple language where parameter names are assigned values. It includes information such as details of the original file transfer request, date and time at which the request is processed by the CDDS, production inform-

```
SPACECRAFT_NAME = Cluster_1;
DATA_SOURCE = EDI;
DATA_TYPE = BSD;

REQUEST_FILENAME = REQUEST.DDS;

SINCE_TIME_SPECIFIED = "1997-08-12Z";
SINCE_TIME_EXPANDED = "1997-08-12T00:00:00Z";

BEFORE_TIME_SPECIFIED = "1997-08-13Z"
BEFORE_TIME_EXPANDED = "1997-08-13T00:00:00Z"

MAXIMUM_BYTES_SPECIFIED = UNLIMITED;
BYTES_DELIVERED = 7666;
COMPRESSION = ZIP;

TARGET_FILENAME_SPECIFIED = "111.222.333.444:[EDI.DDS]BURST.DAT";
TARGET_FILENAME_EXPANDED = "111.222.333.444:[EDI.DDS]BURST.DAT";

TIME_PROCESSED = 1997-08-14t12:33:44Z

COMPUTER = "DEC VAXStation 4000/105A"
OPERATING_SYSTEM = "VAX VMS 5.5.2";
CDDS_SOFTWARE_VERSION = "3.1"
DISTRIBUTION_METHOD = NETWORK;

ERROR_MESSAGE = "NO ERROR";
```

Figure 16. Example of the Content of the Acknowledgement Part of a CDDS Response.

ation indicating the environment that this SFDU is created in, an error message string (this contains the string 'NO ERRORS' for nominal processing with no errors). If there is an error in the target file specification supplied in the request file, or the request file is for any reason not readable, the CDDS can still inform the user of the error by using the default file specification as the target file. The default file specification is another piece of information that each authorised CDDS user must supply upon registration to the on-line delivery service. An example of the acknowledgement part is given in Figure 16.

As it can be seen, the acknowledgement format is very clear for the human reader and also very easy to process by software.

The *Catalogue Entry* contains information specific to the actual data transferred. Clearly, this can be different to the requested data depending upon what time-stamped telemetry or auxiliary data are available. It uses the PVL and contains information such as the spacecraft name to which the data are relevant, the type of data delivered, the SFDU Authority and Description Identifier (ADID) for the included type of data, the time stamps of the earliest and latest packets in the delivered data and the number of packets that are included. The PVL statements are grouped in a catalogue entry using the PVL aggregation construct (BEGIN_OBJECT ... END_OBJECT). An example of the catalogue entry is given in Figure 17.

The *Requested Data* can be from three sources: telemetry data, auxiliary data and Master Catalogue data. Telemetry data is delivered in raw binary format as a set of housekeeping or science data telemetry packets and arranged in time-ascending order. The packets are time-stamped with Spacecraft Event Time (SCET), formatted using the CCSDS Day Segmented time code (Time Code Formats 1990). Auxiliary

```
BEGIN_OBJECT = CATALOGUE_ENTRY

 SPACECRAFT_NAME = Cluster_1;
 DATA_SOURCE = EDI;
 DATA_TYPE = BSD;

 ADID = ECLUB101;

 FIRST_PACKET_TIME = 1997-08-12T08:08:11Z;
 LAST_PACKET_TIME = 1997-08-12T08:28:07Z;

 NUMBER_OF_PACKETS = 1342

END_OBJECT = CATALOGUE_ENTRY;
```

Figure 17. Example of the Content of the Catalogue Entry Part of a CDDS Response.

data are treated in the same way as telemetry data, the only difference being that they are encoded in ASCII as opposed to binary format. Each packet for these two data sources has a standard 14-byte header, the CDDS header, that provides information such as SCET, data type, packet length, spacecraft and ground station identifications, data stream, time quality and telemetry acquisition sequence identification. A non-applicable code is used if a particular piece of information in the header is not relevant for a given data type. The Master Catalogue provides summary information for all the data currently available on-line from the CDDS (i.e., the last ten days). When a user asks for a Master Catalogue, he receives catalogue entries for all the available data types. Although the CDDS supports the concept of private data, no restriction on the access of data for any user is required for Cluster, e.g., a user may access data from any payload. The format of a Master Catalogue is simply all the individual data stream catalogues concatenated together.

The only time that the CDDS user does not receive an SFDU of the structure shown above is when an error has been detected during the processing of the request. When this occurs, an SFDU containing only the acknowledgement section is transferred. Errors due, for instance, to illegal command specification, exceeding of the user-specific quota or too many outstanding file transfers, generate this type of acknowledgement containing details of the particular error message.

6.2. Off-line Data Delivery Service

The off-line data delivery service is totally separate and asynchronous to the on-line data delivery service, although both use the same data pool that is available on the CDDS. This service is wholly controlled by ESOC and does not require any input from external users. As schematically shown in Figure 13, the CDDS stores all housekeeping and science telemetry data and auxiliary data on Recordable Compact Disks (CD-R) that are shipped by fast courier to an external CD manufacturer where they are used as masters and duplicated in about 100 copies on Read-Only Memory Compact Disks (CD-ROM). The CD manufacturer takes also care of packaging and shipping the CD-ROMs to the authorised members of the Cluster community located around the world. On top of this, CD-ROMs are distributed also to the Cluster Co-Is.

One calendar day (from 00:00:00 to 23:59:59) is the basic CD production period. An average of two CD-ROMs per day is foreseen to be produced during the mission operations phase, although the CDDS can support the production of up to four CD-ROMs per day on an exceptional basis. The CDDS generates the CD-R directory structures and populates them with the files on the main VAX computer. Upon completion, the network-connected personal computer is notified and the creation of the CD images and successively the production of the CD-Rs starts automatically. A personal computer is used because during the actual production of a CD-R the data rate to the CD burner must be constant, therefore a completely dedicated and off-line computer is preferred.

The set of CD-ROMs for any one day must be delivered within three weeks from the data generation date. This period includes consolidation of orbit and attitude data (6 days), the production of the CD-Rs at ESOC (1 day), the shipment to the CD-ROM manufacturer (2 days), the CD-ROM mass-production (5 days) and, finally, the delivery of the CD-ROMs directly to the science community members (7 days).

A format compliant to the ISO 9660 level 1 standard is used for CD-ROM so that the disks are readable on all common platforms, e.g., UNIX, VMS, PC, and Mac. In addition to the housekeeping and science telemetry and the auxiliary data for all four spacecraft, the CD-ROMs contain the data descriptions for all included files in support of the SFDU concept. This means that the data on each disk is wholly self-contained.

The data on the CD-ROM are accessible by two methods: either the more conventional direct file access, i.e., each data type is stored under a particular file name in a particular directory, or the whole disk may be viewed as a single logical SFDU product. The SFDU product is formed by having an Exchange Data Unit (EDU) file in the top-level directory of the disk that uses the SFDU referencing technique to logically include all the other data files and data description files on the disk into a single SFDU product. The conventional method of accessing the data presents no problems as long as the user knows what to look for, where it is on the disk, the file naming convention, etc. The SFDU approach has the advantage that nothing needs to be known about the disk in advance except the name of the top level 'index' SFDU file. From this SFDU file the science community member can find all the data on the disk, the corresponding data descriptions and, via the ADIDs (assigned under the authority of ESOC), can link the two together. So, for long-term archive access, when relevant documentation and knowledgeable personnel may no longer be available, this method assures the continued understanding of the data. Additionally, ESOC has developed a fully automatic Control Authority system that implements the CCSDS recommendations and that allows the dissemination, registration and revision of all the Cluster data descriptions using a World Wide

Web interface. At the time in which this article has been written, this system is undergoing testing and for this reason its Internet address can not be provided.

Each CD-ROM has a label that is both printed on its top surface and electronically stored as its volume label. The format of this label is 'yymmdd_n_tv', where 'yy', 'mm' and 'dd' indicate the year, month and day of the data in the CD-ROM, respectively, 'n' the number of the CD-ROM within the particular day, 't' the total number of CD-ROMs within the particular day and 'v' the version number of the CD-ROM, (range A-Z, where A is the initial release).

The CD-ROM has a standard hierarchical directory structure that divides the data into separate directories for each spacecraft and data type, as shown in Figure 18. The directory structure is self-describing. The data generated by the spacecraft and ESOC are split into four sub-directories, one for each spacecraft, in each of these sub-directories there are five further sub-directories. These subdivide each spacecraft's data into the Normal Science Data (NSD), the Burst Science Data (BSD), the Housekeeping Data (HKD), the Auxiliary Data (AUX) and the Housekeeping Parameter Definition Data (HPD) as generated by ESOC. The DATADES sub-directory under the top-root level contains files of all the data descriptions in support of the SFDU concept. Within the DATADES sub-directory there is a sub-directory for each spacecraft, each of these directories contain four sub-directories, each one containing the description files for each of the data types, NSD_DESx for Normal Science Data description files, BSD_DESx for Burst Science Data description files, HKD_DESx for HK Data description files and HPD_DESx for HK Parameter Definition description files. The 'x' indicates the spacecraft that they pertain to, where A is Cluster 1, B is Cluster 2, C is Cluster 3 and D is Cluster 4. There are separate directories for each spacecraft as, due to changes or failures on different spacecraft, the data description information may not be the same for each spacecraft. Also under the DATADES directory there is an AUX_DES directory that contains descriptions of all auxiliary files. These files are always identical in format for all spacecraft as they are centrally generated by ESOC on the ground and not directly by the spacecraft. Furthermore, under the DATADES directory there is an CDDS_DES directory that contains descriptions of the files that are specific to the CDDS system and the CD-ROM, such as the directory listing file and the cumulative index files. The SOFTWARE sub-directory contains software routines delivered by ESOC to assist in processing orbit data.

Note that each CD-ROM contains all files even if the files have zero size. This may occur when there is no data for a particular date for a data source. Zero size files are included, as this makes clear that there was no data for the data source in the time range. If a file is completely missing then it immediately indicates a manufacturing error.

All the files on the CD-ROM follow a particular file-naming convention, this is so that if a file should be copied from the CD-ROM to another media with no knowledge of which directory it came from, then the filename alone is enough to identify the file unambiguously. The file names follow the ISO 9660 level 1

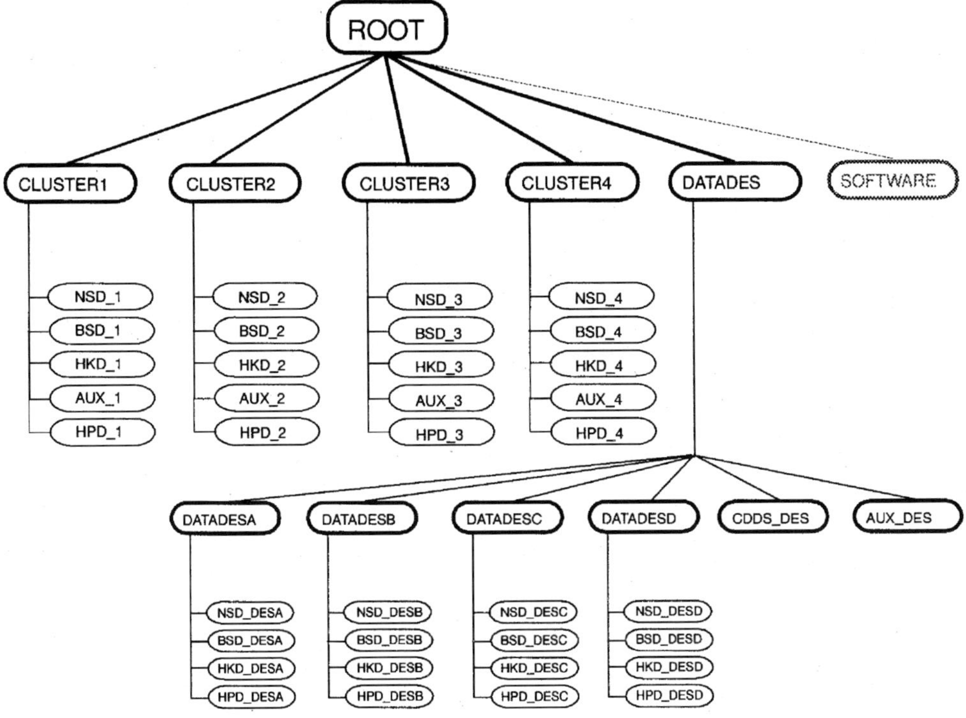

Figure 18. The CD-ROM Directory Structure.

standard for CD-ROMs. This has a basic format of a directory path followed by an '8.3' file name. Whilst this is quite restrictive, it does mean that the files and directory structure can be transferred to any known random-access media with no file-name mapping required. The file names use the convention 'yymmddst.nvc', where 'yy', 'mm' and 'dd' indicate the year, month and day of the first packet in the file, respectively, 'st' the data source and type, 'n' the CD-ROM number within the particular day (range 1–4, first CD-ROM = 1), 'v' the version of the file in case of reissue (range A-Z, initial version = A) and 'c' the spacecraft. Only a small number of files (eight) do not follow this convention because of their particular nature.

For example, normal science data from Cluster spacecraft number 2, from the EDI payload with the first packet from 27 April 1997, would reside in the directory '/CLUSTER2/NSD_2/' relative to the root directory, and would have the file name '970427EN.1A2', assuming that the file resided on the first CD-ROM produced for that day, and was the original CD-ROM, i.e., not a reissue. Similarly, the data description file for this data source (if written on 12 February 1995) would reside in the directory '/DATADES/DATADESB/NSD_DESB/' relative to the root directory, and would have the file name '950212EN.1AB'.

Despite a total capacity of approximately 650 MBytes, each CD-ROM is filled up to about 600 MBytes to leave some margin. Under the current baseline for data production (approximately 1 GByte day^{-1}), two CD-ROMs are sufficient to hold one whole day's worth of data. If, due to the data-reception profile over a single orbit, more than two CD-ROM's worth of data is generated within a single calendar day, then a third and, if necessary, a fourth CD-ROM can be produced to store the data. The splitting of data across several CD-ROMs for a day is done in such a way that each CD-ROM is self-standing.

Due to the fact that raw unprocessed data is being delivered, it is not possible to have a new version of such data, but what must be considered is the possibility of a fault in the manufacture of the CD-ROM, an incorrect selection of files being placed on the CD-ROM or a new release of auxiliary files. Therefore the CDDS has been designed to handle new versions of files in the extraordinary situation, should it arise. The issue of version control is handled by having an index file in the root directory of each CD-ROM. This file lists the full collection of Cluster CD-ROMs since launch. Against each CD-ROM title it indicates the time span that the CD-ROM covers, and also the issue indicator, so that it is easy to see if an older CD-ROM has been replaced by a reissued one. This index file is cumulative, i.e., it contains a record of all issued CD-ROMs up until the present date.

7. Conclusions

This paper has described the Cluster Data Processing System as it has been implemented to meet the Cluster scientific objectives. The Cluster mission is a challenge to ESOC and all the requirements related to the unique characteristics of this mission, as defined by the users, have been carefully analysed, resulting in the design of a distributed computer system. The main motivations for such a decision have been:

– *A distributed architecture reduces the costs of hardware and software*. The costs of procuring and maintaining many small-sized computers (VAXstations) are significantly lower than those for a large computer capable of hosting the whole system. Software licenses are also usually cheaper for small-sized computers. Furthermore, the distributed architecture permits the optimum use of licences throughout the system.

– *In the distributed architecture, each element is independent*. In general, the failure of one element does not affect the others. Furthermore, the global system performance is not compromised in the case of heavy use of resources by one element. This is of particular importance in the case of time-critical tasks (e.g., real-time commanding).

– *In the distributed architecture, off-line tasks are hosted on a dedicated computer*. Important, resource-demanding off-line tasks are totally decoupled from the actual spacecraft operations and other time-critical tasks (e.g., database maintenance).

The experience gained so far in preparing for Cluster has shown that a distributed architecture for science operations provides more flexibility and allows trade-offs to be made between the various elements of the whole system. This optimises the end-to-end data return and reduces the project's costs.

However, distributed systems call for clear definition of the interfaces between the elements. Great attention has to be paid to the definition, the review, and the configuration control of these interfaces. As many of these interfaces involve the science community, it is critical that Cluster community members actively take part in their definition and review. It is also important that these interfaces are tested well in advance of the start of the mission with all parties involved.

ESOC has already achieved great success with the Cluster data-processing system. Our challenge for the future is to build on this experience and implement systems to meet the needs of the future, even more challenging science missions.

References

Cyr, O. C. St., Sáncez-Duarte, L., Martens, P. C. H., Gurman, J. B., and Larduinat, E.: 1995, 'Soho Ground Segment, Science Operations, and Data Products', *Solar Phys.* **162**, 39.

CRID: 1995, 'Command Request Interface Document (CRID)', ESOC, CL-ESC-ID-0003, Issue 2.2, 3 July 1995.

DDID: 1995, 'Data Delivery Interface Document (DDID)', ESOC, CL-ESC-ID-0001, Issue 2.4, 13 February 1995.

ICD: 1995, 'CMCS-CMPS Interface Control Document', ESOC, CL-ESC-ID-0602, Issue 2.0, 25 July, 1995.

MCS-ADD: 1993, 'Cluster MCS Architectural Design Document', ESOC, CL-ESC-RS-0310, Issue 2.4, 31 August, 1993.

MCS-SRD: 1995, 'Cluster MCS Software Requirements Document', ESOC, CL-ESC-RS-0201, Issue 2.1, 28 February, 1995.

MIRD: 1991, 'Cluster Mission Implementation Requirements Document', ESTEC, CL-ESTRSOOO3, Issue 4.0, November 1991.

MPS-ADD: 1994, 'Cluster Mission Planning System Architectural Design Document', ESOC, CL-ESC-RS-0311, Issue 2.1, 30 June, 1994.

PVLS: 1992, 'Recommendation for Space Data System Standards: Parameter Value Language Specification (CCSD0006)', CCSDS 641.0-B-1, Blue Book, Issue 1, Consultative Committee for Space Data Systems, May 1992.

SFDU – Tutorial: 1992, 'Report Concerning Space Data System Standards: Standard Formatted Data Units – A Tutorial', CCSDS 621.0-G-1, Green Book, Issue 1, Consultative Committee for Space Data Systems, May 1992.

SFDU – Structure: 1992, 'Recommendation for Space Data System Standards: Standard Formatted Data Units – Structure and Construction Rules', CCSDS 620.0-B-2, Blue Book, Issue 2, Consultative Committee for Space Data Systems, May 1992.

Time Code Formats: 1990, 'Recommendation for Space Data System Standards: Time Code Formats', CCSDS 301.0-B-2, Blue Book, Issue 2, Consultative Committee for Space Data Systems, April 1990.

Appendix: List of Acronyms

ADID	Authority and Description Identifier
AUX	Auxiliary Data
BSD	Burst Science Data
CCSDS	Consultative Committee for Space Data Systems
CD	Compact Disk
CD-R	Recordable Compact Disk
CD-ROM	Read-Only Memory Compact Disk
CDBS	Cluster Database System
CDCS	Cluster (Spacecraft) Dedicated Control System
CDDS	Cluster Data Disposition System
CDPS	Cluster Data-Processing System
CMCS	Cluster Mission Control System
CMDH	Command History (file)
CMPS	Cluster Mission Planning System
CNCTRS	Cluster Controller and Telemetry Receiver System
Co-I	Co-Investigator
CSDS	Cluster Science Data System
CSDSnet	Cluster Science Data System Network
CVP	Commissioning and Verification Phase
DSF	Detailed Schedule File
DSN	(NASA) Deep-Space Network
EDU	Exchange Data Unit
ESOC	European Space Operations Centre
FDS	Flight Dynamics System
FOP	Flight Operations Procedures
FOPS	Flight Operations Procedure System
FTP	File Transfer Protocol
FTRQ	Fine-Tuning Request (file)
FVSR	File Validation Status Report
GSOP	Ground Segment Operations (file)
HKD	Housekeeping Data
HPD	Housekeeping Parameter Definition (file)
JSOC	Joint Science Operations Centre
LAN	Local Area Network
LEOP	Launch and Early Orbit Phase

LTEF	Long-Term Event File
LTOF	Long-Term Orbit File
MOP	Mission Operations Phase
MOT	Mission Operation Team
NSD	Normal Science Data
OBRQ	Observational Request (file)
OBSM	On-board Software Maintenance
OBT	On-Board Time
OCC	Operations Control Centre
OPRQ	Operations Request (file)
OR	Observational Request
PI	Principal Investigator
PVL	Parameter Value Language
RAM	Random Access Memory
SATT	Spacecraft Attitude and Spin Rates (file)
SCET	Spacecraft Event Time
SCOP	Spacecraft Operations file
SDE	Software Development Environment
SFDU	Standard Formatted Data Unit
SPEVAL	Spacecraft Evaluation System
STEF	Short-Term Event File
STEF	Short-Term Event File
STOF	Short-Term Orbit File
STSP	Solar Terrestrial Science Programme
TC	Telecommand
TCAL	Time Calibration File
TDA	Telemetry Data Acquisition
TM	Telemetry
TOP	Transfer Orbit Phase
TRK	Tracking Data (Ranging and Doppler)
UTC	Universal Time Coordinated
WBD	Wideband Data
XT	X-Terminal

THE CLUSTER SCIENCE DATA SYSTEM (CSDS) – A NEW APPROACH TO THE DISTRIBUTION OF SCIENTIFIC DATA

R. SCHMIDT and C. P. ESCOUBET
Space Science Department of ESA/ESTEC, P.O. Box 299, 2200 AG Noordwijk, The Netherlands
S. J. SCHWARTZ
Astronomy Unit, Queen Mary and Westfield College, Mile End Road, London E1 4 NS, U.K.

Abstract. Cluster is an ESA/NASA four-spacecraft mission designed to study plasma processes in three dimensions using the combined data from eleven instruments on each spacecraft. This mission requires the combination of many measured parameters, and the Cluster community have taken the unprecedented step of establishing a set of high quality data products from all instruments at spin ($\sim$ 4 s) resolution which will be produced and distributed throughout the mission lifetime. The Cluster Science Data System (CSDS) is based on a set of eight data centres which are implemented and funded through national programmes. As part of CSDS, a Joint Science Operations Centre (JSOC) has been established to facilitate the commanding of the 44 instruments. It is co-located with the UK data centre at the Rutherford Appleton Laboratory (RAL), Didcot, United Kingdom. ESA's contribution to CSDS includes the provision of the CSDS User Interface, a dedicated network (CSDSnet) to interconnect the data centres, and the co-ordination of all activities at CSDS level. A wide scientific community wishing to use Cluster data will have differing data rights, experience and means of access. Users will also include those working with data sets from other missions, e.g., Soho, Geotail, Wind, Polar, Interball, and Equator-S. The Cluster Science Data System is primarily designed to support multi-instrument and multi-spacecraft data analysis and it is distributed across six national data centres in Europe, one in the USA, and one in China. CSDSnet will be used to interconnect the European data centres, the Joint Science Operations Centre at Didcot and the spacecraft Operations Control Centre at ESOC in Darmstadt.

1. Introduction

Research on space plasma has reached such a level of maturity that much of the geography of the plasma environment of the Earth is generally understood. However, many of the plasma processes that take place in this environment are vital to our understanding of plasma physics as a whole, but are not yet well understood. One of the reasons for this lack of understanding is that previous space missions have generally had only one, or at best, two simultaneous measurement locations in space. The separation of temporal and spatial variations has thus been extremely difficult. This is particularly true for small-scale structures and waves (where the term small-scale refers to scale lengths ranging from a few 100 to a few 1000 km). Space plasma research uses mainly time series of data that are scalars (e.g. electron number density), vectors (e.g., electric and magnetic fields), tensors (e.g., pressure), particle distribution functions or images. The phenomena being studied cannot usually be located in advance and the generation of catalogues of scientific events is difficult. Space plasma physics investigations normally require data from several instruments.

Space Science Reviews **79:** 557–582, 1997.

In the past, such multi-instrument studies required bi-lateral sharing of usually small data segments between Principal Investigators, or dedicated Co-ordinated Data Analysis Workshops ('CDAW's') to which Investigators brought high-resolution data associated with a particular phenomena or pre-arranged time-interval. On a routine basis, only coarse data products, e.g., ISTP's 'Key Parameters' of 1 min averages or worse, or survey plots, were distributed or available for preliminary analysis and event selection. The many advances made in space physics over the last two decades suggest that such an approach is successful. Yet all researchers know that many more events, statistical analyses, and studies would have been possible were it not for the difficulty in identifying them, in collecting the necessary data sets from the individual Investigators, in re-organising the data into a suitable format, and generally assembling the data products. Exceptions to this approach tend to be smaller, national projects (e.g., *Freja*, AMPTE/IRM) where the bulk of the Principal Investigators are co-located. Occasionally, post-mission archives are established for public access containing relatively high-resolution data (e.g., AMPTE UKS/IRM data at NSSDC and the Geophysical Data Facility at RAL), but these were too late for high-profile science during the intense scientific activity of the mission.

The maturity of space science and the complexity/data volume of the Cluster concept demands that a more open, efficient approach be considered. This large, multi-national mission has taken the unprecedented step of addressing this question and preparing for data dissemination prior to launch. CSDS will provide the entire Cluster community with spin-averaged data from all instruments and all spacecraft within approximately two months of the actual measurements. The data will form a unified database, with common data formatting and access tools. This paper describes the detailed implementation of this system, which will form the underpinning of most, if not all, Cluster science investigations. Of course, detailed scientific studies will still require higher-resolution (in time, dimensionality, or other sophistication) products which will invoke bi-lateral or CDAW methodology. But the quality and commonality of the CSDS databases will significantly enhance the efficiency and productivity of such studies. Many complete detailed scientific studies will probably be accomplished with high-resolution data from one or two instruments coupled with data products drawn from the CSDS databases.

The Cluster mission uses four identical spacecraft that will have a roughly tetrahedral configuration for their orbital positions during crossing of the key regions in geospace. Thus the three-dimensionality of the phenomena being studied will often be resolved, and varying the inter-spacecraft separation distance will enable a range of phenomena to be examined. The polar orbit has an apogee of 19.6 Earth radii and a perigee of 4 Earth radii. The spin axes of the spacecraft will be kept perpendicular to the ecliptic plane. Each spacecraft carries the same set of eleven instruments. The multi-spacecraft, multi-instrument data will provide estimates of field and plasma parameters and gradients thereof, thus enabling, for example, a study of the particle currents which give rise to magnetic field gradients as described via Ampère's law.

CSDS has been designed as a distributed system to make possible the joint scientific analysis of data from all 44 instruments (four sets of eleven). The general approach is to have national data centres located near the Principal Investigators (PIs) and thus near the expertise required for processing the data. One of the major tasks of CSDS is to offer, as a routine, products such as the Summary Parameter Data Base and the Prime Parameter Data Base (see Section 7 of this paper). CSDS also serves to some extent as the infrastructure for the Joint Science Operations Centre (JSOC), which is a staffed facility located at the Rutherford Appleton Laboratory, Didcot (UK), to support the scientific payload operations. Mail and communications services built up for CSDS will be used by the instrument teams to send their commands to JSOC (Hapgood *et al.*, this issue).

The major scientific advances that are likely to result from the Cluster mission will require the use of data from all four spacecraft and generally from more than one instrument. Thus it was essential that the system be designed to make it as easy as possible to combine the various sets of data. This has had several consequences for the designer; for example it means that the data formats must be compatible for all the instruments, and that the electronic data networks must be used wherever possible. For the most effective use of the data it is also important that the user interface be as simple as possible.

2. Implementation of CSDS

At the time of the writing, CSDS is basically ready and awaiting the launch of the Cluster spacecraft. The fact that CSDS, as a distributed system with many interfaces and contributors located in many European and other countries, was ready on time was largely due to a clear management approach. Following a two-year preparation period (1989–1991), implementation started in the second half of 1991. Two management groups were set up with the task of delivering CSDS on time and within the budget allocated by the contributing countries.

The CSDS Steering Committee, with members nominated by the national data centres and by ESA and NASA, carried overall responsibility. In this capacity it scheduled reviews of the status of CSDS at major milestones and monitored the work of the Implementation Working Group. The latter group performed the day-to-day tasks, recommended standards to the Steering Committee, established and monitored the formal interfaces, implemented the system and executed all testing required prior to the final delivery of CSDS.

The implementation of CSDS was broken down into distinctly different phases according to the applicable ESA software engineering standards (ESA-PSS-05, 1991). The CSDS software life cycle is defined as follows.

The preparation phase was completed by the first major review, the Conceptual Design Review in April 1992. The user- and software requirements definition phase culminated in the CSDS Requirements Review held in March 1993 (for JSOC, due to a later start, in August 1993). The successful completion of the architectural design phase was demonstrated in February 1994 and, finally, the review ending the detailed design and production phase was held in March 1995.

The overall readiness of CSDS and JSOC was reviewed in November 1995. The results of an extensive test phase were reported at the same review. CSDS was declared ready for its operations- and maintenance phase, which will formally start about three months after the launch of the Cluster spacecraft.

3. End-to-End Data Flow

The payloads on each spacecraft can be commanded to produce one of three different telemetry streams: 22 kbit/s (normal science data rate plus housekeeping data), 131 kbit s^{-1} (high-speed science data rate plus housekeeping data) and 262 kbit s^{-1}. The highest data rate is only used for the real-time transmission of data from the WBD (Gurnett *et al.*, this issue) instrument to one of the two NASA ground stations located at Goldstone or Canberra. A recent change of the on-board storage from a tape recorder to a solid-state memory which provides twice as much storage capacity, has made it feasible to maintain the nominal bit rate for more than 95% of the orbital period of 57 hours. If the payload were only operated in the normal science data rate then, from the four spacecraft over the two-year period, the mission would produce a data volume of about 4.1×10^{15} bit as an upper limit. In the case of the high-speed data taking, there is a limit in that ESA can only use two ground stations to serve the four spacecraft. This limiting factor is due to the time required to transmit the content of the on-board storage to ground. Extensive studies have shown that, on average, only six hours of high-speed data can be taken per orbit and per spacecraft. The associated total amount of data, if only high-speed data were taken, is 3.6×10^{12} bit as a lower limit. In reality, the Cluster science community will want to use a mixture of normal- and high-speed science data taking, which will result in a final data volume somewhere between the two limits.

The Science Working Team agreed that the Cluster scientists and CSDS data centres should have available the entire data set. The Data Disposition System (DDS), located at the Operations Control Centre at ESOC, Darmstadt, organises the data delivery such that the raw data media contain all data on a per-instrument and per-spacecraft basis, i.e. all science and housekeeping data from the instruments, spacecraft-related data, such as housekeeping data and orbital information are accessible to the community. It was decided to use CD-ROM's as raw data media and, for economy, their shipment will be made to institutions and not to individual scientists (Table I). The daily amount of data will fill, on average, two CD-ROM's and these will be shipped on a weekly basis to 83 institutions in Europe, USA, Japan, China and India. The elapsed time between data taken on the spacecraft and CD-ROM's arriving at the home institutes of the scientists will be about four weeks. The weekly shipment, the postal service and, to a minor extent, the spacecraft visibility by the ground stations are the major factors affecting the delivery time.

A simplified data flow diagram is shown in Figure 1. The Operations Control Centre at ESOC is responsible for gathering and distribution of all science data on CD-ROM's. The organisation of the data on the CD-ROM is described by Sörensen *et al.* (this issue). In addition, ESOC allows remote access to a time limited data set. This service is only available for instrument teams and some of the data centres. After arrival of the raw data at the data centres (Figure 1), a pipeline process extracts the relevant information and produces on a per-day basis a non-validated data set. After validation by the relevant Principal Investigators, the data set becomes available to the other data centres. By fetching data sets from the other data centres a complete data set for a given day is compiled in the form of validated CDF files (Common Data Format; Allen *et al.*, 1996). After ingestion of these files into the data base and the associated update of the catalogues describing the content of the Prime and Summary Parameter Data Base, the files become accessible to the scientific users. The Joint Science Operations Centre (JSOC) provides the contents of their event catalogues (e.g., mission and orbit related information, crossing of boundaries, etc.).

The instrument teams and some of the data centres will have quicker access to mission data by remotely retrieving data from the Data Disposition System located at the Operations Control Centre. This system offers the data soon after they are received by the ground stations but there are restrictions both to the daily data volume requested by a user and the total number of users on the system. This service is intended to provide the instrument teams with a facility that allows quick checks of their instrument's status. It is also important for CSDS, because the timely availability of magnetic field data is essential for the pipeline processing of data from other instruments.

4. National Data Centres and the Joint Science Operations Centre

CSDS is a distributed system consisting of eight nationally funded and operated data centres. In most cases, the data centres produce data products on behalf of the national PI teams. Members of the Cluster science community wishing to access CSDS may contact the manager of their national data centre (*e*-mail addresses are given below). In those countries (Table I) not served by a national data centre, members of the Cluster community may contact the relevant Principal Investigator to determine which data centre should be used. It should be noted that all data centres offer the same data products (see). Scientists from outside the Cluster community will also have access to CSDS, according to the policy on data rights outlined in Sec-

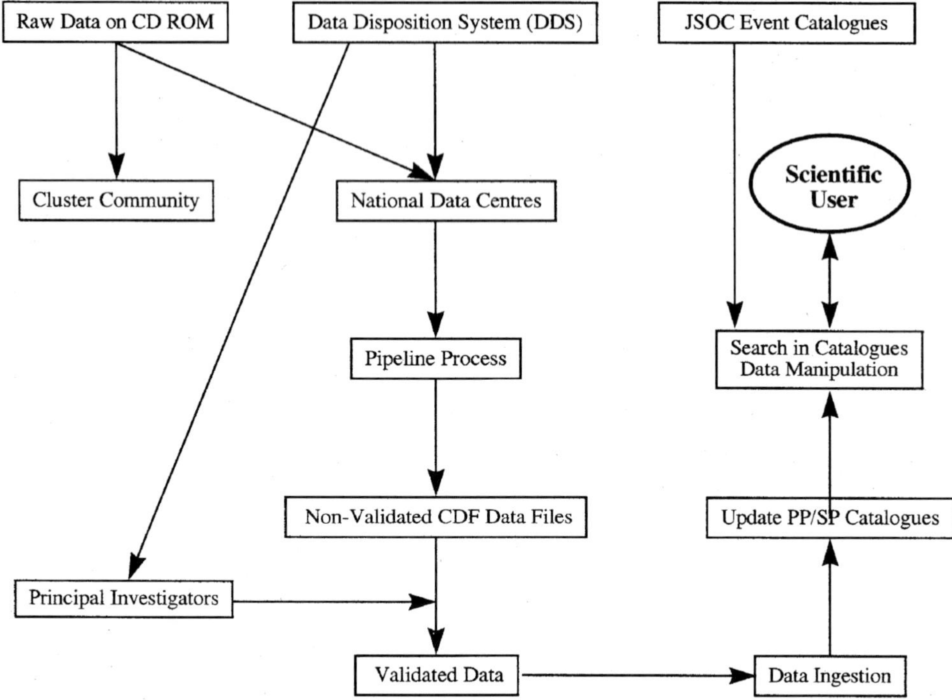

Figure 1. Simplified data flow diagram starting with the availability of raw data at the national data centres. After execution of all steps the scientists can retrieve a complete set of validated Prime and Summary Parameters.

tion 7. Full access can be granted to the Summary Parameters and, possibly during campaigns, to the Prime Parameters but restricted on a time and instrument basis. Inquiries may be made addressed to either the nearest national data centre or to the Cluster Project Scientists (rschmidt@estec.esa.nl or cpescoubet@estec.esa.nl).

The Austrian Cluster Data Centre (ACDC)
The ACDC is hosted by the Department for Experimental Space Research of the Space Research Institute of the Austrian Academy of Sciences (IWF). Its primary tasks are the production of CSDS data products for the Cluster instrument ASPOC (Riedler *et al.*, this issue) and in the provision of access to CSDS and services provided by CSDS for the Austrian community. The operating environment is Open-VMS/Alpha. The ACDC manager can be contacted at the following email address: torkar@fiwf01.dnet.tu-graz.ac.at

The Chinese Cluster Data and Research Centre (CCDRC)
CCDRC is located at the Centre for Space Science and Applied Research, Chinese Academy of Sciences, Beijing. Its task is to serve the Chinese Cluster community currently located at two institutes in the Beijing area. CCDRC is for geographical

Table I

Geographical distribution of the scientific users, distribution of CD-ROMs and location of data centres (status: February 1996)

Country	PIs and/or Co-Is	Number of CD-ROMs	Data Centre
Austria	7	3	yes
Canada	2	2	
China	3	2	yes
Denmark	4	4	
Finland	2	1	
France	34	8	yes
Germany	25	6	yes
Greece	2	2	
Hungary	2	1	yes
India	2	1	
Ireland	1	1	
Israel	1	1	
Italy	7	1	
Japan	3	2	
ESTEC, NL	5	1	after May 1996
Norway	10	3	
Russia	3	2	
Sweden	14	4	yes
Switzerland	1	1	
United Kingdom	20	10	yes
United States	68	27	yes
Total	216	83	

reasons not directly linked to CSDSnet, so the transfer of CSDS data products is organised by ACDC through regular mailings of tapes. Smaller amounts of data will also be exchanged via the Internet. The operating environment is Sun/Solaris and the CCDRC manager can be contacted at: liu@sun20.cssar.ac.cn

The French Cluster Data Centre (CFC)

The services to be provided by the CFC (Centre Français Cluster) fall into two complimentary areas of competence, firstly the operational use of large computing systems and, secondly, the scientific expertise and knowledge of the instruments. The first task is executed by CNES, Toulouse, which also acts as entry point for all users of the national community and international users belonging to French PI teams. The second task is executed by the scientific institutions involved in the development of three scientific instruments. CFC supports the production of data for CIS (Rème *et al.*, this issue), STAFF (Cornilleau *et al.*, this issue) and WHISPER (Décréau *et al.*, this issue). The operating environment is Sun/Solaris. The CFC manager can be contacted at: prado@cst.cnes.fr

The German Cluster Data Centre (GCDC)
The GCDC is located at the Max-Planck-Institut (MPI) für extraterrestrische Physik, Berlin-Adlershof. Its primary purpose is to generate and make available the CSDS data products for the RAPID (Wilken *et al.*, this issue) and EDI (Paschmann *et al.*, this issue) experiments. This is done in close collaboration with the two PI institutes located in Garching and Katlenburg-Lindau. GCDC has also undertaken to produce the Summary Plot Files for all of CSDS; the distribution of the files outside Germany is handled via the other national data centres. The operating environment is Open-VMS/Alpha. The GCDC manager can be contacted at: okp@mpepl.plasma.mpe-garching.mpg.de

The Hungarian Data Centre (HDC)
The HDC is located at the Research Institute for Particle and Nuclear Physics, Budapest. The HDC is one of the three data centres not directly serving a national Principal Investigator team. HDC contributes to CSDS through the provision of the spacecraft auxiliary parameters by retrieving the relevant information from ESOC and offering it in a CSDS-compatible format. Further services concern the distribution of software to produce magnetic field data from the FGM (Balogh *et al.*, this issue) instrument and the associated calibration files for that instrument. The operating environment is Sun/Solaris. The HDC manager of HDC can be contacted at: mariella@rmki.kfki.hu

The Swedish Data Centre (SDC)
The SDC is located at the Alfvén Laboratory, Royal Institute of Technology (KTH), Stockholm. The SDC is responsible for producing data for the EFW (Gustafsson *et al.*, this issue) instrument. The implementation work is shared between research groups at the Swedish Institute of Space Physics, Uppsala, the University of Oslo, Norway, and the University of Oulu, Finland. Its prime customers are the Scandinavian science community and the other data centres. The operating environment is Open-VMS/Alpha. The SDC manager can be contacted at: gh@irfu.se

The United Kingdom Data Centre (UKDC)
The UKDC is located at the Rutherford Appleton Laboratory (RAL), Didcot and it serves the three national PI groups: DWP (Woolliscroft *et al.*, this issue), FGM (Balogh *et al.*, this issue) and PEACE (Johnstone *et al.*, this issue). The complete data base will be made available to the UK Cluster community, including Ireland and to the other data centres. The scientific users interact with UKDC either via the CSDS User Interface or with the UK-specific Science Analysis System. The operating environment is Sun/Solaris. The UKDC manager can be contacted at: p.a.vaughan@rl.ac.uk

The United States Data Centre (USDC)
The USDC is located at Goddard Space Flight Center, Greenbelt, Maryland. It constitutes part of the Command and Data Handling Facility (CDHF) for the Global

Geospace Science (GGS) Project. Although there is one PI team (WBD, Gurnett *et al.*, this issue) located in the United States, USDC is not directly supporting him due to the rather complicated nature of the WBD data. The University of Iowa will produce the relevant data products and forward them directly to CFC, UKDC and SDC. The USDC is linked to the UKDC and by an automatic process its data bases are kept aligned with those of the UKDC. Access to the USDC differs from that with other data centres as it is not based on the Cluster User Interface. The USDC manager can be contacted at: wmish@istp1.gsfc.nasa.gov

The Joint Science Operations Centre (JSOC)
JSOC is co-located with the UKDC at RAL, Didcot (see also Hapgood *et al.*, this issue). The task of JSOC is to support the Cluster Project Scientists in the scientific operations of the payload. To this end, JSOC receives the master science plan from the PI teams and will build the detailed instrument command files. These files are iterated with the instrument teams and then forwarded for final acceptance checking by the Operations Control Centre (OCC, see also Ferri *et al.*, this issue). In case of a rejection by the OCC, e.g., due to a violation of spacecraft constraints, a further iteration with the PI teams is executed. Most of the JSOC catalogues are accessible to CSDS via the CSDS User Interface. The interface with a data centre being operated by the ground based observations community is also implemented by JSOC. The JSOC manager can be contacted at: t.g.dimbylow@rl.ac.uk

5. Architecture of CSDS

The Cluster Science Working Team recommended at its first meeting that a working group be formed to compile a report on the national plans regarding the Cluster data analysis. The working group delivered its report within six months and it concluded that a mission-wide data system might be feasible if it was possible to co-ordinate the essentially independent activities in the ESA Member States. It was proposed to implement a distributed system of nationally funded and operated data centres (Figure 2). Based on this report, ESA released an Announcement of Opportunity (CSDS AO, 1991) to call for the contribution of the individual elements of CSDS. ESA's contribution was restricted to the provision of CSDSnet, the User Interface and the Joint Science Operations Centre (JSOC) in collaboration with the Rutherford-Appleton Laboratory, UK (Hapgood *et al.*, this issue).

Each national data centre would produce data sets on behalf of the Principal Investigator(s) from the same country. Data sets from all other instruments would be retrieved over an electronic network from the relevant data centres. In this way, each data centre could offer the entire data sets to its national user community and avoid duplication of work. From the very beginning, one of the key user requirements was that the system should provide a uniform user interface to ensure that a user, independent of geographic location, could interface with the same system in the same way and visualise the data in the same format.

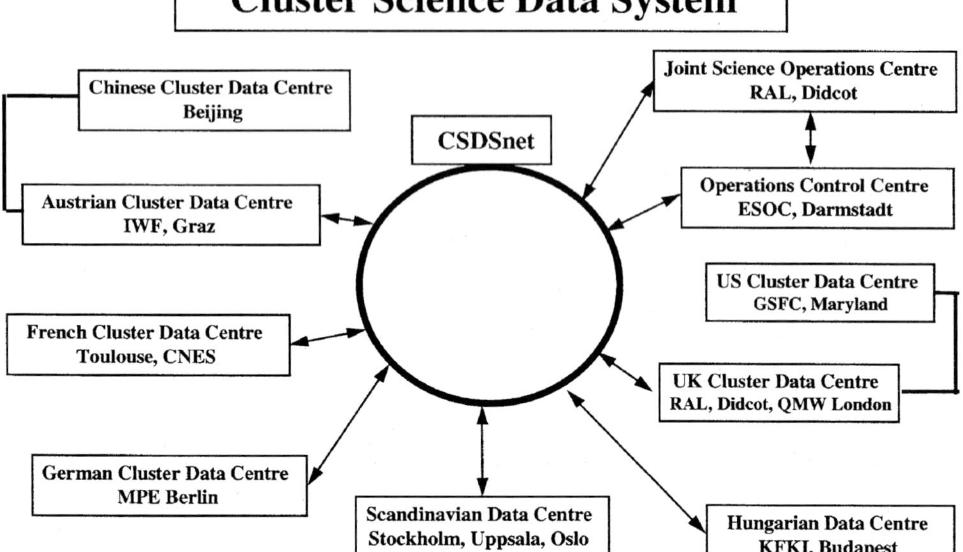

Figure 2. Architecture of the Cluster Science Data System. The users access their national data centre. The data centres themselves are interconnected via CSDSnet, an ESA-provided TCP/IP based network. The American and the Chinese data centres retrieve data via the U.K. and Austrian data centre, respectively.

6. Networking and CSDSnet

It was soon realised that national academic networks offered an acceptable level of service and sufficient bandwidth to transfer large amounts of data. However, the international networking was clearly a bottleneck as far as the connectivity between the CSDS data centres was concerned. A reliable connectivity with sufficient bandwidth is an essential factor in a distributed system. Thus, CSDSnet as a general-purpose, mesh-like ground communications network was established by ESA and designed to provide the following services:

– Retrieval of raw science, housekeeping and auxiliary data available on the Data Disposition System at ESOC.

– Submission of command requests by the PI teams to the Joint Science Operations Centre (JSOC) and from JSOC to the Operations Control Centre (OCC) at ESOC.

– Electronic exchange of processed science data between the individual data centres.

– Access to test data at the sites where the spacecraft integration and test programme is conducted.

It should also be noted that CSDSnet only handles international traffic; national traffic, for instance between a data centre and its user community, is not served

Table II

Committed information rates on frame relay links between the elements of CSDS in kbit s^{-1}. HDC is not connected by frame relay and therefore no committed information rate can be given. Traffic to/from ACDC is routed via UKDC or CFC.

To/from	ESOC	UKDC	CFC	GCDC	SDC	ACDC	ESTEC	HDC
ESOC		8	4	4	4	–	0	N/A
UKDC	48		8	4	4	16	4	N/A
CFC	48	16		16	8	16	4	N/A
GCDC	16	4	4		4	–	4	N/A
SDC	8	8	8	4		–	4	N/A
ACDC	–	16	16	–	–		4	N/A
ESTEC	0	4	4	4	4	–		N/A
HDC	N/A	N/A	N/A	N/A	N/A	N/A	N/A	

by CSDSnet. There is, however, one exception, regarding the traffic between the German Cluster Data Centre and ESOC, which is national in the geographic sense but handled by CSDSnet as it constitutes a critical part of CSDS.

Most of the network design requirements have been developed by a CSDS task group which undertook to model the network traffic in a realistic manner. Additional requirements state that the network must be able to handle the weekly amount of data eight hours per day and five days per week. There is also a need for a high throughput reserve. The available bandwidth per connection shall be at least twice as high as the average data-exchange requirement. The connectivity shall be based exclusively upon the Internet Protocol. The requirement analysis led to a design providing the committed information rates as listed in:

The non-availability of frame relay for the Hungarian Data Centre (HDC) was a certain complication in the sense that the traffic to and from the HDC will be handled via a X.25 connection and routers located both at RAL and CNES.

6.0.1. *Wide area network*

CSDSnet was designed as a network which only supports data and file transfer between data centres (Figure 3). An implicit service is provided to the Principal Investigators by enabling them to retrieve from the data disposition system at ESOC near-real time data over CSDSnet via their national data centres. Email transfer or any other non-CSDS traffic is not supported. CSDSnet is configured as a self-contained TCP/IP inter-network, coexisting with ESA's existing TCP/IP network, ESInet. It has its own addressing, routing and security schemes, to allow it to be tailored to the specific needs of CSDS, and to keep it dedicated to CSDS-related traffic. At the same time, it will maintain full compatibility with the existing ESInet and with all existing European Internet domains to enable seamless interconnection to the latter. CSDSnet is a multi-layered system, compliant with the Open System Interconnection (OSI) 7-layer model, to ensure that each layer can be designed independently.

At each data centre the end systems are interconnected to an Ethernet Local-Area Network segment, which is CSDS-dedicated, i.e., only CSDS hosts and routers attach to it. The LAN-interconnect service provides logical end-to-end connectivity between the distributed hosts at the various data centres, allowing them to run a number of network applications, corresponding to the layer-3 network service of the OSI model implemented on a LAN.

The Wide-Area Network (WAN) access is implemented by attaching routers to the CSDS LAN segment at each data centre site. The routers handle multi-protocol routing, should it be required at a later stage. As can be seen from , there are users in a few countries who are not served by a national data centre (i.e., Switzerland, Greece, Italy, India, Japan, Russia). In order to give these users access to all CSDS services, two different types of data centre have been conceived. Type-1 data centres support full interconnectivity between CSDSnet and both national and international Internet. To this end, a peering relationship has been established between the CSDSnet and a data centre-router to support the connectivity required by the CSDS community to the external users. To become a type-1 data centre, the site router must be capable of exchanging routing information with the external Internet. Type-1 data centres act as entry points for international CSDS users into the CSDS network (UKDC, SDC, HDC and CFC in Figure 3).

Type-2 data centres cannot support connectivity to/from international users. The connectivity is restricted to national users. The ensuing connectivity diagram is shown in Figure 3. It visualises the way CSDSnet interconnects the European national data centres, public Internet and ESA's own IP network, ESINet.

6.1. Network Security and Quality of Service

It is the responsibility of the CSDS end-user to impose security constraints and access-control mechanisms. The provision of data confidentiality services, i.e., encryption, is not required but the CSDS network service will guarantee error-free transmission of data, even across inherently unreliable media. CSDSnet cannot be seen as an isolated entity and it will heavily rely on network services provided by the national Internet networks, i.e., for the access of the Data Disposition System by the PI-teams. A static routing has been implemented whereby the PI uses a national Internet link to the local data centre and from there the traffic is routed onto CSDSnet.

The submission of command requests to JSOC, and subsequently from JSOC to ESOC, requires the highest overall quality of service, particularly in terms of availability, backlog service and operational support. For the exchange of data between data centres the availability and throughput reserve is considered critical. Access to the Data Disposition System at ESOC is very sensitive to backlogs which could build up during network outages or congestion. Adequate throughput reserve must be, therefore, foreseen. In general, the network has to offer high availability and high-throughput reserve.

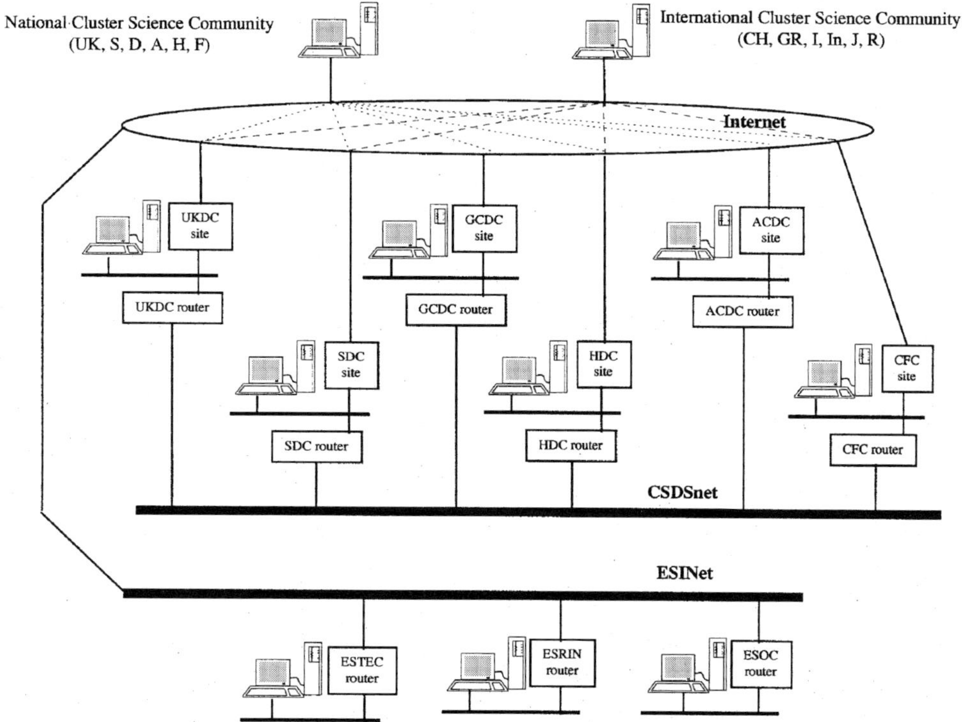

Figure 3. Connectivity diagram of CSDSnet. The keys are explained in Section 4. International users can access the system only via UKDC, SDC, HDC and CFC. The USDC and CCDRC are not connected to CSDSnet.

7. Data Products and Access Rights

A fundamental principle of CSDS is that each data centre generates its own unique data products in a routine manner and makes them available to the other data centres. Thus, the well-organised production and reception of data leads to identical data bases at each data centre. Data products are, in principle, available on-line for retrieval by the users for the duration of the mission. Each of the national data centres participating in CSDS offers three data sets to its users. These data sets are listed below together with the data access rights as formulated by the Cluster Principal Investigators.

An important aspect of CSDS concerns the fact that also scientists not belonging to the Cluster community may get access to some of the data products, i.e. the SPDB and SP as described below. The appropriate procedure will be to contact either a Principal Investigators or one of the addresses listed in Section 4 and an assignment to one of the national data centres will be made. As the national data centres operate under limited budgets the number of external users may be depend on the availability of system resources at the centre.

7.1. Prime parameter data base (PPDB)

The parameters contained in this data base are listed in Table III. The data files are written in the Common Data Format (CDF, see Allen *et al.*, 1995) and are held in physical units with an exhaustive set of ancillary information (or metadata). The PPDB will hold data from all four spacecraft with a time resolution of about 4 s.

7.2. Summary parameter data base (SPDB)

The parameters contained in this data base are listed in . These files are also written in the Common Data Format and are held in physical units with an exhaustive set of ancillary information. The SPDB will only hold data from one of the four spacecraft with a time resolution of 60 s.

7.3. Summary plots (SP)

These are plots of selected summary parameters with one-minute resolution which will be used by the scientists to search for interesting events; it has been agreed that the GCDC will produce these plots as Postscript files centrally and distribute them inside Germany and to the other data centres. The plot information is encoded in a compact form for sending it over the network. The compacted file also includes a detached SFDU (Standard Format Data Unit, containing some metadata). The access to the files will be either through the Cluster User Interface or via 'anonymous FTP'. The Internet addresses of the FTP servers will be defined at a later stage.

An example of a Summary Plot is shown in Figure 4. For a given six-hour time interval there will be no more than four plot pages in A4 format, i.e., a maximum of 24 pages per day. The first plot page gives an overview, while the remaining pages contain plasma composition information, fields and waves and spacecraft configuration information, respectively. shows the first two of a set of four plot pages per six hours. These examples show test pages created as part of the overall CSDS test activity, therefore some of the panels are not filled with data points and the date in the upper right corner is set to 60 September 1996 in order to avoid confusion during the mission.

7.4. Data access rights ('Rules of the road')

The Principal Investigators have agreed upon data access rights which are reproduced below as far as the PPDB and SPDB are concerned:

Prime Parameter Data Base (PPDB). These data sets, generated by software implemented and maintained under PI supervision, validated by PI's, will be the principal means of establishing multi-instrument, co-ordinated data analysis for the Cluster investigator teams. The data consist of spin-averaged parameters (4 s

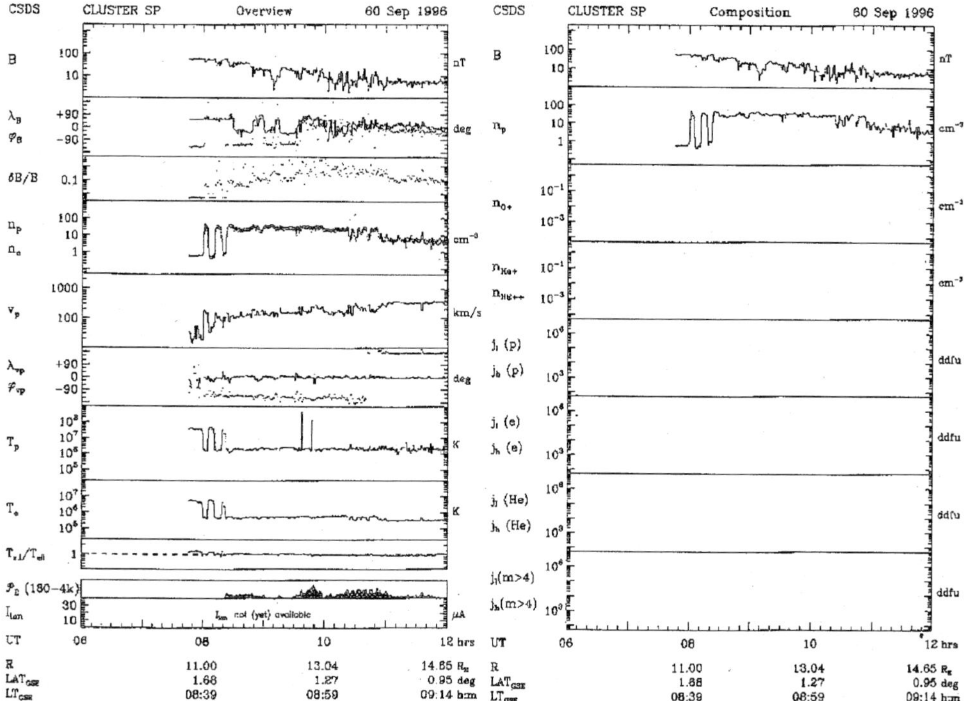

Figure 4. Sample of the first two pages of the summary plot files. These pages cover the overview and the composition data.

resolution) from all instruments on all four Cluster spacecraft as defined, on inputs from the PI's, by the CSDS Implementation Working Group.

These data will be accessible to all Cluster PI's and Co-Is from CSDS data centres. The access by Co-Investigators to the PPDB is through agreed procedures between the PI's, Co-Investigators and the Data Centres.

The purpose of the PPDB is to establish and carry out co-ordinated data analysis within the direct Cluster community only. Neither raw data from the PPDB, nor any data derived from the PPDB can be published without the agreement of the PI responsible for the data. It is the responsibility of the PI of each team to ensure that all Co-Investigators are aware of their obligations under this agreement and that in the event of violation their priviliges may be withdrawn.

Summary Parameters (SPDB). These data sets, generated by software implemented and maintained under PI supervision, validated by PI's, will be available to the Cluster investigator teams and to a much wider community, on a 'best-efforts' basis, for establishing scientific co-operation with other space missions and with ground-based investigators. The data consist of one-minute averaged parameters from all instruments from one of the four Cluster spacecraft as defined, on inputs from the PI's, by the CSDS Implementation Working Group.

These data will be accessible and/or distributed rom the CSDS data centres. Access to these data is not restricted; the implementation of the access and distribution is the responsibility of the Data Centres.

The purpose of the SPDB is to establish co-ordinated data analysis between Cluster scientists and a wide and diverse scientific community. All recipients of the data and the summary plots will be informed that neither raw data from the SPDB, nor any data derived from the SPDB, can be published without the agreement of the Principal Investigator responsible for the data.

Fluxgate Magnetometer Processing Support Data Set (FGM-PSDS). This data set, generated by software implemented and maintained under the supervision of the FGM PI and validated by the FGM PI, will be available in the data centres for supporting the generation of the PPDB/SPDB of those investigations which require FGM data for these purposes. These data consist of 1 s resolution spot values of magnetic field vectors from all four spacecraft. The FGM PSDS will not be used for purposes other than the generation of other instruments' PPDB and SPDB, except with the specific agreement of the FGM PI. Data centres generating the PSDS will be required to ensure conformance with this requirement.

7.5. Detailed Content of the PPDB/SPDB

Table III shows the content of the PPDB and SPDB as tabled in Allen *et al.* (1994). It should be noted that parameters listed under PPDB are available for each spacecraft, thus their actual number is four times higher. The data products are listed on a per-instrument basis. The appearance or absence of a parameter in the SPDB and PPDB is marked by a full and open circle, respectively. A single parameter is stored in a data word of four bytes. Some of the parameters, e.g., plasma flow speed, listed in bold, are vectors and require three data words each. The auxiliary parameters are provided only in the SPDB. This is because the high temporal resolution of the PPDB is not needed for them. This means that access is required to the SPDB when the PPDB data are to be processed with positional information. However, since the data from every instrument will be stored in a file of its own, it does not really matter whether the file containing auxiliary data is designated as belonging to the PPDB or SPDB, as it must in any event be transferred from its source data centre. By assigning it to the SPDB, the time resolution of one minute is emphasised. There are two tetrahedron configuration parameters shown in . The Glassmeier-quality factor Q_G is described in detail in von Stein *et al.*, [1992]. Its mathematical expression is as follows:

$$Q_G = \frac{\text{true volume}}{\text{ideal volume}} + \frac{\text{true total surface area}}{\text{ideal total surface area}} + 1 \, ,$$

where true refers to the actual figures and ideal to a regular tetrahedron with base size equal to the mean separation. For a regular tetrahedron $Q_G = 3$, for a planar configuration $Q_G = 2$, and for a co-linear arrangement $Q_G = 1$.

Other quality factors have been proposed by Robert and Roux [1993], one of which is also included as an auxiliary parameter. Designated Q_R, it is the cube root of the ratio of the volume of the actual figure to the volume of the sphere into which it can be inscribed, normalised to 1 for a regular tetrahedron.

$$Q_R = \left(\frac{\text{true volume}}{\text{volume sphere}} \Big/ \frac{\text{regular volume}}{\text{regular sphere}} \right)^{1/3} .$$

The normalising factor

$$\frac{\text{regular volume}}{\text{regular sphere}} = \frac{2}{3\pi\sqrt{3}} ,$$

Q_R becomes zero when the four satellites are co-planar. Opinions are that both Q_G and Q_R yield useful but different information. The overall size of the spacecraft figure is listed by the minimum and maximum separations, i.e., min $|\Delta\mathbf{R}_{ij}|$ and max $|\mathbf{R}_{ij}|$, respectively.

A rough calculation of the size of the two data bases over a two-year period, assuming a 100% data coverage in the normal science data taking mode, neglecting CDF overhead, but including two time data words per instrument and auxiliary parameters, yields about 1 GBytes for the SPDB. The size of the PPDB will amount to about 26 GBytes. It must be emphasised that these estimates are lower limits.

8. Interface Control, Standardisation and Availability of Documents

A system of distributed data centres poses a large number of problems of organisation and coordination compared to one of a single data centre. Many questions of standardisation and harmonisation have to be addressed. A critical issue in this context is to keep track of all interfaces and control their evolution in a highly reliable manner. This implies that documentation must be easily available and it has to be ensured that all parties work only with its most recent version. Document change control must be handled very accurately. Within CSDS a documentation system was set up whereby the most critical documents (e.g., Requirements Specifications, Interface Control Document) are configuration-controlled by ESA. All underlying documents, as well as data centre-specific documents, are subject to configuration control by the data centres.

8.1. Documentation

The timely availability of documents to the engineers is certainly not a trivial issue. It was felt that documents could be most efficiently stored and distributed by establishing an FTP server with a directory structure according to suppliers of

Table III

Detailed content of the PPDB and SPDB. Bold letters represent vectorial quantities. Descriptions of the instruments and their science parameters cab be found elsewhere in this issue.

Instrument acronym	Parameter	Words	PPDB	SPDB
ASPOC	Ion current	1	•	•
	ASPOC status word	1	•	•
	Epoch time	2	•	•
CIS	$n(\mathrm{p})$, $n(\mathrm{O}^+)$, $n(\mathrm{He}^+)$, $n(\mathrm{He}^{++})$, $n(\mathrm{HIA})$	5	•	•
	$\mathbf{v}(\mathrm{p})$, $\mathbf{v}(\mathrm{HIA})$	6	•	•
	$\mathbf{v}(\mathrm{O}^+)$	3	○	•
	$T(\mathrm{p})$, $T(\mathrm{HIA})$	2	○	•
	$T_\perp(s)$, $T_\parallel(s)$; $s = p$, HIA	4	•	○
	CIS status word	1	•	•
	Epoch time	2	•	•
[n = number density; $\mathbf{v}$ = velocity vector; T = temperature, HIA = Hot Ion Analyser, measures most dominant ion species]				
EDI	$\mathbf{E}$, electron drift velocity	3	•	○
	$\mathbf{v}(\mathrm{ed})$, electron drift velocity	3	•	•
	EDI status word	1	•	•
	Epoch time	2	•	•
FGM	$\mathbf{B}$, magnetic field	3	•	•
	$(\delta B)^2/B^2$, total field variations	1	•	•
	$(\delta\|B\|)^2/B^2$, field magnitude variations	1	•	•
	FGM status word	1	•	•
	Epoch time	2	•	•
PEACE	$n(e)$, $\mathbf{v}(e)$, $T_\parallel(e)$, $T_\perp(e)$, $Q_\parallel(e)$	7	•	•
	PEACE status word	1	•	•
	Epoch time	2	•	•
[Electron density, velocity, temperature, heat flux]				
RAPID	$J_{\mathrm{low}}(s)$, $J_{\mathrm{hi}}(s)$; $s = e$, p, He, mass > 4	8	•	•
	$A_\parallel(e)$, $A_\parallel(\mathrm{p})$	2	•	•
	RAPID status word	1	•	•
	Epoch time	2	•	•
[J = particle flux, 2 energy ranges, 4 species, A = field-aligned flux anisotropy, 2 species]				
DWP	Significance level	1	•	•
	v_{max}, $P(v)$	2	•	○
	DWP status word	1	•	•
	WEC status word	1	•	•
	Epoch time	2	•	•
[Sign. level, freq. v, and energy P of most intense component]				

Table III
(continued)

Instrument acronym	Parameter	Words	PPDB	SPDB
EFW	E_{dusk}, σ_E, V_{PS}, I_{probe}	4	•	•
	E power; 0.3–10, 10–180 Hz	2	•	•
	EFW status word	1	•	•
	WEC status word	1	•	•
	Epoch time	2	•	•
[Dawn-dusk electric field, variations in electric field, probe-s/c voltage, and probe current]				
STAFF	$B_{\|\|}$, $B_{\perp}$, μ, 0.3–10, 10–180, 180–4000 Hz	9	•	•
	B power; 0.3–10, 10–180, 180–4000 Hz	3	•	•
	E power; 10-180, 180–4000 Hz	2	•	•
	STAFF status word	1	•	•
	WEC status word	1	•	•
	Epoch time	2	•	•
WBD	WBD status word	1	•	•
	WEC status word	1	•	•
	Epoch time	2	•	•
WHISPER	$n(e)$, quality	2	•	•
	E power, 4–10, 10–20, 20–80 kHz	3	•	○
	E power, 4–80 kHz	1	○	•
	E variance, 2–80 kHz	1	•	○
	WHISPER status word	1	•	•
	WEC status word	1	•	•
	Epoch time	2	•	•
Auxiliary parameters	$\mathbf{R}_1$, position of s/c 1	3	○	•
	$Delta\mathbf{R}_n$ = 2, 3, 4, position rel. to SC 1	9	○	•
	$\mathbf{V}_1$, velocity vector of SC 1	3	○	•
	Q_G, Q_R, min $\|\Delta\mathbf{R}_{ij}\|$, max $\|\Delta\mathbf{R}_{ij}\|$	4	○	•
[Tetrahedron quality parameters]				
	Θ_{Scn}, Φ_{SCn}, n = 1–4, spin axes in GSE	8	○	•
	GSE $\rightarrow$ GSM conversion angle	1	○	•
	Dipole tilt angle	1	○	•
	Spacecraft status word	1	○	•
	Epoch time	2	○	•
Total number of data words:			108	127

documents. In principle it was agreed that documents be offered as postscript files, in either compressed or uncompressed form. The CSDS document server is open to everyone for retrieving documents. The uploading of documents is, however, restricted.

A user wishing to download documents must make an 'anonymous FTP' call to the server, reachable at ftp.estec.esa.nl or 131.176.1.105. The user will find in /pub/csds the sub-directories named according to the main elements of CSDS (e.g., data centres, user interface, JSOC). Currently, there is a significant amount of documentation on the system that either was required for the design, implementation and test phase. Once CSDS is fully operational, it is expected that documents will be available that provide instructions on how to log into a national data centre and how to get started.

8.2. Software

CSDS must serve a wide and diverse user community and it is well known that a wide variety of computing hardware and software is used in the scientific institutions. For reasons of economy it was necessary to select for CSDS a small number of platforms (i.e., Open-VMS/Alpha and Sun/Solaris) and operating systems. Pipeline processing software at the data centres was developed by the data centres and will also be maintained by them. The Principal Investigator teams have supplied their national data centre with specific codes which enable the centres to open the science packets and produce the PPDB and SPDB. The software to routinely produce SP's was developed by MPE Garching. The Cluster User Interface, developed under an ESA contract with a few scientific institutes, was delivered to the data centres (see Section 9) and it will only be supported on Sun/Solaris and Open-VMS/Alpha. Interfaces to commercial packages such IDL are available where scientists wish to process the data beyond the services offered by the User Interface. ESA will maintain this software until two years after the end of the nominal mission which is currently the end of 1999. Several new releases with enhanced functionalities are planned in the near future. The installation and maintenance at the scientific institutes of the client software is a task of the national data centres. As it is unavoidable that some of the CSDS software must be exchanged among the data centres, it was agreed that software should only be written in FORTRAN 90 or 'C'. In addition, certain coding guidelines had to be followed to ensure portability during the implementation phase and maintainability of the code during the operational phase of CSDS.

8.3. Data files

A critical function of CSDS is the management of large amounts of data products, which need to be computed, ingested, and distributed in an autonomous way by the system. At the outset it was recognised that a self-documenting data format and, wherever possible, the use of existing standards, is desirable. The selection of a standard was a difficult task as standards tend to have a cost impact on the system and their lifetime is normally difficult to predict. Two of the more important data-related standards selected for CSDS concern the Standard Formatted Data Unit (SFDU, 1991–1992) and the Common Data Format (Allen *et al.*, 1996).

The Standard Formatted Data Units (SFDU) have been chosen at the top level of data management. An SFDU file is used to identify and describe each data product generated by CSDS and the file contains information about the data and pointers to the actual data product. The SFDU, normally delivered in form of a detached text file, is a standard adopted and operated under the auspices of the Consultative Committee for Space Data Systems (CCSDS). Its goal is to describe data products by labelling the data objects and each SFDU gets a unique registration number and is 'archived' by a control authority. SFDU's related to CSDS are registered by ESOC, Darmstadt. The use of the SFDU methodology ensures that the description of data format and data content is either contained within the data itself or can be readily obtained from the control authority. This methodology, however, does not resolve the problem of correlating or merging the science data from multiple instruments and missions.

The need of the Cluster community to perform multi-instrument data analyses from the four Cluster spacecraft and to correlate the results into context with data from other space missions, required the selection of a science data format which shares a common data format and a common data dictionary. A common data dictionary provides the means to ensure that identical data items from different space missions are described identically with respect to formats and their units of measurement. A Cluster scientist, fetching spin-plane electric-field measurements from the Cluster wire boom instrument, should be able to call up the data items from the same type of instrument on the NASA/Polar spacecraft by using the same nomenclature.

In order to facilitate this task, the Common Data Format (CDF) was chosen for the storage of the CSDS data base products. However, CDF data sets could have been designed in an almost infinite variety of ways given the flexibility allowed in the CDF conventions. A dedicated CSDS task force under the leadership of the Queen Mary and Westfield College was therefore set up to recommend a Cluster CDF design which meets all the Cluster requirements but maintains compatibility with all the other missions that also generate CDF products. CDF files are produced by data systems in support of numerous missions, such as Polar, Wind, Geotail, Interball or SOHO (particle instruments). The organisation of a CDF file is based on metadata (i.e., data describing the science data part of the file) and the science data. Metadata can be broken down into global metadata, providing identification and information about the data set as a whole, and variable metadata. While global metadata can be used to understand the data set and populate data bases, the variable metadata are defined to provide labels, dependencies, uncertainties and offsets. Variable metadata are defined to allow for later searching, plotting, merging and subsetting. The science data content of a CDF file, i.e., variables, can be organised in a variety of ways, such as scalar and multi-dimensional representation.

9. The Cluster User Interface (CUI)

The main purpose of the Cluster User Interface (CUI) is to provide the scientific community with uniform access to the CSDS data centres. The server parts of CUI are installed at all data centres except the USDC. The users will receive the clients from their data centres. For the individual data centres, the CUI offers local file handling, distribution of validated data files to other data centres, data ordering, user administration, catalogue browsing and data manipulation functions. The CUI allows the registered users to:

- browse the CSDS catalogues,
- fetch prime and/or summary data,
- manipulate and display prime and summary parameters,
- retrieve summary plot files.

These functions are available to the user, from the user account on a client machine, through the CUI session manager. shows the session manager interface. The user is offered access to the catalogue browser, simple display, data manipulation client (ISDAT), change password function and the user information.

The Catalogue Browser opens a graphical user interface (GUI) to access and browse the CSDS catalogues. Its main form is a catalogue menu from where the user can choose the catalogues to be browsed. The following catalogue browsers are available:

- PPDB/SPDB catalogue browser.
- Orbit geometry catalogue browser.
- Catalogue browser for each of the following JSOC catalogues (predicted solar cycle, predicted geometric spacecraft positions, predicted science events, predicted positions of regions of scientific interest and final scientific events).

The Simple Display function provides a Motif-based graphical user interface to perform a quick visualisation of scientific data and textual display of metadata extracted from the CDF files local to the user machine. The user will call this function to verify that the data file content meets expectations. The application enables the user to:

- access to data and metadata in the CDF files,
- display metadata in textual form,
- display one or more parameters versus time,
- display one parameter versus another parameter.

The ISDAT client is a powerful tool to display and manipulate the content of the SPDB and PPDB. The application is based on the client/server architecture. The server normally runs at the data centre but, at a later stage, will also be portable to a user site. The servers tasks are to:

- access a data base according to the user request,
- select data according to the user request,
- convert data into units, time resolution, etc. according to the user request,
- communicate the data to the client,

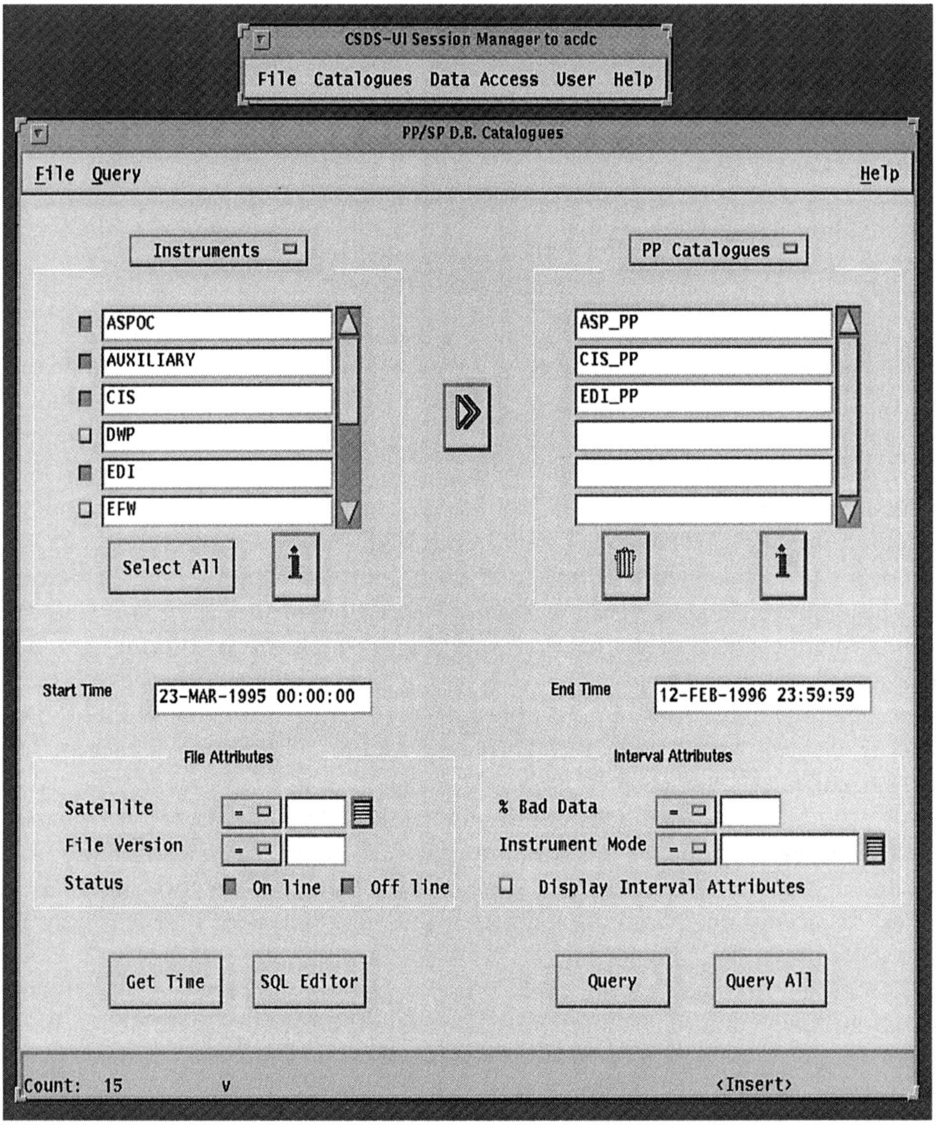

Figure 5. Session manager (above) and catalogue browser (below) of the Cluster User Interface.

– check the access rights of the user.

The ISDAT client, located at the end-user machine, is an analysis and display tool with a graphical user interface which interacts with the ISDAT server at the data centre. Three classes of client are currently implemented: (a) time managers, (b) general clients, and (c) specific clients. A feature of (a) and (b) to be mentioned here is the possibility of establishing direct links to commercial software packages used in the scientific community (e.g., IDL). ISDAT, as part of the CUI, is not intended to be a comprehensive data analysis package to satisfy all the wishes of the scientific community, but it does include the most commonly used applications for the manipulation of scientific data. The user can easily extract data and use commercial packages with more elaborate data analysis tools.

10. Summary

The Cluster community has recognised the need to distribute high quality multi-instrument, multi-spacecraft data in order to maximise the scientific return of the mission. The implementation phase of CSDS is complete and the system awaits the launch of the Cluster spacecraft which is currently planned for the morning of 15 May, 1996. Currently, emphasis is placed on the establishment of operational procedures for each data centre and for the system as a whole. The Cluster mission baseline stipulates that the final orbit is reached about three weeks after launch. Upon arrival in this orbit, all booms carrying parts of the scientific instruments will be deployed and the commissioning of the payload will be initiated. The commissioning phase of the payload will last about ten weeks, during which the instruments will be sequentially switched on and tested. Thus, science data taking only starts just under four months after launch. The first shipment of CD-ROM's, holding data from the first week of Cluster operations, will arrive at the data centres five months after launch.

In view of the sudden arrival of significant amounts of data and the fact that, after five months of elapsed mission time, the more complicated instruments might not have been fully characterised, a certain backlog in the production of SPDB and PPDB may build up. Careful planning for the initial phase of CSDS is therefore extremely important.

The production over a long time of good scientific products requires that the quality of the data products be verified on a regular basis by the Principal Investigator teams. The procedure for the formal release of data files requires that an instrument expert, e.g., the Principal Investigator, actively validates each data file and defines its quality by setting a quality flag and, optionally, explains any further applicable restrictions in the use of the data in form of caveats contained within the file metadata or via a separate caveat file. This approach is considered critical to achieve the best possible quality within the given financial constraints. It is quite obvious that a certain amount of reprocessing of data will be required. The data centres have agreed to undertake a system-wide reprocessing of all data products about three times during the mission, e.g., after six months of CSDS operations and during and after the end of the Cluster mission. The intention is that the last reprocessing activity will form the basis of the final 'Cluster archive'. Although not yet finally agreed, the final archive could be a set of CD-ROMs holding all products produced by CSDS, enhanced with high(er) resolution data and all relevant documentation.

CSDS has also been designed to aid the scientific community in the execution of data analysis campaigns in collaboration with other missions, for instance, in the IACG campaigns (Escoubet *et al.*, this issue). The exchange of data between ISTP missions and CSDS is facilitated by choosing the Common Data Format to organise the data. For this service to work well in the future, it is important that a co-ordinated upgrade philosophy be maintained between CSDS and ISTP-related data systems in case of new releases of the CDF standards.

The CSDS user privileges can be set, provided the Cluster PI's agree, to enable access to a subset of the PPDB for a user who is normally not authorised to access the PPDB. This subset can either be specified by a time period (e.g., only data gathered during a campaign) or by instrument.

Latest news: At the time of the final editing of this paper the following changes have been implemented:

(a) GCDC has been moved from Berlin to Garching. The email address of the GCDC manager in Section 4 was updated accordingly.

(b) The connectivity to HDC has been changed from X25 to frame relay.

(c) The ACDC connectivity was expanded to allow direct traffic from ACDC to all other data centres.

Acknowledgements

The successful delivery of CSDS and JSOC was only made possible by the enthusiastic support of the scientific and technical staff of the national data centres, JSOC and the representatives of the national funding authorities. All parties contributed in an extremely productive manner to the work of the CSDS Steering Committee and the CSDS Implementation Working Group. J. Credland, the ESA Cluster Project Manager, made available expert manpower for planning, schedule- and configuration control and testing. The development and implementation of CSDS User Interface and the CSDSnet was also managed under his supervision.

References

Allen, A. J., Schwartz, S. J., and Burgess, D.: Reference Document for CSDS CDF Implementation, DS-QMW-TN-0003, Jan. 1996. More general information on CDF's can be found at http://nssdc.gsfc.nasa.gov/cdf/html/docs.html.

Balogh *et al.*: 1996, 'The Cluster Magnetic Field Investigation', *Space Sci. Rev.*, this issue.

Cluster Science Data System – Announcement of Opportunity, Released by ESA Head Office Paris, France, SCI(90)/1, July 1990.

Cornilleau *et al.*: 1996, 'The Cluster Spatio-Temporal Analysis of Field Fluctuations (STAFF) Experiment', *Space Sci. Rev.*, this issue.

CSDS-UI: 1995, Scientific User's User Manual, DS-ESR-SM-0002, Issue 2, Revision 0, Issue date 29 June, 1995.

Décréau *et al.*: 1996, 'WHISPER, A Resonance Sounder and Wave Analyser: Performances and Perspectives for the Cluster Mission', *Space Sci. Rev.*, this issue.
ESA-PSS-05 (1991), ESA software engineering standards, Issue 2, ESA PSS-05-0, Issue 2.
Escoubet *et al.*: 1996, 'Cluster – Science and Mission Overview', *Space Sci. Rev.*, this issue.
Gurnett *et al.*: 1996, 'The Wide-Band Plasma Wave Investigation', *Space Sci. Rev.*, this issue.
Gustafsson *et al.*: 1996, 'The Electric Field and Wave Experiment for the Cluster Mission', *Space Sci. Rev.*, this issue.
Hapgood *et al.*: 1996, 'The Joint Science Operations Centre', *Space Sci. Rev.*, this issue.
Johnstone *et al.*: 1996, 'PEACE: A Plasma Electron and Current Experiment', *Space Sci. Rev.*, this issue.
Rème *et al.*: 1996, 'The Cluster Ion Spectrometry (CIS) Experiment', *Space Sci. Rev.*, this issue.
Riedler *et al.*: 1996, 'Active Spacecraft Potential Control', *Space Sci. Rev.*, this issue.
Paschmann *et al.*: 1996, 'The Electron Drift Instrument for Cluster', *Space Sci. Rev.*, this issue.
Robert, P. and Roux, A.: 1993, 'Influence of the shape of the Tetrahedron on the Accuracy of the Estimate of the Current Density', *Proceedings of the Conference on Spatio-Temporal Analysis for Resolving Plasma Turbulence (START)*, Aussois, France, published as ESA-WPP-047.
Sörensen *et al.*: 1996, 'The Cluster Data Processing System: A Distributed System in Support of a Challenging Scientific Mission', *Space Sci. Rev.*, this issue.
Standard Formatted Data Units – Structure and Construction Rules; CCSDS 620.0-R-1.1c, May 1992.
SFDU – Standard Formatted Data Units – A Tutorial; CCSDS 620.0-G-1, May 1992.
SFDU – Parameter Value Language Specification (ccsd0006); CCSDS 641.0-R-0.2; June 1991.
SFDU – Parameter Value Language – A Tutorial; CCSDS 641.0-6-1.0; May 1992.
von Stein, Glassmeier, K.-H., and Dunlop, M.: 1992, *A Configuration Parameter for the Cluster Satellites*, Technical Report 2/1992, Institut für Geophysik und Meteorologie der Technischen Universität Braunschweig.
Wilken *et al.*: 1996, 'RAPID, The Imaging Energetic Particle Spectrometer on Cluster', *Space Sci. Rev.*, this issue.
Woolliscroft *et al.*: 1996, 'The Digital Wave-Processing Experiment on Cluster', *Space Sci. Rev.*, this issue.

EUROPEAN NETWORK FOR THE NUMERICAL SIMULATION OF SPACE PLASMAS

G. CHANTEUR and A. ROUX
Centre d'étude des Environnements Terrestre et Planétaires CETP/CNRS, Vélizy, France

Abstract. The Cluster mission of the European Space Agency (ESA) will allow, for the first time three-dimensional measurements in key regions of the Earth's magnetosphere to be carried out. The European Numerical Simulation Network (ENSN) aims at providing a theoretical support to the mission. We describe the achievements of the ENSN during its first period of activity 1991–1994, during which the network was funded by the European Union. In particular, the ENSN has set up (i) thematic Working Groups on the prime scientific goals of the mission, (ii) a code development Working Group to develop numerical simulation codes specifically adapted to studying magnetospheric boundaries and the corresponding scale mixing, and (iii) software models of Cluster instruments to test in a numerical simulation what the set of four instruments will measure.

Key words: CLUSTER mission – Space Plasmas – Theory – Numerical Simulation

1. Overview of the Network

ESAS's Cluster mission involves four identical spacecraft coordinated in space and tailored to investigate in situ key regions of the Earth's environment, especially the boundary layers. The prime targets are: (i) the collisionless bow shock in front of the Earth's magnetosphere, (ii) the magnetopause and the exterior cusp which bound the planetary magnetic field, and (iii) the geomagnetic tail where explosive release of magnetic energy takes place. The European Network for Numerical Simulation of Space Plasmas aims at providing a theoretical support to the Cluster mission. Owing to the rapid development of supercomputers, it is now possible to carry out two- and even three-dimensional simulations of shocks, for instance, with hybrid and fully kinetic codes. Thus numerical simulations of magnetospheric boundaries with realistic conditions can now be run, permitting the spatial scales parallel and perpendicular to the boundaries to be estimated, a critical issue for the preparation of Cluster. In order to enhance the scientific collaboration with the Cluster community and to contribute to the scientific preparation of the mission, the European Numerical Simulation Network (ENSN) has identified the following tasks as essential:

(1) to set up thematic working groups to direct specific efforts on the prime objectives of the Cluster mission,

(2) to develop new numerical simulation codes with emphasis on the tools that will be necessary to investigate the physics of the boundary layers that are the prime scientific targets of Cluster,

Space Science Reviews **79:** 583–598, 1997.

(3) to share the development of the visualisation software necessary for the display of numerical simulation results on work stations,

(4) to develop software models of Cluster instruments and implement these software models into numerical simulation codes. This will facilitate (i) testing the analysis tools currently being developed within the experimental groups involved in Cluster, and (ii) optimising the distance between the Cluster spacecraft in order to improve the chances of successfully achieving the scientific aims of the mission.

The present report is organised around these four tasks and describes what has been done during the contract in the framework of the SCIENCE programme of the European Union. This contract covered the period March–April 1991 to June 1994. The list of participating groups and their main activities are described in the appendix. ENSN's activities have been approved and are supported by the Cluster Science Working Team (SWT), by the Project Scientist (R. Schmidt), and the Project Manager (J. Credland), who have been regularly informed and consulted. A wide discussion between representatives of the Cluster and Soho communities and members of the ENSN took place in Aussois, early in February 1993.

The network activities were presented during this international conference held in Aussois (1993) and a workshop held in Toulouse (1994). In both cases, the proceedings have been published by ESA and can be obtained upon request.

2. Thematic Working Groups

Even with four spacecraft, the characterisation of three-dimensional structures developing within magnetospheric boundary layers remains a formidable task. Recent progresses in code development (see Section 3) make it possible to run two and even three-dimensional simulations that can greatly help in understanding physical processes developing within magnetospheric boundary layers and in estimating the relevant spatial scales. The ENSN has set up three thematic working groups, to orient the scientific activity of the network toward the prime goals of the Cluster mission, namely:

(1) the collisionless shocks group, led by B. Lembège (CETP)

(2) the magnetopause/cusp/boundary layer group, led by M. Scholer (MPIE)

(3) the plasma sheet group, led by R. Dendy (AEA Culham).

The main results obtained by the participating groups are described below

2.1. Collisionless Shocks and Turbulence Group

Electrostatic turbulence is observed upstream of the bow shock. Numerical simulations have been carried out by Bingham *et al.* (1994) to investigate the role of the inhomogeneity in preventing the stabilisation of electron beams reflected at the shock. In order to model the development of plasma turbulence upstream of quasi-parallel shocks, one-dimensional hybrid simulations have been carried out by Scholer and Burgess (1992), who studied the interaction of ions reflected at the shock with incoming solar wind. They showed that this interaction leads to the generation of Ultra-Low Frequency (ULF) waves, that grow and steepen as time evolves. The same problem was investigated by Dubouloz and Scholer (1993) who also performed one-dimensional hybrid simulations, but took into account a very hot population of reflected ions and explained the formation of 'Short Large Amplitude Magnetic Structures' (SLAMS) with amplitudes reaching a few times the ambient magnetic field. The generation of the pulsating structure in front of the shock was also investigated, via a 2D simulation, by Dubouloz and Scholer (1995). A 2D simulation carried out with an implicit electromagnetic code has been performed to study the role of the whistler-mode waves that develop in front of the bow shock (Pantellini *et al.* 1992). Thanks to these numerical studies, we know that quasi-parallel shocks are continuously 'reformed' and that this reformation process is due to the steepening of the upstream turbulence convected toward the shock.

Two-dimensional fully kinetic simulations have also been carried out by Lembège and Savoini (1992) who studied the self-reformation process in the case of a quasi-perpendicular supercritical shock. Savoini and Lembège (1994) also studied supercritical collisionless shocks but took into account a finite angle between the shock normal and the solar wind magnetic field. In the case of an oblique shock, they were able to reproduce the change from a drifting electron population to a plateau. This numerical study helps to interpret the plateau observed in the electron distribution by spacecraft crossing the Earth's bow shock. The two-dimensional aspects were also investigated by McClements and Dendy (1993a, b), who studied the interaction of a core (solar wind) plasma flowing with respect to a ring simulating ions reflected at the shock. Particle acceleration by shocks and wave emissions in the foreshock have also been subject to cooperative studies involving experimental and numerical works. These studies are important in the context of preparing Cluster measurements.

The nature of plasma turbulence, and the identification of the plasma modes involved (Mirror, Alfvénic) has been the subject of a detailed investigation carried out on the basis of ISEE-1 and -2 results by Lacombe *et al.* (1992). The phase space signature of linear instabilities developing as a consequence of the interaction between the solar wind and the magnetosphere has been studied by Pantellini *et al.* (1994). Hybrid simulations have been used to study ion acceleration in a self-consistent way in the case of quasi-parallel shocks (Giacalone *et al.*, 1992). These results have been compared with existing theories (Giacalone *et al.*, 1993).

One important activity of the group has been the organisation of three European workshops on collisionless shocks in Paris, in March 1991, October 1992 and March 1994, respectively. All details concerning both the advances in the study of collisionless shocks and the collaboration/coordination between the various European teams can be found in the Proceedings of these European workshops. Exchanges of scientists are also worth mentioning; for instance M. Scholer (from

Garching to QMWC), F. Pantellini (from Meudon to QMWC) and N. Dubouloz (from CETP to Garching).

2.2. Magnetopause and Exterior Cusp

In view of the Cluster objectives, the Network decided to concentrate first on understanding local processes that are important for the transfer of mass, energy and momentum from the solar wind to the magnetosphere. These are magnetic reconnection, direct plasma entry and Kelvin–Helmholtz instability. Furthermore, it was decided that it is important to study the four-spacecraft aspect of Cluster in the numerical simulations of such processes. 2D MHD simulations have been performed in order to investigate the Kelvin-Helmholtz instability at the magnetopause. They have shown the inverse cascading toward long wavelengths for the surface waves excited by the instability. These simulations have also demonstrated how the inner edge of the magnetopause boundary layer gets highly corrugated by surface waves. 2D and 3D local MHD simulations have been performed in order to understand the temporal and spatial patterns of magnetopause reconnection (Scholer and Otto, 1992; Otto, 1992). Simulated time series of plasma and magnetic field data at four spatial positions have been constructed similar to those that will be obtained by the Cluster mission. These time series have subsequently been analysed by the same methods that will be applied to the real data (Otto *et al.*, 1993). This is described in Section 5. In order to understand the processes that lead to plasma entry through the magnetopause 2D hybrid simulations of impulsive penetration of plasma density enhancements from the magnetosheath through a tangential discontinuity have been performed. These simulations demonstrate the importance of ion kinetic effects for the understanding of the penetration process (Savoini *et al.* 1994).

The workshops on Magnetopause/Cusp organised by the Network in 1991 and 1992 also brought in European experimenters and data analysts from previous spacecraft missions. This was important to identify the outstanding questions and pending problems. The Network has also contributed to training and mobility. For instance, P. Savoini, a postdoctoral scientist from CETP, spent two years (14 months funded by the Network) at the Max-Planck Institute in Garching, to work on magnetopause problems.

2.3. Plasma Sheet and Substorms

The role played by single-particle dynamics in the destabilisation or reconfiguration of the magnetotail current sheet associated with substorms has been investigated using numerical orbit integration in evolving magnetic geometry. The critical parametric dependence that determines the transitions between different regimes of regular and stochastic behaviour has been identified (Chapman and Watkins, 1993; Chapman, 1994). Cluster will be able to measure the key parameters directly, so that immediate correlations will be inferred between observed substorm processes and the regimes of particle dynamics.

Excitation of ELF waves in the plasma sheet has been studied, using GEOS-1 and -2 observations to anticipate scenarios likely to be encountered by Cluster. The role of different models of the energetic ion distributions exciting the ELF waves was examined, and it was found that the GEOS-1 and -2 measurements of broadband turbulence were more likely to be driven by shell-type than by ring-type distributions (McClements and Dendy, 1993a, b). Cluster measurements of the pitch-angle distribution of energetic ions in the plasma sheet will be important in this context.

Field and ion energisation signatures of substorms have been investigated using 1D and 2D hybrid codes (Richardson and Chapman, 1994). Simulations show distinct signatures that should be observable by Cluster in-situ measurements.

Recent experimental and theoretical results raise questions as to the role of tearing modes in triggering substorms. Since the MHD approximation cannot describe the dissipation required to destabilise tearing modes, the Working Group decided to develop new (non-MHD) codes that could describe large-scale phenomena and still retain kinetic effects, at least on ions. Along this line, 2D and 3D hybrid, gyrokinetic and guiding-centre codes have been, or are currently being developed. These new tools will be very important for interpreting Cluster results. The development of 2D and 3D hybrid, gyrokinetic and guiding-centre codes is a joint undertaking involving several groups (see section on code development). The group has also contributed to training and mobility: A. Matthews worked successively in Meudon, QMWC and MPIE on 2-D and 3-D hybrid codes, and F. Mottez (postdoctoral scientist from CETP) worked for one year in Culham (supported by an ESA Fellowship) to start developing a guiding-centre code. Two workshops, one at RAL in December 1991 and one at QMWC in November 1992 were organised by the Working Group.

3. Code Development Working Group

In agreement with the terms of reference of this European programme, most laboratories of the Network took part in code development activity. The availability of new simulation codes and the improved skill of the involved teams are the most patent results of this three-year activity. The aims of the different projects have been different, hence their present status differ greatly. Briefly speaking, a first kind of developement, using known numerical schemes, targeted operational codes to do physical simulations in the short term; projects of the second kind were more exploratory and aimed at the design of new numerical schemes, especially tailored to investigating the physics of boundary layers where the large (MHD) and the small (kinetic) scales interact.

3.1. Kinetic Codes

3.1.1. *Fokker–Planck Code*

The Culham group has designed a two-dimensional and finite difference Fokker-Planck code to investigate the diffusive acceleration of electrons by resonant lower hybrid waves. The main application is the simulation of processes producing the nonthermal velocity distributions of the auroral electrons (Dendy *et al.*, 1994).

3.1.2. *Hybrid Code*

A new numerical scheme for hybrid simulations has been designed by A. Matthews at Meudon Observatory, where one- and two-dimensional implementations were made and tested (Matthews, 1994). The three-dimensional version has been completed during a long stay of A. Matthews at the Max-Planck Institute for Extra-terrestrial Physics in Garching. The ability of this code to handle efficiently multiple ion species makes it a very valuable tool to investigate the physics of collisionless shocks in the solar wind, low-frequency micro-instabilities at the magnetopause and the associated diffusion of ions through the magnetospheric frontier. This new scheme, named the Current Advance Method and Cyclic Leapfrog (CAM-CL) is distinct from other hybrid schemes due to the following features:

(1) Multiple ion species may be treated with only a single computational pass through the particle data.

(2) While the moment method advances the fluid velocity, CAM advances the ionic current density and therefore computes more efficiently the dynamics of multiple ion species.

(3) This current advance is made easier by the collection of a free streaming ionic current density.

(4) CL is a leapfrog scheme for advancing the magnetic field, adapted from the modified mid-point method; it allows sub-stepping of the magnetic field.

3.1.3. *Darwin Gyrokinetic Code*

The group at Rutherford Appleton Laboratory designed an explicit electromagnetic code, in the Darwin approximation of Maxwell's equations, making use of guiding-centre electrons and ring-shaped ions. Simulations of tearing modes and drift waves are among potential applications of this operational two-dimensional code.

3.1.4. *Guiding Centre Code*

A collaboration of three groups (CETP, Ecole Polytechnique and Culham) extended an implicit electromagnetic code designed at Ecole Polytechnique to simulate the dynamics of the guiding centres of the electrons instead of the electrons themselves. This new scheme does not require projections on the local magnetic field and hence should be more efficient than other similar codes. Additional work is needed to quench a numerical instability.

3.2. Fluid and MHD Codes

3.2.1. *Coupled Neutral Gas and MHD Code*

The team at Bochum University designed a 3D explicit and finite difference code to investigate the coupling of a neutral gas and an MHD fluid through ionisation, recombination and friction (Birk *et al.*, 1993). The code is operational to study the interaction of the ionospheric plasma with the neutral atmosphere. It should help towards understanding the relationships between space observations made by Cluster and ground-based observations.

3.2.2. *MHD in Co-Mobile Coordinates*

In order to study the radial evolution of MHD turbulence in the solar wind, a three-dimensional MHD code has been designed at Meudon Observatory (Grappin *et al.* 1993a). The use of co-mobile coordinates allows a given part of the solar plasma following its outward expansion to be simulated (Grappin *et al.* 1993b).

3.2.3. *Finite Volume Method for MHD*

In order to get rid of the constraint of more or less uniform grids imposed by finite difference methods, a few groups have recently designed numerical schemes for the integration of the MHD equations on irregular grids; one of these has been developed at CETP, in collaboration with the Laboratoire d'Analyse Numérique d'Orsay (LANOR). We were enticed to design Riemann solvers for MHD, because they proved to be very efficient for gas dynamics but, due to the great complexity of the MHD Riemann problem, they usually result in complex and slow schemes. Croisille *et al.* (1995) designed a finite volume method with a kinetic flux splitting which generalises schemes successfully applied to the Euler equations of gas dynamics. The principle of the finite volume method is the following: for any physical quantity satisfying a conservation equation, the variation of its integral over a given cell of the mesh, during a time step, is equal to the integral of the associated flux through the border of the cell during that time step. The kinetic method associates a repartition function of particles to a given macroscopic state and the flux escaping from the cell is related to the particles leaving the cell. Any segment joining two vertices of the mesh separates two cells with different states: the flux through this oriented segment is the algebraic sum of the fluxes escaping from the left and right cells, i.e. a function of these two states. This splitting enforces the positivity of the mass density without any supplementary condition. The scheme is both robust and simple and is consistent with $\operatorname{div} \mathbf{B} = 0$ (Khanfir, 1995). In order to validate this new code Khanfir *et al.* (1995) have made a series of tests including: MHD Riemann problems, linear propagation of the MHD modes in uniform plasma, rotational discontinuities, alfvénic vortices, and an MHD extension of Sedov's strong explosion. A $2\frac{1}{2}$-D implementation of this code is presently operational in cartesian coordinates; an axisymmetric version in cylindrical coordinates is under development.

4. Graphic Software Development Working Group

From the early definition of this European collaboration, graphic software was recognised as very important in helping to display simulation results. It appeared necessary to undertake jointly the development of specific programs which were not available at that time (1989–1990) as commercial products. Many programs have been designed using the Application Visualisation System (AVS). Routines were written to do real-time animation of a computed time-dependent field, but they usually require a very powerful computer to achieve anything worthwhile. The most valuable activity has been the development of various field-line integrators, written in Fortran language, combined with AVS modules to display interactively the results in three dimensions.

Rapid changes occurred with the emergence of de facto standards like Interactive Data Language (IDL) or PV-Wave and AVS, which became available on many workstations. The Network decided to rely upon these commercial products which fulfilled most of its needs. This reorientation benefitted to the development of simulation codes.

5. Software Instrument (SWI) Working Group

The concept of Cluster is based upon four spacecraft coordinated in space, which gives access to 'new' parameters (i.e., not accessible via single-point measurements). In order to take advantage of these new possibilities, new diagnostic tools have to be developed and tested (Burgess, 1994). One of the main goals of the ENSN was to provide Cluster experimenters with the possibility of testing beforehand their multi-point diagnostic tools, with the help of Software models of Cluster Instruments (SWI). The Network has also contributed to the preparation of the Cluster mission by suggesting new, or more easily tractable diagnostic tools described below.

5.1. Software models of cluster instruments (SWI)

A software package that provides simulated spacecraft data flowing through a self-consistent simulation has been developed by the group at Queen Mary and Westfield College. This software can mimic field and particle experiments and produces, as an output, time series of physical parameters at selected points (the spacecraft positions). This software is available to the Cluster community and to the Network and can be used in any type of simulation; it has been tested in the case of four spacecraft crossing a simulated bow shock (Burgess *et al.*, 1993; Giacalone *et al.*, 1994). Emphasis has been placed on the effect of the distance between the spacecraft upon the interpretation of the measurements.

The group at Sussex University has undertaken a close collaboration with the Cluster Fluxgate Magnetometer (FGM) team (PI: A. Balogh). A spectral analysis tool has been designed to determine whether the time series from a given time interval of four spacecraft data can lead to a suitable characterisation of a quasi-static moving structure (Chapman and Dunlop, 1993). A number of simulations using both electrostatic and electromagnetic particle codes have been performed to test various strategies for the detection of space- and phase-bunching wave-particle interactions in velocity space; The performance of one and several bit 'detectors' in both particle auto-correlation and wave-particle cross-correlation modes have been tested with promising results (Mouikis *et al.*, 1993). This work is highly relevant to the Cluster Wave Experiment Consortium Data-Processing Unit (WEC DPU) resident software correlator.

The group at the Swedish Institute of Space Physics (Umea group) is involved in the development of Software Instrument (SWI) for the Wave Experiment Consortium, with special emphasis on the needs of the Electric Field and Wave (EFW) experiment. The work has concentrated on the integration of the SWI with the standard data-analysing packages used for the Cluster experiment. It has been possible to integrate the SWI into the ISDAT data-handling package at a rather low level. The advantage of this is that, from the users' point of view, the SWI is treated as one of the real Cluster experiments, and also that data can be displayed graphically by the same routines as those used for the Cluster data. This saves a lot of work since the SWI output format becomes identical to that used for the real Cluster experiment. Simulation data are also more accessible, since experimenters are already familiar with the graphic interface. The integration of the SWI to ISDAT has been tested in case of a 1D numerical simulation, where time series of the density and electric field were 'measured' and processed through ISDAT to give evidence for electrostatic structures such as double layers (Oscarsson and Holmgren, 1993).

5.2. Tests of New Diagnosis Tools

5.2.1. *Spatio-Temporal Analysis of Electromagnetic Fields*

As regards the investigation of electromagnetic fields by Cluster, each spacecraft will measure two spin-plane components of **E** and **B** and the spin-axis component of **B**. The spectral density matrix is the Fourier transform of the correlation matrix of these measurements, and the field energy distribution is the trace of the former. Given that no more than four spacecraft are generating data for correlation studies, it is essential to choose algorithms that maximise the use of the available time-series. J. L. Pinçon and F. Lefeuvre (STAFF/WEC) have developed a new method, based on the generalisation of a method proposed by Capon, to estimate the spatial and temporal power spectrum from the four-point measurements of the electromagnetic field. To test this method, simulated data were generated by a 2D and a half compressible MHD code. A 'cluster' of three 'spacecraft' measured local values of **E**, **B**, and the density. Initial conditions were chosen, so that slow, fast and torsional Alfvén waves were present. Processing of the resulting simulated data recovered these modes, together with a further mode which was not present in the initial condition and difficult to identify from classical methods. These results demonstrated that the new mode arises from mode coupling between one of the torsional waves and the first harmonic of the slow wave. This numerical simulation has made it possible to test the ability of the method developed by Pinçon and Lefeuvre to identify unambiguously, from a limited set of spacecraft/probe, a nonlinear coupling of waves through independent checks of the selection rules for frequencies and wave vectors, without making any assumption about dispersion relations of the modes (Pinçon *et al.*, 1994).

5.2.2. *Test of Magnetic Reconnection*

The group in Bochum has run 3D MHD simulations with a localized resistivity, to generate signatures of localized magnetic reconnection (Otto *et al.*, 1993). In order to show that data from simulations and satellites are similar, a simple model equilibrium for the dayside magnetopause was chosen. Then the system was allowed to evolve in the presence of localized resistivity, which gives rise to reconnection. When comparing with the observational signatures of a real flux transfer event, it appeared that there was a plausible match to the corresponding traces emerging from the simulation. More systematic data analysis, for example minimum variance studies of field quantities, a search for the de Hoffmann-Teller frame of reference, test of Whalen relation, were also applied to time series extracted from 3D MHD simulations. This study has led to a number of conclusions about the influence of the averaging procedure and about the optimum duration of the data set to be used for these determinations.

5.2.3. *Estimate of Differential Quantities from Four-Point Measurements*

Several studies have been carried out (i) to estimate the accuracy of the determination of curl **B** from four distant points (at Imperial College and at CETP), and (ii) to construct geometrical quality factors indicating where, along the orbit, this estimate is correct (Technische Universitat Braunschweigh and CETP) (Robert and Roux, 1993). Two recent studies in this domain have been made by Coeur-Joly *et al.* (1995) and Robert *et al.* (1995), who (i) tested the relation between the quality of the estimate of the current density and the above-mentioned quality factors, (ii) showed that the estimated value of div**B** cannot be directly related to the uncertainty in estimating curl**B** via four-point measurements, and (iii) studied the evolution along the orbit of the uncertainty in the estimate of curl **B**. Since the geometry of the tetrahedron formed by the four Cluster spacecraft changes along the orbit, these methods should not depend upon the geometry. In the case of the determination of the current density via curl**B**, contour integrals – which are not geometry-dependent – can be used. For other differential operators (div **B**, grad **B**, etc.), another geometry-independent method must be sought. It is described below.

5.2.4. *Barycentric Coordinates*

A new, simple, and elegant method has been designed to estimate the gradient of any physical field, scalar or vectorial, measured by Cluster (Chanteur and Mottez, 1993). Making use of barycentric coordinates for any tetrahedron, the linear spatial interpolation of four simultaneous field samples provides estimators of all the components of the gradient as linear forms involving the four spacecraft in a symmetric way. Applied to the DC magnetic field data, this method gives instantaneous and independent estimates of the nine components of grad **B**. The sum of the diagonal terms is known to be equal to div **B**; a significant deviation from zero could be indicative of the breakdown of the linearity assumption of the field versus space coordinates in the tetrahedron. Any component of the mean current density **J** through the cluster is equal to the difference of two non-diagonal components of grad **B**. The current density **J** may also be estimated, via Ampère's theorem, by computing the circulation of **B** along the edges of three of the four faces of the tetrahedron. The latter method, like the former one, is based upon the linearity assumption with regard to space coordinates, hence it has the same domain of applicability. Both methods give similar results which could be wrong in regions of strong gradients even when the linearity assumption does not break down. The origin of this failure is the same for both methods. Each method computes any component of the current density as a sum of a few terms (two components of grad **B**, or the circulations of **B** along the three edges of a triangular face of the tetrahedron): in a current-free region all these terms should compensate for each other, but in fact they do not compensate exactly when the gradient is strong, which results in a false current detection. A criterion exists to eliminate these false current detections for both methods: when there is a real current through the Cluster tetrahedron, the product of the two components of grad **B** involved in a given component of **J** is negative and, similarly, the three contributions to the circulation of **B** along a given triangular loop have the same sign. This simple idea has been successfully tested in the following way: we have simulated the DC magnetic-field data recorded by a fleet of four spacecraft along their orbits, similar to a typical Cluster orbit, in a dipolar field onto which we superposed a current tube crossed by the cluster near its apogee on the night side. Therefore this simple magnetic model is exactly current-free except for the current tube with a radius equal to 2000 km and an assumed parabolic current density which produces a proper field of 10 nT at its edge. In this simple model the true current is accurately detected and the false current discarded.

5.2.5. *Surface Crossing by Cluster*

A sharp plasma transition layer encountered by a single spacecraft is usually identified by inspection of the electromagnetic field and particle data. Various techniques have been designed in the past to estimate the normal to the 'discontinuity' from the DC magnetic field records; the direction of minimum variance of **B** is generally accepted as the best normal to the crossed surface. But it has been demonstrated

(Paschmann *et al.*, 1986; Sonnerup *et al.*, 1987, 1990) that the direction of maximal variance of the convection electric field **E** provides a better determination of the plane tangent to the discontinuity. The impossibility, without a model, to disentangle spatial and temporal variations from a single spacecraft data, burdens the conclusions. A multi-spacecraft mission, like Cluster, will give new insights provided methods of data analysis taking advantage of the four-point measurements are available. In this respect, a theoretical tool has been designed at CETP (Mottez and Chanteur, 1994) to investigate the local geometry of a surface crossed by Cluster and to determine its velocity in the reference frame of the cluster. The knowledge, for each spacecraft, of the crossing time of the 'discontinuity' and the normal at the crossing point, permits the estimation of the principal curvature radii of the encountered surface by assuming that this surface does not deform or rotate during the time interval necessary for the whole cluster to cross it. The relative velocity between the cluster and the surface is simultaneously determined, the normal component being the most reliable. The method is based on the discretisation of the differential system satisfied by Frenet's vectors and makes use of Euler's formula relating locally the curvature of a normal section to the principal curvatures. Using all the available information in a symmetric way leads to an overdetermined system of linear and nonlinear equations which is solved in a least square sense.

A numerical investigation of the errors affecting the results shows, as could be expected, that they mainly result from the uncertainties concerning the normal directions at the crossing points. When the principal radii of curvature of the surface are much larger than the size of the cluster, the normals differ slightly and the component of the relative velocity along the mean normal direction is rather well estimated. The estimation of the tangential velocity is usually not reliable and may critically affect the estimation of the curvature when this velocity is parallel to a principal section of the surface. The reliability of the estimations is greatly enhanced when the tangential velocity is known.

6. Conclusion

The ENSN has been very successful in developing a fruitful collaboration between the participating groups. This involves joint scientific work, participation of several groups in the development of new codes, exchange of numerical simulation codes and exchange of senior and postdoctoral scientists. Several workshops were organised, about one each year, by the various participating groups. The interaction with the Cluster community took different forms: direct contact between some of the PI groups and ENSN groups, joint studies and meetings such as the international conference held in Aussois, where most of the PI groups and ENSN groups participated actively. Organising an effective collaboration between 10 PI groups and 10 Numerical Simulation groups was a challenge. We have not reached all our objectives, but a very significant amount of important work has been done, as acknowledged during the Cluster-oriented theory workshop held in Toulouse in November 1995. It is also encouraging to see that several groups in Oslo, Orleans, Berlin, and Braunschweig, have expressed their wish to participate in the Network. Given this willingness, we have submitted a new proposal to the European community. The hardware phase being now almost completed, it is essential to focus on the methods that we need to develop to take full advantage of the spatial resolution of the four Cluster spacecraft. To take this need into account, the Network has been opened to groups that have a broad experience in the multi-point analysis of turbulence.

Acknowledgements

The European Network for Numerical Simulation of Space Plasmas has been funded by the European Union in the framework of the SCIENCE Programme (contract number SCI-0468M). The authors acknowledge the financial support of the European Space Agency who granted the Network two contracts (one to Culham and one to QMWC) and one Fellowship (F. Mottez spent a post-doctoral year at Culham Laboratory). French groups acknowledge CCVR and IDRIS Supercomputer Centres for computing resources. The authors are indebted to their colleagues of the different teams participating in this Network for their contributions.

Appendix. List of Participating Groups with a Brief Description of Their Contributions

1. Centre d'étude des Environnements Terrestre et Planétaires (CETP)
   a) ensured the overall coordination of the Network, as well as the coordination of the working groups on shocks, code development, software instrument and graphics.
   b) developed numerical models to optimise Cluster orbits and to test new diagnostic tools for multi-point measurements.
   c) developed a new MHD code based on a finite volume method

   Contact person: Gérard Chanteur, e-mail gerard.chanteur@cetp.ipsl.fr
2. Centre de Physique Theorique (CPT)
   a) made available to the Network its expertise in implicit methods (saving computing time). Codes developed at QMWC, DESPA and CETP have benefitted from this expertise.

   Contact person: Jean-Claude Adam, e-mail adam@orphee.polytechnique.fr
3. Swedish Institute of Space Physics (SISP)
   a) developed analysis tools for space projects (ISDAT).
   b) tested these analysis tools via numerical simulation thanks to the interaction with the Network, in particular with the Bochum and QMWC groups.

Contact person: Tord Oscarsson, e-mail tord.oscarsson@physics.umu.se

4. Max-Planck Institut fur Extraterrestrische Physik (MPIE)
   a) was responsible for the thematic group on magnetopause and cusp.
   b) has developed a strong collaboration with the groups at QMWC and CETP (on shocks) and the Bochum group (on 3D simulations).
   c) has upgraded to 3D the 2D hybrid code developed by DESPA. by DESPA, Meudon.

   Contact person: Manfred Scholer, e-mail mbs@mpe-garching.mpg.de
5. Queen Mary and Westfield College (QMWC)
   a) has developed an interface between (Cluster) particle experiments and numerical simulation codes.
   b) concluded joint studies, in particular with MPIE, on shocks.

   Contact person: David Burgess, e-mail d.burgess@qmw.ac.uk
6. Rutherford Appleton Laboratory (RAL)
   a) has provided expertise in developing codes, in particular hybrid and gyrokinetic.

   Contact person: Robert Bingham, e-mail rbi@vk.rl.ac.uk
7. Royal Institute of Technology (RIT), Stockholm
   a) has investigated possible use of Connection Machines by the Network.

   Contact person: Michael Raadu, e-mail raadu@plasma.kth.se
8. Sussex University (Sussex)
   a) interaction with the Cluster FGM team (Fluxgate Magnetometer).
   b) simulation of stochastic processes in the tail.
   c) the subcontractor at Culham Laboratory has provided expertise on code development and coordinated the Working Group on tail and substorms.

   Contact person: Sandra Chapman, e-mail s.c.chapman@warwick.ac.uk
9. DESPA (Meudon Observatory)
   a) has developed a new hybrid simulation code, presently used by QMWC, MPIE and CETP.

   Contact person: Andre Mangeney, e-mail mangeney@megasg.obspm.fr
10. Bochum University (Bochum)
    a) has developed 3D MHD codes with appropriate ionospheric conditions.
    b) has tested the analysis tools used for analysing boundary crossings via 3D MHD simulations.

References

Bingham, R., Nairn, C. M. C., Shapiro, V. D., and Hall, D. S.: 1994, 'Generation of Upstream Waves by Inhomogeneous Electron Streams', *Phys. Scripta* **50**, 60.

Birk, G. T., Otto, A., and Ziegler, H. J.: 1993, in A. Roux, F. Lefeuvre, and D. LeQueau (eds.), 'A 3D Three-Fluid Code to Study the Magnetosphere-Ionosphere-Thermosphere System', *Spatio-Temporal Analysis for Resolving Plasma Turbulence*, ESA Proceedings WPP-047, ESA Publications Division, Noordwijk, p. 357.

Burgess, D.: 1994, in N. Baker, V. O. Papitashvili, and M. Teague (eds.), 'Mission-Oriented Modelling in Europe', *Solar-Terrestrial Energy Program, COSPAR Colloquia Series*, Vol. 5, Pergamon Press, London, p. 723.

Burgess, D., Giacalone, J., Kucharek, H., and Scholer, M.: 1993, in A. Roux, F. Lefeuvre, and D. LeQueau (eds.), 'Using Software Particle Detectors in Self-Consistent Simulations of Quasi-Parallel Shocks', *Spatio-Temporal Analysis for Resolving Plasma Turbulence*, ESA Proceedings WPP-047, ESA Publications Division, Noordwijk, p. 311.

Chanteur, G. and Mottez, F.: 1993, in A. Roux, F. Lefeuvre, and D. LeQueau (eds.), 'Geometrical Tools for Cluster Data Analysis', *Spatio-Temporal Analysis for Resolving Plasma Turbulence*, ESA Proceedings WPP-047, ESA Publications Division, Noordwijk, p. 341.

Chapman, S. C.: 1994, 'Properties of Single Particle Dynamics in a Parabolic Magnetic Reversal With General Time Dependence', *J. Geophys. Res.* **99**, 5977.

Chapman, S. C. and Dunlop, M. W.: 1993, 'Some Consequences of the Shift Theorem for Multispacecraft Measurements', *Geophys. Res. Letters* **20**, 2023.

Chapman, S. C. and Watkins, B. W.: 1993, 'Parametrization of Chaotic Particle Dynamics in a Simple Time-Dependant Field Reversal', *J. Geophys. Res.* **98**, 165.

Coeur-Joly, O., Robert, P., Chanteur, G., and Roux, A.: 1995, in C. P. Escoubet and R. Schmidt (eds.), 'Simulated Daily Summaries of Cluster Four-Point Magnetic Field Measurements', *Physical Measurements and Mission-Oriented Theory*, ESA Proceedings SP-371, ESA Publications Division, Noordwijk, p. 223.

Croisille, J. P., Khanfir, R., and Chanteur, G.: 1995, 'Numerical Simulation of the MHD Equations by a Kinetic-Type Method', *J. Sci. Comput.* **10**, 81.

Dendy, R. O., Harvey, B. M., O'Brien, M., and Bingham, R.: 1994, 'Fokker-Planck Modelling of Auroral Wave-Particle Interactions', *J. Geophys. Res.*, to appear.

Dubouloz, N. and Scholer, M.: 1993, 'On the Origin of the Short Large-Amplitude Magnetic Structures Upstream of Quasi-Parallel Collisionless Shocks', *Geophys. Res. Letters* **20**, 547.

Dubouloz, N. and Scholer, M.: 1995, 'Two-Dimensional Simulations of Magnetic Pulsations Upstream of the Earth's Bow Shock', *J. Geophys. Res.* **100**, 9461.

Giacalone, J., Burgess, D., and Schwartz, S. J.: 1992, 'Hybrid Simulations of Protons Strongly Accelerated by a Parallel Collisionless Shock', *Geophys. Res. Letters* **19**, 433.

Giacalone, J., Burgess, D., Schwartz, S. J., and Ellison, D. C.: 1993, 'Ion Injection and Acceleration at Parallel Shocks: Comparisons of Self-Consistent Plasma Simulations With Existing Theories', *Astrophys. J.* **402**, 550.

Giacalone, J., Schwartz, S. J., and Burgess, D.: 1994, 'Artificial Spacecraft in Hybrid Simulations of the Quasi-Parallel Earth's Bow Shock: Analysis of Time Series Versus Spatial Profiles and a Separation Strategy for Cluster', *Ann. Geophys.* **12**, 591.

Grappin, R., Mangeney, A., and Velli, M.: 1993a, in A. Roux, F. Lefeuvre, and D. LeQueau (eds.), 'MHD Simulations of Solar Wind Turbulence in Co-Mobile Coordinates', *Spatio-Temporal Analysis for Resolving Plasma Turbulence*, ESA Proceedings WPP-047, ESA Publications Division, Noordwijk, p. 325.

Grappin, R., Velli, M., and Mangeney, A.: 1993b, 'Nonlinear Wave Evolution in the Expanding Solar Wind', *Phys. Rev. Letters* **70**, 2190.

Khanfir, R.: 1995, PhD Thesis, Université Paris XI, Orsay.

Khanfir, R., Chanteur, G., and Croisille, J. P.: 1995, 'A Method of Finite Volumes for Ideal MHD: Algorithm and Tests', *Comp. Phys. Comm.*, to appear.

Lacombe, C., Pantellini, F. G. E., Hubert, D., Harvey, C. C., Mangeney, A., Belmont, G., and Russell, C. T.: 1992, 'Mirror and Alfvénic Waves Observed by ISEE-1and -2 During Crossings of the Earth Bow Shock', *Ann. Geophys.* **10**, 772.

Lembège, B. and Savoini, P.: 1992, 'Non-Stationarity and Stationarity of a 2-D Quasi-Perpendicular Supercritical Collisionless Shock by Self-Reformation', *Phys. Fluids* **4**, 3533.

Matthews, A. P.: 1994, 'Current Advance Method and Cyclic Leapfrog for 2D Multispecies Hybrid Simulations', *J. Comp. Phys.* **112**, 102.

McClements, K. G. and Dendy, R. O.: 1993a, 'Ion Cyclotron Harmonic Wave Generation by Ring Protons in Space Plasmas', *J. Geophys. Res.* **98**, 11689.

McClements, K. G. and Dendy, R. O.: 1993b, 'Ion Cyclotron Wave Emission at the Quasi-Perpendicular Bow Shock', *J. Geophys. Res.* **98**, 15531.

Mottez, F. and Chanteur, G.: 1994, 'Surface Crossing by a Group of Satellites: a Theoretical Study', *J. Geophys. Res.* **99**, 13499.

Mouikis, C. G., Christiansen, P. J., Chapman, S. C., and Watkins, N. W.: 1993, in A. Roux, F. Lefeuvre, and D. LeQueau (eds.), 'The University of Sussex Wave-Particle Correlator: Computer Simulation of Instrument Response', *Spatio-Temporal Analysis for Resolving Plasma Turbulence*, ESA Proceedings WPP-047, ESA Publications Division, Noordwijk, p. 409.

Oscarsson, T. and Holmgren, G.: 1993, in A. Roux, F. Lefeuvre, and D. LeQueau (eds.), 'A Software Instrument for the Cluster Electric Fields and Waves Experiment', *Spatio-Temporal Analysis for Resolving Plasma Turbulence*, ESA Proceedings WPP-047, ESA Publications Division, Noordwijk, p. 377.

Otto, A.: 1992, 'Three-Dimensional Magnetohydrodynamic Simulations of Processes at the Earth's Magnetopause', *Geophys. Astrophys. Fluid Dynamics* **62**, 69.

Otto, A., Ziegler, H. J., and Birk, G. T.: 1993, in A. Roux, F. Lefeuvre, and D. LeQueau (eds.), 'Plasma and Magnetic Signatures Generated by Three-Dimensional MHD Simulations', *Spatio-Temporal Analysis for Resolving Plasma Turbulence*, ESA Proceedings WPP-047, ESA Publications Division, Noordwijk, p. 337.

Pantellini, F. G. E., Burgess, D., and Schwartz, S. J.: 1994, 'Phase-Space Evolution in Linear Instabilities', *Phys. Plasmas* **1** 3784.

Pantellini, F. G. E., Heron, A., Adam, J. C., and Mangeney, A.: 1992, 'The Role of the Whistler Precursor During the Cyclic Reformation of a Quasi-Parallel Shock Wave' *J. Geophys. Res.* **97**, 1303.

Paschmann, G., Papamastorakis, I., Baumjohann, W., Sckopke, N., Carlson, C. W., Sonnerup, B. U. Ö, and Lühr, H.: 1986, 'The Magnetopause for Large Magnetic Shear: AMPTE/IRM Observations', *J. Geophys. Res.* **91**, 11099.

Pinçon, J. L., Lefeuvre, F., and Chanteur, G.: 1994, in N. Baker, V. O. Papitashvili, and M. Teague (eds.), 'Multipoint Analysis of a Simulated Bi-Dimensional MHD Turbulence: Application to the Cluster Mission', *Solar-Terrestrial Energy Program, COSPAR Colloquia Series*, Vol. 5, Pergamon Press, London, p. 785.

Richardson, A. and Chapman, S. C.: 1994, '1-D Hybrid Code Simulations of the Self-Consistent Evolution of a Dipolarizing Field Line', *J. Geophys. Res.* **99**, 17391.

Robert, P. and Roux, A.: 1993, in A. Roux, F. Lefeuvre, and D. LeQueau (eds.), 'Influence of the Shape of the Tetrahedron on the Accuracy of the Estimate of the Current Density', *Spatio-Temporal Analysis for Resolving Plasma Turbulence*, ESA Proceedings WPP-047, ESA Publications Division, Noordwijk, p. 289.

Robert, P., Roux, A., and Coeur-Joly, O.: 1995, C. P. Escoubet and R. Schmidt (eds.), 'Validity of the Estimate of the Current Density Along Cluster Orbit With Simulated Magnetic Data', *Physical Measurements and Mission-Oriented Theory*, ESA Proceedings SP-371, ESA Publications Division, Noordwijk, p. 229.

Savoini, P. and Lembège, B.: 1994, 'Electron Dynamics in 2-D and 1-D Oblique Supercritical Collisionless Shocks', *J. Geophys. Res.* **99**, 6609.

Savoini, P., Scholer, M., and Fujimoto, M.: 1994, 'Two-Dimensional Hybrid Simulations of Impulsive Plasma Penetration Through a Tangential Discontinuity', *J. Geophys. Res.* **99**, 19377.

Scholer, M. and Burgess, D.: 1992, 'The Role of Upstream Waves in Supercritical Quasi-Parallel Shock Reformation', *J. Geophys. Res.* **97**, 8319.

Scholer, M. and Otto, A.: 1991, 'Magnetotail Reconnection, Current Diversion and Field-Aligned Currents', Geophys. Res. Letters **18**, 733.

Sonnerup, B. U. Ö., Papamastorakis, I., Paschmann, G., and Lühr, H.: 1987, 'Magnetopause Properties From AMPTE/IRM Observations of the Convection Electric Field: Method Development', *J. Geophys. Res.* **92**, 12137.

Sonnerup, B. U. Ö., Papamastorakis, I., Paschmann, G., and Lühr, H.: 1990, 'The Magnetopause for Large Magnetic Shear: Analysis of Convection Electric Fields From AMPTE/IRM', *J. Geophys. Res.* **95**, 10541.

OPPORTUNITIES FOR MAGNETOSPHERIC RESEARCH WITH COORDINATED CLUSTER AND GROUND-BASED OBSERVATIONS

H. J. OPGENOORTH
Swedish Institute of Space Physics, Uppsala Division, S-79551 Uppsala, Sweden; Visiting Professor, the Finnish Meteorological Institute, Geophysical Research Division, Box 503, FIN-00101 Helsinki, Finland

M. LOCKWOOD
Rutherford Appleton Laboratory, Chilton, Didcot, OX11 OQX, U.K.

Abstract. ESA's first multi-satellite mission Cluster is unique in its concept of 4 satellites orbiting in controlled formations. This will give an unprecedented opportunity to study structure and dynamics of the magnetosphere. In this paper we discuss ways in which ground-based remote-sensing observations of the ionosphere can be used to support the multipoint *in-situ* satellite measurements. There are a very large number of potentially useful configurations between the satellites and any one ground-based observatory; however, the number of ideal occurrences for any one configuration is low. Many of the ground-based instruments cannot operate continuously and Cluster will take data only for a part of each orbit, depending on how much high-resolution ('burst-mode') data are acquired. In addition, there are a great many instrument modes and the formation, size and shape of the cluster of the four satellites to consider.

These circumstances create a clear and pressing need for careful planning to ensure that the scientific return from Cluster is maximised by additional coordinated ground-based observations. For this reason, the European Space Agency (ESA) established a working group to coordinate the observations on the ground with Cluster. We will give a number of examples how the combined spacecraft and ground-based observations can address outstanding questions in magnetospheric physics. An online computer tool has been prepared to allow for the planning of conjunctions and advantageous constellations between the Cluster spacecraft and individual or combined ground-based systems. During the mission a ground-based database containing index and summary data will help to identify interesting datasets and allow to select intervals for coordinated studies. We illustrate the philosophy of our approach, using a few important examples of the many possible configurations between the satellite and the ground-based instruments.

1. Introduction

Long before satellites were even considered as feasible diagnostic tools for near-Earth space physics, networks of ground-based instruments have been utilized to gain an initial understanding of our closest space environment. Being remote-sensing observations, they are of lower resolution than *in-situ* data, and can at times be more difficult to interprete. Nevertheless, many of the still-discussed processes within the solar wind - magnetosphere - ionosphere system were discovered and initially understood with the help of distributed networks of ground-based instrumentation. Two of the most outstanding discoveries (predictions) of truely magnetospheric processes from the ground were the claim for the existence of

Space Science Reviews **79:** 599–637, 1997.

field-aligned currents by Birkeland (1913), and the deduction of the magnetopause currents by Chapman and Ferraro (1931). Both of these were based on ground-based magnetometer recordings. Also the existence of the ionosphere, the basic morphology of the polar magnetic and auroral substorms, and other features of magnetospheric energy storage and release have been clarifed by ground-based observations, long before *in-situ* satellite measurements allowed us to understand the details of the underlying physical mechanisms in the magnetosphere.

Much of the progress in space science was made due to a well balanced parallel development of ground-based and satellite-borne instrumentation; series of individual discoveries cross-fertilized both methodological approaches. An early example of this was the discovery of the cusp/cleft region of magnetosheath-like particle precipitation, now one of the main objectives of te Cluster mission. A region of dayside auroral luminosity, dominated by red line (630.0 nm) emissions, was first reported by Sandford (1964) using ground-based optical observations. Eather and Mende (1971) used the ratio of the emission intensities at different wavelengths to show that this was caused by relatively soft electron precipitation. In the same year, Heikkila and Winningham (1971) and then Frank (1971) reported *in-situ* satellite observations of magnetosheath-like plasma precipitation in the magnetosphere, the association with the red-dominant aurora being first made by Heikkila (1972). In the recent past many new results have been achieved by planned coordinated observations both from the ground and in space during the same event or within the same spatial structure (see below for many more examples of such types of studies).

Most of the classical ground-based observational techniques are still actively in use today, however, spatial coverage, temporal resolution, instrument sensitivity and accessibility of data have been improved in phase with the technical revolution of space-borne instrumentation. Dense networks of magnetometers, standard and imaging riometers, digital ionosondes and optical cameras and photometers are operated in key regions of the northern hemisphere (with extreme concentrations in Fenno-Scandinavia, Canada, Alaska, and Greenland) and all over the Antarctic continent. All instruments in these networks provide data almost continuously. Optical instruments, e.g., can reveal transient events and track evolving boundaries, and the other network instruments have important applications, including monitoring the latitude of the auroral oval as well as the extent and intensity of disturbances along it (see examples below).

Modern radar technology has opened a wide and exciting field of sophisticated remote sensing measurements of the ionosphere. Radar systems operating at HF and VHF frequencies are sensitive to auroral backscatter from ionospheric irregularities, which drift under the influence of magnetospheric convection electric fields. Bi-static multibeam or scanning coherent radar systems can provide vital 2-dimensional snapshots of the convective flow in the ionospheric F and E region (Hanuise *et al.*, 1993). The SuperDARN network of HF radars, which is now to large part constructed and under operation (Greenwald *et al.*, 1995), can image such flow patterns over a very large fraction of the high-latitude region in the northern hemisphere. Also in the southern hemisphere, a tri-static HF system covering most of Antarctica is being built. The combined system will be unique in its possibility for conjugate studies and will provide an ideal monitor for the dynamical development of magnetospheric assymmetries, during different states of solar wind coupling. In Figure 1 (from Greenwald *et al.*, 1995) we show what spatial coverage of coherent radar systems can be expected for the years of the Cluster mission.

In terms of the number of geophysical parameters measured, the most powerful of the ground-based observatories are the incoherent scatter (IS) radars, which, for example, can be used to measure ion drifts (electric fields), ion and electron temperatures and plasma density throughout all ionospheric layers. With models and complex processing, these radars indirectly yield much more information, including conductivities, neutral winds and precipitating electron spectra. By the time Cluster data-taking commences in 1996, the EISCAT Svalbard Radar, ESR (Cowley *et al.*, 1990) will be in operation on the island of Spitsbergen and this will add to the existing high-latitude IS facilities at Söndre Strömfjord, Millstone Hill, and EISCAT.

The range and flexibility of these radars allows detailed measurements to be made which will be valuable complements to the Cluster observations. However, while most of the earlier mentioned instrument networks operate more or less on a permanent basis (data availibility will only be limited by instrument failure, cloud coverage, and excitation level of ionospheric instabilities, respectively), incoherent radars have limited hours of operation. IS radars require much maintenance, and hardware degradation and power consumption are too expensive to allow operation on a continuous basis. The complexity of the radar systems and in addition the complicated transmission and reception schemes require real-time operator monitoring and supervision of both hardware and software. Furthermore, IS radars usually require large antenae, which are generally steered mechanically. This limits the speed with which the beam can scan across the ionosphere, and calls for careful experiment design to balance the requirements of spatial coverage and time resolution.

Therefore detailed planning is required to ensure that the best opportunities for combined studies with Cluster are exploited. To start with it will be important to match the radar operating modes to the satellite observations, such that the radars genuinely add to the information that the satellites obtain. For example, in order to achieve the right balance between spatial coverage and temporal resolution, antenna scanning patterns appropriate to different conjunctions between the spacecraft and the radars will have to be designed. Similarly, the right balance between spatial and temporal resolution will need to be struck by the pulse coding scheme. For example, for most applications two major antenna pointing geometries of the combined EISCAT UHF, VHF, and ESR radar systems appear to be sufficient and effective. When Cluster crosses field lines connected to the

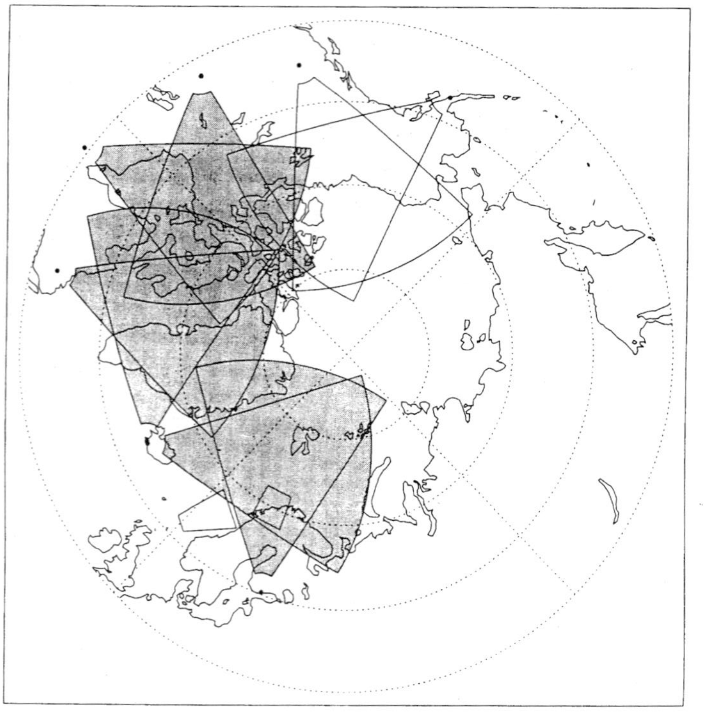

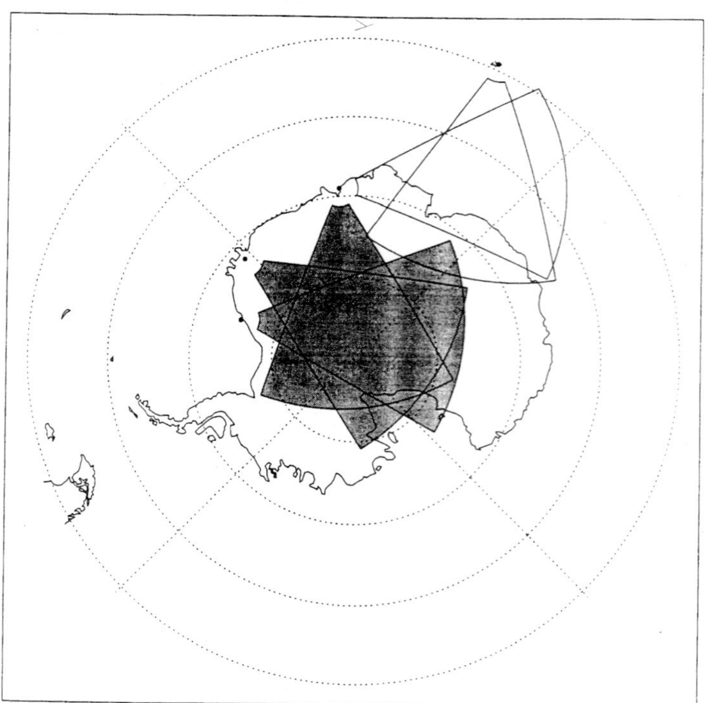

Figure 1. Overview of the operative (shaded) and planned (unshaded) fields-of-view of the SuperDARN network of bistatic HF-radar systems, in the northern and southern polar regions (left and right panel, respectively). In the northern hemisphere even the VHF systems STARE and SABRE will contribute to the SuperDARN coverage (see sketches of their smaller fields-of-view in Scandinavia, left panel; from Greenwald *et al.*, 1995).

cusps or other high latitude regions, a field-aligned pointing direction of the ESR radar, and a north-looking EISCAT VHF dual beam experiment combined with a field-aligned UHF pointing direction (as sketched in Figure 2(a)) will allow us to monitor many ionospheric plasma parameters directly under the Cluster satellite and at the same time monitor the temporal and spatial development of convection flow and particle precipitation in a region between the very high latitude polar cap and the auroral zone, equatorward of the region of key interest. When Cluster crosses field lines connected to the auroral zone above the radars, a field-aligned UHF, vertical VHF, and south-looking ESR beam-swing experiment (as depicted in Figure 2(b)) will allow us to monitor precipitation associated events within the auroral zone, and at the same time monitor the region poleward of the auroral zone (e.g., the open/closed field-line boundary) with the help of the ESR radar. A second antenna on Svalbard, which might become available during the later part of the Cluster mission, would be extremely beneficial for this particular experiment mode, allowing us to exchange the ESR beamswing experiment in Figure 2(b) by a permanent dual-beam southlooking mode (see below for a more detailed discussion of the advantages of such a mode over simple beamswinging experiments). Two antennas on Svalbard would also add interesting possibilities to the Cusp experiment in Figure 2(a)). The EISCAT Scientific Advisory Committee has decided that the combined EISCAT/ESR modes shown in Figure 2 will be the basic modes employed during periods of special interest in relation to the Cluster observations. They have been defined for the first year of coordinated observations and may eventually be upgraded or revised as first experience is gained and evaluated.

In this report we do not wish to review the many capabilities and specialities of all above mentioned instruments, this will instead be done in a source-book for Cluster/Ground-based (CGB) Coordination, which is in press as an ESA Special Publication (SP). Rather we wish to review here in more general sense what kind of studies can be conveyed with combined Cluster and ground-based observations, and what kind of observations we wish to coordinate in order to maximize the scientific outcome of the Cluster mission. Finally we will also describe how the existence of interesting data can be monitored with the help of a ground-based database, containing new indices and general summary data.

As mentioned earlier, ground-based observations of the magnetosphere-ionosphere-thermosphere system are obviously a form of remote sensing. As such they are generally less precise than *in-situ* satellite observations. What therefore do they provide which will enhance the scientific return of the Cluster mission? There are a number of important answers to this question. Firstly, ground-based observatories are spread around the globe and so can be used to simultaneously sense widely-separated regions of the magnetosphere system. Indeed, because the magnetic field lines converge with decreasing altitude, a single ground-based station can monitor a very extensive region of the magnetosphere. However, there are more subtle advantages to ground-based studies. In recent years, the study of transient events and rapid temporal changes in the magnetosphere-ionosphere system has been one

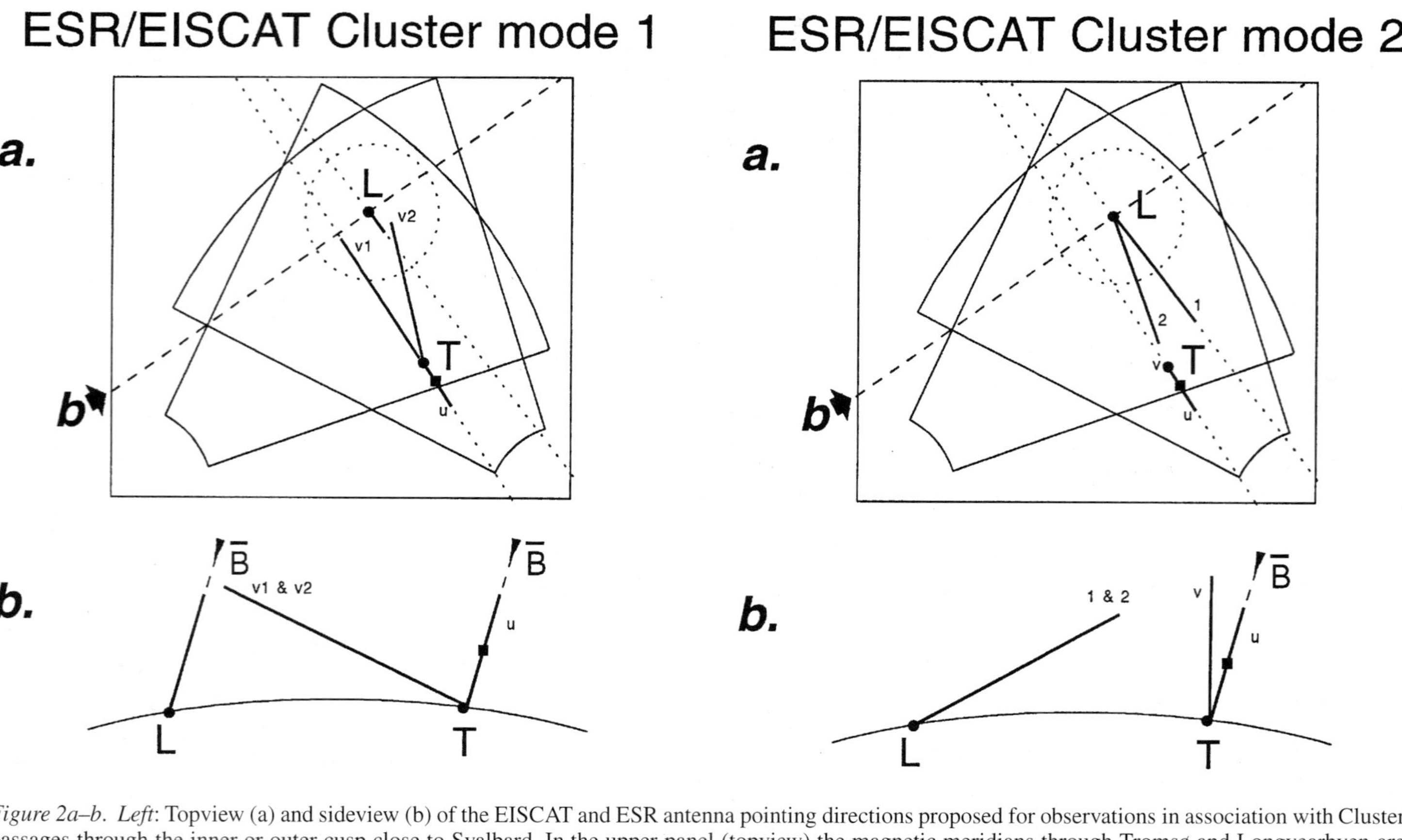

Figure 2a–b. Left: Topview (a) and sideview (b) of the EISCAT and ESR antenna pointing directions proposed for observations in association with Cluster passages through the inner or outer cusp close to Svalbard. In the upper panel (topview) the magnetic meridians through Tromsø and Longyearbyen are indicated by dotted lines. The dotted circle marks the field-of- view of an All-Sky-Camera at Longyearbyen, for 630,0 nm emissions at 250 km altitude. The broken line indicates the viewing direction for the sideview in the bottom panel. We also indicate the combined field-of-view of the European pair of SuperDARN radars, CUTLASS. The filled square marks the altitude, resp. location, of the UHF tristatic electric field measurement. *Right*: as the left but for Cluster observations within the nightside auroral oval and magnetotail regions.

of the most active areas of solar-terrestrial physics research. Ground-based remote-sensing observations have a vital role to play in these studies because they are unique in covering a range of invariant latitudes, at high time resolution and for an extended period of time. *In-situ* satellite observations, on the other hand, provide much higher resolution data but suffer from spatial-temporal ambiguity and from limited spatial coverage. For both remote sensing and *in-situ* measurements, there is a trade-off to be made between spatial coverage, time resolution and the length of the continuous data sequence in any one region of the coupled magnetosphere-ionosphere system. For *in-situ* observations this compromise is set by the orbital dynamics of the satellite; for spatially-integrating ground-based instruments (like magnetometers) it is set by the rotation period of the Earth; but for instruments like radars and imaging riometers, with multiple or steerable beams, this choice can be varied within broad limits set by the rotation of the Earth and the scanning capabilities of the instrument.

Not only are there problems of distinguishing spatial structure from temporal changes in data from a lone spacecraft, but also we cannot determine the motion nor the orientation of observed structures and boundaries. These problems will be addressed in three dimensions for the first time by the four Cluster spacecraft, flying in known but variable configurations. However, they will only answer questions on certain temporal and spatial scales, depending on their separation and altitude (and hence velocity). Ground-based observations can be used in a number of ways to provide important support for Cluster data and greatly enhance the mission's scientific return.

From a survey of the literature Lockwood and Opgenoorth (1995) have defined four classes of scientific investigations, where simultaneous satellite and ground-based observations will be advantageous over single instrument data. We do not attempt to review all such measurements in this paper, but give selected examples to illustrate the classes of application and to look at their particular potential for combined Cluster and ground-based observations.

1.1. Resolution of Spatial and Temporal Variations

The ground-based data can be used to extend the range of time scales of temporal variations which can be studied and can also be used to interpolate between data taken at different times by different Cluster craft at a given point in space. Recent examples of this kind of application (with lone satellites) have included studies of precipitation of magnetosheath-like plasma in what we now know to be transient events, called travelling convection vortices (TCVs), as detected by conjugate arrays of ground-based magnetometers and radars (Potemra *et al.*, 1992; Heikkila *et al.*, 1989). A second example of such an application is the resolution of spatial and temporal variations of the magnetopause reconnection rate (which give cusp ion 'steps' in satellite data) by using simultaneous incoherent scatter observations (Lockwood *et al.*, 1993a; Lockwood, 1995a). A related study by Pinnock *et al.*

(1993) showed that the region of cusp precipitation, as seen by a low-altitude satellite, was co-incident with a longitudinal flow channel seen by an HF backscatter radar: longitudinal flows were also detected by the satellite, but only the radar could resolve that this flow channel was elongated and that it was one of a sequence of transient flow events. Both transient longitudinal flow channels and cusp ion steps are predicted ionospheric signatures of magnetopause reconnection bursts (i.e., flux transfer events or FTEs): the flow channels are expected when the magnitude of the dawn/dusk component of the magnetosheath field is large, the cusp ion steps will be more common when it is small. Such predictions for these, and other, transient events will be ideally tested by combined ground-based observations of the cusp ionosphere while Cluster is at the dayside magnetopause or crossing the dayside auroral oval.

1.2. Placing Satellite Observations in Context

We can also use ground-based observations to place the Cluster observations in context, in both time and space. For example, ground-based data can be used to define boundaries (e.g., convection reversal boundaries, the auroral electrojets, the locations of arcs, the zero potential contour between flow cells), which can give information about which regions of the magnetosphere-ionosphere system the spacecraft were in. Using ground-based instrumentation, sequences of magnetospheric changes can be monitored and certain Cluster observations can be put into the right context of the development stage of the disturbance, as e.g., substorms. It will also be possible to monitor the overall disturbance state of the magnetosphere for individual limited periods of Cluster data.

An example of placing a spatial structure in context of the larger-scale spatial distribution is the recent work on the dayside field-aligned currents and magnetosheath-like plasma precipitations by de la Beaujardère *et al.* (1993). They used radar observations of the convection pattern to resolve an ambiguity of which convection cell a satellite passed through. Likewise, ground-based data can determine when a feature was seen by a satellite in a sequence of events. This is particularly important for studies of the evolution of the magnetosphere-ionosphere system during substorms. Opgenoorth *et al.* (1989) employed ground-based data to investigate the evolution of a westward-travelling surge and showed that the satellite data were within the surge head which had recently ceased moving. Pellinen *et al.* (1992) used ground-based data with auroral images from satellites to show that the recovery phase is much more complex than a simple global return to quiet conditions. Even from multi-satellite data both observations would not have been clearly attributable to the corresponding substorm phases.

Using ground-based data to place satellite measurements in a sequence of events has sometimes produced results which appear to conflict with the conclusions of other studies, which place them in a certain region of the magnetosphere-ionosphere system. There is much to be gained from resolving such conflicting evidence. For example, a major question in recent substorm research has been when and where substorm onset is located and, a related question, when and where the open lobe flux built up in the growth phase is destroyed by tail reconnection. McPherron *et al.* (1993) used ground-based observations of substorm onset to show that tail lobe field strengths begin to decay at onset, implying that enhanced tail reconnection causes onset and that the poleward expansion of the aurora is due to the closure of open flux. On the other hand, Lopez *et al.* (1993) compared particle and field data from the tail plasma sheet with observations by ground magnetometers and auroral imagers and have provided evidence that the tailward expansion of activity in the near-Earth tail is related to the poleward expansion of the aurora, implying that onset is located Earthward of, and occurs before significant closure of open flux by tail reconnection. It is clear that the resolution of these conflicting observations will require combinations of ground-based and satellite data (see review by Lockwood, 1995b). It may eventually help to resolve the general debate between the 'classical near-Earth neutral line' model and the 'Kiruna conjecture', which included a simultaneous or initial disruption of the cross tail current at 8 to 10 R_E in the substorm onset mechanism (see Kennel *et al.*, 1992). The orbit and separation of the Cluster spacecraft will be ideal to study the causal sequence of substorm features in the near Earth magnetotail, as it allows to determine the direction of propagation of various disturbances in the magnetospheric plasma. This may finally resolve the temporal sequence of the processes involved and clarify the location of the initial disturbance onset.

1.3. Providing Ionospheric Boundary Conditions to Studies of Magnetosphere-Ionosphere Coupling

The ionosphere is not just a passive mirror of magnetospheric processes, but an active part of a genuinely coupled system. In modelling the magnetospheric observations, it is vital to know the prevailing boundary conditions in the ionosphere. In particular, ionospheric conductivities are of importance and can be derived from altitude profiles from incoherent scatter radars or by comparison of electric and magnetic field values (e.g., Kirkwood *et al.*, 1988; Buchert *et al.*, 1988; Brekke *et al.*, 1989; Kirkwood, 1994). Observed conductivities can be used in a wide variety of ways to add crucial information to a number of studies. These include: studying Alfvén wave reflection at the ionosphere, for example in TCV events (Glassmeier, 1992); testing theories of magnetosphere-ionosphere interaction, for example in substorms (Kan, 1993) and, in particular, using numerical models (Hesse and Birn, 1991a); calculating inductive time-constants for non-steady convection (Sanchez *et al.*, 1991); estimating Ohmic heat dissipation (Foster *et al.*, 1983; Heelis and Coley, 1988; Weiss *et al.*, 1992) and deriving snapshots of the convection pattern by magnetometer inversion techniques (Richmond, 1992; Knipp *et al.*, 1993).

1.4. Quantitative Estimates from Combined Data

Ground-based data can also be used with satellite data to gain information which cannot be obtained from either on their own. Obvious examples of this type of application would include the recognition of structures and sequences of events such that the mapping of convection-dispersed particle populations, waves, magnetic and electric fields from the magnetosphere to the ionosphere is revealed (for example, (Elphic *et al.*, 1990)). However, there are other less obvious applications: Lockwood *et al.* (1993a) have recently used a combination of satellite and radar data to compute the distance from the magnetopause reconnection site to the satellite. This measurement is not possible from either of the two data sets in isolation. Another example is the comparison of electron spectra seen at a satellite with that inferred at low altitudes on the same field line by an IS radar, giving evidence for field-aligned particle acceleration at heights between the two (Kirkwood *et al.*, 1989; Kirkwood and Eliasson, 1990). At low altitudes the consecutive passage of four satellites through virtually one and the same region of space will help to reveal spatial and temporal variations of field-aligned acceleration processes.

2. Conjunctions and Constellations

In order to plan coordinated observations using Cluster and ground-based facilities, the European Space Agency ESA established a working group (Opgenoorth, 1993) for which the authors act as chairman (HJO) and the representative for incoherent scatter facilities (ML). The working group has met several times and organised workshops in Orleans, France, in March 1994, and Rome, Italy, in April 1995. As an initial basis for planning coordinated observations, the working group has followed ESA's Cluster Science Plan by classifying the orbits, such that apogee falls into one of four magnetic local time (MLT) sectors, namely 6 hours around 0, 6, 12, and 18 MLT (i.e. the midnight, dawn, noon and dusk sectors). We also consider the ground-based station or meridian chain of stations to be simultaneously with Cluster in one of the same four MLT sectors, which divides the possibilities into a total of 16 combinations. For each of the 16 there are a number of points on the Cluster orbit near which coordinated observations with a certain ground observatory are of special scientific interest. Thus far, we have defined 67 such conjunctions and configurations. Note that in this paper, we refer to 'configurations' between any one ground-based observatory and the group of four Cluster spacecraft: this should not be confused with the configurations of the four craft, relative to each other, which is variable and an important and complex part of the operations planning for the Cluster mission (Rodriguez-Canabal *et al.*, 1993). The numbers in Figure 3 refer to those configurations or conjunctions which we have identified for periods when the Cluster orbit plane is close to the noon-midnight (GSE XZ) plane, whereas those in Figure 4 are when the orbit plane is closer to the dawn-dusk (GSE YZ) plane.

Figure 3 views the Earth and the Cluster orbit (thick line) from dusk and the small arrow shows the location of a considered key ground-based observatory or local network of stations. The thin lines show a typical magnetopause location, along with geomagnetic field lines which thread the dayside low-latitude boundary layer (LLBL), the high latitude boundary layer (HLBL) or mantle and the tail neutral sheet. Figure 4 views the Earth and the Cluster orbit from the Sun and the thin lines show a typical magnetopause and field lines which pass through the low-latitude boundary layer on the dawn and dusk flanks of the magnetosphere.

To understand what is meant here by a configuration, consider the segment of the orbit marked 1 in the top left part of Figure 3. For this configuration, the satellites are near apogee in the central current sheet of the tail, while the ground-based station in question makes observations of the midnight sector auroral oval. This is an example of a near-conjugate configuration. However, we also consider many non-conjugate configurations to be important. Configuration 2, on the same plot, is one such case, allowing ground-based observations of the development of the substorm aurora and electrojets in the midnight sector while Cluster makes simultaneous observations in the tail lobe. In Figures 3 and 4 we label configurations where the ground-based observatory and Cluster are in opposite hemispheres with an asterix. In many of these cases, much of the same science can be addressed as when the two are in the same hemisphere; however, the interpretation of such data is often likely to be more difficult and, unless there are specific reasons to the contrary, the opposite-hemisphere configurations are considered to be of lower priority. However, we note that in cases where the satellite and radar data can be considered to be of similar type and quality, we may sometimes be able to use opposite-hemisphere observations to test for conjugate and non-conjugate phenomena (e.g., Greenwald *et al.*, 1990; Rodger *et al.*, 1994b).

A comprehensive list of potential scientific objectives for each numbered conjunction or configuration in Figure 3 and 4 is presented in the paper by Lockwood and Opgenoorth (1995) and the reader is referred to that paper or, better still, to an on-line interactive implementation of Figures 3 and 4 which can be acessed via World-Wide Web on Internet, with a graphics-handling browser such as Mosaic or Netscape (the relevant URL is http://www.gbdc.rl.ac.uk/). One of the main purposes of this list is to provoke thought about the potential uses of a ground-based observatory when in a certain configuration with the Cluster spacecraft. Configuration 1, for example, is when Cluster is near apogee and near the tail neutral sheet, while the ground-based observatory is near midnight, configuration 2 has Cluster in the tail lobe (on a pass with apogee near midnight), with the ground-based observatory near midnight. Note that these configurations will not occur in the order 1, 2, 3, ..., for any one observatory because as the satellites move along the orbit, the ground-based station rotates with the Earth. Note also that one configuration for one observatory will simultaneously be a different number configuration for another station. For example, the EISCAT and Söndre Strömfjord incoherent scatter radars are separated by approximately 6 hours of MLT. Thus,

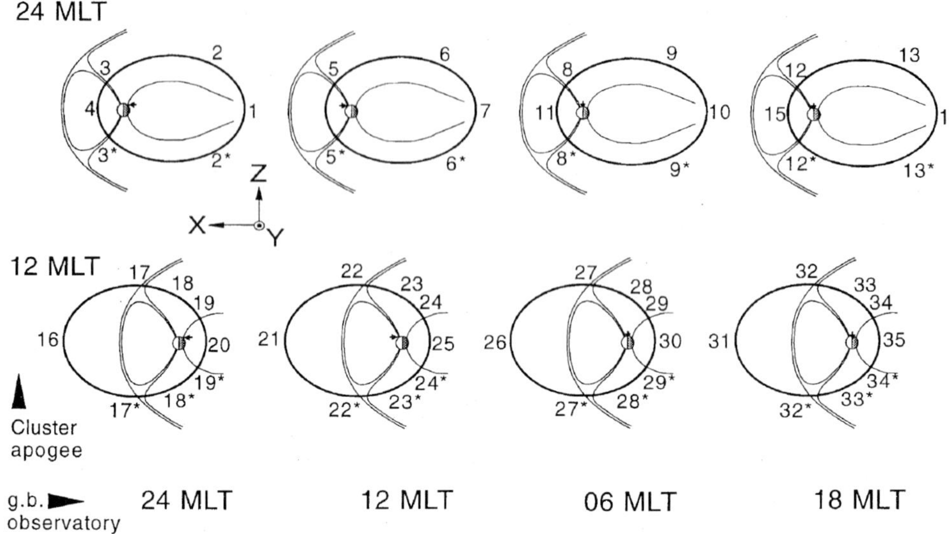

Figure 3. Cluster orbits for when the orbit plane is close to the noon-midnight (GSE XZ) plane. The Earth and the Cluster orbit (thick line) are viewed from dusk and the small arrows show the location of a ground observatory (in this case EISCAT/ESR). The thin lines show a typical magnetopause location along with geomagnetic field lines which thread the dayside low-latitude boundary layer (LLBL), the high latitude boundary layer (HLBL or mantle) and the plasma sheet boundary layer. The numbers refer to satellite locations for configurations/conjunctions with the ground station which we have identified to be of particular scientific interest (see text). The upper row of four figures are all for satellite apogee in the midnight sector and the lower row are for apogee in the noon sector. The vertical columns are for the sector in which the ground-based observatory is situated (i.e., from left to right at midnight, noon, dawn and dusk), when the satellite is at the numbered location. Note that because the ground observatory rotates as the satellite moves along the orbit, the numbered configurations occur in a complex sequence. Configurations where the ground station and Cluster are in opposite hemispheres with an asterix (from Lockwood and Opgenoorth, 1995).

for example configuration 5 (with EISCAT in the noon auroral oval and Cluster flying through the mid-altitude cusp) would simultaneously be a configuration 8 for Söndre Strömfjord, which would then be in the dawn auroral oval. It is important to consider all the possible useful configurations between any one station and Cluster, because the number of ideal configurations for each individual station is limited. Furthermore, the separation of the 4 Cluster craft will be different in one year from when in the same location one year later. This means that particular configurations, which were considered useful during the first year of Cluster observations, will not exactly reoccur during the second year.

3. Examples of Scientific Problems to be Adressed for a Few Direct Conjunctions

The permutations of science topics and satellite-ground configurations identified in the Cluster/Ground-Based (CGB) Online Planning Tool are far too numerous to

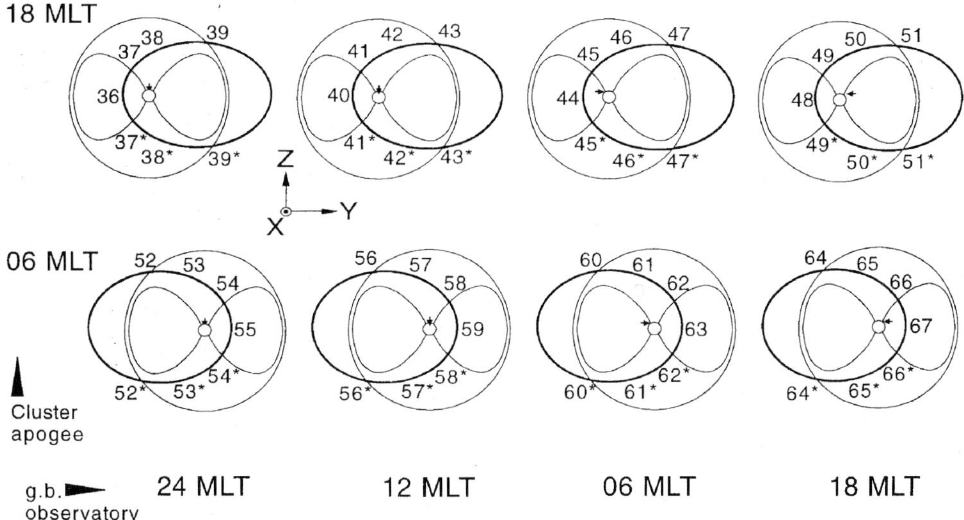

Figure 4. Corresponding plots to Figure 3 for when the orbit plane is close to the dawn-dusk (GSE YZ) plane, so that satellite apogee is in the dusk sector (upper panel) or the dawn sector (lower panel). The Earth and the Cluster orbit (thick line) are viewed from the Sun and the thin lines show a typical magnetopause and field lines which pass through the low-latitude boundary layer on the dawn and dusk flanks of the magnetosphere (from Lockwood and Opgenoorth, 1995).

discuss here in detail. However, to illustrate the choice of scientific objectives and the priorities, we will here shortly illustrate the potential possibilities for mainly two cases, giving outlooks to closely related conjunctions. We chose configuration 1, one of the most important of many novel possibilities for substorm studies, and configuration 5, which will give extremely exciting new possibilities for studies of the dayside boundary layers and cusp. These two cases exemplify the kind of arguments and thinking we have used in the compilation of the topical priority lists displayed in the planning tool at the CGB database.

3.1. Studies of Nightside Magnetospheric Disturbances when both Cluster Apogee and a Concentration of Ground-Based Instruments are close to Magnetic Midnight (Configuration 1)

Configuration 1, with Cluster in the nightside magnetotail conjugated with a concentrated ground-based network area such as Fenno-Scandinavia, North-America, or Greenland is ideal for studies of the substorm cycle with its many, still unsolved problems.

Even during times of extreme magnetic quiescence, often associated with northward IMF conditions, the magnetosphere does not remain undisturbed. Kamide *et al.* (1975, 1977) and Lui *et al.* (1976) have shown that substorm-like magnetic activity can occur, even on a highly contracted auroral oval. To study such phenomena we require observations of the contracted oval, by stations and radars with a

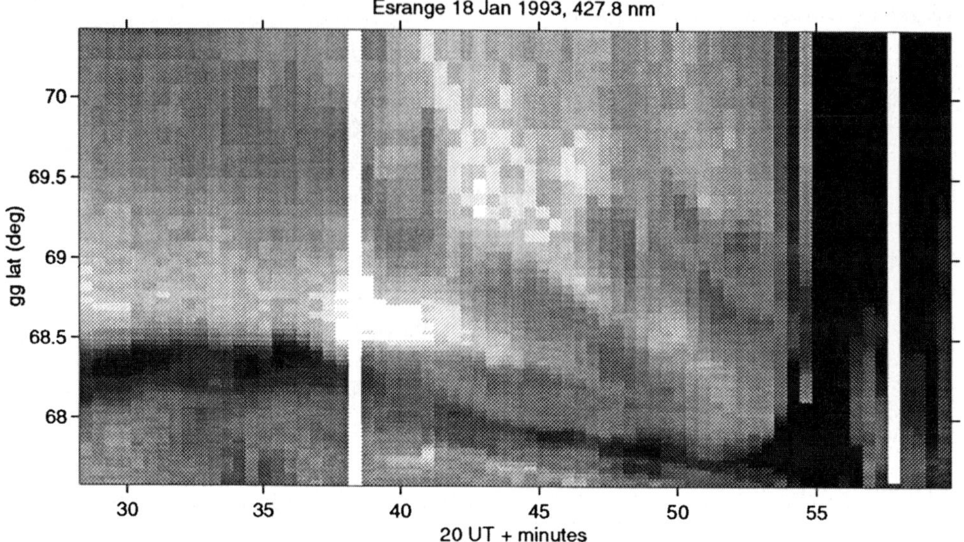

Figure 5. Measurements versus latitude and time of the 427.8 nm auroral emission with a scanning photometer at Esrange, Sweden. The grey scale is logarithmic, with white and black corresponding to intensities below 700 R and above 7000 R, respectively. The two white vertical bars are due to backgrond measurements. Grey structures correspond to southward drifting auroral arcs, and the black structure after 2052 UT to the northward expansion of an auroral break-up (from Persson *et al.*, 1994b).

field-of-view at very high-latitudes, along with simultaneous Cluster measurements in the tail lobe or plasma sheet. Since disturbances during quiet times are by definition localized and small in amplitude, the multi-spacecraft concept of the Cluster mission will improve our chances to (a) observe and (b) identify such events. Other features of the relatively quiet magnetosphere during northward IMF are the so-called 'Theta' auroras and Sun-aligned arcs (Murphree and Cogger, 1981; Frank *et al.*, 1982; Murphree *et al.*, 1989) and auroral structures within the polar cap, such as in the 'teardrop' or 'horse-collar' aurora (Hones *et al.*, 1989). Detailed studies of such very high latitude phenomena will be possible with the new ESR and SuperDARN radars as well as Söndre Strömfjord and EISCAT. Understanding the possible underlying magnetospheric processes and topology will be considerably improved by multisatellite observations in the magnetospheric tail lobes and plasma sheet.

The early development of a substorm is often characterized by equatorward-drifting auroral arcs. They are seen in the late growth phase and early expansion phase, poleward of where onset will later occur, respectively has already occurred. A typical example of such multiple southward drifting arcs prior to the poleward substorm expansion is shown in Figure 5. This data is from a latitudinally scanning photometer and was earlier published by Aikio *et al.* (1996) and Persson *et al.* (1994b). The dark striations in the figure correspond to multiple southward drifting arcs in the substorm growth phase, which starts at about 20:30 UT. The equatorward drift motion of these arcs was observed to be unaffected by a substorm onset at lower latitudes, starting from the equatormost auroral arc close to the isotropic trapping boundary. Only after several minutes the arcs were engulfed by the expanding substorm aurora. This observation supports the concept of substorm onset occurring in the near-Earth central plasma sheet, such that the boundary plasma sheet (BPS) and plasma sheet boundary layer (PSBL) remain unaffected at least for the first ten minutes of the expansion phase after onset (see also Persson *et al.* (1994a, b) and Gazey *et al.*, 1995). With a single satellite it is hard to identify such structures, but they are clearly seen in auroral images and IS radar data and the combination of such ground-based facilities and the four Cluster satellites will be invaluable for the identification of the magnetospheric sources of these arcs and their associated field-aligned current systems (Fukunishi *et al.*, 1993).

The mechanism that leads to the onset of a magnetospheric substorm is still the prime unsolved question of magnetospheric research. Recently, the scenario of the so-called 'Kiruna Conjecture' of substorms (Kennel, 1992) has increasingly gained acceptance. This is because evidence has accumulated that substorm onset usually takes place around $X = -8\ R_E$, which is relatively close to the Earth in the central tail current sheet (see Lui, 1991, and references therein). In contrast to such an inferred near-Earth location of onset, signatures of reconnection (specifically the direction of accelerated plasma flows and the polarity of the magnetic field across the tail current sheet) place the near-Earth neutral line (NENL), which is a competing mechanism for the initial substorm onset, somewhere beyond $X = -20\ R_E$ (Baumjohann *et al.*, 1991; Baumjohann, 1993). This conclusion is very important, because it may indicate that the observed disruption of the cross-tail current at substorm onset is not co-located with the NENL but happens considerably Earthward of it. Important questions will then have to be raised as to whether the NENL gives rise to the current disruption, or *vice versa* (see review by Lockwood, 1995b). From recent modelling studies (Hesse and Birn, 1991b) it appears as if the near-Earth neutral line formation would be responsible for the production of vorticity and flux pile-up in the plasmasheet Earthward of it, which could as a secondary effect lead to the formation of a current wedge there.

In a very recent study Nakamura *et al.* (1994) have used data from the IMP 6, 7, and 8 satellites to study the characteristics plasma flow and magnetic field characteristics close to the neutral sheet. They found strong evidence (fast tailward streaming plasma flows, and negative B_z) for near-Earth neutral line formation in association with substorms. Because of the orbital characteristics of the used satellites they could identify the probable region for NENL formation to lie between 18 and 24 R_E, which is closer than earlier studies based on Ampte/IRM would place it (see above).

These recent results are very promising for the Cluster mission. With an apogee at 19.5 R_E Cluster will in configuration 1 have a reasonable chance to pass through the likely region of NENL formation, and should at the same time, according to the Kiruna conjecture, be able to see signatures of the substorm current disruption expanding tailward. It will be of crucial importance for the understanding of the causality of substorm processes to have four-satellite observations in this region at the right moment of substorm development. On the ground the IS and HF radars, surrounding magnetometer arrays, and other ground-based instrumentation such as optical and riometer imagers, will allow us to monitor the onset and spreading of the current disruption, thereby identifying when such Cluster observations are being made. It is crucial to understand when in the evolution of the substorm (as seen from the ground) do the signatures of tail reconnection (as seen by Cluster) commence and how do they relate in time and space to the near-Earth current disruption features. In the later development of the substorm we also need to investigate when the NENL starts to reconnect open lobe flux, thereby detaching the plasmoid from the Earth (Moldwin and Hughes, 1992; Slavin *et al.*, 1992). Note that by the time that this pinching off takes place, the plasmoid may well be already moving down the tail (Owen and Slavin, 1992). Opgenoorth *et al.* (1994) have recently shown that there are clear auroral intensifications even in the later phases of substorm development (in the recovery phase), and these may be associated with the relatively late detachment of a plasmoid. As a characteristic sign of such enhanced recovery phase precipitation they found a strong, possibly purely adiabatic acceleration to energies between 20 and 30 keV. The spatial dynamics of near-Earth recovery of the magnetotail are still completely unknown, but could be studied if Cluster is at the right time at the right place. Again only ground-based data can help to identify those intervals of Cluster data, which would be valuable for a study on this particular topic.

Ground-based data often show a substorm onset, which subsequently fails to develop into a full substorm (Lui *et al.*, 1976; Untiedt *et al.*, 1978; Koskinen, 1992). At other times substorms fail to develop at all over a prolonged perid of time, despite favourable IMF condititons (Yahnin *et al.*, 1994; Sergeev *et al.*, 1994). During multiple substorm intensifications the appearance and strength of the disturbance does not always agree in the ionosphere and the near-Earth space (Yeoman *et al.*, 1994; Grande *et al.*, 1992, 1994). To understand the relative energy release during real substorms, incomplete substorms (pseudo-breakups) or multiple-onset substorms it will be important to detect and track the development of the open/closed field-line boundary with the multiple Cluster spacecraft, and appropriate ground-based instrumentations such as incoherent scatter radars.

3.2. Studies of Dayside Boundary Layers, Cusp and Cleft Using Configuration 5

The previous sub-section considered measurements by Cluster near apogee on the nightside of the Earth. By way of contrast, we here will concentrate on the special aspects of mid-altitude Cluster measurements, on dayside auroral field-lines. Specifically, we look at configuration 5 from Figure 3, to illustrate the

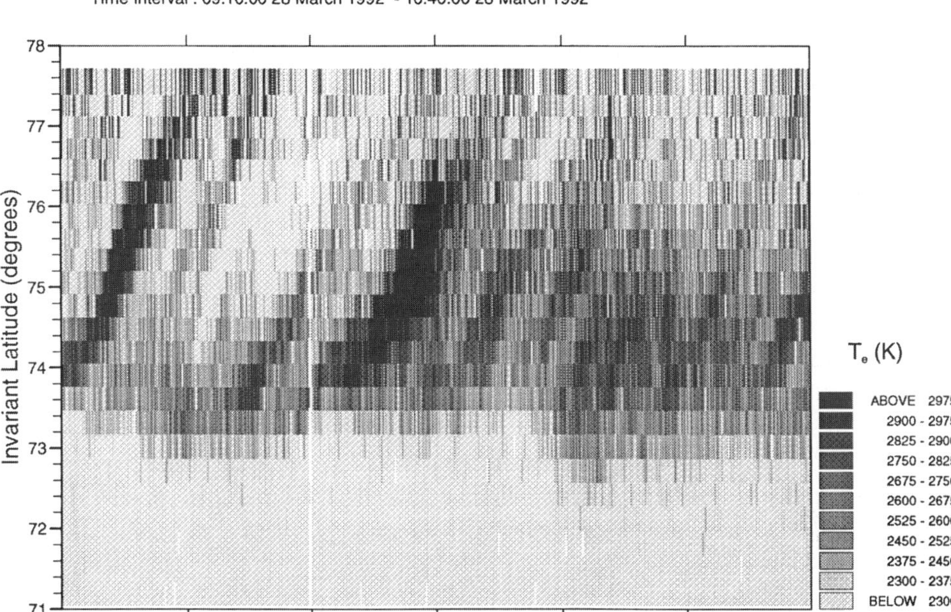

Figure 6. Measurements versus latitude and time of the F-region electron temperature with a north-looking EISCAT UHF experiment at low elevation. Dark structures correspond to northward moving regions of high electron temperature (from Lockwood *et al.*, 1993a).

benefits of the Cluster four-satellite concept, when not in tetrahedron configuration, for coordinated measurements with ground-based instruments. Close to perigee the satellite configuration will be more or less linear along the Cluster orbit (a 'string of pearls'), which means that the satellites will consecutively pass through almost the same regions of space. Depending on the original, near-apogee tetrahedron size, the time span for repetitive sounding of any one region will vary between several to several tens of minutes, which may sometimes allow us to monitor the full temporal development of dayside transient events. Ground-based optical instruments in the dayside cusp/cleft region have revealed at least two classes of transient events. The first are poleward moving events, which either break away from the background cusp/cleft aurora, or may even collectively make up that background (e.g., Sandholt *et al.*, 1985). These are often associated with longitudinal flow channels (Lockwood *et al.*, 1989) and cusp ion steps (Lockwood *et al.*, 1993a). These contrast with the second class of optical events which are associated with travelling convection vortices (TCV's) (Friis-Christensen *et al.*, 1988; Glassmeier *et al.*, 1989). These move longitudinally around the the oval, away from noon and at latitudes just equatorward of the background cusp/cleft (Jacobsen *et al.*, 1991; Heikkila *et al.*, 1989; Potemra *et al.*, 1992; Lühr *et al.*, 1995).

Much recent interest has focused on how and where reconnection takes place at the dayside magnetopause (see reviews by Lockwood, 1995a; Crooker and Toffeletto, 1995). Lockwood and Smith (1994) have recently made predictions of the cusp ion dispersion signatures, as would be seen by the Cluster craft during mid-altitude cusp crossings, when the rate of reconnection at the dayside magnetopause is pulsed. The predicted poleward migration of the cusp ion steps (caused by the periods of low reconnection rate between pulses) could be detected by comparing the ion data from the different Cluster spacecraft. The theory predicts this motion to be associated with poleward-moving ionospheric electron temperature enhancements, 630 nm (red-line) auroral transients and transient bursts of longitudinal flow. These have already been detected by EISCAT and optical instruments on Svalbard (Lockwood *et al.*, 1993a; Sandholt *et al.*, 1990). Figure 7 shows examples of poleward-moving events seen by the EISCAT UHF radar, as reported by Lockwood *et al.* (1993a). The plot shows electron temperature enhancements in the F-region, consistent with the effects of soft, magnetosheath-like electron precipitation.

Figure 7 is an example of a common form of data presentatiom for incoherent scatter radars, showing a measured parameter (in this case the electron temperature) as a function of time (x axis) and range from the radar (y axis). The plot reveals that the enhancements were propagating away from the radar. This demonstrates the kind of ambiguity that can arise in such radar data, because a fixed beam was used which slants at low elevations through the ionosphere. Thus the observed features could have resulted either from poleward-moving latitudinal structures, or from rising thin altitudinal structures (layers). Fortunately, data were also simultaneously recorded by the VHF radar, pointing at a different elevation and these showed the events were indeed poleward-moving latitudinal structures. Gaining this information could have been achieved with a single radar, but only by scanning the beam and consequently reducing the time resolution to twice the duration of each scan cycle (the Nyqvist limit). Depending on the scan speed of the radar, this could well have caused a failure to detect the poleward-moving events. The data shown are continuous measurements at 10-s resolution, but an analysis of the effects of antenna scanning have shown that the region of enhanced temperatures would have appeared continuous if a scan were used, even if the cycle time was a short as 4 min. The example serves to illustrate how carefully an experiment must be designed if a certain feature (in this case the poleward-moving events) is to be detected. In addition, it illustrates the importance of information on, and understanding of, the radar mode when looking at such data presentations.

The data shown in Figure 7 are very similar in form to keograms of optical emissions from the cusp region, which also show such poleward-moving events when the IMF is southward. Indeed, there are good theoretical reasons to expect the electron temperature enhancements to be associated with red-line (630 nm)

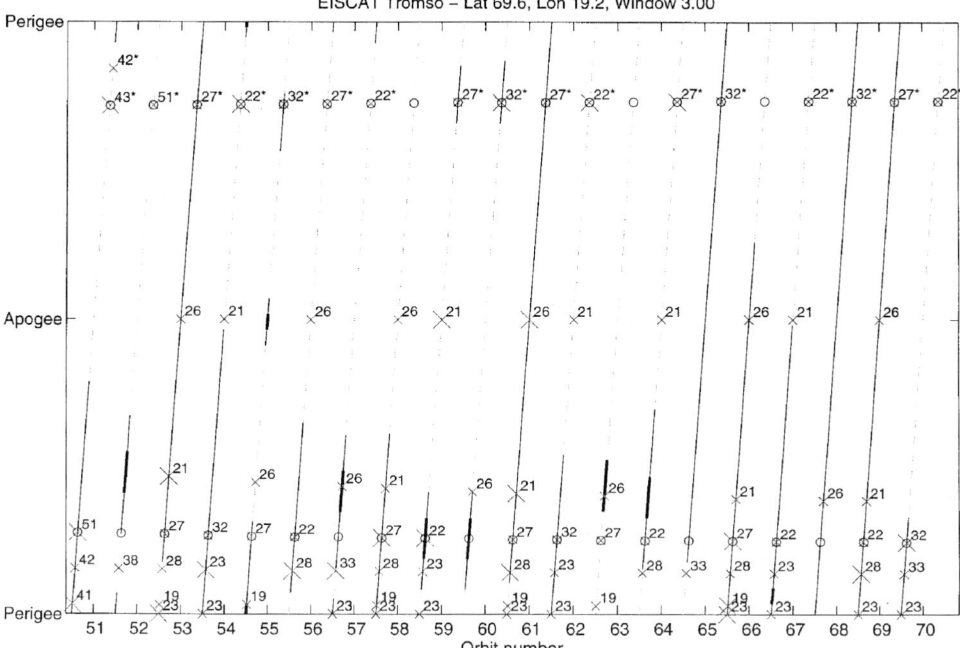

Figure 7. Example of a so called 'Bryant-Plot' used for identification of coinciding Cluster and ground-based measurements. Such plots, indicating the periods of Cluster data taking along a sequence of orbits and the times and types of conjunctions with a user selected ground-based instrument, can be created with a WWW tool at the GBDC. The full lines indicate Normal Mode Cluster data taking, the heavier lines Burst Mode periods. The crosses indicate conjunctions or configurations (according to the numbers given in Figures 3 and 4) with the chosen station, and the size of the crosses roughly indicates the quality of the conjunction (see text for more details).

dominant emissions. A key question about the red-line auroral light is the altitude profile of the emissions, as this influences our estimates of the size of the events, as derived from all-sky images (Lockwood *et al.*, 1993b). The emission profiles of 630 nm light are determined by the altitude profiles of ionospheric electron density and temperature, both of which are enhanced by the precipitating magnetosheath plasma in the cusp region. In addition, the transient patches of 630 nm emission are associated with small regions of dominant 557.7 nm (green-line) emission. These were shown, in one case at least, to be coincident with the upward field-aligned current of the oppositely directed matched pair which transmit the longitudinal motion into the ionosphere (Sandholt *et al.*, 1990; Lockwood *et al.*, 1993b). However, the causes of the required electron acceleration are not known. The new ESR IS radar will be ideal for studying these processes and the emission profiles because Svalbard offers optical observations of the cusp in darkness, as well as field-aligned radar measurements of ionospheric parameters. In particular, if reconnection pulses are confirmed to be the origin of poleward-moving events in the radar data, it becomes crucial to measure their area because this gives an estimate of the total flux opened by each reconnection pulse, and hence the contribution to the average transpolar voltage (see Lockwood *et al.*, 1993b).

Thus the combination of the EISCAT, ESR and Cutlass radars with optical observations will be and ideal complement for Cluster electron and ion precipitation observations for configuration 5. From these combined observations we should be able to determine where the reconnection takes place, how the reconnection rate varies and what conditions prevail at the reconnection site (see Lockwood, 1995b; Lockwood *et al.*, 1994).

However, a major complication in this area of research has been that transient flows, aurorae and field-aligned currents are also key features of travelling convection vortices (TCV's) (Friis-Christensen *et al.*, 1988; Glassmeier *et al.*, 1989). These are thought to result from solar wind pressure pulses but a variety of different mechanisms have been proposed (Kivelson and Southwood, 1991; Lysak and Lee, 1992). Thus the origin, as well as the propagation and lifetime, of TCVs is still not known. In addition, they appear to be associated with soft precipitation equatorward of the background cusp/cleft (Heikkila *et al.* 1989; Potemra *et al.*, 1992; Jacobsen *et al.*, 1991; Lühr *et al.*, 1995) which is not predicted by the current theories of their generation. For these studies it might be beneficial to have configurations between Cluster and ground-based radars which are not exactly conjugated, but rather offset by some longitudinal difference of $\leq 90°$ (see, e.g., configurations 8, 12, 27, 32 in Figure 3). In these configurations Cluster might monitor the TCV formation process while ground-based stations could follow the drift along the late morning or early afternoon auroral oval. The large scale magnetometer networks and semiglobal coverage of the SuperDARN radars will add important additional information.

As on the nightside, magnetic mapping is uncertain in the cusp/cleft region, and in addition is likely to be highly dependent on the amount of open flux threading the dayside magnetopause (Crooker et al., 1991; Crooker and Tofelletto, 1995). Induction effects mean that the voltage pulses (i.e., flux transfer events) in the magnetopause are decoupled from the ionosphere where they may cause only smoothed poleward flow unless the magnetosheath B_y component is large [see review by Lockwood, 1995a). Comparisons between radar flow observations and Cluster data, when in close conjunction in the cusp/cleft region will help answer the vexed questions of how both magnetopause magnetic and electric fields map into the ionosphere. Much attention has been given to the cusp when the IMF is southward and relatively little to its behaviour when it is northward, which can often be complex (e.g., Weiss *et al.*, 1995). Configuration 5 would be valuable for studying how the northward IMF cusp relates to transpolar arcs and sunward convection in the lobe.

Lastly, the cusp/cleft region is known to be a major source of ionospheric plasma for the polar magnetosphere in the cleft ion fountain (Lockwood *et al.*, 1985). The IS radars and digital ionosondes could be used to detect the upflows in the cusp ionosphere while the Cluster spacecraft observe them in the dayside auroral oval and their dispersion by convection into the near-Earth lobe. Thus the combined Cluster-ground-based data can yield information about the location and causes of the cleft ion fountain.

4. An Operations Scenario

At the time of writing, the exact Cluster orbit is unknown and hence also the UT at which the spacecraft are in any one location. As it is this UT which determines the location of a ground-based facility, this information is vital for planning coordinated measurements with any one ground-based observatory. Consequently, detailed plans on when favourable configurations or conjunctions will occur cannot yet be made. However, to gain an idea for the likely operating schedules we here consider the nominal Cluster orbit of 57 hours (i.e., 2 days 9 hours). Table I illustrates the evolution of the relative locations of the satellites and one ground-based station, as e.g., incoherent scatter radars or other campaign instruments, following an ideal occurrence of just one configuration (the example chosen here is configuration number 5). This conjunction is said to be ideal if the satellites are at noon when crossing the dayside auroral oval, and the considered ground-based observatory is also at magnetic noon (which is at a UT of roughly 9 hr at, e.g., Svalbard). This can be seen to be the case for orbit 1 because the difference in longitude between the ideal and actual radar sites, δL, is zero; as is the difference between the ideal and actual MLT of the satellite, δMLT. At the same point of the next orbit (2), the radar location is far from ideal, with $\delta L = 9$ hr. For orbit 3, δL is -6 hr when Cluster is in the interior cusp. Note that although this is not a usable occurrence of the very important configuration 5, satellite data on the cusp could still be of use for ground-based coordinaton, because it is an ideal occurrence of configuration 8 (which as mentioned above could be used for TCV studies). Orbit 5 gives a configuration 3 when Cluster is in the interior cusp, and would allow for studies of dayside trigger of nightside disturbances. The interior cusp crossing on orbit 7 yields configuration 12, which is of smilar use as configuration 8, but for TCVs drifting along the dusk oval. Configuration 5 is regained on orbit 9. Note, however, that in this 8-orbit cycle, the satellite has drifted by 1.25 hrs of MLT (because the satellite orbit plane moves through 0.156 hours of MLT per orbit, covering 24 hours in a year), and therefore the ideal conjunction of Orbit 1 does not exactly re-occur. If, for example, we wish the satellite to be within 2 hours of the ideal conjunction for any one configuration, we will only have 2 or 3 such configuration re-occurrences per year of the mission. This is true for the nominal orbit, however, any deviation from the planned orbit can alter these numbers in either direction.

A corresponding analysis can be applied to each of the many useful configurations or conjunctions which occur at other portions of the Cluster orbit. The key point for operations planning is that different ideal configurations will occur during the same orbits. For example, the (very high priority) configuration 1, for

Table I
The repetition of a configuration for an orbit period of 57 hours

Oribt number	Day number	UT (hr)	MLT difference of GB station from satellite δL (hr)	Deviation of satellite ML from ideal, δMLT (hr)
1	1	9	0	0
2	3	18	9	0.156
3	6	3	−6	0.312
4	8	12	3	0.468
5	10	21	12	0.642
6	13	6	−3	0.780
7	15	15	6	0.936
8	18	0	−9	1.092
9	20	9	0	1.248

e.g., EISCAT, requires Tromsø at 24 MLT, so that the satellites are at apogee in the tail at about 19:30-23:30 UT. This is roughly achieved during orbit 2 in Table I on day 4 at 23:30 UT and during orbit 7 on day 16 at 20:30 UT. However, other important configurations with other ground-based instruments will also occur in the same period. We have also noticed that often two areas equipped with key ground-based instruments (as e.g., Greenland and Scandinavia) are offset in local time by about 90°, which means that when one local network encounters configuration 1 or 5 the other will encounter configuration 8 or 12 (see Figure 3). Naturally such conjunctions require coordinated operations of the US and European incoherent radars. The complexity of the planning is yet further increased by the choices for operations made by the Cluster SWT, their selection of data-gathering periods, satellite separation strategy and the instrument modes.

This planning cycle may sound extremely complicated, but is actually more easily understandable when presented on an interactive computer system. Therefore we have prepared the so-called Cluster/Ground-Based conjunction planning tool at our data center at the Rutherford Appleton Laboratory (RAL). We make this system available for the STP community on the World-Wide-Web, via Internet. It allows the user to compile a list of the predicted occurrences of one or more of the configurations for a user-specified ground-based observatory, to within a δL tolerance which is also set by the user.

For planning of combined observations, it is vital to know in advance which of the configurations of the satellites with the ground stations will take place during proposed data taking periods. In order to let scientists (using instruments both on Cluster and the ground) see the implications of a proposed data-gathering sequence, the WWW planning tool at RAL provides so-called 'Bryant Plots' of the orbit. These were used to great effect during the AMPTE mission and an example for the nominal Cluster orbit is here presented in Figure 7. The plot shows time elapsed since the last occurrence of satellite perigee (y axis) as a function of time (x axis). Each orbit thus appears as a straight line (of unity slope) from on perigee to the next, perigee being at both the bottom and the top of the plot. The number of each orbit is labelled along the x axis at the time when the satellite moves through apogee (at the centre of the x axis). Figure 7 is for a period when Cluster apogee is near noon and in the upstream solar wind. At any one time, one of three line types is used to show the orbit: the thick solid line segments show periods when it is currently planned that Cluster will be taking data in burst mode, the thin line segments are where it is to be in normal mode and the dashed line shows where no data-taking is planned. Also on each orbit we mark the occurrences of certain events, based on statistical magnetosphere models: e.g., in this example magnetopause crossings are denoted by open circles. Figure 7 is based on the version of the Cluster Science Plan of March 4, 1996, using data on the orbit and the data taking plans supplied to the CGBDC at RAL by the Cluster JSOC (Joint Operations Science Centre). The planning tool will always give predictions based on the most up-to-date data from JSOC who will provide such data weekly.

Figure 7 also marks a series of points for each orbit, at which a desirable configuration/conjunction with a user-selectable ground-based observatory takes place. The plot shown in Figure 7 is for the EISCAT site at Tromsø but could be for any site, selected by name or coordinates by the WWW user. Beside each cross, the configuration/conjunction number (as given in Figures 3 and 4) is given. Note that in this period there are several occurrences of configuration 22 (Cluster at the magnetic cusp on the magnetopause, with EISCAT at noon) and configuration 19 (Cluster crossing the nightside auroral oval near perigee while EISCAT is close to conjugate with it). If the configuration/conjunction is a 'good' one, the cross is larger in size: 'good' defined as the ground station being 1 hour off an ideal location (e.g., for configuration 22, this means the ground station is in the MLT range of 11–13 hr) or the satellite MLT is the same as that of the satellites to within 1 hour. The figure can be used to study the implications of the proposed data taking for the ground-station in question. The planning tool can then be used to get further information of any case in tabular form. For example, data are taken during all of the occurrences of configuration 22 during the 20 orbits shown, with burst mode on orbits 72 and 85, both of which are 'good' occurrences. On the other hand, of the 7 occurrences of configuration 19, data are only taken on 3, with only 1 in burst mode. Only on orbit 79 the configuration 19 is classed as 'good'.

It can also be seen that on orbit 77 burst mode data are taken when the satellites pass through the nightside auroral zone, but this occurs at a time when EISCAT or other ground-based networks are not anywhere close to configuration 19. This is a typical example where the ground-based community (particularly those scientists interested in using EISCAT to study substorms and the nightside auroral oval) could ask for the Cluster burst mode data taking to be moved to, for example, orbit 79 when a good configuration 19 occurs with EISCAT. However such requests of change in the Cluster master science plan will only be requested when the final Cluster orbit is known, and they will have to be related to a whole combination of other change requests for the benefit of other ground-based instrumentation, or general satellite requirements. It will be necessary for the working group to make clear priority choices for change requests.

A Bryant plot of each individual orbit can be obtained from the WWW planning tool, from the lists of configuration(s) selected by he user. For longer-term planning, the user can select the range of orbits required on each plot (Figure 7 shows 20) when he selects the ground-based observatory. It is expected that this planing tool will remain of great use even after the data have been acquired and are being exploited. It will then be used to define periods when data were taken with the satellites and a ground station in a given configuration.

With the help of an image handling browser such as Mosaic or Netscape the planning procedure is easily explorable under the URL **http://www.gbdc.rl.ac.uk/** at the World Data Center C1 at RAL (see also Lockwood and Opgenoorth (1995) for more information on the WWW planning tool.)

5. Ground-Based Summary and Index Data

In order to carry out truely coordinated studies between Cluster and the many ground-based instruments, it will not only be necessary to predict ideal conjunctions and coordinate the actual operation of the measurement. When selecting data from the huge database that is going to be produced, we need also fast and easily accessible criteria to judge, whether the magnetosphere was in an interesting or suitable disturbance state, or whether certain events did or did not take place. To that end the ground-based community will provide the users of Cluster data with a database containing quicklock or summary data from the various instruments and networks. Every instrument type has its own procedure of presenting key observations in quicklook form, and the exact layout of the data will be described in detail in an ESA Special Publication on Cluster Ground-Based Coordination. Here we only show a few examples of how such daily overview plots of certain key instruments can look like. Figure 5, which was discussed above in a more scientific context, is a typical example of a latitude versus time plot, derived from optical instruments such as photometers and All-Sky cameras (ASC). Such plots allow the determination of location and dynamics of precipitation regions in various key areas of the polar regions; dense instrument networks of optical stations do exist in Scandinavia, Greenland, Canada, Alaska, and Antarctica. Similar latitude versus time plots of electrojet positions and strength can be derived from data of magnetometer chains, as they exist e.g., in Canada, Scandinavia and Greenland. No example plots of such data are given here, but the interested reader can find many nice examples on the WWW home pages of the Canopus and Image magnetometer networks. The CGB WWW-pages at RAL will contain direct links to these overview data.

In addition all individual radars of the SuperDARN network will provide daily overviews in the form of range versus time diagrams of backscatter power, line of sight velocity and spectral widths, from the central beam direction. Again, the combined overviews from all radars cannot only inform about the existence and quality of data, but also at the same time provide an initial indication about the spatial and temporal development of ionospheric convection boundaries. A typical example of such a plot from the Finnish SuperDARN station in Hankasalmi (the eastern part of the European CUTLASS system) is given in Figure 8.

Finally, the incoherent scatter radars such as EISCAT, Söndre Strömfjord and Millstone Hill will provide (whenever operative) overview data of the temporal development of certain ionospheric parameters such as electron density, electron and ion temperature and plasma velocity versus altitude or latitude (or both, in case of antenna scans). Figures 9(a) and 9(b) gives a typical example of how the EISCAT multi radarsystem can provide combined information of plasma parameters versus altitude and time, along a field-aligned position as depicted in Figure 2, and at the same monitor the plasma conditions versus range (latitude) and time inside the polar cap for an oblique viewing direction of another antenna (see even Section 3.2 for a more detailed discussion of such type of data). This simultaneous dataset can provide information on both the local precipitation and the expansion and contraction of boundaries and auroral features over an adjacent area further north. As explained above the future combined EISCAT ESR facility, will cover a wide range of the high latitude region with such information.

In many cases the CGBDC is organised in such a way that the actual data for the overview plots will not reside at the CGBDC at RAL, but the users will instead be provided with links to the homepages of the original instrumentations. Apart from the above described optical stations, ionosondes, riometers, incoherent and coherent radars, and of course, magnetometers around the world from both hemispheres, even geostationary satellite data overviews will be available in a similar manner.

To help browsing such key instrumentation another WWW facility at the CGB-DC at RAL provides an (as complete as possible) list of ground-based instruments from which data is (or may become) available for coordinated Cluster studies. Only instruments relevant to magnetosphere-ionosphere studies have been included. At present the list contains 850 instruments at over 450 sites. We would welcome any corrections and additions to this list which can be returned to us using the WWW facility We plan to constantly update his list during the Cluster mission. Table II gives the first entries in an alphabetical listing of the instruments as an example for the information contained. The list does not give magnetic coordinates because these are available on the Web pages, using a coordinate system and magnetic field model of the user's choice. This will enable users to obtain all magnetic coordinates in a common and self- consistent frame. The facility will also allow the user to

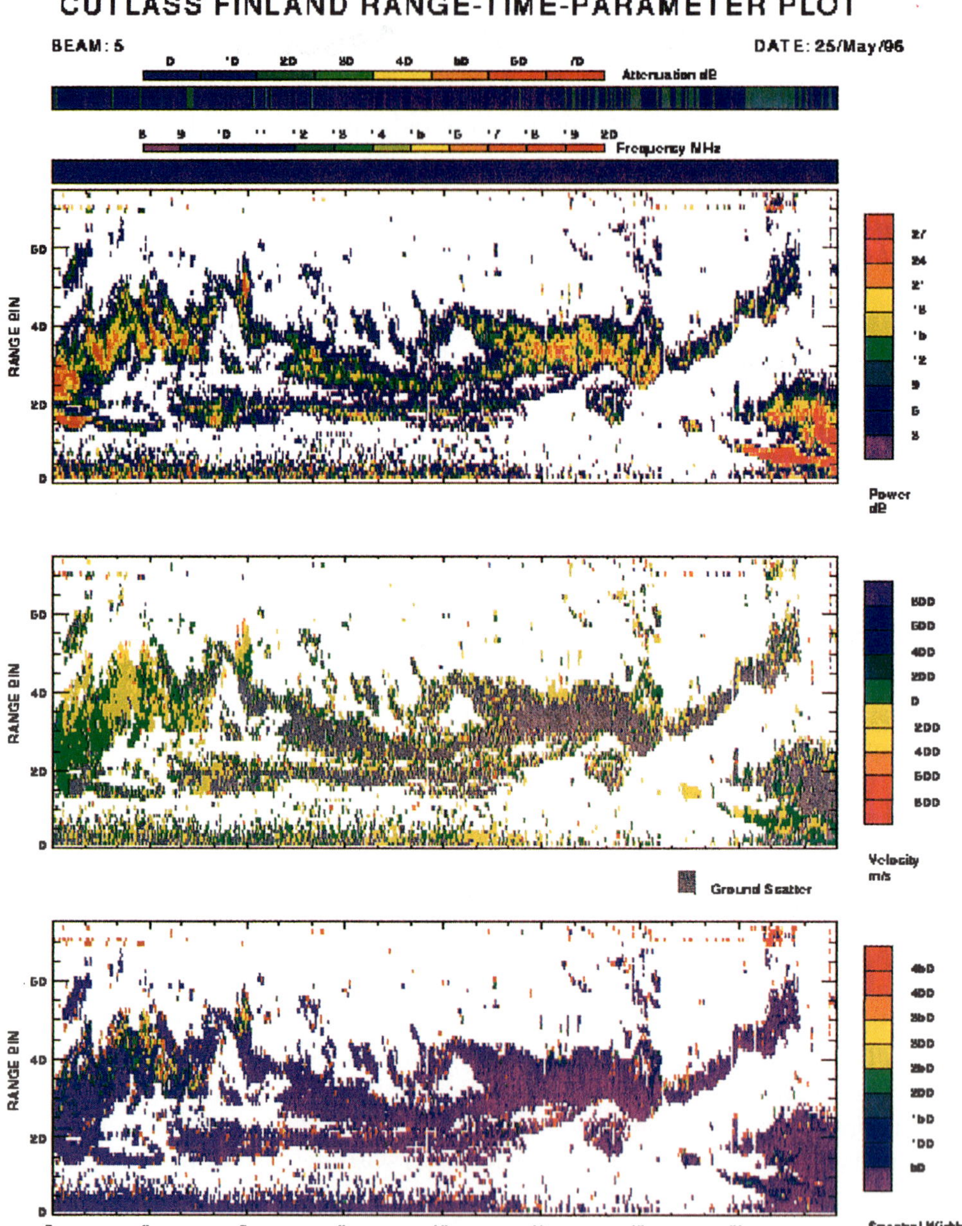

Figure 8. Example of a daily overview plot from an individual SuperDARN radar station, giving a summary of range versus time observations of backscatter power (*top panel*), line-of-sight velocity (*central panel*) and spectral width (*bottom panel*). This dataexample comes from the western station (Hankasalmi) of the European radar pair (CUTLASS), overlooking Svalbard.

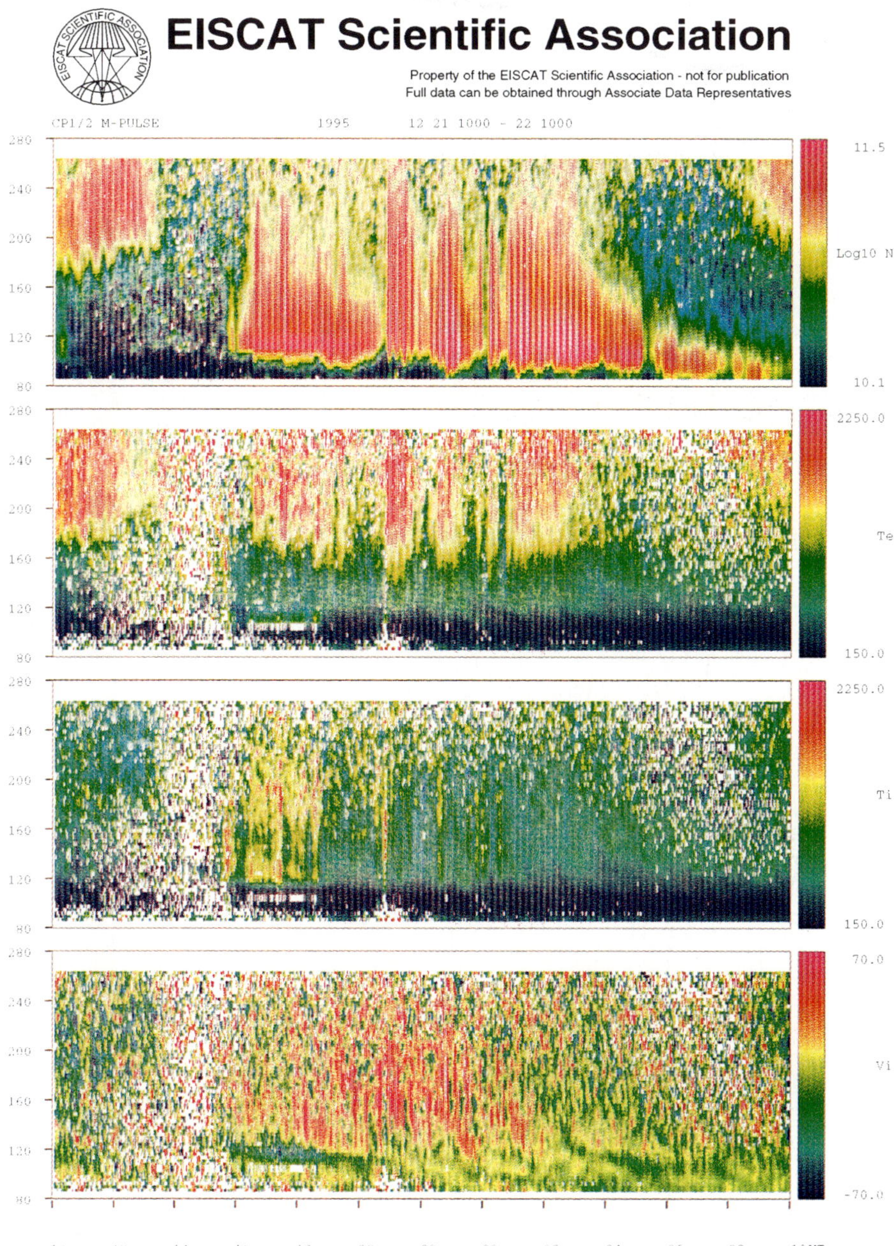

Figure 9a. Daily overview plot of EISCAT measurements of plasmaparameters (electron density, electron and ion temperature and ion velocity) as a function of altitude and time for a field-aligned pointing experiment. Times of precipitation can easily be recognized in the electron density data. (Reproduced with special permission of the EISCAT Scientific Association.)

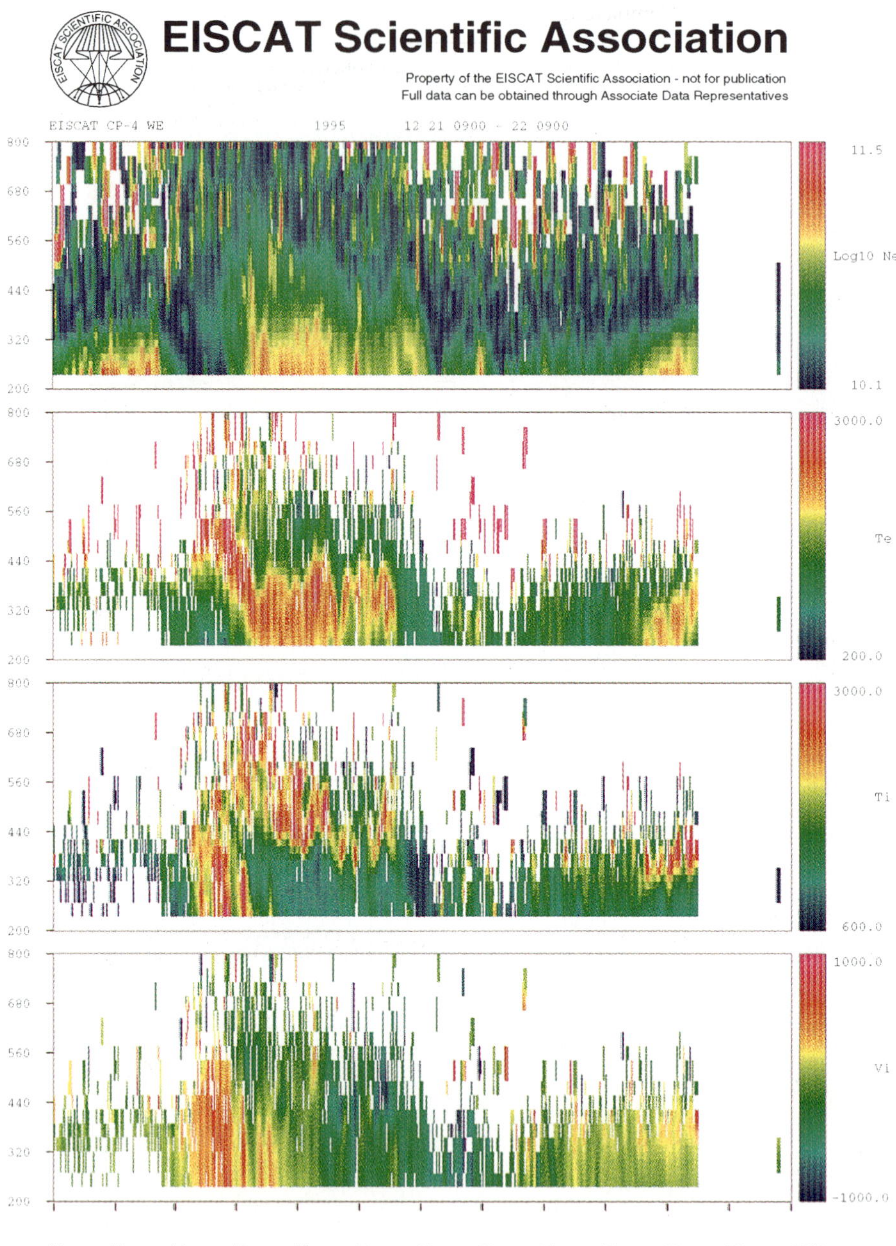

Figure 9b. Daily overview plot of simultaneous EISCAT measurements of plasma parameters (electron density, electron and ion temperature and ion velocity) as a function of range (latitude) and time for an experiment with a north-pointing antenna. In these kinds of plots the expansion of precipitation regions can be identified as regions of enhanced electron temperature. Regions of enhanced ion temperature indicate strong electric fields. (Reproduced with special permission of the EISCAT Scientific Association.)

Table II

Complete list of ground-based observing stations

This is a preliminary list of all ground–based observing stations that we know about. In due course you will be able to select the information from this list using various criteria. Until then, each line in the station list corresponds to one instrument, and thus sites equipped with many instruments appear on several lines. An individual instrument may also appear more than once if it belongs to more than one group of instruments. The entries are arranged alphabetically by station name, and the various codes and acronyms in the list are explained in associated tables:

- Instrument codes (R for riometer, M for magnetometer, ... etc)
- Group acronyms used for chains and networks of stations, for data centres and for instruments.
- Station synonyms where a station is known by more than one name.
- Institution acronyms and abbreviations used to describe the institutes which operate the instrument in question.

The last column of the list is the e–mail address of a useful contact person for the station.

Station	Code	Geographic coordinates	Instr-ument	Chain	Operated by	Contact e-mail
AGO A77	B1	77.58 S 336.63 E	M	AGONET	BAS	J.Dudeney@bas.ac.uk
AGO A77	B1	77.58 S 336.63 E	R	AGONET	BAS	J.Dudeney@bas.ac.uk
AGO A80	B2	80.75 S 339.60 E	LF-HF	AGONET	U.Stan	jameslabelle@dartmouth.edu
AGO A80	B2	80.75 S 339.60 E	M	AGONET	BAS	J.Dudeney@bas.ac.uk
AGO A80	B2	80.75 S 339.60 E	MP	AGONET	ACM	engebret@augsburg.edu
AGO A80	B2	80.75 S 339.60 E	R	AGONET	BAS	J.Dudeney@bas.ac.uk
AGO A80	B2	80.75 S 339.60 E	VLF	AGONET	BAS	J.Dudeney@bas.ac.uk
AGO A81	B3	81.50 S 3.00 E	LF-HF	AGONET	U.Stan	jameslabelle@dartmouth.edu
AGO A81	B3	81.50 S 3.00 E	M	AGONET	BAS	J.Dudeney@bas.ac.uk
AGO A81	B3	81.50 S 3.00 E	MP	AGONET	ACM	engebret@augsburg.edu
AGO A81	B3	81.50 S 3.00 E	R	AGONET	BAS	J.Dudeney@bas.ac.uk
AGO A81	B3	81.50 S 3.00 E	VLF	AGONET	BAS	J.Dudeney@bas.ac.uk
AGO A84	B4	84.0 S 335.0 E	LF-HF	AGONET	U.Stan	jameslabelle@dartmouth.edu
AGO A84	B4	84.0 S 335.0 E	M	AGONET	BAS	J.Dudeney@bas.ac.uk
AGO A84	B4	84.0 S 335.0 E	MP	AGONET	ACM	engebret@augsburg.edu
AGO A84	B4	84.0 S 335.0 E	R	AGONET	BAS	J.Dudeney@bas.ac.uk
AGO A84	B4	84.0 S 335.0 E	VLF	AGONET	BAS	J.Dudeney@bas.ac.uk
AGO I1	I1	74.07 S 164.12 E	M	AGONET	–	–
AGO J1	J1	70.0 S 39.5 E	M(p)	AGONET	NIPR	–
Abisko	ABK	68.35 N 18.82 E	ASC(4freq)	ALIS	IRFK	steen@irf.se
Abisko	ABK	68.35 N 18.82 E	R	–	SGO	Hilkka.Ranta@csc.fi
Abisko	ABK	68.36 N 18.82 E	M	–	–	–
Adelaide	ADL	34.67 S 138.65 E	M	–	–	–
Adelaide	–	34.7 S 138.6 E	ID	ULCAR	–	reinisch@cae.uml.edu
Agua Verde	–	25.4 S 290.0 E	ID(p,c)	ULCAR	–	reinisch@cae.uml.edu
Aire Sur L'Adour	–	43.4 N 0.0 W	BLS	–	–	–
Akita	AK 539	39.70 N 140.10 E	I	WDC-C1	–	M.Wild@rl.ac.uk
Alert	ALE	82.45 N 297.7 E	ID	–	UWO	macdougal@uwo.ca
Alert	ALE	82.50 N 297.50 E	M	–	GSC	vanbeek@geolab.emr.ca
Alfred Faure	–	46.43 S 51.87 E	M	–	ULP	schlich@??
Alibag	ABG	18.64 N 72.87 E	M	Intermagnet	–	d.kerridge@bgs.ac.uk
Alice Springs	ASP	23.77 S 133.88 E	M	NGDC	BMR	stewart@lodestone.bmr.gov.au
Alice Springs	–	24.0 S 143. E	ID	ULCAR	DSTO	wardbd@hfrd.srl.dsto.gov.au
Alma Ata	AA 343	43.20 N 77.00 E	I	WDC-C1	–	M.Wild@rl.ac.uk
Alma Ata	AAA	43.25 N 76.92 E	M	–	–	–
Almeria	ALM	36.85 N 357.54 E	M	–	–	–
Amderma	AM 269	69.46 N 61.41 E	I	WDC-C1	–	M.Wild@rl.ac.uk
Amderma	AMD	69.40 N 60.20 E	M	–	AARI	olegtro@geophys.spb.su
Amderma (Calgary)	AMD	69.47 N 61.42 E	M	–	–	–
Ammassalik	AMK	65.60 N 322.37 E	M	GreenlandE	DMI/U.Mich	efc@dmi.min.dk
Anchorage	AMU	61.24 N 210.13 E	M	–	U.Alaska	jvo@maxwell.gi.alaska.edu
Andøya	–	69.295N 16.030E	ASC	–	ARR	kolbjorn@arr.nsc.no
Andøya	–	69.295N 16.030E	ATV	–	ARR	kolbjorn@arr.nsc.no
Andøya	–	69.295N 16.030E	ELF	–	ARR	kolbjorn@arr.nsc.no
Andøya	–	69.295N 16.030E	ID(p)	–	IAP Kuehlungsborn	??
Andøya	–	69.295N 16.030E	M	–	ARR	kolbjorn@arr.nsc.no
Andøya	–	69.295N 16.030E	MF(p)	–	IAP Kuehlungsborn	??
Andøya	–	69.295N 16.030E	MP	–	ARR	kolbjorn@arr.nsc.no
Andøya	–	69.295N 16.030E	MSP	–	ARR	kolbjorn@arr.nsc.no
Andøya	–	69.295N 16.030E	R	–	ARR/DMI	kolbjorn@arr.nsc.no
Andøya	–	69.295N 16.030E	RR	–	ARR	kolbjorn@arr.nsc.no
Andøya	–	69.295N 16.030E	SIR	–	UWO	??
Andøya	–	69.295N 16.030E	SM	–	MPI Lindau	??
Andøya	–	69.295N 16.030E	VLF	–	ARR	kolbjorn@arr.nsc.no

generate maps of selected stations and to select stations from maps. How to make full use of the planning tool is described within the system itself, and it would be beyond the scope of this article to go into more detail.

While quicklook data often is, strictly speaking, only usable for event selection or classification, one set of ground-based quicklook data has during a long time enjoyed very frequent use in publications and is considered as a direct monitor of the magnetospheric state of energy input or release. These are the magnetic AE indices (Davis and Sugiura, 1966), which represent the maximum horizontal magnetic deflection along a ring of selected magnetometers around the northern hemisphere auroral oval. Useful (and much used) as they are, the AE indices nevertheless have several shortcomings for Cluster and ground-based operations planning, of which the most critical one is relatively slow availibility.

Acknowledging the success and long-term usability of the $AE(12)$ indices we decided to provide a similar index, using data with one minute time resolution from 10 magnetometer stations along the standard auroral oval. The used stations are roughly the same (or at equivalent locations) as the original AE stations. The rationale for deviations from the standard AE procedure was the need for fast and easy data access. A survey of the literature revealed that substorms were usually identified by AE perturbations dominated by large negative AL-values, or simply using the AL-index alone. It was therefore decided that for Cluster mission-oriented tasks AL-type (lower envelope of the magnetic H-disturbance) of indices were required.

Since the basic index is derived from stations along the standard auroral oval, it has been named Standard Oval (SO) index. One of the basic differences in the production of the SO-index is that it will be made available on-line, as soon as data from a few stations have come into the database at RAL. The dataset will then automatically be updated as more data are received. The index will always be quoted with the number of contributing stations in parentheses and in addition, the summary plots will show explicitly the longitude of the stations used. This means that immediate event recognition and selection can be carried out for many coordinated studies within certain key regions of the auroral zone, even if not all data have yet been collected; collaborative studies can be started without delay. Later, while the scientific work is proceeding, the magnetic index will be automatically updated to improve coverage and quality. The user will always be supplied with the most up-to-date index values.

This procedure is a considerable improvement of the accessibility of magnetic indices. There is no immediate drawback in the quality, unless one wants to carry out a study over a region from where data has yet to be received. Recently Kauristie *et al.* (1996) have shown, that local magnetic indices contain much of the information of the full $AE(12)$ index for stations within several hours around magnetic midnight. As a result, many event studies involving ground-based networks in highly instrumented regions need not be delayed by incomplete AE index coverage.

Thus the new index will overcome the problems of timely availability, that the AE index suffers from. The other main shortcoming of the standard $AE(12)$ index is its restriction to stations along the location of the standard nightside oval. During very quiet and very disturbed conditions $AE(12)$ is known not to represent the actual disturbance state of the magnetosphere, but to underestimate it

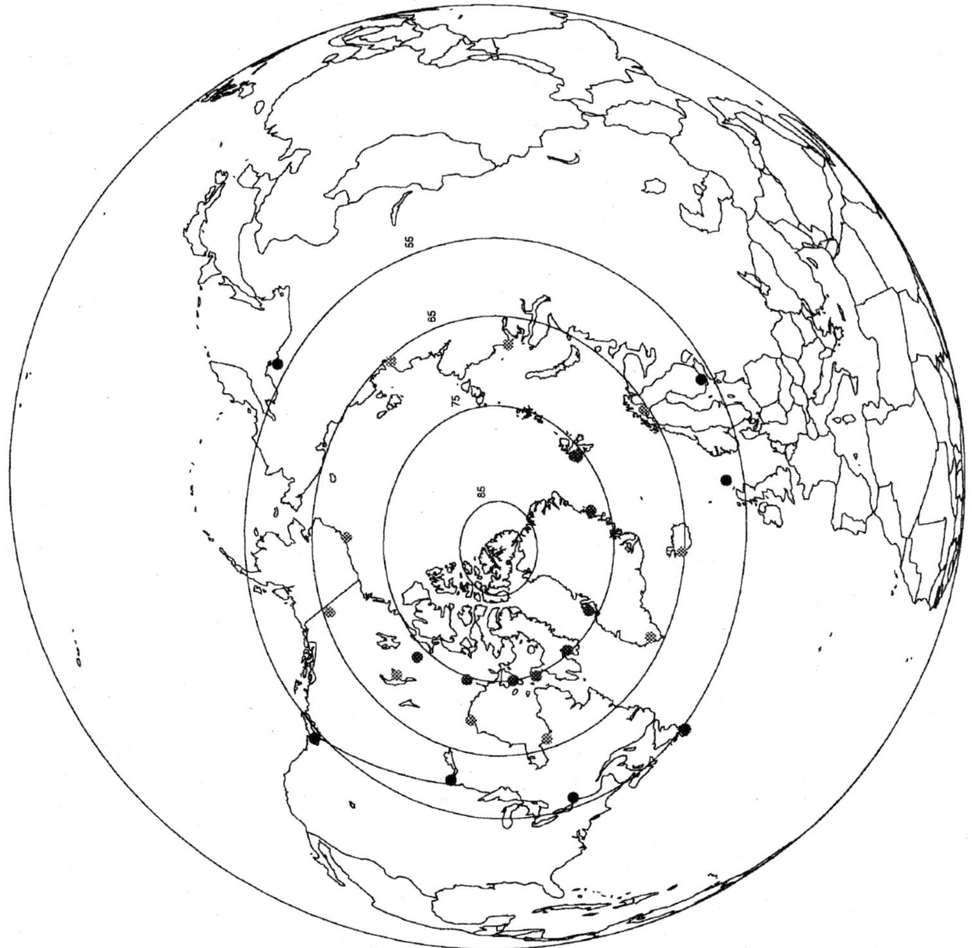

Figure 10. Map of northern hemisphere magnetometer stations which are planned to be used for the derivation of the Contracted, Standard, and Expanded Oval indices, as indicated by the three oval rings.

(Kamide and Akasofu, 1983). We plan to tackle this drawback by using two other (though less complete) rings of magnetometer stations in (i) very high and (ii) sub-auroral latitudes for the construction of complementary indices. These two rings are located at latitudes, where the auroral oval is typically observed when it is either more contracted or more expanded than the standard average oval. Consequently we name the complete new family of indices SO-, CO-, and EO-index for the Standard, Contracted, and Expanded Oval, respectively. The locations of stations which will be used for these three indices are given Figure 10.

For all three indices we will use the same procedure as described above, i.e., construct a preliminary local index as soon as data are available, and update as more data are received. Using this procedure it may eventually be even possible to add data from additional or new stations to these indices, thereby improving the longitudinal sensitivity, at least in regions where enough stations are available. As described above, small substorms, multiple substorm intensifications, and small-scale dayside magnetic disturbances are prime topics for coordinated Cluster and ground-based studies, and a denser index network will eventually allow the recognition of many more such events.

Using the CGB-DC facilities on WWW (see above), it will be possible to view the current Standard, Contracted and Expanded Oval Indices, and decide whether they are useful for the intended study. In order to improve the understanding and information content of the datasets we will not only indicate the actual stations from which data have been included, but also indicate which station produces the maximum envelope portion of the index curve, allowing the user to locate events and follow their motion in longitude or magnetic local time (MLT). An example of such a preliminary summary index plot is given in in Figure 11. In the panels below the index curves, the stations from which data have been included are denoted by thin lines, and those contributing to the envelope function of the final indices are marked with thicker bars. In plots like the one in Figure 11, the relative size of the three indices will be useful for the analysis of the energy state of the magnetosphere, as characterised by the latitudinal location of the maximum electrojet activity at any one MLT. It will also be possible to follow a substorm expansion in latitude, for example when starting off in the standard oval (SO-index) and expanding into the contracted oval index (CO). In our example of December 24, 1995 such an event can be seen from 18:00 to 22:00 UT. Another event on the same day at 10:00 UT shows a disturbance in the standard oval which does not expand to the contracted oval in the same longitude sector. It will thus be possible to discriminate multiple onsets in the standard oval, from expansion phase intensifications of the poleward edge of the substorm bulge. During extremely disturbed times (i.e., under major magnetic storms) much of the variability of the magnetosphere will become visible in the EO-index (no example available yet), and during more quiet times the CO-index will allow the recognition of small substorms or alternatively dayside magnetic disturbances.

We believe that both the more timely availability and better latitudinal coverage of this new family of geomagnetic disturbance indices will be of value for the Cluster ground-based coordination. Event selection for prime orbits can commence roughly one month after its occurrence, and the index used in the scientific evaluation of the event will eventually improve in quality – but not necessarily change in its principal content – as the study converges towards publishable results.

Dayside magnetic disturbances have recently been recognized as important signatures of magnetospheric reconfigurations caused by changes in the solar wind/magnetosphere coupling (see Section 3.2 above). Usually these disturbances are much smaller than typical nightside disturbances – a few 100 nT as compared to substorm disturbances of up to 1000 nT. Consequently they will either disap-

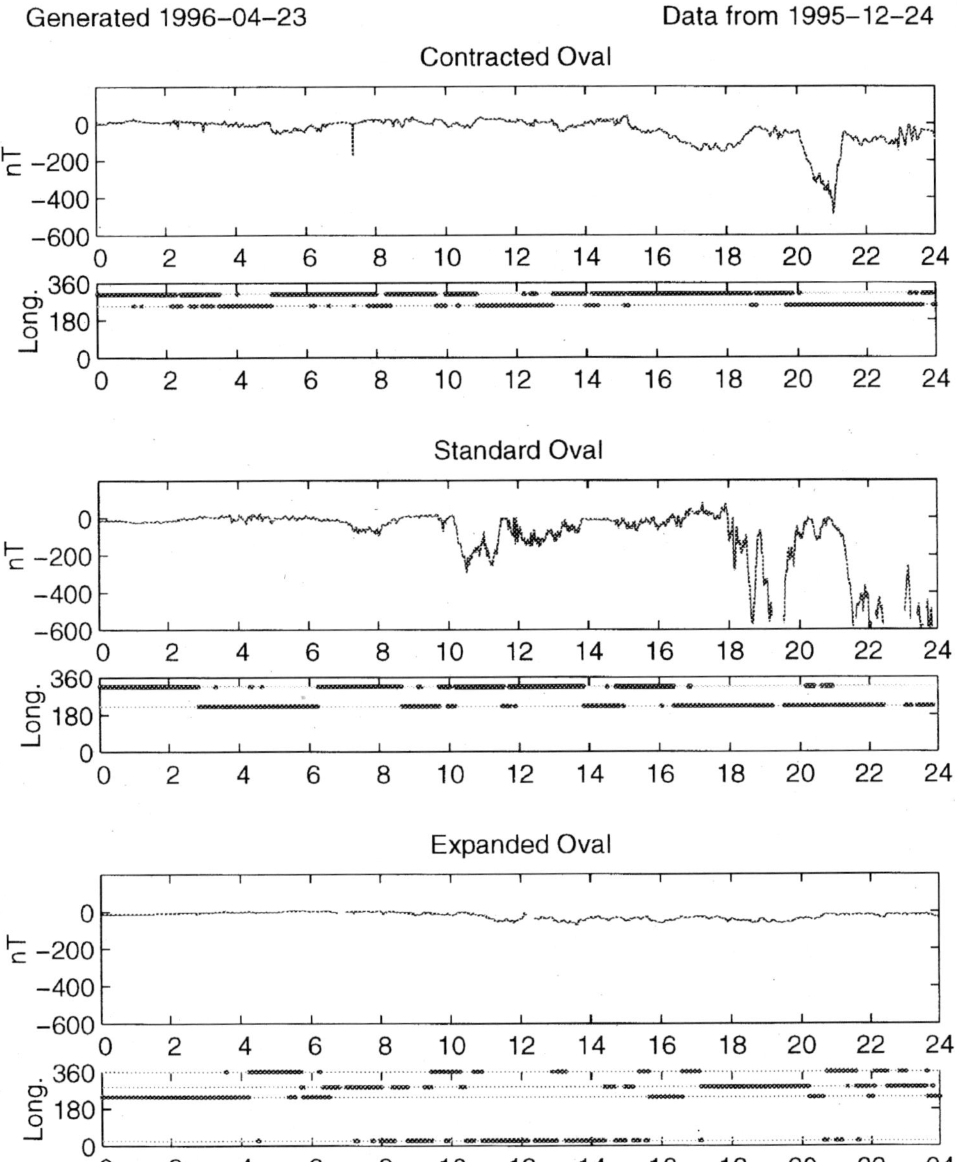

Figure 11. Example plot of the Contracted, Standard, and Expanded Oval indices. Below the index plots all stations contributing to the index are indicated. The station that contributes to the envelope function of the actual index is marked with a heavier bar. Plots based on roughly this number of stations will be available one month after the date in question and will subsequently be upgraded as data from further stations are received.

pear within the normal index envelope or not affect Standard Oval index stations because of their occurrence at cusp latitudes, which corresponds to our new CO-index latitude. The new CO-index, with the station contributing to the envelope clearly marked, will allow the identification of such dayside disturbances and discriminate them from nightside disturbances. This development would thus widen the applicability of the new indices for event selection during the Cluster mission.

In conclusion, we consider this new family of magnetospheric disturbance indices not only to be particularly useful for the Cluster mission, but also for similar coinciding and future missions or dedicated ground-based measurement campaigns. We would like to stress, however, that our new family of indices is not meant to be a substitute for the existing AE-index. They are derived in far too crude a manner (mainly due to lack of time and manpower) to replace AE in detailed quantitative studies. For example, because of the need for rapid compilation with minimal manual intervention we use an automated background subtraction algorithm. We would like to point out that the new CGBDC indices are basically intended for the rapid evaluation of conditions which may have significant implications for:

(a) event selection for coordinated Cluster and ground-based studies,

(b) the operations planning of coordinated Cluster / ground-based observations, and

(c) the operations planning of the Cluster mission itself.

For example, at some instance it might be important to have an estimate of how many substorms were seen while Cluster was at apogee in the tail, when deciding on future operation modes for the following passes.

6. Conclusions

We have reviewed a small fraction of opportunities, where ground-based observations can be used to support the Cluster mission to maximum effect. We have also briefly reviewed some combined satellite and ground observations carried out in the past, and suggested objectives to stimulate thinking about the variety of measurements which could be carried out with a true multi-satellite mission concept. This review is far from complete, but examples have been selected to illustrate the range of uses of ground-based data and the potential to support Cluster observations. Again it should be noted that from the many selected configurations many examples are not even close to magnetic conjugacy between the satellites and key ground-based observatories or networks. Instead often an offset is planned to allow for special coordinated measurements of travelling phenomena, to allow cause and effect studies, as e.g., the dayside activity monitored by Cluster, and possibly associated nightside disturbances monitored by ground-based equipment on the nightside.

We have also illustrated how data from global networks, e.g. magnetometers and large scale HF radarsystems, can be useful not only for dedicated conjugated studies, but to provide a background information on the spatial and temporal large scale state of the magnetosphere, thereby allowing to interprete the Cluster observations within the correct context. This line of thoughts has lead to the development of new magnetic disturbance indices, which are more readily available and more adaptive to different states of the magnetosphere then the widely used AE (or AU and AL indices). Such index and other ground-based summary data will be available for the Cluster mission from the Cluster Ground-Based Data Center at RAL.

Acknowledgements

We are grateful to M. A. Hapgood, project scientist of the Cluster Joint Satellite Operation Center (JSOC) at RAL, for assistance in establishing the required interfaces with JSOC, and to M. N. Wild and R. Stamper for the planning and database software design, and the implementation of the Cluster Ground-Based Data Centre (GBDC) at RAL. We are grateful to the other members of the ESA Cluster Ground-Based Working Group for many discussions. Much input to the above considerations was received at the two international workshops organized by the Cluster Ground-Based Working Group in Orleans and Rome, and we would particularly like to thank J.-P. Villain and M. Candidi, for the local organisation of these workshops. Another more informal workshop on magnetic data handling, held at the Finnish Meteorological Institute in Helsinki, Finland, has lead to the development of the magnetic disturbance indices that we will provide for the Cluster mission. Lastly we are greatly indebted to the many other scientists, who are already, or will be, contributors to the summary data in the GBDC.

References

Aikio, A. T. and Kaila K. U.: 1996, 'A Substorm Observed by EISCAT and Other Ground-Based Instruments, Evidence for Near Earth Substorm Initiation', *J. Atmospheric Terrest. Phys.* **58**, 5.

Baumjohann, W.: 1993, 'The Near-Earth Plasma Sheet, an Ampte/IRM Perspective', *Space Sci. Rev.* **64**, 141.

Baumjohann, W., Paschmann, G., Nagai, T., and Lühr, H.: 1991, 'Superposed Epoch Analysis of the Substorm Plasma Sheet', *J. Geophys. Res.* **96**, 11 605.

Birkeland, K.: 1913, *The Norwegian Aurora Polaris Expedition, 1902–1903*, Vol. II, Aschehoug, Nygaard, Christiania.

Brekke, A., Hall, C. and Hansen, T. L.: 1989, 'Auroral Ionospheric Conductances During Disturbed Conditions', *Ann. Geophys.* **7**, 269.

Buchert, S., Baumjohann, W., Haerendel, G., La Hoz, C., and Lühr, H.: 1988, 'Magnetometer and Incoherent Scatter Observations of an Intense Ps 6 Pulsation Event', *J. Atmospheric. Terrest. Phys.* **50**, 357.

Chapman, S. and Ferraro, V. A. C.: 1931, 'A New Theory of Magnetic Storms, Part 1, The Initial Phase', *Terrest. Mag. Atmos. Elect.* **36**, 77.

Cowley, S. W. H.: 1982, 'The Causes of Convection in the Earth's Magnetosphere: A Review of Developments During IMS', *Rev. Geophys.* **20**, 531.

Cowley, S. W. H., Van Eyken, A. P., Thomas, F. C., Williams, D. J. S., and Willis, D. M.: 1990, 'The Scientific and Technical Case for a Polar Cap Radar', *J. Atmospheric Terrest. Phys.* **52**, 645.

Crooker, N. U. and Toffoletto, F. R.: 1995, in B. U. O Sonnerup (ed.), *Physics of the Magnetopause*, Proc. Chapman Conf. on the Magnetopause, American Geophysical Union Monograph, in press.

Crooker N. U., Toffoletto, F. R., and Gussenhoven, M. S.: 1991, 'Opening the Cusp', *J. Geophys. Res.* **96**, 3497.

Davis, T. N. and Sugiura, M.: 1966, 'Auroral Electrojet Index AE and Its Universal Time Variations', *J. Geophys. Res.* **71**, 785.

de la Beaujardère, O., Newell, P. T., and Rich, R.: 1993, 'Relationship Between Birkeland Current Regions, Particle Participation, and Electric Fields', *J. Geophys. Res.* **98**, 7711.

de la Beaujardère, O., Lyons, L. R., Ruohoniemi, J. M., Friis-Christensen, E., Danielsen, C., Rich, F. J., and Newell, P. T.: 1994, 'Quiet Time Intensifications Along the Poleward Auroral Boundary Near Midnight', *J. Geophys. Res.* **99**, 287.

Eather, R. H. and Mende, S. B.: 1971, 'Airborne Observations of Auroral Precipitation Patterns', *J. Geophys. Res.* **76**, 1746.

Elphic, R. C., Lockwood, M., Cowley, S. W. H., and Sandholt, P. E.: 1990, 'Flux Transfer Events at the Magnetopause and in the Ionosphere', *Geophys. Res. Letters* **17**, 2241.

Foster, J. C., St. Maurice, J.-P., and Abrev, V. J.: 1983, 'Joule Heating at High Latitudes', *J. Geophys. Res.* **88**, 4885.

Frank, L. A.: 1971, 'Plasmas in the Earth's Polar Magnetosphere', *J. Geophys. Res.* **76**, 5202.

Frank, L. A., Craven, J., Burch, J. L., and Winningham, J. D.: 1982, 'Polar Views of the Earth's Aurora with Dynamics Explorer', *Geophys. Res. Letters* **9**, 1001.

Friis-Christensen, E., McHenry, M. A., Clauer, C. S., and Vennerstrom, S.: 1988, 'Ionospheric Travelling Convection Vortices Observed Near the Polar Cleft: A Triggered Response to Sudden Changes in the Solar Wind', *Geophys. Res. Letters* **15**, 253.

Fukunishi, H. Y., Takahashi, T., Nagatsuma, T., Mukai, T., and Machida, S.: 1993, 'Latitudinal Structures of of Nightside Field-Aligned Currents and Their Relationships to the Plasma Sheet Region', *J. Geophys. Res.* **98**, 11 235, 1993

Gazey, N. G. J., Lockwood, M., Smith, P. N., Coles, S., Bunting, R. J., Lester, M., Aylward, A. D., Yeoman, T. Y., and Lühr, H.: 1995, 'The Development of Substorm Cross-Tail Current Disruption as Seen from the Ground', *J. Geophys. Res.*, in press.

Glassmeier, K-H.: 1992, 'Travelling Magnetospheric Convection Twin-Vortices: Observations and Theory', *Ann. Geophys.* **10**, 547.

Glassmeier, K.-H., Hoenisch, M., and Untied, J.: 1989, 'Ground-Based and Satellite Observations of Travelling Magnetospheric Convection Twin Vortices', *J. Geophys. Res.* **94**, 2520.

Grande, M., Perry, C. H., Hall, D. S., Wilken, B., Livi, S., Søraas, F., and Fennel, J. F.: 1992, 'Composition Signatures of Substorm Injections', in *Proceedings of the International Conference on Substorms (ICS-1)*, Kiruna, Sweden, March 1992, ESA SP-335, p. 111.

Greenwald, R. A., Baker, K. B., Ruohoniemi, J. M., Dudeney, J. R., Pinnock, N., Mattin, N., Leonard, J. M., and Lepping, R. P.: 1990, 'Simultaneous Conjugate Observations of Dynamic Variations in High-Latitude Dayside Convection Due to Changes in IMF B_y, *J. Geophys. Res.* **95**, 8057.

Greenwald, R. A., Baker, K. B., Dudeney, J. R., Pinnock, M., Jones, T. B., Thomas, E. C., Villain, J.-P., Cerisier, J.-C., Senior, C., Hanuise, C., Hunsucker, R. D., Sofko, G., Koehler, J., Nielsen, E., Pellinen, R., Walker, A. M. D., Sato, N., and Yamagishi, H.: 1995, 'DARN/SuperDARN, A Global View of the Dynamics of High Latitude Convection', *Space Sci. Rev.* **71**, 761.

Hanuise, C., Senior, C., Cerisier, J. C., Villain, J.-P., Greenwald, R. A., Ruohoniemi, J. M., and Baker, K. B.: 1993, 'Instantaneous Mapping of High-Latitude Convection with Coherent HF Radars', *J. Geophys. Res.* **98**, 17387.

Heelis, R. A. and Coley, W. R.: 1988, 'Global and Local Joule Heating Effects Seen by DE-2', *J. Geophys. Res.* **93**, 7551.

Heikkila, W. J.: 1972, *The Morphology of Auroral Particle Precipitation in Space Research 12*, Akademie-Verlag, p. 1343.

Heikkila, W. J. and Winningham, J. D.: 1971, 'Penetration of Magnetosheath Plasma to Low Altitudes Through the Dayside Magnetospheric Cusps', *J. Geophys. Res.* **76**, 883.

Heikkila, W. J., Jorgensen, T. S., Lanzerotti, L. J., and Maclennan, C. J.: 1989, 'A Transient Auroral Event on the Dayside', *J. Geophys. Res.* **94**, 15 291.

Hesse, M. and Birn, J.: 1991a, 'Magnetosphere-Ionosphere Coupling During Plasmoid Evolution: First Results', *J. Geophys. Res.* **96**, 11 513.

Hesse, M. and Birn, J.: 1991b, 'On Dipolarization and Its Relationship to the Substorm Current Wegde', *J. Geophys. Res.* **96**, 19 417.

Hones, E. W., Jr., Craven, J. D., Frank, L. A., Evans, D. S., and Newell, P. T.: 1989, 'The Horse-Collar Aurora: a Frequent Pattern of the Aurora in Quiet Times', *Geophys. Res. Letters* **16**, 37.

Jacobsen, B., Sandholt, P. E., Lybekk, B., and Egeland, A.: 1991, 'Transient Auroral Events Near Midday: Relationship with Solar Wind/Magnetosheath Plasma and Magnetic Field Conditions', *J. Geophys. Res.* **96**, 1327.

Kan, J. R.: 1993, 'A Global Magnetosphere-Ionosphere Coupling Model of Substorms', *J. Geophys. Res.* **98**, 17263.

Kamide Y., Akasofu, S. I., Deforest, S. E., and Kisabeth, J. L.: 1975, 'Weak and Intense Substorms', *Planetary Space Sci.* **23**, 579.

Kamide, Y., Perreault, P. D., Akasofu, S.-I., and Winningham, D.: 1977, 'Dependence of Substorm Occurrence Probability on the Interplanetary Magnetic Field and on the Size of the Auroral Oval', *J. Geophys. Res.* **82**, 5521.

Kamide, Y. and Akasofu, S.-I.: 1983, 'Notes on the Auroral Electrojet Indices', *J. Geophys. Res.* **71**, 1647.

Kauristie, K., Pulkkinen, T. I., Pellinen, R. J., and Opgenoorth, H. J.: 1995, 'Comparison Between the Global *AE* Chain and a Local Meridional Magnetometer Chain', *Ann. Geophys.*, submitted.

Kennel, C.: 1992, 'The Kiruna Conjecture: The Strong Version', *Proceedings of the International Conference on Substorms (ICS-1)*, Kiruna, Sweden, March 1992, ESA SP-335, p. 599.

Kirkwood, S.: 1994, 'An Improved Conductivity Model for Substorm Modelling', *Proceedings of the International Conference on Substorms (ICS-2)*, Fairbanks, Alaska, March 1994, p. 33.

Kirkwood, S. and Eliasson, L.: 1990, 'Energetic Particle Precipitation in the Substorm Growth Phase Measured by Eiscat and Viking', *J. Geophys. Res.* **95**, 6025.

Kirkwood, S., Opgenoorth, H. K., and Murphree, J. S.: 1988, 'Ionospheric Conductivities, Electric Fields and Currents Associated with Auroral Substorms Measured by the EISCAT Radar', *Planetary Space Sci.* **36**, 1359.

Kirkwood, S., Eliasson, L., Opgenoorth, H. J., and Pellinen-Wannberg, A. 1989, 'A Study of Auroral Electron Acceleration Using the EISCAT Radar and the Viking Satellite', *Adv. Space Res.* **9**, 49.

Kivelson, M. G. and Southwood, D. J.: 1991, 'Ionospheric Travelling Vortex Generation by Solar Wind Buffeting of the Magnetosphere', *J. Geophys. Res.* **96**, 1661.

Knipp, D., Emery, B. A., Richmond, D., Crooker, N. U., Hairston, M. R., Cumnock, J. A., Denig, W. F., Rich, F. J., de la Beaujardère, O., Ruohoniemi, J. M., Rodger, A. S., Crowley, G., Ahn, B.-H., Evans, D. S., Fuller-Rowell, T. J., Friis-Christiansen, E., Lockwood, M., Kroehl, H., McClennan, C., McEwin, A., Pellinen, R. J., Morris, GR. J., Burns, B., Papitashvili, V., Zaitzev, A., Troshichev, O., Sato, N., Sutcliffe, P., Tomlinson, L.: 1993, 'Ionospheric Convection Response to Strong, Slow Variations in a Northward Interplanetary Magnetic Field: A Case Study for January 14, 1988', *J. Geophys. Res.* **98**, 19 273.

Koskinen, H. E. J., Pulkkinen, T. I., Pellinen, R. J., Bösinger, T., Baker, D. N., and Lopez, R. E.: 1992, 'Characteristics of Pseudobreakups', *Proceedings of the International Conference on Substorms (ICS-1)*, Kiruna, Sweden, March 1992, ESA SP-335, p. 111.

Lockwood, M.: 1995a, in B. U. O. Sonnerup (ed.), 'Ground-Based and Satellite Observations of the Cusp: Evidence for Pulsed Magnetopause Reconnection', *Physics of the Magnetopause*, Proc. Chapman Conf. on the Magnetopause, American Geophysical Union Monograph 90, 417.

Lockwood, M.: 1995b, 'Large-Scale Fields and Flows in the Magnetosphere-Ionosphere System', *Surveys Geophys.* **16**, 389.

Lockwood, M. and Cowley, S. W.: 1992, 'Ionospheric Convection and the Substorm Cycle, Substorms 1', *Proceedings of the First International Conference on Substorms (ICS-1)*, Kiruna, Sweden, March 1992, ESA SP-335, p. 99.

Lockwood, M. and Opgenoorth, H. J.: 1995, 'Opportunities for Magnetospheric Research Using EISCAT/ESR and CLUSTER', *J. Geoelect. Geomag.* **47**, 699.

Lockwood, M. and Smith, M. F.: 1994, 'Low- and Mid-Altitude Cusp Particle Signatures for General Magnetopause Reconnection Rate Variations: I – Theory', *J. Geophys. Res.* **99**, 8531.

Lockwood, M., Chandler, M. O., Horwitz, L., Waite, J. H., Jr., Moore, T. E., and Chappell, C. R.: 1985, 'The Cleft Ion Fountain', *J. Geophys. Res.* **90**, 9736.

Lockwood, M., Sandholt, P. E., Cowley, S. W. H., and Oguti, T.: 1989, 'Interplanetary Magnetic Field Control of Dayside Auroral Activity and the Transfer of Momentum Across the Dayside Magnetopause', *Planetary Space Sci.* **37**, 1347.

Lockwood, M., Denig, W. F., Farmer, A. D., Davda, V. N., Cowley, S. W., and Lühr, H.: 1993a, 'Ionospheric Signatures of Pulsed Reconnection at the Earth's Magnetopause', *Nature* **361**, 424.

Lockwood, M., Carlson, H. C., and Sandholt, P. E.: 1993b, 'The Implications of the Altitude of Transient 630 nm Dayside Auroral Emissions', *J. Geophys. Res.* **98**, 15571.

Lockwood, M., Onsager, T. G., Davis, C. J., Smith, M. F., and Denig, W. F.: 1994, 'The Characteristics of the Magnetopause Reconnection X-Line Deduced from Low-Altitude Satellite Observations of Cusp Ions', *Geophys. Res. Letters* **21**, 2757.

Lopez, R. E., Koskinen, H. E. J., Pulkkinen, T. I., Bösinger, T., McEntire, R. W., and Potemra, T. A.: 1993, 'Simultaneous Observation of the Poleward Expansion of Substorm Electrojet Activity and the Tailward Expansion of Current Sheet Disruption in the Near-Earth Magnetotail', *J. Geophys. Res.* **98**, 9285.

Lui, A. T. Y.: 1991, 'A Synthesis of Magnetospheric Substorm Models', *J. Geophys. Res.* **96**, 1849.

Lui, A. T. Y., Akasofu, S. I., Hones, E. W., Bame, S. J., and McIlwain, C. E.: 1976, 'Observations of the Plasma Sheet During a Contracted Oval Substorm in a Prolonged Quiet Period', *J. Geophys. Res.* **81**, 1415.

Lühr, M., Lockwood, M., Sandholt, P. A., Hansen, T. L., and Moretto, T.: 1995, 'Multi-Instrument Ground-Based Observations of a Travelling Convection Vortex Event', *Ann. Geophys.*, in press.

Lysak, R. L. and Lee, D. H.: 1992, 'Response of the Dipole Magnetosphere to Pressure Pulses', *Geophys. Res. Letters* **19**, 937.

McPherron R. L., Angelopoulos, V., Baker, D. N., and Hones, E. W., Jr.: 1993, 'Is There a Near-Earth Neutral Line?', *Adv. Space Res.* **13** (4), 1730.

Murphree, J. S. and Cogger, L. L.: 1981, 'Observed Connections Between Apparent Polar Cap Features and the Instantaneous Diffuse Auroral Oval', *Planetary Space Sci.* **29**, 1143.

Murphree, J. S., Elphinstone, R. D., Cogger, L. L., and Wallis, D. D.: 1989, 'Short-Term Dynamics of the High-Latitude Auroral Distribution', *J. Geophys. Res.* **94**, 6969.

Moldwin, M. B. and Hughes, W. J.: 1992, 'On the Formation and Evolution of Plasmoids: A Survey of ISEE 3 Geotail Data', *J. Geophys. Res.* **97**, 19 259.

Nakamura, R., Baker, D. N., Fairfield, D. H., Mitchell, D. G., McPherron, R. L., and Hones, E. W., Jr.: 1994, 'Plasma Flow and Magnetic Field Characteristics Near the Midtail Neutral Sheet', *J. Geophys. Res.* **99**, 23 591.

Opgenoorth, H. J.: 1993, 'Coordination of Ground-Based Observations with Cluster', in *Cluster: Mission, Payload and Supporting Activities*, ESA SP-1159, p. 301.

Opgenoorth, H. J., Bromage, B., Fontaine, D., La Hoz, C., Huuskonen, A., Kohl, H., Løvhaug, P., Wannberg, G., Gustavsson, G., Murphree, J. S., Eliasson, L., Marklund, G., Potemra, T. A., Kirkwood, S., Nielsen, E., and Wahlund, J. E.: 1989, 'Coordinated Observations with EISCAT and the Viking Satellite: the Decay of a Westward Travelling Surge', *Ann. Geophys.* **7**, 479.

Opgenoorth, H. J., Persson, M. A. L., Pulkkinen, T. I., and Pellinen, R. J.: 1994, 'The Recovery Phase of Magnetospheric Substorms and Its Association with Morning-Sector Aurora', *J. Geophys. Res.* **99**, 4115.

Owen, C. J. and Slavin, J. A.: 1992, 'Energetic Ion Events Associated with Travelling Compression Regions', in *Proceedings of the First International Conference on Substorms, (ICS-1)*, Kiruna, Swede, March 1992, ESA SP-335, p. 365.

Pellinen, R. J., Koskinen, H. I. J., Pulkinen, T. I., Murphree, J. S., Rostoker, G., and Opgenoorth, H. J.: 1990, 'Satellite and Ground-Based Observations of a Fading Transpolar Arc', *J. Geophys. Res.* **95**, 5817.

Pellinen, R. J., Opgenoorth, H. J., and Pulkkinen, T. I.: 1992, 'Substorm Recovery Phase: Relationship to Next Activation', *Proceedings of the International Conference on Substorms (ICS-1)*, Kiruna, Sweden, March 1992, ESA SP-335, p. 469.

Persson, M. A. L., Opgenoorth, H. J., Pulkkinen, T. I., Ericsson, A. I., Dovner, P. O., Reeves, G. D., Belian, R. D., Andre, M., Blomberg, L. G., Erlandson, R. E., Boehm, M. H., Aikio, A. T., and Häggström, I.: 1994a 'Near-Earth Substorm Onset: A Coordinated Study', *Geophys. Res. Letters* **21**, 1875.

Persson, M. A. L., Aikio, A., and Opgenoorth, H. J.: 1994b, 'Satellite Ground-Based Coordination: Late Growth Phase and Early Expansion Phase of a Substorm', in *Proceedings of the International Conference on Substorms (ICS-2)*, Fairbanks, Alaska, March 1994, in press.

Pinnock, M., Rodger, A. S., Dudeney, J. R., Baker, K. B., Newell, P. T., Greenwald, R. A., and Greenspan, M. E.: 1993, 'Observations of an Enhanced Convection Channel in the Cusp Ionosphere', *J. Geophys. Res.* **98**, 3767.

Pinnock, M., Rodger, A. S., Dudeney, J. R., Rich, F., and Baker, K. B.: 1995, 'High Spatial and Temporal Resolution Observations of the Ionospheric Cusp', *Ann. Geophys.* **13**, 919.

Potemra, T. A., Erlandson, R. E., Zanetti, L. J., Arnold, R. L., Woch, J., and Friis-Christensen, E.: 1992, 'The Dynamic Cusp', *J. Geophys. Res.* **97**, 2835.

Richmond, A. D.: 1992, 'Assimilative Mapping of Ionospheric Electrodynamics', *Adv. Space Res.* **12** (6), 59.

Rodger, A. S., Pinnock, M., Dudeney, J. R., Baker, K. B., and Greenwald, R. A.: 1994a, 'A New Mechanism for Polar Patch Formation', *J. Geophys. Res.* **99**, 6425.

Rodger, A. S., Pinnock, M., Dudeney, J. R., Waterman, J., de la Beaujardère, O., and Baker, K. B.: 1994b, 'Simultaneous Two-Hemisphere Observations of the Presence of Polar Patches in the Night Ionosphere', *Ann. Geophys.* **12**, 642.

Rodriguez-Canabal, J., Warhaut, M., Schmidt, R., and Bello-Mora, M.: 1993, 'The Cluster Orbit and Mission Scenario', in *Cluster: Mission, Payload and Supporting Activities*, ESA SP-1159, p. 259.

Sanchez, E. R., Siscoe, G. L., and Meng, C. I.: 1991, 'Inductive Attenuation of the Transpolar Voltage', *Geophys. Res. Letters* **18**, 1173.

Sandford, B. P.: 1964, 'Aurora and Airglow Intensity Variations with Time and Magnetic Activity at Southern High Latitudes', *J. Atmospheric Terrest. Phys.* **26**, 749.

Sandholt, P. E., Egeland, A., Holtet, J. A., Lybekk, B., Svenes, K., and Asheim, S.: 1985, 'Large- and Small-Scale Dynamics of the Polar Cusp', *J. Geophys. Res.* **90**, 4407.

Sandholt, P. E., Lockwood, M., Oguti, T., Cowley, S. W. H., Freeman, K. S. C., Lybekk, B., Egeland, A., and Willis, D. M.: 1990, 'Midday Auroral Breakup Events and Related Energy and Momentum Transfer from the Magnetosheath', *J. Geophys. Res.* **95**, 1039.

Sergeev, V. A., Elphic, R. C., Mozer, F. S., Saint-Marc, a., and Sauvard, J. A.: 1992, 'A Two-Satellite Study of Nightside Flux Transfer Events in the Plasma Sheet', *Planetary Space Sci.* **11**, 1551.

Slavin, J. A., Smith, M. F., Mazur, E. L., Baker, D. N., Iyemori, T., Singer, H. J., and Greenstadt, E. W.: 1992, 'ISEE 3 Plasmoid and TCR Observations During an Extended Interval of Substorm Activity', *Geophys. Res. Letters* **19**, 825.

Smith, M. F. and Rodgers, D. J.: 1991, 'Ion Distributions at the Dayside Magnetopause', *J. Geophys. Res.* **96**, 11 617.

Untiedt, J., Pellinen, R., Küppers, F., Opgenoorth, H. J., Pelster, W. D., Baumjohann, W., Ranta, R., Kangas, J., Chechowsky, P., and Heikkila, W.: 1978, 'Observations of the Initial Development of an Auroral and Magnetic Substorm at Magnetic Midnight', *J. Geophys.* **45**, 41.

Weiss, L. A., Reiff, P. H., Moses, J. J., Heelis, R. A., and Moore, B. D.: 1992, 'Energy Dissipation in Substorms', in *Proceedings of the International Conference on Substorms (ICS-1)*, Kiruna, Sweden, March 1992, ESA SP-335, p. 309.

Weiss, L. A., Reiff, P. H., Carlson, H. C., Weber, E. J., and Lockwood, M.: 'Flow-Aligned Jets in the Magnetospheric Cusp: Results from the GEM Pilot Programme', *J. Geophys. Res.*, in press.

Yeoman, T. K., Lühr, H., Friedel, R. W. H., Coles, S., Grande, M., Perry, C. H., Lester, M., Smith, P. N., Singer, H. J., and Orr, D.: 'CRRES/Ground-Based Multi-Instrument Observations of an Interval of Substorm Activity', *Ann. Geophys.* **12**, 1158.

DIRECTORY OF CLUSTER COMMUNITY MEMBERS

(24 May, 1996)

1. Brief Guide on Directory

This directory consists of the address, telephone, fax and e-mail of all Cluster Principal investigators and Co- investigators. In addition the persons from the Cluster Science Data System steering committee and implementation working group, the Joint Science Operation Centre and the persons involved in Cluster in the European Space Operation Centre are also included.

Abbreviation guide:

Cluster Experiments

ASPOC	Active Spacecraft Potential Control
CIS	Cluster Ion Spectrometry
DWP	Digital Wave Processor
EDI	Electron Drift Instrument
EFW	Electric Fields and Waves
FGM	Fluxgate Magnetometer
PEACE	Plasma Electron and Current Experiment
RAPID	Research with Adaptive Particle Imaging Detectors
STAFF	Spatio-Temporal Analysis of Field Fluctuations Experiment
WBD	Wide Band Data
WHISPER	Waves of High Frequency and Sounder for Probing of Electron Density by Relaxation

Others

CSDS	Cluster Science Data System
CoI	Co-Investigator
DCM	Data Centre Manager
ESOC	European Space Operations Centre
ESTEC	European Space Technology Centre
IWG	Implementation Working Group (CSDS)

Space Science Reviews **79:** 639–658, 1997.

JSOC	Joint Science Operations Centre
PI	Principal Investigator
SC	Steering Committee (CSDS)
TM	Technical Manager

This directory may be obtained from the World Wide Web at:
ftp :: //ftp.estec.esa.nl/pub/csds/ps/html/directory.htm
or by anonymous FTP from ftp.estec.esa.nl (131.176.1.105):
directory *pub/csds/ps/address* and filenames are
cluster_directory.ps (PostScript), and *cluster_directory.txt* (ASCII).

For further information, please contact:

C. P. Escoubet
Space Science Department
ESA/ESTEC
Postbus 299, Noordwijk, The Netherlands
Tel: +31-71-5653454
Fax: +31-71-5654697
Email: cpescoubet@estec.esa.nl

Acuna, Dr Mario H., NASA, Goddard Space Flight Center, Code 695, Greenbelt, MD 20771, U.S.A.
Tel: 1-301-286-7258, Fax: 1-301-286-1683, Email: lepvax::u2mha
CoI: FGM

Aggson, Dr Thomas, NASA, Goddard Space Flight Center, Code 696, Greenbelt, MD 20771, U.S.A.
Tel: , Fax: , Email:
CoI: EFW

Ahlen, Dr Lennart, Swedish Institute of Space Physics, Uppsala Division, S-75591 Uppsala, Sweden
Tel: 46-18-303631, Fax: +46-18-403100, Email: ala@irfu.se
EFW, TM

Alcayde, Dr Denis, CESR/CNRS, B.P. 4346, 9 Avenue du Colonel Roche, F-31029 Toulouse Cedex, France
Tel: +33-61-556677, Fax: +33-61-556701, Email: alcayde@cesr.cnes.fr
CoI: CIS

Allen, Dr A. J., Queen Mary and Westfield College, Astronomy Unit, Mile End Road, London, E1 4NS, U.K.
Tel: +44-171-775-3174, Fax: +44-181-981-9587, Email: a.j.Allen@qmw.ac.uk
IWG

Alleyne, Dr Hugo, University of Sheffield, Dept. of Physics, Hicks Building, Hounsfield Rd., Sheffield, S3 7RH, U.K.
Tel: +44-114-2824354, Fax: +44-114-2728079, Email: h.alleyne@sheffield.ac.uk
PI: DWP, CoI: WHISPER

Alsop, Dr Chris, Mullard Space Science Laboratory, Univ. College London, Dept. of Physics, Holmbury St. Mary, Dorking, Surrey, RH5 6NT, U.K.
Tel: +44-1483-204167, Fax: +44-1483-278312, Email: ca@mssl.ucl.ac.uk
CoI: PEACE, TM

Alvisi, Mr. Gian Franco, ESA/ESRIN, Via Galileo Galilei, Casella Postale 64, 00044 Frascati, Italy
Tel: +39-6-94180248, Fax: +39-6-94180649, Email: Galvisi@esrin.esa.it
SC
Amata, Dr Ermano, CNR-Istituto Fisica Spazio Interplanetario, Cas. Postale 27, 00044 Frascati, Italy
Tel: 39-694-186233, Fax: 39-694-26814, Email: amata@hp.ifsi.fra.cnr.it
CoI: EDI CIS
Arballo, Dr John K., Jet Propulsion Laboratory, MS 169-506, 4800 Oak Grove Drive , Pasadena, CA 91109, U.S.A.
Tel: +1-818-354-9110, Fax: +1-818-354-8895, Email: jarballo@jplsp.jpl.nasa.gov
Axford, Prof. William .I., Max-Planck-Institut für Aeronomie, Postfach 20, D-37189 Katlenburg-Lindau, Germany
Tel: 49-5556-979-0, Fax: +49-5556-979-240, Email: axford@linmpi.mpae.gwdg.de
CoI: RAPID
Bahnsen, Dr Axel, Danish Space Research Inst., Lundoftevej 7, DK-2800 Lyngby, Denmark
Tel: +45-4588-2277, Fax: , Email:
CoI: STAFF
Baker, Dr Daniel .N., University of Colorado, LASP, Boulder, CO 80309, U.S.A.
Tel: 1-303-492-4509, Fax: 1-303-492-6444, Email: baker@zodiac.colorado.edu
CoI: RAPID
Balikhin, Dr M., LPCE/CNRS, Laboratoire de Physique et Chimie, de l'Environnement, 3A Ave. de la Recherche Scientifique, F-45071 Orleans Cedex 2, France
Tel: , Fax: , Email:
CoI: DWP
Balogh, Prof. Andre, Imperial College, The Blackett Laboratory, Space and Atmospheric Physics Group, Prince Consort Road, London, SW7 2BZ, U.K.
Tel: 44-171-594-7768, Fax: 44-171-594-7772, Email: s.balogh@ic.ac.uk
PI: FGM, SC
Balsiger, Prof. Hans, Universitaet Bern, Physikalisches Institut, Sidlerstrasse, 5, CH-3012 Bern, Switzerland
Tel: +41-31-631-4414, Fax: +41-31-631-4405, Email:
CoI: CIS
Bame, Dr Samuel J., Los Alamos National Laboratory, Space and Atmospheric Sciences (NIS-1), Mail Stop D466, Los Alamos, NM 87545, U.S.A.
Tel: +1-505-667-5308, Fax: +1-505-665-7395, Email: sbame@lanl.gov
CoI: PEACE
Barraclough, Dr Bruce L., Los Alamos National Laboratory, Space and Atmospheric Sciences (NIS-1), Mail Stop D466, Los Alamos, NM 87545, U.S.A.
Tel: +1-505-667-8244, Fax: +1-505-665-7395, Email: bbarraclough@lanl.gov
CoI: PEACE
Bauer, Mr. Otto H., MPI für Extraterrestrische Physik, Postfach 1603, D-85740 Garching, Germany
Tel: +49-89-3299-3591, Fax: +49-89-3299-3569, Email: mpe::ohb
CoI: EDI
Baumjohann, Dr Wolfgang, MPI für Extraterrestrische Physik, Postfach 1603, D-85740 Garching, Germany
Tel: +49-89-3299-3539, Fax: +49-89-3299-3569, Email: bj@mpe-garching.mpg.de
CoI: EDI, SC
Bavassano-Cattaneo, Dr Maria Bice, CNR-Istituto Fisica Spazio Interplanetario, Cas. Postale 27, 00044 Frascati, Italy
Tel: 39-694186231, Fax: 39-69426814, Email: bice@ifsi.fra.cnr.it
CoI: CIS
Bekkum, Mr. Huib v., ESA/ESTEC (PKC), Postbus 299, 2200 AG Noordwijk, The Netherlands
Tel: +31-71-5653427, Fax: +31-71-5656280, Email: hvbekkum@vmprofs.estec.esa.nl
IWG
Belian, Dr Richard D., Los Alamos National Laboratory, MS D438, POB 1663, Los Alamos, NM 87545, U.S.A.

Tel: +1-505-667-9714, Fax: +1-505-667-6937, Email: essdp1::belian
CoI: RAPID
Belluci, Dr G.C., CNR-Istituto Fisica Spazio Interplanetario, Cas. Postale 27, 00044 Frascati, Italy
Tel: , Fax: 39-69426814, Email:
CoI: CIS
Berchem, Dr Jean, University of California at Los Angeles, Inst. Geophys. & Planet. Physics, 405 Hilgard Avenue, Los Angeles, CA 90095-1567, U.S.A.
Tel: +1-310-206-2849, Fax: +1-310-206-3051, Email: jberchem@igpp.ucla.edu
Berthelier, Dr Jean-Jacques., CETP, 10/12 Avenue de l'Europe, 78140 Velizy, France
Tel: +33-1-39254905, Fax: +33-1-39254922, Email: jean-jacques.berthelier@cetp.ipsl.fr
CoI: PEACE
Blake, Dr J. Bernard, Aerospace Corporation, MS M2-259, P.O. Box 92957, Los Angeles, CA 90009, U.S.A.
Tel: +1-310-336-7078, Fax: +1-310-336-1636, Email: dirac2::blake
CoI: RAPID
Blanc, Dr Michel, Observatoire Midi Pyrenees, 14 av. Edouard Belin, F-31400 Toulouse, France
Tel: +33-61-332850, Fax: +33-61-536722, Email: blanc@obs-mip.fr
CoI: PEACE
Blomberg, Dr Lars, Royal Institute of Technology, Alfven Laboratory, Division of Plasma Physics, S-10044 Stockholm, Sweden
Tel: +46-8-7907697, Fax: +46-8-245431, Email: blomberg@plasma.kth.se
CoI: EFW
Borg, Dr H., Swedish Institute for Space Physics, Institutet foer Rymdfysik (IRF), University of Umea, S-90187 Umea, Sweden
Tel: +46-90-165043, Fax: +46-90-166673, Email: hans.borg@physics.umu.se
CoI: RAPID
Bosqued, Dr Jean-Michel, CESR/CNRS, B.P. 4346, 9 Avenue du Colonel Roche, F-31029 Toulouse Cedex, France
Tel: +33-61-556673, Fax: +33-61-556701, Email: Jean-Michel.Bosqued@cesr.cnes.fr
CoI: CIS
Bostrom, Dr Rolf, Swedish Institute of Space Physics, Uppsala Division, S-75591 Uppsala, Sweden
Tel: 46-18-303610, Fax: +46-18-403100, Email: rb@irfu.se
CoI: EFW
Bougeret, Dr Jean-Louis, Observatoire de Paris-Meudon, DESPA, Place Jules Janssen, F-92195 Meudon Principal Cedex, France
Tel: +33-1-4534-7704, Fax: +33-1-4507-2806, Email: meudon::bougeret
CoI: WBD
Bruno, Dr Roberto, CNR-Istituto Fisica Spazio Interplanetario, CP27, 00044 Frascati, Italy
Tel: 39-6-94186235, Fax: 39-6-9426814, Email: bruno@ifsi.fra.cnr.it
CoI: CIS
Buchert, Dr Stephan, TU Braunschweig, Institut für Geophysik und Meteorologie, Mendelssohnstr. 3, D-38106 Braunschweig, Germany
Tel: +49-531-391-5215, Fax: +49-531-391-5220, Email: scb@geophys.nat.tu-bs.de
IWG
Buchner, Dr Joerg, MPI für extraterrestriche Physik, External Branch, Rudower Chaussee 5, D-12489 Berlin-Adlershof, Germany
Tel: +49 (30) 6392 3937, Fax: +49 (30) 6392-3937, Email: jb@mpe.FTA-berlin.de
Burch, Dr James L., Southwest Research Institute, P.O. Drawer 28510, 6220 Culebra road, San Antonio, TX 78284, U.S.A.
Tel: 1-210-522-2526, Fax: +1-210-647-4325, Email: jburch@swri.edu
CoI: PEACE
Burgess, Dr David, Queen Mary and Westfield College, Astronomy Unit, Mile End Road, London, E1 4NS, U.K.
Tel: +44-171-975-55460, Fax: +44-181-981-9587, Email: d.burgess@qmw.ac.uk
CoI: DWP

Canu, Dr Patrick, CETP, 10/12 Avenue de L'Europe, F-78140 Velizy, France
Tel: +33-1-39254888, Fax: +33-1-39254872, Email: patrick.canu@cetp.ipsl.fr
CoI: WHISPER WBD, IWG

Cao, Mr. B.-J., CESR/CNRS, B.P. 4346, 9 Avenue du Colonel Roche, F-31029 Toulouse Cedex, France
Tel: +33-61-556679, Fax: +33-61-556701, Email:
CoI: CIS

Carlson, Dr Charles W., University of California, Space Science Laboratory, Berkeley, CA 94720, U.S.A.
Tel: +1-510-642-1478, Fax: +1-510-643-8302, Email: cwc@ssl.berkeley.edu
CoI: CIS

Carter, Dr Paul, Mullard Space Science Laboratory, Univ. College London, Dept. of Physics, Holmbury St. Mary, Dorking, Surrey, RH5 6NT, U.K.
Tel: +44-1483-204105, Fax: +44-1483-278312, Email: pjc@mssl.ucl.ac.uk
CoI: PEACE

Cattell, Prof. Cynthia A., School of Physics and Astronomy, Tate Laboratory of Physics, 116 Church Street S.E., Minneapolis, MN 55455-0112, U.S.A.
Tel: +1-612-6268918, Fax: +1-612-6262029, Email: cattell@belka.spa.umn.edu
CoI: EFW

Chaizy, Dr Patrick, Rutherford Appleton Laboratory, Chilton, Didcot, Oxon, OX11 0QX, U.K.
Tel: +44-1-235-445437, Fax: +44-1-235-446509, Email: pac@jsoc1.bnsc.rl.ac.uk
JSOC

Chanteur, Dr Gerard, CETP, 10/12 Avenue de L'Europe, F-78140 Velizy, France
Tel: , Fax: +33-1-39254887, Email: gerard.chanteur@cetp.ipsl.fr
STAFF

Chapman, Dr Sandra C., University of Warwick, Physics Department, Coventry, CV4 7AL, U.K.
Tel: +44-1203-523390, Fax: +44-1203-692016, Email: s.c.chapman@warwick.ac.uk
CoI: DWP

Coates, Dr Andrew, Mullard Space Science Laboratory, Univ. College London, Dept. of Physics, Holmbury St. Mary, Dorking, Surrey, RH5 6NT, U.K.
Tel: +44-1483-204145, Fax: +44-1483-278312, Email: ajc@mssl.ucl.ac.uk
CoI: PEACE

Cornilleau-Wehrlin, Dr Nicole, CETP, 10/12 Avenue de L'Europe, F-78140 Velizy, France
Tel: +33-1-39254898, Fax: +33-1-39254887, Email: nicole.cornilleau@cetp.ipsl.fr
PI: STAFF, CoI: EFW WHISPER DWP

Cowley, Prof. Stanley W.H, Department of Physics and Astronomy, University of Leicester, University Road, Leicester, LE1 7RH, U.K.
Tel: 44-116-223-1331, Fax: 44-116-252-3555, Email: swhcl@ion.le.ac.uk
CoI: FGM

Credland, Mr. John, ESA/ESTEC (PK), Postbus 299, 2200 AG Noordwijk, The Netherlands
Tel: +31-71-5653201, Fax: +31-71-5656280, Email: jcredlan@vmprofs.estec.esa.nl
SC

Cros, Dr Alain, CESR/CNRS, B.P. 4346, 9 Avenue du Colonel Roche, F-31029 Toulouse Cedex, France
Tel: +33-61-556674, Fax: +33-61-556701, Email: alain.cros@cesr.cnes.fr
CoI: CIS, TM

Curtis, Dr D., University of California, Space Science Laboratory, Berkeley, CA 94720, U.S.A.
Tel: +1-510-642-5998, Fax: +1-510-643-8302, Email: dwc@ssl.berkeley.edu
CoI: CIS

d'Uston, Dr Claude, CESR/CNRS, B.P. 4346, 9 Avenue du Colonel Roche, F-31029 Toulouse Cedex, France
Tel: +33-61-556672, Fax: +33-61-556701, Email: lionel.duston@cesr.cnes.fr
CoI: CIS

Daly, Dr Patrick W., Max-Planck-Institut für Aeronomie, Postfach 20, D-37189 Katlenburg-Lindau, Germany

Tel: +49-5556-979-279, Fax: +49-5556-979-240, Email: daly@linmpi.mpae.gwdg.de
CoI: RAPID CIS, SC, IWG
Dandouras, Dr Iannis, CESR/CNRS, B.P. 4346, 9 Avenue du Colonel Roche, F-31029 Toulouse Cedex, France
Tel: +33-61-558320, Fax: +33-61-556701, Email: iannis.dandouras@cesr.cnes.fr
CoI: CIS, IWG
de Feraudy, Dr Herve, CETP, 10/12 Avenue de L'Europe, F-78140 Velizy, France
Tel: +33-1-39254903, Fax: +33-1-4529-4872, Email: deferaudy@cetp.ipsl.fr
CoI: WHISPER
de la Porte, Dr Bertrand, CETP, 10/12 Avenue de L'Europe, F-78140 Velizy, France
Tel: +33-1-39254856, Fax: +33-1-39254872, Email: bertrand.delaporte@cetp.ipsl.fr
STAFF
Decreau, Dr Pierrette M.E., LPCE/CNRS, Laboratoire de Physique et Chimie, de l'Environnement, 3A Ave. de la Recherche Scientifique, F-45071 Orleans Cedex 2, France
Tel: +33-38-515281, Fax: +33-38-631234, Email: pdecreau@cnrs-orleans.fr
PI: WHISPER, CoI: DWP EFW STAFF
Di Lellis, Dr A., CNR-Istituto Fisica Spazio Interplanetario, Cas. Postale 27, 00044 Frascati, Italy
Tel: 39-694186244, Fax: 39-69426814, Email: dilellis@ifsi.fra.cnr.it
CoI: CIS
Dimbylow, Dr Trevor, Rutherford Appleton Laboratory, Chilton, Didcot, Oxon, OX11 0QX, U.K.
Tel: +44-1-235-445827, Fax: +44-1-235-446509, Email: t.g.dimbylow@rl.ac.uk
, IWG, JSOC
Donzelli, Mr. Paolo, ESA/ESRIN, Via Galileo Galilei, Casella Postale 64, 00044 Frascati, Italy
Tel: +39-6-94180486, Fax: +39-6-94180649, Email: donzelli@esrin.esa.it
, IWG
Drigani, Mr. Fulvio, ESA/ESTEC (PKC), Postbus 299, 2200 AG Noordwijk, The Netherlands
Tel: +31-71-5654604, Fax: +31-71-5656280, Email: fdrigani@vmprofs.estec.esa.nl
, IWG
Dunford, Dr Erik, Rutherford Appleton Laboratory, Chilton, Didcot, Oxon, OX11 0QX, U.K.
Tel: +44-1-235-445450, Fax: +44-1-235-446667, Email: e.dunford@rl.ac.uk
SC
Egeland, Prof. Alv, University of Oslo, Department of Physics, P.O. Box 1048, Blindern, N-0316 Oslo, Norway
Tel: +47-2285-5672, Fax: +47-2285-5671, Email: alv.egeland@fys.uio.no
CoI: EFW DWP
Eliasson, Dr Lars, Swedish Institute of Space Physics, PO Box 812, S-98128 Kiruna, Sweden
Tel: 46-980-79087, Fax: 46-980-79050, Email: lars@irf.se
CoI: CIS
Ellwood, Mr. John, ESA/ESTEC (PKP), Postbus 299, 2200 AG Noordwijk, The Netherlands
Tel: +31-71-5653507, Fax: +31-71-5656280, Email: jellwood@vmprofs.estec.esa.nl
, SC
Elphic, Dr Richard C., Los Alamos National Laboratory, MS D466, Los Alamos, NM 87545, U.S.A.
Tel: +1-505-665-3693, Fax: +1-505-665-3332, Email: relphic@lanl.gov
CoI: FGM
Escoubet, Dr C. Philippe, ESA/ESTEC (SO), Postbus 299, 2200 AG Noordwijk, The Netherlands
Tel: +31-71-5653454, Fax: +31-71-5654697, Email: cpescoubet@estec.esa.nl
CoI: CIS, IWG
Fairfield, Dr Donald H., NASA, Goddard Space Flight Center, Code 695, Greenbelt, MD 20771, U.S.A.
Tel: 1-301-286-7472, Fax: 1-301-286-3271, Email: u2dhf@lepdhf.gsfc.nasa.gov
CoI: FGM
Falthammar, Prof. Carl-Gunne, Royal Institute of Technology, Alfven Laboratory, Division of Plasma Physics, S-10044 Stockholm, Sweden
Tel: +46-8-7907685, Fax: +46-8-245431, Email: falthammar@plasma.kth.se
CoI: EFW, SC

Fazakerley, Dr Andrew, Mullard Space Science Laboratory, Univ. College London, Dept. of Physics, Holmbury St. Mary, Dorking, Surrey, RH5 6NT, U.K.
Tel: +44-1483-204175, Fax: +44-1483-278312, Email: anf@mssl.ucl.ac.uk
CoI: PEACE
Fehringer, Dr Michael, Institut fur Physik, Forschungszentrum Seibersdorf, A-2444 Seibersdorf, Austria
Tel: +43-2254-780-3137, Fax: +43-2254-74060, Email: fehringer@zdfzsarcs.ac.at
CoI: ASPOC
Fennell, Dr Joseph P., Aerospace Corporation, P.O. Box 92957, Los Angeles, CA 90009, U.S.A.
Tel: +1-310-336-7075, Fax: +1-310-336-1636, Email: joe-fennel@qmail2.aero.org
CoI: RAPID
Ferri, Dr Paolo, ESOC (MOD/SMD/MCS), Robert-Bosch-Str. 5, D-64293 Darmstadt, Germany
Tel: +49-6151-90-2332, Fax: +49-6151-90-485, Email: pferri@esoc.esa.de
, ESOC
Ferron, Mr. Pierre, Rutherford Appleton Laboratory, Chilton, Didcot, Oxon, OX11 0QX, U.K.
Tel: +44-1-235-445437, Fax: +44-1-235-446509, Email: psf@jsoc2.bnsc.rl.ac.uk
, JSOC
Fiala, Dr Vladimir, Institute of Atmospheric Physics, Academy of Science of the Csech Republic, Bocni II-1401, 141 31 Prague 4, Csech Republic
Tel: +42-2-67-103300, Fax: +41-2-76-2528, Email: fiala@ufa.cas.cz
CoI: WHISPER
Fillius, Dr Walker, University of California, San Diego, Center for Astrophysics and Space Sciences, Mail Code 0111, 9500 Gilman Dr, La Jolla, CA 92093-0111, U.S.A.
Tel: +1-619-534-3315, Fax: +1-619-534-0177, Email: wfillius@ucsd.edu
CoI: EDI
Fontaine, Dr Dominique, CETP, 10/12 Avenue de l'Europe, 78140 Velizy, France
Tel: +22-1-39254917, Fax: +33-1-39254922, Email: dominique.fontaine@cetp.ipsl.fr
CoI: PEACE
Formisano, Dr Vittorio, CNR-Istituto Fisica Spazio Interplanetario, Cas. Postale 27, 00044 Frascati, Italy
Tel: 39-6-94186218, Fax: 39-6-9426814, Email: formisan@hp.ifsi.fra.cnr.it
CoI: CIS EDI
Friedel, Dr Reiner, Max-Planck-Institut für Aeronomie, Postfach 20, D-37189 Katlenburg-Lindau, Germany
Tel: +49-5556-979-441, Fax: +49-5556-979-240, Email: friedel@hydra.mpae.gwdg.de
CoI: CIS
Fritz, Dr Theodore A., Boston University, Centre for Space Physics, 725 Commonwealth Avenue, Boston, MA 02215, U.S.A.
Tel: +1-617-353-7446, Fax: +1-617-353-6463, Email: fritz@buasta.bu.edu
CoI: RAPID
Fuselier, Dr Stephen, Lockheed Research Laboratories, Dept. 91-20, Bldg. 252, 3251 Hanover St., Palo Alto, CA 94304, U.S.A.
Tel: +1 (415) 4243334, Fax: 1-415-424-3333, Email: fuselier@space.lockheed.com
Gedalin, Dr Michael, Ben-Gurion University, Department of Physics, P.O.B. 653, Beer-Sheva, Israel
Tel: 972-(0)7-461645, Fax: 972-(0)7-472904, Email: gedalin@bgumail.bgu.ac.il
CoI: DWP
Ghielmetti, Dr Arthur G., Lockheed Research Laboratories, Dept. 91-20, Bldg. 252, 3251 Hanover St., Palo Alto, CA 94304, U.S.A.
Tel: +1 (415) 4243257, Fax: 1-415-424-3333, Email: gmetti@space.lockheed.com
CoI: CIS
Glassmeier, Prof. Karl-Heinz, TU Braunschweig, Institut für Geophysik und Meteorologie, Mendelssohnstr. 3, D-38106 Braunschweig, Germany
Tel: +49-531-391-5214, Fax: +49-531-391-5222, Email: khg@geophys.nat.tu-bs.de
CoI: FGM, IWG
Gliem, Dr Fritz, Inst. f. Datenverarb. Anlagen (IDA), Postfach 3329, Hans Sommer Str. 66, D-38023

Braunschweig, Germany
Tel: +49-531-391-3740, Fax: +49-531-391-4587, Email:
CoI: RAPID
Goldstein, Dr Melvyn L., NASA, Goddard Space Flight Center, Code 692, Greenbelt, MD 20771, U.S.A.
Tel: +1-301-286-7828, Fax: +1-301-286-1683, Email: u2mlg@ruach.gsfc.nasa.gov
CoI: PEACE
Goldstein, Dr Raymond, Jet Propulsion Laboratory, 4800 Oak Grove Drive, Pasadena, CA 91109, U.S.A.
Tel: 1-818 354 0241, Fax: 1-818-354-8895, Email: jplsp::rgoldstein
CoI: ASPOC
Golton, Mr. E., Rutherford Appleton Laboratory, Chilton, Didcot, Oxon, OX11 0QX, U.K.
Tel: +44-1-235-446429, Fax: +44-1-235-446667, Email: e.golton@rl.ac.uk
IWG
Gosling, Dr John T., Los Alamos National Laboratory, Space and Atmospheric Sciences (NIS-1), Mail Stop D466, Los Alamos, NM 87545, U.S.A.
Tel: +1-505-667-5389, Fax: +1-505-665-3332, Email: jgosling@lanl.gov
CoI: PEACE
Gough, Dr M. Paul, University of Sussex, Space Science Centre, School of Engineering, Falmer, Brighton, BN1 9QT, U.K.
Tel: +44-1273-678421, Fax: +44-1273-678399, Email: michaelg@central.sussex.ac.uk
CoI: STAFF DWP
Gramkow, Mr. Bodo A., ESA/ESTEC (PKP), Postbus 299, 2200 AG Noordwijk, The Netherlands
Tel: +31-71-5654564, Fax: +31-71-5656280, Email: bgramkow@vmprofs.estec.esa.nl
IWG
Grande, Dr Manuel, Rutherford Appleton Laboratory, Chilton, Didcot, Oxon, OX11 0QX, U.K.
Tel: +44-1-235-446501, Fax: +44-1-235-445848, Email: m.grande@rl.ac.uk
CoI: PEACE RAPID
Grard, Dr Rejean, ESA/ESTEC (SO), Postbus 299, 2200 AG Noordwijk, The Netherlands
Tel: +31-71-5653596, Fax: +31-71-5654697, Email: rgrard@estec.esa.nl
CoI: ASPOC EFW
Greenwald, Dr Raymond, Aplied Physics Laboratory, John Hopkins University, John Hopkins Rd, Laurel, MD 20723, U.S.A.
Tel: +1-301-953-5408-2184, Fax: +1-301-992-4560, Email: ray.greenwald@jhuapl.edu
CoI: EDI
Gurgiolo, Dr Chris, Southwest Research Institute, P.O. Drawer 28510, 6220 Culebra road, San Antonio, TX 78228-0510, U.S.A.
Tel: +1-210-522-2075, Fax: +1-210-647-4325, Email: chris@bilbo.space.swri.edu
CoI: PEACE
Gurnett, Dr Donald, University of Iowa, Dept. of Physics and Astronomy, Iowa City, Iowa 52242, U.S.A.
Tel: +1-319-335-1697, Fax: +1-319-335-1753, Email: gurnett@iowave.physics.uiowa.edu
PI: WBD, CoI: EFW STAFF WHISPER DWP
Gustafsson, Dr Georg, Swedish Institute of Space Physics, Uppsala Division, S-75591 Uppsala, Sweden
Tel: 46-18-303611, Fax: +46-18-403100, Email: gg@irfu.se
PI: EFW, CoI: STAFF WHISPER WBD DWP
Guttler, Dr Wolfgang, Max-Planck-Institut für Aeronomie, Postfach 20, D-37189 Katlenburg-Lindau, Germany
Tel: 49-5556-979-371, Fax: +49-5556-979-139, Email: wguettler@linmpi.mpae.gwdg.de
RAPID, TM
Haerendel, Dr Gerhard, MPI für Extraterrestrische Physik, Postfach 1603, D-85740 Garching, Germany
Tel: +49-89-3299-3516, Fax: +49-89-3299-3569/-3351, Email: mpe::hae
CoI: EDI WBD

Haernquist, Mr. Gunnar, ESA/ESTEC (PS), Postbus 299, 2200 AG Noordwijk, The Netherlands
Tel: +31-71-5654925, Fax: +31-71-5656280, Email: gharnqui@vmprofs.estec.esa.nl
, IWG
Hall, Dr David S., Rutherford Appleton Laboratory, Building R254, Chilton, Didcot, Oxon, OX11 0QX, U.K.
Tel: +44-1-235-446503, Fax: +44-1-235-445848, Email: d.hall@rl.ac.uk
CoI: RAPID PEACE
Hapgood, Dr Michael, Rutherford Appleton Laboratory, Chilton, Didcot, Oxon, OX11 0QX, U.K.
Tel: +44-1-235-446520, Fax: +44-1-235-446509, Email: m.hapgood@rl.ac.uk
CoI: PEACE, IWG, JSOC
Hardy, Dr D.A., PL/GPSG, Hanscom AFB, MA 01731-3010, U.S.A.
Tel: +1-617-377-3102, Fax: +1-617-377-2491, Email: hardy@zircon.plh.af.mil
CoI: PEACE
Harvey, Dr Christopher C., Observatoire de Paris-Meudon, DESPA, Place Jules Janssen, F-92195 Meudon Principal Cedex, France
Tel: +33-1-4507-2805, Fax: +33-1-4507-2806, Email: harvey@obspm.fr
CoI: STAFF EFW DWP, SC, IWG
Hayakawa, Dr Hajime, Institute of Space and Astronautical Sciences, 3-1-1, Yoshinodai, Sagamihara Kanagawa, 229, Japan
Tel: 81-427513911-2511, Fax: 81-427594236, Email: hayakawa@gtl.isas.ac.jp
CoI: EDI
Helliwell, Dr Robert A., STAR Laboratory, Durand 325, Stanford University, Stanford, CA 94305-4055, U.S.A.
Tel: +1-415-723-3582, Fax: +1-415-723-9251, Email: rah@nova.stanford.edu
CoI: WBD
Hill, Dr Peter, Rutherford Appleton Laboratory, Chilton, Didcot, Oxon, OX11 0QX, U.K.
Tel: +44-1-235-445437, Fax: +44-1-235-446509, Email: pmh@jsoc1.bnsc.rl.ac.uk
, JSOC
Ho, Dr Christian M., Jet Propulsion Laboratory, MS 169-506, 4800 Oak Grove Drive , Pasadena, CA 91109, U.S.A.
Tel: +1-818-354-7894, Fax: +1-818-354-8895, Email: cho@jplsp.jpl.nasa.gov
Holback, Dr Bengt, Swedish Institute of Space Physics, Uppsala Division, S-75591 Uppsala, Sweden
Tel: 46-18-303642, Fax: +46-18-403100, Email: bh@irfu.se
CoI: EFW
Holmgren, Dr Gunnar, Swedish Institute of Space Physics, Uppsala Division, S-75591 Uppsala, Sweden
Tel: 46-18-303631, Fax: +46-18-403100, Email: gh@irfu.se
CoI: EFW DWP, IWG, DCM
Holter, Dr Oivin, University of Oslo, Department of Physics, P.O. Box 1048 Blindern, N-0316 Oslo, Norway
Tel: +47-2285-5636, Fax: +47-2285-5671, Email: oivin.holter@fys.uio.no
CoI: EFW
Holtet, Dr Jan A., University of Oslo, Department of Physics, P.O. Box 1048 Blindern, N-0316 Oslo, Norway
Tel: +47-2285-5669, Fax: +47-2285-5671, Email: j.a.holtet@fys.uio.no
CoI: EFW
Horne, Dr Richard, British Antartic Survey, High Cross, Madingley Road, Cambridge, CB3 0ET, U.K.
Tel: 44-1223-251542, Fax: 44-1223-362616, Email: r.horne@bas.ac.uk
CoI: DWP
Hovestadt, Dr Dieter, MPI für Extraterrestrische Physik, Postfach 1603, D-85740 Garching, Germany
Tel: +49-89-3299-3817, Fax: +49-89-3299-3569, Email: dih@mpesmp.mpe-garching.mpg.de
CoI: CIS
Hubert, Dr Daniel, Observatoire de Paris-Meudon, DESPA, Place Jules Janssen, F-92195 Meudon

Principal Cedex, France
Tel: +33-1-4507-7703, Fax: +33-1-4507-2806, Email: Daniel.Hubert@obspm.fr
CoI: STAFF

Huff, Dr Richard, University of Iowa, Dept. of Physics and Astronomy, Iowa City, Iowa 52242, U.S.A.
Tel: +1-319-335-1934, Fax: +1-319-335-1753, Email: huff@iowave.physics.uiowa.edu
WBD, TM

Hultqvist, Dr Bengt, Swedish Institute of Space Physics, PO Box 812, S-98128 Kiruna, Sweden
Tel: 46-980-79060, Fax: 46-980-15465, Email: hultqv@irf.se
CoI: RAPID

Inan, Prof. Umran S., Stanford University, STAR Laboratory, Durand 321, Stanford, CA 94305, U.S.A.
Tel: +1-415-723-4994, Fax: +1-415-723-9251, Email: inan@nova.stanford.edu
CoI: WBD

Ip, Dr Wing H., Max-Planck-Institut für Aeronomie, Postfach 20, D-37189 Katlenburg-Lindau, Germany
Tel: +49-5556-979-416, Fax: +49-5556-979240, Email: ip@linmpi.mpae.gwdg.de
CoI: RAPID

Iversen, Dr Iver B., Space Physics Workshop, Bakkevej 12, 3460 Birkeroed, 2800 Lyngby, Denmark
Tel: +45 (42) 815642, Fax: +45 (42) 815688, Email:
CoI: WHISPER DWP

Jacquey, Dr Christian, CESR/CNRS, B.P. 4346, 9 Avenue du Colonel Roche, F-31029 Toulouse Cedex, France
Tel: +33-61-556672, Fax: +33-61-556701, Email: christian.jacquey@cesr.cnes.fr
CoI: CIS

Johnstone, Prof. Alan D., Mullard Space Science Laboratory, Univ. College London, Dept. of Physics, Holmbury St. Mary, Dorking, Surrey, RH5 6NT, U.K.
Tel: +44-1483-204144, Fax: +44-1483-278312, Email: adj@mssl.ucl.ac.uk
PI: PEACE,

Kecskemety, Dr Karolyi, KFKI Research Institute, for Particle and Nuclear Physics, P.O. Box 49, H-1525 Budapest, Hungary
Tel: 36-11699-499, Fax: 36-11696-567, Email: kecske@rmki.kfki.hu
CoI: RAPID

Kellogg, Prof. Paul J., University of Minnesota, Dept. of Physics, Minneapolis, Minnesota 55455, U.S.A.
Tel: +1-612-6241669, Fax: +1-612-6262029, Email: waves::kellogg
CoI: EFW

Kerr, Dr Stephen S. , University of California, San Diego, Center for Astrophysics and Space Sciences, Mail Code 0111, 9500 Gilman Dr, La Jolla, CA 92093-0111, U.S.A.
Tel: +1-619-534-0179, Fax: +1-619-534-0177, Email: skerr@ucsd.edu

Kessel, Dr Ramona, NASA, Goddard Space Flight Center, Code 632, Greenbelt, MD 20771, U.S.A.
Tel: +1-301-286-6595, Fax: +1-301-286-1771, Email: kessel@nssdca.gsfc.nasa.gov
CoI: PEACE

Kettmann, Dr Georg, Max-Planck-Institut für Aeronomie, Postfach 20, D-37189
Katlenburg-Lindau, Germany
Tel: +49-5556-979-338, Fax: +49-5556-979-240, Email: kettmann@linmpi.gwdg.de
, IWG

Khurana, Dr Krishan, University of California, Inst. Geophys. & Planet. Physics, Dept. of Earth and Space Sciences, 405 Hilgard Avenue, Los Angeles, CA 90095-1567, U.S.A.
Tel: +1-310-825-8240, Fax: +1-310-206-8042, Email: kkhurana@igpp.ucla.edu

Kintner, Dr Paul, School of Electrical Engineering, Space Plasma Physics Group, 302 Frank H. T. Rhodes Hall, Cornell University, Ithaca, NY 14853, U.S.A.
Tel: +1-607-2555304, Fax: +1-607-2556236, Email: paul@ee.cornell.edu
CoI: EFW

Kistler, Dr Lynn, University of New Hampshire, Space Science Center, Science and Engineering

Research Center, Durham, New Hampshire 03824, U.S.A.
Tel: +1-603-862-1399, Fax: +1-603-862-1915, Email: lynn.kistler@unh.edu
CoI: CIS
Kivelson, Prof. Margaret G., University of California, Inst. Geophys. & Planet. Physics, Dept. of Earth and Space Sciences, 405 Hilgard Avenue, Los Angeles, CA 90095-1567, U.S.A.
Tel: +1-310-825-3435, Fax: +1-310-206-8042, Email: mkivelson@igpp.ucla.edu
CoI: FGM
Klecker, Dr B., MPI für Extraterrestrische Physik, Postfach 1603, D-85740 Garching, Germany
Tel: +49-89-3299-3872, Fax: +49-89-3299-3569, Email: bek@mpe.mpe-garching.mpg.de
CoI: CIS
Klimov, Dr Stanislav, Space Research Institute, Russian Academy of Sciences, Profsoyuznaya 84/32, 117 810 Moscow GSP-7, Russia
Tel: +7-095-331100, Fax: +7-310-7023, Email: klimov@esoc1.iki.rssi.ru
CoI: EFW
Kofman, Dr Wlodek, Ecole Nat. Superieure d'Ingenieurs, Electriciens de Grenoble, Boite Postale 46, F-38402 St. Martin d'Heres Cedex, France
Tel: +33-76-826263, Fax: +33-76-826384, Email: wlodek@cephag.observ-gr.fr
CoI: STAFF DWP
Koons, Dr Harry C., Aerospace Corporation, MS M2-260, P.O. Box 92957, Los Angeles, CA 90009, U.S.A.
Tel: +1-310-336-6519, Fax: +1-310-336-1636, Email: dirac2::koons
CoI: DWP
Korth, Dr Axel, Max-Planck-Institut für Aeronomie, Postfach 20, D-37189 Katlenburg-Lindau, Germany
Tel: +49-5556-979-430, Fax: +49-5556-979240, Email: korth@linmpi.mpae.gwdg.de
CoI: RAPID CIS
Kovrazkhin, Dr Rostislav A., Space Research Institute, Russian Academy of Sciences, Profsoyuznaya 84/32, 117810 Moscow GSP-7, Russia
Tel: , Fax: 7-095-310-7023, Email: kovrazkhin@romance.iki.rssi.ru
CoI: CIS
Krasnosel'skikh, Dr Vladimir V., LPCE/CNRS, Laboratoire de Physique et Chimie, de l'Environnement, 3A Ave. de la Recherche Scientifique, F-45071 Orleans Cedex 2, France
Tel: +33-38-631234, Fax: +33-38-515281, Email: vkrasnos@cnrs-orleans.fr
CoI: DWP WHISPER
Kremser, Prof.. Gerhard, Max-Planck-Institut für Aeronomie, Postfach 20, D-37189 Katlenburg-Lindau, Germany
Tel: +49-5556-979-434, Fax: +49-5556-979-240, Email: kremser@lnkr.mpae.gwdg.de
CoI: RAPID
Kucharek, Dr Harald, MPI für Extraterrestrische Physik, Postfach 1603, D-85740 Garching, Germany
Tel: +49-89-3299-3349, Fax: +49-89-3299-3569, Email: hak@mpe-garching.mpg.de
CoI: CIS
LaBelle, Dr James, Dartmouth College, Dept Physics & Astronomy, Hanover, NH 03755, U.S.A.
Tel: +1-603-646-1394, Fax: +1-603-646-1446, Email: jlabelle@einstein.dartmouth.edu
CoI: DWP
Lacombe, Dr C., Observatoire de Paris-Meudon, DESPA, Place Jules Janssen, F-92195 Meudon Principal Cedex, France
Tel: +33-1-4507-7734, Fax: +33-1-4507-2806, Email: clacombe@megasx.obsmp.fr
CoI: STAFF
Lakhina, Prof. G.S., Indian Institute of Geomagnetism, Colaba, 4000005 Bombay, India
Tel: +91-22-218-9569, Fax: +91-22-218-9568, Email: lakhina@iigmo.ernet.in
CoI: PEACE
Lebreton, Dr Jean-Pierre, ESA/ESTEC (SO), Postbus 299, 2200 AG Noordwijk, The Netherlands
Tel: +31-71-5653596, Fax: +31-71-5654697, Email: jlebreton@estec.esa.nl
CoI: EFW

Lefeuvre, Dr Francois, LPCE/CNRS, Laboratoire de Physique et Chimie de l'Environnement, 3A Ave. de la Recherche Scientifique, F-45071 Orleans Cedex 2, France
Tel: +33-38-515284, Fax: +33-38-631234, Email: lefeuvre@cnrs-orleans.fr
CoI: STAFF
Lennartsson, Dr W., Lockheed Research Laboratories, Dept. 91-20, Bldg. 252, 3251 Hanover St., Palo Alto, CA 94304, U.S.A.
Tel: +1 (415) 4243259, Fax: +1-415-424-3333, Email: lenn@space.lockheed.com
CoI: CIS
Lin, Dr Robert P., University of California, Space Science Laboratory, Berkeley, CA 94720, U.S.A.
Tel: +1-510-6421149, Fax: +1-510-6438302, Email: boblin@ssl.berkeley.edu
CoI: CIS
Lindqvist, Dr Per-Arne, Royal Institute of Technology, Alfven Laboratory, Division of Plasma Physics, S-10044 Stockholm, Sweden
Tel: +46-8-790-7696, Fax: +46-8-245431, Email: lindqvist@plasma.kth.se
CoI: EFW, IWG
Liu, Dr Z.X., Center for Space Science and, Applied Research, PO Box 8701, 100080 Beijing, China
Tel: +86-10-257-6898, Fax: 86-10-257-6921, Email: liu@sun20.cssar.ac.cn
CoI: PEACE, SC, IWG, DCM
Livi, Dr Stefano, Max-Planck-Institut für Aeronomie, Postfach 20, D-37189 Katlenburg-Lindau, Germany
Tel: 49-5556-979-429, Fax: +49-5556-979-139, Email: livi@linmpi.mpae.gwdg.de
CoI: RAPID
Lockwood, Dr Michael, Rutherford Appleton Laboratory, EISCAT Group, Chilton, Didcot, Oxon, OX11 0QX, U.K.
Tel: +44-1-235-446496, Fax: +44-1-235-445848, Email: mike@eiscat.ag.rl.ac.uk
CoI: PEACE
Louarn, Dr Philippe, Observatoire Midi Pyrenees, 14 ev. Edouard Belin, F-31400 Toulouse, France
Tel: +33-61332946, Fax: +33-61-536722, Email: louarn@srvdec.obs-mip.fr
CoI: STAFF
Luehr, Prof. Dr Hermann, TU Braunschweig, Institut für Geophysik und Meteorologie, Mendelssohnstr. 3, D-38106 Braunschweig, Germany
Tel: +49-531-391-5222, Fax: +49-531-391-5220, Email: luehr@geophys.nat.tu-bs.de
CoI: FGM
Lundin, Dr Rickard, Swedish Institute of Space Physics, PO Box 812, S-98128 Kiruna, Sweden
Tel: 46-980-79063, Fax: 46-980-79050, Email: rickard@irf.se
CoI: CIS
Lyons, Dr Lawrance R., Aerospace Corporation, POB 92957, Los Angeles, CA 90009, U.S.A.
Tel: +1-310-336-5470, Fax: , Email: dirac2::lyons
CoI: RAPID
Manning, Mr. Robert, Observatoire de Paris-Meudon, DESPA, Place Jules Janssen, F-92195 Meudon Principal Cedex, France
Tel: +33-1-4507-7686, Fax: +33-1-4507-2806, Email: manning@obspm.fr
CoI: EFW
Marklund, Dr Goran, Royal Institute of Technology, Alfven Laboratory, Division of Plasma Physics, S-10044 Stockholm, Sweden
Tel: +46-8-790-7695, Fax: +46-8-245431, Email: marklund@plasma.kth.se
CoI: EFW
Markwardt, Dr Matthias, MPI für extraterrestriche Physik, External Branch, Rudower Chaussee 5, D-12489 Berlin, Germany
Tel: +49 (30) 6392 3934, Fax: +49 (30) 6392 3939/3929, Email: mm@mpe.fta-berlin.de
, IWG, DCM
Martin, Mr. Warren L., Jet Propulsion Laboratory, MS 303-401, 4800 Oak Grove Drive, Pasadena, CA 91109, U.S.A.
Tel: +1-818-354-5635, Fax: +1-818-393-1692, Email: warren.l.martin@jpl.nasa.gov
CoI: WBD

Matthaeus, Dr William, University of Delaware, Bartol Research Institute, Newark, DE 19717, U.S.A.
Tel: +1-302-831-2780, Fax: +1-302-831-1843, Email: yswhm@bartol.udel.edu
CoI: PEACE
Maynard, Dr Nelson, Mission Research Corporation, 1 Terra Bld., Suit 302, Nashua, New Hampshire 03062-2801, U.S.A.
Tel: +1-603-8910070, Fax: +1-603-8910088, Email: afgl::maynard
CoI: EFW
Mazelle, Dr Christian, CESR/CNRS, B.P. 4346, 9 Avenue du Colonel Roche, F-31029 Toulouse Cedex, France
Tel: +33-61-558320, Fax: +33-61-556672, Email: christian.mazelle@cesr.cnes.fr
McCarthy, Dr M., University of Washington, GEOPHYSICS, Box 351650, Rm. 202, ATG Bldg., Seattle, WA 98195-1650, U.S.A.
Tel: 1-206-685-2543, Fax: 206-685-3815, Email: mccarthy@geophys.washington.edu
CoI: CIS
McComas, Dr David, Los Alamos National Laboratory, Space and Atmospheric Sciences (NIS-1), Mail Stop D466, Los Alamos, NM 87545, U.S.A.
Tel: +1-505-667-0138, Fax: +1-505-665-3332, Email: dmccomas@lanl.gov
CoI: PEACE
McFadden, Dr James P., University of California, Space Science Laboratory, Berkeley, CA 94720, U.S.A.
Tel: +1-510-642-9918, Fax: +1-510-643-8302, Email: mcfadden@ssl.berkeley.edu
CoI: CIS
McIlwain, Prof. Carl E., University of California, San Diego, Center for Astrophysics and Space Sciences, Mail Code 0111, 9500 Gilman Dr, La Jolla, CA 92093-0111, U.S.A.
Tel: +1-619-534-3314, Fax: +1-619-534-0177, Email: cmcilwain@ucsd.edu
CoI: EDI
McKenna-Lawlor, Prof. Susan, St. Patrick's College, Experimental Physics Department, Maynooth, Ireland
Tel: 353-16-286-788, Fax: 353-16-286-470, Email: stil@vax1.may.ie
CoI: RAPID
Merri, Dr Mario, ESOC (FCSD/DPD), Robert-Bosch-Str. 5, D-64293 Darmstadt, Germany
Tel: +49-6151-90-2292, Fax: +49-6151-90-3010, Email: mmerri@esoc.esa.de
, IWG, ESOC
Meyer, Dr Alain, CETP, 10/12 Avenue de L'Europe, F-78140 Velizy, France
Tel: , Fax: , Email:
STAFF, TM
Mish, Dr William, NASA, Goddard Space Flight Center, Code 694, Greenbelt, MD 20771, U.S.A.
Tel: +1-301-286-5444, Fax: +1-301-286-1683, Email: wmish@istp1.gsfc.nasa.gov
SC, IWG, DCM
Moebius, Dr Eberhard, University of New Hampshire, Space Science Center, Science and Engineering Research Center, Durham, New Hampshire 03824, U.S.A.
Tel: +1-603-862-1397, Fax: +1-603-862-1915, Email: moebius@rotor.sr.unh.edu
CoI: CIS
Motschmann, Dr Uwe, TU Braunschweig, Institut für Geophysik und Meteorologie, Mendelssohnstr. 3, D-38106 Braunschweig, Germany
Tel: +49-531-391-5225, Fax: +49-531-391-5220, Email: uwe@geophys.nat.tu-bs.de
Moulin, Mr. Serge, ESA/ESRIN, Via Galileo Galilei, Casella Postale 64, 00044 Frascati, Italy
Tel: +39-6-94180275, Fax: +39-6-94180649, Email: moulin@esrin.esa.it
, IWG
Mozer, Prof. Forrest S., University of California, Space Science Laboratory, Berkeley, CA 94720, U.S.A.
Tel: +1-510-6420549, Fax: +1-510-6437629, Email: fmozer@ssl.berkeley.edu
CoI: EFW DWP
Mursula, Prof. Kalevi, University of Oulu, Department of Physical Sciences, FIN-90570 Oulu, Fin-

land
Tel: 358-81-5531366, Fax: 358-81-5531287, Email: kalevi.Mursula@oulu.fi
CoI: RAPID EFW
Musmann, Dr Gunter, TU Braunschweig, Institut für Geophysik und Meteorologie, Mendelssohnstr. 3, D-38106 Braunschweig, Germany
Tel: +49-531-391-5217, Fax: +49-531-391-8126, Email: musmann@geophys.nat.tu-bs.de
CoI: FGM
Mutel, Dr R.L., University of Iowa, Dept. of Physics and Astronomy, Iowa City, Iowa 52242, U.S.A.
Tel: +1-319-335-1950, Fax: +1-319-335-1753, Email: rlm@astro.physics.uiowa.edu
CoI: WBD
Nakamura, Dr Masato, University of Tokyo, STP Dept. of Earth & Planet. Physics, 7-3-1, Hongo, Bonkyo-ku, Tokyo 113, Japan
Tel: +81-30-815-9215, Fax: +81-3-3818-0745, Email: mnakamur@grl.s.u-tokyo.ac.jp
CoI: EDI
Narheim, Mr. Bjorn, Norwegian Defence Research Establishment, Division of Electronics, P.O. Box 25, N-2007 Kjeller, Norway
Tel: +47-6380-7308, Fax: 47-6380-7212, Email: btn@ffi.no
CoI: ASPOC PEACE
Nesbit, Mr. Mark, ESA/ESTEC (PK), Postbus 299, 2200 AG Noordwijk, The Netherlands
Tel: +31-71-5655191, Fax: +31-71-5656280, Email: mdn@cluster.estec.esa.nl
, IWG
Neubauer, Prof. Fritz M., Inst. für Geophysik & Meteorologie, Universitaet zu Koeln, Albertus-Magnus-Platz, D-50923 Koeln, Germany
Tel: +49-221-470-2310, Fax: +49-221-470-5198, Email: neubauer@geo.Uni-Koeln.de
CoI: FGM
Noyes, Mr. Jeff, ESA/ESTEC (PKP), Postbus 299, 2200 AG Noordwijk, The Netherlands
Tel: +31-71-5653433, Fax: +31-71-5656280, Email: jnoyes@vmprofs.estec.esa.nl
, IWG
Olsen, Dr Richard C., Naval Postgraduate School, Physics Department, Monterey, CA 93943, U.S.A.
Tel: 1-408-646-2019, Fax: 1-408-646-2834, Email: olsen@physics.nps.navy.mil
CoI: ASPOC
Opgenoorth, Dr Hermann, Swedish Institute of Space Physics, Uppsala Division, S-75591 Uppsala, Sweden
Tel: 46-18-303661, Fax: +46-18-403100, Email: opg@irfu.se
SC
Papamastorakis, Prof. Ioannis, University of Crete, Physics Department, Leof Knossou, P.O. Box 2208, 71409 Iraklion, Greece
Tel: 30-81-239757, Fax: 30-81-239735, Email: papamast@iesl.forth.gr
CoI: CIS
Parks, Prof. George K., University of Washington, GEOPHYSICS, Box 351650, Rm. 202, ATG Bldg., Seattle, WA 98195-1650, U.S.A.
Tel: 206-543-0953, Fax: 206-685-3815, Email: parks@geophys.washington.edu
CoI: CIS
Parrot, Dr Michel, LPCE/CNRS, Laboratoire de Physique et Chimie
de l'Environnement, 3A Ave. de la Recherche Scientifique, F-45071 Orleans Cedex 2, France
Tel: +33-38-515291, Fax: +33-38-631234, Email: mparrot@cnrs-orleans.fr
CoI: STAFF DWP, IWG
Paschmann, Dr Goetz, MPI für Extraterrestrische Physik, Postfach 1603, D-85740 Garching, Germany
Tel: +49-89-3299-3868, Fax: +49-89-3299-3569, Email: gep@mpe-garching.mpg.de
PI: EDI, CoI: CIS
Pecseli, Prof. Hans L., University of Oslo, Department of Physics, P.O. Box 1048 Blindern, N-0316 Oslo, Norway
Tel: +47-2285-5637, Fax: +47-2285-5671, Email: hans.pecseli@fys.uio.no
CoI: EFW

Pedersen, Dr Arne, ESA/ESTEC (SI), Postbus 299, 2200 AG Noordwijk, The Netherlands
Tel: +31-71-5653594, Fax: +31-71-5654699, Email: apederse@estec.esa.nl
CoI: ASPOC STAFF EFW EDI
Pedersen, Dr Brit M., Observatoire de Paris-Meudon, DESPA, Place Jules Janssen, F-92195 Meudon Principal Cedex, France
Tel: +33-1-4507-7809, Fax: +33-1-4507-7959, Email: meudon::pedersen
CoI: WBD
Pellat, Prof. Rene, CNES, 2 Place Maurice Quentin, F-75039 Paris Cedex 01, France
Tel: +33-1-44767500, Fax: +33-1-44767676, Email:
CoI: CIS
Pellinen, Prof. Risto, Finish Meteorological Institute, Box 503, FIN-00101 Helsinki, Finland
Tel: 358-01929500, Fax: 358-01929539, Email: risto.pellinen@fmi.fi
CoI: STAFF
Perraut, Dr Sylvaine, CETP, 10/12 Avenue de L'Europe, F-78140 Velizy, France
Tel: +33-1-39254891, Fax: +33-1-39254872, Email: sylvaine.perraut@cetp.ipsl.fr
CoI: STAFF
Pfaff, Dr Robert, NASA, Goddard Space Flight Center, Code 696, Greenbelt, MD 20771, U.S.A.
Tel: +1-301-286-6328, Fax: +1-301-286-1648, Email: rob.pfaff@gsfc.nasa.gov
CoI: EFW
Pincon, Dr Jean-Louis, LPCE/CNRS, Laboratoire de Physique et Chimie, de l'Environnement, 3A Ave. de la Recherche Scientifique, F-45071 Orleans Cedex 2, France
Tel: +33-38-517820, Fax: +33-38-631234, Email: jlpincon@cnrs-orleans.fr
CoI: STAFF
Poussin, Dr Herve, CNES, bpi1501, 18 Avenue Edouard Belin, 31055 Toulouse Cedex, FRANCE
Tel: +33-61-274967, Fax: +33-61-273084, Email: poussin@cst.cnes.fr
IWG
Prado, Mr. Jean-Yves, CNES, CFC, 18 Avenue Edouard Belin, F-31055 Toulouse Cedex, France
Tel: +33-61273704, Fax: +33-61282995, Email: prado@cst.cnes.fr
SC, IWG, DCM
Primdahl, Mr. Fritz, Danish Technical University, Department of Electrophysics, Bldg. 322, DK-2800 Lyngby, Denmark
Tel: +45-45-253445, Fax: +45-45-887133, Email: fp@iris.aef.dtu.dk
CoI: FGM
Prokopiu, Dr Konstantin, MPI für Extraterrestrische Physik, Postfach 1603, D-85740 Garching, Germany
Tel: +49-89-3299-3520, Fax: +49-89-3299-3554,
Email: okp@mpepl.plasma.mpe-garching.mpg.de
IWG
Pu, Dr Zuyin, Peking University, Department of Geophysics, 100871 Beijing, China
Tel: +86-1-2642-..., Fax: +86-1-2642-635, Email:
CoI: RAPID
Quinn, Dr Jack M., Space Science Center, Morse hall, University of New Hampshire, Durham, NH 03824, U.S.A.
Tel: +1-603-862-2976, Fax: +1-603-862-0311, Email: jack.quinn@unh.edu
CoI: EDI
Rajaram, Prof. R., Indian Institute of Geomagnetism, Colaba, 4000005 Bombay, India
Tel: , Fax: +91-22-218-9568, Email:
CoI: PEACE
Randriamboarisson, Dr Orelien, LPCE/CNRS, Laboratoire de Physique et Chimie, de l'Environnement, 3A Ave. de la Recherche Scientifique, F-45071 Orleans Cedex 2, France
Tel: +33-38-517648, Fax: +33-38-631234, Email: randriam@cnrs-orleans.fr
CoI: WHISPER
Rathje, Dr R., Inst. f. Datenverarb. Anlagen (IDA), Postfach 3329, Hans Sommer Str. 66, D-38023 Braunschweig, Germany
Tel: +49-531-391-3798, Fax: +49-531-391-4587, Email: rathje@ida.ing.tu-bs.de

CoI: RAPID
Reiff, Prof. Patricia, Rice University, Dept. Space Physics & Astronomy, MS 108, POB 1892, Houston, TX 77251-1892, U.S.A.
Tel: +1-713-527-4634, Fax: +1-713-285-5143, Email: reiff@alfven.rice.edu
CoI: PEACE
Reme, Prof. Henri, CESR/CNRS, B.P. 4346, 9 Avenue du Colonel Roche, F-31029 Toulouse Cedex, France
Tel: +33-61-556665, Fax: +33-61-556701, Email: henri.reme@cesr.cnes.fr
PI: CIS, , SC
Rezeau, Dr Laurence, CETP, 10/12 Avenue de L'Europe, F-78140 Velizy, France
Tel: +33-1-39254910, Fax: +33-1-39254922, Email: laurence.rezeau@cetp.ipsl.fr
CoI: STAFF
Reznikov, Dr A., IZMIRAN, Academy of Sciences, 142092 Troitsk, Moscow Region, Russia
Tel: , Fax: , Email:
CoI: DWP
Riedler, Prof. Willi, Institut für Weltraumforschung, Oesterr. Akademie der Wissenschaften, Inffeldgasse 12, A-8010 Graz, Austria
Tel: +43-316-463696-14, Fax: +43-316-463697, Email: riedler@fiwf01.tu-graz.ac.at
PI: ASPOC, CoI: FGM, SC
Robert, Dr Patrick, CETP, 10/12 Avenue de L'Europe, F-78140 Velizy, France
Tel: +33-1-39254909, Fax: +33-1-39254922, Email: patrick.robert@cetp.ipsl.fr
CoI: STAFF
Rodriguez–Canabal, Dr Jose, ESOC (MOD-MAS), Robert-Bosch-Str. 5, D-64293 Darmstadt, Germany
Tel: +49-6151-90-2205, Fax: +49-6151-90-2625, Email: jrodrigu@esoc.esa.de
ESOC
Rosenbauer, Dr Helmut, Max-Planck-Institut für Aeronomie, Postfach 20, D-37189 Katlenburg-Lindau, Germany
Tel: +49-5556-979-425, Fax: +49-5556-979-148, Email: rosenbauer@linmpi.mpae.gwdg.de
CoI: CIS
Roth, Dr Ilan, University of California, Space Science Laboratory, Berkeley, CA 94720, U.S.A.
Tel: +1-510-642-1327, Fax: +1-510-643-8302, Email: ilan@ssl.berkeley.edu
CoI: EFW
Roux, Dr Alain, CETP, 10/12 Avenue de L'Europe, F-78140 Velizy, France
Tel: +33 -1-3925-4899, Fax: +33-1-39254887, Email: Alain.roux@cetp.ipsl.fr
CoI: FGM STAFF EFW WBD DWP
Ruedenauer, Prof. Fritz, Institut für Physik, Forschungszentrum Seibersdorf, A-2444 Seibersdorf, Austria
Tel: +43-2254-780-3141, Fax: +43-2254-74060, Email: ruedenauer@zdfzs.arcs.ac.at
CoI: ASPOC
Sandahl, Dr Ingrid, Swedish Institute of Space Physics, PO Box 812, S-98128 Kiruna, Sweden
Tel: 46-980-79084, Fax: 46-980-79050, Email: ingrid@irf.se
CoI: RAPID
Sarris, Prof. Emmanuel T., Demokritos University of Thrace, Division of Telecommunications and Space Science, 67100 Xanthi, Greece
Tel: +30-541-26948, Fax: +30-541-27264, Email: sarris@xanthi.cc.duth.gr
CoI: RAPID
Sauvaud, Dr Jean-Andre, CESR/CNRS, B.P. 4346, 9 Avenue du Colonel Roche, F-31029 Toulouse Cedex, France
Tel: +33-61-556676, Fax: +33-61-556701, Email: jean-andre.sauvaud@cesr.cnes.fr
CoI: CIS
Schmidt, Dr Rudolf, ESA/ESTEC (SO), Postbus 299, 2200 AG Noordwijk, The Netherlands
Tel: +31-71-5653603, Fax: +31-71-5654697, Email: rschmidt@estec.esa.nl
CoI: ASPOC EFW, SC, IWG
Scholer, Dr Manfred, MPI für Extraterrestrische Physik, Postfach 1603, D-85740 Garching, Ger-

many
Tel: +49-89-3299-3821, Fax: +49-89-3299-3569, Email: mbs@mpe-garching.mpg.de
CoI: RAPID CIS
Schwartz, Dr Steve J., Queen Mary and Westfield College, Astronomy Unit, Mile End Road, London, E1 4NS, U.K.
Tel: +44-171-975-5449, Fax: +44-181-981-9587, Email: s.j.schwartz@qmw.ac.uk
CoI: PEACE, IWG
Schwetterle, Mr. Marc, ESA/ESTEC (PKP), Postbus 299, 2200 AG Noordwijk, The Netherlands
Tel: +31-71-5654847, Fax: +31-71-5656280, Email: mschwett@vmprofs.estec.esa.nl
Schwingenschuh, Dr Konrad, Institut für Weltraumforschung, Oesterr. Akademie der Wissenschaften, Inffeldgasse 12, A-8010 Graz, Austria
Tel: +43-316-468730, Fax: +43-316-463697, Email: 29134::schwingen
CoI: FGM
Sckopke, Dr Norbert, MPI für Extraterrestrische Physik, Postfach 1603, D-85740 Garching, Germany
Tel: +49-89-3299-3870, Fax: +49-89-3299-3569, Email: nos@mpe-garching.mpg.de
CoI: EDI CIS, IWG
Scudder, Dr Jack, University of Iowa, Dept. of Physics and Astronomy, Iowa City, Iowa 52242, U.S.A.
Tel: +1-319-335-0804, Fax: +1-319-335-1753, Email: iowasp::scudder
CoI: CIS
Sene, Dr Francois X., LPCE/CNRS, Laboratoire de Physique et Chimie
de l'Environnement, 3A Ave. de la Recherche Scientifique, F-45071 Orleans Cedex 2, France
Tel: +33-38-515277, Fax: +33-38-631234, Email: fsene@cnrs-orleans.fr
WHISPER, TM
Sharber, Dr James, Southwest Research Institute, P.O. Drawer 28510, San Antonio, TX 78284, U.S.A.
Tel: 1-210-522-3853, Fax: +1-210-647-4325, Email: sharber@swri.space.swri.edu
CoI: PEACE
Shelley, Dr Edward G., Lockheed Research Laboratories, Dept. 91-20, Bldg. 252, 3251 Hanover St., Palo Alto, CA 94304, U.S.A.
Tel: +1 (415) 4243253, Fax: +1 (415) 4243333, Email: shelley@space.lockheed.com
CoI: CIS
Singer, Dr Howard J., NOAA R/E/SE, 325 Broadway, Boulder, CO 80303-3328, U.S.A.
Tel: +1-303-4976959, Fax: +1-303-4973645, Email: hsinger@sel.noaa.gov
CoI: EFW
Skogvold, Mr. Stein, ESA/ESTEC (WGE), Postbus 299, 2200 AG Noordwijk, The Netherlands
Tel: +31-71-5654110, Fax: +31-71-5656280, Email: sskogvol@estec.esa.nl
IWG
Slavin, Dr James A., NASA, Goddard Space Flight Center, Code 696, Greenbelt, MD 20771, U.S.A.
Tel: 1-301-286-5839, Fax: 1-301-286-1648, Email: LEPVAX::U6JAS
CoI: FGM
Smiddy, Dr Michael, AFGL/PHG, Hanscom Air Force Base, Hanscom, MA 01731, U.S.A.
Tel: +1-617-3772431, Fax: +1-617-3774498, Email: afgl::smiddy
CoI: EFW
Sonnerup, Dr Bengt U., Darmouth College, Thayer School of Engineering , Hanover, NH 03755, U.S.A.
Tel: +1-603-646-2883, Fax: +1-603-646-3856, Email: sonnerup@dartmouth.edu
CoI: CIS
Soraas, Dr Finn, University of Bergen, Dept. of Physics, Allegaten Garten 55, N-5007 Bergen, Norway
Tel: 47-5521-2714, Fax: 47-5531-8334, Email: finn.soraas@fi.uib.no
CoI: RAPID
Southwood, Prof. David J., Imperial College, The Blackett Laboratory, Space and Atmospheric Physics Group, Prince Consort Road, London, SW7 2BZ, U.K.

Tel: 44-171-594-7500, Fax: 44-171-594-7772, Email: d.southwood@ic.ac.uk
CoI: FGM
Spangler, Dr Steven R., University of Iowa, Dept. of Physics and Astronomy, Iowa City, Iowa 52242, U.S.A.
Tel: +1-319-335-1948, Fax: +1-319-335-1753, Email: srs@astro.physics.uiowa.edu
CoI: WBD
Stasiewicz, Dr Kristof, Swedish Institute of Space Physics, Uppsala Division, S-75591 Uppsala, Sweden
Tel: 46-18-303643, Fax: +46-18-403100, Email: ks@irfu.se
CoI: EFW
Steiger, Dr F., Institut für Physik, Forschungszentrum Seibersdorf, A-2444 Seibersdorf, Austria
Tel: +43-2254-780-0, Fax: +43-2254-74060, Email: aconet::zdvaxa::steiger
CoI: ASPOC
Svenes, Dr Knut Ragnar, Norwegian Defence Research Establishment, Division of Electronics, P.O. Box 25, N-2007 Kjeller, Norway
Tel: 47-6380-7326, Fax: 47-6380-7212, Email: knut.svenes@ffi.no
CoI: PEACE
Sweeney, Dr Mark, ESOC (MOD/SMD/MCS), Robert-Bosch-Str. 5, D-64293 Darmstadt, Germany
Tel: +49-6151-90-2733, Fax: +49-6151-90-485, Email: msweeney@esoc.esa.de
ESOC
Szego, Dr Karolyi, KFKI Research Institute, for Particle and Nuclear Physics, P.O. Box 49, H-1525 Budapest, Hungary
Tel: 36-1-155-1682, Fax: 36-1-169-6567, Email: szego@rmki.kfki.hu
SC
Tanskanen, Prof. Pekka, University of Oulu, Department of Physical Sciences, FIN-90570 Oulu, Finland
Tel: 358-81-5531284, Fax: 358-81-5531287, Email: pekka.tanskanen@oulu.fi
CoI: RAPID EFW
Tatrallyay, Dr Mariella, KFKI Research Institute, for Particle and Nuclear Physics, P.O. Box 49, 1525 Budapest, HUNGARY
Tel: +36-1-155-3494, Fax: +36-1-169-6567, Email: mariella@rmki.kfki.hu
CoI: FGM, IWG, DCM
Temerin, Dr Michael, University of California, Space Science Laboratory, Berkeley, CA 94720, U.S.A.
Tel: +1-510-643-9859, Fax: +1-510-643-8302, Email: temerin@ssl.berkeley.edu
CoI: EFW
Thomlinson, Dr John G., Imperial College, The Blackett Laboratory, Space and Atmospheric Physics Group, Prince Consort Road, London, SW7 2BZ, U.K.
Tel: 44-171-594-6756, Fax: 44-171-594-7772, Email: j.thomlinson@ic.ac.uk
FGM, TM
Thomsen, Dr Michelle, Los Alamos National Laboratory, Space and Atmospheric Sciences (NIS-1), Mail Stop D466, Los Alamos, NM 87545, U.S.A.
Tel: +1-505-667-1210, Fax: +1-505-665-7395, Email: mthomsen@lanl.gov
CoI: PEACE
Thouvenin, Dr Jean-Pierre, CNES/DP/ST, 18 Avenue Edouard Belin, 31055 Toulouse, France
Tel: +33-61274524, Fax: +33-61274013, Email: thouvenin@cnesta.span.cnes.fr
SC
Thrane, Prof. Eivind V., Norwegian Defence Research Establishment, Division of Electronics, P.O. Box 25, N-2007 Kjeller, Norway
Tel: +47-63-807330, Fax: +47-63-807212, Email: eivind.thrane@ffi.no
CoI: EFW
Torbert, Dr Roy B., University of New Hampshire, Space Science Center, Science and Engineering Research Center, Durham, New Hampshire 03824, U.S.A.
Tel: +1-603-862-2899, Fax: +1-603-862-1915, Email: 6866::torbert
CoI: ASPOC EDI

Torkar, Dr Klaus, Institut für Weltraumforschung, Oesterr. Akademie der Wissenschaften, Inffeldgasse 12, A-8010 Graz, Austria
Tel: +43 (316) 463696-25, Fax: +43 (316) 463697, Email: torkar@fiwf01.tu-graz.ac.at
CoI: ASPOC, SC, IWG, DCM, TM
Treumann, Dr Rudolf A., MPI für Extraterrestrische Physik, Postfach 1603, D-85740 Garching, Germany
Tel: +49-89-3299-3641, Fax: +49-89-3299-3569, Email: 28773::tre
CoI: EDI
Troim, Mr. Jan, Norwegian Defence Research Establishment, P.O. Box 25, N-2007 Kjeller, Norway
Tel: +47-63-807000, Fax: +47-63-807212, Email:
CoI: ASPOC
Trotignon, Dr Jean-Gabriel, LPCE/CNRS, Laboratoire de Physique et Chimie, de l'Environnement, 3A Ave. de la Recherche Scientifique, F-45071 Orleans Cedex 2, France
Tel: +33-38-515263, Fax: +33-38-631234, Email: jgtrotig@cnrs-orleans.fr
CoI: WHISPER
Tsuruda, Dr Koichiro, Institute of Space and Astronautical Sciences, 3-1-1, Yoshinodai, Sagamihara Kanagawa, 229, Japan
Tel: 81-427-513940, Fax: 81-427-594236, Email: tsuruda@gtl.isas.ac.jp
CoI: EDI
Tsurutani, Dr Bruce T., Jet Propulsion Laboratory, MS 169-506, 4800 Oak Grove Drive , Pasadena, CA 91109, U.S.A.
Tel: +1-818-354-7559, Fax: +1-818-354-8895, Email: btsurutani@jplsp.jpl.nasa.gov
CoI: FGM
Ullaland, Dr Stein, University of Bergen, Dept. of Physics, Allegaten 55, N-5007 Bergen, Norway
Tel: 47-55212749, Fax: 47-55318334, Email: stein.ullaland@fi.uib.no
CoI: RAPID
Ungstrup, Dr Eigil, Geophysical Institute, Haraldsgade 6, DK-2200 Copenhagen N, Denmark
Tel: 45-31-835694, Fax: 45-35-822565, Email: ungstrup@nbivax.nbi.dk
CoI: WBD
Vaisberg, Prof. Oleg L., Space Research Institute, Russian Academy of Sciences, Profsoyuznaya 84/32, 117810 Moscow GSP-7, Russia
Tel: 7-095-333-3456, Fax: 7-095-310-7023, Email: olegv@afed.iki.rssi.ru
CoI: PEACE
Valavanoglou, Dr Nicolaas, Institut für Weltraumforschung, Oesterr. Akademie der Wissenschaften, Inffeldgasse 12, A-8010 Graz, Austria
Tel: +43 (316) 8737437, Fax: +43 (316) 463697, Email:
CoI: ASPOC
Vasyliunas, Dr Vitenis M., Max-Planck-Institut für Aeronomie, Postfach 20, D-37189 Katlenburg-Lindau, Germany
Tel: 49-5556-979-435, Fax: 49-5556-979-240, Email: vasyliunas@linax1.dnet.gwdg.de
CoI: RAPID CIS
Vaughan, Dr Peter, Rutherford Appleton Laboratory, Building R68, Chilton, Didcot, Oxon, OX11 0QX, U.K.
Tel: +44-1235-446269, Fax: +44-1235-446667, Email: p.a.vaughan@rl.ac.uk
IWG, DCM
Villain, Dr Jean-Paul, LPCE/CNRS, Laboratoire de Physique et Chimie,
de l'Environnement, 3A Ave. de la Recherche Scientifique, F-45071 Orleans Cedex 2, France
Tel: +33-38-515287, Fax: +33-38-631234, Email: jvillain@cnrs-orleans.fr
CoI: WHISPER
Walton, Dr D., Mullard Space Science Laboratory, Univ. College London, Dept. of Physics, Holmbury St. Mary, Dorking, Surrey, RH5 6NT, U.K.
Tel: +44-1483-204143, Fax: +44-1483-278312, Email: dmw@mssl.ucl.ac.uk
CoI: PEACE
Warhaut, Dr Manfred, ESOC (MOD-SMD), Robert-Bosch-Str. 5, D-64293 Darmstadt, Germany
Tel: +49-6151-90-2791, Fax: +49-6151-90-3409, Email: mwarhaut@esoc.esa.de

SC, IWG, ESOC

Wenzel, Dr K. P., ESA/ESTEC (SO), Postbus 299, 2200 AG Noordwijk, The Netherlands
Tel: +31-71-5653573, Fax: +31-71-5654697, Email: kpwenzel@estec.esa.nl
Whipple, Dr Elden, University of Washington, Geophysics Department, c/o Space Physics Group, Seattle, WA 98195, U.S.A.
Tel: 206-402-0821, Fax: , Email: whipple@geophys.washington.edu
CoI: ASPOC EDI
Whitman, Mr. Rusty, Computer Sciences Corporation, 7700 Hubble Drive, Lanham-Seabrook, MD 20706, U.S.A.
Tel: +1-301-794-2361, Fax: +1-301-794-8355, Email: whitman@istp2.gsfc.nasa.gov
Wilken, Dr Berend, Max-Planck-Institut für Aeronomie, Postfach 20, D-37189 Katlenburg-Lindau, Germany
Tel: 49-5556-979-431, Fax: +49-5556-979-139, Email: wilken@linmpi.mpae.gwdg.de
PI: RAPID,
Winningham, Dr J. David, Southwest Research Institute, P.O. Drawer 28510, 6220 Culebra road, San Antonio, TX 78238- 0510, U.S.A.
Tel: +1-210-522-3075, Fax: +1-210-647-4325, Email: david@dews1.space.swri.edu
CoI: PEACE
Woch, Dr Joachim, Max-Planck-Institut für Aeronomie, Postfach 20, D-37189 Katlenburg-Lindau, Germany
Tel: 49-5556-979-447, Fax: +49-5556-979-139, Email: woch@linmpi.mpae.gwdg.de
CoI: RAPID
Wygant, Dr John, School of Physics and Astronomy, Tate Laboratory of Physics, 116 Church Street S.E., Minneapolis, MN 55455-0112, U.S.A.
Tel: +1-612-6241335, Fax: +1-612-6262029, Email: wygant@ham.spa.umn.edu
CoI: EFW
Young, Dr David T., Southwest Research Institute, P.O. Drawer 28510, 6220 Culebra road, San Antonio, TX 78228-0510, U.S.A.
Tel: +1-210-522-2743, Fax: +1-210-647-4325, Email: dave@swri.space.swri.edu
CoI: PEACE
Zhao, Dr Hua, Center for Space Science and, Applied Research, PO Box 8701, 100080 Beijing, China
Tel: , Fax: 86-10-257-6921, Email: hzhao@sun20.cssar.ac.cn
CoI: ASPOC